Fundamentals of PWM Dc-to-Dc Power Conversion

Third Edition

Byungcho Choi
Kyungpook National University
Daegu, Korea

가디언북
GuardianBook Publishing

ISBN 979-11-87021-06-3

이 도서의 국립중앙도서관 출판예정도서목록(CIP)은 서지정보유통지원시스템 홈페이지(http://seoji.nl.go.kr)와 국가자료종합목록 구축시스템(http://kolis-net.nl.go.kr)에서 이용하실 수 있습니다. (CIP제어번호 : CIP2019026643)

Published by GuardianBook Publishing, Daegu, Korea.

Printed in the Republic of Korea.

10 9 8 7 6 5 4 3 2 1

To my family –
Meeleeyah,
Jieyeon, David,
Ann, Luke,
Yeonsoo, and Sunghyuk

PREFACE

Most industrial and consumer electronics are powered from a dc source whose voltage level is fixed in advance. For examples, telecommunication electronics usually receive their input power from a 48 V dc source and cellular phones operate with 3 V rechargeable batteries. Meanwhile, the circuits inside such equipment require several dc voltages that are different from the voltage level of the input source. Accordingly, each equipment contains specially-engineered circuit blocks capable of changing the input voltage to another dc level, as required by the various internal circuits.

The process of changing the voltage level of a dc power source into another value is referred to as the dc-to-dc power conversion. The dc-to-dc power conversion is performed in many different ways, each with a distinctive circuit technique. The most popular scheme is the dc-to-dc power conversion circuit employing the pulsewidth modulation (PWM) technique. The dc-to-dc power conversion based on the PWM technique is called the PWM dc-to-dc power conversion. Circuitries designed for the PWM dc-to-dc power conversion are called PWM dc-to-dc converters

This book is primarily intended to be a textbook for undergraduate and first-year postgraduate students who are beginning to study power electronics focusing on the PWM dc-to-dc power conversion. The book supplements existing textbooks with more dedicated treatments on the PWM dc-to-dc power conversion. The book is also written as a reference book for engineers working in the area of modeling, analysis, and control of PWM dc-to-dc converters.

This book is divided into three parts based on technical contents and targeted readers. The first four chapters cover the static characteristics of PWM dc-to-dc converters, concentrating on the steady-state time-domain operation. This part is mainly for undergraduate students exposed to power electronics for the first time. Experienced engineers or postgraduate students may quickly review or skip some sections in this part.

The next three chapters constitute the second part of this book, which deals with the small-signal dynamic characteristics of PWM dc-to-dc converters. This part covers the modeling, small-signal transfer functions, and dynamic performance of PWM dc-to-dc converters. While most materials are adequate for junior or senior students with reasonable academic backgrounds, some advanced topics can be omitted for inexperienced undergraduate students.

The last three chapters are devoted to the control design of PWM dc-to-dc converters. This part presents the functional basics, dynamic modeling and analysis, compensation design, and applications of both voltage mode control and current mode control of PWM dc-to-dc converters. One chapter fully investigates the sampling effects of current mode control. This chapter is aimed for experienced engineers and postgraduate students. Engineers may acquire in-depth knowledge about current mode control. For postgraduate students, this part could serve as a foundation to start out their researches on relevant topics.

This book is suited for a textbook for one-semester power electronics classes for senior undergraduate students or first-year postgraduate students. A typical syllabus for an undergraduate class or postgraduate class will consist of as follows.

Undergraduate Class
Chapter 1: PWM Dc-to-Dc Power Conversion
Chapter 2: Power Stage Components
Chapter 3: Buck Converter
Chapter 4: Dc-to-Dc Power Converter Circuits
Chapter 5: Modeling PWM Dc-to-Dc Converters
Chapter 6: Power Stage Transfer Functions
Chapter 7: Dynamic Performance of PWM Dc-to-Dc Converters
Chapter 8: Feedback Compensation and Closed-Loop Performance

Postgraduate Class
Chapter 3: Buck Converter
Chapter 5: Modeling PWM Dc-to-Dc Converters
Chapter 6: Power Stage Transfer Functions
Chapter 7: Dynamic Performance of PWM Dc-to-Dc Converters
Chapter 8: Feedback Compensation and Closed-Loop Performance
Chapter 9: Current Mode Control (I) – Functional Basics and Classical Analysis
Chapter 10: Current Mode Control (II) – Sampling Effects and New Control Design Procedures

In writing this book, special efforts have been made in the two aspects listed below.

1) Computer simulations are used as much as possible, as an instrumental tool to demonstrate the validity of theoretical developments and accuracy of analytical predictions.

2) Technical contents are presented in a format directly adaptable to practical applications. Dc-to-dc converter design examples are given with all engineering details so that each design can immediately be transformed into working hardware.

The following materials are prepared to assist students and lecturers who will study and teach the PWM dc-to-dc power conversion using this book.

End-of-Chapter Problems: Each chapter contains a number of problems to reinforce the technical contents of the text. Among the problems, the ones marked '*' are more important and essential problems. The answers to the selected problems are given in the appendix.

On-Line Teaching and Learning Aid: PDF® files for lecture slides are available at `http://pwmconverters.bchoi`.

The author is indebted to many individuals who helped to improve editorial and technical presentations of this book. The author expresses special respects and thanks to Prof. Sang Dae Yu at Kyungpook National University (KNU) for his vast knowledge about LaTex and wholehearted help, all embedded into the typesetting of this book. The author also acknowledges the contribution of Dr. Syam Kumar Pidaparthy at KNU in deriving some equations and preparing illustrations in the book. Author's warmest appreciation goes to Jieyeon, the author's eldest daughter, for becoming a knowledgeable power electronics engineer herself and assisting with numerous help for the publication of this book.

This book has evolved from the class materials taught by the author over the last 15 years at Kyungpook National University (KNU) in Daegu, Korea. The author expresses affections and thanks to the former and current students at KNU, who inspired the author to undertake the task of writing this book. Some materials in this book were presented by the author in short courses at several power electronics industries. The author is also grateful for the engineers who participated in the short courses and provided valuable feedback.

BYUNGCHO CHOI

Daegu, Korea
August 2019

CONTENTS IN BRIEF

CONTENTS

CHAPTER 1

PWM DC-TO-DC POWER CONVERSION

The dc-to-dc power conversion is broadly referred to as the process of generating a desired dc voltage using a dc source whose voltage level is different from the desired value: namely, changing the voltage level of a dc source into another value. The dc-to-dc power conversion based on the pulsewidth modulation (PWM) technique is called the PWM dc-to-dc power conversion.

This introductory chapter presents an overview of the PWM dc-to-dc power conversion. We discuss basic principles and unique natures of dc-to-dc power conversion circuits, along with the concept of the PWM technique. This chapter also presents the features and issues of PWM dc-to-dc power conversion systems employed to modern electronic equipment and systems. Finally, this chapter outlines the contents of forthcoming chapters.

1.1 PWM DC-TO-DC POWER CONVERSION

The PWM dc-to-dc power conversion is described as the process of changing the voltage level of a dc source using the PWM technique. However, a more definitive and precise description is necessary to comprehend the natures and features of the PWM dc-to-dc power conversion circuit.

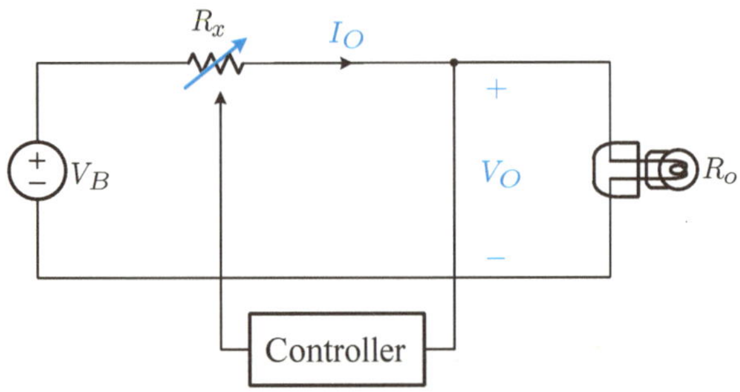

Figure 1.1 Conventional approach to lighting electric bulb.

1.1.1 Dc-to-Dc Power Conversion

To formulate an accurate description of the dc-to-dc power conversion, this section discusses two different approaches to operating an electric bulb using a dc voltage sourced from a battery. It is presumed that the electric bulb requires a strict 12 V for operation, while the battery voltage is varied between 18 V and 30 V depending on its charging status and operational conditions. Figure 1.1 shows the first approach where a variable resistor and controller are employed between the battery and electric bulb. The controller is assumed to only draw a negligible current.

In Fig. 1.1, the controller adjusts the resistance of the variable resistor R_x to meet the following relationship

$$\boxed{V_O = \frac{R_o}{R_x + R_o} V_B = 12\,\text{V}} \tag{1.1}$$

where V_O is the voltage across the electric bulb, R_o denotes the resistance of the bulb, and V_B is the battery voltage that varies between $18\,\text{V} < V_B < 30\,\text{V}$. Figure 1.1 certainly fulfills the goal of providing a fixed dc voltage from a variable voltage source, however, it has one critical problem that renders this approach impractical.

The variable resistor is accompanied by an ohmic power loss

$$\boxed{P_{loss} = P_{in} - P_{out} = I_O V_B - I_O V_O = I_O (V_B - V_O)} \tag{1.2}$$

where P_{in} is the input power drawn from the battery, P_{out} is the output power delivered to the electric bulb, and I_O is the current flowing from the battery to the electric bulb. The power loss is given by the product of the load current and the difference between the battery voltage and bulb voltage. This power loss easily becomes significant. For example, when an electric bulb that consumes a 60 W power at 12 V voltage level is connected to a 30 V battery, the power loss is as large as $P_{loss} = (60/12)\,(30-12) = 90\,\text{W}$. This loss is even larger than the power consumed in the bulb, $P_{out} = 60$ W.

The power loss is always transformed into heat and the resulting heat must be removed using an appropriate cooling system. The cooling system usually employs

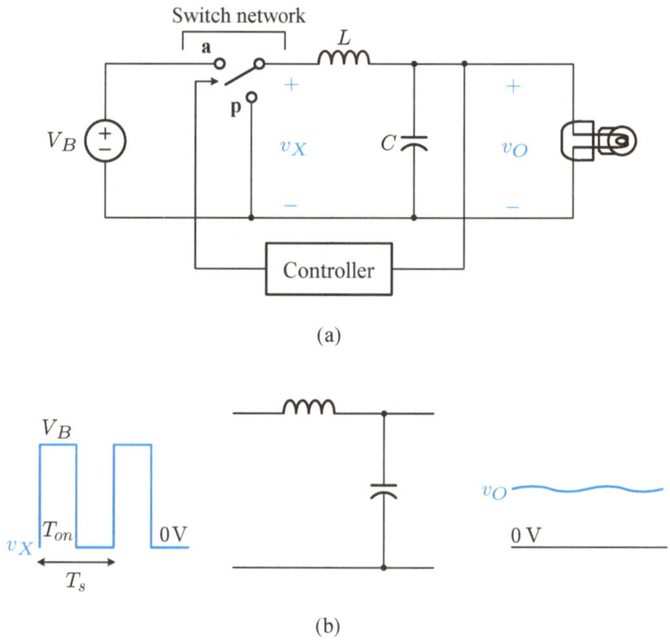

Figure 1.2 Dc-to-dc power conversion applied to power electric bulb. (a) Circuit diagram. (b) Input and output waveforms of LC filter.

bulky heat sinks and noisy fans, consequently increasing the size and weight of the entire system. Accordingly, Fig. 1.1 cannot be used for applications where the dimension and weight should be limited, which is usually the case for most modern electronic equipment and systems.

Figure 1.2 shows an alternative approach where a *switch network* and LC filter are inserted between the battery and electric bulb. The switch network periodically changes its connection. Within each switching period T_s, the switch network maintains position **a** for T_{on} and position **p** for the remaining part of the switching period, $T_s - T_{on}$. This switch network is called the *single-pole double-throw (SPDT) switch* because it contains one pole which is always connected to one of the two contacts, the throw **a** and the throw **p**. With the switching action of the SPDT switch, the battery voltage is transformed into a rectangular waveform at the output of the SPDT switch, v_X in Fig. 1.2. The rectangular waveform is then applied to the LC filter. The LC filter alters the rectangular waveform into a smoothly-filtered continuous voltage waveform, v_O in Fig. 1.2.

If the LC filter provides sufficient filtering, the output voltage nearly becomes a dc waveform, corresponding to the average value of v_X

$$\boxed{v_O(t) \approx V_O = \bar{v}_X(t) = \frac{T_{on}}{T_s} V_B} \tag{1.3}$$

To maintain $V_O = 12$ V at the presence of the battery voltage variation, the controller adjusts the ratio T_{on} to T_s. With a fixed T_s, the controller changes T_{on} to meet the condition

$$\frac{T_{on}}{T_s} V_B = 12 \text{ V} \tag{1.4}$$

For example, with a battery voltage $V_B = 24$ V and switching period $T_s = 10\,\mu$s, the controller generates $T_{on} = 5\,\mu$s to produce $V_O = (5 \times 10^{-6}/10 \times 10^{-6})\,24 = 12$ V. If the battery voltage is increased to $V_B = 30$ V, the controller reduces T_{on} to $4\,\mu$s to regulate V_O at 12 V: $V_O = (4 \times 10^{-6}/10 \times 10^{-6})\,30 = 12$ V.

Although Fig. 1.1 and Fig. 1.2 both achieve the same goal, a crucial difference exists between them. Figure 1.2 presumes a lossless operation because the SPDT switch and reactive components in the LC filter do not consume any power. The lossless operation eliminates all the problems associated with the power loss. Because no heat management is required, the circuit can be packaged with a smaller size and lighter weight, thereby making it fully compatible with modern electronic systems.

A more definitive description of the dc-to-dc power conversion is now established as the process of changing the voltage level of a dc source, while eliminating or minimizing power loss. In this perspective, Fig. 1.2 is a typical example of the dc-to-dc power conversion circuit, while the conventional circuit illustrated in Fig. 1.1 is not classified so.

1.1.2 PWM Technique

The concept of PWM technique can be envisaged from the operation of Fig. 1.2, where the ratio T_{on} to T_s of the SPDT switch is adjusted to keep the output voltage constant. By changing the T_{on}/T_s ratio, the pulsewidth of the rectangular voltage waveform, which passes through the LC filter to yield its average value as the output voltage, is adaptively modulated so that the output voltage remains constant despite the input voltage variation. This control scheme is called the *pulsewidth modulation (PWM) technique* and the dc-to-dc conversion circuit based on the PWM scheme is known as the *PWM dc-to-dc converter*. The PWM dc-to-dc converter is widely adapted to modern industrial and consumer electronics, thereby becoming the most prevailing dc-to-dc power conversion circuit.

1.2 DC-TO-DC POWER CONVERSION SYSTEM

The basic concept illustrated in Fig. 1.2 is generalized into dc-to-dc power conversion systems whose block diagram representation is shown in Fig. 1.3. The system consists of the dc source, dc-to-dc converter, and load. The dc source provides an arbitrary dc voltage to the dc-to-dc converter. The dc-to-dc converter then converts the level of the given dc voltage to the value required by the load and delivers it to the load. The load is an application system that operates with a fixed voltage and eventually consumes electrical power. This section presents the characteristic features of the dc source, dc-to-dc converter, and load.

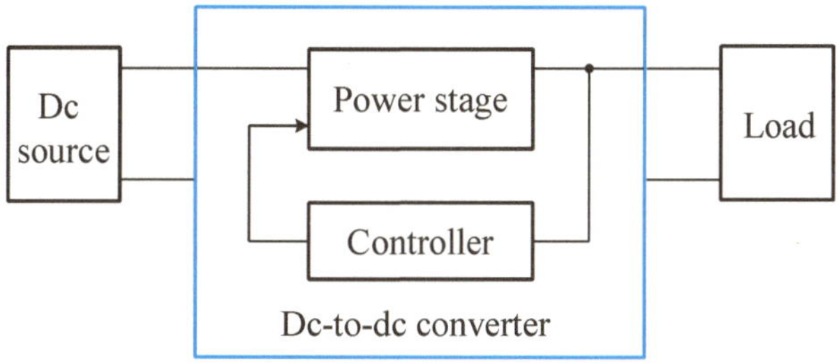

Figure 1.3 Dc-to-dc power conversion system.

Dc Source with Non-Ideal Characteristics

The practical dc source falls short of the characteristics of an ideal voltage source in several aspects. First, the voltage level of the dc source could vary with time, as is the case with batteries, fuel cells, and other standalone dc sources. The change in the voltage could occur either gradually or abruptly, depending on the characteristics and status of the dc source.

Second, a rectified ac source is often used as a substitute for the dc source. For this case, the rectified ac source could contain a considerable amount of ac components, known as *ac ripple*. In addition, the output of the rectified ac source could be corrupted with various noises. Accordingly, the dc source represents an arbitrary non-ideal source whose voltage can be varied, polluted with ac ripple and noises, and switched from one value to another.

Dc-to-Dc Converter as Voltage Source

The dc-to-dc converter receives an arbitrary voltage from the non-ideal source and is required to provide a fixed dc voltage for the load. Thus, in addition to altering the voltage level, the dc-to-dc converter should have the capacity of maintaining its output voltage constant at the presence of the variation, ac ripple component, and abrupt change in the input voltage. Ideally, the dc-to-dc converter should function as an *ideal* voltage source, powered by a non-ideal voltage source and programmed to produce the required dc voltage for the load, regardless of the status of the voltage source.

Although practical dc-to-dc converters are more complicated in their structure and operation than those of Fig. 1.2 in the previous section, they still can be divided into two functional blocks: the power stage and the controller. The power stage alters the level of the input voltage into a desired value using various circuit components, while the controller provides the necessary signals for the power stage to execute its function.

Dc-to-dc converters come with numerous variations in their power stage configuration and each dc-to-dc converter is named differently after its power stage structure. In spite of the diversity in structure, all the power stages employ the common electronic components to perform the dc-to-dc power conversion. The power stage utilizes 1) semiconductor devices to implement the function of the SPDT switch, 2) energy storage components to perform filtering, and 3) transformers to change the voltage and current levels of circuit variables while transferring electrical energy.

The controller also varies widely in its structure and functionality. Nonetheless, all the controllers perform the same role of providing the control signals that are required for the power stage to generate a fixed output voltage, regardless of variations in the input voltage and other operational circumstances. In PWM dc-to-dc converters, this important function is executed in a closed-loop fashion using the PWM technique. The closed-loop PWM controller uses various analog and digital ICs, as well as discrete circuit components, to generate the required control signals.

Load as Dynamic Current Sink

The load of a dc-to-dc converter can be any electronic equipment or system operating with a fixed dc voltage. The load draws the current from the dc-to-dc converter to meet its power requirement. Thus, the load current could fluctuate depending on operational situations of the load system. In particular, when high-frequency digital systems are connected to a dc-to-dc converter, the load current change could occur frequently and rapidly, including step changes between two different values. Accordingly, the load system presents a dynamic current sink to the dc-to-dc converter, whose current level could change widely and abruptly.

1.3 FEATURES AND ISSUES OF PWM DC-TO-DC CONVERTER

The PWM dc-to-dc converter is a closed-loop controlled dc-to-dc converter which operates based on the principle of PWM. The PWM dc-to-dc converter is intended to function as an efficient and reliable voltage source, interfacing with the non-ideal dc source and dynamic load system. Accordingly, there are specific features demanded for the dc-to-dc converter and issues involved with implementing such features in a PWM dc-to-dc converter.

Power Stage Components

A dc-to-dc converter employs semiconductor devices, reactive components such as inductors and capacitors, and transformers in the power stage; in contrast, the power stage never contains any resistive component in order to avoid power loss. The semiconductor devices operate as the switches that losslessly alternate on-state and off-state at very high frequency, up to several MHz range in some applications. Due to this switching action, all the power stage components are subjected to periodic voltage and current excitations. The switching action and periodic operation are the characteristic features of the power stage components employed in dc-to-dc converters. Circuit analysis skills beyond the standard linear circuit theory are required for understanding the behaviors of the power stage components under periodic switching operations.

Power Stage Configuration

A dc-to-dc converter is required to accept an arbitrary input voltage and to generate a predetermined output voltage. The ratio between the input voltage and output voltage could be either very large or considerably small. Also, a dc-to-dc converter should provide an arbitrary load current, as required by the load. Accordingly, there are demanding requirements for the converter power stage in voltage and current ratings, input-to-output voltage ratio, and power handling capacity. In addition, dc-to-dc converters are frequently needed to provide galvanic isolation between the source and load. To meet these demands, numerous power stage configurations have been developed, each with a different complexity and functionality. The power stage configuration occupies a large and important portion of the dc-to-dc power conversion technology.

Dynamic Modeling and Analysis

A dc-to-dc converter should function as a voltage source which holds its output voltage constant at the desired value, irrespective of any potential changes in the input voltage, load current, and other operational environments. This vital function is achieved by the closed-loop feedback controller operating under the principles of PWM.

It is well known that a closed-loop controlled system becomes unstable if the system is not properly designed. Stability can be assessed by investigating the dynamic properties of the closed-loop controlled system. There are many analytical methods to judge whether a closed-loop controlled system is stable or not. However, these methods are mainly intended for linear time-invariant systems. The PWM dc-to-dc converter falls into the category of the nonlinear time-variant system to which the aforementioned stability analysis methods cannot be directly applied.

The dynamic modeling refers to the analytical process of describing the dynamic characteristics of the nonlinear PWM dc-to-dc converter in a special format to which all the classical analysis methods, originally aimed to linear systems, can be applied. Therefore, the dynamic modeling allows us to investigate the stability and performance of the nonlinear time-variant PWM dc-to-dc converter using the familiar classical control theory. The dynamic modeling and ensuing analysis using the resultant model are collectively called as the *dynamic modeling and analysis*. The dynamic modeling and analysis plays an important role in the PWM power conversion technology and deserves rigorous treatment.

Dynamic Performance and Control Design

The performance of a dc-to-dc converter will be divided into two categories in this book: the static performance and dynamic performance. The *static performance* specifies the dc-to-dc converter as a static voltage source. The static performance metrics include the input-to-output voltage conversion ratio, current/voltage stress of semiconductor switches, and power handling capacity. The static performance is solely determined by the power stage and is irrelevant to the feedback controller.

The second category is the *dynamic performance* which depicts the dc-to-dc converter as a closed-loop controlled dynamic system. The most important dynamic performance metric is stability. The dc-to-dc converter should establish a periodic steady-state operation to produce the desired output voltage. When a certain disturbance is introduced, the converter could temporarily deviate from its steady-state operation. However, the converter should always return to the original operation as the disturbance disappears. This essential feature is possible only when the converter meets the stability criterion.

Another important dynamic performance metric is the step load response. When a step change occurs in the load current, the output voltage would show a transitional excursion before it returns to its steady-state value. The transitional output voltage response is called the *step load response* in this book. The step load response is of particular concern when digital logic circuits are employed as the load. Modern logic circuits operate with a very tightly-regulated low voltage, for example 2.1 ± 0.02 V, and draw a large pulsating current. These logic circuits naturally and frequently incur substantial step changes in the load current. For this case, the output voltage excursion should be controlled within certain limits, in order to avoid the potential failure of digital logic circuits due to an excessive transitional overshoot or undershoot in the supply voltage.

The dynamic performance is solely determined by the feedback controller. For a given power stage configuration, the controller should be designed for stability and good dynamic performance. While the controller design is primarily based on the dynamic modeling and analysis, it also requires broad knowledge about the control theory, linear system theory, and feedback compensation design.

1.4 CHAPTER HIGHLIGHTS

The purpose of this book is to cover the PWM dc-to-dc power conversion, while focusing on the features and issues addressed in the previous section. The book consists of three parts, which individually underscore the issues of PWM dc-to-dc conversion and collectively address the dynamics and design of complete PWM dc-to-dc converters.

Part I: Circuits for Dc-to-Dc Power Conversion

Part I covers the power stage components, circuit configuration, and operation of PWM dc-to-dc converters. Circuit properties of semiconductor switches and energy storage elements are all presented. The circuit topologies and operation of most widely-used PWM dc-to-dc converters are discussed.

Chapter 2 deals with the power stage components. This chapter presents the circuit behavior of semiconductor switches, inductors, capacitors, and transformers, all operating under periodic excitations. First, Chapter 2 describes the operation of MOSFETs as an active switch, diodes as a passive switch, and MOSFET-diode pairs as a single-pole double-throw (SPDT) switch. Then, this chapter discusses the basic circuit equations of inductors and capacitors. Lastly, this chapter describes the operation of transformers and introduces a circuit model for practical transformers.

Chapter 3 presents the simplest PWM dc-to-dc power conversion circuit, known as the buck converter. Theoretical basics and functional details of the buck converter are both presented. This chapter also illustrates circuit analysis techniques that are commonly applicable to all other forthcoming PWM dc-to-dc converters. In addition, Chapter 3 discusses the underlying basics of the PWM technique and closed-loop control of dc-to-dc converters.

Chapter 4 deals with the topology and operation of an important class of PWM dc-to-dc converters. This chapter first introduces two important non-isolated converters, boost converter and buck/boost converter. A number of isolated converters, derived from the non-isolated converter, are also presented. For each converter, the origin of the circuit topology is first illustrated and the steady-state operation is then investigated using the analysis techniques established in Chapter 3.

Part II: Modeling, Dynamics, and Performance of PWM Dc-to-Dc Converters

Part II starts from the modeling of PWM dc-to-dc converters and proceeds with the analysis of the power stage dynamics of the PWM converters studied in Chapters 3 and 4. This part investigates the closed-loop dynamics of PWM dc-to-dc converters and presents the techniques to evaluate stability and performance of dc-to-dc converters.

Chapter 5 covers the dynamic modeling of PWM dc-to-dc converters. This chapter illustrates the procedures of describing the dynamic characteristics of nonlinear time-variant PWM dc-to-dc converters, using the terms and formats that have been used for linear time-invariant systems. As the ultimate outcome of the dynamic modeling, this chapter provides a linear circuit model for PWM dc-to-dc converters, which allows us to investigate the nonlinear converter dynamics via conventional s-domain techniques.

Chapter 6 presents the dynamic analysis of PWM dc-to-dc converters using the linear circuit model developed in Chapter 5. This chapter describes the dynamic characteristics of an important class of PWM dc-to-dc converters, thus providing theoretical foundations for the control design and closed-loop analysis.

In Chapter 7, the dynamic performance of the closed-loop controlled PWM dc-to-dc converter is discussed. The implication and significance of the performance metrics are demonstrated with examples. The method of stability analysis and performance evaluation, based on the Nyquist stability criterion and Bode plot technique, are presented.

Part III: Control Schemes and Converter Performance

Part III explores two control schemes of PWM dc-to-dc converters, voltage mode control and current mode control, focusing on the frequency- and time-domain performance of PWM converters. The principles, dynamics, and implementation of the two control schemes are presented in a comparative manner. Special emphases are placed on the design and performance of the respective control scheme.

Chapter 8 is devoted to the closed-loop performance and feedback compensation of PWM dc-to-dc converters employing voltage mode control. This chapter first introduces a graphical analysis method which facilitates the dynamic analysis and compensation design. Using this graphical method, the chapter presents theoretical details of compensation design. The impacts of the compensation design on the converter performance are addressed. Practical techniques to correlate the compensation parameters with the frequency- and time-domain performance are demonstrated with several examples.

Chapter 9 presents functional basics and dynamic analyses of PWM dc-to-dc converters employing current mode control, which utilizes an additional feedback from the inductor current. The motivation and benefits of current mode control are described. The converter dynamics under current mode control are investigated, leading to general procedures for the closed-loop analysis and control design.

Current mode control is executed by sampling and holding the error signal produced by the fast-varying inductor current waveform. Due to this sampling and holding mechanism, current mode control exhibits characteristics of a sampled-data system. This unique feature of the control scheme is known as the sampling effects of current mode control. Chapter 10 extensively analyzes the sampling effects and the outcomes are converted into alternative design equations for current mode control. Several design examples are given to substantiate the alternative design equations.

REFERENCES

1. R. W. Erickson and D. M. Maksimović, *Fundamentals of Power Electronics*, 2nd Ed., Kluwer Academic Publishers, 2001.
2. Byungcho Choi, *Dynamics and Control of Switchmode Power Conversions in Distributed Power Systems*, Ph. D. Dissertation, Virginia Polytechnic Institute and State University, Blacksburg, VA, May 1992.

PART I

CIRCUITS FOR PWM DC-TO-DC POWER CONVERSION

The circuits intended for the PWM dc-to-dc power conversion employ semiconductor devices, inductors, capacitors, and transformers: all commonly found in other conventional linear/nonlinear circuits. What makes the PWM power conversion circuits unique is their distinctive operational circumstances. The semiconductor devices function as the switches and the other components are subjected to periodic voltage/current excitations. The harmonious operation of the inductors, capacitors, and transformers, in synchronization with the active/passive switches, produces the functionalities and circuit waveforms needed for the PWM dc-to-dc power conversion.

Part I deals with the circuits for the PWM dc-to-dc power conversion, ranging from the circuit equations of individual components to the operations of numerous PWM dc-to-dc converters. In Chapter 2, the responses of the inductors, capacitors, and transformers under periodic excitations are described. Conditions and constraints of circuit variables of the individual components are given. The flux linkage balance condition for inductors and the charge balance condition for capacitors are presented as the fundamental principles that determine circuit waveforms in steady state.

Chapter 3 considers the buck converter, which performs the step down dc-to-dc power conversion. The basic principles and functional details of the buck converter are analyzed. During the analysis, several general techniques are established, which are universally applicable to all other dc-to-dc converters. The closed-loop feedback control using PWM scheme is presented. The steady-state and transient responses of the closed-loop controlled buck converter are illustrated via PSpice® simulations. This chapter provides all the information necessary to comprehend the functions of closed-loop controlled PWM converters.

Chapter 4 presents topologies and operations of an important class of PWM dc-to-dc converters. This chapter provides a systematic view of the evolution and diversification of converter topologies. The circuit waveforms of the boost converter and buck/boost converter are investigated using the analysis techniques established in Chapter 3. The analyses are extended to a number of isolated PWM converters including the flyback converter, bridge-type converters, and various forward converters.

CHAPTER 2

POWER STAGE COMPONENTS

Power stages of PWM dc-to-dc converters contain semiconductor switches, inductors, capacitors, and transformers. This chapter describes the operations of these power stage components under periodic excitations. The current chapter also presents several circuit theorems that will be used in later chapters for analyses of dc-to-dc power conversion circuits. Finally, this chapter discusses two practical switching circuits, the solenoid drive circuit and capacitor charging circuit, in order to demonstrate operations of power stage components in real applications.

2.1 SEMICONDUCTOR SWITCHES

Dc-to-dc power conversion circuits extensively use active and passive semiconductor switches. The active switch refers to the three-terminal semiconductor device whose *on/off* status is actively controlled by exciting one of the device terminals. On the other hand, the passive switch is the two-terminal device whose on/off status is passively

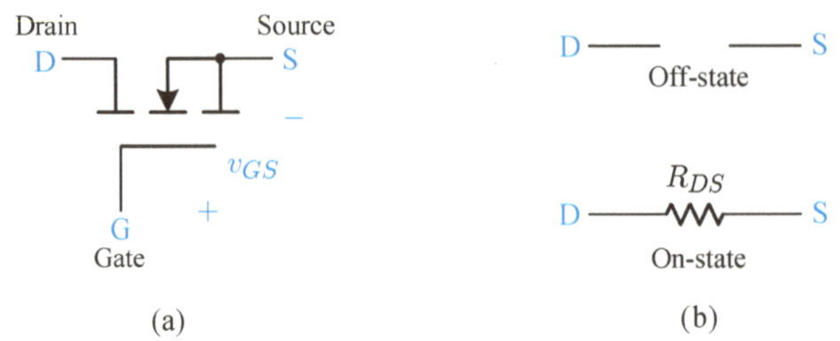

Figure 2.1 MOSFET. (a) Circuit symbol. (b) Circuit representations.

determined by the conditions of the application circuit. The switching action of active and passive semiconductor switches allows the voltage and current waveforms of the application circuit to be altered, as required to perform the desired dc-to-dc power conversion.

As illustrated in Fig. 1.2 in Section 1.1.1, the dc-to-dc power conversion is performed by converting the dc voltage into a rectangular waveform and filtering the resulting rectangular voltage into another dc voltage. The conversion from the dc input to the intermediate rectangular waveform is performed using the single-pole double-throw (SPDT) switch. It will be shown that the SPDT switch is implemented with a pair of active and passive semiconductor switches.

There are numerous semiconductor switches available for dc-to-dc conversion circuits. The selection of switching devices depends on both how well the existing devices perform and what the application circuits require. In many dc-to-dc conversion circuits, MOSFETs are commonly used for active switches because of their fast switching characteristics compared with other alternatives. For passive switches, fast recovery diodes or Schottky diodes are used due to their excellent switching characteristics. This section describes the function of MOSFETs as an active switch, diodes as a passive switch, and MOSFET-diode pairs as an SPDT switch. Emphases are placed on the functional behavior of these devices in dc-to-dc power conversion circuits, rather than on their physical or operational principles.

2.1.1 MOSFETs

When employed as an active switch, MOSFETs can only have either an off-state or on-state. Figure 2.1 shows the symbol of an n-channel MOSFET and its circuit representations for the off-state and on-state. When the gate drive signal, denoted by v_{GS} in Fig. 2.1(a), is below the threshold voltage, the MOSFET is turned-off. At the off-state, the drain-source terminal simply behaves open-circuited because the conduction channel is not created. When the gate drive signal is larger than the threshold voltage, the conduction channel is formed and the MOSFET is turned-on. Once turned-on, the drain and source terminals are connected through the resistance of the conduction channel, denoted by R_{DS} in Fig. 2.1(b). The channel resistance R_{DS} is usually so small that a turned-on MOSFET can be viewed as a short circuit.

Figure 2.2 Diode. (a) Circuit symbol. (b) Circuit representations.

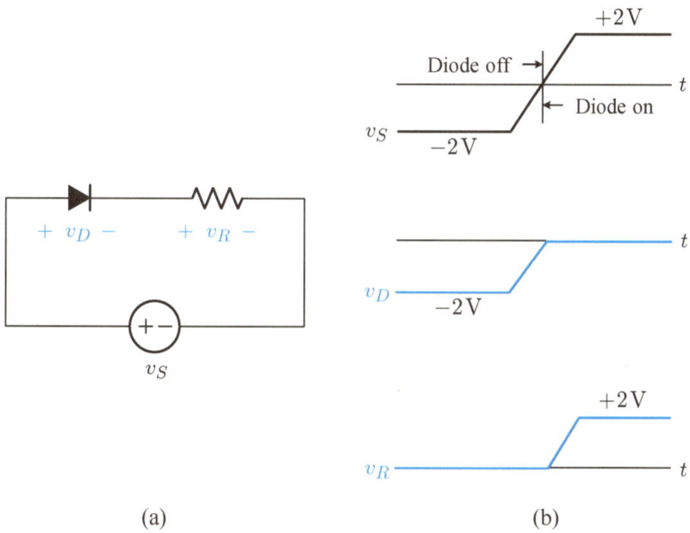

Figure 2.3 Diode switching with voltage source. (a) Circuit diagram. (b) Voltage waveforms.

2.1.2 Diodes

Diodes always presume one of two possible states, the off-state or on-state. Figure 2.2 depicts the symbol of a diode and its circuit representations for off-state and on-state. At an off-state, the diode becomes open-circuited. At an on-state, a practical diode is viewed as a series connection of a voltage source V_D, resistor, and *ideal* diode. The voltage source and resistor are necessary to approximate the nonlinear $v - i$ characteristics of the diode to a piecewise linear function. The ideal diode ensures the *unidirectional* current flow from the anode terminal to cathode terminal. The values of the voltage source and resistor are usually negligibly small. Accordingly, the on-state circuit representation is approximated to a short circuit for practical circuit analyses. Yet, the current flows only from the anode terminal to cathode terminal.

The state of a diode is determined by the condition of the application circuit. When the application circuit forces a positive voltage from the anode to cathode terminal, the diode establishes the on-state. Conversely, when the application circuit imposes a negative voltage, the diode is turned-off. As an illustration, Fig. 2.3(a) shows a simple circuit consisting of a diode, resistor, and time-varying voltage source, v_S.

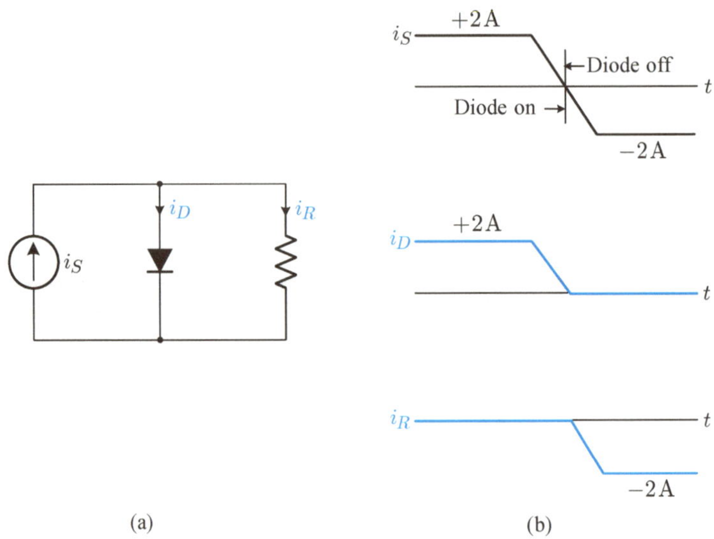

Figure 2.4 Diode switching with current source. (a) Circuit diagram. (b) Current waveforms.

The diode is turned-off when v_S is negative and turned-on when v_S is positive. The transition from the off-state to on-state occurs at the moment the negative voltage v_S increases to hit the *zero* voltage. The voltage waveforms across the diode and resistor are determined as shown in Fig. 2.3(b), based on the circuit behavior of the diode at the on-state and off-state.

The state of a diode can also be judged based on the direction of the current flow. When the application circuit forces a *positive* current flow from the anode terminal to cathode terminal, the diode is turned-on and retains the on-state as long as the current remains positive. On the other hand, when the application circuit forces a *negative* current flow from the cathode terminal to anode terminal, the diode becomes turned-off.

Figure 2.4 is a simple circuit which illustrates the diode switching using a time-varying current source, i_S. When i_S is positive, the diode is turned-on and carries the entire current. Conversely, when i_S becomes negative, the diode is turned-off and the current diverts to the resistor. The transition from the on-state to off-state happens at the instant the positive current i_S reduces to the *zero* current.

2.1.3 MOSFET-Diode Pair as SPDT Switch

When a MOSFET and diode are paired together, they function as an SPDT switch. This section illustrates the implementation of the SPDT switch with a MOSFET-diode pair. Figure 2.5(a) shows a simple switching circuit where a voltage source is connected to a current source via an SPDT switch. The SPDT switch periodically toggles its position as illustrated in Fig. 2.5(a). Figure 2.5(b) shows representations of

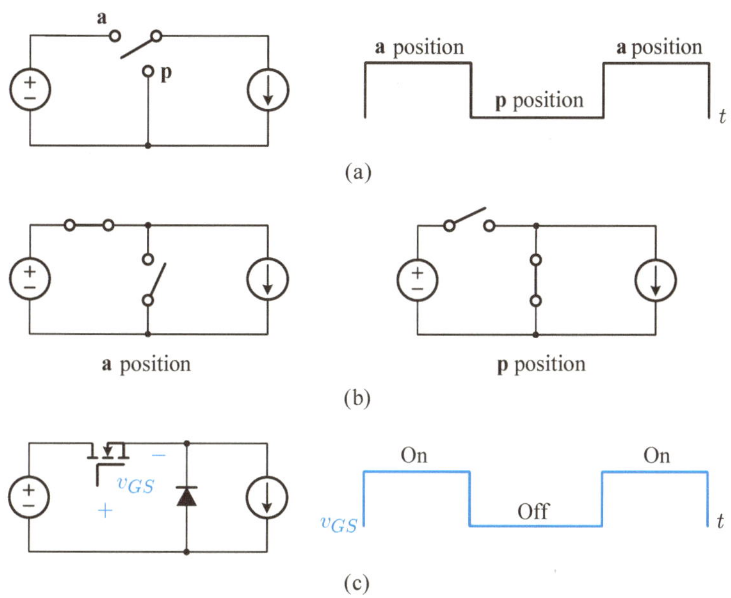

Figure 2.5 Simple switching circuit. (a) Original circuit with SPDT switch. (b) Equivalent circuits. (c) Implementation of SPDT switch using semiconductor switches.

the switching circuit with two different SPDT switch positions: one is at the **a** position and the other is at the **p** position. Finally, Fig. 2.5(c) depicts an implementation of the SPDT switch using MOSFET and diode.

The MOSFET is turned-on/off by the gate drive signal v_{GS}. When the MOSFET is turned-on by the *on signal*, the diode is turned-off because the voltage source forces a negative voltage across the diode. Conversely, when the MOSFET is turned-off by the *off signal*, the current source forces the diode to turn on. The equivalence between Fig. 2.5(a) and Fig. 2.5(c) now becomes apparent. As shown in this example, the SPDT switch is usually constructed using a MOSFET-diode pair.

2.2 ENERGY STORAGE AND TRANSFER DEVICES

Dc-to-dc power conversion circuits utilize energy storage devices to perform filtering and energy transfer devices to transfer electrical energy while altering the magnitude of voltage and current waveforms. Energy storage and transfer devices include inductors, capacitors, and transformers. This section analyzes the circuit behavior of these devices under periodic excitations.

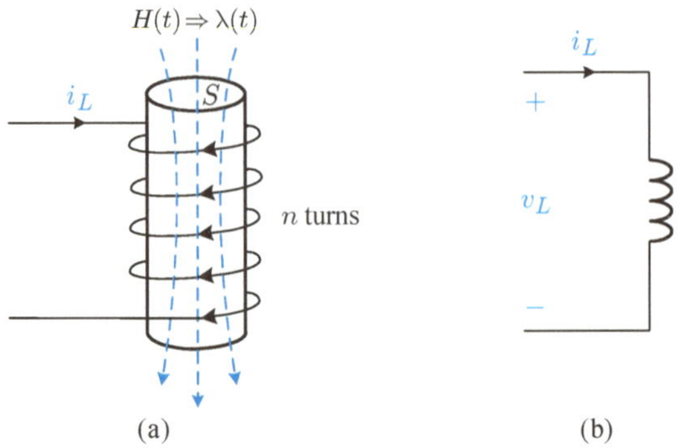

Figure 2.6 Inductor and its terminal properties. (a) Electro-magnetic process within inductor. (b) Polarity/direction of inductor voltage/current waveform.

2.2.1 Inductors

The current section deals with circuit properties of inductors. In addition to basic circuit equations, the operations of inductors under periodic excitations are discussed. This section also presents important circuit theorems for inductors.

Circuit Equations

An inductor is typically constructed by winding a copper wire on a magnetic core. When the current i_L flows into the inductor as shown in Fig. 2.6(a), a series of electro-magnetic phenomena occur inside the magnetic core. First, the current passing through the copper winding creates magnetic field intensity H. The magnetic field intensity in turn produces magnetic flux density: $B = \mu H$ with μ representing the permeability of the core. The total magnetic flux inside the core is then calculated as $\phi = BS$ where S is the cross-sectional area of the core. Finally, the magnetic flux linkage is given by $\lambda = n\phi$ where n is the number of the copper winding turns. The cause-and-effect of the aforementioned process is summarized as $i_L(t) \Rightarrow H(t) \Rightarrow B(t) \Rightarrow \phi(t) \Rightarrow \lambda(t)$. When the current i_L is considered as the input of the electro-magnetic process, the magnetic flux linkage λ becomes the output of the process. The ratio of the output variable, λ , to the input variable of the process, i_L, is defined as the inductance of the inductor

$$\boxed{L \equiv \frac{\lambda(t)}{i_L(t)}} \tag{2.1}$$

According to Faraday's law, the time-varying magnetic flux linkage induces a voltage across the inductor terminals

$$v_L(t) = \frac{\mathrm{d}\lambda(t)}{\mathrm{d}t} \tag{2.2}$$

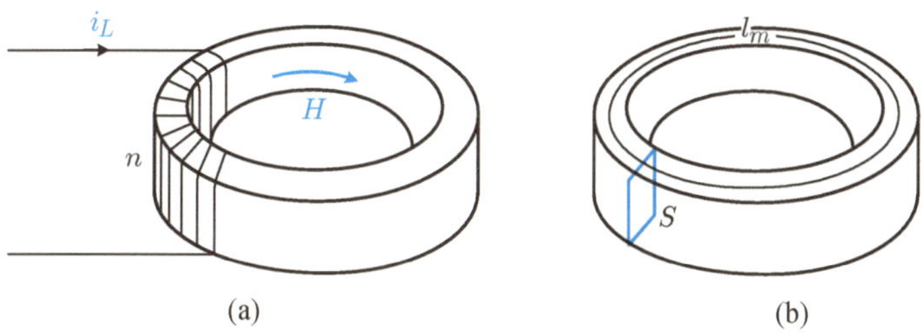

Figure 2.7 Toroidal inductor. (a) Inductor structure. (b) Geometry of toroidal core.

Using the definition of the inductance, Faraday's law is rewritten as

$$\boxed{v_L(t) = L\frac{\mathrm{d}i_L(t)}{\mathrm{d}t}} \tag{2.3}$$

thus establishing the $i - v$ relationship of the inductor. The polarity of the induced voltage is defined as shown in Fig. 2.6(b).

Equation (2.3) is integrated to yield an alternative circuit equation for the inductor

$$\boxed{i_L(t) = \frac{1}{L}\int v_L(t)\,\mathrm{d}t} \tag{2.4}$$

When a dc voltage V_S is applied across the inductor, the inductor current is given by

$$\boxed{i_L(t) = \frac{V_S}{L}t} \tag{2.5}$$

Expression (2.5) indicates that the inductor current continues to increase linearly without bounds, as long as the dc voltage is present. In practice, however, a real inductor cannot sustain an excessively large current and eventually becomes saturated. The saturation of an inductor will be discussed later in this section.

Inductance of Toroidal Inductor — EXAMPLE 2.1

This example evaluates the inductance of an inductor fabricated using a toroidal core, as depicted in Fig. 2.7. Referring to Fig. 2.7, the inductance is given by

$$L = \frac{\lambda(t)}{i_L(t)} = \frac{n\phi(t)}{i_L(t)} = \frac{nSB(t)}{i_L(t)} = \frac{\mu_r\mu_o nS}{i_L(t)}H(t) \tag{2.6}$$

where μ_r is the relative permeability of the core material, μ_o is the permeability of free space, n denotes the turns of the inductor winding, and S is the cross-sectional area of the core.

Ampere's law leads to the relationship

$$H(t)\,l_m = n\,i_L(t) \qquad \Rightarrow \qquad H(t) = \frac{n\,i_L(t)}{l_m} \tag{2.7}$$

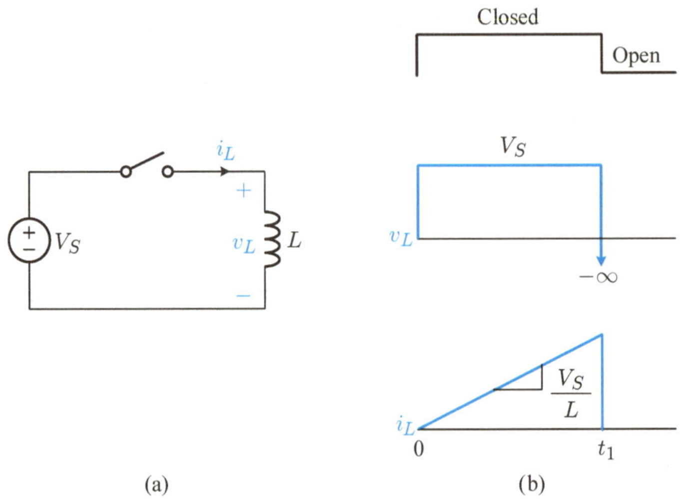

Figure 2.8 Inductive switching circuit. (a) Circuit diagram. (b) Circuit waveforms.

where l_m is the length of the magnetic path along the toroidal core. From (2.6) and (2.7), it follows that

$$L = \frac{\mu_r \mu_o n S}{i_L(t)} \frac{n\, i_L(t)}{l_m} \tag{2.8}$$

The final expression for the inductor is given by

$$\boxed{L = \mu_r \mu_o \frac{S}{l_m} n^2} \tag{2.9}$$

EXAMPLE 2.2 Inductive Switching Circuit

This example illustrates the behavior of the inductor excited by a pulse voltage waveform. Figure 2.8 shows a simple switching circuit, along with its circuit waveforms. Assume that the inductor is initially unenergized and the switch is opened. When the switch is closed at $t = 0$, a dc voltage V_S is applied across the inductor and i_L starts to increase linearly with a slope V_S/L. The switch is then opened at $t = t_1$, thereby forcing the inductor current to collapse. The change in the inductor current in turn induces a voltage across the inductor. Equation (2.3) indicates that a negative voltage, whose magnitude is proportional to the decaying slope of the inductor current, will be induced across the inductor.

When the switch is assumed to open instantaneously, the magnitude of the induced voltage is infinite. In practice, however, the magnitude of the negative voltage is finite because a real switch can only react with a finite response time. Even so, the magnitude is still large enough to destroy any practical semiconductor switch.

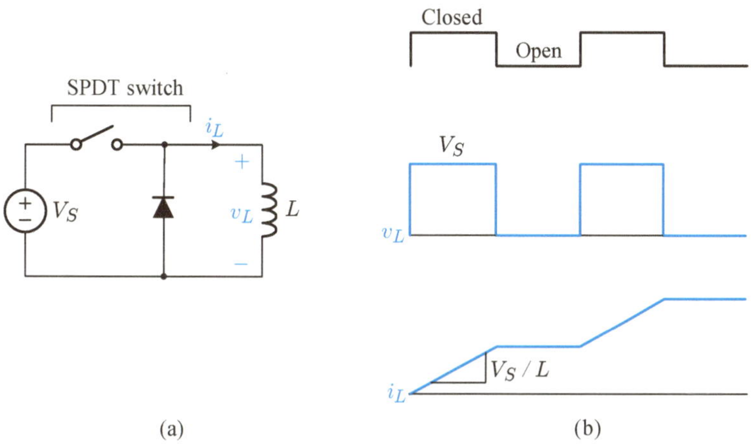

Figure 2.9 Inductive switching circuit. (a) Circuit diagram. (b) Circuit waveforms.

The behavior of the inductor can also be explained using the energy conservation principle. As the inductor current ramps up, the magnetic energy stored inside the inductor continues to increase. When the inductor current is collapsed at $t = t_1$, the total energy $E_m = 0.5Li_L^2(t_1)$ is instantaneously released in the shape of a voltage spike. As illustrated in this example, a sudden interruption of the inductor current incurs a *destructive* voltage spike and therefore should be avoided, unless the circuit is intentionally designed to operate so.

Inductive Switching Circuit — EXAMPLE 2.3

This example shows the operation of another inductive switching circuit. In the circuit in Fig. 2.9(a), a dc voltage source is connected to an inductor through an active-passive switch pair. The on/off status of the active switch is controlled by the switch drive signal shown in Fig. 2.9(b). As explained earlier, the active-passive switch pair functions as an SPDT switch. When the active switch turns on, the passive switch is turned-off; conversely, when the active switch turns off, the passive switch is turned-on. With the active switch turned-on, the dc voltage V_S is applied across the inductor L and the inductor current thus ramps up with a slope V_S/L.

When the active switch turns off, the diode is turned-on because the current-carrying inductor behaves as a current source. Once the diode is turned-on, the voltage across the inductor becomes zero and is clamped at zero thereafter. Referring to (2.3), the inductor current remains unchanged when its voltage stays constant; the inductor current thus maintains the value that has developed before the switch is turned-off. When the switch is turned-on in the next operational cycle, the inductor current again starts to increase. By repeating this process, the inductor current proceeds as depicted in Fig. 2.9(b).

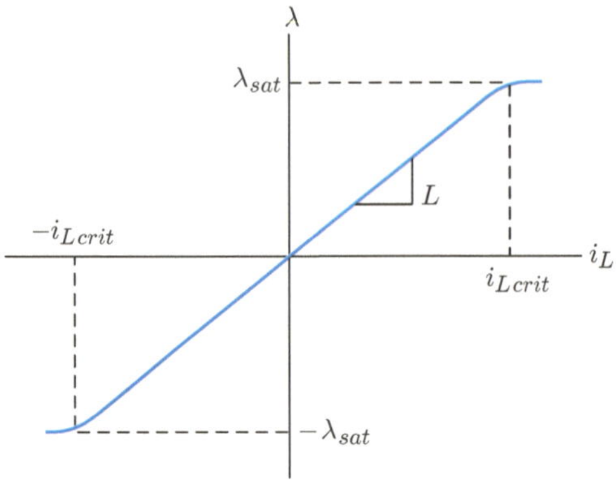

Figure 2.10 i_L–λ characteristics of inductor.

Saturation of Inductor

Inductors invariably *saturate* when they are forced to carry an excessively large current. Figure 2.10 illustrates the relationship between the inductor current i_L and magnetic flux linkage λ. As i_L is increased from zero, so is λ. However, the linear relationship between i_L and λ is valid only within a limited range. When the inductor current exceeds the critical values, $\pm i_{L\,crit}$ in Fig. 2.10, the magnetic flux linkage remains the same at $\pm \lambda_{sat}$, regardless of the magnitude of the inductor current; this phenomenon is referred to as the magnetic saturation.

Because the slope of the i_L–λ curve represents the inductance, the saturation implies a *zero* inductance for the inductor. Thus, the saturation causes a profound impact on the circuit operation. For example, when a dc voltage V_S is applied to an inductance L, the current increases towards $i_{L\,crit}$ by the equation $i_L(t) = (V_S/L)\,t$. When the inductor current reaches $i_{L\,crit}$, magnetic saturation occurs. Upon magnetic saturation, the inductor current shoots up boundlessly because the inductance becomes *null* at the instant of saturation. A real circuit cannot support such an excessive inductor current, and magnetic saturation therefore should be avoided to prevent a catastrophic failure of the circuit.

Magnetic saturation is attributed to the properties of the core material used for the inductor. Numerous magnetic dipoles† exist inside the core material. When no external current is present, the magnetic dipoles are randomly oriented. For this case, the core material does not exhibit any magnetic properties because the effects of individual magnetic dipoles are canceled by each other.

When the inductor current begins to flow, the magnetic field is developed inside the core and some magnetic dipoles start to align in parallel with the magnetic field, thereby generating additional magnetic flux linkage; this effect is known as the magnetic induction. As the current continues to grow, more magnetic dipoles

† Very small conceptual magnets that are assumed to exist inside magnetic materials.

align with the magnetic field, thereby increasing the magnetic flux linkage. When the inductor current reaches the critical value $i_{L\,crit}$, all the magnetic dipoles line up to the direction of the magnetic field and the magnetic flux linkage attains its maximum value, λ_{sat}. The core now becomes saturated and the magnetic flux linkage remains at λ_{sat} even if the inductor current is further increased. The inductance of the saturated inductor is zero because the slope of the $i_L - \lambda$ curve is zero.

Flux Linkage Balance Condition or Volt-Sec Balance Condition

In most switching circuits, inductors are operated in such a way that the magnetic flux linkage is increased during one part of a switching period and then decreased during the other part of the same switching period. The *flux linkage balance condition* asserts that the increase in the flux linkage, $\Delta\lambda_{inc}$, in one switching period should be equal to the decrease in the flux linkage, $\Delta\lambda_{dec}$, in that switching period

$$\boxed{\Delta\lambda_{inc} = \Delta\lambda_{dec}} \tag{2.10}$$

or equivalently the net change in the magnetic flux linkage over one switching period must be zero.

The justification for the flux linkage balance condition is self-explanatory. If the net magnetic flux linkage change within each switching period is *not* balanced at zero, the magnetic flux linkage will continue to grow towards the positive direction or negative direction, eventually encountering the magnetic saturation illustrated in Fig. 2.10. Therefore, all the inductors in a properly-designed inductive switching circuit should satisfy the flux linkage balance condition.

The flux linkage balance condition is paraphrased into an alternative form, which is more convenient for circuit analysis purposes. Faraday's law states that

$$v_L(t) = \frac{\mathrm{d}\lambda(t)}{\mathrm{d}t} \tag{2.11}$$

where v_L is the inductor winding voltage and λ denotes the magnetic flux linkage across the inductor winding. Equation (2.11) is rearranged as

$$\mathrm{d}\lambda(t) = v_L(t)\mathrm{d}t \tag{2.12}$$

to express the flux linkage change as a function of the inductor voltage. Now assume that the inductor is excited by a rectangular voltage waveform in Fig. 2.11. The positive voltage V_1 increases the magnetic flux linkage, whereas the negative voltage $-V_2$ decreases the flux linkage. The flux linkage increase during T_1 is given by

$$\Delta\lambda_{inc} = V_1 T_1 \tag{2.13}$$

and the flux linkage decrease during T_2 becomes

$$\Delta\lambda_{dec} = V_2 T_2 \tag{2.14}$$

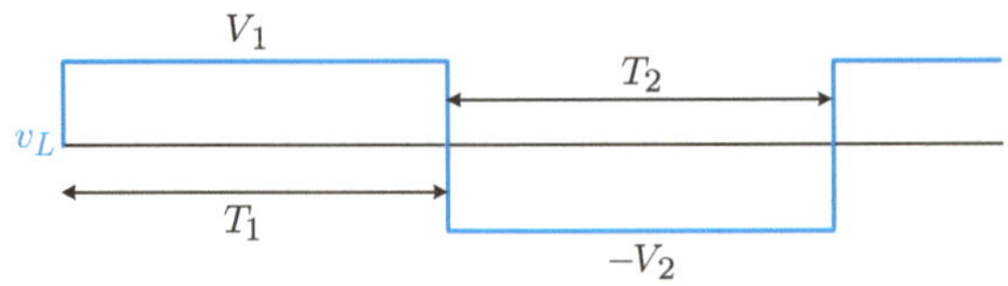

Figure 2.11 Volt-sec balance condition: $V_1T_1 = V_2T_2$.

By equating (2.13) and (2.14) based on the flux linkage balance condition, it follows that

$$\boxed{\Delta\lambda_{inc} = \Delta\lambda_{dec} \quad \Rightarrow \quad V_1T_1 = V_2T_2} \tag{2.15}$$

which states that the product of the voltage and time interval during which the inductor voltage is positive should be equal to that calculated when the inductor voltage is negative, or equivalently *the average value of the inductor voltage over one switching period must be zero.* This principle is referred to as the *volt-sec balance condition.* The volt-sec balance condition is generalized as follows. The average value of the inductor voltage calculated over integer multiples of the switching period is zero. Furthermore, the average value of the inductor voltage in general can be considered to be zero, assuming that the averaging is performed over a sufficiently longer period than the switching period.

A properly designed inductive switching circuit will establish the steady-state equilibrium where all the circuit variables are settled down into the values that satisfy the volt-sec balance condition on inductors. Thus, the volt-sec balance condition can be used as a circuit theorem in evaluating steady-state values of circuit waveforms.

Freewheeling Path and Freewheeling Diode

As illustrated in Example 2.2, the inductor current should *not* be discontinued abruptly because a sudden interruption of the inductor current generates a destructive high voltage spike. Therefore, the circuit path, which is connected to an inductor and to be open-circuited during the circuit operation, should always be accompanied by an alternative circuit path. The alternative path shall be operative only provided that the original inductor current path is broken. This alternative path for the inductor current is called the *freewheeling path.* The freewheeling path is usually constructed with a diode, which is normally turned-off and only becomes turned-on when the original inductor current path is disrupted. The diode employed to provide a freewheeling path is called a *freewheeling diode.*

EXAMPLE 2.4 Inductive Switching Circuit with Freewheeling Path

This example illustrates the operation of an inductive switching circuit with a freewheeling path. Figure 2.12 shows the circuit diagram and waveforms of an inductive switching circuit. If the freewheeling diode located in the middle of Fig. 2.12(a) is

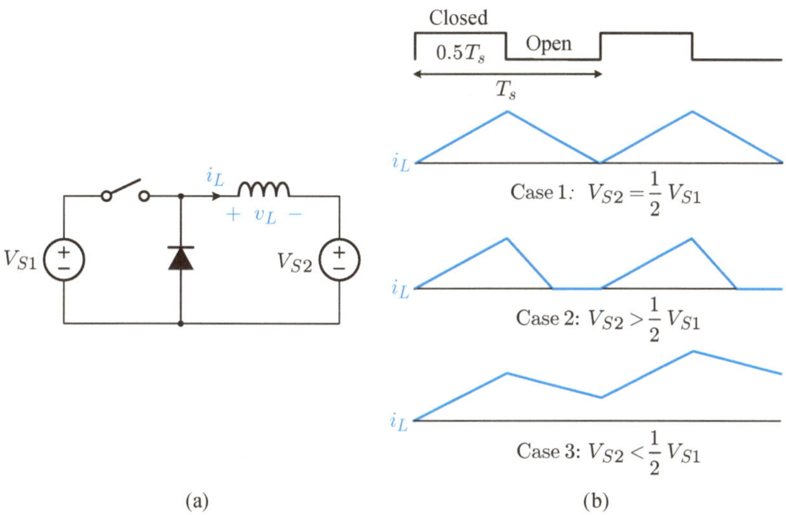

Figure 2.12 Inductive switching circuit with freewheeling path. (a) Circuit diagram. (b) Inductor current waveforms.

not present, a high voltage spike will be generated when the switch is opened. The freewheeling diode prevents such a voltage spike. When the switch is turned-off, the freewheeling diode conducts, thereby creating a freewheeling path for the inductor current.

When the switch is turned-on, the inductor current increases according to the equation

$$i_L(t) = \frac{v_L}{L}t = \frac{V_{S1} - V_{S2}}{L}t \tag{2.16}$$

On the other hand, when the switch is turned-off and the freewheeling diode conducts, the inductor current decreases as

$$i_L(t) = -\frac{V_{S2}}{L}t \tag{2.17}$$

Now, it is assumed that the active switch is turned-on for half the switching period and turned-off for the remaining switching period, as shown in Fig. 2.12(b). With this assumption, the following three different cases are considered:

- Case 1: $V_{S2} = \frac{1}{2}V_{S1}$
- Case 2: $V_{S2} > \frac{1}{2}V_{S1}$
- Case 3: $V_{S2} < \frac{1}{2}V_{S1}$

The inductor current waveforms i_L for these three cases are shown in Fig. 2.12(b). For Case 1, the increasing slope of the inductor current is identical to the decreasing slope, resulting in a periodic triangular waveform. For Case 2, the falling rate is faster

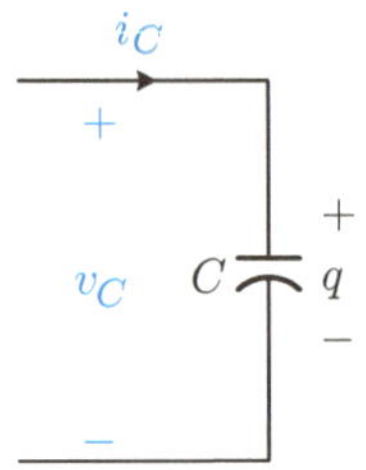

Figure 2.13 Capacitor and polarity/direction of terminal voltage/current waveforms.

than the rising rate, as such, the inductor current i_L is reduced to zero before the onset of the next operational period. When i_L is reduced to zero, the diode turns off and remains off for the remaining switching period, because the diode cannot carry the current in the reverse direction.

Finally, for Case 3, the rate of the inductor current falling is slower than the rate of the inductor current rising. At each switching period, the final value of the inductor current will be larger than the initial value. Accordingly, the inductor current will respond as shown in the bottom plot in Fig. 2.12(b). Case 3 violates the flux linkage balance condition and the circuit never reaches steady state. This circuit will eventually encounter inductor saturation and ensuing catastrophic failure. Readers are urged to sketch the inductor voltage waveforms for the three cases, in order to confirm the compliance with or violation of the volt-sec balance condition on the inductor, particularly to understand how Case 2 meets the volt-sec balance condition.

2.2.2 Capacitors

Along with inductors, capacitors are widely used in dc-to-dc power conversion circuits as an energy storage component. This section discusses the circuit properties of capacitors and investigates their operation under periodic excitations. The current section also presents important circuit theorems for capacitors.

Circuit Equations

Capacitors are typically fabricated by placing a pair of conductor plates in parallel and filling the gap with a dielectric material. When a voltage v_C appears across the parallel plates, the electric charge is accumulated inside the dielectric material, as shown in Fig. 2.13. The capacitance of a capacitor is defined as the ratio of the accumulated electric charge q to the voltage v_C across the capacitor

$$\boxed{C \equiv \frac{q(t)}{v_C(t)}} \tag{2.18}$$

On the other hand, the current through the capacitor is defined as

$$i_C(t) \equiv \frac{\mathrm{d}q(t)}{\mathrm{d}t} \tag{2.19}$$

Using the definition of the capacitance in (2.18), (2.19) is rearranged as

$$\boxed{i_C(t) = C\frac{\mathrm{d}v_C(t)}{\mathrm{d}t}} \tag{2.20}$$

leading to the $v-i$ relationship for capacitors. The polarity/direction of the voltage/current waveform associated with a capacitor is defined as shown in Fig. 2.13. Equation (2.20) is integrated to yield an alternative circuit equation for capacitors

$$\boxed{v_C(t) = \frac{1}{C}\int i_C(t)\,\mathrm{d}t} \tag{2.21}$$

When a capacitor is connected to a dc current I_S, the voltage across the capacitor rises linearly

$$\boxed{v_C(t) = \frac{I_S}{C}t} \tag{2.22}$$

Equation (2.22) indicates that the capacitor voltage continues to increase boundlessly at the presence of the dc current. However, a practical capacitor will be damaged or destroyed when the capacitor voltage is increased excessively.

Capacitive Switching Circuit — EXAMPLE 2.5

This example illustrates the operation of a simple capacitive switching circuit. Figure 2.14(a) shows a capacitive switching circuit where an ideal current source is connected to an uncharged capacitor. The switch in the middle of the circuit is initially closed. When the switch is opened at $t = 0$, the dc current I_S flows into the capacitance C, thereby raising the capacitor voltage as $v_C(t) = (I_S/C)t$. When the switch is closed at $t = t_1$, the capacitor voltage is forced to collapse. A sudden collapse of the capacitor voltage in turn induces an infinitely large current according to (2.20).

Certainly, a real semiconductor switch cannot carry such a large current and will be permanently damaged. In terms of the energy conservation principle, the energy accumulated in the capacitor at $t = t_1$, $E_e = 0.5Cv_C^2(t_1)$, is instantaneously released when the switch is closed. This instantaneous energy discharge will destroy the semiconductor switch.

Capacitive Switching Circuit — EXAMPLE 2.6

The operation of another capacitive switching circuit is shown here. Figure 2.15(a) shows a capacitive switching circuit in which the current source is connected to a capacitor through a diode. The switch in the middle of the circuit is initially closed.

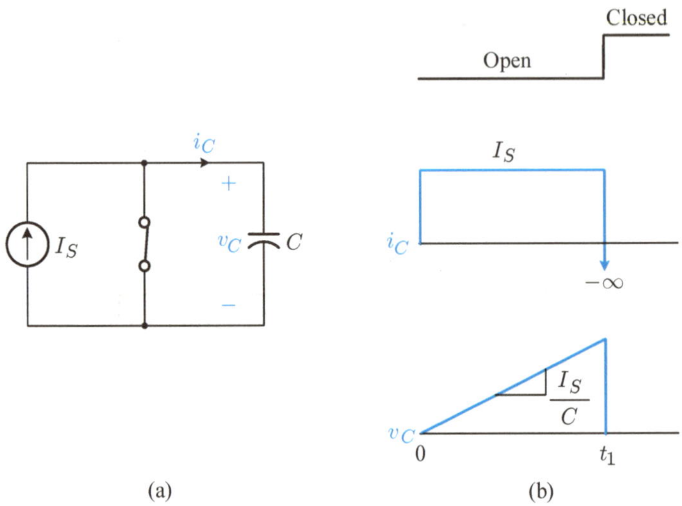

Figure 2.14 Capacitive switching circuit. (a) Circuit diagram. (b) Circuit waveforms.

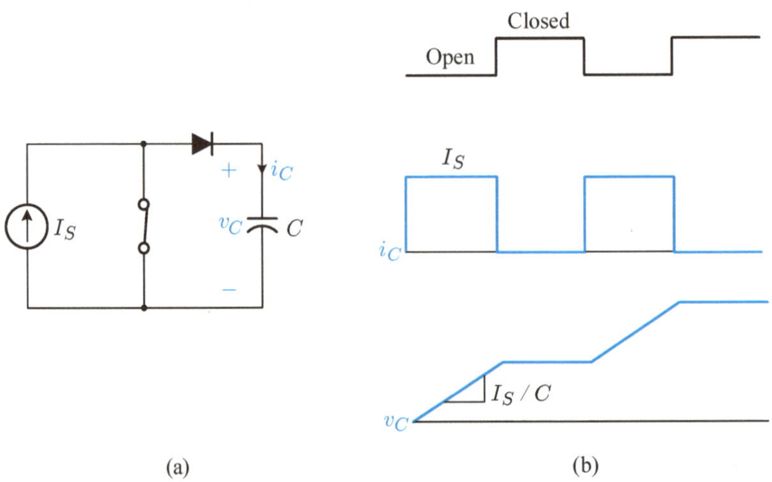

Figure 2.15 Capacitive switching circuit. (a) Circuit diagram. (b) Circuit waveforms.

When the switch is opened, the current source turns on the diode and charges the capacitor, thereby raising the capacitor voltage by the equation $v_C(t) = (I_S/C)t$. When the switch is closed, the diode is reverse-biased by the elevated capacitor voltage. The capacitor voltage remains constant because the capacitor is isolated from the current source by the turned-off diode. The diode, which isolates the charged capacitor from the current source, is called an *isolation diode*. When the switch is opened in the next switching period, the capacitor voltage starts to rise again. By repeating this process, the circuit produces the waveforms shown in Fig. 2.15(b).

Insulation Breakdown

As the capacitor voltage continues to rise, it eventually reaches the critical value that breaks the insulation of the dielectric material inside the capacitor. This insulation breakdown causes a permanent damage to the capacitor and application circuit. Accordingly, all capacitors are specified with the highest voltage they can sustain. The circuit shown in Fig. 2.15 is not practical because the capacitor voltage will eventually rise to the critical value that triggers the insulation to break down.

Charge Balance Condition or Amp-Sec Balance Condition

The *charge balance condition* states that the net change in the charge accumulation in a capacitor should be balanced at zero for each switching period

$$\boxed{\Delta q_{inc} = \Delta q_{dec}} \tag{2.23}$$

where Δq_{inc} is the charge increase during one switching period and Δq_{dec} is the charge decrease in that switching period. As with the case of the flux linkage balance condition, the charge balance condition is a prerequisite to assume steady-state operation for a capacitive switching circuit. Because the incremental charge in a capacitor, Δq, is given by the product of the capacitance C and incremental capacitor voltage, Δv_C

$$\Delta q = C\Delta v_C \tag{2.24}$$

the violation of the charge balance condition implies that the capacitor voltage will continue to rise in the positive or negative polarity until the capacitor encounters insulation breakdown. Because this is unacceptable to any application circuits, all capacitors should meet the charge balance condition.

The charge balance condition is converted into an alternative form using the definition of the current

$$i_C(t) = \frac{\mathrm{d}q(t)}{\mathrm{d}t} \tag{2.25}$$

where i_C is the capacitor current. Equation (2.25) is written as

$$\mathrm{d}q(t) = i_C(t)\mathrm{d}t \tag{2.26}$$

to describe the charge balance condition as a function of the capacitor current, i_C.

When a capacitor is excited by a periodic rectangular current shown in Fig. 2.16, the charge balance condition implies

$$\boxed{I_1T_1 = I_2T_2} \tag{2.27}$$

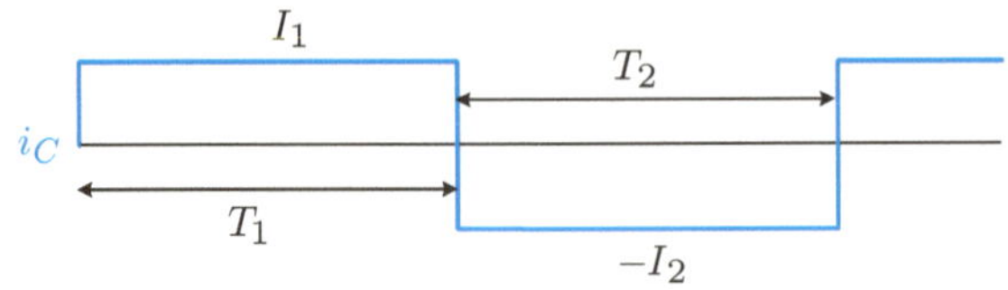

Figure 2.16 Amp-sec balance condition: $I_1T_1 = I_2T_2$.

because the positive current I_1 accumulates the electric charge, whereas the negative current $-I_2$ depletes the electric charge. Equation (2.27) is referred to as the *amp-sec balance condition* as it places a constraint on the product of the magnitude and the period of the capacitor current.

As a generalization of the charge balance condition, the average value of the capacitor current in general is considered zero. This generalized charge balance condition could simplify the analysis of capacitive switching circuits.

EXAMPLE 2.7 Capacitive Switching Circuit

This example illustrates the operation of another capacitive switch circuit. Figure 2.17(a) shows a capacitive switching circuit, consisting of two current sources, active switch, diode, and capacitor. As initial conditions, the capacitor is uncharged and the active switch is closed. Under this situation, I_{S1} flows through the active switch while I_{S2} circulates via the diode. For this case, the capacitor current i_C is zero, as shown in Fig. 2.17(a). Now, the circuit begins its operation by opening the active switch.

When the active switch is opened, I_{S1} flows into the capacitor while I_{S2} still runs through the diode, as shown in Fig. 2.17(b). Accordingly, the capacitor voltage increases linearly. When the active switch is closed, the diode is reverse-biased by the elevated capacitor voltage and I_{S2} now flows into the capacitor, in the *opposite* direction to I_{S1}, as shown in Fig. 2.17(c). Now it is assumed that the switch is periodically opened and closed, with an equal time length for the open and closed periods. The following three cases are considered:

- Case 1: $I_{S1} = I_{S2}$
- Case 2: $I_{S1} < I_{S2}$
- Case 3: $I_{S1} > I_{S2}$

The capacitor voltage waveforms v_C for these three cases are shown in Fig. 2.17(d). For Case 1, the capacitor voltage v_C becomes a symmetric triangular waveform with the identical rising and falling slopes. For Case 2, the falling slope is steeper than the rising slope. Thus, v_C reduces to zero before the onset of the next operational period. When v_C becomes zero, the diode is turned-on. Under this condition, I_{S2} diverts from the capacitor to the diode and v_C is thus held zero for the remaining operational period.

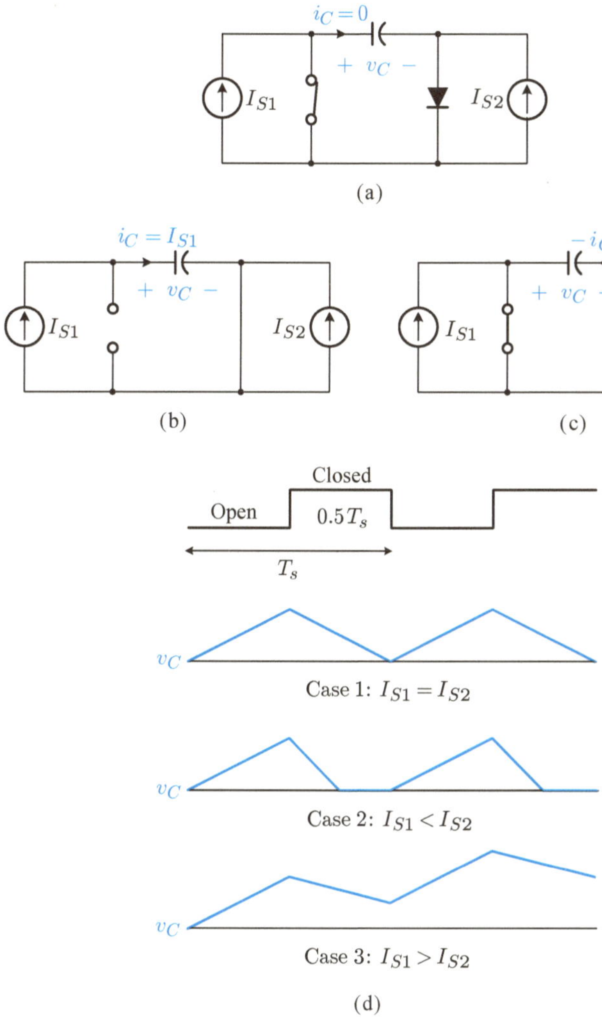

Figure 2.17 Capacitive switching circuit. (a) Circuit diagram. (b) Current flow with switch open. (c) Current flow with switch closed. (d) Capacitor voltage waveforms.

For Case 3, the rate of the capacitor voltage decreasing is slower than the rate of the capacitor voltage increasing. Accordingly, the capacitor voltage v_C will increase cycle-by-cycle, as illustrated in Fig. 2.17(d). In fact, Case 3 breaks the charge balance condition and the circuit never reaches steady state. It would be informative to sketch the capacitor current waveform i_C, in order to comprehend how each of the three cases does or does not meet the amp-sec balance condition.

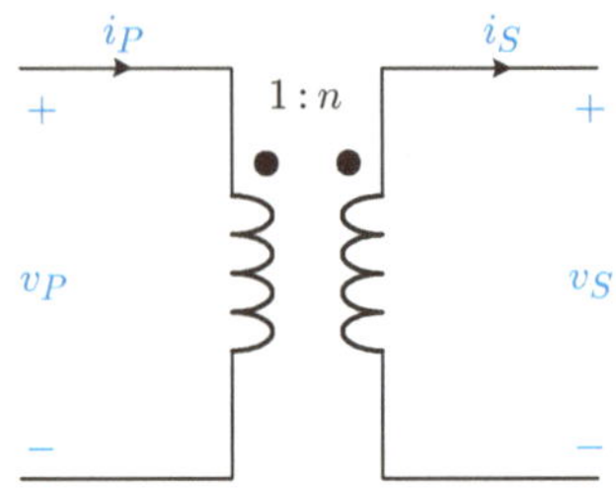

Figure 2.18 Symbol and polarity/direction of voltage/current waveforms of ideal transformer.

2.2.3 Transformers

Transformers are widely used in dc-to-dc power conversion circuits to change the levels of voltage and current waveforms while transferring electrical energy. Although the electro-magnetic process inside the transformer is rather complicated, the external circuit properties can readily be described by a circuit model. This section discusses the circuit model for practical transformers and illustrates its use.

First, the concept of the *ideal* transformer is discussed. Then, the electro-magnetic process of practical transformers is described using the equations of the ideal transformer and inductor, leading to the circuit model of practical transformers.

Ideal Transformer

The ideal transformer is a conceptual device that satisfies predefined relationships among its circuit variables. Figure 2.18 shows the symbol and polarity/direction of terminal voltage/current waveforms of an ideal transformer. The following terms are defined for the circuit variables and parameter associated with the ideal transformer:

- $v_P(t)$: primary voltage $i_P(t)$: primary current
- $v_S(t)$: secondary voltage $i_S(t)$: secondary current
- n: turns ratio

The symbol *dot* • in Fig. 2.18 is involved with the definitions for the polarity/direction of the primary and secondary voltage/current waveforms. In other words, the polarity and direction of the circuit variables are determined in reference to the position of •. Details about • will be given in the next section which deals with non-ideal practical transformers. The turns ratio n is the key parameter that defines the relationships between the terminal voltage and current waveforms of the ideal transformer

$$\boxed{v_S(t) = n\,v_P(t)} \tag{2.28}$$

$$\boxed{i_P(t) = n\,i_S(t)} \tag{2.29}$$

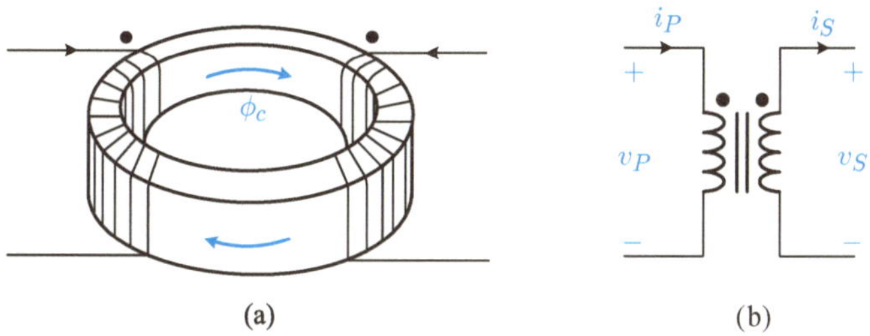

Figure 2.19 Structure, symbol, and polarity of practical transformer. (a) Structure and dot convention. (b) Symbol and polarity of practical transformer.

The ideal transformer is an imaginary device whose terminal circuit variables are governed by (2.28) and (2.29). For example, if a dc value $v_P = V_1$ is applied as the primary voltage, the secondary voltage is also a dc, given by $v_S = nV_1$. It can be inferred from (2.28) and (2.29) that, when one port of an ideal transformer is short-circuited or open-circuited, the other port becomes the same. Although defined as a conceptual device, the ideal transformer plays a critical role in characterizing the circuit properties of practical transformers. In fact, the ideal transformer is conceived as a means of describing the circuit properties of non-ideal practical transformers. Now, we start the discussions about practical transformers.

Practical Transformers

Practical transformers are fabricated by wrapping two or more copper windings around a magnetic core. Figure 2.19(a) shows a simplified structure of a two-winding transformer built on a toroidal magnetic core.† If one winding is referred to as the primary winding, the other winding becomes the secondary winding. To determine the terminal circuit characteristics of the practical transformer, it is necessary to know *how the primary and secondary windings are wound.* In other words, the polarity/direction of the terminal circuit waveforms depends on the internal pattern of the transformer windings. The following *dot convention* has been used to specify the winding pattern of transformers.

Dot Convention: For both primary and secondary windings, one end of the winding is marked with a *dot* • as a means of indicating the *polarity* of the transformer windings. The location of • is determined as follows. When the primary current flows into the *dotted* end of the winding and the secondary current also enters the dotted end of the winding, two magnetic fluxes, generated by the two currents running into the respective windings, are directed to be additive, as illustrated

† In practical transformers. the two windings are not separated as shown in Fig. 2.19(a). The two windings are alternatively and evenly wound all around the magnetic core, in order to maximize the magnetic coupling between them.

in Fig. 2.19(a). Namely, the primary and secondary currents flowing into the winding terminals marked • produce mutually additive magnetic flux. Readers are urged to examine Fig. 2.19(a) to confirm the dot convention, pattern of the transformer windings, and direction of the magnetic flux, all illustrated in the figure.

The operation of the transformer is based on the coupling between the primary winding and the secondary winding via the electro-magnetic induction. According to Lenz's law, the electro-magnetic induction occurs in such a way that the induced magnetic flux, produced as the outcome of the induction, opposes the first magnetic flux that initiated the induction process. In conjunction with the dot convention, Lenz's law is reiterated as follows. *When the initiating current enters the dotted end of the primary winding, the induced current should leave from the dotted end of the secondary winding.* The two magnetic fluxes, the initiating flux and the induced flux, then oppose each other, thereby complying with Lenz's law.

Figure 2.19(b) shows the symbol of a practical transformer together with the polarity/direction of its terminal voltage/current waveforms. The directions of the terminal currents are defined by Lenz's law and dot convention. The polarities of the winding voltages accord with the directions of the winding currents. The polarity/direction of the terminal circuit variables, along with the location of • in Fig. 2.19(b), is consistent with that of the ideal transformer in Fig. 2.18. The vertical bars between the primary and secondary windings signify the presence of a magnetic core in the practical transformer. The vertical bars also symbolically differentiate the practical transformer from the ideal transformer.

The polarity of the transformer can be defined in other way around. In other words, the position of • on each winding can be swapped, along with the reversal of the polarity/direction of voltage/current waveforms associated with each winding. This, of course, does not alter the circuit properties of the transformer.

Circuit Model for Practical Transformers

Unlike to the cases of inductors and capacitors, it is not straightforward to directly set up the circuit equations for practical transformers. One alternative approach is to develop a specific circuit model, which describes the electrical behavior of transformers using standard circuit components. In this section, we develop a circuit model for practical transformers by combining an inductor and ideal transformer. Once the circuit model is obtained, the equations for practical transformers can readily be written using the circuit equations of the inductor and ideal transformer.

A circuit model for practical transformers can be created based on Faraday's law, Ampere's law, and properties of the magnetic core. Figure 2.20 shows the configuration used in developing a circuit model for the practical transformer. To simplify the model derivation, a *perfect* magnetic coupling is presumed for the practical transformer. In other words, the common magnetic flux, ϕ_c in Fig. 2.20(a), entirely passes through both the primary and secondary windings without any leakage component.

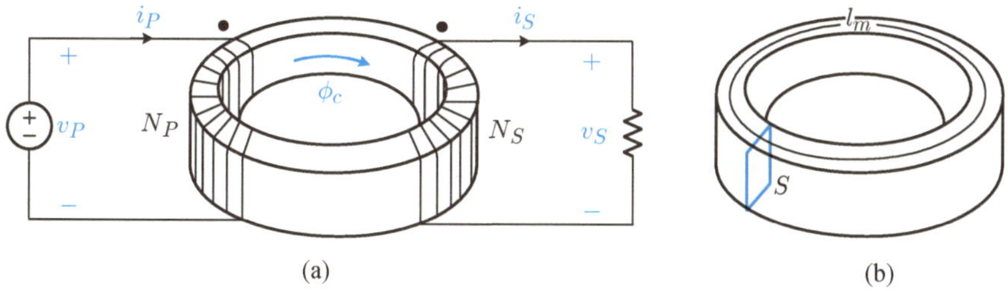

Figure 2.20 Simple circuit using practical transformer. (a) Circuit configuration. (b) Geometry of toroidal core.

According to Faraday's law, the terminal voltage of each winding is given by

$$v_P(t) = \frac{\mathrm{d}\lambda_P(t)}{\mathrm{d}t} = N_P \frac{\mathrm{d}\phi_c(t)}{\mathrm{d}t} \tag{2.30}$$

$$v_S(t) = \frac{\mathrm{d}\lambda_S(t)}{\mathrm{d}t} = N_S \frac{\mathrm{d}\phi_c(t)}{\mathrm{d}t} \tag{2.31}$$

where λ_P and λ_S represent the magnetic flux linkages across the primary and secondary windings due to the common magnetic flux ϕ_c, while N_P and N_S denote the number of the primary and secondary winding turns. From (2.30) and (2.31), the relationship between the terminal voltages is given as

$$\boxed{v_S(t) = \frac{N_S}{N_P} v_P(t)} \tag{2.32}$$

The connection between the terminal currents is formulated by applying Ampere's law along the magnetic path of the transformer

$$N_P i_P(t) - N_S i_S(t) = H_c l_m \tag{2.33}$$

where H_c is the magnetic field intensity associated with ϕ_c, and l_m is the length of the magnetic path. The $-$ sign in the second term of the left-hand side of (2.33) indicates that i_S is coming out of the • terminal while i_P is entering the • terminal. Equation (2.33) is rearranged as

$$\boxed{i_P(t) = \frac{N_S}{N_P} i_S(t) + \frac{H_c l_m}{N_P}} \tag{2.34}$$

The last term in (2.34) is referred to as the magnetizing current

$$i_m(t) \equiv \frac{H_c l_m}{N_P} \tag{2.35}$$

The magnetic flux linkage across the primary winding is given by

$$\lambda_P(t) = N_P\, \phi_c(t) = N_P\, \mu_r\, \mu_o H_c S \tag{2.36}$$

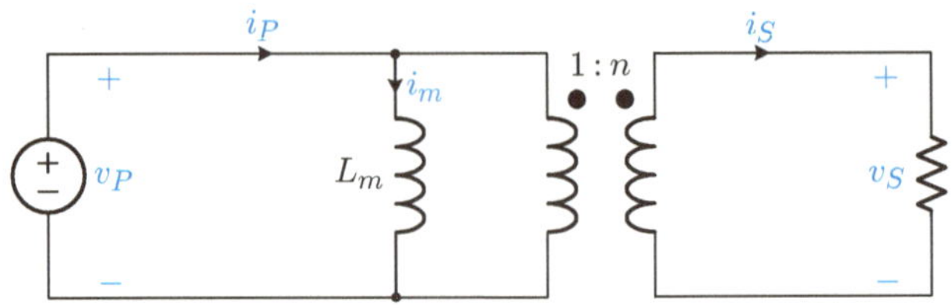

Figure 2.21 Equivalent circuit representation of Fig. 2.20.

with μ_r the relative permeability of the core material, μ_o the permeability of free space, and S the cross-sectional area of the core. The inductance, associated with the magnetizing current of (2.35) and magnetic flux linkage of (2.36), is now defined as

$$L_m = \frac{\lambda_P(t)}{i_m(t)} = \frac{N_P\,\mu_r\,\mu_o H_c S}{\dfrac{H_c l_m}{N_P}} = \mu_r\,\mu_o \frac{S}{l_m} N_P^2 \tag{2.37}$$

This inductance is termed as the magnetizing inductance because it is associated with the magnetizing current. Equation (2.37) indicates that the magnetizing inductance in fact corresponds to the inductance that is evaluated at the primary side of the transformer with the secondary winding removed or unaccounted for, as can be confirmed from Example 2.1.

The circuit behavior of Fig. 2.20, given by (2.32) and (2.34), is represented by the circuit model in which the practical transformer is replaced with an ideal transformer plus magnetizing inductance. Figure 2.21 shows such a circuit model, where the turns ratio of the ideal transformer is determined as

$$n = \frac{N_S}{N_P} \tag{2.38}$$

and the magnetizing inductance L_m is given by (2.37). The validity of Fig. 2.21 is confirmed by showing that the circuit model reproduces the original equations of (2.32) and (2.34). Referring to Fig. 2.21 and the circuit equation of the $1:n$ ideal transformer, the relationship between the terminal voltages is given as

$$v_S(t) = n v_P(t) \tag{2.39}$$

and the terminal currents are related as

$$i_P(t) = n i_S(t) + i_m(t) \tag{2.40}$$

The magnetizing current i_m is evaluated as

$$i_m(t) = \frac{\lambda_P(t)}{L_m} = \frac{N_P\,\mu_r\,\mu_o H_c S}{\mu_r\,\mu_o \dfrac{S}{l_m} N_P^2} = \frac{H_c l_m}{N_P} \tag{2.41}$$

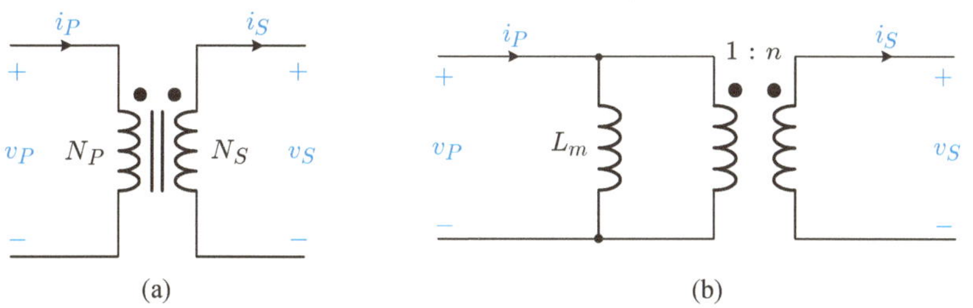

Figure 2.22 Symbol and circuit model for practical transformer. (a) Symbol. (b) Circuit model.

Now, the current equation (2.40) becomes

$$i_P(t) = \frac{N_S}{N_P} i_S(t) + \frac{H_c l_m}{N_P} \tag{2.42}$$

which is the original current equation of (2.34).

To summarize the model derivation, Fig. 2.22 shows the symbol and circuit model of the practical transformer. The turns ratio of the ideal transformer is given by

$$\boxed{n = \frac{N_S}{N_P}} \tag{2.43}$$

and the magnetizing inductance is determined as

$$\boxed{L_m = \mu_r \mu_o \frac{S}{l_m} N_P^2} \tag{2.44}$$

The magnetizing inductance represents the *non-ideal property* of practical transformers. Thus, it is good engineering practice to maximize the magnetizing inductance while complying with other design constraints. The larger the magnetizing inductance, the closer the practical transformer is to the ideal transformer. The magnetizing inductance becomes infinitely large when the permeability of the core material is assumed to be infinite. For such a case, the magnetizing inductance disappears and the circuit model reduces to the ideal transformer.

Simple Transformer Circuit — EXAMPLE 2.8

This example illustrates the operation of an ideal transformer in a simple circuit. Referring to Fig. 2.23, assume the permeability of the core material is infinite and the transformer thus behaves as an ideal transformer. A sinusoidal voltage $v_P(t) = \sin t$ is applied to the primary winding of the two winding transformer, whose second winding is terminated with a 0.5 Ω resistor. *From the structure and turns of the windings,* the secondary winding voltage is given by $v_S(t) = -2\sin t$ and the secondary current is $i_S(t) = -2\sin t/0.5 = -4\sin t$. The primary current becomes $i_P(t) = -2 \cdot i_S(t) = 2 \cdot 4\sin t = 8\sin t$.

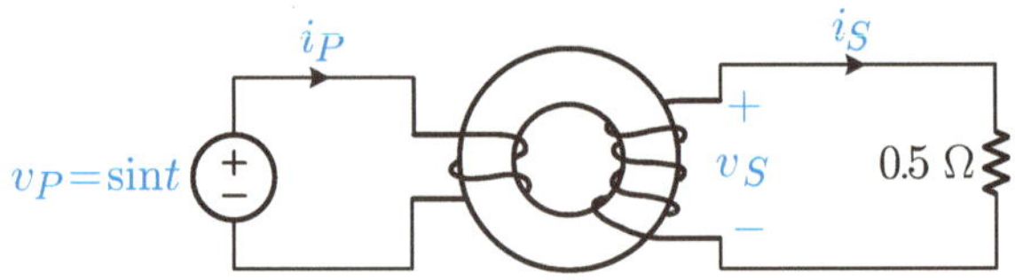

Figure 2.23 Simple transformer circuit.

Figure 2.22(b) is the simplest circuit model for practical transformers. The model becomes more complex when the imperfect magnetic coupling, parasitic circuit components, and detailed core properties are incorporated. However, the operations of most dc-to-dc power conversion circuits can duly be described using the transformer model in Fig. 2.22(b).

EXAMPLE 2.9 Simple Circuit with Practical Transformer

This example illustrates the use of the circuit model for practical transformers. Figure 2.24(a) shows a simple circuit in which a practical transformer is connected to a voltage source v_P and current source i_S. The waveforms for the voltage source v_P and current source i_S are shown in Fig. 2.24(a). The transformer is built using a toroidal core with $\mu_r = 5000$, $S = 1\ \text{cm}^2$, and $l_m = 4\pi\times10^{-1}$ cm. The turns of the primary winding is $N_P = 10$ and that of the secondary winding is $N_S = 20$. Figure 2.24(b) shows the circuit model of Fig. 2.24(a). The magnetizing inductance is evaluated as

$$\begin{aligned} L_m &= \mu_r\,\mu_o \frac{S}{l_m} N_P^2 \\ &= 5\times10^3\,4\pi\times10^{-7}\,\frac{10^{-4}}{4\pi\times10^{-3}}\,10^2 = 5\ \text{mH} \end{aligned}$$

and the turns ratio is

$$n = \frac{N_S}{N_P} = \frac{20}{10} = 2$$

The circuit variables are evaluated as

$$v_S(t) = n v_P(t) = 2\,v_P(t) \tag{2.45}$$

and

$$i_P(t) = n i_S(t) + i_m(t) = 2\,i_S(t) + i_m(t) \tag{2.46}$$

with

$$i_m(t) = \frac{1}{L_m}\int v_P(t)\,\text{d}t \tag{2.47}$$

The average value of i_m is assumed zero in this example. Figure 2.24(c) shows the steps of constructing v_S and i_P, based on the expressions (2.45) through (2.47) along with v_P and i_S waveforms given in Fig. 2.24(a).

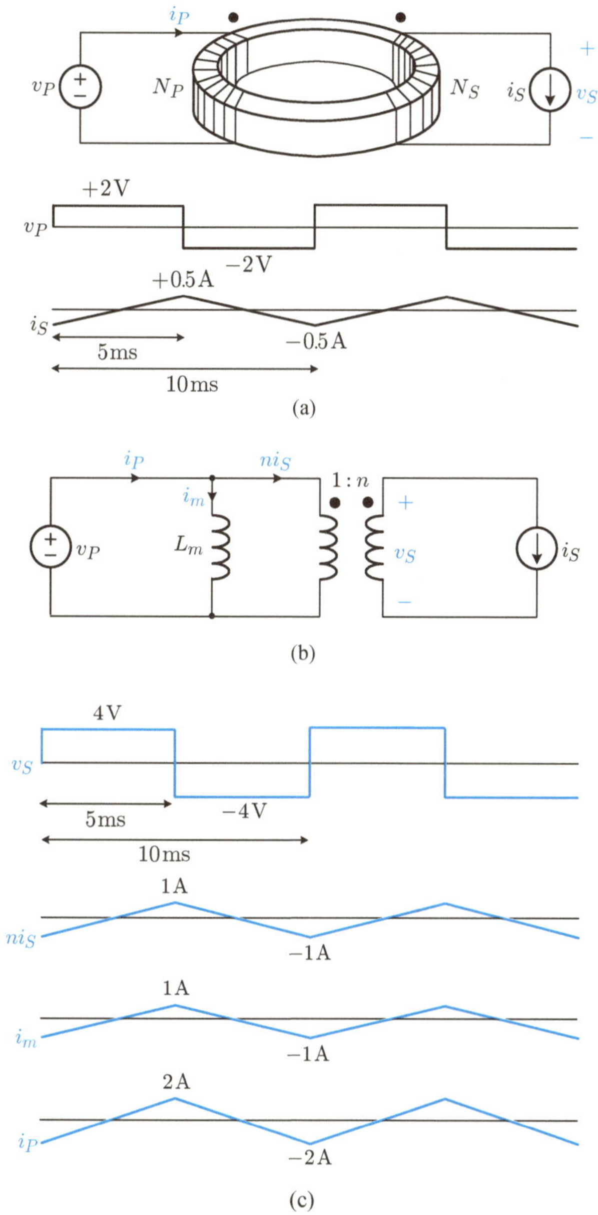

Figure 2.24 Simple transformer circuit. (a) Circuit configuration. (b) Circuit model. (c) Circuit waveforms.

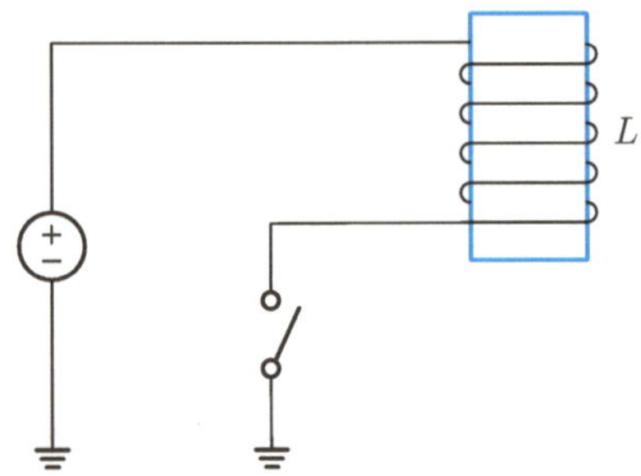

Figure 2.25 Conceptual solenoid drive scheme.

2.3 SWITCHING CIRCUITS IN PRACTICE

Switching circuits, constructed with the power stage components discussed in this chapter, have been used in many industrial and consumer electronics. In these applications, semiconductor switches and energy storage/transfer elements are appropriately combined to operate as a practical switching circuit. This section presents a couple of such examples.

2.3.1 Solenoid Drive Circuits

An inductor fabricated by winding a copper coil around an iron core is called a solenoid. The solenoid is often used as an actuator in industrial electronics. When a solenoid is connected to a voltage source by turning on a semiconductor switch, a linearly-increasing current is established in the solenoid inductance and energy is thus accumulated inside the solenoid. Although some energy is dissipated during the operation as an actuator, most of the accumulated energy will still remain in the solenoid after operation. The solenoid drive circuit must be properly designed so that the remnant energy is safely removed from the solenoid inductance without damaging the semiconductor switch. In this section, different solenoid drive circuits are analyzed focusing on their efficiency. To simplify ensuing discussions, the energy consumed during the actuator operation is assumed negligible and the solenoid is thus represented by a pure inductor.

A conceptual solenoid drive circuit is shown in Fig. 2.25. It becomes immediately apparent that this drive circuit is *not* workable due to the lack of the freewheeling path. When the switch is closed, the inductor current increases linearly, thereby storing energy in the solenoid. When the switch is opened, the inductor current suddenly loses its path and the stored energy is instantaneously released in an uncontrolled manner. This abrupt energy discharge, occurring in the form of a high voltage spike, would destroy the semiconductor switch. This section discusses two different solenoid drive circuits that operate without damaging the semiconductor switch.

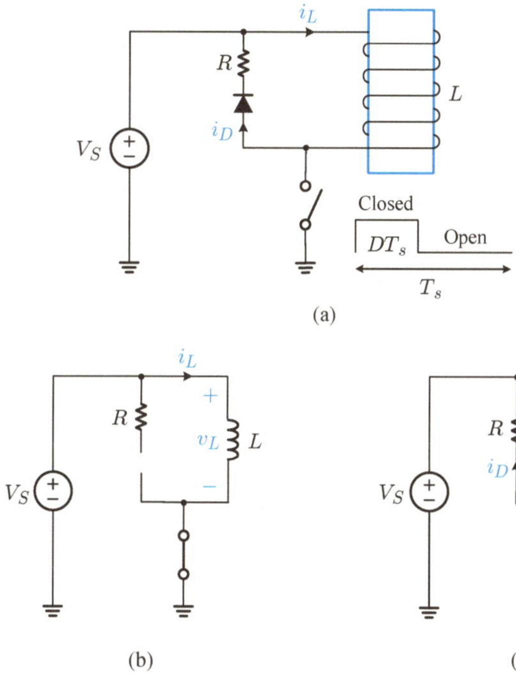

Figure 2.26 Dissipative solenoid drive circuit. (a) Circuit diagram. (b) Energy build-up period. (c) Energy removal period.

Dissipative Solenoid Drive Circuit

Figure 2.26(a) shows the first workable solenoid drive circuit where a freewheeling path, consisting of a diode and resistor, is employed in parallel with the solenoid. When the switch is closed, the diode is reverse-biased and does not affect the circuit operation. However, when the switch is opened, the diode conducts to create a freewheeling path for the inductor current. This prevents the instantaneous energy discharge and protects the switch. Detailed circuit operations are explained below.

Energy Build-up Period: When the switch is closed, the freewheeling diode is reverse-biased and the circuit is represented by Fig. 2.26(b). The inductor current increases linearly

$$i_L(t) = \frac{V_S}{L}t \tag{2.48}$$

thereby piling up energy in the solenoid inductance.

Energy Removal Period: When the switch is opened at $t = DT_s$,[†] the diode is turned-on and the freewheeling path is closed. The equivalent circuit in this period is shown in Fig. 2.26(c). The freewheeling current is given by

$$\boxed{i_D(t) = i_L(t) = i_L(DT_s)\, e^{-t/\tau}} \tag{2.49}$$

with $\tau = L/R$. As the freewheeling current circulates through the resistor and solenoid inductor, the energy stored in the solenoid inductance is gradually dissipated at the resistor. As the freewheeling current reduces to a negligible value, the stored energy is practically all removed by dissipation. Although the solenoid drive circuit functions properly without damaging the switch, efficiency of the drive circuit will be low because the stored energy is dissipated in the resistor. This solenoid drive circuit is referred to as the dissipative solenoid drive circuit.

EXAMPLE 2.10 Dissipative Solenoid Drive Circuit

In this example, the operation of the dissipative solenoid drive circuit is illustrated using PSpice®simulations. Figure 2.27(a) is the circuit diagram of a dissipative solenoid drive circuit. Figure 2.27(b) illustrates the major circuit waveforms with $V_S = 90\,\text{V}$, $L = 180\,\text{mH}$, $R = 20\,\Omega$, $T_s = 50\,\text{ms}$, and $D = 0.2$. When the switch is closed, the inductor current grows linearly and reaches the peak value at $t = DT_s = 0.2 \cdot 50 \times 10^{-3}$ s

$$i_{L\,peak} = \frac{V_S}{L} DT_s = \frac{90}{180 \times 10^{-3}}\, 0.2 \cdot 50 \times 10^{-3} = 5\ \text{A}$$

The total energy

$$E_m = \frac{1}{2} L i^2_{Lpeak} = \frac{1}{2} 180 \times 10^{-3}\, 5^2 = 2.25\ \text{J}$$

is stored in the inductor.

When the switch is opened, the energy removal period starts and the inductor current i_L freewheels through the loop created by the diode, resistor, and solenoid inductor. The diode current i_D decays exponentially while dissipating the energy at the resistor. When the stored energy is all dissipated in the resistor, i_D converges to zero. The operation of the circuit is repeated four times in Fig. 2.27. The waveforms of the source current i_S and inductor voltage v_L show details of the circuit operation.

Non-Dissipative Solenoid Drive Circuit

Efficiency of the solenoid drive circuit will be improved if the stored energy is recovered by the drive circuit, rather than wasted in the resistor. Figure 2.28(a) shows a non-dissipative solenoid drive circuit. The circuit employs a pair of synchronized

[†] D denotes the ratio of the switch-on period to the operational period and is called the duty ratio of the switch.

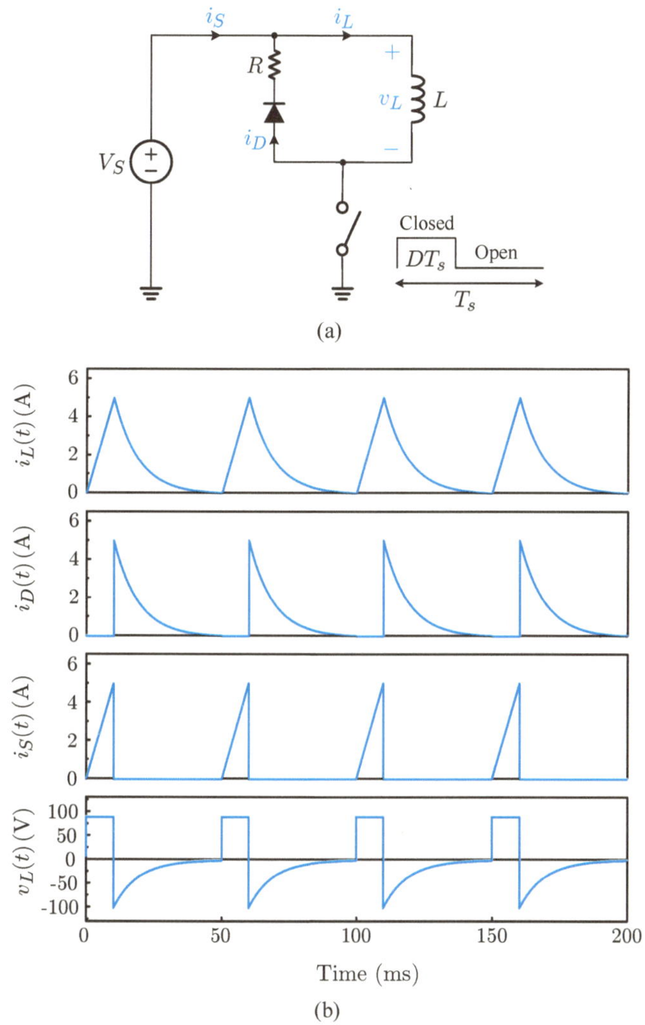

Figure 2.27 Dissipative solenoid drive circuit. (a) Circuit diagram. (b) Circuit waveforms.

switches and two diodes. The synchronized switches are turned-on and turned-off simultaneously, while the two diodes are used to create a freewheeling path for the inductor current. By turning on the two synchronized switches, a linearly-increasing inductor current is produced. When the synchronized switches are turned-off, the inductor current flows through the two diodes and the stored energy is transferred back to the voltage source. Details of the drive circuit are described as follows.

Energy Build-up Period: When the synchronized switches are closed, the two diodes are individually reverse-biased by V_S. The equivalent circuit in this period is

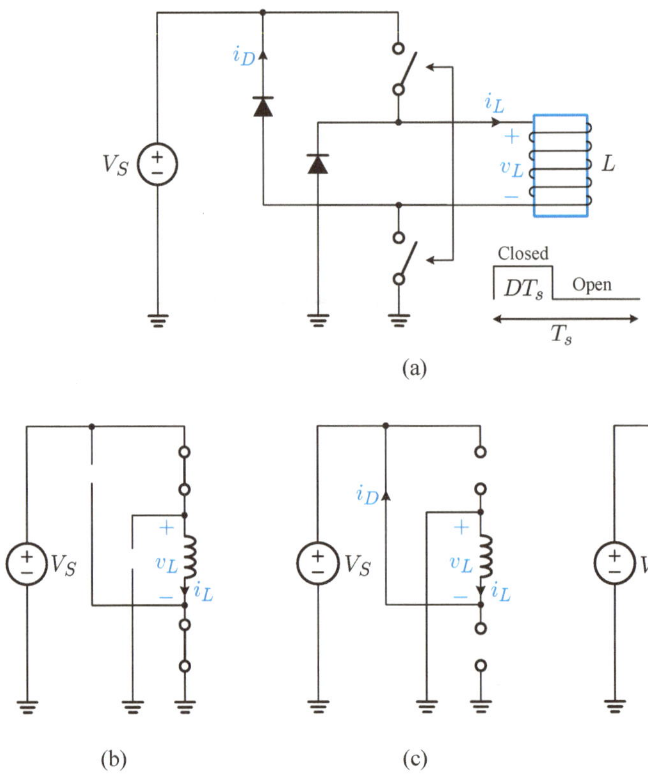

Figure 2.28 Non-dissipative solenoid drive circuit. (a) Circuit diagram. (b) Energy build-up period. (c) Energy recovery period. (d) Recess period.

shown in Fig. 2.28(b). During this period, the solenoid inductor current increases linearly

$$\boxed{i_L(t) = \frac{V_S}{L}t} \tag{2.50}$$

and energy is delivered from the source to solenoid.

Energy Recovery Period: When the switches are opened at $t = DT_s$, the two diodes conduct simultaneously and a freewheeling path is created as shown in Fig. 2.28(c). The solenoid voltage now becomes $-V_S$ and the freewheeling current ramps down from the peak with a slope $-V_S/L$

$$\boxed{i_D(t) = i_L(t) = i_L(DT_s) - \frac{V_S}{L}t} \tag{2.51}$$

During this period, the freewheeling current runs into the voltage source and the energy stored in the solenoid is transferred back to the voltage source. The stored energy is thus recovered by the drive circuit, rather than dissipated in the circuit.

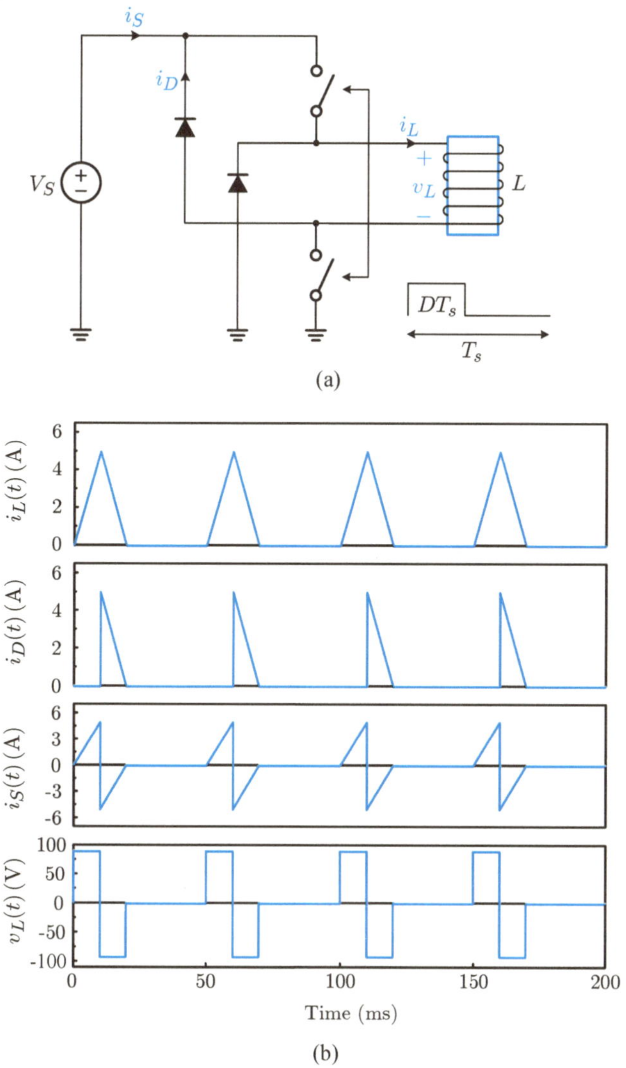

Figure 2.29 Non-dissipative solenoid drive circuit. (a) Circuit diagram. (b) Circuit waveforms.

Recess Period: When the freewheeling current is reduced to zero, the two diodes are turned-off and remain off thereafter. Because the solenoid is isolated from the source, all the circuit variables are zero, as shown in Fig. 2.28(d).

EXAMPLE 2.11 Non-Dissipative Solenoid Drive Circuit

This example illustrates the operation of the non-dissipative solenoid drive circuit. Figure 2.29 shows the circuit diagram and simulated waveforms of a non-dissipative solenoid drive circuit. The operational conditions and circuit parameters are $V_S = 90\,\text{V}$, $L = 180\,\text{mH}$, $T_s = 50\,\text{ms}$, and $D = 0.2$.

Referring to Fig. 2.29(b), the circuit operation is explained as follows. During the energy build-up period, the circuit waveforms are the same as the previous dissipative drive circuit. However, during the energy recovery period, the inductor current i_L flows back to the source through the freewheeling diodes, thereby returning the accumulated energy to the source. The source current i_S becomes negative in this period: $i_S = -i_D = -i_L$. The negative current signifies the retrieval of the energy transferred to the solenoid drive circuit during the energy build-up period.

2.3.2 Capacitor Charging Circuit

As another example of industrial applications, Fig. 2.30(a) shows a switching circuit consisting of a practical transformer, active switch, diode, and capacitor. Here, the transformer windings are arranged such that • appears at the opposite terminals of the primary and secondary windings. This circuit is capable of charging the capacitor to a desired voltage level. A simple application will be a high voltage generator, which is used in flash lamp driving circuits for hand-held cameras. Figure 2.30(b) shows the circuit model of Fig. 2.30(a), where the practical transformer is replaced with a $1:n$ ideal transformer and magnetizing inductance L_m. The operation of the capacitor charging circuit is explained using the circuit model in Fig. 2.30(b).

Energy Build-up Period: When the switch is closed, the energy build-up period starts. The circuit model for this period is shown in Fig. 2.31(a). The source voltage V_S is applied across the magnetizing inductance and primary winding of the ideal transformer. Due to the polarity of the transformer windings, the diode is turned-off by the reverse voltage of $nV_S + v_C$, where v_C is the voltage across the capacitor. A linearly-increasing current flows through the magnetizing inductance and active switch

$$\boxed{i_m(t) = i_Q(t) = \frac{V_S}{L_m}t} \qquad (2.52)$$

In this period, the primary winding does not carry any current because the secondary winding is open-circuited; for this case, the primary and secondary winding currents are both zero to meet the current equation of the $1:n$ ideal transformer. With the increasing magnetizing current, energy is piled up in the magnetizing inductance of the transformer.

Energy Transfer Period: When the switch is opened at $t = DT_s$, the energy transfer period begins. As shown in Fig. 2.31(b), the magnetizing current now diverts from the active switch to the primary winding of the ideal transformer, and

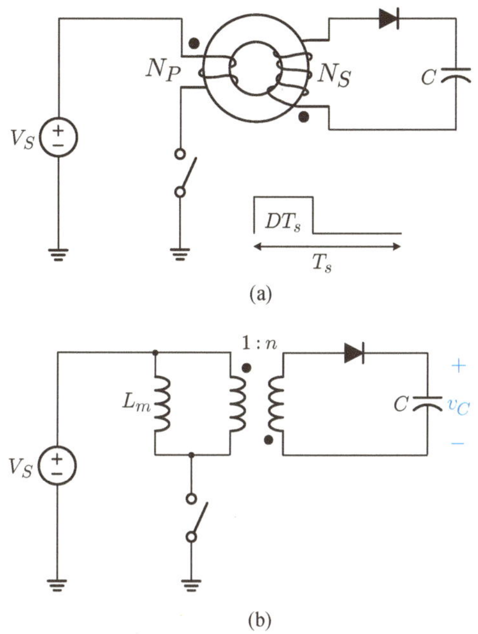

Figure 2.30 Capacitor charging circuit and its circuit model. (a) Capacitor charging circuit. (b) Circuit model: L_m is the magnetizing inductance and $n = N_S/N_P$ is the turns ratio of transformer.

circulates through the loop formed by the magnetizing inductance and the primary winding. This current in turn forces the diode to conduct, thus complying with the current equation of the $1 : n$ ideal transformer. Under this situation, the capacitor voltage is reflected to the primary winding and applied across the magnetizing inductance in the *negative* polarity. The magnetizing current now declines as

$$i_m(t) = -\frac{1}{L_m}\int \frac{1}{n} v_C(t)\,\mathrm{d}t \tag{2.53}$$

As the magnetizing current i_m continues to decrease, the accumulated energy is transferred to the capacitor, thereby increasing the capacitor voltage. The diode current is determined by the circuit equation of the $1 : n$ ideal transformer

$$i_D(t) = \frac{1}{n} i_m(t) \tag{2.54}$$

Recess Period: When i_m is reduced to zero, so is the diode current and the accumulated energy is completely transferred to the capacitor. Now, the diode is turned-off and the capacitor voltage is held constant. The circuit model in this recess period is shown in Fig. 2.31(c).

By repeating the above operational cycle, the charge storage in the capacitor will be gradually increased, along with the elevation in the capacitor voltage.

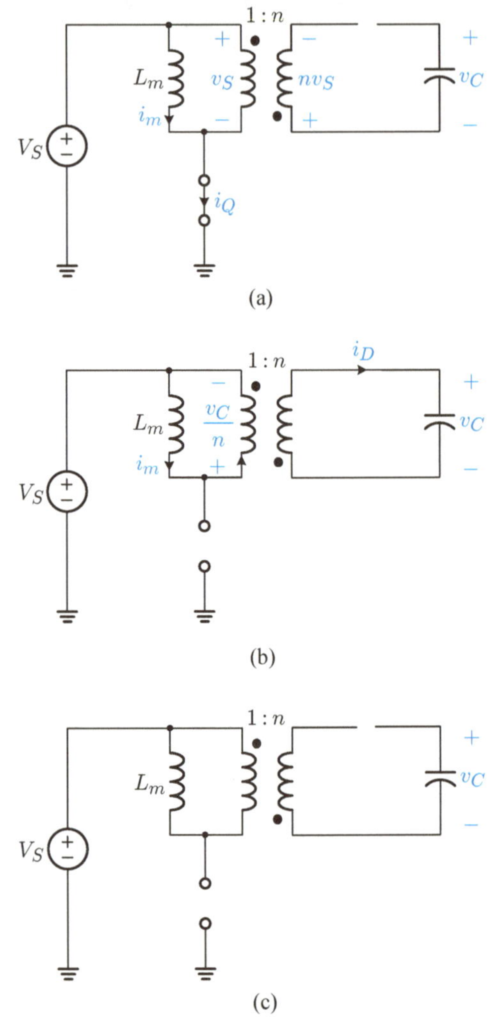

Figure 2.31 Operation of capacitor charging circuit. (a) Energy build-up period. (b) Energy transfer period. (c) Recess period.

EXAMPLE 2.12 Capacitor Charging Circuit

Operational details of the capacitor charging circuit are illustrated in this example. Figure 2.32 shows the circuit diagram and simulated waveforms of a capacitor charging circuit with $V_S = 24$ V, $L_m = 120\ \mu$H, $C = 5\ \mu$F, $n = 1$, $T_s = 100\ \mu$s, and $D = 0.4$.

Figure 2.32(b) shows the major waveforms of the capacitor charging circuit. The circuit operation is explained as follows. The magnetizing current i_m, which is iden-

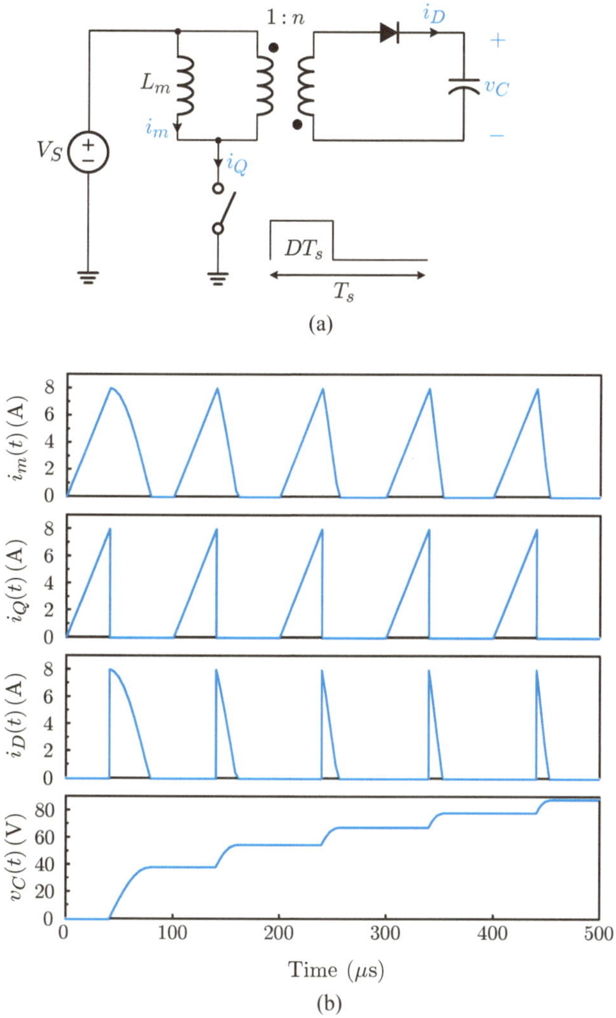

Figure 2.32 Capacitor charging circuit. (a) Circuit diagram. (b) Circuit waveforms.

tical to the switch current i_Q during the energy build-up period, increases up to the peak value

$$i_{m\,peak} = \frac{V_S}{L_m} DT_s = \frac{24}{120 \times 10^{-6}} 0.4 \cdot 100 \times 10^{-6} = 8 \text{ A}$$

thereby accumulating the total energy

$$E_m = \frac{1}{2} L_m i_{m\,peak}^2 = \frac{1}{2} 120 \times 10^{-6}\, 8^2 = 3.84 \text{ mJ} \tag{2.55}$$

in the transformer. This energy is delivered to the capacitor by the diode current during the energy transfer period. Thus, E_m represents the amount of the energy transfer from the voltage source to capacitor during one operational period. The capacitor voltage v_C continues to increase in the period in which the diode current i_D exists. When i_D reduces to zero, the recess period starts and v_C is held constant until the next energy transfer period.

Figure 2.32(b) reveals important features of the circuit operation. First, although the magnetizing current i_m increases linearly during the energy build-up period, it decays in a nonlinear fashion in the energy transfer period. The voltage across the magnetizing inductance is a constant V_S during the energy build-up period. However, in the energy transfer period, the time-varying capacitor voltage is reflected through the ideal transformer and applied to the magnetizing inductor in the negative polarity. Accordingly, the magnetizing current declines in a nonlinear manner.

Second, the duration of the energy transfer period becomes shorter as the operational cycle proceeds. The decaying slope of the magnetizing current is proportional to the magnitude of the capacitor voltage, which increases cycle-by-cycle. Consequently, the rate of the energy transfer becomes progressively faster, resulting in a continuous reduction in the energy transfer period.

Finally, the incremental growth in the capacitor voltage becomes smaller as the operational cycle proceeds. For each operational period, a fixed amount of energy is transferred to the capacitor, resulting in the cycle-by-cycle growth in the capacitor voltage. Because the energy stored in the capacitor is a quadratic function of the capacitor voltage, the increase in the capacitor voltage becomes smaller in proportion to the initial voltage at each operational cycle. More precisely, the following relationship is written from the circuit operation

$$\frac{1}{2}C(v_C(k-1)+\Delta v_C(k))^2 - \frac{1}{2}Cv_C^2(k-1) = E_m \tag{2.56}$$

where $v_C(k-1)$ is the initial capacitor voltage at the kth period and $\Delta v_C(k)$ is the increase in the capacitor voltage during the kth period, and E_m is the amount of the energy transfer in one operational period. From (2.56), the capacitor voltage increase in Fig. 2.32(b) is given by

$$\Delta v_C(k) = \sqrt{v_C^2(k-1) + \frac{2E_m}{C}} - v_C(k-1) \tag{2.57}$$

which is a decreasing function of the initial capacitor voltage $v_C(k-1)$.

The capacitor voltage at the end of the kth operational period, $v_C(kT_s)$, is found from the energy balance relationship

$$kE_m = \frac{1}{2}Cv_C^2(kT_s) \quad \Rightarrow \quad v_C(kT_s) = \sqrt{\frac{2E_m}{C}k} \tag{2.58}$$

where E_m given by (2.55). For example, the capacitor voltage at the end of the 5th operational period is given by

$$v_C(5T_s) = v_C(500\,\mu\text{s}) = \sqrt{\frac{2\cdot 3.84\times 10^{-3}}{5\times 10^{-6}}5} = 87.6\text{ V}$$

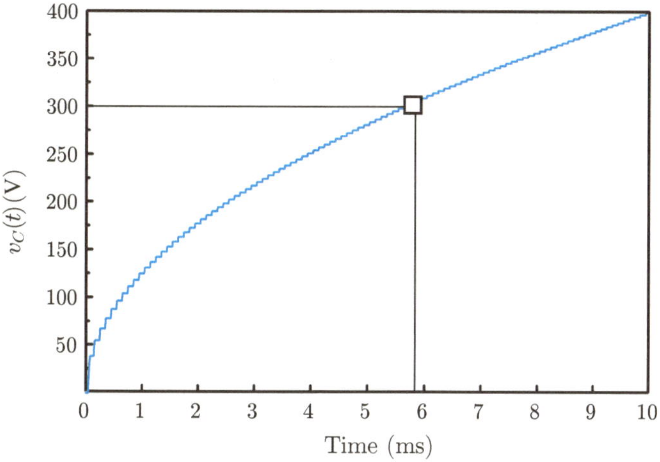

Figure 2.33 Capacitor voltage waveform.

Figure 2.33 shows the capacitor voltage v_C for a longer time period. The time interval required to charge the capacitor at a fixed V_C is determined as follows. First, the required number of the charing operation, k, is found from

$$\frac{1}{2}CV_C^2 = kE_m \quad \Rightarrow \quad k = \frac{1}{2}CV_C^2\frac{1}{E_m} \tag{2.59}$$

Now the total time required for charging V_C is given by

$$T_{charge} = kT_s = \underbrace{\frac{1}{2}CV_C^2\frac{1}{E_m}}_{k}T_s \tag{2.60}$$

where T_s is the operational period. As an example, the time needed to charge $V_C =$ 300 V is given by

$$\begin{aligned} T_{charge} &= \frac{1}{2}CV_C^2\frac{1}{E_m}T_s \\ &= \frac{1}{2}5\times10^{-6}\,300^2\frac{1}{3.84\times10^{-3}}100\times10^{-6} \\ &= 5.86 \text{ ms} \end{aligned}$$

Figure 2.33 shows a close agreement with this prediction.

2.4 SUMMARY

This chapter investigated the circuit behavior of power stage circuit components. Operations of semiconductor switches, inductors, capacitors, and transformers were presented with several examples. One critical functional component in dc-to-dc power conversion circuits is the single-pole double-throw (SPDT) switch, which toggles its connection periodically. In all the switching circuits covered in this chapter and other dc-to-dc conversion circuits to be studied in later chapters, the SPDT switch is implemented using the MOSFET-diode pair. In particular, in the capacitor charging circuit in Section 2.3.2, the MOSFET and diode collectively function as an SPDT switch even though they are physically separated by the transformer.

The energy storage and transfer devices include inductors, capacitors, and transformers. The freewheeling path and flux linkage balance condition are critical to understanding operations of inductive switching circuits. For capacitive switching circuits, the charge balance condition can be used to analyze the circuit waveforms. The practical transformer is modeled as a combination of the ideal transformer and magnetizing inductance. The operation of the practical transformer is then analyzed using the circuit equations of the ideal transformer and inductor.

This chapter analyzed two practical switching circuits – the solenoid drive circuit and capacitor charging circuit. Computer simulations are given to illustrate the operations of the switching circuits. PSpice®codes for these simulations are available at `http://pwmconverters.bchoi`, along with those of other forthcoming simulation examples.

REFERENCES

1. R. D. Middlebrook and S. Ćuk, *Advances in Switch-Mode Power Conversion*, TESLAco, Pasadena, CA, 1983.
2. D. W. Hart, *Power Electronics*, McGraw-Hill, New York, NY, 2011.

PROBLEMS

2.1 Assume that the inductor and capacitor in Fig. P2.1 are initially unenergized.

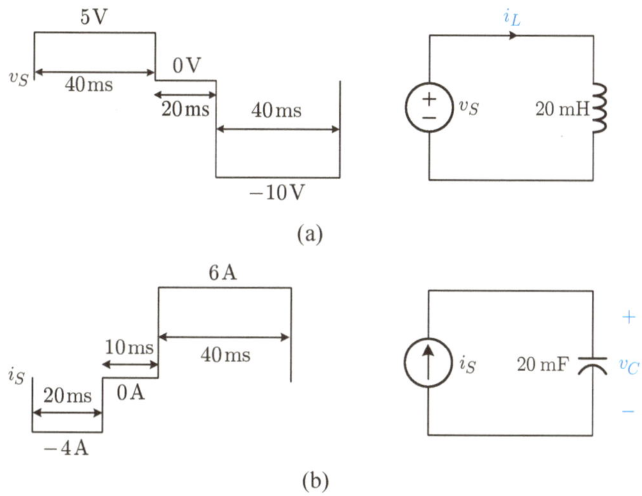

Fig. P2.1

a) For Circuit (a), sketch the inductor current i_L for $0 < t < 100$ ms.
b) For Circuit (b), sketch the capacitor voltage v_C for $0 < t < 70$ ms.

2.2* A switch drive signal shown in Fig. P2.2 is commonly applied to the following three circuits.

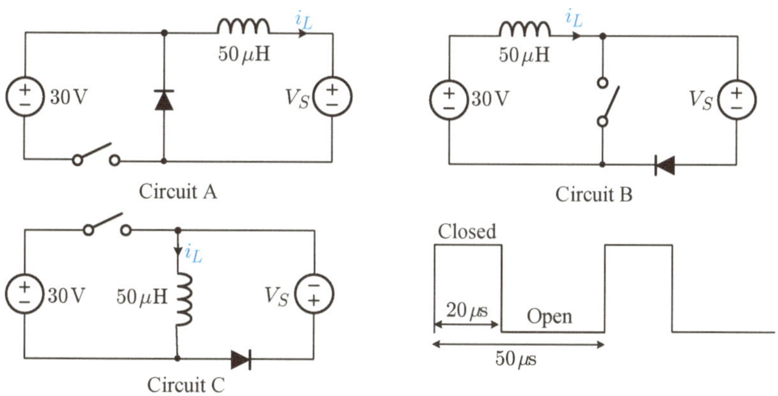

Fig. P2.2

a) For each circuit, determine the magnitude of V_S such that the inductor current i_L settles into a continuous periodic triangular waveform.

b) Assume the inductor is initially unenergized. Using the V_S values found in a), sketch i_L of each circuit for the two operational periods. Show the maximum value of i_L.

2.3* Shown in Fig. P2.3 are four different switching circuits along with their respective inductor voltage or capacitor current waveform.

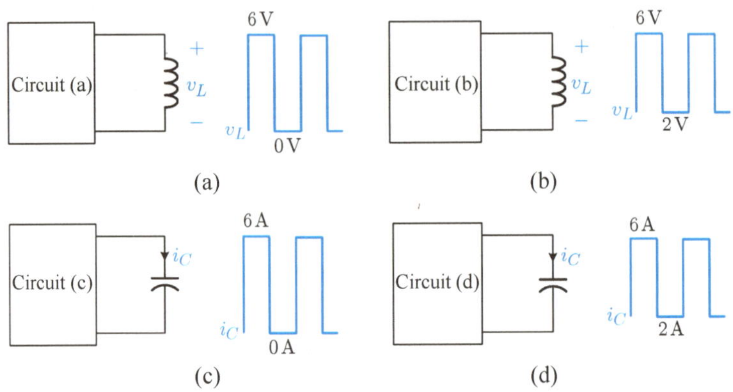

Fig. P2.3

a) Construct Circuit (a) using a 4 V voltage source, 2 V voltage source, active switch, and diode so that the circuit generates the given inductor voltage waveform v_L.
b) Build Circuit (b) using the same circuit components as those of a) so that the circuit produces the given inductor voltage waveform v_L.
c) Construct Circuit (c) using a 4 A current source, 2 A current source, active switch, and diode so that the circuit generates the given capacitor current waveform i_C.
d) Build Circuit (d) using the same circuit components as those of c) so that the circuit produces the given capacitor current waveform i_C.

2.4* Figure P2.4 shows a switching circuit, along with the switch drive signal and i_L–λ curve of the inductor L.
a) Find the inductance of L.
b) Assume $V_X = 16$ V and sketch i_L for the first three operational periods. Show the peak value of i_L.
c) Find the time instant at which the inductor would saturate.
d) Find the minimum value for the V_X that avoids the inductor saturation.

2.5* Consider the circuits shown in Fig. P2.5 and answer the questions.
a) For Circuit (a), assume that the inductor is initially unenergized. Sketch the i_L waveform for the first two operational periods for the following two cases.

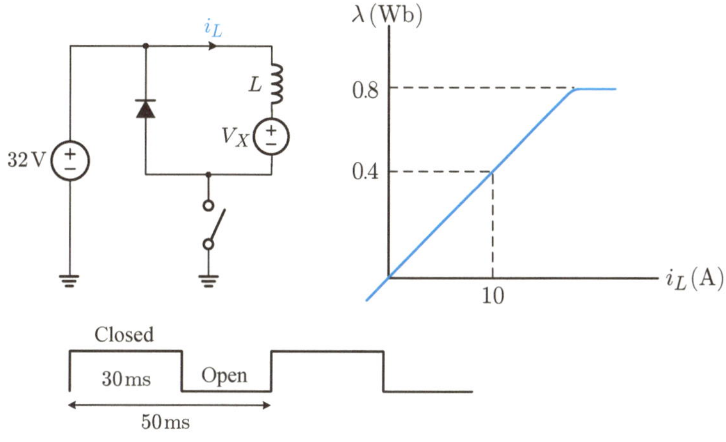

Fig. P2.4

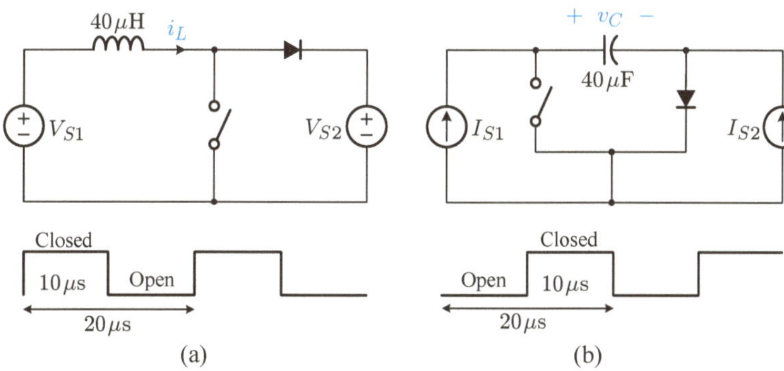

Fig. P2.5

i) $V_{S1} = 4$ V and $V_{S2} = 6$ V
ii) $V_{S1} = 2$ V and $V_{S2} = 6$ V
Show the maximum and minimum values of i_L.

b) Assume that the capacitor is initially uncharged in Circuit (b). Draw the v_C waveform for the first two operational periods for the following two cases.
i) $I_{S1} = 4$ A and $I_{S2} = 2$ A
ii) $I_{S1} = 2$ A and $I_{S2} = 4$ A
Show the maximum and minimum values of v_C.

2.6 In the circuit in Fig. P2.6, the SPDT switch holds **x** position for DT_s and holds **y** position for $(1 - D)T_s$.

a) For Circuits A, B, C, find the expression of v_2/v_1.
b) For Circuit D, find the relationship of i_2/i_1.

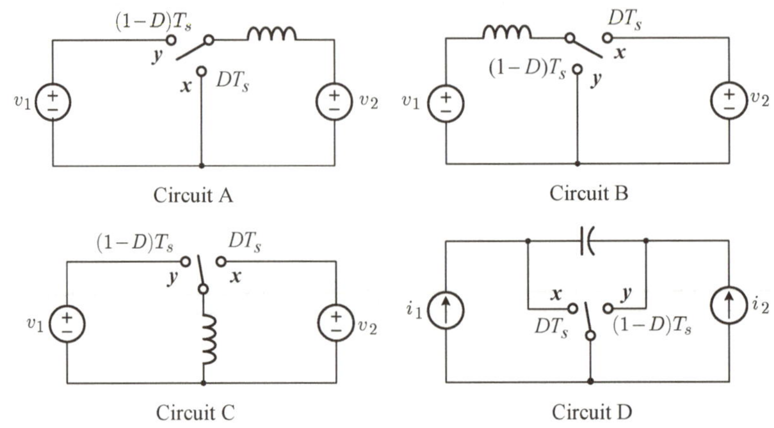

Fig. P2.6

2.7 In the four switching circuits shown in Fig. P2.7, the SPDT switch holds position **a** for DT_s and position **p** for $(1 - D)T_s$. Answer the following questions.

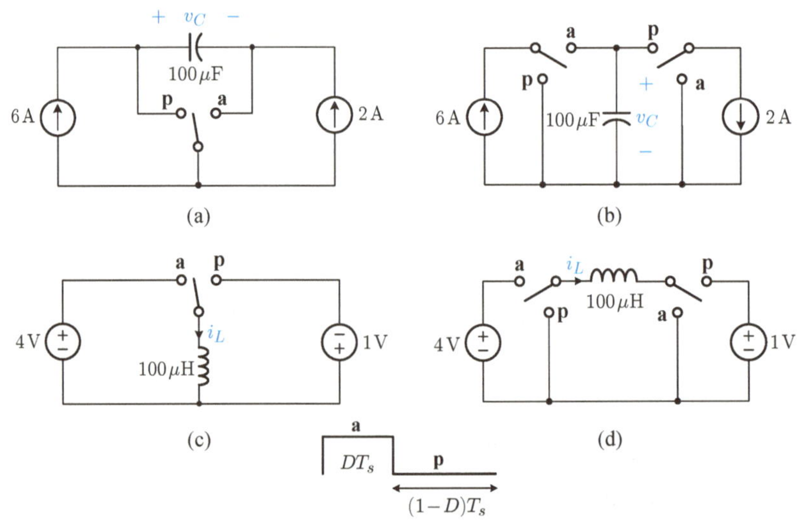

Fig. P2.7

a) For Circuits (a) and (b), determine the values for D and T_s so that each circuit settles in a steady-state operation with an 1.8 V peak value of the capacitor voltage, $v_{C\,peak} = 1.8$ V. Assume the capacitor is initially uncharged and the capacitor voltage does not stay at 0 V.

b) For Circuits (c) and (d), determine the values for D and T_s so that each circuit establishes a steady-state operation with $i_{L\,peak} = 1.2$ A. The inductor is initially unenergized and the inductor current does not stay at 0 A.

2.8* A switching circuit shown in Fig. P2.8 is configured using an SPDT switch and other circuit components. The SPDT switch periodically changes its position, as illustrated in Fig. P2.8. Assume the average value of i_{L2} is 2 A and answer the questions.

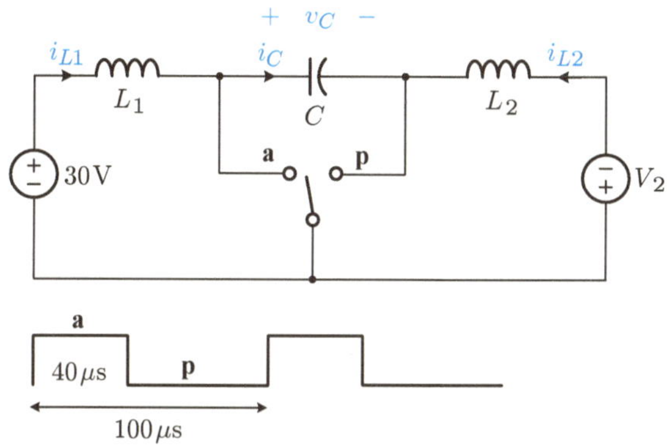

Fig. P2.8

a) Use the flux linkage balance condition on L_1 to evaluate the average value of v_C.
b) Apply the flux linkage balance condition to L_2 to find the value for V_2.
c) Use the charge balance condition on C to evaluate the average value of i_{L1}. The average value of i_{L2} is 2 A.
d) Assume $L_1 = 4.8$ mH, $L_2 = 1.2$ mH, and $C = 120\ \mu$F. Sketch i_{L1}, i_{L2}, i_C, and v_C for the two operational periods. Show the maximum and minimum values of each waveform.

2.9* As shown in Fig. P2.9, a switching circuit configured using an SPDT switch and other circuit components. The SPDT switch periodically changes its structure as illustrated in Fig. P2.9.

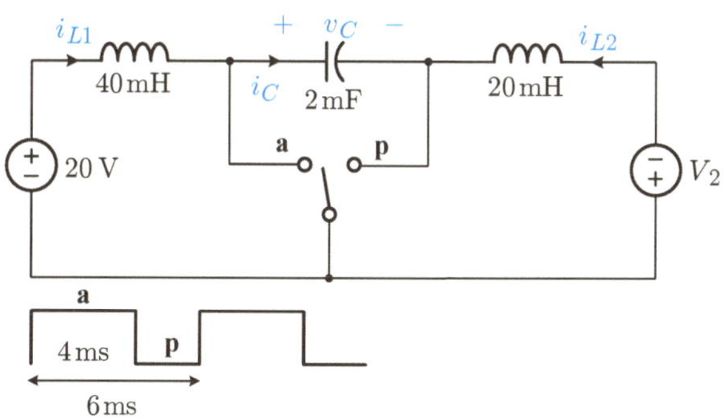

Fig. P2.9

a) Evaluate the average value of v_C and V_2.
b) Assume that the average value of i_{L1} is 4 A. Sketch i_{L1} and i_{L2} for the two operational periods. Show the maximum and minimum values of the waveforms.
c) Using a) and b), sketch the v_C waveform for the two operational periods. Show the maximum and minimum values.

2.10 Figure P2.10 shows a switching circuit configured using a MOSFET and diode along with other linear circuit components. Assume the MOSFET switch is activated with a duty ratio of $D = 0.5$ at $T_s = 6$ ms.

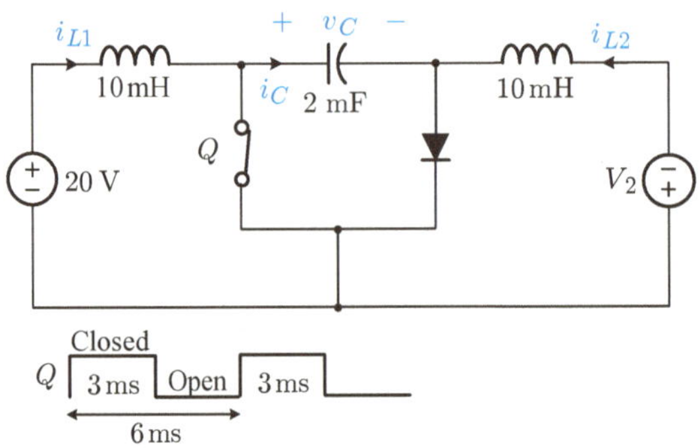

Fig. P2.10

a) Evaluate the average value of v_C and V_2.
b) Assume that the average value of i_{L1} is 0 A. Sketch i_{L1}, i_{L2}, i_C, and v_C for the two operational periods. Show the maximum and minimum values of the waveforms.

2.11* The common switch drive signal is applied to four switching circuits shown in Fig. P2.11. For each circuit, sketch i_S, v_T, and i_{D2} for the two operational periods. Assume the transformer is ideal.

2.12 A transformer is built using a toroidal core with $\mu_r = 5000$, $S = 2\ \text{cm}^2$, and $l_m = 4\pi \times 10^{-1}$ cm. The turns of the primary winding are 200 and the turns of the secondary winding are 400. The permeability of the free space is $\mu_o = 4\pi \times 10^{-7}$ H/m.

a) A simple circuit shown in Fig. P2.12(a) is constructed using the transformer discussed above. Find the expressions for i_P and v_S.
b) Now assume that the circuit is modified as shown in Fig. P2.12(b) in which the two identical transformers, each constructed as described above, are used. Find the expressions for i_P, i_m, v_m, and v_O.

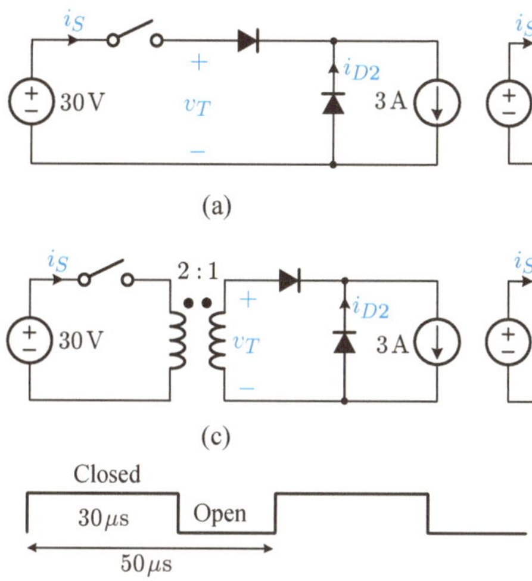

Fig. P2.11

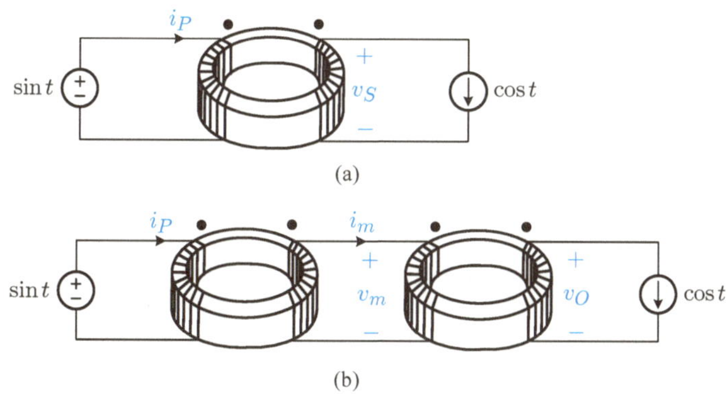

Fig. P2.12

2.13 A couple of experiments are performed to extract the circuit parameters of a practical transformer.

- A 0.5 H inductance is measured at the primary winding with the secondary winding open-circuited.
- When $v_S(t) = 20 \sin t$ is connected to the primary winding, $v_O(t) = 5 \sin t$ is measured at the open-circuited secondary winding.

a) Draw the circuit model for the transformer. Show all the circuit parameters in your model.

b) Five different circuits in Fig. P2.13 are configured using the transformer discussed above. For each circuit, find an expression for the primary winding current, i_P.

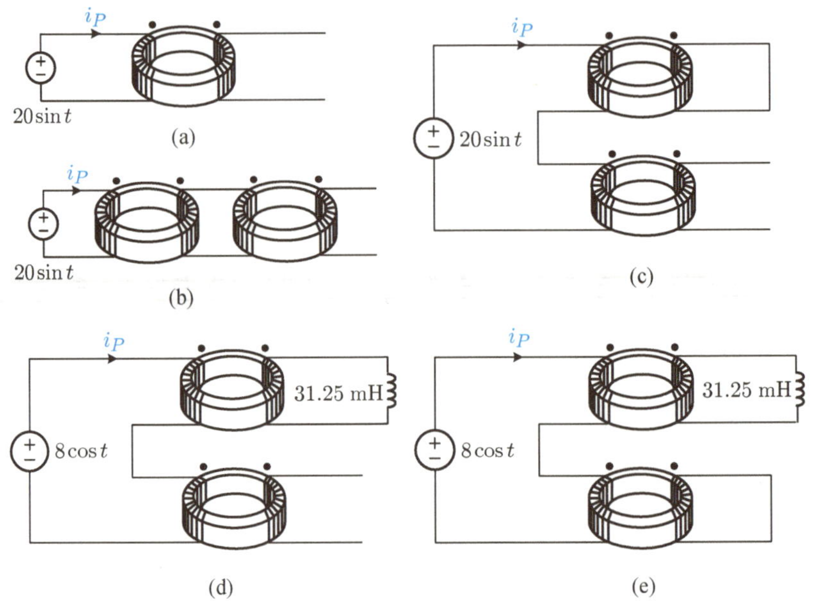

Fig. P2.13

2.14* An inductor is built using a toroidal core with $\mu_r = 5000$, $S = 2\ \text{cm}^2$, and $l_m = 4\pi \times 10^{-1}$ cm.

a) Figure P2.14(a) shows the i_L–λ curve of the inductor.
   i) Evaluate the inductance of the inductor.
   ii) Find the number of the turns of the inductor winding.

b) A switching circuit shown in Fig. P2.14(b) is constructed using the inductor discussed in a). In the following problems, the average value of i_P is zero: $\bar{i}_P(t) = 0$.
   i) Assume $T_s = 1$ ms and find the maximum value of V_S for a proper operation of the circuit.
   ii) Now assume $V_S = 30$ V and find the maximum value for T_s.
   iii) For the case with $T_s = 1$ ms and $V_S = 30$ V, sketch i_P for the two operational periods. Show the maximum and minimum values of the waveform.

c) A secondary winding is added to the inductor discussed above, thereby yielding a two-winding transformer with the turns ratio of 1 : 0.5. As shown in Fig. P2.14(c), a 3.75 mH inductor is terminated across the secondary winding of the transformer. For the case with $T_s = 1$ ms, $V_S = 30$ V, and $\bar{i}_P(t) = 0$, sketch i_S and i_P for the two operational periods. Show the maximum and minimum values of the waveforms.

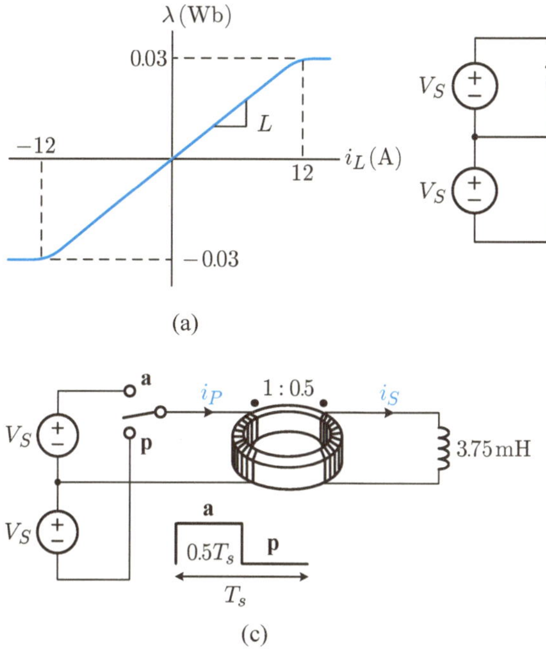

Fig. P2.14

2.15* Figure P2.15 shows a conceptual circuit configured with two inductors and one ideal transformer. The following experiments are performed to determine the values of the inductances and the turns ratio of the ideal transformer.

- A 2 mH inductance is measured at the terminal x–x′ with the terminal y–y′ short-circuited.
- A 7 mH inductance is measured at the terminal x–x′ with the terminal y–y′ open-circuited.
- When a voltage source $v_S(t) = \sin t$ is applied across the terminal x–x′, the same voltage $v_O(t) = \sin t$ is measured across the terminal y–y′.

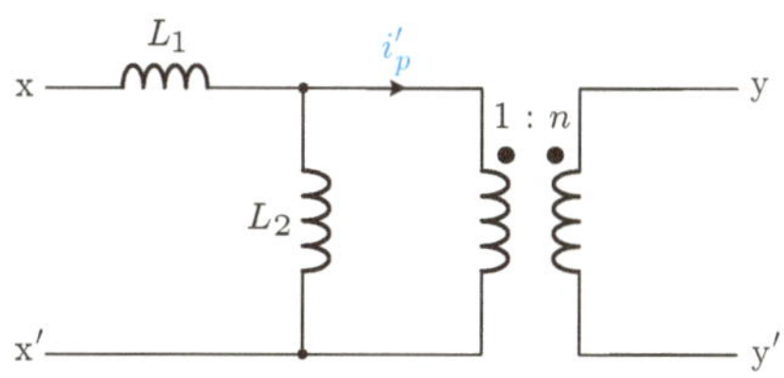

Fig. P2.15

a) Find the values for the inductances, L_1 and L_2, and the turns ratio n of the ideal transformer.

b) Find the expression for i_p' when a voltage source $v_S(t) = \sin t$ is connected across the x–x′ terminal and the y–y′ terminal is shorted.

2.16 Referring to the two circuits shown in Fig. P2.16, answer the following questions. The transformer is built using a toroidal core with $\mu_r = 5000$, $S = 2\text{cm}^2$, and $l_m = 4\pi\times10^{-1}$ cm. The turns of the primary winding are 10 and the turns of the secondary winding are 20.

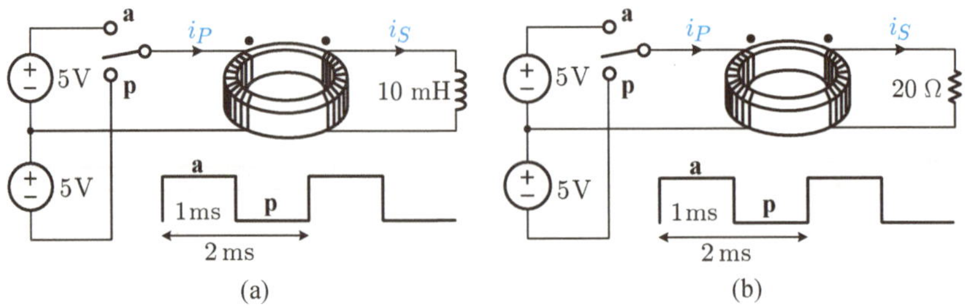

Fig. P2.16

a) For Circuit (a), sketch the steady-state waveforms of i_P and i_S for the two switching periods. Show the maximum and minimum values of each waveform. Assume that the average values of i_P and i_S are zero.

b) Repeat a) for Circuit (b).

2.17* In the circuit shown in Fig. P2.17, the transformer is built by winding $N_P = 2$ and $N_S = 4$ turns of copper coils around a toroidal core with $\mu_r = 5 \times 10^4$, $\mu_o = 4\pi \times 10^{-7}$ H/m, $S = 2$ cm^2, and $l = 4\pi \times 10^{-1}$ cm.

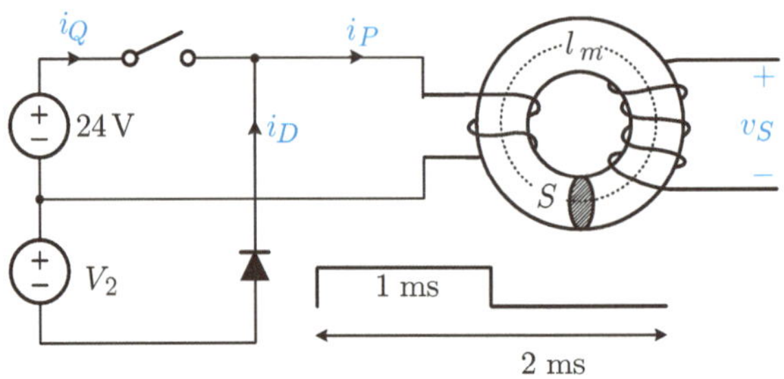

Fig. P2.17

a) Assume $V_2 = 0$ and sketch i_P, i_D, v_S, and i_Q for the initial two operational periods. Show the maximum and minimum values of each waveform. Assume the three reactive components in the circuit are initially unenergized.

b) Assume $V_2 = 24$ V and repeat a).

c) Now assume $V_2 = 48$ V and repeat a).

2.18 Referring to Fig. P2.18, the transformer is built by winding $N_P = 2$ and $N_S = 4$ turns of copper coils around a toroidal core with $\mu_r = 5 \times 10^4$, $\mu_o = 4\pi \times 10^{-7}$ H/m, $S = 2$ cm^2, and $l = 4\pi \times 10^{-1}$ cm. Sketch i_S, i_P, i_Q, and i_D for the two initial operational periods. Assume the reactive components are initially unenergized. Show the peak value of each waveform.

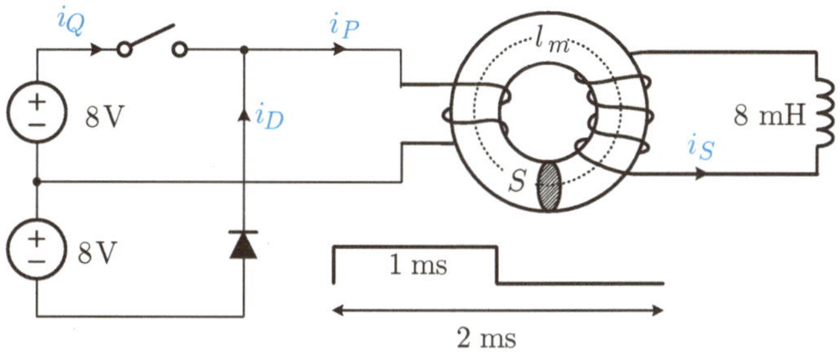

Fig. P2.18

2.19* Figure P2.19 shows two solenoid drive circuits, along with their switch drive signal and/or inductor current waveform.

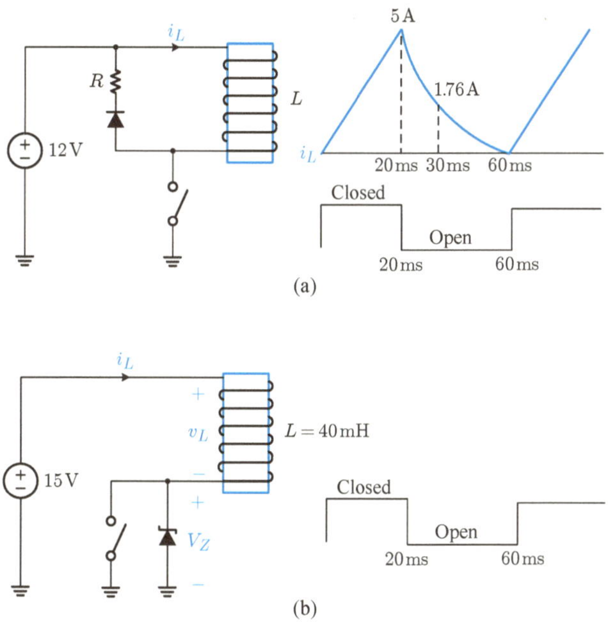

Fig. P2.19

a) Referring to Fig. P2.19(a), find the inductance L of the solenoid and the resistance of R.

b) Use the results of a) to calculate
   i) the peak energy stored in the solenoid,
   ii) the energy consumed at the resistor in each switching period, and
   iii) the average power supplied by the source.

c) An alternative solenoid drive circuit is configured as shown in Fig. P2.19(b). Referring to the solenoid drive circuit, answer the questions.
i) Assuming $V_Z = 25$ V for the Zener diode, sketch i_L and v_L for the initial two operational periods. Label for the maximum and minimum values of each waveform.
ii) Now assume $V_Z = 20$ V for the Zener diode and sketch i_L for the initial two operational periods. Show the maximum value and the final value of i_L. Do you think this circuit is workable? Justify your answer.

2.20* Three different solenoid drive circuits are shown in Fig. P2.20. Referring to the common switch drive signal, answer the questions.

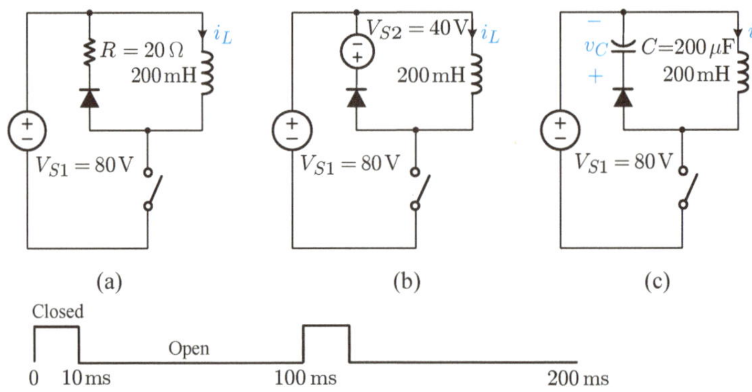

Fig. P2.20

a) Circuit (a) uses a diode-resistor branch as a freewheeling path.
i) Find the expression for $i_L(t)$ for 10 ms $< t <$ 100 ms.
ii) Find the power consumed at the resistor $R = 20\ \Omega$.

b) Circuit (b) employs an independent voltage source, $V_{S2} = 40$ V, in the freewheeling path.
i) Sketch i_L for $0 < t < 100$ ms. Label the value of i_L at $t = 20$ ms on your sketch.
ii) When the switch is turned-off, what would happen to the energy stored in the inductor?

c) Circuit (c) uses a practical capacitor $C = 200\ \mu$F in place of V_{S2}.
i) Find the value of v_C at $t = 99$ ms. Assume the capacitor is initially uncharged and the inductor current i_L becomes zero at $t = 99$ ms.
ii) Sketch the general shapes of i_L and v_C for $0 < t < 200$ ms.
iii) What would eventually happen to the circuit as time elapses?

2.21 A transformer is built using a toroidal core with $\mu_r = 5000$, $S = 2$ cm^2, and $l_m = 4\pi \times 10^{-1}$ cm. The turns of the primary winding are 10 and the turns of the secondary winding are 20. The two different switching circuits shown in Fig. P2.21 are built using the transformer. Referring to the common switch drive signal, sketch i_S and i_T of each circuit for the first two operational periods. Show the maximum and minimum values of the waveforms on your sketch.

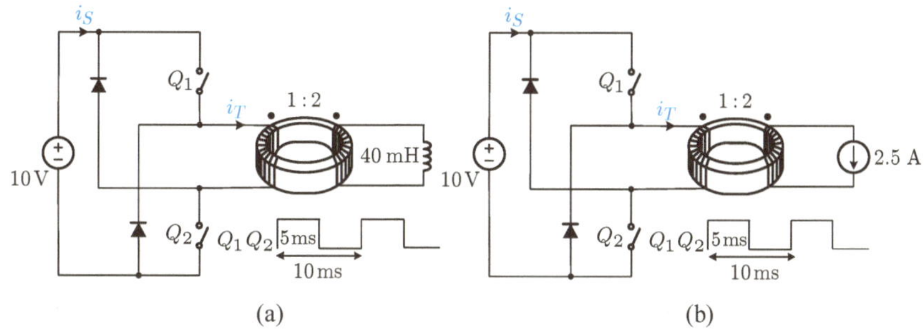

Fig. P2.21

2.22 Consider the solenoid drive circuit shown in Fig. P2.22 and answer the questions.

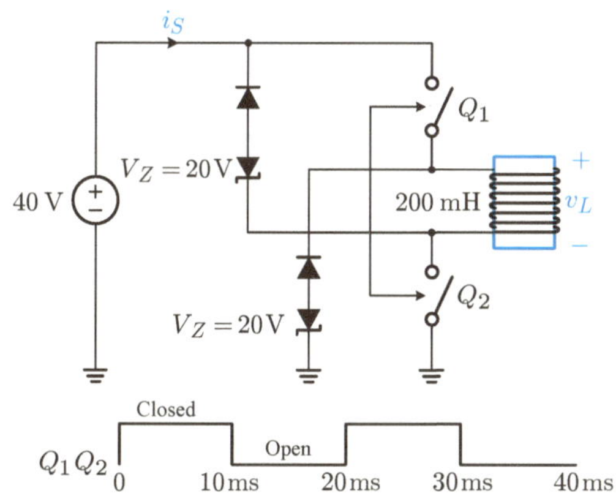

Fig. P2.22

a) Sketch the waveforms of i_S and v_L for $0 < t < 40$ ms. Show the maximum and minimum values on your sketch.
b) Calculate the following items evaluated during $0 < t < 20$ ms:
   i) the energy transferred from the source to the solenoid drive circuit,
   ii) the energy returned from the solenoid drive circuit to the source, and
   iii) the energy dissipated in the solenoid drive circuit.

2.23 The non-dissipative solenoid drive circuit discussed in Section 2.3.1 is modified as shown in Fig. P2.23. Referring to the switch drive signal, sketch i_L, i_{D1}, i_{D2}, i_{Q1}, and v_X for $0 < t < 6$ ms. Label the maximum and minimum values of each waveform.

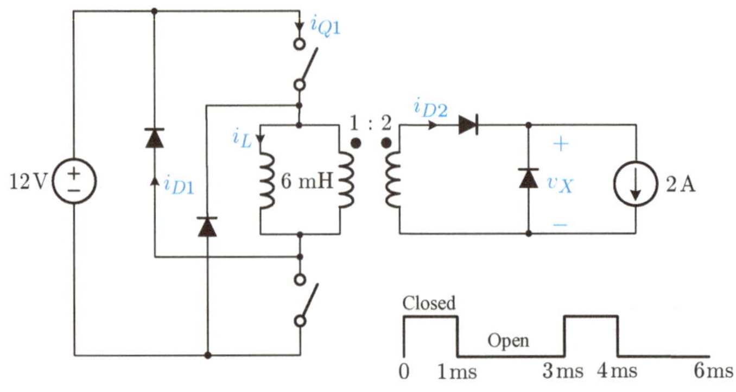

Fig. P2.23

2.24 Consider the capacitor charging circuit and its major circuit waveforms shown in Fig. P2.24.

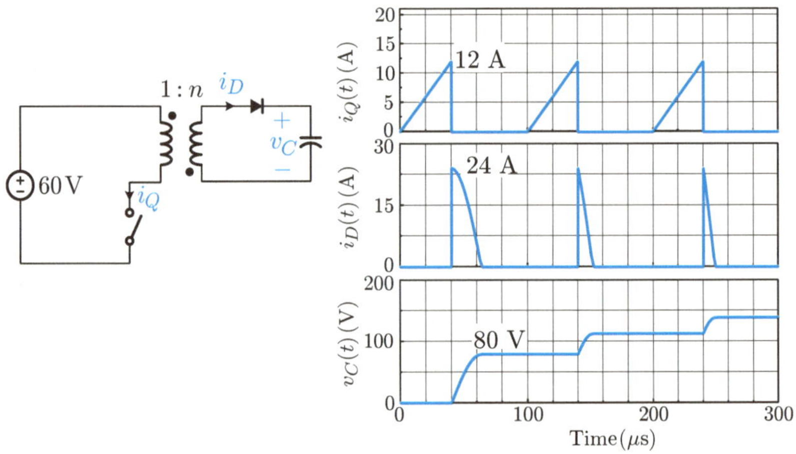

Fig. P2.24

a) Referring to the circuit waveforms, calculate the following circuit parameters and operational conditions of the solenoid driving circuit:
i) L_m: magnetizing inductance of the transformer,
ii) n: turns ratio of the transformer,
iii) D: duty ratio of the switch, and
iv) C: capacitance of the output capacitor.

b) Evaluate the following items based on the circuit waveforms:
i) the energy transferred from the voltage source to the capacitor during the first 10 switching periods,

ii) the capacitor voltage at the end of the 10th switching period, and
iii) the time instant when the capacitor voltage elevates approximately to 300 V.

c) Suggest a new switch drive scheme that would minimize the total time required to charge the capacitor to a specified voltage.

2.25* Consider the circuit shown in Fig. P2.25 along with its switch drive signal. Assume that the inductor and capacitor are initially unenergized.

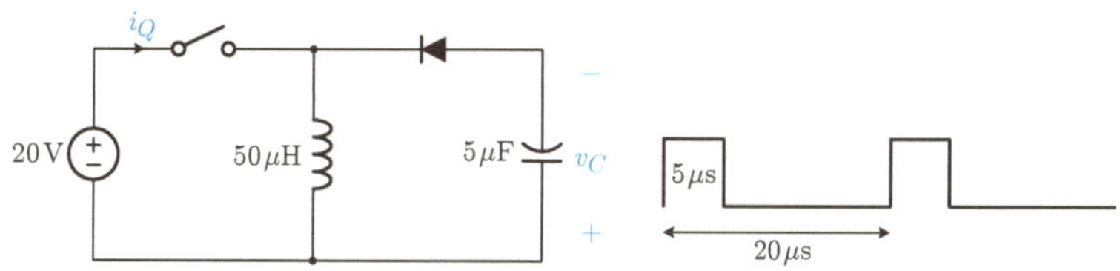

Fig. P2.25

a) Sketch i_Q and v_C for the first two operational periods. Show all the characteristic features of the waveforms including their maximum and minimum values.

b) Evaluate v_C at $t = 2.0$ ms.

2.26* In the circuit shown in Fig. P2.26, the transformers are built by winding $N_P = 2$ and $N_S = 4$ turns of copper coils around a toroidal core with $\mu_r = 5 \times 10^4$, $\mu_o = 4\pi \times 10^{-7}$ H/m, $S = 2$ cm^2, and $l = 4\pi \times 10^{-1}$ cm. The patterns of the transformer windings are not identical. For each circuit, sketch the primary winding current i_P and the secondary winding voltage v_S for the first two operational periods. Show the maximum and minimum values. Assume the reactive components in the circuits are initially unenergized.

2.27 Figure P2.27 shows the two switching circuits are constructed using the transformers discussed in Problem 2.26.

a) For Circuit A, evaluate the following items based on the circuit operation.
i) the energy transferred from the 24 V voltage source to the switching circuit during one operational period.
ii) the energy delivered to the capacitor during one operational period.
iii) the energy dissipated in the switching circuit during one operational period.
iv) the energy returned from the switching circuit to the 24 V voltage source during one operational period.

b) Repeat a) for Circuit B.

2.28* Figure P2.28 shows the two switching circuits along with their switch drive signals.

a) For Circuit (a), evaluate the following items based on the circuit operation:
i) the energy transferred from the 24 V voltage source to the switching circuit during one operational period,

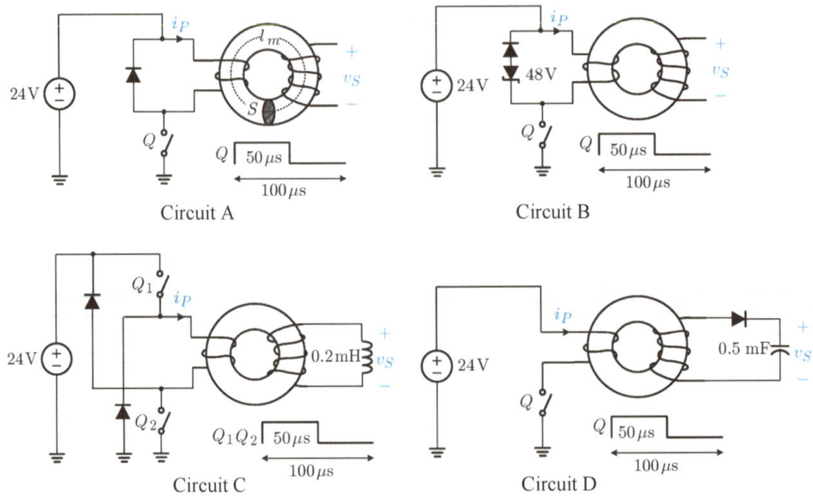

Fig. P2.26

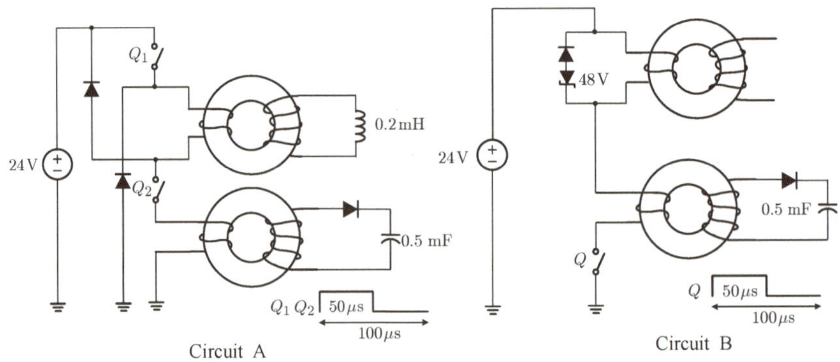

Fig. P2.27

ii) the energy delivered to the capacitor during one operational period,
iii) the energy dissipated in the switching circuit during one operational period, and
iv) the capacitor voltage v_C at the end of the first five operational periods.

b) For Circuit (b), evaluate the following items:
i) the energy transferred from the 24 V voltage source to the switching circuit during one operational period,
ii) the energy delivered to the capacitor during one operational period,
iii) the energy returned from the switching circuit to the 24 V voltage source during one operational period, and
iv) the capacitor voltage v_C at the end of the first five operational periods.

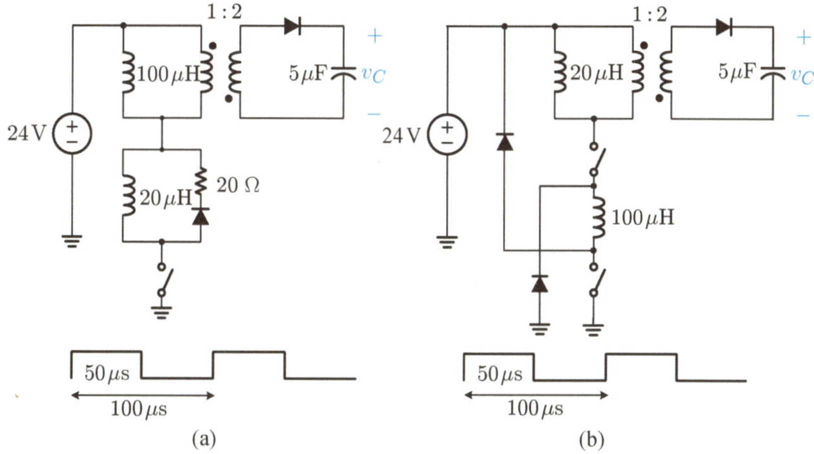

Fig. P2.28

2.29* A transformer is built using a toroidal core with $\mu_r = 5000$, $S = 2\ \text{cm}^2$, and $l_m = 4\pi \times 10^{-1}$ cm. The turns of the primary winding are 5 and those of the secondary winding are 10. The switching circuit shown in Fig. P2.29 is fabricated using the transformer described above.

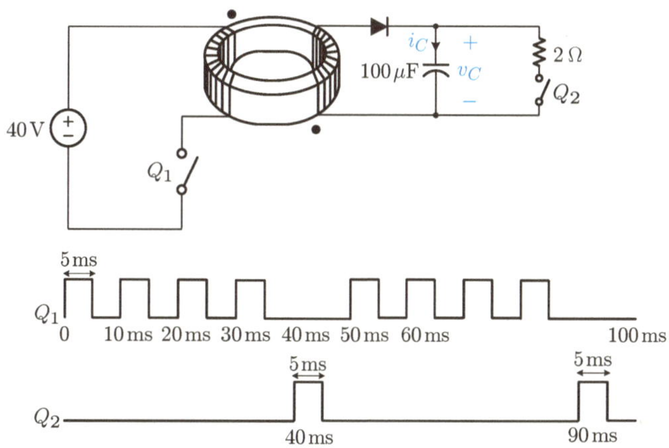

Fig. P2.29

a) Referring to the switch drive signals, sketch the waveforms v_C and i_C for $0 < t < 100$ ms. Show all the important information on your sketch so that your knowledge about the circuit operation is fully exposed.

b) Calculate the average power consumed in the circuit.

2.30* Assume all the circuit components are ideal in the circuits in Fig. P2.30.

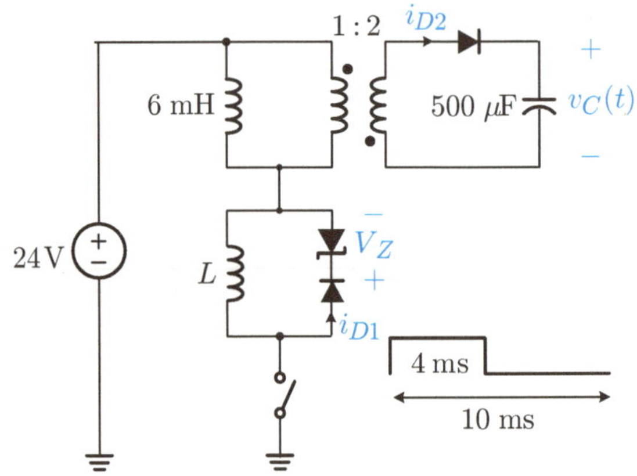

Fig. P2.30

a) Determine the value of inductance L such that a 25% of the total energy delivered from the voltage source is dissipated in the circuit.
b) Assume the circuit operates with the inductance L determined in a). Find the minimum value of the breakdown voltage of the Zener diode, V_{Zmin}, which ensures a proper operation of the circuit.
c) Assume the circuit operates with the inductance L determined in a) and with the Zener diode with $V_Z = 2V_{Zmin}$ where V_{Zmin} is determined in b). Sketch i_{D1}, i_{D2}, and v_C for the first two operational periods. Show the maximum and minimum values of the waveforms. The reactive components are initially unenergized.

2.31 Consider the circuit shown in Fig. P2.31 and answer the question.

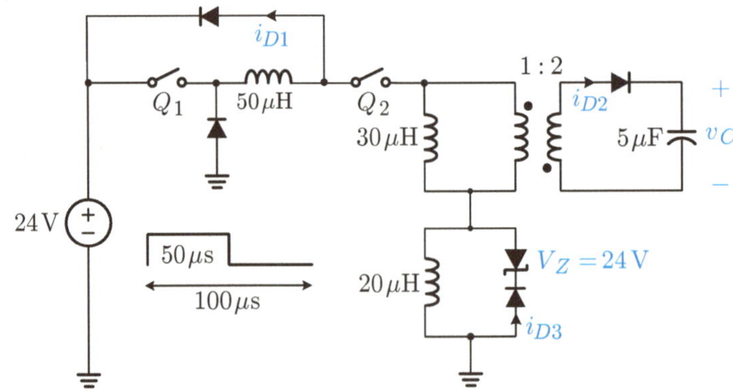

Fig. P2.31

a) The two synchronized switches, Q_1 and Q_2, operate with the duty ratio of $D = 0.5$ at $T_s = 100$ μs. Evaluate the following items based on the circuit operation.

i) the energy transferred from the 24 V voltage source to the switching circuit during one operational period.
ii) the energy delivered to the capacitor during one operational period.
iii) the energy dissipated in the switching circuit during one operational period.
iv) the energy returned from the switching circuit to the 24 V voltage source during one operational period.
v) the capacitor voltage v_C at the end of the first five operational periods.

b) Sketch i_{D1}, i_{D2}, and i_{D3} for $0 < t < 200$ μs. Specify the peak value and show all characteristic features of the current waveforms.
c) Evaluate the average value of i_{D1} and i_{D3}.

2.32* In the circuit shown in Fig. P2.32, the transformer is built by winding $N_P = 2$ and $N_S = 4$ turns of copper coils around a toroidal core with $\mu_r = 5 \times 10^4$, $\mu_o = 4\pi \times 10^{-7}$ H/m, $S = 2$ cm^2, and $l = 4\pi \times 10^{-1}$ cm. The two synchronized switches Q_1 and Q_2 operate with a duty ratio of 0.5 at $T_s = 4$ ms.

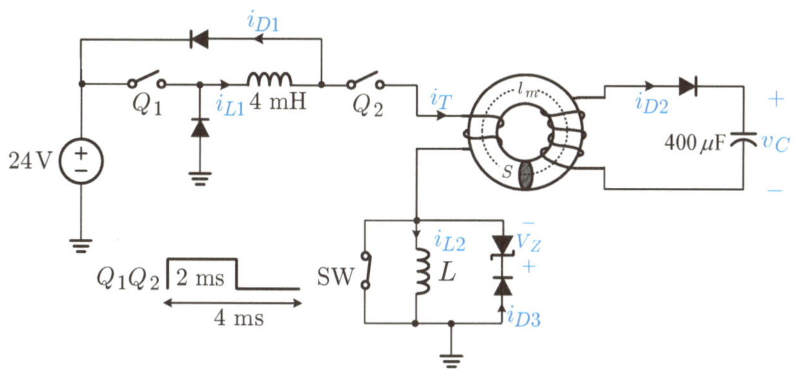

Fig. P2.32

a) Assume the switch SW is closed and sketch i_{L1}, i_{D1}, i_T, i_{D2}, and v_C for the first two operational periods. Assume the reactive components are initially unenergized. Show the peak value of each waveform.
b) Now assume the switch SW is opened and answer the questions.
i) Determine the value of the inductance L such that a 20% of the total energy delivered from the voltage source is dissipated in the circuit.
ii) Assume the circuit operates with the inductance L determined in i). Find the minimum of the breakdown voltage of the Zener diode, V_{Zmin}, which ensures a proper operation of the circuit.
iii) Assume the circuit operates with the inductance L determined in i) and with a Zener diode with $V_Z = 2V_{Zmin}$ where V_{Zmin} is found in ii). Sketch i_{L2} and i_{D3} for the initial two operational periods. Show the peak value of each waveform.

CHAPTER 3

BUCK CONVERTER

There are numerous PWM dc-to-dc converters with different power processing functions and respective application areas. Among these, the buck converter has a very simple structure and is straightforward in operation, while also possessing all the essential features commonly found in other PWM dc-to-dc converters. The output voltage of the buck converter is always lower than the input voltage. For this reason, the buck converter is also called the step-down converter.

There is another type of dc-to-dc converters that invariably provides a higher output voltage than the input voltage, called the boost converter or step-up converter. Furthermore, a dc-to-dc converter that either steps down or steps up the input voltage can be synthesized by combining the step-down converter and step-up converter. The resulting converter is called the buck/boost converter or up-down converter.

This chapter deals with the buck converter, while the succeeding chapter covers the boost and buck/boost converters, as well as other PWM dc-to-dc converters. Although focused on the buck converter, theoretical basis and analysis techniques presented in this chapter can readily be extended to other PWM dc-to-dc converters including the boost and buck/boost converters.

As discussed in Chapter 1, the source of a PWM dc-to-dc converter could be any practical standalone dc source or non-ideal dc source obtained by rectifying an ac voltage source. In addition, the load can be any electrical device, equipment, or system operating with a constant voltage. In this chapter, however, an ideal dc source and resistive load will be assumed to concentrate on the study of the dc-to-dc power conversion itself. The issues involved with the non-ideal characteristics of the source and load systems will be addressed in a later publication.

This chapter discusses the functional basics and operational details of the buck converter. Several important analysis techniques are introduced and used to reveal the operation and properties of the buck converter. This chapter also illustrates the closed-loop control of the buck converter using the PWM scheme.

3.1 IDEAL STEP-DOWN DC-TO-DC POWER CONVERSION

The buck converter is the simplest circuit configuration that can perform the step-down dc-to-dc power conversion. The concept of the step-down dc-to-dc conversion was introduced in Section 1.1.1 in conjunction with the electric bulb driving circuit in Fig. 1.2. The current section now presents the theoretical aspects of the step-down power conversion.

The step-down dc-to-dc power conversion can be explained using a conceptual diagram, as shown in Fig. 3.1(a), consisting of two functional blocks: a single-pole double-throw (SPDT) switch and ideal low pass filter. Within one switching period T_s, the SPDT switch holds position **a** for T_{on} and position **p** for $T_{off} = T_s - T_{on}$. The time period T_{on} is defined as the *on-time period*, while T_{off} is denoted as the *off-time period*. The ratio T_{on} to T_s is defined as the *duty ratio* or *duty cycle* D of the SPDT switch

$$\boxed{D \equiv \frac{T_{on}}{T_s}} \tag{3.1}$$

Similarly, the ratio T_{off} to T_s is defined as

$$D' \equiv \frac{T_{off}}{T_s} = \frac{T_s - T_{on}}{T_s} = 1 - D \tag{3.2}$$

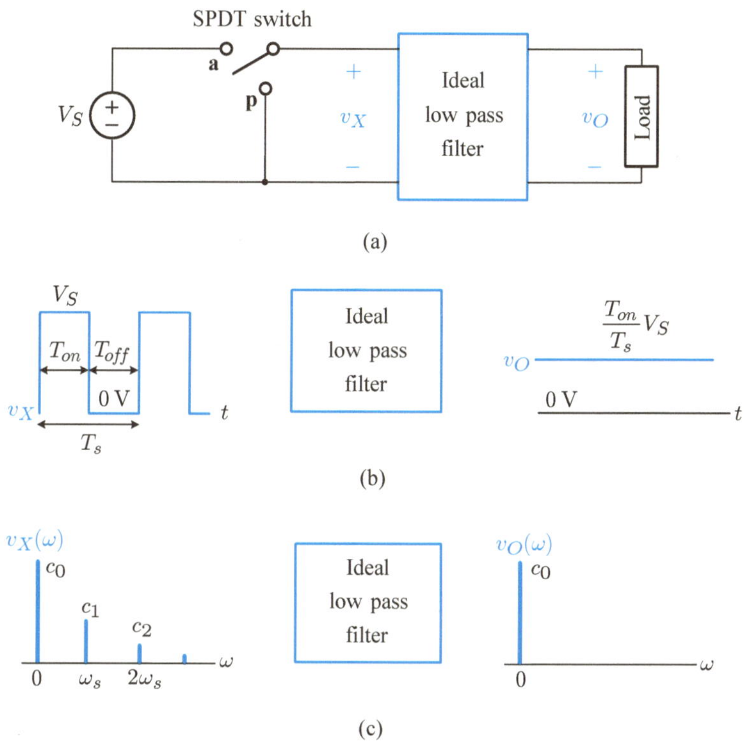

Figure 3.1 Ideal step-down dc-to-dc power conversion. (a) Block diagram representation. (b) Time-domain description. (c) Frequency-domain interpretation.

The SPDT switch transforms the input voltage V_S into a rectangular waveform v_X, as shown in Fig. 3.1. The rectangular waveform v_X is then applied to the input of the ideal low pass filter.

Using Fourier series expansion, v_X is expressed as the sum of dc component and harmonic sinusoids

$$v_X(t) = c_0 + \sum_{n=1}^{\infty} c_n \sin(n\omega_s t + \theta_n) \tag{3.3}$$

where c_0 represents the dc component and the switching frequency $\omega_s = 2\pi/T_s$ is the fundamental frequency of v_X. The dc component c_0 is simply the average value of v_X

$$c_0 = \bar{v}_X(t) = \frac{T_{on}}{T_s} V_S = DV_S \tag{3.4}$$

If the cut-off frequency of the ideal low pass filter, ω_c, is lower than the fundamental frequency of v_X, $\omega_c < \omega_s$, all the harmonic components are completely blocked and only the dc component appears as the output of the low pass filter

$$\boxed{v_O(t) = c_0 = DV_S} \tag{3.5}$$

Figure 3.1(b) shows the time-domain input-to-output description of the ideal low pass filter, while Fig. 3.1(c) depicts the frequency-domain interpretation.

The dc-to-dc power conversion illustrated in Fig. 3.1 has the following properties.

1) The circuit provides a pure dc voltage for the load, due to the ideal characteristics of the low pass filter.

2) The voltage gain of the circuit, the ratio v_O to V_S, is simply the duty ratio D of the SPDT switch. Thus, the output voltage can be adjusted by controlling the duty ratio of the SPDT switch.

3) Because $0 < D < 1$, the output voltage is always lower than the input voltage.

For above reasons, the power conversion illustrated in Fig. 3.1 is called the ideal step-down dc-to-dc power conversion.

3.2 BUCK CONVERTER: STEP-DOWN DC-TO-DC CONVERTER

Although the block diagram in Fig. 3.1 is a conceptual illustration of the ideal step-down dc-to-dc power conversion, there exists a practical converter which is very similar to Fig. 3.1 in structure and operation. This converter is called the buck converter and is one most widely-used dc-to-dc converter. This section explains the circuit configuration and theoretical basics of the buck converter, while succeeding sections describe operational details.

3.2.1 Evolution to Buck Converter

The functional diagram in Fig. 3.1 is transformed into the buck converter by implementing the SPDT switch with semiconductor switches and replacing the ideal low pass filter with an LC filter. Figure 3.2 shows the circuit diagram of the buck converter along with its major waveforms. The SPDT switch is configured with the MOSFET-diode pair. The MOSFET switch is turned-on/off by the gate drive signal v_{GS}, meanwhile the status of the diode is determined by the condition of the MOSFET switch. When the MOSFET switch is turned-on by v_{GS}, the diode is turned-off because the input voltage V_S reverse-biases the pn junction. Conversely, when the MOSFET switch is turned-off, the inductor current forces the diode to conduct, thereby creating a freewheeling path.

A second-order LC filter is used as a functional substitute for the ideal low pass filter. The LC filter, despite its far from ideal characteristics, provides more than adequate filtering for most applications. It will be shown later that the consequences of the non-ideal filtering property of the LC filter are indeed negligible. A pure resistor is assumed for the load to simplify ensuing discussions.

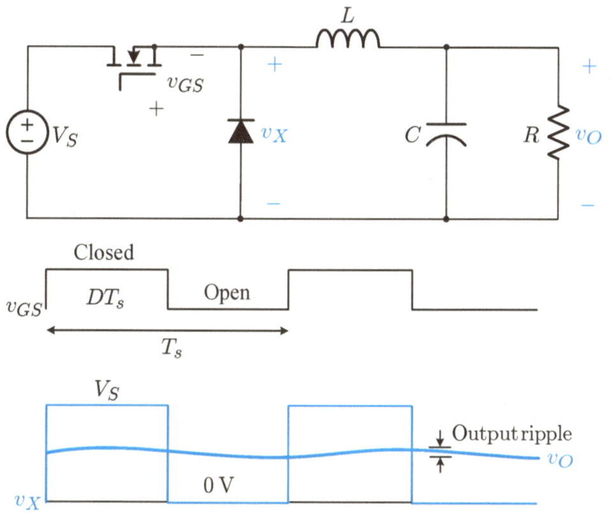

Figure 3.2 Buck converter and major waveforms.

3.2.2 Frequency-Domain Analysis

The LC filter is unable to completely remove the high-frequency harmonics from v_X and the output voltage v_O invariably contains an ac component, as illustrated in Fig. 3.2. The ac component contained in the output voltage is referred to as the *output ripple* or *switching ripple*.

A simple frequency-domain analysis is performed to determine the output ripple due to the non-ideal characteristics of the LC filter. The input-to-output transfer function of the LC filter with a load resistor R is given by

$$F_f(s) = \frac{v_o(s)}{v_x(s)} = \frac{\frac{1}{sC} \parallel R}{sL + \frac{1}{sC} \parallel R} = \frac{1}{1 + \frac{s}{Q\omega_o} + \frac{s^2}{\omega_o^2}} \tag{3.6}$$

where ω_o is the pole frequency

$$\omega_o = \frac{1}{\sqrt{LC}} \tag{3.7}$$

and Q is the quality factor of the filter circuit

$$Q = R\sqrt{\frac{C}{L}} \tag{3.8}$$

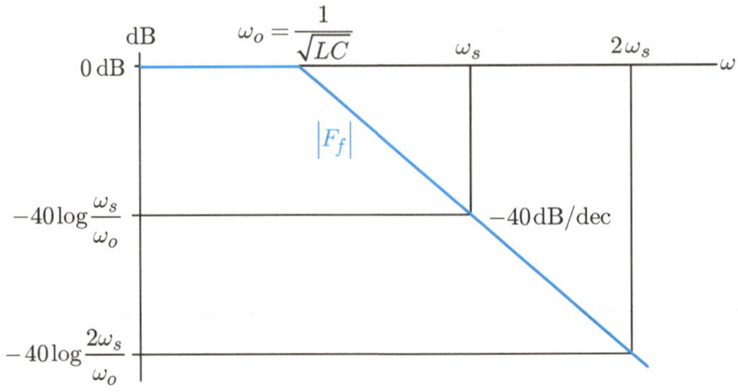

Figure 3.3 Asymptotic plot for transfer function of second-order LC filter.

Figure 3.3 shows the asymptotic plot of $|F_f|$, drawn with the assumptions that the switching frequency is higher than the pole frequency of the filter: $\omega_s > \omega_o$. The LC filter passes the dc component unchanged, yet attenuates the harmonic components of v_X. The input of the LC filter is represented by Fourier series

$$\boxed{v_X(t) = c_0 + \sum_{n=1}^{\infty} c_n \sin(n\omega_s t + \theta_n)} \tag{3.9}$$

The coefficients for the harmonic sinusoids are determined as

$$c_n = \frac{\sqrt{2}V_S}{n\pi}\sqrt{1 - \cos\left(n2\pi D\right)} \tag{3.10}$$

from the Fourier series expansion. Referring to Fig. 3.3, the nth order harmonic is attenuated by $-40\log(n\omega_s/\omega_o)$, while the dc component remains unaltered. Accordingly, the output of the LC filter is expressed as

$$\boxed{v_O(t) = c_0 + \sum_{n=1}^{\infty} c_n' \sin(n\omega_s t + \theta_n')} \tag{3.11}$$

where

$$\begin{aligned} c_n' &= c_n 10^{\frac{-40\log\left(\frac{n\omega_s}{\omega_o}\right)}{20}} \\ &= \underbrace{\frac{\sqrt{2}V_S}{n\pi}\sqrt{1 - \cos\left(n2\pi D\right)}}_{c_n}\, 10^{\frac{-40\log\left(\frac{n\omega_s}{\omega_o}\right)}{20}} \end{aligned} \tag{3.12}$$

The second term on the right-hand side of (3.11), representing the total sum of the harmonic sinusoids, collectively constitutes the output ripple. Equation (3.12) shows that the output ripple is reduced to a negligible level if the condition $\omega_s \gg \omega_o$ is met. This can be achieved by either lowering the pole frequency with larger filter components or increasing the switching frequency.

Output Ripple Estimation — EXAMPLE 3.1

This example demonstrates the accuracy of the previous frequency-domain ripple analysis. A buck converter operating with $V_S = 16$ V, $L = 40\ \mu$H, $C = 470\ \mu$F, $R = 1\ \Omega$, $T_s = 20\ \mu$s, and $D = 0.25$ is considered. The pole frequency of the filter is given by

$$\omega_o = \frac{1}{\sqrt{LC}} = \frac{1}{\sqrt{40 \times 10^{-6}\, 470 \times 10^{-6}}} = 7.29 \times 10^3 \text{ rad/s}$$

and the switching frequency of the converter is

$$\omega_s = \frac{2\pi}{T_s} = \frac{2\pi}{20 \times 10^{-6}} = 3.14 \times 10^5 \text{ rad/s}$$

The dc component of the output is determined as

$$c_O = DV_S = 0.25 \cdot 16 = 4 \text{ V}$$

and the magnitude of the fundamental sinusoid is calculated as

$$\begin{aligned} c_1' &= \underbrace{\frac{\sqrt{2}\,V_S}{1 \cdot \pi}\sqrt{1 - \cos(1 \cdot 2\pi D)}}_{c_1}\ 10^{\frac{-40\log\left(\frac{1\,\omega_s}{\omega_o}\right)}{20}} \\ &= \frac{\sqrt{2}\ 16}{1 \cdot \pi}\sqrt{1 - \cos(1 \cdot 2\pi\, 0.25)}\ 10^{\frac{-40\log\left(\frac{1 \cdot 3.14\times10^5}{7.29\times10^3}\right)}{20}} \\ &= 3.88 \times 10^{-3} \end{aligned}$$

Similarly, the magnitudes of the higher-order harmonics are determined as $c_2' = 6.86 \times 10^{-4}$, $c_3' = 1.44 \times 10^{-4}$, $c_4' = 0$, and $c_5' = 3.10 \times 10^{-5} \cdots$.

Two important observations are made from the previous analysis.

1) The coefficients of the harmonic components of the output voltage are negligibly small compared to the dc component. This indicates that the LC filter provides sufficient filtering and the output of the converter is practically $c_O = 4$ V.

2) Because the coefficient of the fundamental sinusoid, c_1', is much larger than the coefficients of the higher-order harmonics, the amplitude of the output ripple is estimated as $\Delta v_O(t) \approx 2\,c_1' = 2 \cdot 3.88 \times 10^{-3} = 7.76$ mV.

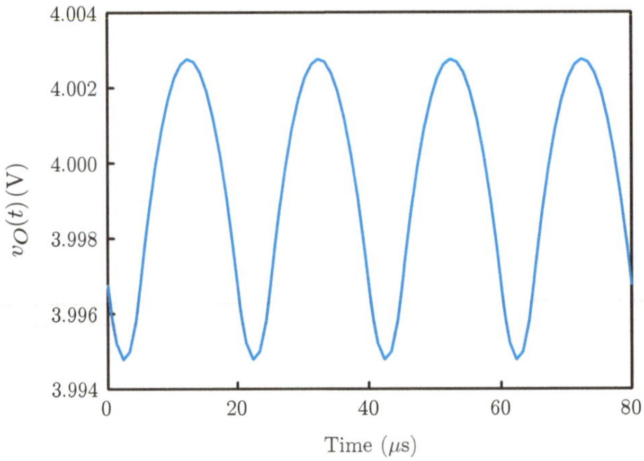

Figure 3.4 Output voltage waveform of buck converter.

Figure 3.4 shows the output voltage of the buck converter obtained from PSpice® simulation. The magnitude of the output ripple is very close to the analytical prediction of 7.76 mV.†

As demonstrated in Example 3.1, a simple second-order LC filter readily attenuates the output ripple to a negligible level. The second-order LC filter is just one specific example among many possible filter configurations. If a larger output ripple is allowed, the capacitor can be removed from the second-order filter, resulting in a first-order LR filter. Conversely, if further ripple attenuation is required, a higher-order filter or multi-stage filter could be employed.

EXAMPLE 3.2 Output Filter Configuration and Output Ripple

This example illustrates the structure and ripple attenuation of the three buck converters configured with different output filters. Figure 3.5 shows the buck converters considered in this example: the first with a first-order LR filter, the second with a second-order LC filter, and the last with a two-stage LC filter. All the three buck converters operate with the input voltage of $V_S = 6$ V and the duty ratio of $D = 0.5$ at the switching frequency of 50 kHz. The circuit parameters of the output filters are shown in Fig. 3.5.

The three converters would produce the same 3 V output. However, the magnitude of the ripple component will vary due to the difference in the output filter configuration. PSpice® simulations are performed to investigate the output filter characteristics and output ripple attenuation.

† The average value of the output voltage is slightly less than the theoretical value 4.0 V. The offset is due to the non-ideal characteristics of the diode used in the simulations.

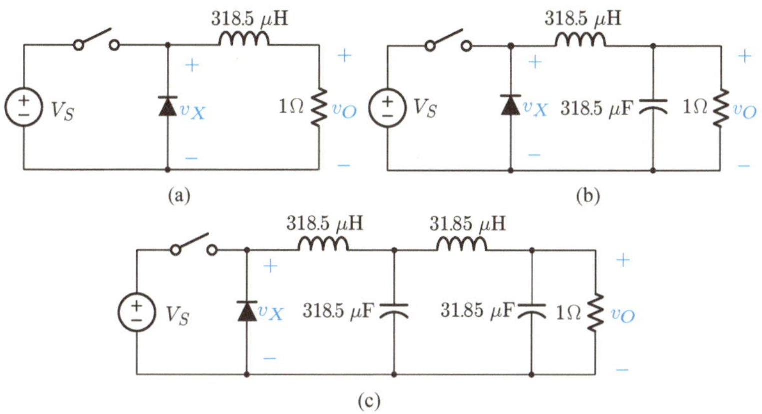

Figure 3.5 Buck converters with different output filter configurations. (a) Buck converter with first-order output filter. (b) Buck converter with second-order output filter. (c) Buck converter with two-stage output filter.

Figure 3.6(a) shows the simulated Bode plots of the filter transfer functions, $F_f(s) = v_o(s)/v_x(s)$, of the three buck converters. The Bode plots indicate that the first-order filter provides only a 40 dB attenuation for the fundamental component, while the second-order output filter offers a 80 dB attenuation. The two-stage output filter attenuates the fundamental component by 120 dB. Figure 3.6(b) shows the output voltage waveforms of the three buck converters. The first-order output filter generates a large triangular-shaped output ripple. The second-order filter yields a small sinusoidal output ripple. The two-stage output filter results in a nearly undetectable output ripple due to the much enhanced attenuation.

3.3 BUCK CONVERTER IN START-UP TRANSIENT

This section explains the operation of the buck converter during the start-up process. The concept of piecewise linear analysis is first introduced and the start-up response is then explained using the piecewise linear analysis method.

3.3.1 Piecewise Linear Analysis

The time-domain analysis of a dc-to-dc converter is not straightforward mainly because the switching action of the semiconductor devices forces the converter to periodically change its topological structure. One standard approach to analyzing the time-domain behavior of dc-to-dc converters is the piecewise linear analysis, where a dc-to-dc converter is considered as a combination of several linear circuits; each of them is valid for a specific time interval within one switching period.

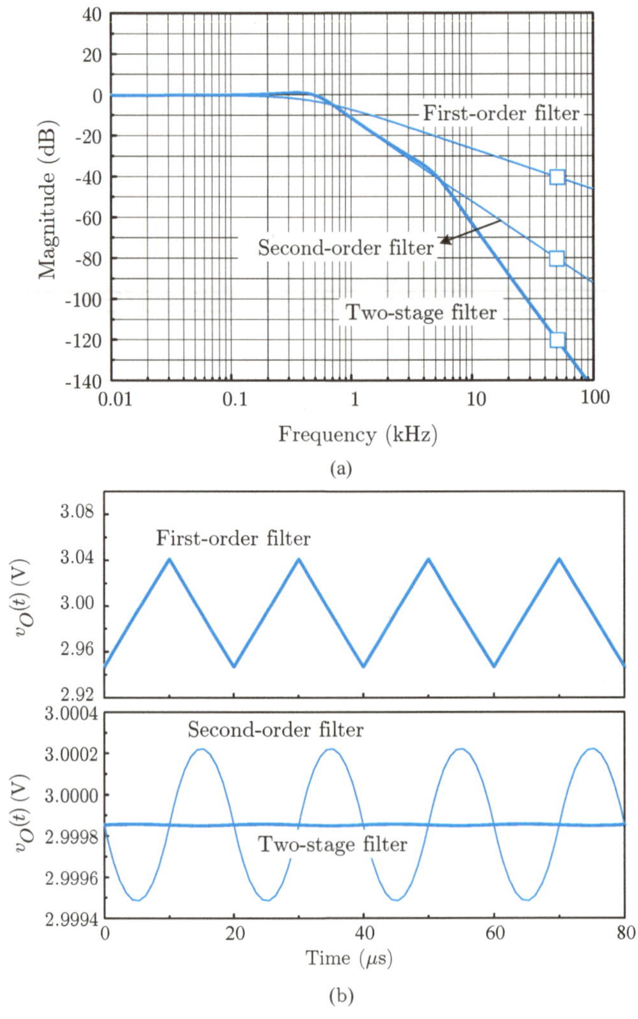

Figure 3.6 Filter transfer functions and output voltage ripple. (a) Bode plots of filter transfer functions. (b) Output voltage of three buck converters

The circuit that is effective when its switch drive signal is *on* (thus, the active switch is turned-on and the diode is turned-off) is defined as the *on-time subcircuit*. Likewise, the effective circuit when the switch drive signal is *off* (thus, with the active switch off and diode on) is denoted as the *off-time subcircuit*. The operation of a dc-to-dc converter is analyzed by first examining each subcircuit individually and later considering the circuit behavior of the two subcircuits collectively. Figure 3.7 shows a buck converter and its on-time and off-time subcircuits. The piecewise linear analysis using these subcircuits simplifies the analysis process and, more importantly, provides significant insights on the converter operation.

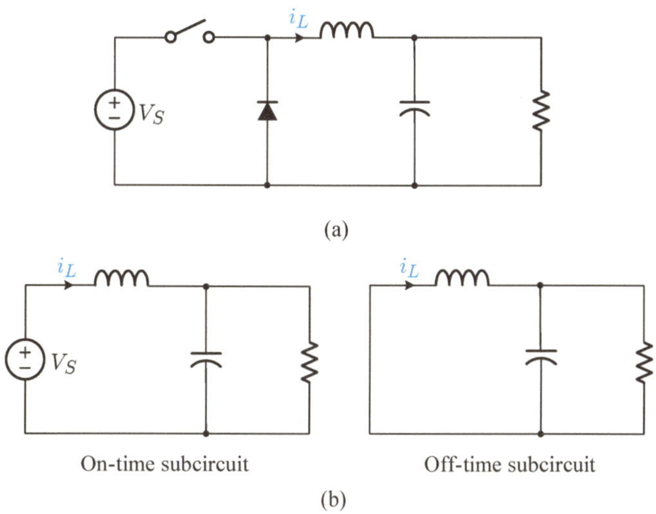

Figure 3.7 Buck converter and on-time and off-time subcircuits. (a) Buck converter. (b) On-time and off-time subcircuits.

The piecewise linear analysis will be used in the this chapter, to investigate the transient and steady-state responses of the buck converter. Furthermore, this analysis method will be employed throughout the book, as an underlying basis to portray the steady-stage waveforms of various PWM dc-to-dc converters.

3.3.2 Start-up Response

The time-domain response of dc-to-dc converters can be analyzed by solving the circuit equations for the on-time and off-time subcircuits successively. For example, the response of the inductor current during the start-up process can be calculated as outlined below. First, the circuit equations for the on-time subcircuit are solved for $0 < t < DT_s$ with zero initial conditions. The expression for the inductor current is found from the resulting solutions. Second, the circuit equations for the off-time subcircuit are solved for $DT_s < t < T_s$ considering the circuit variables evaluated at $t = DT_s$ as the new initial conditions. By repeating this process, the transient response of the inductor current for the entire start-up period is obtained.

As an alternative to the previous iterative analysis, a qualitative method can be used to predict the behavior of the inductor current. For the time period $0 < t < DT_s$ during which the converter retains the on-time subcircuit, the voltage source transfers energy to the inductor. As a result, the inductor current is increased as illustrated in Fig. 3.8. For the time interval $DT_s < t < T_s$ in which the converter holds the off-time subcircuit, the energy stored in the inductor is released to the load, and consequently the inductor current is decreased.

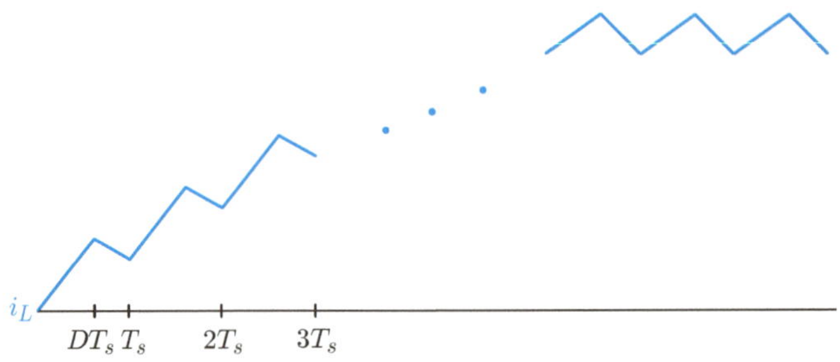

Figure 3.8 Qualitative behavior of inductor current.

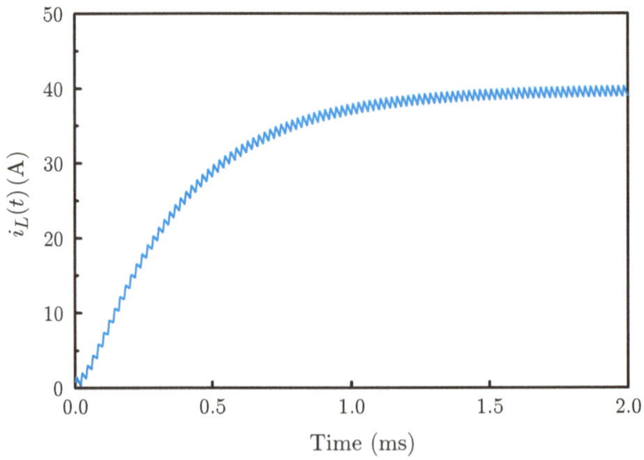

Figure 3.9 Start-up response of inductor current.

During the early stage of the start-up process, the energy transferred from the source is larger than the energy released to the load. Accordingly, the net change in the energy storage in the inductor for each switching period is positive. This implies that the inductor current progresses as illustrated in Fig. 3.8. As time passes, the difference between the amounts of the transferred energy and the released energy lessens. When the on-time energy transfer becomes the same as the off-time energy release, the converter establishes steady state and the inductor current settles into a periodic triangular waveform.

EXAMPLE 3.3 Start-up Response of Inductor Current

This example illustrates the start-up response of the buck converter. Figure 3.9 shows the inductor current during the start-up process of the buck converter introduced in Example 3.1. A load resistance $R = 0.1\ \Omega$ is used in this simulation, while other parameters are the same as those of Example 3.1.

3.4 BUCK CONVERTER IN STEADY STATE

A properly designed dc-to-dc converter should reach steady state. When a dc-to-dc converter is in steady state, the inductor current settles into a periodic triangular waveform and the output voltage becomes nearly constant with a small ripple component. This section first introduces several circuit analysis techniques and later analyzes the circuit waveforms of the buck converter in steady state.

3.4.1 Circuit Analysis Techniques

In addition to standard circuit equations, several specific analysis techniques are used for dc-to-dc power conversion circuits. These include the piecewise linear analysis, small-ripple approximation, flux linkage balance condition on inductors, and charge balance condition on capacitors.

Piecewise Linear Analysis

The piecewise linear analysis technique is a very efficient method to analyze time-variant dc-to-dc power conversion circuits. When coupled with the forthcoming small-ripple approximation, the piecewise linear analysis allows us to study dc-to-dc converters only using very simple circuit equations. In most cases, the circuit waveforms are accurately predicted by quick graphical constructions based on basic circuit equations.

Small-Ripple Approximation

Another useful technique for the steady-state analysis of dc-to-dc converters is the *small-ripple approximation*. This analysis postulates that *the ripple component contained in the output of a converter is so small that the output voltage can be considered a pure dc*. This assumption greatly simplifies the circuit analysis without compromising accuracy.

In practice, the ripple component in the output voltage is actually very small compared to the dc component, and its effects on the converter operation are negligible. Consequently, an exact circuit analysis including the ripple component is usually unnecessary and the simplified analysis assuming a constant output is fully adequate for most cases. The streamlined analysis based on the small-ripple approximation quickly yields simple circuit equations, and the resulting solutions retain sufficient accuracy for all engineering purposes.

With the small-ripple approximation, we perform the simplified analysis assuming the output voltage is a pure dc without any ripple component. Nonetheless, the actual output voltage ripple can always be accurately estimated from the results of the simplified analysis based on the small-ripple approximation. Consequently, the small-ripple approximation does not cause any real loss in the analysis accuracy. Discussions about the output ripple estimation are given in Section 3.4.3.

Flux Linkage Balance Condition and Charge Balance Condition

Inductors used in dc-to-dc conversion circuits should satisfy the flux linkage balance condition; the flux linkage increase during an on-time period should be the same as the flux linkage decrease during an off-time period. As demonstrated in Section 2.2.1, the flux linkage balance condition is transformed into the volt-sec balance condition. The volt-sec balance condition states that *the product of the voltage level and time interval, evaluated over the period when the inductor voltage is positive, should be the same as the voltage-time product calculated when the inductor voltage is negative*. Furthermore, as the generation of the volt-sec balance condition, the average value of the inductor voltage is considered to be zero in general.

Capacitors are subjected to the charge balance condition; the net change in the charge storage in a capacitor should be balanced at zero for each switching period. The charge balance condition is converted into the amp-sec balance condition. The amp-sec balance condition indicates that *the product of the positive current value (coming into the capacitor) and the current-flowing period should be the same as the product of the negative current value (leaving from the capacitor) and the corresponding current-flowing time*. More generally, the average value of the capacitor current is considered to be zero, as is the case for the inductor voltage.

The volt-sec balance condition and amp-sec balance condition place constraints on the circuit variables of inductors and capacitors in dc-to-dc power conversion circuits. When dc-to-dc converters reach steady state, the circuit variables settle down to the values that satisfy the volt-sec and amp-sec balance conditions. Thus, these conditions could be used as the circuit theorems that ease the steady-state analysis of dc-to-dc converters.

3.4.2 Steady-State Analysis

The steady-state analysis of the buck converter is now performed based on the piecewise linear analysis, along with the small-ripple approximation, volt-sec balance condition, amp-sec balance condition, and other standard circuit analysis techniques. Figure 3.10 shows a buck converter, on-time and off-time subcircuits, and major circuit waveforms. Based on the small-ripple approximation, the output of the converter is considered as a constant V_O for both the on-time and off-time subcircuits. From Fig. 3.10(b), the voltage across the inductor is determined as

$$v_L(t) = \begin{cases} V_S - V_O & \text{for on-time subcircuit} \\ -V_O & \text{for off-time subcircuit} \end{cases} \tag{3.13}$$

By applying the volt-sec balance condition to the inductor, it follows that

$$(V_S - V_O)DT_s = V_O(1 - D)T_s \tag{3.14}$$

which is simplified to

$$\boxed{V_O = DV_S} \tag{3.15}$$

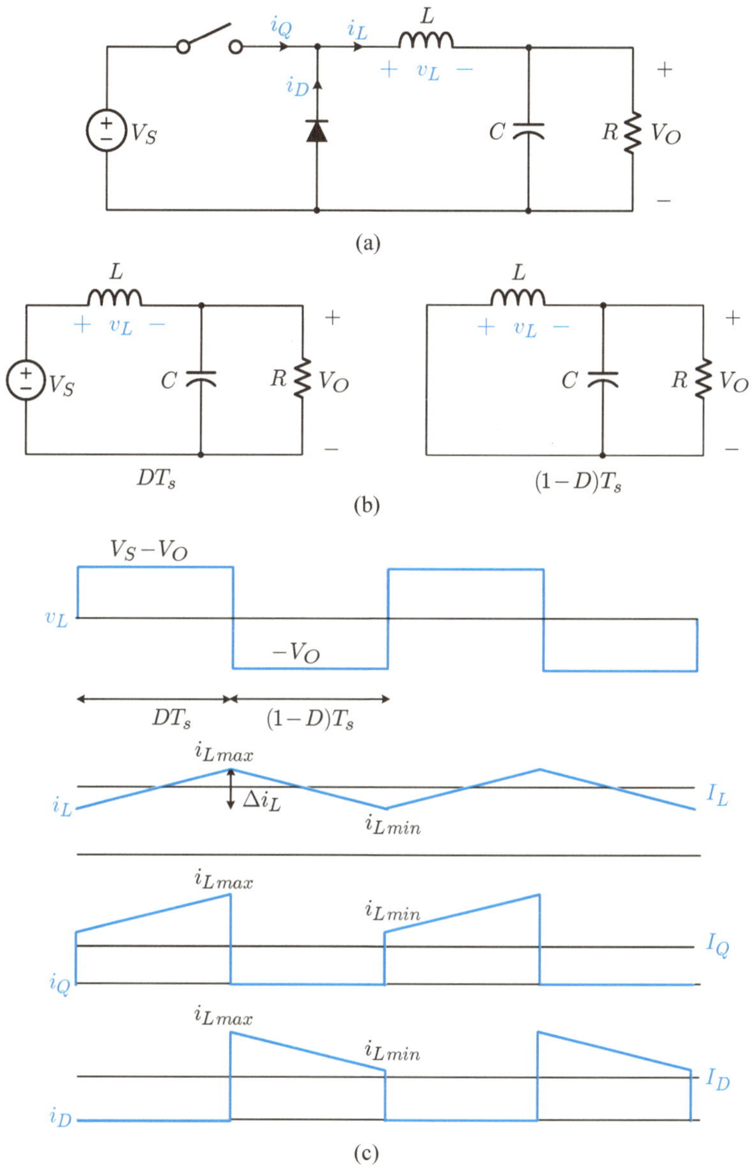

Figure 3.10 Steady-state analysis of buck converter. (a) Buck converter. (b) On-time and off-time subcircuits. (c) Major waveforms.

yielding the voltage gain expression. This voltage gain is identical, as it should be, to the result of the previous frequency-domain analysis.

The circuit equation for the inductor current is given by

$$i_L(t) = \frac{v_L(t)}{L}t = \frac{V_S - V_O}{L}t \tag{3.16}$$

for on-time period DT_s and

$$i_L(t) = -\frac{V_O}{L}t \tag{3.17}$$

for off-time period $(1-D)T_s$. It is evident from (3.16) and (3.17) that the inductor current ramps up during on-time period and ramps down during off-time period, resulting in a periodic triangular waveform shown in Fig. 3.10(c).

The average value of the inductor current is determined as follows. The charge balance condition assures that the average value of the current flow through the capacitor is zero. This implies that the average value of the inductor current should be identical to the dc current passing through the load resistor

$$\boxed{\bar{i}_L(t) = I_L = \frac{V_O}{R}} \tag{3.18}$$

On the other hand, it can also be inferred from (3.16) and (3.17) that the difference between the maximum and minimum values of the inductor current is given by

$$\boxed{\Delta i_L = \frac{v_L}{L}\Delta t = \frac{V_S - V_O}{L}DT_s = \frac{V_O}{L}(1-D)T_s} \tag{3.19}$$

The maximum and minimum values of the inductor current then become

$$i_{L\,max} = I_L + \frac{1}{2}\Delta i_L \tag{3.20}$$

and

$$i_{L\,min} = I_L - \frac{1}{2}\Delta i_L \tag{3.21}$$

The inductor current flows through the active switch during on-time period, and then freewheels through the diode during off-time period. Figure 3.10(c) also shows the switch current i_Q and diode current i_D, whose average values are given by

$$\bar{i}_Q(t) = I_Q = DI_L \tag{3.22}$$

and

$$\bar{i}_D(t) = I_D = (1-D)I_L \tag{3.23}$$

Table 3.1 Steady-State Analysis of Buck Converter

Circuit variable	Expression
V_O	$V_S D = 16 \cdot 0.25 = 4\ \text{V}$
$v_{L\,max}$	$V_S - V_O = 16 - 4 = 12\ \text{V}$
$v_{L\,min}$	$-V_O = -4\ \text{V}$
I_L	$\frac{V_O}{R} = \frac{4}{1} = 4\ \text{A}$
Δi_L	$\frac{V_S - V_O}{L} DT_s = \frac{16-4}{40 \times 10^{-6}} 0.25 \cdot 20 \times 10^{-6} = 1.5\ \text{A}$
$i_{L\,max}$	$I_L + \frac{1}{2}\Delta i_L = 4 + \frac{1.5}{2} = 4.75\ \text{A}$
$i_{L\,min}$	$I_L - \frac{1}{2}\Delta i_L = 4 - \frac{1.5}{2} = 3.25\ \text{A}$

Steady-State Analysis of Buck Converter **EXAMPLE 3.4**

This example illustrates the accuracy of the steady-state analysis based on the small-ripple approximation. The parameters of the buck converter are $V_S = 16$ V, $L = 40\ \mu$H, $C = 470\ \mu$F, $R = 1\ \Omega$, $T_s = 20\ \mu$s, and $D = 0.25$. The important values for the voltage and current waveforms of the buck converter in Example 3.1 are evaluated in Table 3.1. Figure 3.11 shows the simulated converter waveforms.

3.4.3 Evaluation of Output Voltage Ripple

The output of the buck converter contains a ripple component due to the non-ideal filtering characteristics of an LC filter. An exhaustive analysis could be performed to yield an accurate expression for the output voltage including the ripple component. In practice, however, a quick estimation of the magnitude of the output ripple is more practical and useful than the exact analysis. This section presents a simple output ripple analysis.

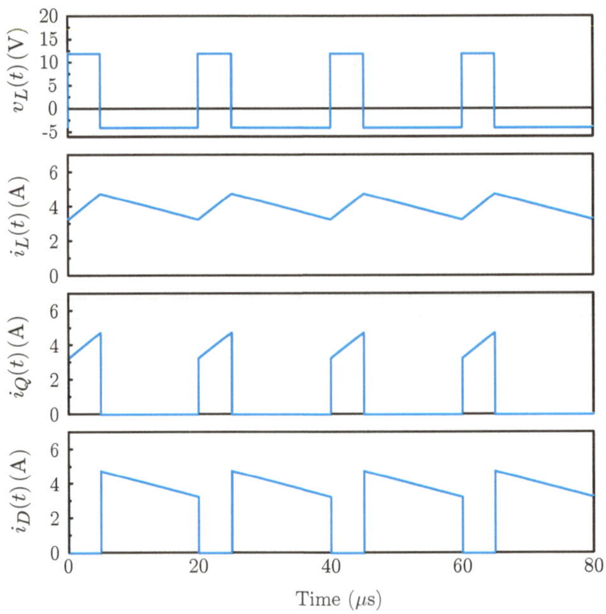

Figure 3.11 Buck converter circuit waveforms.

Evaluation with Ideal Capacitor

Figure 3.12 shows the circuit diagram and waveforms, associated with the ripple analysis. As shown in the earlier analysis, the inductor current is a triangular waveform consisting of dc and ac components. The dc component passes entirely through the load resistor because the capacitor presents an infinitely large impedance to the dc current.

On the other hand, the ac component, the triangular portion of the inductor current, would flow through both the capacitor and the load resistor. However, a practical dc-to-dc converter usually employs a large capacitor to provide sufficient filtering. As a result, the reactance of the capacitor evaluated at the switching frequency is much smaller than the load resistance. Thus, it is highly realistic to assume that the triangular portion of the inductor current fully passes through the capacitor. This situation is illustrated in Fig. 3.12(b).

The triangular current passing through the capacitor develops an ac voltage, thereby creating a ripple component on top of the dc output voltage. The ac voltage across the capacitor is given by

$$v_{ac}(t) = \frac{1}{C} \int i_C(t)\, \mathrm{d}t \tag{3.24}$$

where i_C represents the capacitor current. It can be deduced from (3.24) that v_{ac} grows during the period the capacitor current i_C is positive and then decays when i_C

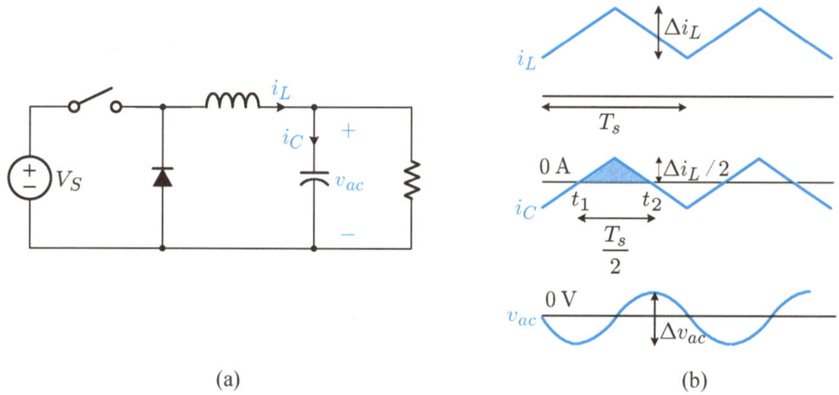

Figure 3.12 Estimation of output ripple. (a) Circuit diagram. (b) Current waveforms and output ripple.

is negative, as illustrated in Fig. 3.12(b). The peak-to-peak value of v_{ac} is then found by integrating i_C over the time period during which i_C remains positive

$$\boxed{\Delta v_{ac} = \frac{1}{C}\int_{t_1}^{t_2} i_C(\tau)\,\mathrm{d}\tau} \tag{3.25}$$

The integration on the right-hand side of (3.25) in effect corresponds to the area of the triangle enclosed by the positive inductor current and time axis, as highlighted in Fig. 3.12(b). By evaluating the area of the shaded triangle with the incorporation of the fact $|t_2 - t_1| = T_s/2$, the expression (3.25) becomes

$$\Delta v_{ac} = \frac{1}{C}\left(\frac{1}{2}\right)\left(\frac{1}{2}\Delta i_L\right)\left(\frac{1}{2}T_s\right) \tag{3.26}$$

Incorporating (3.19) into (3.26) leads to the expression for the peak-to-peak variation of v_{ac}

$$\boxed{\Delta v_{ac} = \frac{1}{8}\frac{V_O}{LC}(1-D)T_s^2} \tag{3.27}$$

The voltage swing Δv_{ac}, given by (3.27), corresponds to the magnitude of the ripple component, Δv_O, superimposed on the dc value of the output voltage, V_O.

Current Waveforms and Output Ripple — EXAMPLE 3.5

This example illustrates the accuracy of the preceding output ripple analysis. Figure 3.13 shows the simulated waveforms for the inductor current i_L, capacitor current i_C, and output voltage v_O of the buck converter used in the previous example. Based

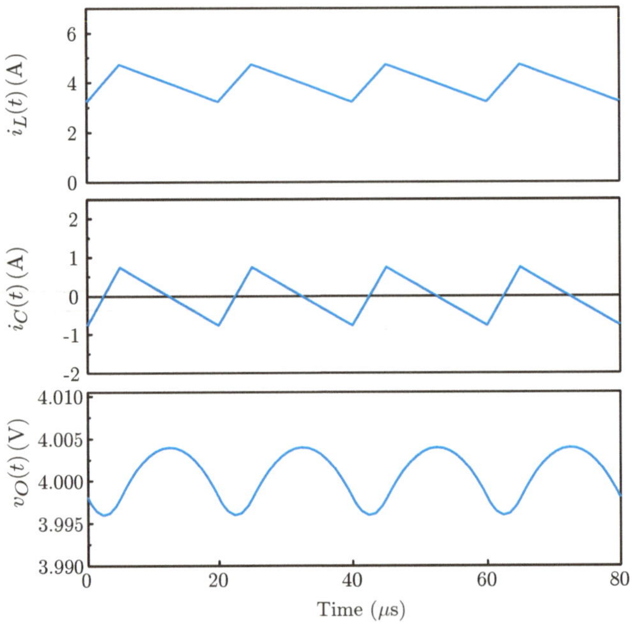

Figure 3.13 Current waveforms and output ripple of buck converter.

on (3.27), the amplitude of the output ripple is estimated as

$$\begin{aligned}\Delta v_O = \Delta v_{ac} &= \frac{1}{8}\frac{V_O}{LC}(1-D)T_s^2\\ &= \frac{1}{8}\frac{4}{40\times10^{-6}\,470\times10^{-6}}(1-0.25)\left(20\times10^{-6}\right)^2\\ &= 7.98\ \text{mV}\end{aligned}$$

The result of this time-domain analysis is very close to the outcome of the previous frequency-domain analysis in Example 3.1. The current analysis shows $\Delta v_O =$ 7.98 mV, whereas the previous frequency-domain analysis predicted $\Delta v_O = 7.76$ mV. The exact PSpice® simulation in Fig. 3.13 shows $\Delta v_O = 7.95$ mV.

Effects of Parasitic Resistance of Capacitor

Earlier discussions about the steady-state operation of the buck converter are based on the assumption that all the circuit components are ideal. In practice, however, the performance of the converter is influenced by non-ideal characteristics of real circuit components. The most noticeable among them is the impact of the parasitic resistance of a real capacitor.

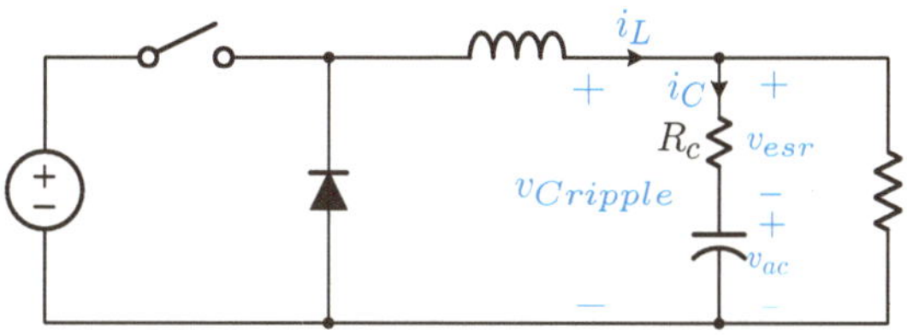

Figure 3.14 Buck converter with practical capacitor.

Real capacitors contain an internal parasitic resistance, referred to as the *equivalent series resistance* (*esr*), due to the non-ideal characteristics of the dielectric material. Consequently, the circuit model of a real capacitor should include the esr, labeled as R_c in Fig. 3.14; a real capacitor is considered as a series connection of the esr and capacitance. With the presence of the esr, the voltage ripple within a practical capacitor, $v_{Cripple}$, is divided into two components — the ripple voltage produced by the current-carrying capacitance, v_{ac}, and the ripple voltage due to the voltage drop at the esr, v_{esr}, as shown in Fig. 3.14

$$v_{Cripple}(t) = v_{ac}(t) + v_{esr}(t) = \frac{1}{C}\int i_C(t)\,\mathrm{d}t + i_C(t)R_c \tag{3.28}$$

where i_C denotes the capacitor current, which corresponds to the triangular portion of the inductor current. From (3.28), the magnitude of the output voltage ripple is expressed as

$$\boxed{\Delta v_O = \Delta v_{Cripple} \approx \frac{1}{C}\int_{t_1}^{t_2} i_C(\tau)\,\mathrm{d}\tau + \Delta i_C R_c} \tag{3.29}$$

Now, using the fact $\Delta i_C = \Delta i_L$, the magnitude of the output ripple is given by

$$\Delta v_O \approx \frac{1}{C}\int_{t_1}^{t_2} i_C(\tau)\,\mathrm{d}\tau + \Delta i_L R_c \tag{3.30}$$

The effect of the esr on the output voltage ripple is rather substantial. The magnitude of the ripple component at the esr is usually much larger than the ripple component produced by the capacitance itself. Accordingly, the magnitude of the output voltage ripple is approximated as

$$\boxed{\Delta v_O \approx \Delta i_L R_c} \tag{3.31}$$

considering the fact

$$\frac{1}{C}\int_{t_1}^{t_2} i_C(\tau)\,\mathrm{d}\tau \ll \Delta i_L R_c \tag{3.32}$$

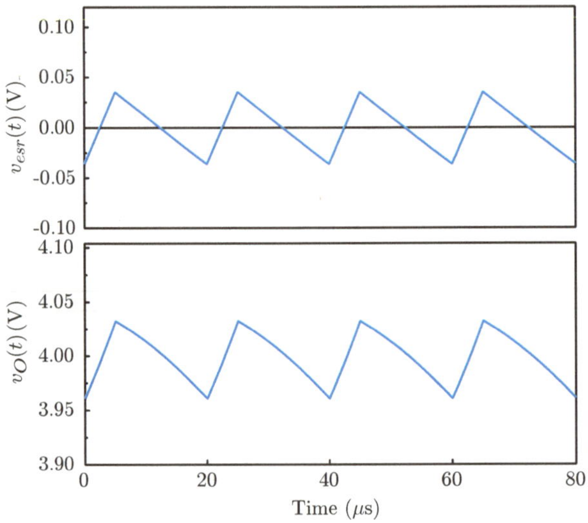

Figure 3.15 Capacitor esr voltage v_{esr} and output voltage v_O.

EXAMPLE 3.6 Output Ripple with Capacitor Esr

This example illustrates the effects of the capacitor esr on the output voltage ripple. The esr of the 470 μF output capacitor used in the buck converter examples is assumed as $R_c = 0.05\ \Omega$. Figure 3.15 shows the simulated waveforms of the voltage across the esr, v_{esr}, and the output voltage, v_O, of the buck converter. As predicted, the output ripple is similar to the triangular-shaped voltage drop at the capacitor esr. The amplitude of the output ripple is also close to the analytical prediction of (3.31): $\Delta v_O \approx \Delta i_L R_c = 1.5 \cdot 0.05 = 0.075$ V.

As illustrated above, the ripple component due to the current-carrying capacitance is negligible and the output voltage ripple is practically determined by the esr of the output capacitor. However, this does not imply that the previous ripple analysis with an ideal capacitor is pointless. The method of ripple calculation in (3.26) is still useful in the sense that the method itself can be applied to many other cases of dc-to-dc converter analyses.

3.5 BUCK CONVERTER IN DISCONTINUOUS CONDUCTION MODE

The operation of the buck converter is relatively simple and easy to understand. The piecewise linear analysis, based on the small-ripple approximation and flux linkage/charge balance conditions, yields an accurate solution for the steady-state operation. However, there are also cases where the buck converter does not follow the aforementioned operational principles and exhibits complex circuit behaviors. This section deals with a new operational mode, known as the discontinuous conduction mode operation.

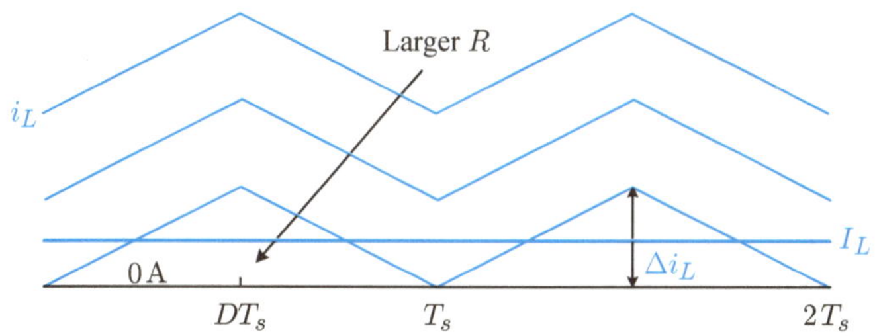

Figure 3.16 Inductor current of buck converter as load resistance increases.

3.5.1 Origin of Discontinuous Conduction Mode Operation

The existence of the new operational mode is explained using Fig. 3.16 which shows a series of inductor currents for a buck converter. Each inductor current waveform is generated with a different load resistance but with the same duty ratio. As the load resistance becomes larger, the inductor current shifts downwards while retaining the same shape. This is because the average value of the inductor current is inversely proportional to the load resistance

$$I_L = \frac{V_O}{R} \tag{3.33}$$

yet the slope of the inductor current remains the same regardless of the change in the load resistance — the inductor voltage that determines the slope of the inductor current will not change as long as the duty ratio remains unaltered. As the load resistance continues to increase, an instant emerges where the minimum value of the inductor current becomes zero. As shown in Fig. 3.16, this situation occurs when

$$I_L = \frac{V_O}{R} = \frac{1}{2}\Delta i_L \tag{3.34}$$

If the load resistance is further increased beyond the value that satisfies (3.34), while the converter obeys the same operational principles, the inductor current will shift further down, thereby becoming negative for a certain interval in the switching period. This situation implies that the inductor current should change its direction. However, the current reversal is impossible because the inductor current flows through the diode which is unable to deliver current in the opposite direction.

In reality, when the load resistance becomes large enough to push the average inductor current below the critical value, $I_L = \Delta i_L/2$, the converter no longer follows the operational principles discussed earlier and enters a new operational mode. This new operational mode is called the *discontinuous conduction mode* (*DCM*) where the inductor current vanishes for a certain time interval within each switching period, thereby becoming discontinuous.

The inductor current in DCM operation is illustrated in Fig. 3.17. The qualitative behavior of the inductor current is explained below.

1) During an on-time period, the inductor voltage, $v_L = V_S - V_O$, is always positive and this forces the inductor current to grow linearly. However, as illustrated in Fig. 3.17, the slope of the on-time inductor current varies when the load

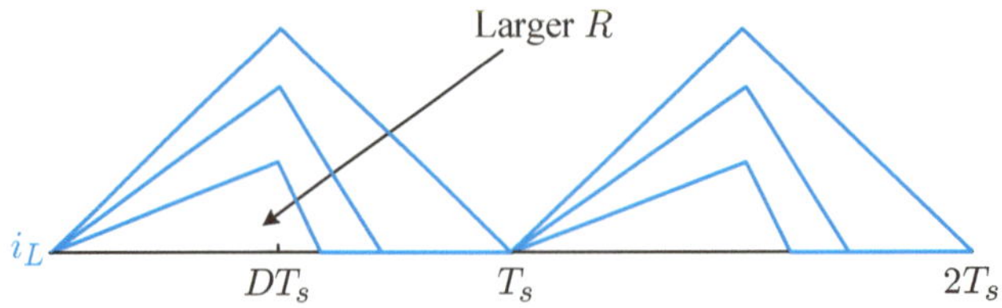

Figure 3.17 Inductor current in DCM operation.

resistance is changed in DCM – the larger the load resistance, the more gradual the slope. The reason for this behavior is that *the voltage gain in DCM operation depends on not only the duty ratio but also the load resistance*. As will be demonstrated in Section 3.5.3, the DCM voltage gain is proportional to the load resistance; namely, as the load resistance becomes larger, so does the voltage gain. Accordingly, a greater load resistance produces a larger output voltage, thereby lowering the ascending slope of the inductor current, $(V_S - V_O)/L$.

2) During an off-time period, the negative inductor voltage, $v_L = -V_O$, forces the inductor current to decay linearly. The condition for DCM is $I_L < \Delta i_L/2$, and this condition implies that the inductor current reduces to zero before the next switching period begins. When the inductor current becomes zero, the diode is turned-off and remains off for the remaining part of the switching period. As addressed above, the output voltage in DCM increases in proportion to the load resistance. Accordingly, as the load resistance grows larger, the decaying slope becomes steeper. As the decaying slope becomes steeper, the period during which the inductor current is absent grows wider, as illustrated in Fig. 3.17.

In contrast to the DCM operation, the case where the inductor current is present all the time is called *continuous conduction mode* (*CCM*) operation. We implicitly assumed CCM operation in the converter analyses of the previous sections. It should be noted that the DCM operation is as equally important and practical as the CCM operation, because all the dc-to-dc converters, even intended for CCM operation at normal conditions, will encounter DCM operation when the load current becomes smaller than the critical value, $I_L = \Delta i_L/2$.

3.5.2 Conditions for DCM Operation

It is evident from the previous analysis that the operational mode of the converter is determined as

$$\begin{aligned} I_L &> \frac{1}{2}\Delta i_L : \quad \text{CCM} \\ I_L &= \frac{1}{2}\Delta i_L : \quad \text{borderline between CCM and DCM} \\ I_L &< \frac{1}{2}\Delta i_L : \quad \text{DCM} \end{aligned} \tag{3.35}$$

Referring to (3.18) and (3.19), the condition for the borderline between CCM and DCM operation is expressed as

$$I_L = \frac{1}{2}\Delta i_L \quad \Rightarrow \quad \frac{V_O}{R} = \frac{1}{2}\frac{V_O}{L}(1-D)T_s \tag{3.36}$$

Equation (3.36) is used to find the critical value for the load resistance or filter inductance that places the converter at the CCM/DCM borderline

$$R_{crit} = \frac{2L}{(1-D)T_s} \tag{3.37}$$

$$L_{crit} = \frac{(1-D)RT_s}{2} \tag{3.38}$$

Equations (3.37) and (3.38) enable us to determine the operational mode, based on the value of the specific circuit component. When the load resistance is larger than R_{crit}, the converter establishes DCM operation. Likewise, when the inductance is reduced to a value smaller than L_{crit}, the converter enters DCM operation.

Operational Mode Change with Resistance Variation — EXAMPLE 3.7

This example illustrates the change in the operational mode as the load resistance is varied. The critical resistance R_{crit} of the buck converter in the preceding examples is evaluated as

$$R_{crit} = \frac{2L}{(1-D)T_s} = \frac{2 \cdot 40 \times 10^{-6}}{(1-0.25)\,20 \times 10^{-6}} = 5.33\ \Omega$$

The inductor current waveforms of the buck converter, whose load resistance is varied between $1\ \Omega < R < 150\ \Omega$, are shown in Fig. 3.18. The converter is on the boundary of CCM operation and DCM operation when $R = 5.33\ \Omega = R_{crit}$ and enters DCM when the load resistance is further increased.

Operational Mode Change with Inductance Variation — EXAMPLE 3.8

This example shows the operational mode change when the inductance is altered. The critical inductance L_{crit} of the previous buck converter with a load resistance $R = 1\ \Omega$ is given by

$$L_{crit} = \frac{(1-D)RT_s}{2} = \frac{(1-0.25)1 \cdot 20 \times 10^{-6}}{2} = 7.5\ \mu\text{H}$$

Figure 3.19 shows the inductor current waveforms of the converter where the inductance is varied between $4\ \mu\text{H} < L < 40\ \mu\text{H}$. As the inductance becomes smaller, the inductor current swing Δi_L increases, thereby causing the converter to operate on the borderline when $L = 7.5\ \mu\text{H} = L_{crit}$, and in DCM when $L < L_{crit}$.

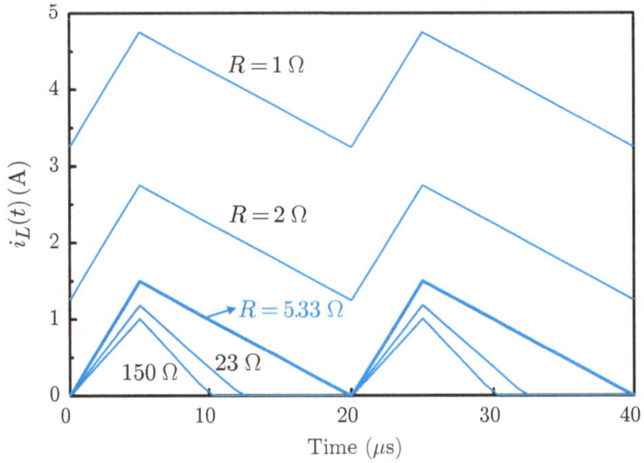

Figure 3.18 Inductor current waveforms with different load resistances.

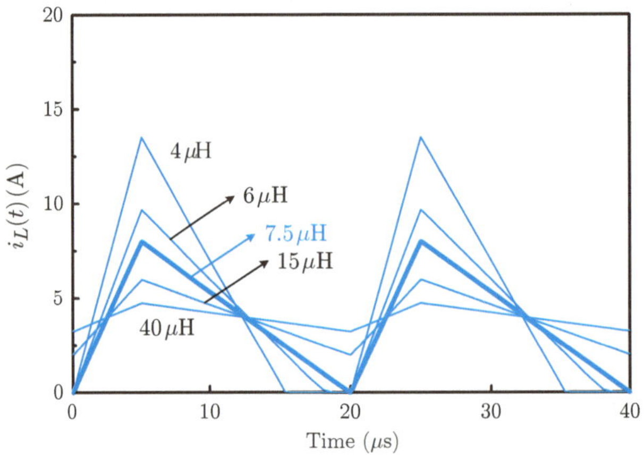

Figure 3.19 Inductor current waveforms with different filter inductances.

3.5.3 Steady-State Operation in DCM

In DCM operation, three topological modes exist within one switching period, as illustrated in Fig. 3.20. In addition to the on-time and off-time subcircuits found in CCM operation, a new subcircuit appears as the inductor current disappears during an off-time period. This third subcircuit is referred to as the DCM subcircuit in Fig. 3.20(a).

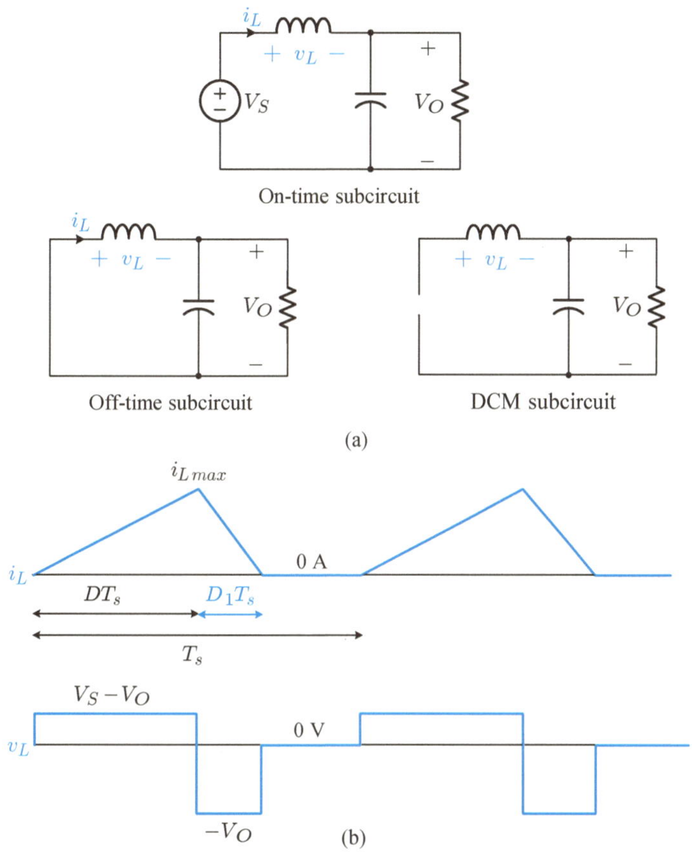

Figure 3.20 Operation of buck converter in DCM. (a) Three subcircuits. (b) Inductor current i_L and inductor voltage v_L.

Figure 3.20(b) depicts typical DCM waveforms for the inductor current and inductor voltage. The notation D_1 in Fig. 3.20(b) is defined as

$$D_1 \equiv \frac{\text{a part of off-time period in which inductor current exists}}{\text{switching period}}$$

When the converter is reduced to the DCM subcircuit, the inductor current and voltage both become zero, as shown in Fig. 3.20(b): $i_L = 0 \Rightarrow \Delta i_L = 0 \Rightarrow v_L = L(\Delta i_L/\Delta t) = 0$. By applying the volt-sec balance condition to the inductor, it follows that

$$(V_S - V_O)DT_s = V_O D_1 T_s \tag{3.39}$$

which is simplified to

$$\frac{V_O}{V_S} = \frac{D}{D + D_1} \tag{3.40}$$

The unknown variable D_1 should be eliminated from (3.40) to result in the complete DCM voltage gain. For the elimination of D_1, an additional equation is formulated as follows. The average value of the inductor current should be the same as the load current, because the average capacitor current is zero due to the charge balance condition. From the geometry of the inductor current, the average inductor current is given by

$$I_L = \frac{\frac{1}{2} i_{L\,max}(D + D_1)T_s}{T_s} \tag{3.41}$$

Based on the facts that

$$I_L = I_O = \frac{V_O}{R} \tag{3.42}$$

and

$$i_{L\,max} = \frac{V_O}{L} D_1 T_s \tag{3.43}$$

the expression (3.41) is rewritten as

$$\frac{V_O}{R} = \frac{1}{2}\frac{V_O}{L} D_1 T_s (D + D_1) \tag{3.44}$$

which is simplified to

$$D_1^2 + DD_1 - \frac{2L}{RT_s} = 0 \tag{3.45}$$

By solving (3.45) for a positive D_1

$$D_1 = \frac{1}{2}\left(-D + \sqrt{D^2 + \frac{8L}{RT_s}}\right) \tag{3.46}$$

and finally incorporating (3.46) into (3.40), the DCM voltage gain is determined as

$$\boxed{\frac{V_O}{V_S} = \frac{2D}{D + \sqrt{D^2 + \dfrac{8L}{RT_s}}}} \tag{3.47}$$

The voltage gain is a nonlinear function of the circuit parameters and operational conditions.

A couple of informative observations can be made from the previous analysis. First, it is evident from (3.47) that the voltage gain increases as the load resistance grows larger, as addressed earlier in conjunction with the behavior of the inductor current in DCM. Second, the following relationship usually holds

$$\frac{2D}{D + \sqrt{D^2 + \dfrac{8L}{RT_s}}} > D \tag{3.48}$$

Table 3.2 Steady-State Analysis in DCM

Circuit variable	Expression
D_1	$\frac{1}{2}\left(-D+\sqrt{D^2+\frac{8L}{RT_s}}\right)$
	$=\frac{1}{2}\left(-0.25+\sqrt{0.25^2+\frac{8\cdot 40\times 10^{-6}}{12\cdot 20\times 10^{-6}}}\right)=0.47$
V_O	$\frac{2D}{D+\sqrt{D^2+\frac{8L}{RT_s}}}V_S$
	$=\frac{2\cdot 0.25}{0.25+\sqrt{0.25^2+\frac{8\cdot 40\times 10^{-6}}{12\cdot 20\times 10^{-6}}}}16=5.59\text{ V}$
$i_{L\,max}$	$\frac{V_O}{L}D_1T_s=\frac{5.59}{40\times 10^{-6}}0.47\cdot 20\times 10^{-6}=1.31\text{ A}$

for the properly selected buck converter parameters. Equation (3.48) indicates that the DCM voltage gain is larger than the CCM counterpart when the same duty ratio is assumed.

Steady-State Analysis in DCM — EXAMPLE 3.9

This example illustrates the circuit waveforms of a buck converter in DCM. The load resistance of the buck converter used in the previous examples is increased to $R = 12\ \Omega$, while the other parameters retain the original values: $V_S = 16$ V, $L = 40\ \mu$H, $C = 470\ \mu$F, $T_s = 20\ \mu$s, and $D = 0.25$. Because $R = 12\ \Omega > R_{crit} = 5.33\ \Omega$, the converter is in DCM region. Theoretical predictions for the steady-state response of the converter are listed in Table 3.2. The accuracy of the theoretical predictions is confirmed with the simulated waveforms in Fig. 3.21.

The operational mode of a dc-to-dc converter can also go through the CCM/DCM transition when its duty ratio is varied. The condition for the DCM operation

$$\frac{V_O}{R} < \frac{1}{2}\frac{V_O}{L}(1-D)T_s \tag{3.49}$$

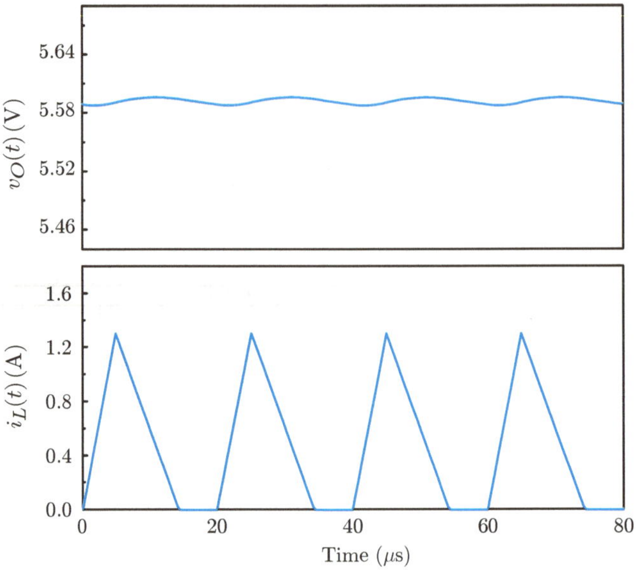

Figure 3.21 Output voltage v_O and inductor current i_L of buck converter in DCM.

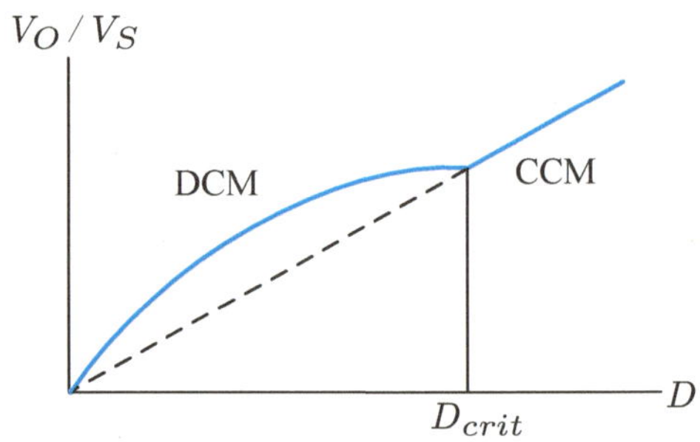

Figure 3.22 Voltage gain of buck converter as function of duty ratio.

is rearranged as

$$D < \left(1 - \frac{2L}{RT_s}\right) \tag{3.50}$$

By defining the right-hand side of (3.50) as

$$1 - \frac{2L}{RT_s} \equiv D_{crit} \tag{3.51}$$

it is concluded that the converter operates in DCM when $0 < D < D_{crit}$ and in CCM with $D_{crit} < D < 1$. Figure 3.22 depicts the voltage gain of a buck converter.

The gain curve follows the DCM gain formula given by (3.47) until the duty ratio increases to D_{crit}, and tracks the CCM formula, $V_O/V_S = D$, thereafter.

It is worthwhile to note that the converter remains in CCM operation for all $0 < D < 1$, if the condition

$$\frac{2L}{RT_s} > 1 \tag{3.52}$$

is satisfied. This condition is derived from

$$D_{crit} = 1 - \frac{2L}{RT_s} < 0 \tag{3.53}$$

which negates the existence of a positive D_{crit}.

Buck Converter Example **EXAMPLE 3.10**

This example shows the case where the converter retains the CCM operation for $0 < D < 1$. For the buck converter used in the previous examples with the circuit parameters of $L = 40\ \mu\text{H}$, $R = 1\ \Omega$, and $T_s = 20\ \mu\text{s}$, it follows that

$$\frac{2L}{RT_s} = \frac{2 \cdot 40 \times 10^{-6}}{1 \cdot 20 \times 10^{-6}} = 4 > 1$$

indicating that the converter will remain in CCM for all $0 < D < 1$.

As demonstrated in this section, the DCM operation causes significant changes in the operational principles and steady-state characteristics of the buck converter. Furthermore, the DCM operation also alters the dynamic characteristics of the converter. While detailed analyses of DCM dynamics will be covered later in Chapter 6, it should be reminded that dc-to-dc converters frequently cross the CCM/DCM boundary and therefore should be designed with both CCM and DCM operation in consideration.

3.6 CLOSED-LOOP CONTROL OF BUCK CONVERTER

In real applications, dc-to-dc converters are powered by a non-ideal voltage source rather than a constant dc source. In addition, converters are loaded with general electrical equipment, not a resistor. Accordingly, dc-to-dc converters experience certain variations in the input voltage and load current.

As addressed in Chapter 1, a dc-to-dc converter is intended to be a voltage source and, as such, the converter should maintain its output voltage at the desired value, regardless of any changes in the input voltage or load current. This feature is referred to as the output voltage regulation or simply *dc regulation.* The term dc regulation implies *regulating the output voltage of a converter at a fixed dc value in steady state.*

To implement the dc regulation, a functional connection must be made between the output voltage and duty ratio of the converter. More specifically, a closed-loop

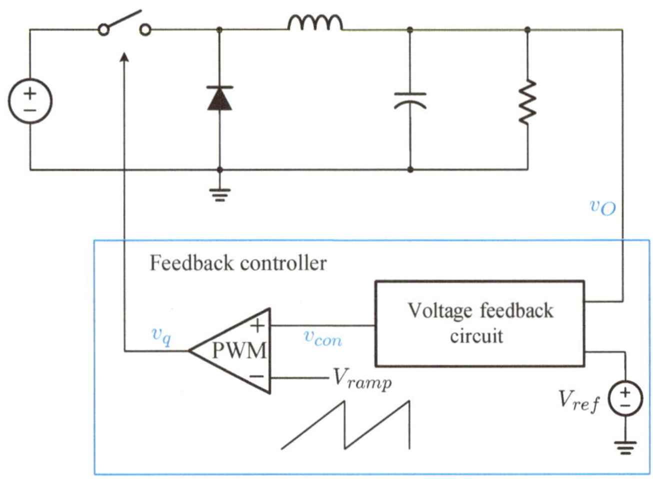

Figure 3.23 Closed-loop controlled buck converter.

feedback control should be employed around the dc-to-dc converter, which adaptively changes the duty ratio of the active switch at the presence of the input voltage and load current variations. This section deals with the dc regulation of a closed-loop controlled buck converter.

Closing a feedback loop is not an easy task and thus requires bread knowledge about modeling, analysis, and control of dynamic systems. Yet, because the interest of the current section is limited to the issues related to the dc regulation, the dynamic modeling, analysis, and control design of dc-to-dc converters are postponed to later chapters.

3.6.1 Closed-Loop Feedback Controller

Figure 3.23 shows a simplified diagram of the buck converter equipped with a closed-loop feedback controller. The feedback controller consists of two functional blocks: the pulsewidth modulation (PWM) block and voltage feedback circuit. The PWM block controls the duty ratio of the active switch using the output of the voltage feedback circuit, labeled as the control voltage v_{con} in Fig. 3.23, and the ramp signal V_{ramp} generated inside the feedback controller. The output of the PWM block is denoted as v_q in Fig. 3.23.

The voltage feedback circuit utilizes the output voltage v_O and reference voltage, denoted as V_{ref} in Fig. 3.23, to generate the control voltage v_{con}. The voltage feedback circuit and PWM block collectively enforce the output of the converter to track the reference voltage in steady state, thereby achieving dc regulation. This section describes the functional details of the closed-loop dc regulation using the PWM scheme.

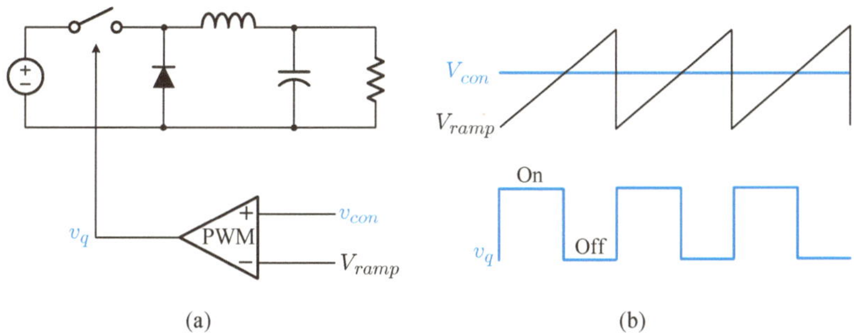

Figure 3.24 PWM block. (a) Block diagram. (b) Timing diagram.

Pulsewidth Modulation

The output of the PWM block is the switch drive signal v_q, whose pulsewidth is modulated to yield the desired duty ratio for the active switch. As shown in Fig. 3.24(a), the PWM block utilizes the control voltage v_{con} and ramp signal V_{ramp}, in order to issue the switch drive signal v_q.

Figure 3.24(b) shows a timing diagram for the PWM waveforms, where the control voltage is assumed to remain constant at $v_{con} = V_{con}$. First, a periodic ramp signal V_{ramp} is generated inside the feedback controller. The period of the ramp signal is in fact the switching period of the converter. The switch drive signal becomes *on* at the instant the ramp signal starts ramping up and resets *off* when the ramp signal intersects the control voltage. By repeating this process, the PWM block generates a periodic switch drive signal whose pulsewidth is modulated in proportion to the magnitude of the control voltage V_{con}.

Figure 3.25 depicts the outputs of the PWM block for two different control voltage values, V_{con} and V'_{con}. The modulation scheme generates a switch drive signal whose trailing edge is modulated to change the duty ratio within a fixed switching period. As such, the modulation scheme is called a *constant-frequency trailing edge modulation.* While alternative modulation schemes are also available for PWM dc-to-dc converters, this scheme is most widely used and thus adopted as the standard PWM in this book.

Voltage Feedback Circuit

Figure 3.26 shows a simplified diagram of a closed-loop controlled buck converter. The output voltage v_O is fed to a voltage feedback circuit, consisting of an op amp, reference voltage V_{ref}, and two impedance blocks, $Z_1(s)$ and $Z_2(s)$. The output of the voltage feedback circuit is the control voltage v_{con}, which is used as the input signal for the PWM block.

The voltage feedback circuit operates based on the principle of the negative feedback. When the output voltage v_O grows larger than the nominal value, the output of the op amp, v_{con}, becomes lower than the previous value. With the lowered v_{con},

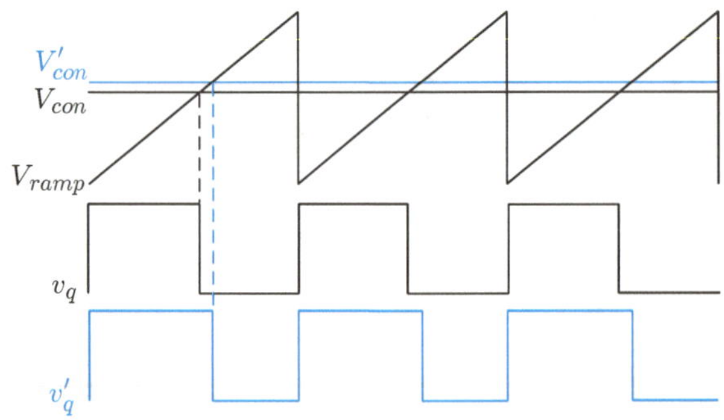

Figure 3.25 Constant-frequency trailing edge modulation.

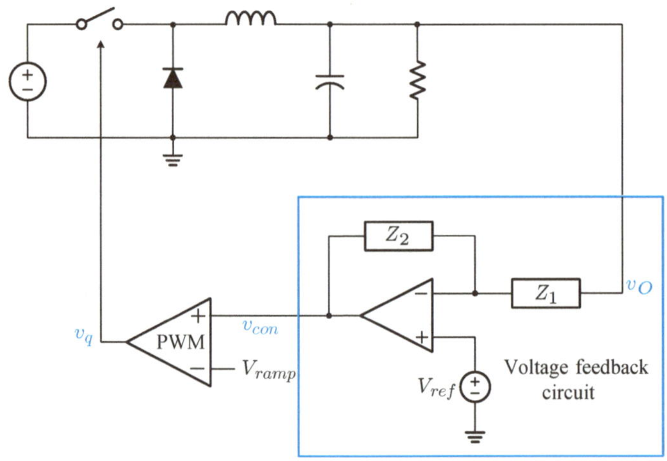

Figure 3.26 Voltage feedback circuit and closed-loop control.

the PWM block produces the switch drive signal v_q whose duty ratio is reduced. The reduced duty ratio in turn brings down the output voltage to the nominal value.

The mechanism of the dc regulation is explained using the node equation formulated at the inverting terminal of the op amp. Using the *virtual short* between the two input terminals of the op amp, the node equation is written as

$$\frac{v_O - V_{ref}}{Z_1(s)} = \frac{V_{ref} - v_{con}}{Z_2(s)} \tag{3.54}$$

which is rearranged as

$$\frac{Z_2(s)}{Z_1(s)}(v_O - V_{ref}) = V_{ref} - v_{con} \tag{3.55}$$

The nature of the dc regulation is deduced from (3.55) as follows.

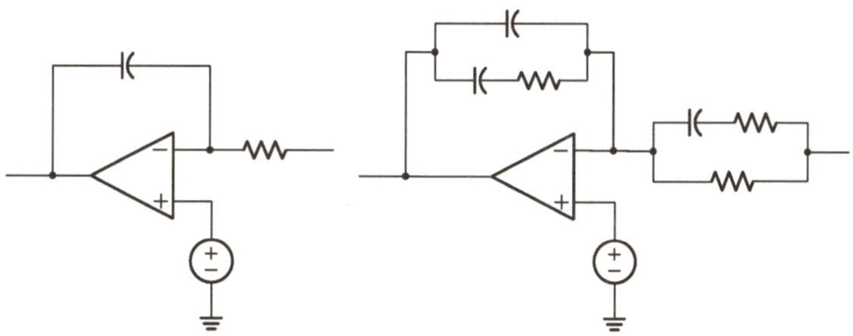

Figure 3.27 Voltage feedback circuits. (a) Miller integrator. (b) Three-pole two-zero circuit.

1) For a proper operation of the converter, the output of the feedback circuit, v_{con}, should have a finite value within the lower and upper bounds of the op amp output voltage.

2) Because both the two variables in the right-hand side of (3.55), V_{ref} and v_{con}, are finite in their magnitude, the term $(|Z_2|/|Z_1|)(v_O - V_{ref})$ in the left-hand side of the equation should be also finite. Otherwise, the expression (3.55) becomes inconsistent.

3) The steady state is an equilibrium where all the time-varying components of circuit variables settle down to zero and the circuit responds only to dc components. As such, the impedances evaluated at dc, $Z(j0)$, are the factor that determines the steady-state operation of the voltage feedback circuit.

4) If the impedance ratio $|Z_2(j0)|/|Z_1(j0)|$ is selected to be infinite, the variable $(V_O - V_{ref})$ should converge to zero in order to make their product finite

$$\frac{|Z_2(j0)|}{|Z_1(j0)|}(V_O - V_{ref}) = \infty \cdot 0 \quad \Rightarrow \quad \text{a finite constant}$$

This condition implies $V_O = V_{ref}$ in steady state, thereby achieving dc regulation.

In conclusion, a voltage feedback circuit that satisfies the condition

$$\boxed{\frac{|Z_2(j0)|}{|Z_1(j0)|} = \infty} \tag{3.56}$$

forces the output voltage to track the reference voltage in steady state. Accordingly, any circuit that meets the condition (3.56) provides dc regulation, thus eligible for a voltage feedback circuit. The simplest among these is Miller integrator shown in Fig. 3.27(a). However, a voltage feedback circuit employing Miller integrator in its original form does not provide good dynamic performance for reasons that will be

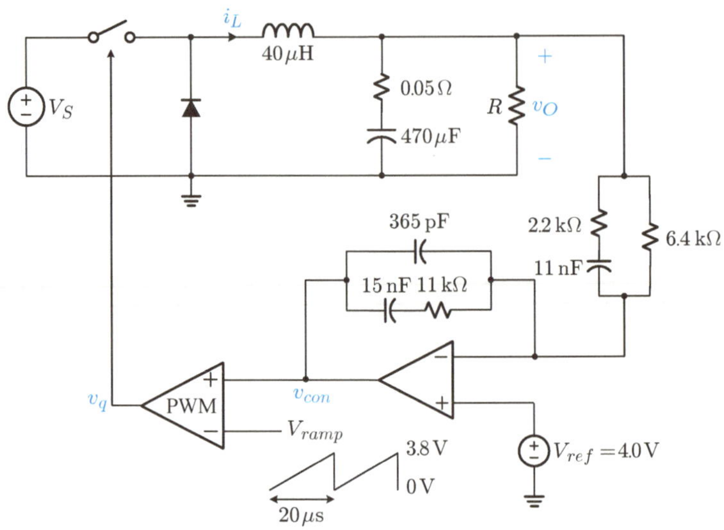

Figure 3.28 Closed-loop controlled buck converter.

discussed in Chapter 8. Consequently, a simple Miller integrator is rarely used for high performance design.

As will be demonstrated in Chapter 8, it has been established from the past researches that a three-pole two-zero circuit shown in Fig. 3.27(b) is the optimal feedback circuit for the buck converter. The three-pole two-zero circuit, named so after its transfer function, provides excellent dynamic performance as well as tight dc regulation. It is easy to confirm that the three-pole two-zero circuit meets the requirement of (3.56). Because the design of voltage feedback circuit requires comprehensive knowledge about dynamic modeling and frequency-domain analysis, this topic will be separately treated in Chapter 8.

3.6.2 Responses of Closed-Loop Controlled Buck Converter

This section presents the time-domain response of a closed-loop controlled buck converter obtained from PSpice® simulations. Figure 3.28 shows the schematic diagram of a closed-loop controlled buck converter. The power stage parameters are identical to those of the buck converter used in the previous examples, however, the input voltage V_S and load resistance R are allowed to change during the converter operation, in order to investigate transient responses of the converter.

The output of the converter is regulated at $V_O = V_{ref} = 4$ V and the switching frequency is $f_s = 1/(20 \times 10^{-6}) = 50$ kHz. The ramp signal varies from 0 to 3.8 V. A three-pole two-zero circuit is employed as the voltage feedback circuit. The circuit parameters for the voltage feedback circuit are optimally selected for good closed-loop performance, as will be demonstrated in Chapter 8.

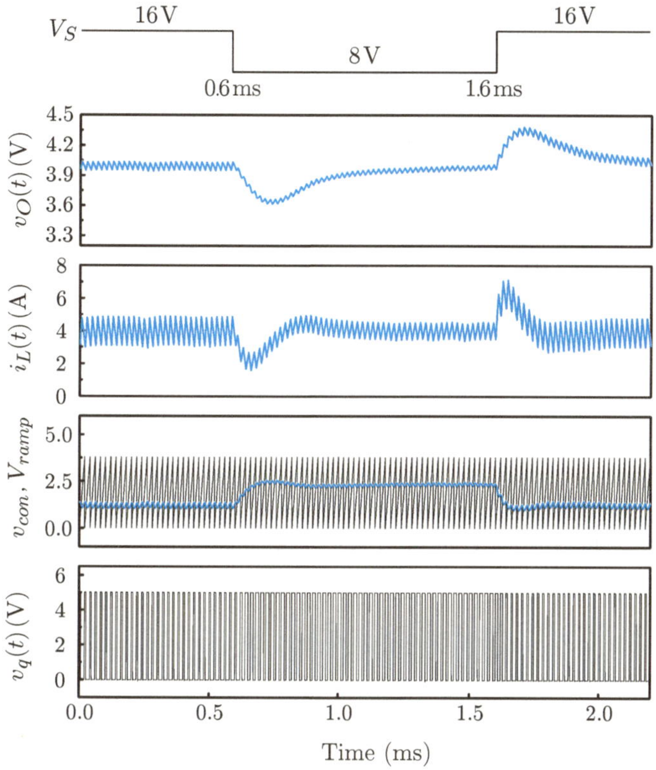

Figure 3.29 Step input response of buck converter: output voltage v_O, inductor current i_L, control voltage v_{con} and ramp signal V_{ramp}, and switch drive signal v_q.

Step Input Response

The buck converter undergoes a series of step changes in its input voltage, whereas the load resistance is fixed at $R = 1\ \Omega$. As shown in Fig. 3.29, the input voltage is changed as $V_S = 16\ \text{V} \Rightarrow 8\ \text{V} \Rightarrow 16\ \text{V}$ during the converter operation. First, the steady-state waveforms of the converter with $V_S = 16$ V and $R = 1\ \Omega$ are analyzed. The output voltage v_O is regulated at 4 V with a small ripple component. The inductor current i_L is a triangular waveform. The average value of the inductor current equals to the load current

$$I_L = \frac{V_O}{R} = \frac{4}{1} = 4\ \text{A}$$

The ripple component of the inductor current is given by

$$\Delta i_L = \frac{V_S - V_O}{L} DT_s = \frac{16 - 4}{40 \times 10^{-6}} \frac{4}{16} 20 \times 10^{-6} = 1.5\ \text{A}$$

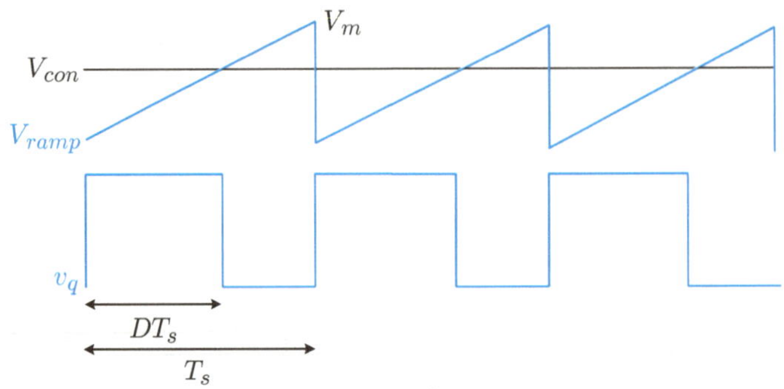

Figure 3.30 PWM block waveforms.

The control voltage v_{con} and ramp signal V_{ramp} are simultaneously shown in the third plot in Fig. 3.29, while the switch drive signal v_q is displayed in the bottom plot. The average value for the control voltage is evaluated from the PWM waveforms shown in Fig. 3.30. From the PWM waveforms, it can be seen that

$$T_s : DT_s = V_m : V_{con} \tag{3.57}$$

where V_m represents the magnitude of the ramp signal and V_{con} is the steady-state control voltage. The above expression is simplified to $V_{con} = DV_m$. With the given operational conditions, the initial value for the control voltage is evaluated as

$$V_{con} = DV_m = \frac{V_O}{V_S}V_m = \frac{4}{16}3.8 = 0.95 \text{ V}$$

The control voltage contains a ripple component on top of the dc value evaluated above. The switch drive signal v_q shows the duty ratio of the active switch. The ramp signal, control voltage, and switch drive signal altogether demonstrate the principles of the PWM control.

When the input voltage steps down from 16 V to 8 V at $t = 0.6$ ms, several changes occur in the converter waveforms. First, the output voltage exhibits a transitional undershoot before it returns to the nominal value $V_O = 4$ V. Upon a sudden decrease in V_S, the feedback controller responds to the change with a certain response time. Before the feedback controller establishes a new steady state, the change in the input voltage influences the output voltage as follows. When the input voltage is reduced, the energy delivered from the input source to the output capacitor is also reduced, incurring an energy deficit in the output capacitor. This energy deficit causes the capacitor voltage to drop, resulting in a transitional undershoot in the output voltage. As time elapses, the feedback controller establishes a new steady state and the output voltage tracks back to the nominal value. The behavior of the inductor current can be similarly explained.

With the reduced input voltage $V_S = 8$ V, the ripple component of the inductor current is decreased to

$$\Delta i_L = \frac{V_S - V_O}{L} DT_s = \frac{8-4}{40 \times 10^{-6}} \frac{4}{8} 20 \times 10^{-6} = 1 \text{ A}$$

The control voltage gradually increases to produce a larger duty ratio, required to regulate the output voltage with the reduced input voltage. The new value for the control voltage is determined as

$$V_{con} = \frac{V_O}{V_S} V_m = \frac{4}{8} 3.8 = 1.9 \text{ V}$$

The duty ratio of the switch drive signal also gradually grows based on the PWM principle.

The transient behavior of the converter due to a step increase in the input voltage is interpreted in the opposite way to the step-down case. The instantaneous increase of the input voltage causes the output voltage and inductor current to go through a transitional overshoot. As the control signal gradually decreases, thereby reducing the duty ratio to $D = 4/16 = 0.25$, the output voltage returns to the nominal value.

The buck converter exhibits very stable and well-controlled transient behavior. In fact, the feedback circuit parameters are carefully selected to yield optimal transient behavior. Discussions about the feedback circuit parameters will be given in Chapter 8.

Step Load Response

Now the converter goes through a series of step changes in the load resistance, $R = 1\ \Omega \Rightarrow 2\ \Omega \Rightarrow 1\ \Omega$, while the input voltage is fixed at $V_S = 16$ V. The pattern of the step changes and the simulated waveforms are shown in Fig. 3.31. With a step load change from $R = 1\ \Omega$ to $2\ \Omega$ at $t = 0.6$ ms, the energy release from the capacitor to the load resistor suddenly becomes smaller than the previous value, resulting in an energy surplus at the output capacitor. Accordingly, the capacitor voltage momentarily exceeds the nominal value, thus producing an output voltage overshoot. The inductor current indicates a decrease in the load current. The inductor current shows that the converter still remains in CCM operation with $R = 2\ \Omega$. The resistance that places the converter on the borderline between CCM and DCM operation is given by

$$R_{crit} = \frac{2L}{(1-D)T_s} = \frac{2 \cdot 40 \times 10^{-6}}{(1-0.25)\, 20 \times 10^{-6}} = 5.33\ \Omega$$

Figure 3.31 also shows the transient waveforms of the control voltage and switch drive signal. After passing through a short transition period, the control voltage and switch drive signal return to their original waveforms — the duty ratio is invariant to the load resistance value, as far as the converter operates in CCM while maintaining the output voltage in regulation.

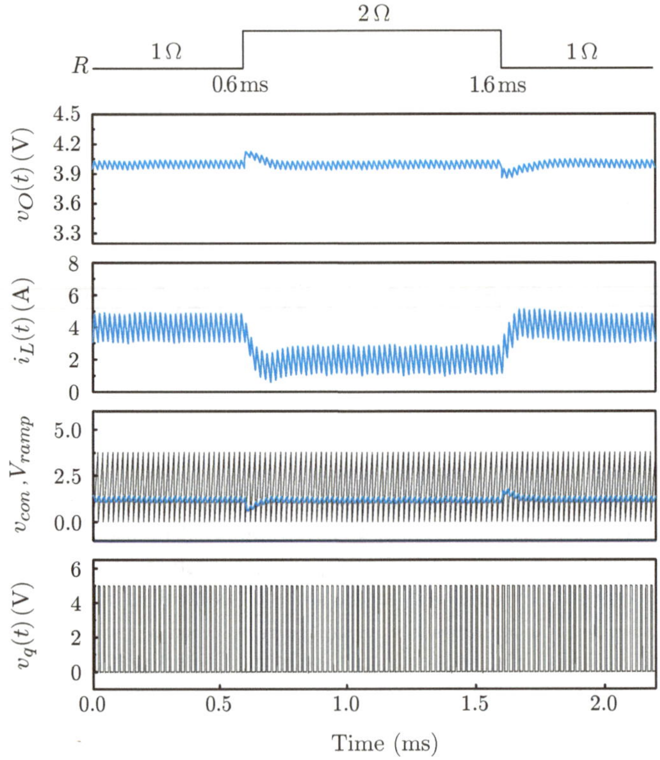

Figure 3.31 Step load response of buck converter.

The transient behavior in response to the step load change from $R = 2\ \Omega$ to $1\ \Omega$ can be understood in the opposite way to the previous analysis. In this case, the output voltage exhibits an undershoot due to the transitional energy deficit.

Operational Mode Change Response

For this case, the load resistance temporarily steps up to a higher value, $R = 8\ \Omega$, well above the critical value of $R_{crit} = 5.3\ \Omega$, while the input voltage remains constant at $V_S = 16$ V. The converter is expected to undergo transitions in the operational mode because it crosses the CCM/DCM boundary. The simulated transient responses of the converter are shown in Fig. 3.32.

With the step increase from $R = 1\ \Omega$ to $8\ \Omega$ at $t = 0.6$ ms, the output voltage exhibits an overshoot and at the same time the converter enters a DCM operation within a few switching cycles, as demonstrated by the inductor current waveform. The output voltage then gradually settles to the target value $V_O = 4$ V, because *the feedback controller guarantees dc regulation regardless of the change in the operational mode.* It is important to notice that, unlike the previous step load response

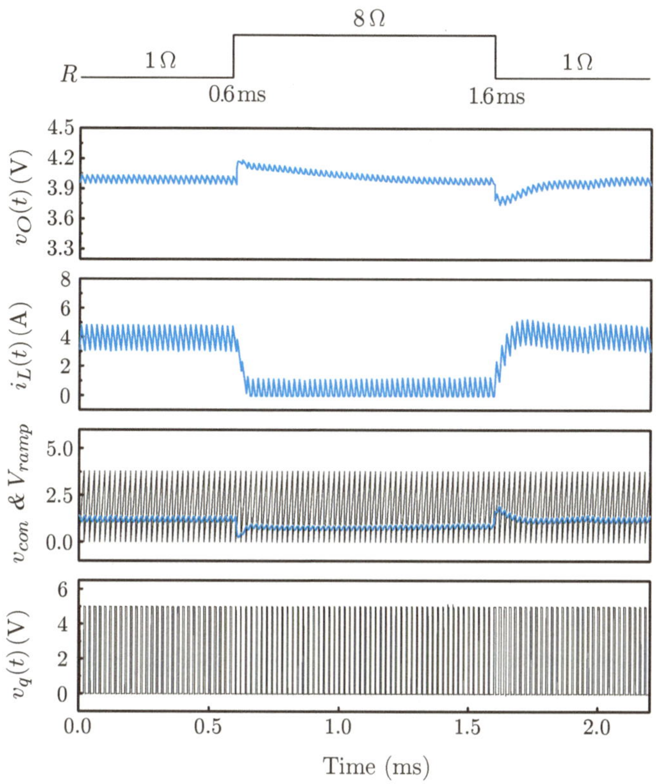

Figure 3.32 Operational mode change response of buck converter.

case where the converter remains in CCM operation all the time and the control voltage retains the same value in steady state, the values of both the control voltage and duty ratio decrease when the converter enters DCM operation. This is because the gain formula in DCM operation differs from the CCM case. As such, the converter must settle in a new duty ratio pertinent to the DCM operation. In fact, the DCM gain is larger than the CCM gain for the same duty ratio. Thus, when the converter enters DCM operation, the duty ratio should be lessened to produce the same output voltage, as shown in Fig. 3.32. With the step decrease in the load resistance from 8 Ω to 1 Ω at $t = 1.6$ ms, the converter returns to CCM operation after a short transition period.

3.7 CHAPTER SUMMARY

A step-down dc-to-dc power conversion is achieved by altering a dc input into a rectangular waveform and filtering the resulting rectangular waveform into another dc value. The buck converter is the simplest circuit that can perform the step-down dc-to-dc power conversion.

The buck converter employs an active-passive switch pair and LC low pass filter. The output of a buck converter invariably contains a ripple component. However, the magnitude of the ripple component is so small that the output voltage can be considered as a pure dc, given by the product of the input voltage and the duty ratio of the active switch — this presumption is called the small-ripple approximation. The small-ripple approximation is very useful for steady-state analysis.

The piecewise linear analysis is adapted to describe the behavior of the buck converter during the start-up transient and in steady state. In the piecewise linear analysis, the converter is decomposed into the on-time and off-time subcircuits. The operation of the converter is investigated by studying each subcircuit individually and later considering the behavior of the two subcircuits collectively. The piecewise linear analysis, when applied along with the small-ripple approximation, readily offers accurate predictions for steady-state converter waveforms. Readers will further appreciate the usefulness of this analysis method in the next chapter which deals with various dc-to-dc converter circuits.

The performance of a dc-to-dc converter is influenced by the non-ideal characteristics of the circuit components. In particular, the equivalent series resistance (esr) of the output capacitor determines the magnitude of the output ripple. Furthermore, it will be shown in Chapter 6 that the esr of the output capacitor has rather significant impacts on the dynamic characteristics of the converter.

Most dc-to-dc converters employ unidirectional switches which only deliver the current in one direction. Accordingly, the switch current cannot alter its direction regardless of any changes in operational conditions. This constraint forces dc-to-dc converters to move into a new operational mode, referred to as a discontinuous conduction mode (DCM), when the operational conditions cross certain boundaries. In DCM operation, there exist time intervals in which the inductor current stays at zero value. The converter operation in DCM differs from the continuous conduction mode (CCM) operation in which the inductor current is present all the time. This operational difference causes notable changes in the steady-state characteristics and transient response of the converter. Details about the DCM operation are addressed in Section 3.5.

A dc-to-dc converter should provide a fixed output voltage for all the operational conditions. For this purpose, the feedback control is employed to adaptively adjust the duty ratio so that the converter produces a desired output voltage regardless of the changes in the input voltage, load current, or operational mode – CCM or DCM. The feedback controller consists of the PWM block and voltage feedback circuit. The PWM block generates a pulsewidth-modulated switch drive signal, using the control voltage provided by the voltage feedback circuit. To regulate the output voltage at a fixed value, the voltage feedback circuit should satisfy the specific condition given by (3.56), $|Z_2(j0)|/|Z_1(j0)| = \infty$, as is the case with the three-pole two-zero circuit shown in Fig. 3.27(b).

PSpice® simulations have proved very useful in visualizing converter operations and substantiating theoretical predictions. The simulations shown in Section 3.6.2 provide valuable insights on both steady-state and transient responses of a closed-loop controlled PWM dc-to-dc converter.

REFERENCES

1. Byungcho Choi, *Pulsewidth Modulated Dc-to-Dc Power Conversion*, Wiley, New Jersey, 2013.
2. R. W. Erickson and D. Maksimović, *Fundamentals of Power Electronics*, Kluwer Academic Publishers, 2001.
3. L. Dixon, "Spice Simulation of Switching Power Supply Performance," Unitrode Power Supply Design Seminar Manual: SEM-800, pp.1-1–1-14, 1991.

PROBLEMS

3.1* Listed below are several useful circuit theorems or analysis techniques that are frequently applied to dc-to-dc power conversion circuits. For each item, give a brief description, mathematical expression, or illustrative example.

i) volt-sec balance condition ii) amp-sec balance condition
iii) small-ripple approximation iv) piecewise linear analysis
v) flux linkage balance condition vi) charge balance condition

3.2* A secondary filter stage is added to a conventional buck converter, yielding a buck converter with a two-stage output filter. The resulting buck converter is shown in Fig. P3.2. Circuit analyses reveal that the transfer function of the two-stage filter can be approximated as

$$\frac{v_o(s)}{v_x(s)} \approx \frac{1}{\left(1+\frac{s}{Q_1\omega_{o1}}+\frac{s^2}{\omega^2_{o1}}\right)\left(1+\frac{s}{Q_2\omega_{o2}}+\frac{s^2}{\omega^2_{o2}}\right)}$$

$$Q_1 = R\sqrt{\frac{C_1}{L_1}} \qquad \omega_{o1} = \frac{1}{\sqrt{L_1C_1}}$$

$$Q_2 = R\sqrt{\frac{C_2}{L_2}} \qquad \omega_{o2} = \frac{1}{\sqrt{L_2C_2}}$$

when the conditions $L_1 \gg L_2$ and $C_1 \gg C_2$ are met.

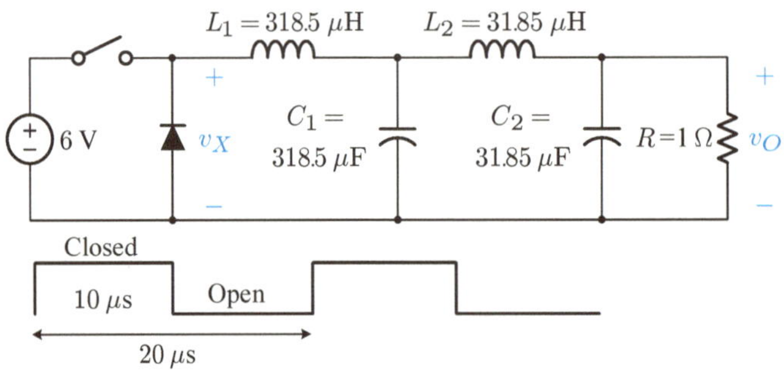

Fig. P3.2

a) Perform the frequency-domain analysis to predict the magnitude of the output voltage ripple. Consider only the first harmonic component of the output voltage.

b) Now assume that the secondary filter stage is removed, $L_2 = C_2 = 0$. For this case, find new values for L_1 and C_1 so that the converter produces the same output voltage ripple as that of a). Assume $Q_1 = R\sqrt{C_1/L_1} = 1$ for this problem.

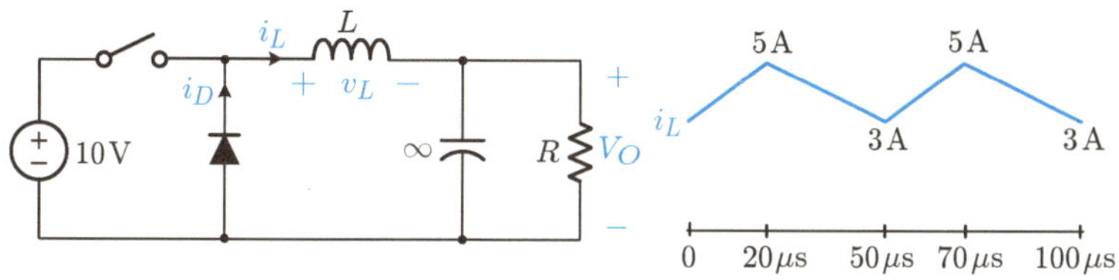

Fig. P3.3

c) Based on the results of a) and b), state the advantage of using a two-stage output filter rather than a single-stage output filter.

3.3 A buck converter and its inductor current waveform are shown in Fig. P3.3.

a) Evaluate the output voltage V_O.

b) Sketch the steady-state waveforms of the inductor voltage v_L and diode current i_D for the two operational periods. Label the maximum and minimum values on your sketch.

c) Evaluate the inductance L and the load resistance R.

3.4* Consider the dc-to-dc conversion system shown in Fig. P3.4. Referring to the switch drive signals for Q_1 and Q_2, answer the questions.

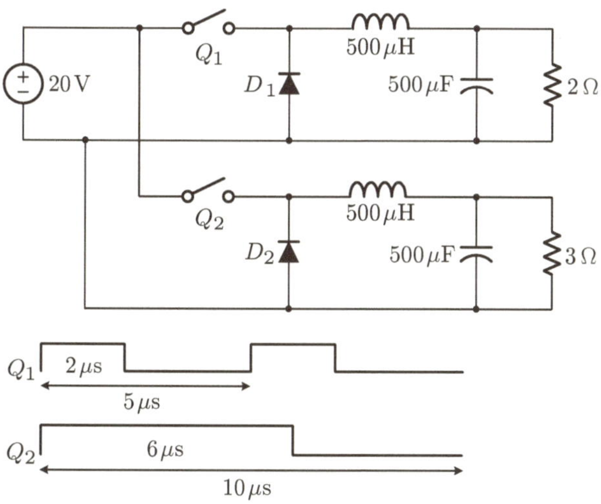

Fig. P3.4

a) First assume that all the circuit components are ideal and the power conversion is performed losslessly. Evaluate the average current flowing through each semiconductor switch, $\bar{i}_{Q1}$, $\bar{i}_{D1}$, $\bar{i}_{Q2}$, and $\bar{i}_{D2}$.

b) Now, assume the following practical devices for the semiconductor switches:
Q_1 and Q_2: MOSFET with the channel resistance of $R_{DS} = 0.5\ \Omega$
D_1 and D_2: Schottky diode with the turn-on voltage of $V_D = 0.5$ V
Calculate the average power dissipated at each practical switch.

c) Lastly, calculate the overall efficiency of the system using the following formula

$$\eta = \frac{P_{out}}{P_{out} + P_{loss}}$$

where P_{out} is the power delivered to the output resistors and P_{loss} is the sum of the power losses, each evaluated in b) for the respective practical switch.

3.5 Shown in Fig. P3.5 is a buck converter along with its switch drive signal.

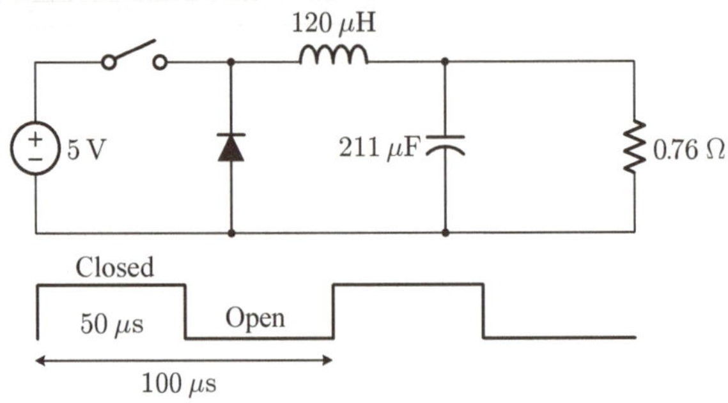

Fig. P3.5

a) Perform the frequency-domain analysis illustrated in Example 3.1, in order to predict the magnitude of the output voltage ripple.
b) Use the equation (3.27) in the text to estimate the size of the output voltage ripple. Compare the result with the outcome of a).

3.6* Figure P3.6 shows two buck converters with different output filter configurations. Referring to the circuit diagrams, answer the questions.

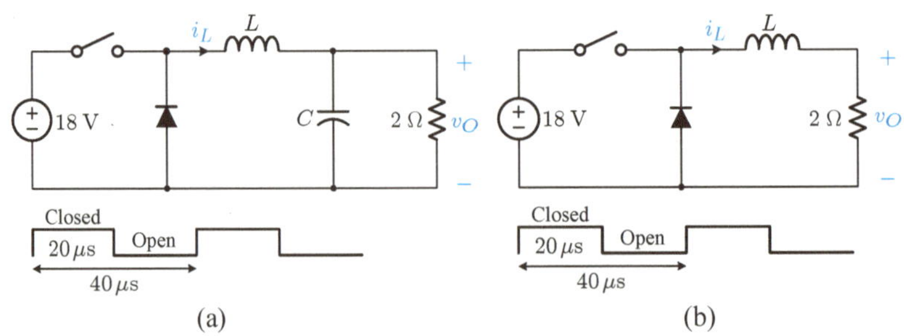

Fig. P3.6

a) For Converter (a), determine the values for L and C to meet the design specifications of $\Delta i_L/I_L = 0.2$ and $\Delta v_O/V_O = 0.02$.
b) For Converter (b), determine the value for L to comply with the design specification of $\Delta v_O/V_O = 0.02$.

3.7* Shown in Fig. P3.7(a) are a buck converter and the Bode plot of its output filter transfer function, $|F_f| = |v_o|/|v_x|$.

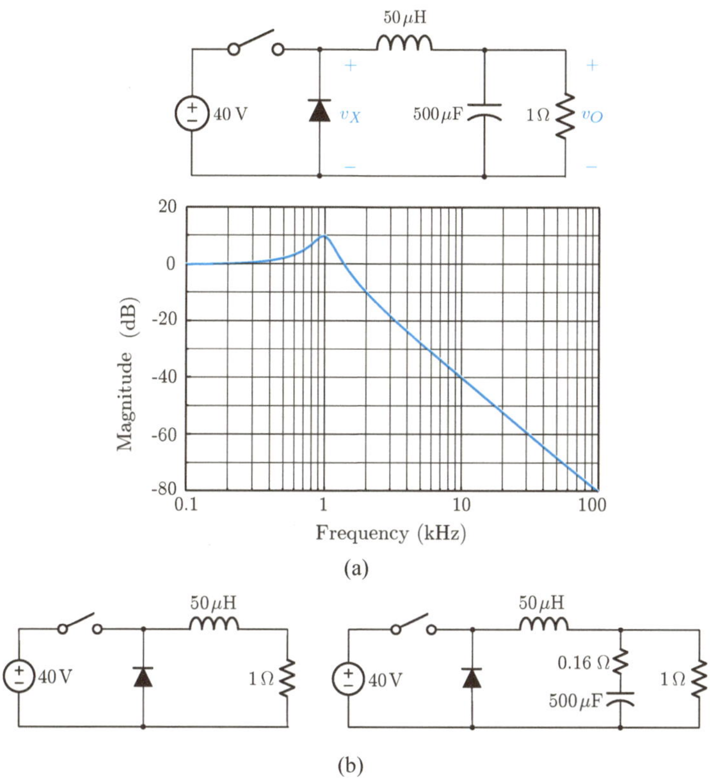

Fig. P3.7

a) Referring to Fig. P3.7(a), estimate the magnitude of the output voltage ripple when the buck converter operates at 30 kHz with $D = 0.5$. Use the information given in the Bode plot.

b) Now assume the buck converter is modified into the two converter circuits shown in Fig. P3.7(b). For each of these two converter circuits, repeat a).

3.8* Referring to Fig. P3.8, answer the following questions.

a) Figure P3.8(a) shows the three buck converters. The common switch drive signal is applied to the three converters. Perform the time-domain analysis to estimate the output voltage ripple of each converter.

b) Shown in Fig. P3.8(b) are the Bode plots of the output filter transfer functions of the three converters. Referring to the Bode plots, perform the frequency-domain analysis to predict the magnitude of the output voltage ripple.

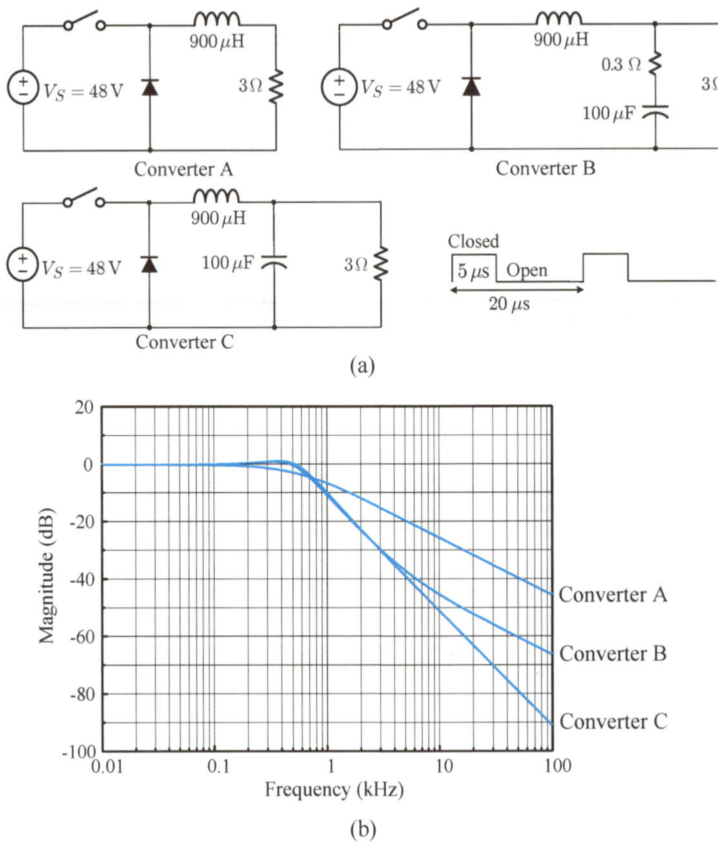

Fig. P3.8

3.9* Figure P3.9 shows the inductor current of a buck converter that operates at the boundary between the continuous conduction mode (CCM) and discontinuous conduction mode (DCM).

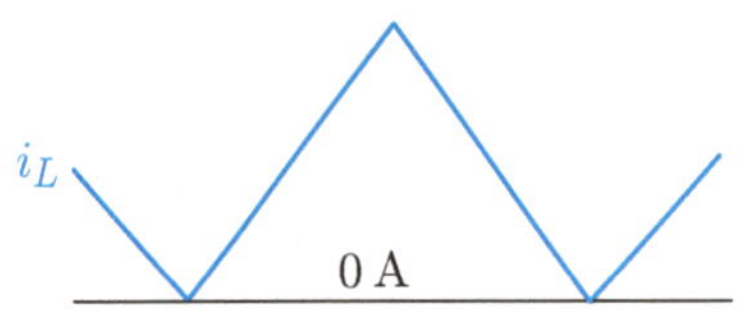

Fig. P3.9

a) Assume that the converter operates in an open-loop condition with a fixed duty ratio. The load resistance is increased beyond the critical value that places the converter at the CCM/DCM borderline, while the input voltage remains the same. Sketch a family of inductor currents, each with a different load resistance, in order to illustrate the transition pattern of the inductor current as the load resistance is increased.

b) The converter is now closed-loop controlled, thereby achieving the output voltage regulation, and again operates at the CCM/DCM boundary. Assuming the load resistance is increased beyond the critical value, while the input voltage remains the same, sketch a family of inductor currents to show the transition pattern of the inductor current.

c) Again assume the closed-loop control and the borderline operation for the converter. For this case, the input voltage is increased beyond the critical value for the CCM/DCM boundary, while the load resistance remains the same. Sketch a family of the inductor currents that exhibits the transition of the inductor current as the input voltage is increased.

3.10* Consider the converters shown in Fig. P3.10 and answer the questions.

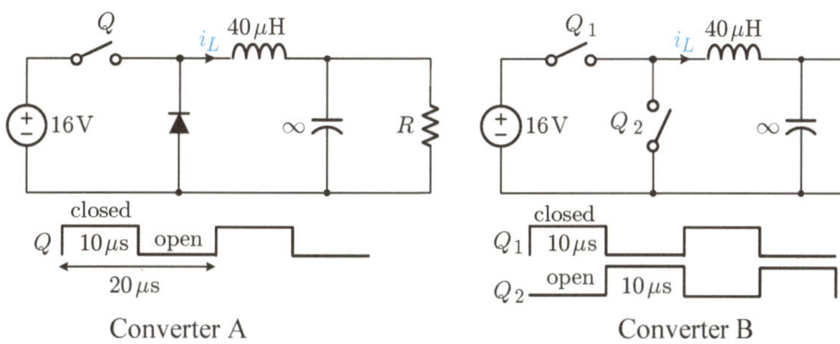

Fig. P3.10

a) In Converter A, a unidirectional switch Q is driven with $D = 0.5$ for $T_s = 20\ \mu s$. Answer the following questions.
i) Sketch the steady-state waveforms of i_L for $R = 2\ \Omega$, $R = 8\ \Omega$, and $R = \infty\ \Omega$. Show the three i_L waveforms simultaneously on the same plot for the two operational periods. Show the maximum and minimum values of the waveforms.
ii) On the plot obtained in i), sketch i_L waveforms for the case of $2\ \Omega < R < 8\ \Omega$ and for the case of $8\ \Omega < R < \infty$, in order to show the evolution pattern of i_L as R varies as $2\ \Omega < R < \infty$.

b) For Converter B, assume Q_1 and Q_2 are ideal bidirectional switches which can deliver the current in both directions. The two bidirectional switches are complementary driven with $D = 0.5$ for $T_s = 20\ \mu s$. Repeat i) and ii) in a) for Converter B.

c) For Converter B, sketch i_L for the case of $R = 18\ \Omega$ for the two operational periods. Show the maximum and minimum values.

3.11 Figure P3.11 shows a buck converter and its inductor current waveform. Based on the given information, evaluate the output voltage V_O, load current I_O, inductance L, and the load resistance R. You are not allowed to use the DCM gain formula in this problem.

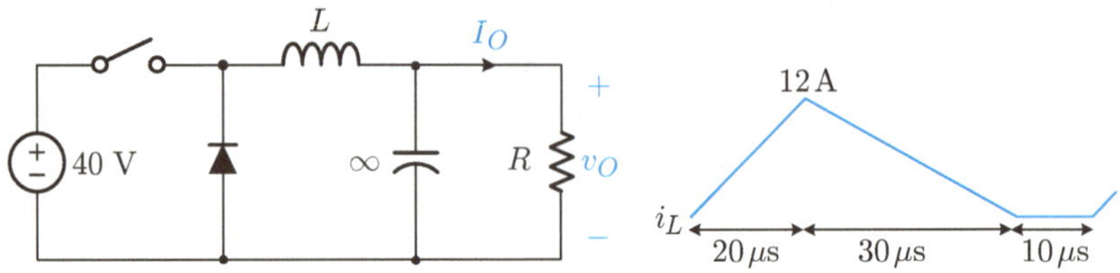

Fig. P3.11

3.12* Shown in Fig. P3.12 are the circuit diagram of a buck converter and its inductor current waveform.

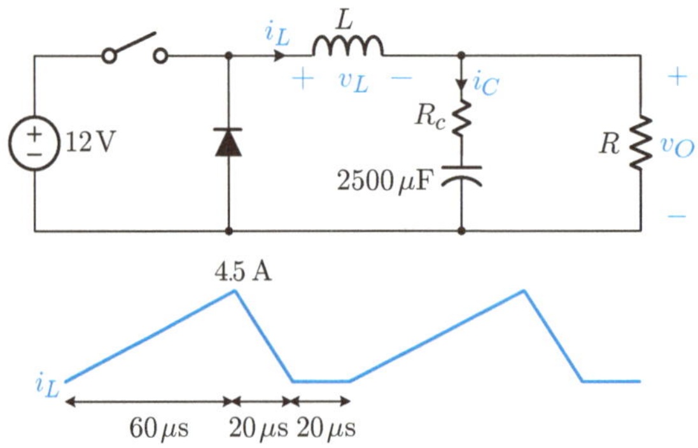

Fig. P3.12

a) Referring to the inductor current i_L, sketch the inductor voltage v_L for the two switching periods. Show the maximum and minimum values on your sketch.

b) Find values for the inductance L and resistance R.

c) Sketch the capacitor current i_C for the two switching periods. Show the maximum and minimum values.

d) Evaluate the output voltage ripple Δv_O when $R_c = 0.12\ \Omega$. Assume $C = 2500\ \mu\text{F}$ is sufficiently large so that the ripple component across the capacitance can be ignored.

e) Evaluate Δv_O when $R_c = 0$.

3.13 Figure P3.13 shows a family of voltage conversion curves of a buck converter, as a function of the duty ratio D and dimensionless parameter $\tau = 2L/(RT_s)$. The converter operates with $T_s = 20\ \mu\text{s}$.

a) Sketch the steady-state waveform of i_L when the operational condition of the converter is located at Point A. Show the maximum, minimum, and average values of the waveform on your sketch.

b) Assume that the operational condition is located at Point B and repeat a).

c) Repeat a) for Point C.

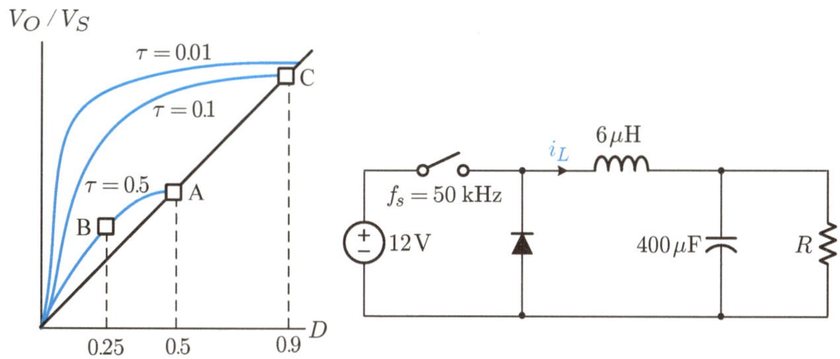

Fig. P3.13

3.14 Referring to the buck converter in Fig. P3.14, answer the following questions.

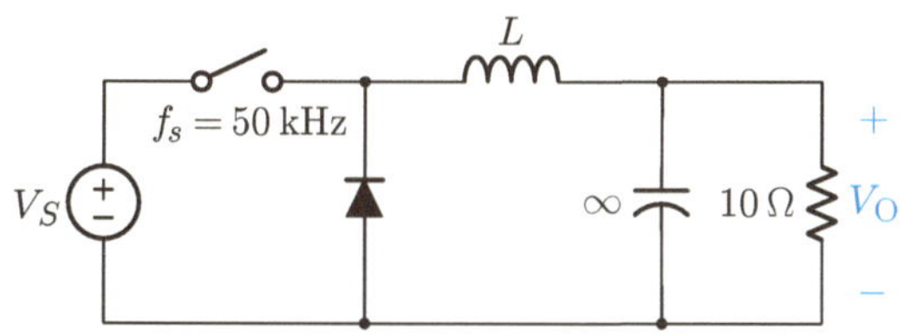

Fig. P3.14

a) Determine the inductance L so that the converter operates at the CCM/DCM borderline with $D = 0.6$.

b) Find the minimum inductance L that ensures CCM operation of the converter for all $0 < D < 1$.

c) Now assume $L = 60\ \mu$H and sketch the general shape of the dc voltage gain curve, V_O/V_S vs D, for all $0 < D < 1$.

3.15 Figure P3.15 shows two different buck converters. Converter (a) operates in an open-loop fashion, while Converter (b) operates with a closed-loop control.

a) For Converter (a), answer the following questions.

i) Assume $R_x = \infty$ and sketch the steady-state waveforms of i_L and v_{con} for the two operational periods.

ii) Now assume $R_x = 10$ kΩ and repeat i).

b) Repeat a) for Converter (b).

3.16* Figure P3.16 shows four buck converters. Assume the circuit components are all ideal. For each converter, calculate the average values of the control voltage v_{con} and output voltage v_O in CCM operation.

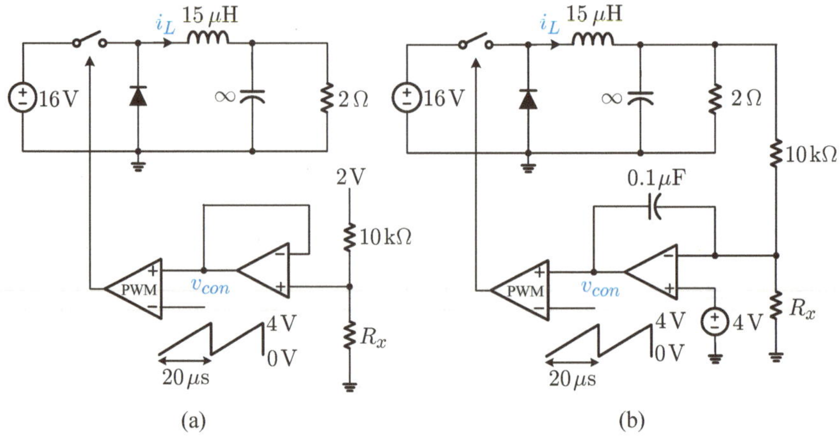

Fig. P3.15

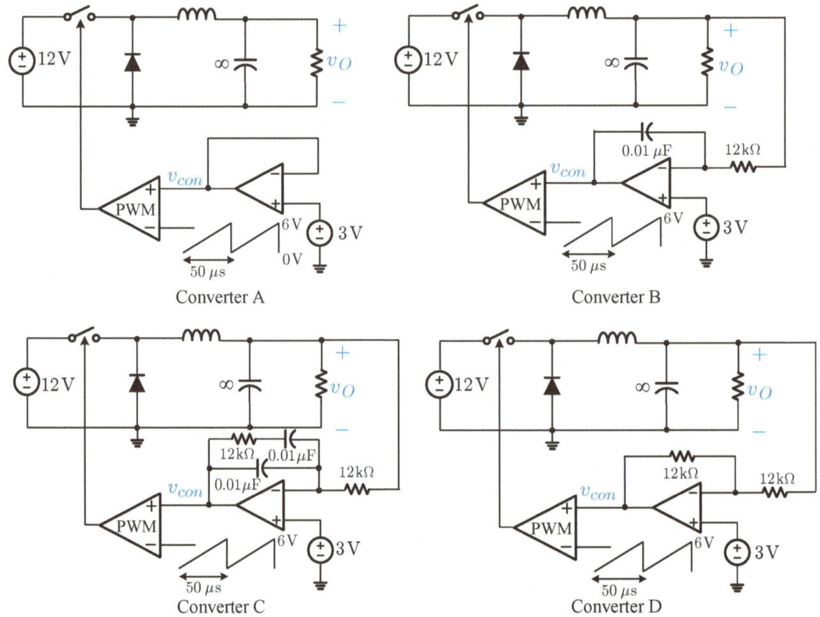

Fig. P3.16

3.17* Assume all the circuit components are ideal in the two converter circuits shown in Fig. P3.17.

a) For Converter A, draw i_L waveforms for the two operational periods when $R = 5\ \Omega$, $R = 8\ \Omega$, and $R = 12\ \Omega$. Show the average value, ascending slope, and transition pattern of i_L in a comparative manner. Show the peak value of i_L when $R = 5\ \Omega$.

b) Repeat a) for Converter B.

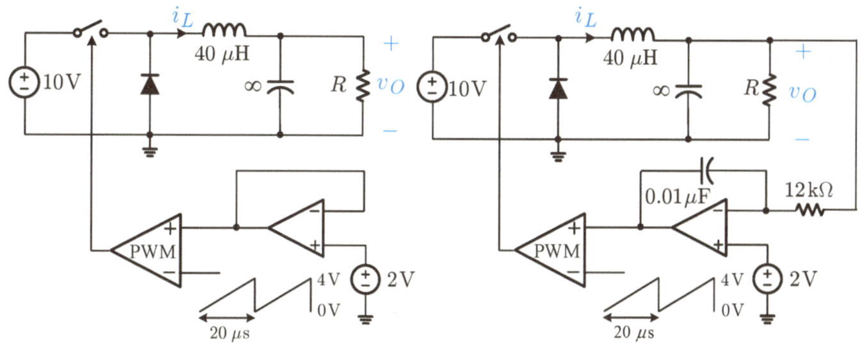

Converter A　　　　Converter B

Fig. P3.17

3.18 Consider the closed-loop controlled buck converter shown in Fig. P3.18 and answer the questions. Assume $R_1 = \infty$ in the voltage feedback circuit for Problems a) and b).

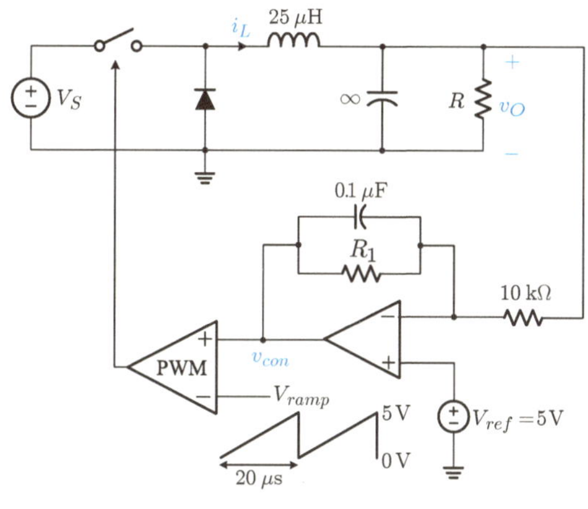

Fig. P3.18

a) With $V_S = 20\text{V}$, $R = 1\Omega$, and $R_1 = \infty$, sketch the steady-state waveforms of the inductor current i_L and control voltage v_{con} for the two switching periods. Label the maximum and minimum values on your sketch.

b) When the operational conditions of the converter are varied as $15\text{ V} < V_S < 25\text{ V}$ and $0.5\ \Omega < R < 2\ \Omega$ with $R_1 = \infty$, find the range of the control voltage v_{con} for the entire operating range: $(\quad) < v_{con} < (\quad)$.

c) With $R_1 = 10\text{ k}\Omega$, the converter fails to achieve dc regulation. Explain the reason for this.

d) When $V_S = 20\text{ V}$ and $R = 1\ \Omega$ with $R_1 = 10\text{ k}\Omega$, find the steady-state values of the output voltage v_O and control voltage v_{con}.

3.19 Figure P3.19 shows a closed-loop controlled buck converter and its voltage gain curve. Answer the following questions.

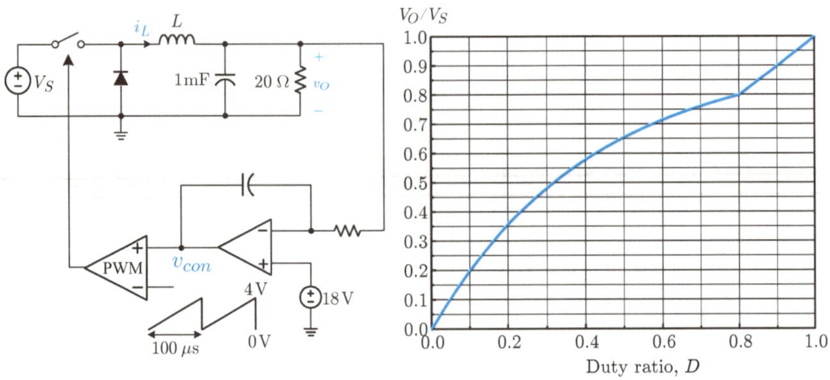

Fig. P3.19

a) Find the ranges for the duty ratio D and control voltage v_{con}, when the input voltage is varied between $20\text{ V} < V_S < 60\text{ V}$.

b) Find the inductance of the filter inductor L.

c) Sketch the inductor current i_L for the two switching periods for the three cases of $V_S = 20$ V, $V_S = 22.5$ V, and $V_S = 60$ V. Show the maximum and minimum values of i_L.

3.20 Consider the buck converter shown in Fig. P3.20 and answer the following questions.

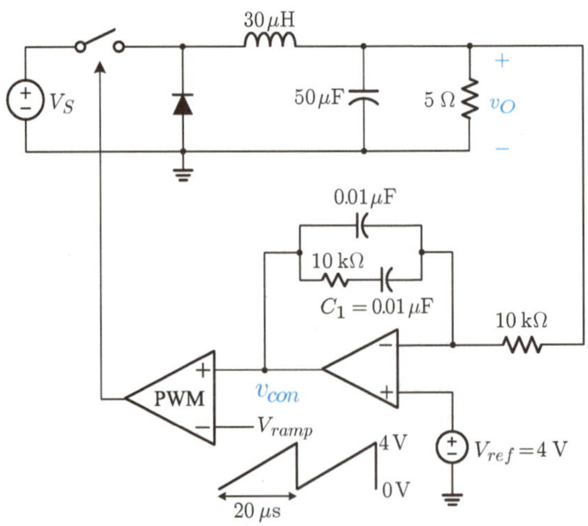

Fig. P3.20

a) Assume that the input voltage is varied between 8 V $< V_S <$ 20 V. For this condition, the converter could operate in either CCM or DCM, depending on the input voltage range. Fill in the blanks in the following descriptions for operation of the buck converter.
   i) The converter operates in CCM for ( ) $< V_S <$ ( ) and the control voltage varies between ( ) $< v_{con} <$ ( ) in this CCM operation.
   ii) The converter operates in DCM for ( ) $< V_S <$ ( ) and the control voltage varies between ( ) $< v_{con} <$ ( ) in this DCM operation.

b) Now assume $V_S = 10$ V and answer the questions.
   i) Find the steady-state values of the output voltage v_O and control voltage v_{con}.
   ii) Assume that the capacitor C_1 is accidentally open-circuited. Do you expect any change in the steady-state waveform of v_O or v_{con}? If you do not expect any change, give the reason for your claim. If you do predict a change in v_O or v_{con}, evaluate the average value of the corresponding waveform in the new steady state.
   iii) Assume the compensation capacitor C_1 is accidentally short-circuited and repeat ii).

3.21 Figure P3.21 shows a closed-loop controlled buck converter. Answer the questions.

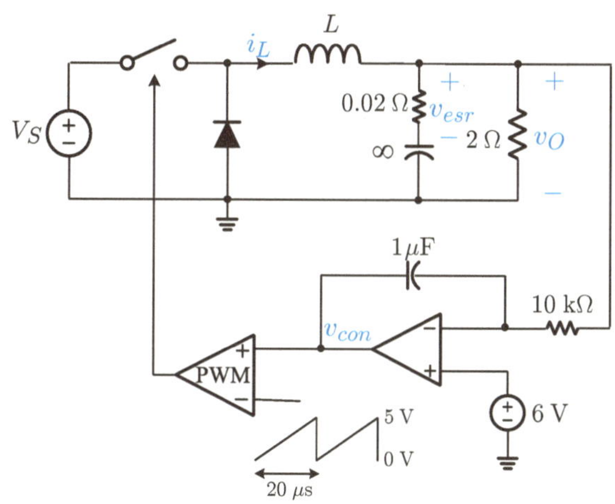

Fig. P3.21

a) Determine the inductance L such that the converter operates at the borderline between CCM and DCM with the input voltage of $V_S = 24$ V. For this case, sketch the inductor current i_L and the ripple component of the output voltage, $\tilde{v}_O = v_O - V_O$, for the two operational periods. Show the maximum and minimum values.

b) Assume the input voltage is varied between 12 V $< V_S <$ 24 V, while the inductance found in a) is used in the output filter. Under these conditions, find the range of the following operational condition and circuit variables of the converter.

i) ( ) $< D <$ ( ) ii) ( ) $< v_{con} <$ ( )
iii) ( ) $< \Delta i_L <$ ( ) iv) ( ) $< \Delta v_O <$ ( )

3.22* Shown in Fig. P3.22 are the transient responses of the major circuit waveforms of a closed-loop controlled buck converter. Case (a) is the transient response of the inductor current i_L and control voltage v_{con} in response to the step changes in the input voltage, $V_{S1} \Rightarrow V_{S2} \Rightarrow V_{S1}$ with $V_{S1} > V_{S2}$. Case (b) is the transient responses due to the step changes in the load resistor, $R_1 \Rightarrow R_2 \Rightarrow R_1$ with $R_1 < R_2$.

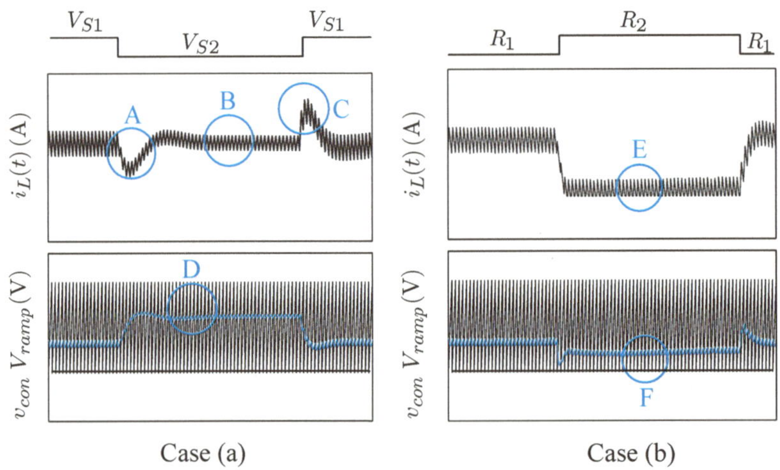

Fig. P3.22

Table P3.22

Circled period	Observed behavior	Cause/origin of behavior
A	Undershoot in i_L	( )
B	Decrease in the ripple of i_L	( )
C	Overshoot in i_L	( )
D	Increase in v_{con}	( )
E	Discontinuity in i_L	DCM operation with $R > R_{crit}$
F	Decrease in v_{con}	( )

Table P3.22 summarizes i) the prominent behavior of the inductor current i_L and control voltage v_{con}, observed in the period circled on the corresponding circuit waveform, and ii) a brief explanation for the cause/origin of the respective circuit behavior. Based on the given information and operational principle of the converter, fill in the blanks in Table P3.22.

CHAPTER 4

DC-TO-DC POWER CONVERTER CIRCUITS

Numerous dc-to-dc converters are currently being employed in various electronic equipment and systems. These converters outwardly look so varied in their topology and operation that the diversity of dc-to-dc converters is seemingly amazing, even mysterious.

Despite the variety and dissimilarity in topological structures, most dc-to-dc converters have evolved from the *three basic converters* known as the buck converter, boost converter, and buck/boost converter. Furthermore, among these three converters, the buck converter is the forerunner of the other two converters — specifically, the boost converter is derived, at least in its functional abstraction, from the buck converter and the buck/boost converter is created by combining the buck converter and boost converter. The buck converter was studied in the previous chapter.

In this chapter, we study an important class of dc-to-dc converters including the boost converter, buck/boost converter, and other converters derived from the three basic converters. For each converter, the origin of the circuit topology is first illustrated by showing how the converter circuit has evolved from its respective forerunning converter. This chapter then investigates the operational details of the converter and presents steady-state waveforms for both continuous conduction mode (CCM) and discontinuous conduction mode (DCM) operation. Because the feedback controller covered in Section 3.6.1 is commonly applicable to all the PWM dc-to-dc converters, the current chapter only deals with the power stage operation.

There are many ways to classify dc-to-dc converters, each with a different criterion for classification. One way is to divide PWM dc-to-dc converters into non-isolated converters and isolated converters. In a non-isolated converter, the input port of the converter shares a common ground with its output port. The buck converter is an example of the non-isolated converter. On the other hand, in an isolated converter, the input and output ports of the converter are electrically isolated. A transformer is commonly used to provide galvanic isolation. In this chapter, we study the important classes of non-isolated and isolated dc-to-dc converters. Discussions start with the non-isolated boost and buck/boost converters and later extend to isolated converters.

4.1 BOOST CONVERTER

The boost converter is a step-up dc-to-dc power conversion circuit that invariably produces a higher output voltage than the input voltage. In this section, the topological structure of the boost converter is first discussed. The steady-state operation in both CCM and DCM is then investigated using the analysis techniques established in Chapter 3.

4.1.1 Evolution to Boost Converter

The boost converter is viewed as a modification of the buck converter, specially configured to offer a larger output voltage than the input voltage. Figure 4.1 illustrates the procedure of transforming the buck converter into the boost converter through steps of circuit manipulations. The buck converter is shown in Fig. 4.1(a), where **a**

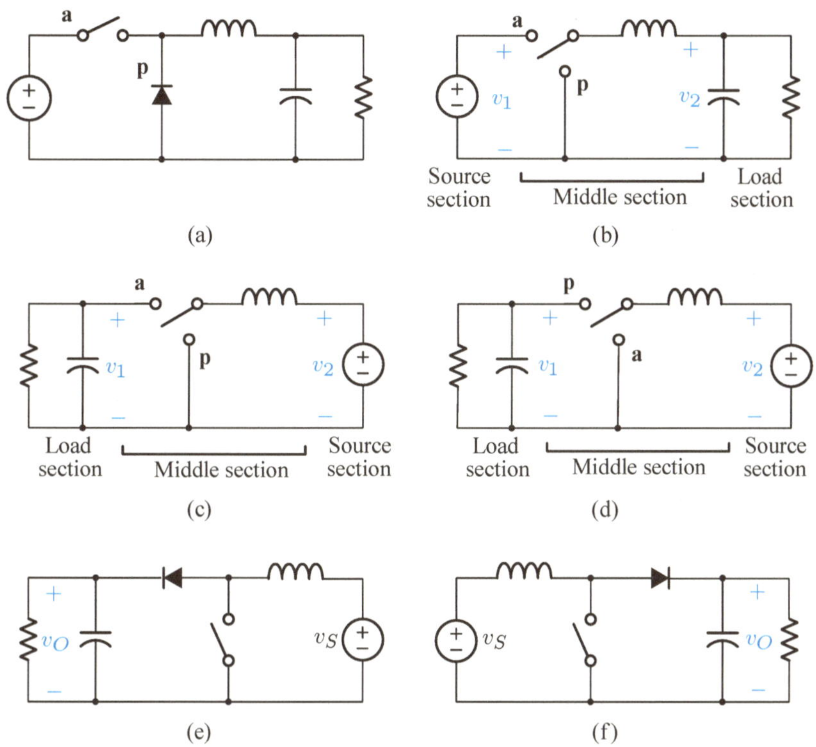

Figure 4.1 Derivation of boost converter. (a) Buck converter. (b) Alternative representation. (c) Interchange of source section and load section. (d) Interchange of passive switch and active switch. (e) Realization of SPDT switch. (f) Boost converter.

represents the node to which the active switch is connected, while **p** is the node where the passive switch is present. Figure 4.1(a) is transformed to Fig. 4.1(b), by replacing the active-passive switch pair with a single-pole double-throw (SPDT) switch. The SPDT switch is connected to **a** for DT_s and **p** for $(1-D)T_s$. As shown in Fig. 4.1(b), the circuit is partitioned into the source section, middle section, and load section. The middle section includes the SPDT switch and inductor. Referring to the voltage across the left-hand side of the middle section as v_1 and that of the right-hand side as v_2, the volt-sec balance condition is applied to the inductor

$$(v_1 - v_2)DT_s = v_2(1 - D)T_s \tag{4.1}$$

resulting in

$$\boxed{v_2 = v_1 D} \tag{4.2}$$

As the first step of the circuit manipulation, one can interchange the location of the source section and load section, while keeping the middle section unchanged, as shown in Fig. 4.1(c). With this modification, v_2 can be viewed as the input voltage associated with the source section, while v_1 becomes the output voltage across the load section. Now, the following two facts need to be kept in mind before continuing the circuit manipulation.

1) The relation (4.2), derived from the volt-sec balance condition on the inductor, is also valid in Fig. 4.1(c), regardless of the position change between the source and load sections.

2) The parameter D in (4.2) denotes the fraction of the switching period in which the SPDT switch is connected to the positive side of v_1.

As the second step, for the purpose of allowing energy flow from the source section to load section, the locations of the active and passive switches are interchanged,[†] resulting in the circuit shown in Fig. 4.1(d). Here, the fraction of the switching period in which the SPDT switch resides on the positive side of v_1 is $(1 - D)$ rather than D. Accordingly, the parameter D in (4.2) should be replaced with $(1 - D)$, resulting in a new relationship

$$\boxed{v_2 = v_1(1 - D)} \tag{4.3}$$

Figure 4.1(d) is then transformed to Fig. 4.1(e), where the SPDT switch is represented by an active-passive switch pair, v_1 is renamed as v_O for the output voltage, and v_2 is renamed as v_S for the input voltage. Finally, Fig. 4.1(e) is rearranged into the standard form shown in Fig. 4.1(f). This new converter circuit is called the boost converter because its voltage gain

$$\boxed{\frac{v_O}{v_S} = \frac{1}{1 - D}} \tag{4.4}$$

is always larger than unity for all $0 < D < 1$, thereby boosting the input voltage to a higher value.

As illustrated above, the boost converter is viewed as an evolution from the buck converter, created by interchanging the source and load connections and rearranging the switches in such a way that allows energy flow from the source section to the load section. During this modification, the step-up feature of the boost converter has naturally arisen.

4.1.2 Steady-State Analysis in CCM

The previous section described the evolution of the boost converter from the buck converter. The steady-state analysis of the boost converter in CCM operation is now performed using the analysis techniques established in Chapter 3.

[†]For proper operation as a dc-to-dc converter, the circuit should transfer energy from the source section to the inductor when the active switch is turned-on, and release the transferred energy to the load section when the active switch is turned-off. Although the on-time energy transfer is achieved in Fig. 4.1(c), the off-time energy release is not feasible. The on-time energy transfer and off-time energy release become both possible when the locations of the active and passive switches are interchanged, as shown in Fig. 4.1(d). Additional discussions about this feature are given in Section 4.3.

Steady-State Operation in CCM

Figure 4.2 shows the boost converter, on-time and off-time subcircuits, and major circuit waveforms in CCM operation. The standard techniques used in Chapter 3, such as the small-ripple approximation, flux linkage balance condition, and charge balance condition, are employed in Fig. 4.2. This figure also incorporates that the output voltage is always larger than the input voltage, $V_O > V_S$. The volt-sec balance condition is now applied to the inductor

$$V_S DT_s = -(V_S - V_O)(1 - D)T_s \tag{4.5}$$

resulting in the voltage gain expression

$$\boxed{\frac{V_O}{V_S} = \frac{1}{1 - D}} \tag{4.6}$$

which is identical to (4.4). The gain expression indicates that the voltage gain increases from unity to infinity when the duty ratio is varied between $0 < D < 1$.

During an on-time period, the inductor current ramps up with a slope V_S/L. Meanwhile, during an off-time period, the inductor current ramps down with a slope $(V_S - V_O)/L$. The excursion of the inductor current is determined as

$$\boxed{\Delta i_L = \frac{V_S}{L} DT_s = \frac{V_O - V_S}{L}(1 - D)T_s} \tag{4.7}$$

As illustrated in Fig. 4.2(c), the inductor current flows through the active switch during the on-time period and freewheels through the diode during the off-time period. Accordingly, the following relationships hold among the average value of the switch current, I_Q, average value of the diode current, I_D, and average value of the inductor current, I_L

$$I_Q = DI_L \tag{4.8}$$

$$I_D = (1 - D)I_L \tag{4.9}$$

In the boost converter, the diode is connected to the load, whereas the inductor is located at the source side. For this case, the average value of the inductor current is determined as explained below. Due to the charge balance condition, the average value of the capacitor current is zero. Accordingly, the average value of the diode current is to be equal to the load current, which is given by the ratio of the output voltage to load resistance

$$I_D = I_O = \frac{V_O}{R} \tag{4.10}$$

Using (4.9) and (4.10), the average value of the inductor current is found as

$$\boxed{I_L = \frac{1}{1 - D} I_D = \frac{1}{1 - D} \frac{V_O}{R}} \tag{4.11}$$

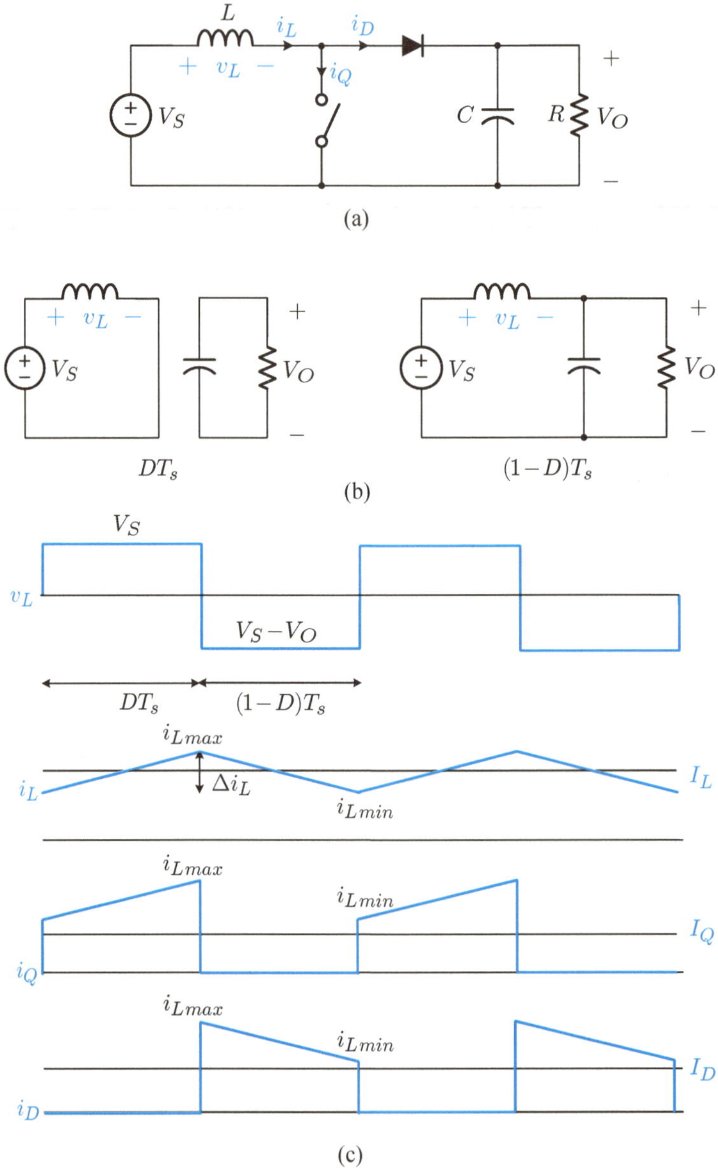

Figure 4.2 Steady-state analysis of boost converter in CCM. (a) Boost converter. (b) On-time and off-time subcircuits. (c) Major waveforms.

Table 4.1 Steady-State Analysis of Boost Converter

Circuit variable	Expression
V_O	$\frac{1}{1-D}V_S = \frac{1}{1-0.4}12 = 20\text{ V}$
$v_{L\,max}$	$V_S = 12\text{ V}$
$v_{L\,min}$	$V_S - V_O = 12 - 20 = -8\text{ V}$
I_L	$\frac{1}{1-D}\frac{V_O}{R} = \frac{1}{1-0.4}\frac{20}{5} = 6.67\text{ A}$
Δi_L	$\frac{V_S}{L}DT_s = \frac{12}{160\times10^{-6}}0.4\cdot 20\times10^{-6} = 0.6\text{ A}$
$i_{L\,max}$	$I_L + \frac{1}{2}\Delta i_L = 6.67 + \frac{0.6}{2} = 6.97\text{ A}$
$i_{L\,min}$	$I_L - \frac{1}{2}\Delta i_L = 6.67 - \frac{0.6}{2} = 6.37\text{ A}$

The maximum and minimum values of the inductor current are expressed as

$$i_{L\,max} = I_L + \frac{1}{2}\Delta i_L \tag{4.12}$$

and

$$i_{L\,min} = I_L - \frac{1}{2}\Delta i_L \tag{4.13}$$

where I_L is given by (4.11) and Δi_L is determined by (4.7).

Steady-State Operation of Boost Converter — EXAMPLE 4.1

This example shows the steady-state analysis and circuit waveforms of a boost converter. The operational conditions and circuit parameters are $V_S = 12$ V, $L = 160\ \mu$H, $C = 400\ \mu$F, $T_s = 20\ \mu$s, and $D = 0.4$. The steady-state values of the major voltage and current waveforms are evaluated as shown in Table 4.1. The simulated steady-state circuit waveforms are given in Fig. 4.3.

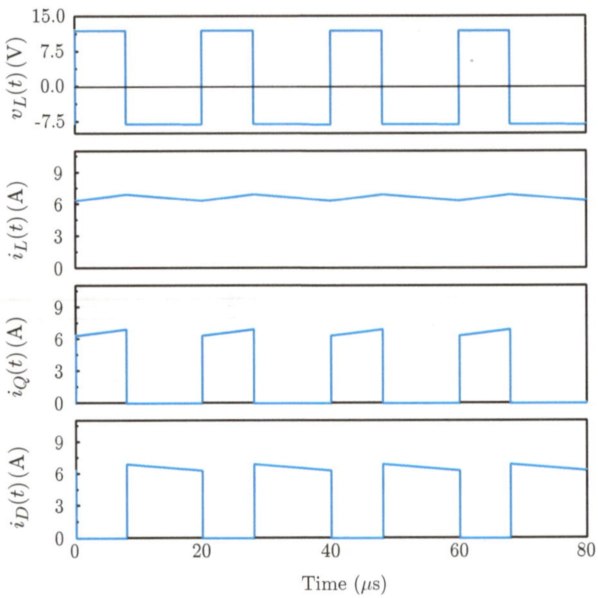

Figure 4.3 Circuit waveforms of boost converter.

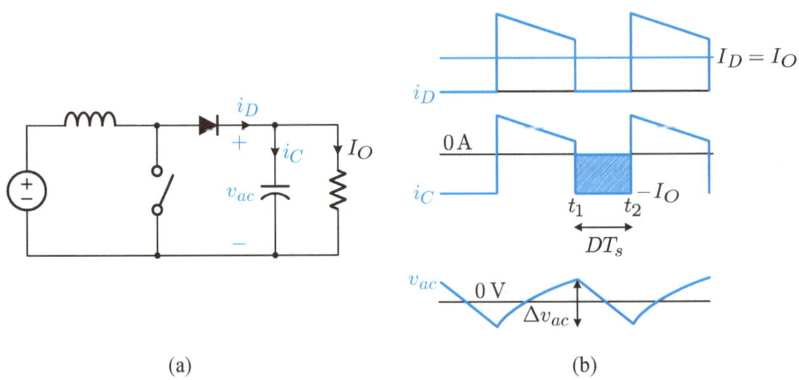

Figure 4.4 Estimation of output ripple. (a) Circuit diagram. (b) Current waveforms and output ripple.

Estimation of Output Voltage Ripple

The magnitude of the output voltage ripple is evaluated from the waveforms shown in Fig. 4.4. The diode current i_D is split into the capacitor current i_C and load current I_O. As discussed in Section 3.4.3, the load current I_O carries the dc component of i_D while the capacitor current i_C absorbs the ac component of i_D: $I_O = I_D$ and $i_C = i_D - I_D$. This situation is illustrated in Fig. 4.4(b).

The peak-to-peak value of the capacitor voltage, Δv_{ac}, or the magnitude of the output voltage ripple, Δv_O, can be found by integrating i_C over the time period in which i_C remains negative and by dividing the resulting value by the capacitance

$$\Delta v_{ac} = \Delta v_O = \frac{1}{C}\int_{t_1}^{t_2} i_C(\tau)\,\mathrm{d}\tau \tag{4.14}$$

The operation of (4.14) is equivalent to evaluating the area of the highlighted rectangle in Fig. 4.4(b) and dividing the area by the capacitance. By noting that $|t_2-t_1| = DT_s$, the magnitude of the output voltage ripple is given by

$$\boxed{\Delta v_O = \frac{1}{C}I_O DT_s = \frac{1}{C}\frac{V_O}{R}DT_s} \tag{4.15}$$

It should be noted that (4.15) is only valid when the boost converter is operating in CCM and the output capacitor does not contain an equivalent series resistance (esr). When the esr R_c is present, the output ripple is given by

$$\boxed{\Delta v_O \approx \Delta i_D\, R_c = i_{D\,max}\, R_c = i_{L\,max}\, R_c} \tag{4.16}$$

for the boost converter operating in either CCM or DCM. The output ripple of the boost converter, given by $\Delta v_O \approx i_{Lmax}\, R_c$ in (4.16), is significantly larger than that of the buck converter, given by $\Delta v_O \approx \Delta i_L R_c$.

4.1.3 Steady-State Analysis in DCM

The boost converter operates at the borderline between CCM and DCM when the condition $I_L = \Delta i_L/2$ is satisfied. The CCM/DCM boundary condition is expressed in terms of circuit parameters and operating conditions

$$\frac{1}{1-D}\frac{V_O}{R} = \frac{1}{2}\frac{V_S}{L}DT_S \tag{4.17}$$

based on (4.11) and (4.7). The expressions (4.17) and (4.6) is used to find the critical value for the load resistance

$$\boxed{R_{crit} = \frac{2L}{D(1-D)^2 T_s}} \tag{4.18}$$

that determines the operational mode. When the load resistance is larger than R_{crit}, the converter enters DCM operation. Otherwise, the converter remains in CCM operation.

Figure 4.5 shows the major circuit waveforms of a boost converter operating in DCM. Application of the volt-sec balance condition to the inductor yields

$$V_S DT_s = (V_O - V_S)D_1 T_s \tag{4.19}$$

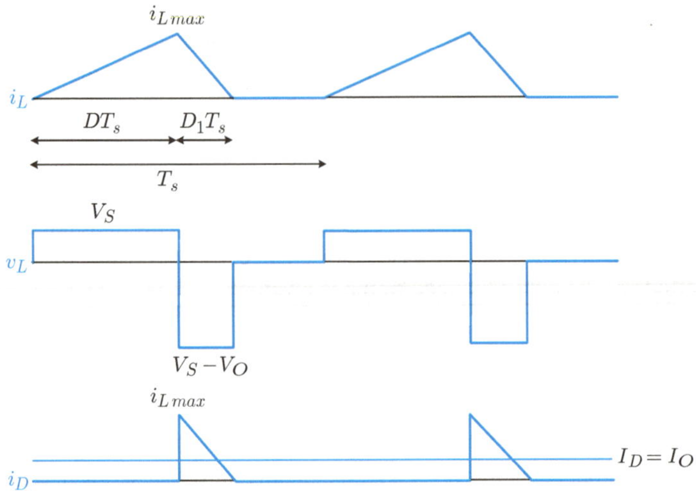

Figure 4.5 Circuit waveforms of boost converter in DCM operation.

which indicates that

$$\frac{V_O}{V_S} = \frac{D + D_1}{D_1} \tag{4.20}$$

The unknown variable D_1 should be removed from (4.20) to yield a complete voltage gain expression.

For the boost converter, the average value of the diode current should be the same as the load current, $I_D = I_O$, because the diode is connected to the output section of the converter. From the waveform of the diode current i_D, the following equation is set forth

$$\frac{\frac{1}{2} i_{L\,max} D_1 T_s}{T_s} = I_O \tag{4.21}$$

By incorporating the fact

$$i_{L\,max} = \frac{V_S}{L} DT_s \tag{4.22}$$

and

$$I_O = \frac{V_O}{R} \tag{4.23}$$

the equation (4.21) is rearranged as

$$\frac{1}{2}\frac{V_S}{L} DT_s D_1 = \frac{V_O}{R} \tag{4.24}$$

to yield the expression for D_1

$$D_1 = \frac{V_O}{V_S}\frac{2L}{RDT_s} \tag{4.25}$$

Now, the expressions (4.20) and (4.25) are combined together, resulting in a quadratic equation for V_O/V_S

$$\left(\frac{V_O}{V_S}\right)^2 - \left(\frac{V_O}{V_S}\right) - \frac{D^2RT_s}{2L} = 0 \tag{4.26}$$

The non-negative solution to (4.26) becomes the DCM gain of the boost converter

$$\frac{V_O}{V_S} = \frac{1}{2}\left(1 + \sqrt{1 + \frac{2D^2RT_s}{L}}\right) \tag{4.27}$$

The DCM voltage gain is a nonlinear function of circuit parameters and operational conditions. The following two observations, previously found in the buck converter case, also hold true for the boost converter.

1) The voltage gain grows larger as the load resistance is increased.
2) For most practical converter parameters, the DCM voltage gain is larger than the CCM counterpart when the same duty ratio is assumed.

4.1.4 Effects of Parasitic Resistance on Voltage Gain

The voltage gain of the ideal boost converter in CCM operation, $V_O/V_S = 1/(1-D)$, predicts that the output voltage increases without limits as the duty ratio D approaches unity, thereby yielding an infinite output voltage with $D = 1$. However, the operational principle of the boost converter contradicts this scenario. With $D = 1$, the active switch remains closed all the time and the diode never conducts. This implies that the output section of the converter is isolated from the source. For this case, the output voltage is zero because no energy is delivered from source to load.

The conflict between the ideal voltage gain and the operational principle of the converter is resolved when the non-ideal characteristics of the circuit components are incorporated. Figure 4.6 shows a boost converter along with its on-time and off-time subcircuits, where the winding resistance R_l is included to a practical inductor, while the other circuit components are still assumed ideal. The voltage gain expression of the converter with the presence of R_l is now derived to investigate the behavior of the *non-ideal* boost converter.

The volt-sec balance condition of the inductor is formulated as

$$(V_S - \bar{i}_L(t)R_l)DT_s = (V_O - \bar{i}_L(t)R_l - V_S)(1-D)T_s \tag{4.28}$$

where $\bar{i}_L$ represents the average value of the inductor current: $\bar{i}_L = I_L$. The average inductor current $\bar{i}_L$ has the same effect as the original inductor current i_L, as far as the volt-sec balance condition is concerned. The expression (4.28) is simplified to

$$V_S = I_LR_l + (1-D)V_O \tag{4.29}$$

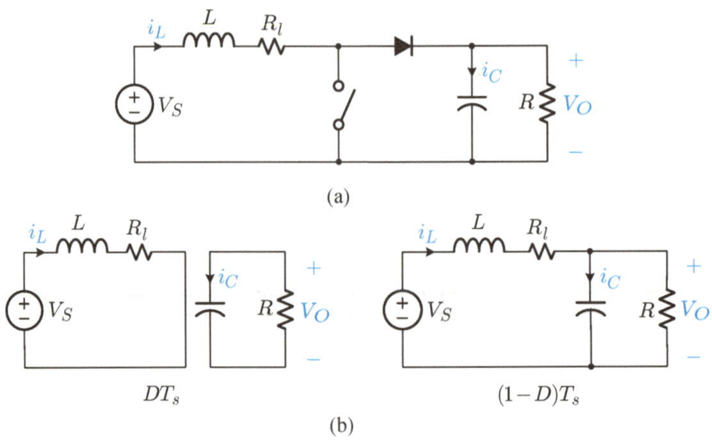

Figure 4.6 Boost converter with inductor winding resistance. (a) Circuit diagram. (b) On-time and off-time subcircuits.

On the other hand, the amp-sec balance condition on the output capacitor is written as

$$-\frac{V_O}{R}DT_s = \left(\frac{V_O}{R} - \bar{i}_L(t)\right)(1-D)T_s \tag{4.30}$$

which is simplified to

$$\frac{V_O}{R} = (1-D)I_L \tag{4.31}$$

The desired expression for the voltage gain is now obtained by simultaneously solving (4.29) and (4.31) to eliminate I_L. The resulting voltage gain is arranged as the product of the voltage gain of the ideal boost converter and the correction factor that accounts for the effect of the winding resistor R_l

$$\boxed{\frac{V_O}{V_S} = \frac{1}{1-D}\left(\frac{1}{1+\frac{1}{(1-D)^2}\frac{R_l}{R}}\right)} \tag{4.32}$$

The key parameter in the correction factor is the ratio of the winding resistance to the load resistance, R_l/R. Figure 4.7 shows the voltage gain curves, calculated using (4.32) with different values for R_l/R. The voltage gain curves reveal substantial deviations from the ideal case. The deviation intensifies as the R_l/R ratio increases, showing a wide gap between the actual voltage gain and the ideal voltage gain when the duty ratio is large. In particular, all the voltage gain curves merge to *zero* as the duty ratio approaches unity, rather than growing boundlessly. This phenomenon is consistent with the operation of the boost converter. With $D = 1$, the output of the converter is never connected to the source, and the output capacitor thus remains uncharged producing *zero* output voltage.

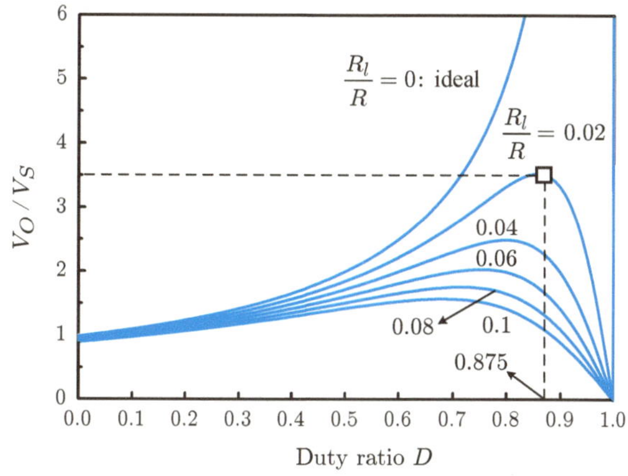

Figure 4.7 Voltage gain of non-ideal boost converter.

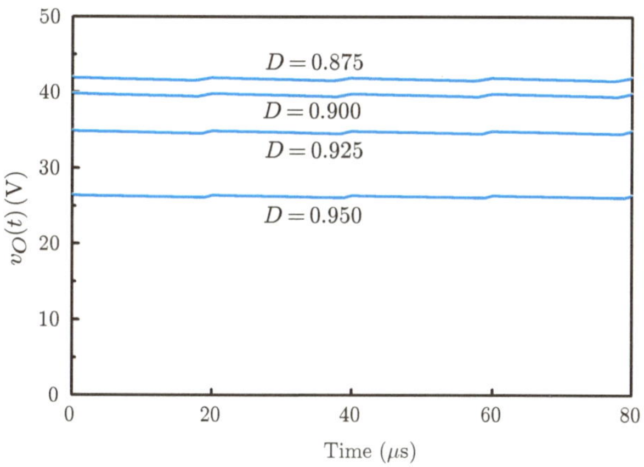

Figure 4.8 Output voltage waveforms of boost converter with $R_l/R = 0.02$.

It is noteworthy that the R_l/R ratio limits the maximum voltage gain that a boost converter can produce. For example, with $R_l/R = 0.02$, Fig. 4.7 indicates that the maximum voltage gain is approximately 3.5 at the duty ratio of $D = 0.875$. If a voltage gain larger than 3.5 is required, R_l/R should be reduced to be smaller than 0.02.

Output Voltage with Inductor Winding Resistance — EXAMPLE 4.2

This example substantiates the results of the preceding gain analysis. The boost converter introduced in Example 4.1 has the parameters of $V_S = 12\,\text{V}$, $L = 160\,\mu\text{H}$, $C = 400\mu\text{F}$, and $R = 5\Omega$. Now, a 0.1Ω winding resistance is included in the inductor,

Table 4.2 Output Voltage Analysis with Inductor Winding Resistance

Duty ratio	Output voltage
$D = 0.875$	$V_O = \dfrac{1}{1-D}\left(\dfrac{1}{1+\dfrac{1}{(1-D)^2}\dfrac{R_l}{R}}\right)V_S = \dfrac{1}{1-0.875}\left(\dfrac{1}{1+\dfrac{1}{(1-0.875)^2}\dfrac{0.1}{5}}\right)12 = 42\text{ V}$
$D = 0.900$	40 V
$D = 0.925$	35 V
$D = 0.950$	27 V

resulting in $R_l/R = 0.1/5 = 0.02$. Under this condition, a series of simulations is performed with four different values for the duty ratio: $D = 0.875$, 0.900, 0.925, and 0.950. The output voltage waveforms of the boost converter are shown in Fig. 4.8. As expected from Fig. 4.7, the output voltage attains its peak value at $D = 0.875$ and decreases as the duty ratio is further increased. The theoretical predictions of the output voltage based on (4.32) are given in Table 4.2.

4.2 BUCK/BOOST CONVERTER

The buck/boost converter is a dc-to-dc power conversion circuit that can either step-up or step-down the input voltage. In this sense, the buck/boost converter is also called the up-down converter. This section discusses the circuit topology and steady-state operation of the buck/boost converter.

4.2.1 Evolution to Buck/Boost Converter

The buck/boost converter is created by cascading a buck converter and boost converter and simplifying the cascaded circuit. Figure 4.9(a) shows the buck converter connected to a boost converter downstream. Figure 4.9(a) is changed to Fig. 4.9(b) by replacing the active-passive switch pairs with SPDT switches. It is now assumed

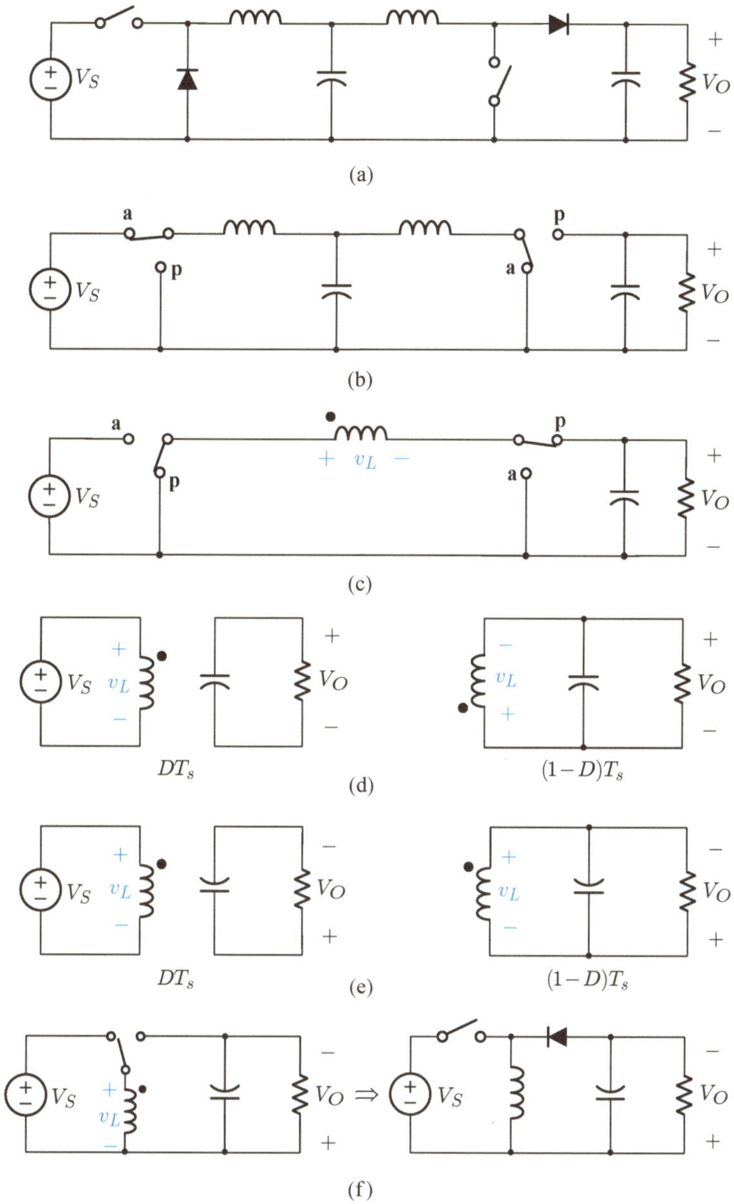

Figure 4.9 Derivation of buck/boost converter. (a) Cascaded connection of buck converter and boost converter. (b) Alternative representation using two synchronized SPDT switches. (c) Non-inverting buck/boost converter. (d) Subcircuits of non-inverting buck/boost converter. (e) Modification of subcircuits. (f) Buck/boost converter.

that the two SPDT switches are synchronized so that both the SPDT switches are held at position **a** for DT_s and position **p** for $(1 - D)T_s$.

It should be noted that the upstream buck converter in Fig. 4.9(b) will be functioning even if its output filter capacitor is removed. The output filter capacitor, employed to enhance the filtering performance in a standalone buck converter, becomes redundant when the buck converter is connected to a boost converter which has its own output capacitor.

Once the output capacitor of the buck converter is removed, the inductor of the buck converter and that of the boost converter can be merged together, resulting in the circuit shown in Fig. 4.9(c). In Fig. 4.9(c), the left-hand side of the inductor terminal is marked • to highlight the polarity of the inductor voltage. This circuit is referred to as the non-inverting buck/boost converter for reasons that will become clear shortly.

Application of the volt-sec balance condition to the inductor in Fig. 4.9(c) yields

$$V_S DT_s = V_O(1 - D)T_s \tag{4.33}$$

which is simplified to result in the voltage gain of the buck/boost converter

$$\boxed{\frac{V_O}{V_S} = D\frac{1}{1 - D}} \tag{4.34}$$

The first term D originates from the buck converter upstream, while the second term $1/(1 - D)$ comes from the boost converter downstream.

The on-time and off-time subcircuits of the non-inverting buck/boost converter are shown in Fig. 4.9(d). The subcircuits in Fig. 4.9(d) are redrawn into Fig. 4.9(e) keeping the circuit properties unchanged. The vertical flipping of the secondary side of the on-time subcircuit and the off-time subcircuit does not affect the circuit operation but the polarity of the output voltage in Fig. 4.9(e) is inverted from that of Fig. 4.9(d). A new circuit is now synthesized from the two subcircuits in Fig. 4.9(e). The resulting circuit is shown in Fig. 4.9(f) in two different forms; one is with the SPDT switch and the other is with an active-passive switch pair. It is easy to see that the new circuit reduces to the two subcircuits in Fig. 4.9(e) with the two different conditions of the SPDT switch.

The converter circuit shown in Fig. 4.9(f) is named as the buck/boost converter because the converter is derived from the connection of the buck converter and the boost converter. The polarity of the output voltage of the buck/boost converter is opposite to that of the input voltage. Meanwhile, in the converter circuit in Fig. 4.9(c), the output voltage has the same polarity as the input voltage, as such, the converter is referred to as the non-inverting buck/boost converter.

4.2.2 Steady-State Analysis in CCM

The steady-state analysis of the buck/boost converter is performed in the same manner as that of the buck converter and boost converter cases. The piecewise linear analysis is merged with the small-ripple approximation and volt-sec balance condition.

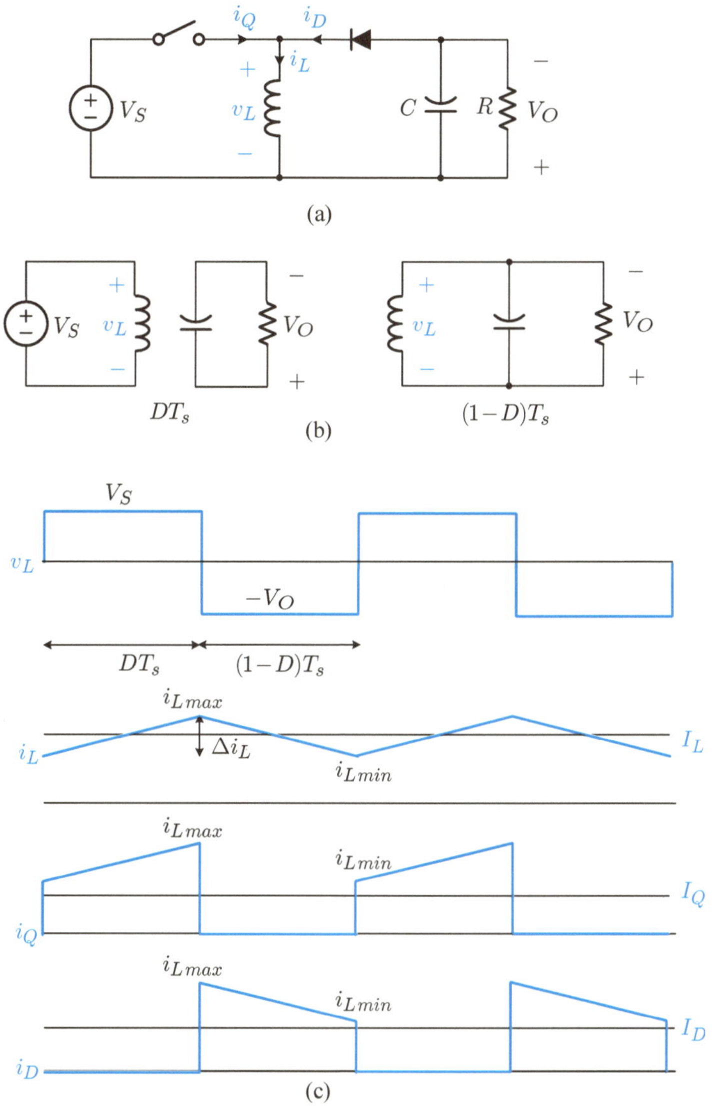

Figure 4.10 Steady-state analysis of buck/boost converter in CCM. (a) Buck/boost converter. (b) On-time and off-time subcircuits. (c) Major circuit waveforms.

Steady-State Operation in CCM

Figure 4.10 shows the buck/boost converter, on-time and off-time subcircuits, and major circuit waveforms. The steady-state operation of the converter is illustrated with the circuit waveforms in Fig. 4.10(c). Application of the volt-sec balance condition to the inductor

$$V_S D T_s = V_O (1 - D) T_s \tag{4.35}$$

yields the voltage gain of the buck/boost converter

$$\boxed{\frac{V_O}{V_S} = \frac{D}{1 - D}} \tag{4.36}$$

The gain formula predicts that the output voltage varies from zero to infinity when the duty ratio is varied between $0 < D < 1$. However, as is the case with the boost converter, the voltage gain deviates from the ideal prediction when non-ideal circuit components are considered. It can be shown that the voltage gain is modified as

$$\boxed{\frac{V_O}{V_S} = \frac{D}{1 - D} \left(\frac{1}{1 + \dfrac{1}{(1 - D)^2} \dfrac{R_l}{R}} \right)} \tag{4.37}$$

at the presence of the inductor winding resistance R_l. The structure of (4.37) is the same as the voltage gain structure of the practical boost converter given in (4.32). Accordingly, the actual buck/boost converter will resemble the behavior of the practical boost converter discussed in Section 4.1.4.

The excursion of the inductor current is given by

$$\boxed{\Delta i_L = \frac{V_S}{L} D T_s = \frac{V_O}{L} (1 - D) T_s} \tag{4.38}$$

The average values of the switch current and diode current are determined as

$$I_Q = D I_L$$

$$I_D = (1 - D) I_L \tag{4.39}$$

Same as the boost converter case, the load current is supported by the diode current. Thus, the average diode current becomes the load current: $I_D = I_O = V_O/R$. Accordingly, the average value of the inductor current is given as

$$\boxed{I_L = \frac{1}{1 - D} I_D = \frac{1}{1 - D} \frac{V_O}{R}} \tag{4.40}$$

The maximum value of the inductor current then becomes

$$i_{L\,max} = I_L + \frac{1}{2} \Delta i_L \tag{4.41}$$

The minimum value is given by

$$i_{L\,min} = I_L - \frac{1}{2} \Delta i_L \tag{4.42}$$

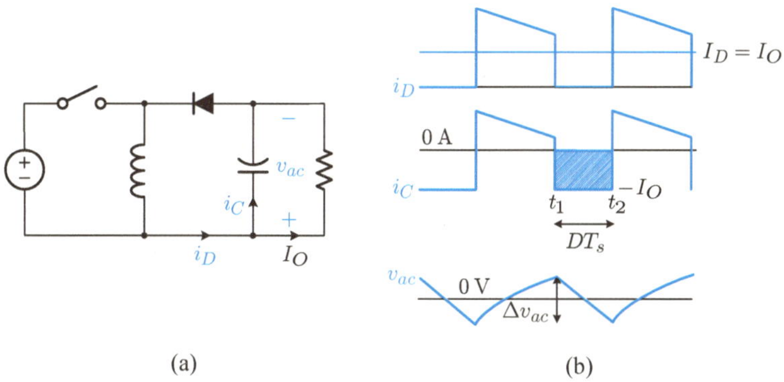

Figure 4.11 Estimation of output ripple. (a) Circuit diagram. (b) Current waveforms and output ripple.

Estimation of Output Voltage Ripple

Figure 4.11 shows the circuit diagram and waveforms that are used in predicting the magnitude of the output voltage ripple. The waveforms are identical to those of the boost converter shown in Fig. 4.4. The output stage of the buck/boost converter retains the circuit properties of the boost converter, because the converter is evolved from the connection of the upstream buck converter and the downstream boost converter. Consequently, the output voltage ripple of the buck/boost converter is identical to the boost converter case

$$\boxed{\Delta v_{ac} = \Delta v_O = \frac{1}{C}\frac{V_O}{R}DT_s} \tag{4.43}$$

This expression only holds true when the converter is in CCM operation and no esr is present in the output capacitor. When the esr of the output capacitor is considered, the output ripple is approximated to

$$\boxed{\Delta v_O \approx \Delta i_D R_c = i_{D\,max} R_c = i_{L\,max} R_c} \tag{4.44}$$

for the buck/boost converter operating in either CCM or DCM.

Steady-State Operation of Buck/Boost Converter — EXAMPLE 4.3

This example demonstrates the circuit waveforms of a buck/boost converter. The circuit parameters of the buck/boost converter are $V_S = 12$ V, $L = 160\ \mu$H, $C = 400\ \mu$F, $R = 5\ \Omega$, $T_s = 20\ \mu$s, and $D = 0.4$. Table 4.3 summarizes the steady-state values of major circuit variables. Figure 4.12 shows the simulated waveforms of important circuit variables.

Table 4.3 Steady-State Analysis of Buck/Boost Converter

Circuit variable	Expression
V_O	$\frac{D}{1-D}V_S = \frac{0.4}{1-0.4}12 = 8\ \text{V}$
$v_{L\,max}$	$V_S = 12\ \text{V}$
$v_{L\,min}$	$-V_O = -8\ \text{V}$
I_L	$\frac{1}{1-D}\frac{V_O}{R} = \frac{1}{1-0.4}\frac{8}{5} = 2.67\ \text{A}$
Δi_L	$\frac{V_S}{L}DT_s = \frac{12}{160\times10^{-6}}0.4\cdot 20\times10^{-6} = 0.6\ \text{A}$
$i_{L\,max}$	$I_L + \frac{\Delta i_L}{2} = 2.67 + \frac{0.6}{2} = 2.97\ \text{A}$
$i_{L\,min}$	$I_L - \frac{\Delta i_L}{2} = 2.67 - \frac{0.6}{2} = 2.37\ \text{A}$

4.2.3 Steady-State Analysis in DCM

For the buck/boost converter, the CCM/DCM borderline condition, $I_L = \Delta i_L/2$, is written as

$$\frac{1}{1-D}\frac{V_O}{R} = \frac{1}{2}\frac{V_S}{L}DT_s \tag{4.45}$$

using (4.40) and (4.38). From (4.45) and (4.36), the critical value for the load resistor is given by

$$\boxed{R_{crit} = \frac{2L}{(1-D)^2T_s}} \tag{4.46}$$

The converter enters DCM operation with the load resistance larger than R_{crit}.

Figure 4.13 shows the major DCM waveforms of the buck/boost converter. Application of the volt-sec balance condition to the inductor yields

$$V_S DT_s = V_O D_1 T_s \tag{4.47}$$

which is simplified to

$$\frac{V_O}{V_S} = \frac{D}{D_1} \tag{4.48}$$

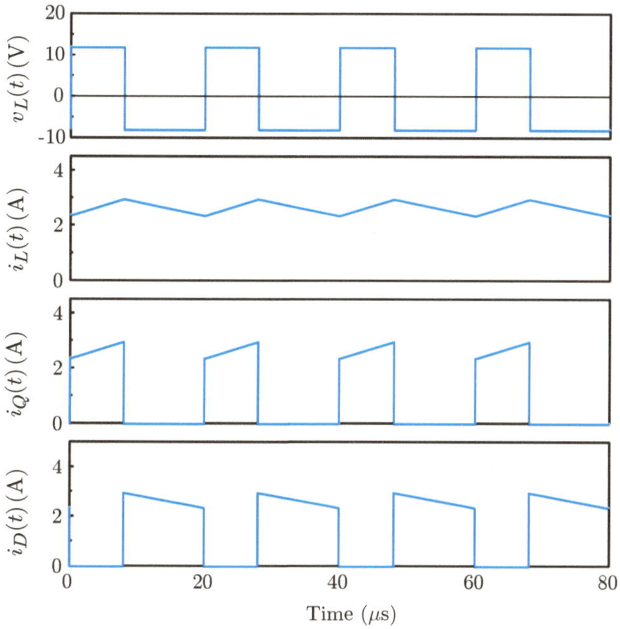

Figure 4.12 Circuit waveforms of buck/boost converter.

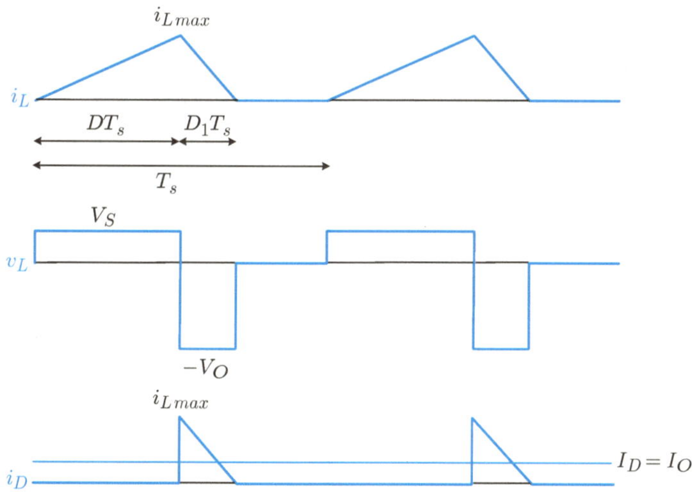

Figure 4.13 Circuit waveforms of buck/boost converter in DCM operation.

By equating the average diode current I_D to the load current, it follows that

$$\frac{1}{2}\frac{V_S}{L}DT_sD_1 = \frac{V_O}{R} \tag{4.49}$$

which provides an expression for D_1

$$D_1 = \frac{V_O}{V_S}\frac{2L}{RDT_s} \tag{4.50}$$

Finally, by combining (4.48) and (4.50), the DCM voltage gain is given by

$$\boxed{\frac{V_O}{V_S} = D\sqrt{\frac{RT_s}{2L}}} \tag{4.51}$$

As is the case with the buck and boost converters, the voltage gain is proportional to the load resistance. Also, the DCM voltage gain is usually larger than the CCM counterpart when the same duty ratio is assumed.

4.3 THREE BASIC CONVERTERS

We have studied the three basic dc-to-dc converter topologies: the buck converter, the boost converter, and the buck/boost converter. The structure, operational principles, and voltage gain of these three converters are now reviewed from a more general perspective.

Structure and Operation of Three Basic Converters

The three basic converters contain a common circuit block, consisting of an SPDT switch and inductor. This common circuit block is shown in Fig. 4.14(a). This circuit block is referred to as the *three-terminal cell.* The node **a** denotes the active switch terminal and **p** indicates the passive switch terminal. The node **c** stands for the common terminal to which the filter inductor is connected. The other end of the filter inductor is denoted as the inductor terminal **i**.

The three-terminal cell is inserted between the source and load in three different ways for the three basic converters. As shown in Fig. 4.14, the inductor terminal **i** is directed to the output port in the buck converter, while the inductor terminal is linked to the input port in the boost converter. In the buck/boost converter, the inductor terminal is connected to the ground.

Despite the different configurations, the three converters have a common aspect that enables each circuit to function as a dc-to-dc power converter. In each converter, energy is transferred from the source to inductor when the common terminal is connected to the active terminal. On the other hand, when the common terminal is tied to the passive terminal, energy is released from the inductor to load. Thus, the switching action of the SPDT switch provides an energy transfer from the source to

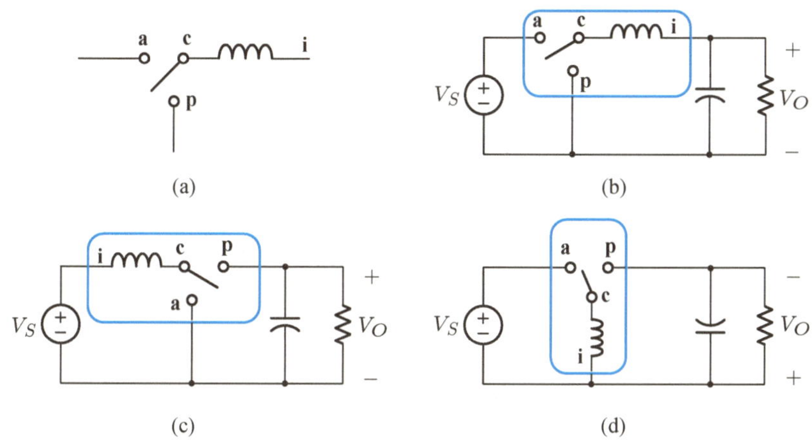

Figure 4.14 Three-terminal cell and basic converters. (a) Three-terminal cell. (b) Buck converter. (c) Boost converter. (d) Buck/boost converter.

load. The amount of the energy transfer is determined by the duty ratio of the active switch. The inductor is used as an intermediate energy storage component.

To summarize, in the three basic dc-to-dc converters, the three-terminal cell is positioned to link the source and the load, while allowing a controlled energy transfer between them by using the inductor as the transitional energy reservoir. In fact, the buck converter, boost converter, and buck/boost converter are the only three possible topological structures that can perform the dc-to-dc power conversion using the three-terminal cell shown in Fig. 4.14(a).

Voltage Gain of Three Basic Converters

The three basic dc-to-dc converters offer the different input-to-output voltage gains. Figure 4.15 shows the voltage gain characteristics of the three converters. For each converter, the voltage gain is displayed for both CCM operation and DCM operation using D_{crit} as the borderline. The converter operates in DCM when its duty ratio is less than D_{crit} and moves to CCM operation when the duty ratio grows larger than D_{crit}. The expressions for the DCM voltage gain and D_{crit} of the three converters are summarized in Table 4.4.

4.4 FLYBACK CONVERTER: TRANSFORMER-ISOLATED BUCK/BOOST CONVERTER

For many dc-to-dc converter applications, electrical isolation between the input and output ports is required by law for practical reasons such as safety. One simple way to provide such an input-output isolation is to insert an isolation transformer in the middle section of dc-to-dc converters. Indeed, many isolated converter topologies

Table 4.4 DCM Voltage Gain and D_{crit} of Three Basic Converters

	DCM voltage gain	D_{crit}
Buck converter	$\dfrac{2D}{D+\sqrt{D^2+\dfrac{8L}{RT_s}}}$	$\dfrac{1}{2}(1-D_{crit})=\dfrac{L}{RT_s}$
Boost converter	$\dfrac{1}{2}\left(1+\sqrt{1+\dfrac{2D^2RT_s}{L}}\right)$	$\dfrac{1}{2}D_{crit}(1-D_{crit})^2=\dfrac{L}{RT_s}$
Buck/boost	$D\sqrt{\dfrac{RT_s}{2L}}$	$\dfrac{1}{2}(1-D_{crit})^2=\dfrac{L}{RT_s}$

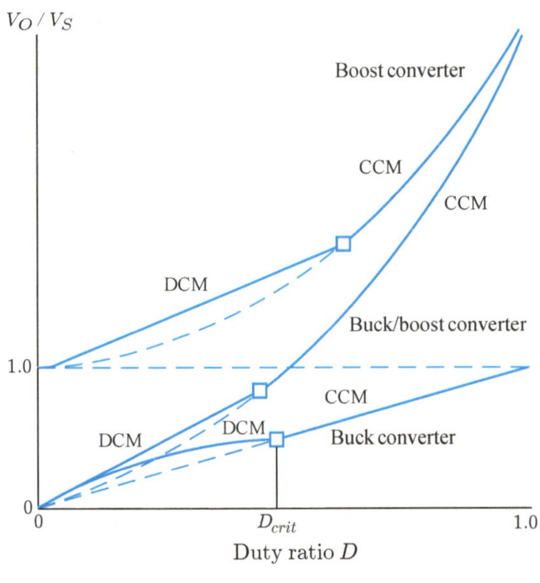

Figure 4.15 Voltage gain curves of three basic converters.

are proliferated by adding a transformer to the three basic dc-to-dc converters. The flyback converter is a typical example of them.

The flyback converter is evolved from the buck/boost converter by adding an isolation transformer and simplifying the resulting circuit. The flyback converter has a very simple structure with minimal component count, while providing the desired input-output isolation. Accordingly, the flyback converter is widely used as a cost-

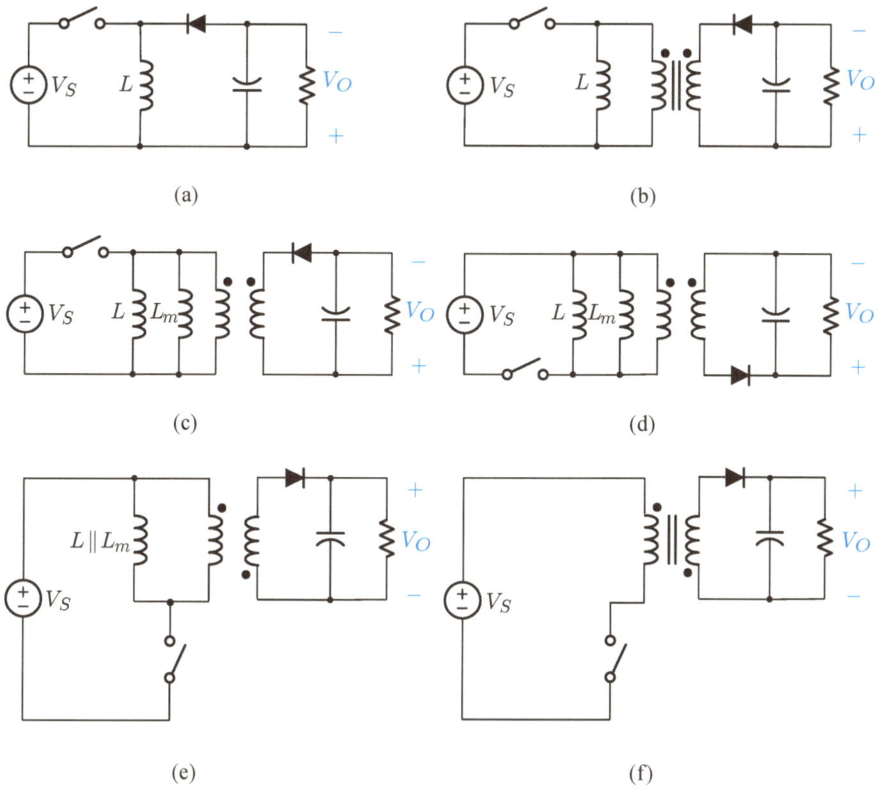

Figure 4.16 Evolution to flyback converter. (a) Buck/boost converter. (b) Insertion of isolation transformer. (c) Circuit model. (d) Equivalent circuit model. (e) Modified circuit model. (f) Flyback converter.

effective dc-to-dc conversion circuit for consumer electronics. This section presents the topological origin and steady-state operation of the flyback converter.

4.4.1 Evolution to Flyback Converter

Figure 4.16 shows the evolution of the flyback converter from the forerunning buck/boost converter. The buck/boost converter is shown in Fig. 4.16(a). A transformer is inserted in the middle section of the buck/boost converter, resulting in the circuit shown in Fig. 4.16(b). After replacing the transformer with a combination of the magnetizing inductance L_m and ideal transformer, as shown in Fig. 4.16(c), the active switch and diode are relocated without affecting the converter operation, as shown in Fig. 4.16(d). Figure 4.16(d) is then changed to Fig. 4.16(e) by combining the inductor of the buck/boost converter and the magnetizing inductance of the transformer, and vertically flipping the circuitry located in the secondary side of the ideal transformer. The final form of the flyback converter is shown in Fig. 4.16(f), where the composite inductance, $L\|L_m$, is considered as the magnetizing inductance of the isolation transformer.

The flyback converter utilizes the magnetizing inductance of the isolation transformer as its functional inductance. Therefore, the isolation transformer should be fabricated in such a way that could offer a controllable magnetizing inductance. One easy solution is to create a gap in the magnetic path of the transformer. This is commonly implemented by introducing an air gap between the magnetic cores, which effectively determines the magnetizing inductance of the isolation transformer.

4.4.2 Steady-State Analysis in CCM

Functionally speaking, the flyback converter is a buck/boost converter with transformer isolation. Thus, the flyback converter closely mimics the buck/boost converter in its operation. Figure 4.17 illustrates the operation of the flyback converter in CCM, where a $1\!:\!n$ turns ratio is assumed for the transformer. The transformer is represented by the parallel connection of magnetizing inductance and ideal transformer.

During an on-time period in which the active switch is closed, the diode is reverse-biased by the sum of the output voltage and the voltage across the secondary transformer winding: $V_O + nV_S$. Accordingly, the on-time subcircuit for DT_s is resulted as shown in Fig. 4.17(b).

The operation of the converter during the on-time period DT_s is explained as follows. The primary winding of the ideal transformer does not carry any current because the secondary winding is open-circuited. The voltage across the magnetizing inductance, v_m in Fig. 4.17, is identical to the input voltage V_S. The current through the magnetizing inductance, i_m, thus ramps up with a slope V_S/L_m. The magnetizing current i_m flows through the switch, $i_m = i_Q$, while both the primary and secondary currents of the ideal transformer are zero. During this period, energy is transferred from the source to the magnetizing inductance of the transformer. As the magnetizing current i_m increases linearly, so does the energy stored in the transformer.

When the active switch is turned-off, the magnetizing current i_m directs towards the primary winding of the transformer and this in turn pushes a current flow from the *undotted* terminal of the secondary winding into the diode at the secondary side of the transformer. Accordingly, the diode is turned-on and the circuit now presents the off-time subcircuit in Fig. 4.17(b) for $(1-D)T_s$. In this period, the magnetizing current i_m runs into the *undotted* terminal of the primary winding, while the diode current i_D leaves from the *undotted* terminal of the secondary winding. The diode current in this period is given by $i_D = i_m/n$, according to the current equation of the $1:n$ ideal transformer.

During the off-time period, the output voltage V_O is reflected through the $1:n$ transformer and applied to the magnetizing inductance L_m in the negative polarity. Thus, the magnetizing current i_m ramps down with a slope $(-V_O/n)/L_m$. As i_m linearly decreases, the energy piled up in the transformer during the on-time period is delivered to the load resistor and output capacitor. Figure 4.17(c) shows the circuit waveforms of the flyback converter, constructed based on the operational principles explained above.

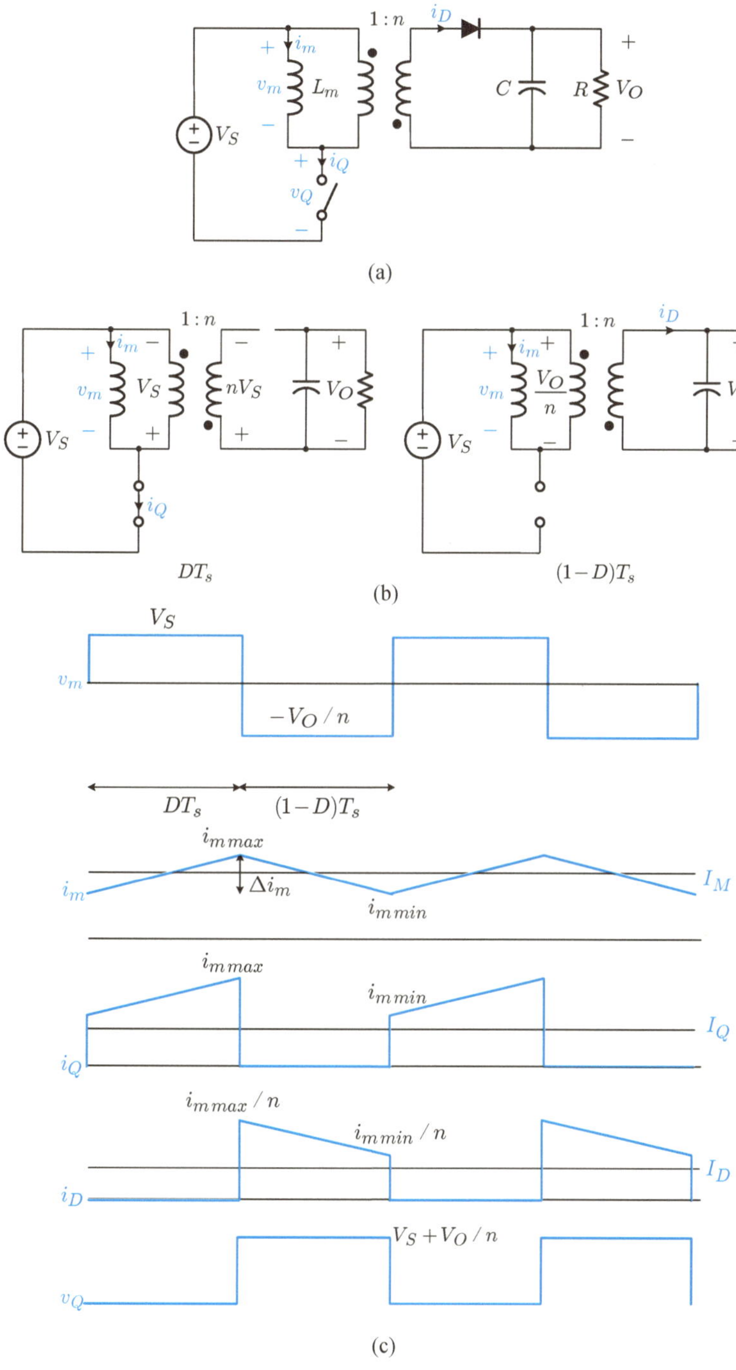

Figure 4.17 Steady-state analysis of flyback converter with $1:n$ transformer. (a) Flyback converter. (b) On-time and off-time subcircuits. (c) Major circuit waveforms.

Application of the volt-sec balance condition to the magnetizing inductance L_m yields

$$V_S D T_s = \frac{V_O}{n}(1-D)T_s \tag{4.52}$$

which leads to the voltage gain for the flyback converter

$$\boxed{\frac{V_O}{V_S} = \frac{D}{1-D}n} \tag{4.53}$$

The voltage gain is given by the product of the voltage gain of the buck/boost converter and the turns ratio of the transformer.

The magnetizing current swing is given by

$$\boxed{\Delta i_m = \frac{V_S}{L_m}DT_s = \frac{V_O/n}{L_m}(1-D)T_s} \tag{4.54}$$

The average value of the diode current is equal to the load current because the diode is connected to the load

$$I_D = \frac{V_O}{R} \tag{4.55}$$

The average diode current I_D is related to the average magnetizing current I_M by the following equation

$$I_D = (1-D)I_M\frac{1}{n} \tag{4.56}$$

From (4.55) and (4.56), the average magnetizing current is given by

$$\boxed{I_M = \frac{1}{1-D}nI_D = \frac{1}{1-D}n\frac{V_O}{R}} \tag{4.57}$$

The maximum value of the magnetizing current is determined as

$$i_{m\,max} = I_M + \frac{1}{2}\Delta i_m \tag{4.58}$$

and the minimum value is given by

$$i_{m\,min} = I_M - \frac{1}{2}\Delta i_m \tag{4.59}$$

The voltage across the active switch, v_Q in Fig. 4.17(a), is zero during the on-time period. During the off-time period, v_Q is given by the sum of the input voltage and the reflected output voltage

$$\boxed{v_Q = V_S + \frac{V_O}{n}} \tag{4.60}$$

The operation of the flyback converter is very similar to that of the buck/boost converter, yet providing an input-to-output isolation, positive output voltage, and more flexible voltage gain.

Table 4.5 Steady-State Analysis of Flyback Converter

Circuit variable	Expression
V_O	$\frac{D}{1-D}nV_S = \frac{0.4}{1-0.4}0.5 \cdot 24 = 8\text{ V}$
$v_{m\,max}$	$V_S = 24\text{ V}$
$v_{m\,min}$	$-\frac{V_O}{n} = -\frac{8}{0.5} = -16\text{ V}$
I_M	$\frac{1}{1-D}n\frac{V_O}{R} = \frac{1}{1-0.4}0.5\frac{8}{2.5} = 2.67\text{ A}$
Δi_m	$\frac{V_S}{L_m}DT_s = \frac{24}{160 \times 10^{-6}}0.4 \cdot 20 \times 10^{-6} = 1.2\text{ A}$
$i_{m\,max}$	$I_M + \frac{1}{2}\Delta i_m = 2.67 + \frac{1.2}{2} = 3.27\text{ A}$
$i_{m\,min}$	$I_M - \frac{1}{2}\Delta i_m = 2.67 - \frac{1.2}{2} = 2.07\text{ A}$
$i_{Q\,max}$	$i_{mmax} = 3.27\text{ A}$
$i_{D\,max}$	$\frac{i_{mmax}}{n} = \frac{3.27}{0.5} = 6.54\text{ A}$
$v_{Q\,max}$	$V_S + \frac{V_O}{n} = 24 + \frac{8}{0.5} = 40\text{ V}$

Steady-State Operation of Flyback Converter — EXAMPLE 4.4

This example shows the steady-state analysis and circuit waveforms of a flyback converter. The operational conditions and power stage parameters of the flyback converter are $V_S = 24\,\text{V}$, $L_m = 160\,\mu\text{H}$, $n = 0.5$, $C = 400\,\mu\text{F}$, $R = 2.5\,\Omega$, $T_s = 20\,\mu\text{s}$, and $D = 0.4$. The results of the steady-state analysis are summarized in Table 4.5. The simulated circuit waveforms are shown in Fig. 4.18.

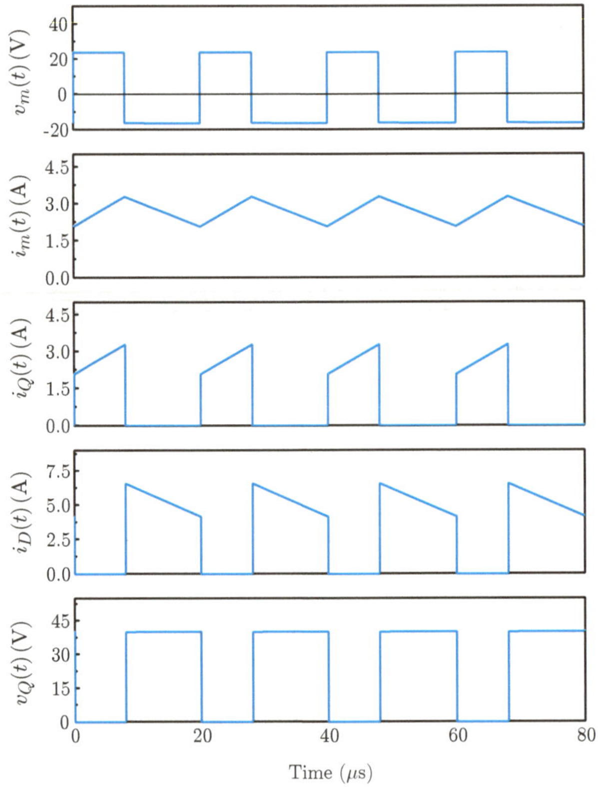

Figure 4.18 Circuit waveforms of flyback converter.

4.4.3 Steady-State Analysis in DCM

For the flyback converter, the CCM/DCM borderline condition $I_M = \Delta i_m/2$ is expressed as

$$\frac{1}{1-D} n \frac{V_O}{R} = \frac{1}{2} \frac{V_S}{L_m} DT_s \tag{4.61}$$

based on (4.57) and (4.54). By incorporating (4.53) into (4.61), the critical load resistance is found as

$$\boxed{R_{crit} = 2n^2 \frac{L_m}{(1-D)^2 T_s}} \tag{4.62}$$

If the load resistance is increased beyond R_{crit}, the converter encounters DCM operation.

Figure 4.19 shows the major waveforms of the flyback converter in DCM operation. When i_m becomes zero, the diode at the secondary side of the transformer is turned-off. Under this condition, i_m remains at zero and thus v_m also becomes zero. Accordingly, the voltage across the active switch equals to the input voltage, $v_Q = V_S$.

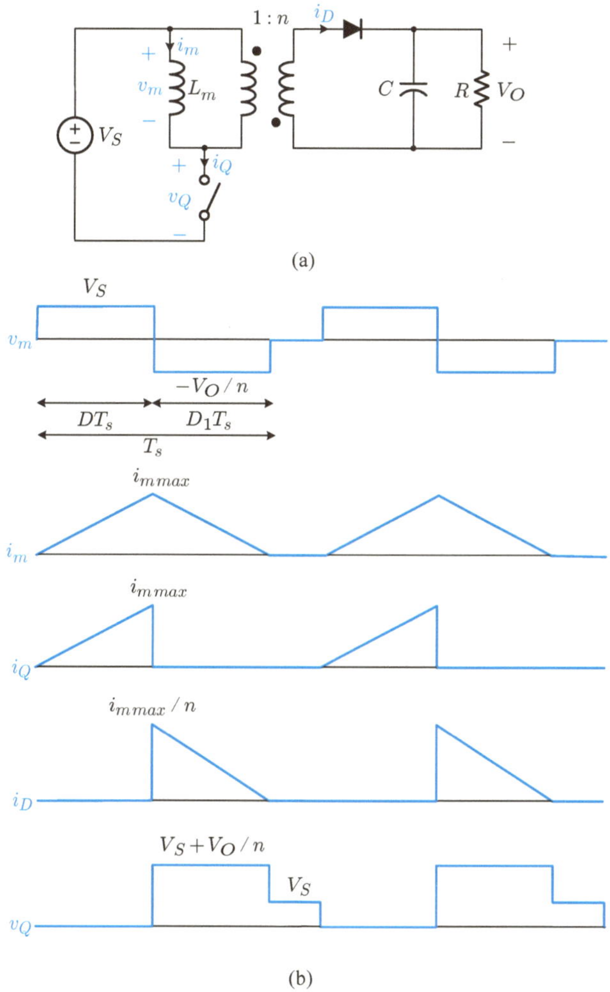

Figure 4.19 Steady-state analysis of flyback converter in DCM. (a) Circuit diagram. (b) Major circuit waveforms.

Application of the volt-sec balance condition to the magnetizing inductance yields

$$V_S D T_s = \frac{V_O}{n} D_1 T_s \tag{4.63}$$

which leads to

$$\frac{V_O}{V_S} = \frac{D}{D_1} n \tag{4.64}$$

By equating the average value of the diode current to the load current

$$\frac{\frac{1}{2}i_{mmax}\frac{1}{n}D_1T_s}{T_s} = \frac{\frac{1}{2}\left(\frac{V_S}{L_m}DT_s\right)\frac{1}{n}D_1T_s}{T_s} = \frac{V_O}{R} \tag{4.65}$$

an expression for D_1 is obtained

$$D_1 = \frac{V_O}{V_S}\frac{2L_m}{RDT_s}n \tag{4.66}$$

Finally, by combining (4.64) and (4.66), the DCM voltage gain is given by

$$\boxed{\frac{V_O}{V_S} = D\sqrt{\frac{RT_s}{2L_m}}} \tag{4.67}$$

The voltage gain expression is independent of the transformer turns ratio n, and in fact is the same as the DCM gain of the buck/boost converter.

4.5 BRIDGE-TYPE BUCK-DERIVED ISOLATED DC-TO-DC CONVERTERS

Most isolated dc-to-dc converters evolved from the three basic converters. We studied the flyback converter, derived from the buck/boost converter by adding an isolation transformer.

Among isolated converter topologies, the dc-to-dc converters derived from the buck converter are widely used in medium-to-high power applications. In this section, a family of buck-derived isolated dc-to-dc converters, called bridge-type converters due to their circuit structure, will be studied. Another family of buck-derived isolated converters will be studied in the next section, which is referred to as forward converters based on their operational principles.

The derivation of an isolated dc-to-dc converter from the buck converter is illustrated in Fig. 4.20, where the buck converter is divided into three functional blocks: the source, SPDT switch, and low-pass filter circuit in Fig. 4.20(a). The functions of the SPDT switch and low pass filter were explained in Section 3.2.

Isolated dc-to-dc converters are derived from the buck converter shown in Fig. 4.20(a), by replacing the SPDT switch with a *switch network* which performs the same function while providing an input-output isolation. This concept is illustrated in Fig. 4.20(b). The requirements of the switch network are summarized below.

1) The switch network should provide a rectangular voltage waveform, v_N in Fig. 4.20(b), as the input voltage to the low-pass filter circuit. The top value v_N is proportionate to the source voltage V_S and the bottom value is zero. The low-pass filter circuit smooths v_N, thereby producing the average value of v_N as the output voltage of the converter: $\bar{v}_N(t) = V_O$.

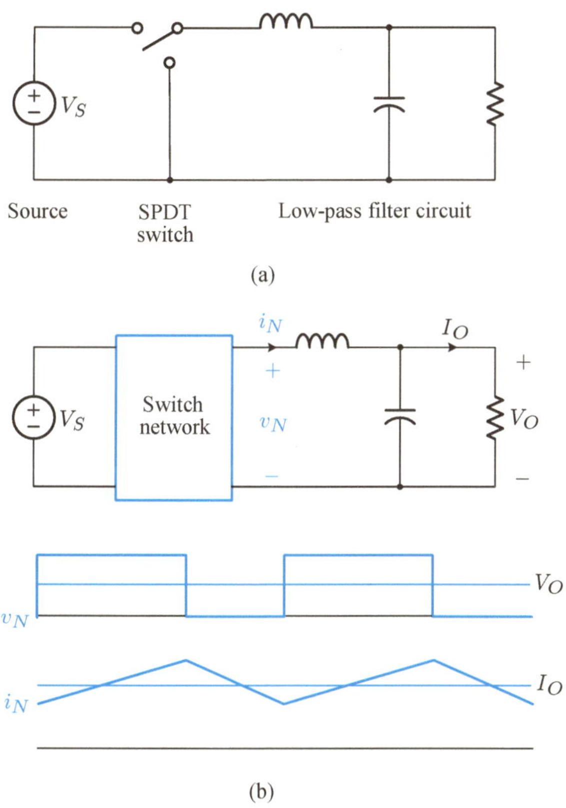

Figure 4.20 Derivation of isolated dc-to-dc converter from buck converter. (a) Buck converter. (b) Buck-derived isolated dc-to-dc converter and switch network waveforms.

2) The switch network should deliver a triangular current waveform, i_N in Fig. 4.20(b), to the input terminal of the low-pass filter circuit. In particular, the switch network should provide a freewheeling path for i_N when v_N is zero. The average value of i_N is the output current of the converter: $\bar{i}_N(t) = I_O$.

3) The switch network should offer the input-output isolation. Most switch networks employ a transformer for the input-output isolation.

In this section, we first investigate the circuit configurations of the switch network. Then, this section reviews the circuit models for multi-winding transformers which are frequently employed as an isolation transformer in the switch network. Finally, we study the operation of three important buck-derived isolated converters: the full-bridge converter, half-bridge converter, and push-pull converter. It will be shown that, although named differently due to the difference in the switch network, the operation of these three buck-derived isolated converters is all functionally equivalent to that of the buck converter.

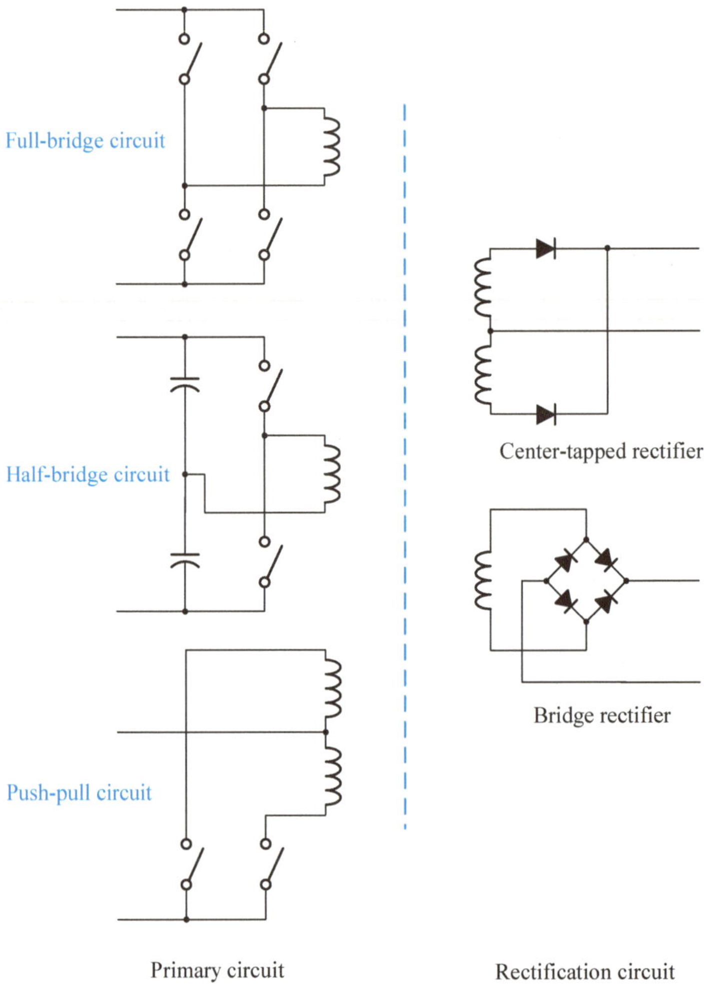

Figure 4.21 Structure of switch network.

4.5.1 Switch Network and Multi-Winding Transformer

This section introduces circuit configurations for the switch network. Because most switch networks contain a multi-winding transformer, the circuit models for multi-winding transformers are also discussed.

Switch Network Structure

Figure 4.21 shows the generic structure of the switch network. The switch network is divided into two circuits using the isolation boundary as the borderline. The circuit located at the left-hand side of the isolation boundary is referred to as the primary circuit, while the circuit appearing at the other side is denoted as the rectification circuit.

There are several circuit configurations that are employed as the primary circuit or rectification circuit. If a particular circuit configuration is selected each for the primary circuit and rectification circuit, a specific switch network will be created. Figure 4.21 illustrates the formation of the switch network, where three primary circuits and two rectification circuits are shown across the isolation boundary. An arbitrary pair of these circuits, one from the primary circuits and the other from the rectification circuits, can be merged to function as a switch network for a transformer-isolated buck-derived dc-to-dc converter.

The primary circuits are named as the full-bridge circuit, half-bridge circuit, and push-pull circuit. The term *full-bridge* stems from the structure of the circuit, which has a pair of vertical legs consisting of two active switches connected in series. In the *half-bridge* circuit, only one active switch leg exists while the other leg is replaced with two capacitors connected in series. On the other hand, the *push-pull* circuit is named after its operational principle; one switch pushes the current into the transformer, while the other switch pulls the current out of the transformer.

The two rectifications circuits in Fig. 4.21 are called the center-tapped rectifier and bridge rectifier. It will be shown that these two rectifiers are equivalent in their function and thus can be used interchangeably. A switch network could employ a two-, three-, or four-winding transformer, depending on the topologies of the primary and rectification circuits.

Circuit Models for Multi-Winding Transformers

The switch network usually contains a multi-winding transformer. Thus, it is needed to review the circuit models for transformers before investigating the operations of transformer-isolated dc-to-dc converters. Figure 4.22(a) shows a two-winding transformer and its circuit model. The circuit model contains the magnetizing inductance and ideal transformer. Details about the transformer model were covered in Section 2.2.3. From the transformer circuit model, the following relationships are written

$$\boxed{v_P(t) : v_S(t) = 1 : n} \tag{4.68}$$

$$\boxed{i_P(t) = i_m(t) + n\, i_S(t)} \tag{4.69}$$

with

$$\boxed{n = \frac{N_S}{N_P}} \tag{4.70}$$

where N_P is the turns of the primary winding and N_S is that of the secondary winding. The magnetizing current i_m is given by

$$i_m(t) = \frac{1}{L_m} \int v_P(t)\, \mathrm{d}t \tag{4.71}$$

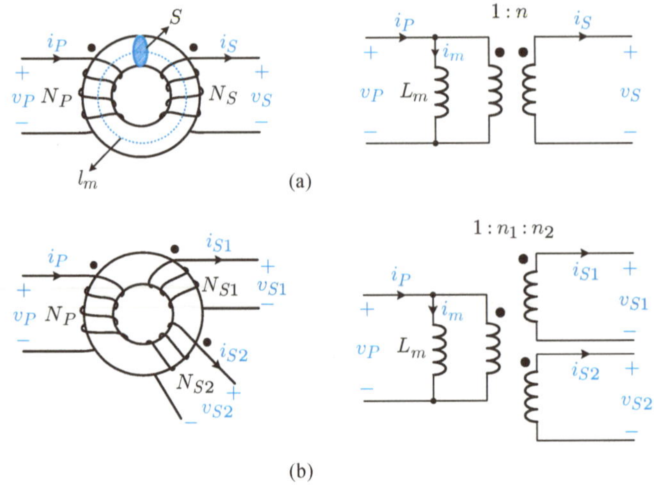

Figure 4.22 Practical transformers and circuit models. (a) Two-winding transformer. (b) Three-winding transformer.

The magnetizing inductance L_m is expressed as

$$L_m = \mu_r \mu_o \frac{S}{l_m} N_P^2 \tag{4.72}$$

in terms of the transformer parameters shown in Fig. 4.22(a).

Figure 4.22(b) shows a three-winding transformer and its circuit model. The circuit model consists of the magnetizing inductance and ideal three-winding transformer. While the circuit model can be derived from the electro-magnetic phenomena inside the three-winding transformer, it also can be considered as an extension of the two-winding transformer model. Equations (4.68) and (4.69) are modified for the three-winding transformer

$$\boxed{v_P(t) : v_{S1}(t) : v_{S2}(t) = 1 : n_1 : n_2} \tag{4.73}$$

$$\boxed{i_P(t) = i_m(t) + n_1 i_{S1}(t) + n_2 i_{S2}(t)} \tag{4.74}$$

with

$$\boxed{n_1 = \frac{N_{S1}}{N_P} \quad \text{and} \quad n_2 = \frac{N_{S2}}{N_P}} \tag{4.75}$$

where N_P, N_{S1}, and N_{S2} represent the respective turns of each transformer winding. Expressions for the magnetizing current i_m and magnetizing inductance L_m are the same as those of the previous two-winding transformer. Polarity/direction of the winding voltage/current is consistent with Lenz's law and *dot convention* discussed in Section 2.2.3. The circuit model and circuit equations for the three-winding transformer can readily be extended for the transformers with four or more windings.

When the magnetizing inductance L_m is assumed to be infinite, the circuit model reduces to an ideal three-winding transformer. The voltage equation of the ideal three-winding transformer is the same as (4.73). On the other hand, the current equation is modified to

$$\boxed{i_P(t) = n_1 i_{S1}(t) + n_2 i_{S2}(t)} \tag{4.76}$$

because i_m in (4.71) becomes zero with $L_m = \infty$.

4.5.2 Full-Bridge Converter

As the first example of bridge-type isolated converters, Fig. 4.23 shows a circuit diagram of the full-bridge converter along with its major waveforms. The full-bridge circuit is employed as the primary circuit, as the name indicates. A three-winding transformer is used for the isolation transformer and the center-tapped rectifier is selected for the rectification circuit. The operation of the full-bridge converter is now discussed with the aid of the circuit model of the three-winding transformer.

Operation with Ideal Transformer

The three-winding transformer in the middle of the switch network is assumed to be an ideal three-winding transformer with an $N_P : N_{S1} : N_{S2} = 1 : n : n$ turns ratio. This assumption simplifies the converter operation. The effects of non-ideal characteristics of a practical transformer will be later discussed. With the assumption of the $1 : n : n$ turns ratio, the transformer equations are expressed as

$$v_P(t) : v_{S1}(t) : -v_{S2}(t) = 1 : n : n \tag{4.77}$$

$$i_P(t) = n\,(i_{D1}(t) - i_{D2}(t)) \tag{4.78}$$

based on the polarity/direction of the voltage/current waveforms shown in Fig. 4.23.

Four active switches in the primary circuit, $Q_1 - Q_4$, are grouped into the two switch pairs, $\{Q_1\,Q_2\}$ and $\{Q_3\,Q_4\}$. Two individual switches in each switch pair are driven synchronously while the two switch pairs are operated complementary, as illustrated by the switch drive signals in Fig. 4.23(b). The time interval in which either the switches $\{Q_1\,Q_2\}$ or $\{Q_3\,Q_4\}$ are closed is referred to as the on-time period, T_{on}, while the time interval in which none of the switches is conducting is called the off-time period.

During an on-time period, the primary winding of the transformer is connected to the source, and the other two windings develop their terminal voltages according to the circuit equation of the ideal three-winding transformer. During an off-time period, the primary winding is isolated from the source, thereby making the terminal voltages of the three transformer windings identically zero.

The rectification circuit performs a full-wave rectification, thus providing the desired switch network voltage waveform, v_N, for the low pass filter circuit. The value of v_N during on-time periods is given by the product of the input voltage and

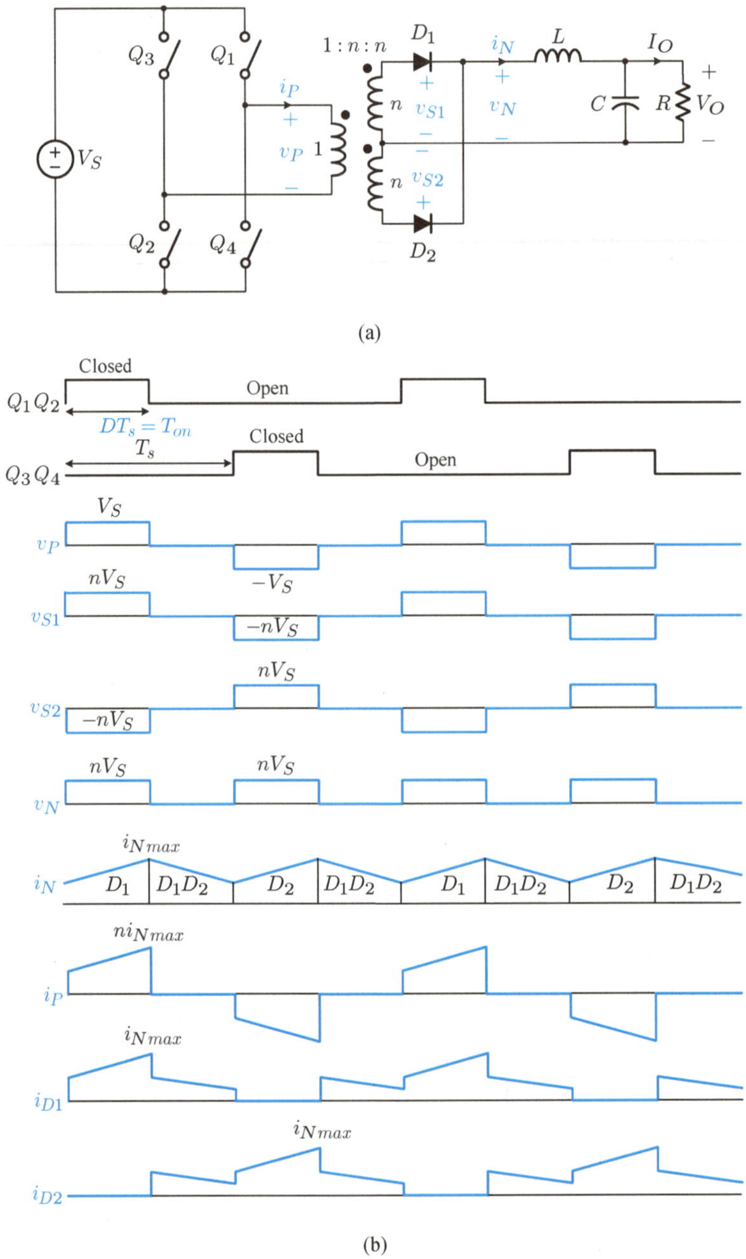

Figure 4.23 Full-bridge converter. (a) Circuit diagram. (b) Major circuit waveforms.

transformer turns ratio. The width of v_N is determined by the *effective* duty ratio[†] of the active switches, represented by D in the switch drive signals. Accordingly, the output voltage of the converter is given by

$$\boxed{V_O = \bar{v}_N(t) = nDV_S} \tag{4.79}$$

During an on-time period, the output current of the switch network, i_N, linearly increases with a slope $(nV_S - V_O)/L$. When $\{Q_1\, Q_2\}$ are closed in the on-time period, the primary current i_P flows into the *dotted* end of the primary winding and the secondary current i_N circulates through the upper diode D_1, while the lower diode D_2 is reverse-biased by the voltage of $2nV_S$. Conversely, when $\{Q_3\, Q_4\}$ are closed, i_P reverses its direction and i_N now circulates through D_2, while D_1 is reverse-biased. The specific diode, which is conducting during the respective operational period, is labeled on the waveform of i_N.

During an off-time period in which all four switches are open and the primary current i_P is zero, the filter inductor current i_N turns on D_1 and D_2 simultaneously to create a freewheeling path. The transformer winding voltages are zero during the off-time period. Accordingly, i_N ramps down with a slope $-V_O/L$. The declining freewheeling current i_N is equally divided into i_{D1} and i_{D2}. Because i_{D1} comes out of the *dotted* terminal and i_{D2} enters the *dotted* terminal, while i_P remains zero, the current equation for the ideal three-winding transformer, given by (4.78), is satisfied. The current waveforms are shown in Fig. 4.23(b).

The switch network produces the desired waveforms for v_N and i_N, and the ideal three-winding transformer satisfies the circuit equations defined by (4.73) and (4.76). The average value of i_N becomes the load current

$$\boxed{I_O = \bar{i}_N(t) = \frac{V_O}{R}} \tag{4.80}$$

Two underlying assumptions are employed in the previous analysis. First, CCM operation is assumed, which is true only when the condition $I_O > \Delta i_N/2$ is satisfied. Yet, Fig. 4.23(b) can readily be modified to yield the circuit waveforms for DCM operation. In fact, the DCM operation of the buck converter, covered in Section 3.5.3, can be extended to the full-bridge converter with minor modifications. The second assumption is the ideal characteristics of the three-winding transformer. When a practical transformer is employed, the converter operation will be altered due to the presence of the magnetizing inductance. The operation with a magnetizing inductance is covered in the next section.

Effects of Magnetizing Inductance

Figure 4.24(a) shows the circuit model of a practical three-winding transformer. The practical transformer is modeled as a parallel connection of the magnetizing

[†]The effective duty ratio is defined as $T_{on}/(T_s'/2)$ where T_s' is the switching period of each individual switch.

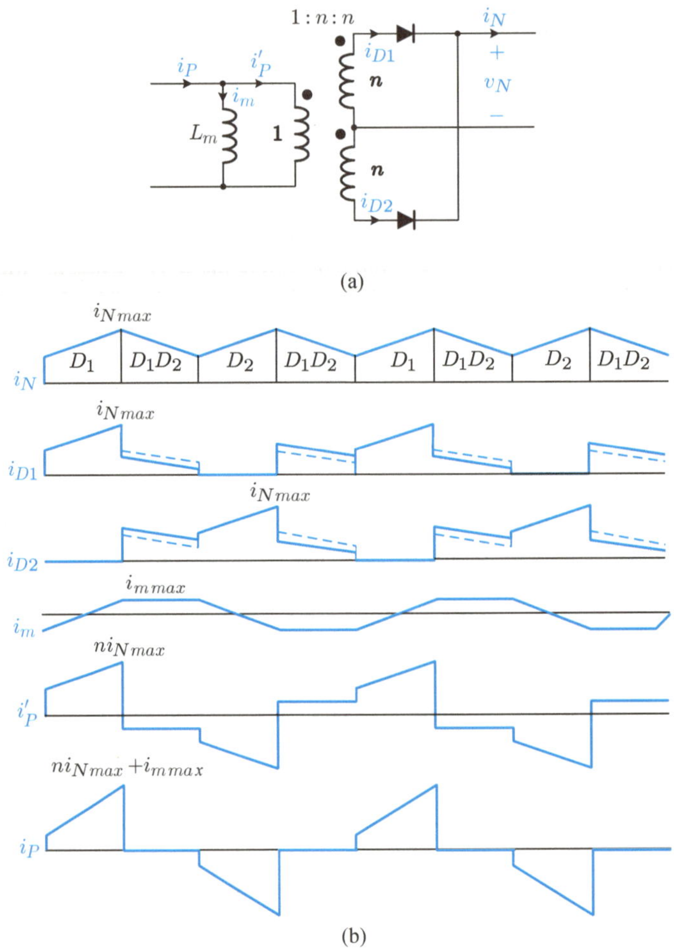

Figure 4.24 Switch network with practical transformer. (a) Practical three-winding transformer. (b) Major waveforms.

inductance L_m and ideal three-winding transformer. The primary winding current of the ideal transformer is renamed as i'_P. The current i_P now denotes the input current to the practical transformer. It can easily be deduced that the filter inductor current i_N remains the same because the switch network voltage v_N is the same as the previous case of ideal transformer. On the other hand, the other current waveforms will be altered due to the magnetizing current i_m running through L_m.

The voltage waveform across the magnetizing inductance L_m naturally meets the volt-sec balance condition and the magnetizing current i_m thus is symmetrical. During an on-time period, i_m increases or decreases depending on the polarity of the voltage across L_m. The magnetizing current, commutating via the closed primary-side switches, runs through L_m and does not enter the primary winding of the ideal transformer. Accordingly, during the on-time period, the winding currents of the ideal transformer, i'_P, i_{D1}, and i_{D2}, are identical to those of the previous case.

During an off-time period in which all the active switches are open, the magnetizing current i_m now circulates through the primary winding of the ideal transformer. In this period, i_m remains constant because the voltage across the magnetizing inductance is clamped at zero. With the presence of i_m in the primary winding, the secondary currents, i_{D1} and i_{D2}, are altered to meet the current equation of the ideal three-winding transformer

$$\boxed{i'_p(t) = -i_m(t) = n(i_{D1}(t) - i_{D2}(t))} \tag{4.81}$$

while maintaining their sum unchanged

$$\boxed{i_N(t) = i_{D1}(t) + i_{D2}(t)} \tag{4.82}$$

Figure 4.24(b) illustrates the current waveforms under this situation. The diode currents, i_{D1} and i_{D2}, are *unevenly* distributed during the off-time period to satisfy the equations (4.81) and (4.82) at the presence of the magnetizing current. The dashed lines show the original i_{D1} and i_{D2} waveforms at the absence of the magnetizing current. The current waveform i'_P is constructed using the fact $i'_P = ni_{D1}$ or $i'_P = ni_{D2}$ for the on-time period, and $i'_P = -i_m$ for the off-time period. The i_P is obtained using the fact $i_P = i'_P + i_m$ for the on-time period and $i_P = 0$ for the off-time period.

When the magnetizing inductance L_m approaches infinity, i_m reduces to zero and the current waveforms return to those of the previous ideal transformer case. For most transformer fabrications, the magnetizing inductance is maximized within practical limits to enhance efficiency and performance of transformers. Accordingly, the magnetizing inductance is sufficiently large and the magnetizing current is negligibly small. For this reason, it is common practice to assume the ideal characteristics for transformers employed in bridge-type isolated dc-to-dc converters.

Another justification for not considering the magnetizing inductance is that the input voltage of the switch network is symmetric and automatically meets the volt-sec balance condition on L_m. If this is not the case, the magnetizing inductance L_m, however large it might be, will eventually saturate to cause a fatal failure of the converter. This transformer saturation will not happen to the bridge-type isolated converters. The assumption of the ideal transformer will be implicitly used in the upcoming discussions about bridge-type isolated converters.

Steady-State Operation of Full-Bridge Converter — EXAMPLE 4.5

This example shows the steady-state analysis and circuit waveforms of a full-bridge converter. The operational conditions and circuit parameters of the converter are $V_S = 64\,\text{V}$, $n = 0.5$, $L = 40\,\mu\text{H}$, $C = 400\,\mu\text{F}$, $R = 1\,\Omega$, $T_s = 20\,\mu\text{s}$, and $D = 0.25$. For the given operational conditions, the steady-state values for major circuit variables are calculated in Table 4.6, while the corresponding PSpice® simulations are shown in Fig. 4.25.

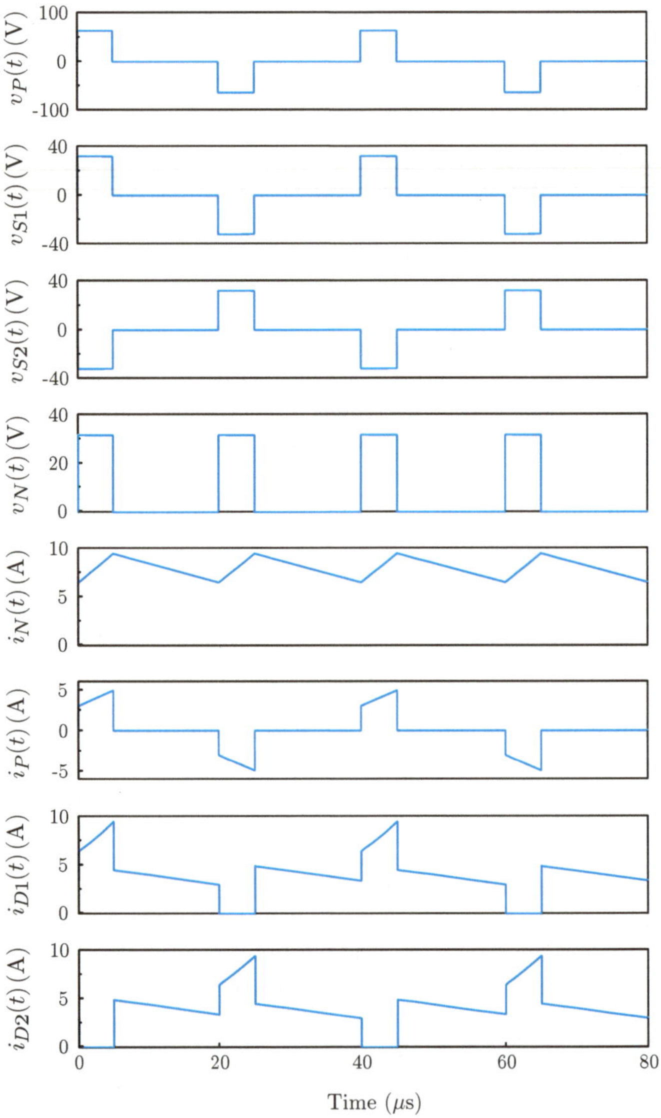

Figure 4.25 Circuit waveforms of full-bridge converter.

Table 4.6 Steady-State Analysis of Full-Bridge Converter

Circuit variable	Expression
V_O	$nDV_S = 0.5 \cdot 0.25 \cdot 64 = 8\text{ V}$
$v_{N\,max}$	$nV_S = 32\text{ V}$
Δi_N	$\frac{nV_S - V_O}{L}DT_s = \frac{0.5 \cdot 64 - 8}{40 \times 10^{-6}}0.25 \cdot 20 \times 10^{-6} = 3\text{ A}$
$i_{N\,max}$	$\frac{V_O}{R} + \frac{1}{2}\Delta i_N = \frac{8}{1} + \frac{1}{2}3.0 = 9.5\text{ A}$
$i_{N\,min}$	$\frac{V_O}{R} - \frac{1}{2}\Delta i_N = \frac{8}{1} - \frac{1}{2}3.0 = 6.5\text{ A}$
$i_{D1\,max}$	$i_{Nmax} = 9.5\text{ A}$
$i_{D2\,max}$	$i_{Nmax} = 9.5\text{ A}$

4.5.3 Half-Bridge Converter

The half-bridge converter can be configured by combining the half-bridge circuit with either the bridge rectifier or center-tapped rectifier. The bridge rectifier is used for the half-bridge converter in Fig. 4.26(a). The two capacitors, C_1 and C_2 in Fig. 4.26(a), function as a voltage divider and each capacitor develops a dc voltage of $V_S/2$ at its terminals.

The switch drive signals and voltage waveforms across the primary and secondary windings of the two-winding transformer are shown in Fig. 4.26(b). When Q_1 is turned-on, v_P and v_S both become positive. Conversely, when Q_2 is closed, v_P and v_S are negative. During an off-time period in which both Q_1 and Q_2 are open, the transformer is functionally isolated from the switch network, thereby yielding the conditions $v_P = v_S = 0$ and $i_P = i_S = 0$.

As shown in Fig. 4.26(b), the bridge rectifier circuit generates v_N that alternates between $nV_S/2$ and zero. The output voltage of the converter then becomes

$$\boxed{V_O = \bar{v}_N(t) = \frac{nDV_S}{2}} \qquad (4.83)$$

The shape of i_N is the same as the previous full-bridge converter case, yet the commutation of the diodes is different. When Q_1 is closed, D_1 and D_2 carry i_N.

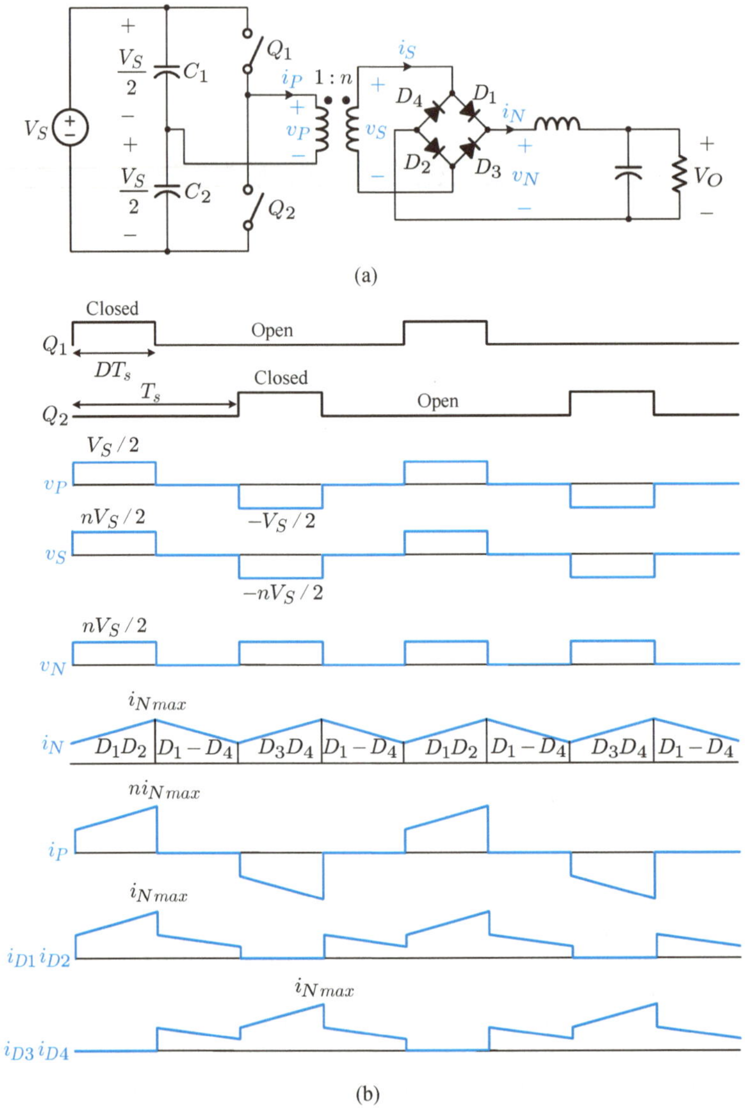

Figure 4.26 Half-bridge converter. (a) Circuit diagram. (b) Major circuit waveforms.

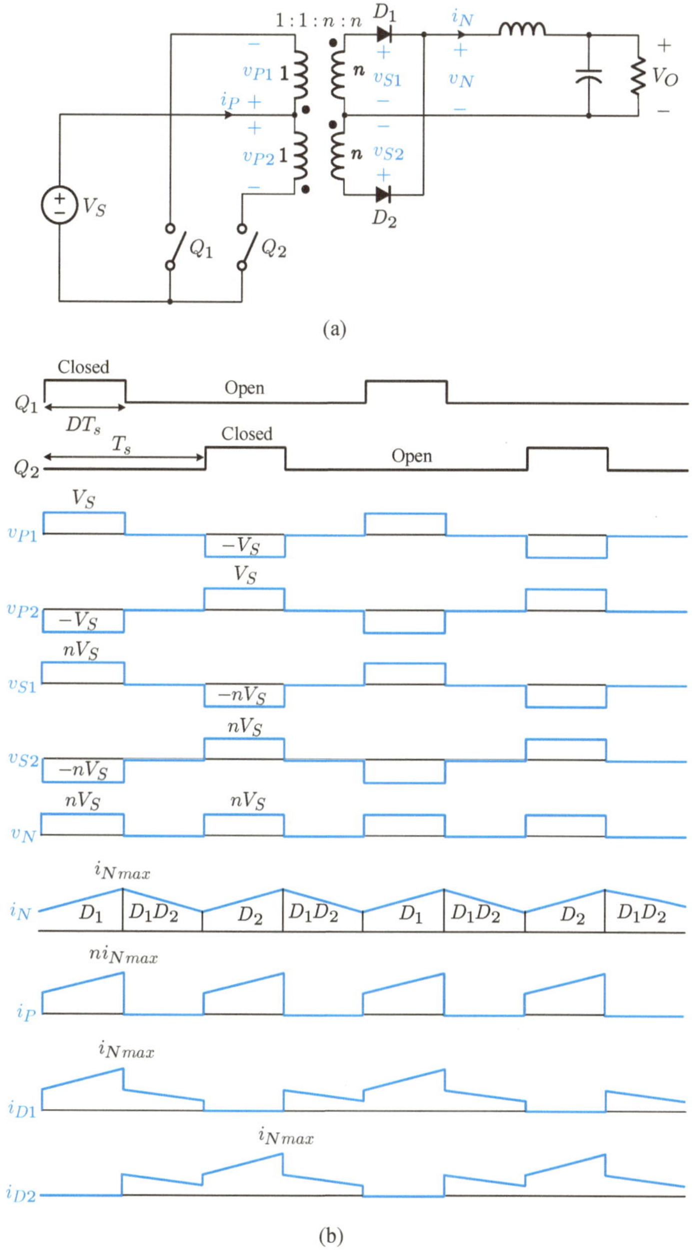

Figure 4.27 Push-pull converter. (a) Circuit diagram. (b) Major circuit waveforms.

Conversely, when Q_2 is closed, D_3 and D_4 conduct. During an off-time period in which the transformer is functionally separated from the other part of the circuit, i_N turns on all the four diodes simultaneously and freewheels through them. It is apparent that the converter could employ a center-tapped rectifier instead of the bridge rectifier. When a center tapped rectifier is used, the current waveforms are identical to those of Fig. 4.23.

4.5.4 Push-Pull Converter

Another variation of the bridge-type converter is the push-pull converter shown in Fig. 4.27. The converter uses two active switches and one four-winding transformer. While a center-tapped rectifier is used in Fig. 4.27, the bridge rectifier could be employed as an alternative, provided that the four-winding transformer is replaced with a three-winding transformer.

When the switch Q_1 is closed, v_{P1} and v_{S1} become positive while v_{P2} and v_{S2} are negative. Conversely, when Q_2 is turned-on, v_{P2} and v_{S2} now become positive while the other two windings develop a negative voltage. The terminal voltages of the four windings are shown in Fig. 4.27. The rectification circuit performs a full-wave rectification, thus yielding the desired v_N waveform. The current waveforms associated with the four-winding transformer are also shown in Fig. 4.27. The voltage and current waveforms in Fig. 4.27 satisfy the circuit equations for the four-winding transformer.

The operation of the push-pull converter is functionally the same as that of the full-bridge converter. In fact, the push-pull converter replicates the operation of the full-bridge converter using two active switches and a transformer with two primary windings.

4.6 FORWARD CONVERTERS

The forward converter is another class of buck-derived isolated dc-to-dc converters. While also derived from the buck converter, the forward converter has unique circuit characteristics, not found in the previous bridge-type converters. Distinctive features of the forward converter will become apparent as we study the operation of the converter.

4.6.1 Basic Operational Principles

Functional basics of the forward converter are first explained using a conceptual converter configured with an ideal transformer. Figure 4.28(a) is the circuit diagram of a conceptual forward converter that employs a 1 : 1 ideal transformer. The converter possesses configuration of the buck-derived isolated converter, shown in Fig. 4.20(b). However, the structure of the switch network differs from that of the previous bridge-type converters. The switch network contains one active switch, one ideal two-

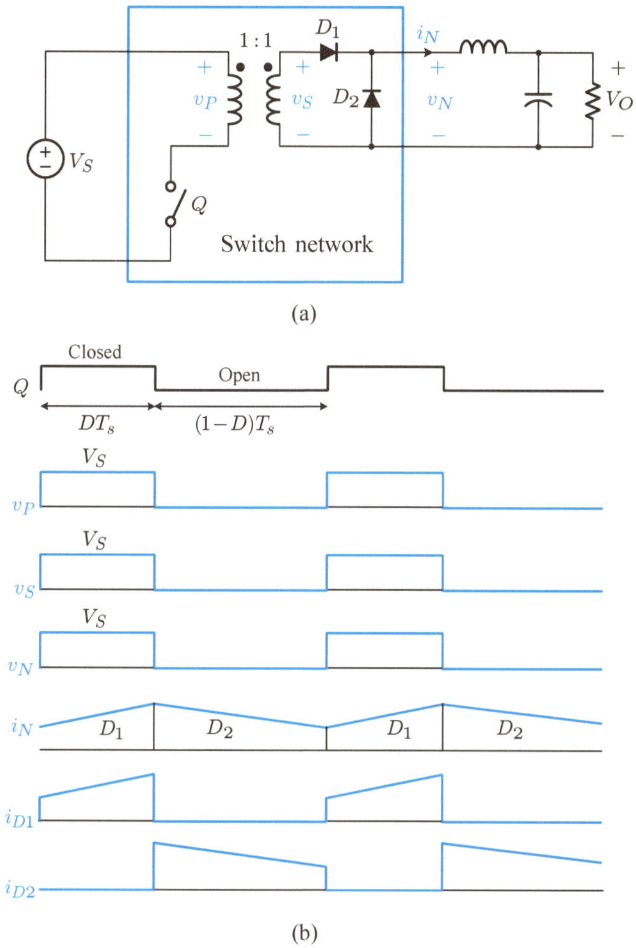

Figure 4.28 Conceptual forward converter with ideal transformer. (a) Circuit diagram. (b) Major circuit waveforms.

winding transformer, and two diodes. This switch network is the simplest among all the switch networks used in buck-derived isolated converters.

The operation of the conceptual forward converter is illustrated in Fig. 4.28(b). When the switch is closed, the positive voltage appears at the *dotted* terminal of the secondary transformer winding. During this period, D_1 is turned-on and D_2 is reverse-biased by the secondary winding voltage v_S. Therefore, the switch network current i_N flows through D_1 and the switch network voltage v_N equals to the input voltage V_S. During this period, energy is delivered in the *forward* direction through the ideal transformer.

When the switch is opened, both the primary current and secondary current are identically zero according to the current equation of the ideal two-winding trans-

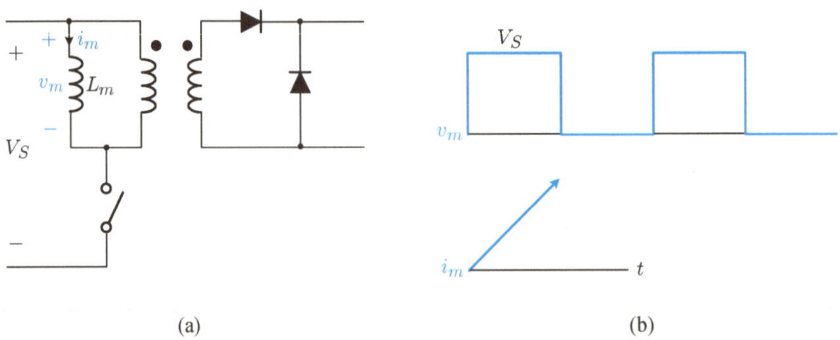

Figure 4.29 Switch network with practical transformer. (a) Circuit diagram. (b) Reset problem.

former. In this period, the ideal transformer is functionally isolated from the converter circuit. The switch network current i_N now freewheels through D_2 and v_N thus becomes zero. The switch network produces the required waveforms for v_N and i_N, and the output voltage is given by

$$\boxed{V_O = \bar{v}_N(t) = DV_S} \tag{4.84}$$

and the output current is given by

$$\boxed{I_O = \bar{i}_N(t) = \frac{V_O}{R}} \tag{4.85}$$

The operation of the conceptual forward converter is identical to that of the buck converter.

Reset Problem and Reset Circuit

When the ideal transformer is replaced with a practical transformer, the conceptual forward converter in Fig. 4.28 immediately becomes inoperative. Figure 4.29(a) shows the switch network including the magnetizing inductance of a practical transformer. Now, the problem of the converter is clear with the presence of the magnetizing inductance L_m. As shown in Fig. 4.29(b), the voltage across the magnetizing inductance v_m does not meet the volt-sec balance requirement. More seriously, the magnetizing current, elevated by the equation $i_m(t) = (V_S/L_m)t$ during an on-time period, instantly loses its path when the active switch is opened. The only way for i_m to maintain its continuity is to divert into the *undotted* terminal of the primary winding of the ideal transformer. If this is the case, the secondary current must run into the *dotted* terminal of the secondary winding. However, this current flow is not possible because the diode cannot deliver any current from cathode to anode.

The problem associated with the volt-sec balance condition on L_m or the continuity in i_m is termed as the reset problem, indicating the problem of resetting the magnetic flux inside the core to its original state.

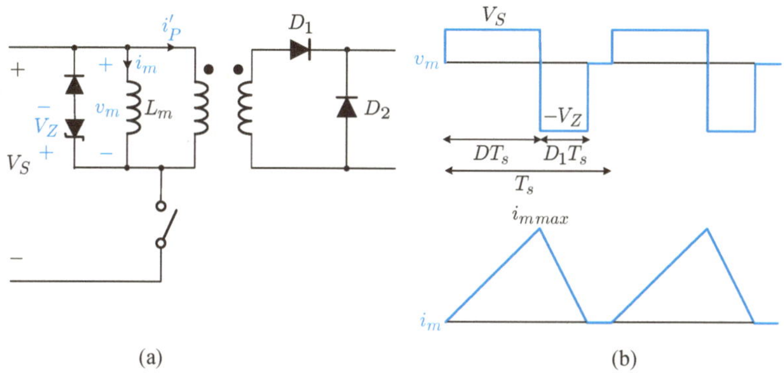

Figure 4.30 Switch network with Zener diode reset. (a) Circuit diagram. (b) Major circuit waveforms.

A variety of auxiliary circuits are employed as a means of meeting the volt-sec balance condition on L_m while providing a continuous path for i_m. The auxiliary circuits, applied to practical forward converters to solve the reset problem, are called reset circuits.

Switch Network with Zener Diode Reset

The Zener diode is often employed to resolve the reset problem. Figure 4.30 shows the Zener diode reset circuit which utilizes a pair of Zener diode and regular diode across the primary winding of the transformer. During an on-time period, the regular diode is reverse-biased and the Zener diode reset circuit does not interfere with the circuit operation. The magnetizing current linearly increases with a slope V_S/L_m during this period.

During an off-time period, the magnetizing current i_m activates the Zener diode and circulates through the loop formed by the magnetizing inductance, Zener diode, and regular diode. In this period, the primary and secondary currents of the ideal transformer are both zero: $i'_P = i_{D1} = 0$. The breakdown voltage of the Zener diode, V_Z, is applied to L_m in the negative polarity and i_m thus linearly decreases with a slope $-V_Z/L_m$. When i_m is reduced to zero, the regular diode turns off and remains off until the onset of the next off-time period. The waveforms for v_m and i_m are shown in Fig. 4.30(b). The volt-sec balance condition on L_m yields

$$V_S D T_s = V_Z D_1 T_s \tag{4.86}$$

where $D_1 T_s$ is the fraction of the off-time period in which the magnetizing current is present. The magnetizing current i_m attains its peak value at the end of the on-time period and returns to zero during the off-time period, thus resetting the magnetic flux to its initial state.

While the Zener diode reset circuit is simple in structure and operation, it only yields a lossy reset. The energy transferred to L_m during the on-time period is dissipated at the Zener diode during the off-time period. Accordingly, the Zener diode reset circuit is rarely used for applications where efficiency is an important consideration.

Many alternative reset circuits are available for forward converters. Some reset circuits result in a lossy reset while others operate losslessly. All these reset circuits do not alter the operation of the forward converter, yet only function during the off-time period to meet the volt-sec balance condition on L_m while maintaining the continuity in i_m. In other words, the forward converter follows the simple operation of the conceptual converter shown in Fig. 4.28, while bypassing the reset problem with an appropriate reset circuit.

Switch Network with Tertiary Winding Reset

It is also possible to reset the transformer using an additional winding. This alternative reset scheme is shown in Fig. 4.31. The reset is performed by the tertiary winding wound on the same magnetic core on which the original transformer is built. Figure 4.31(a) shows a switch network employing the tertiary winding reset. The structure of the three-winding transformer is also shown. For simplicity of discussions, a turns ratio of $N_P:N_S:N_T = 1:1:1$ is assumed for the three-winding transformer. Figure 4.31(b) shows the major circuit waveforms of the switch network. Referring to Fig. 4.31, the operational cycle of the switch network is divided into three periods.

On-Time Period DT_s: This period corresponds to the interval in which the active switch Q is closed. The input voltage V_S appears at all the three transformer windings in the positive polarity — with the positive voltage at the *dotted* winding terminal. Accordingly, the diode D_1 is forward-biased while the other two diodes are reverse-biased. In particular, the diode at the tertiary winding, the *reset diode* D_C, is reversed-biased by $2V_S$ — the input voltage V_S plus the voltage across the tertiary winding, which is V_S for the $1:1:1$ transformer turns ratio. The output current of the switch network, i_N, flows out of D_1 while the same current runs into the *dotted* terminal of the primary winding. The magnetizing current i_m increases linearly with a slope V_S/L_m. The sum of the magnetizing current and primary winding current passes through Q_1. The energy in the magnetizing inductance gradually increases and reaches the peak value at the end of this period.

Reset Period D_1T_s: This period starts when the active switch is opened, and ends when the magnetizing current i_m is reduced to zero. *Opening the active switch triggers the reset diode D_C to conduct and deliver i_C through the tertiary winding.* The tertiary winding current i_C flows into the voltage source V_S connected across the left-hand side of the switch network. This enables the magnetizing current i_m to run continuously while satisfying the current equation of the $1:1:1$ ideal transformer, as explained above.

The magnetizing current i_m now circulates through the loop formed by the magnetizing inductance and primary transformer winding. This current is

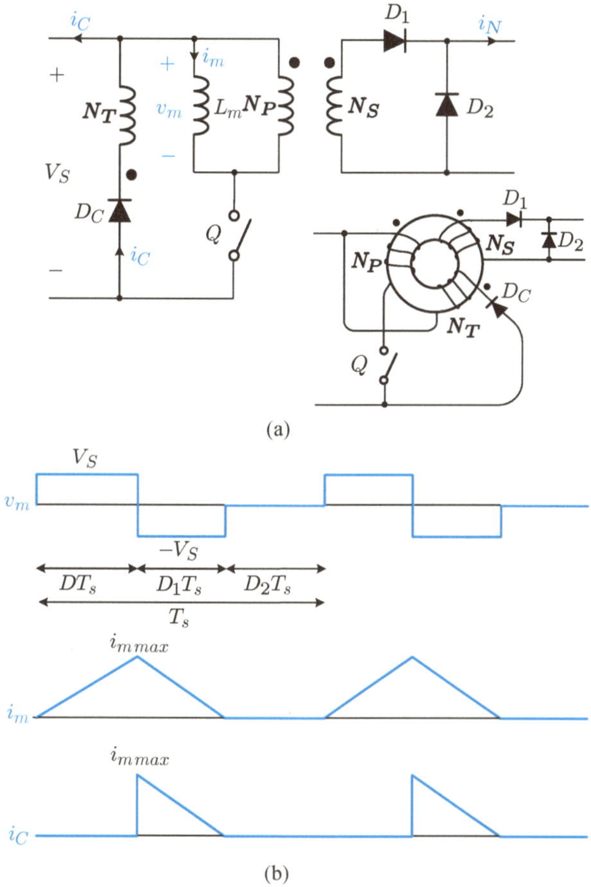

Figure 4.31 Switch network with tertiary winding reset. (a) Circuit diagram. (b) Major circuit waveforms.

coming out of the *dotted* terminal of the primary winding. The continuous flow of i_m is possible because the same current flows into the *dotted* terminal of the tertiary winding, $i_C = i_m$, while the secondary winding carries *zero* current, as illustrated in Fig. 4.31(b). The turn-on of D_C also forces all the three windings to take V_S in the negative polarity – with the negative voltage at the *dotted* terminal. Accordingly, D_1 is reverse-biased and i_N now freewheels through D_2.

The input voltage V_S is applied to the magnetizing inductance L_m in the negative polarity and i_m thus decreases linearly with a slope $-V_S/L_m$. As i_m decays from its peak, the energy stored in L_m is transferred back to the voltage source located at the left-hand side of the switch network. Thus, the circuit provides a lossless reset—all the energy delivered to the magnetizing inductance L_m during the on-time period is fully recovered by the voltage source during the reset period.

Recess PeriodD_2T_s: When the condition $i_C = i_m = 0$ is reached, the reset diode D_C turns off and remains off until the next reset period. All the currents and voltages associated with the switch network are zero in this period.

With the $1:1:1$ transformer turns ratio, the rising slope of i_m during the on-time period is the same as the decaying slope during the reset period. This implies that the maximum duty ratio of the active switch should be limited to 0.5 $(0 < D < 0.5)$ for proper operation of the converter. When the duty ratio exceeds 0.5, the magnetizing current will increase cycle-by-cycle, and the magnetizing inductance will eventually saturate, leading to a fatal circuit failure.

4.6.2 Tertiary-Winding Reset Forward Converter

A forward converter is configured by combining the switch network discussed in the previous section and low-pass filter circuit. The resulting converter is shown in Fig. 4.32, which is referred to as the tertiary-winding reset forward converter. In the tertiary-winding reset circuit in Fig. 4.32(a), the primary and tertiary windings retain the $N_P : N_T = 1:1$ turns ratio, yet the primary and secondary windings have an $N_P : N_S = 1:n$ turns ratio. The tertiary winding is called the *reset winding*. Figure 4.32(a) also shows the typical magnetizing current i_m. Figure 4.32(b) depicts three subcircuits of the converter, each representing the converter circuit during the on-time, reset, or recess period. Expressions of major circuit waveforms, derived from these three subcircuits, are listed in Table 4.7.

Figure 4.33 shows the circuit waveforms, constructed using the equations in Table 4.7. The waveforms indicate that the operation of the forward converter is identical to that of the conceptual converter in Fig. 4.28, except for the waveforms associated with the magnetizing inductance and reset winding. The output voltage of the converter is given by

$$\boxed{V_O = \bar{v}_N(t) = nDV_S} \tag{4.87}$$

with the $1:n$ turns ratio between the primary and secondary windings.

The peak value of i_m is given by

$$i_{m\,max} = \frac{V_S}{L_m} DT_S \tag{4.88}$$

The maximum value for the the switch network current i_N is given by

$$i_{N\,max} = \frac{V_O}{R} + \frac{1}{2}\frac{nV_S - V_O}{L} DT_s \tag{4.89}$$

The minimum value of i_N is given by

$$i_{N\,min} = \frac{V_O}{R} - \frac{1}{2}\frac{nV_S - V_O}{L} DT_s \tag{4.90}$$

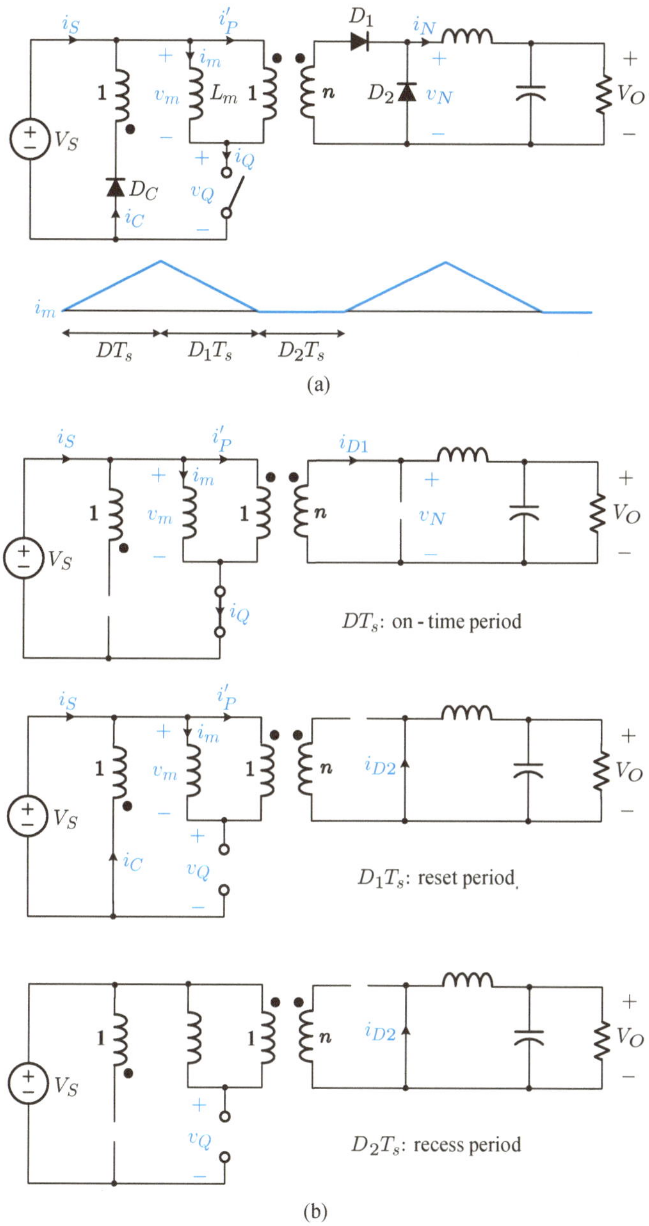

Figure 4.32 Tertiary-winding reset forward converter. (a) Circuit diagram and magnetizing current. (b) Three subcircuits.

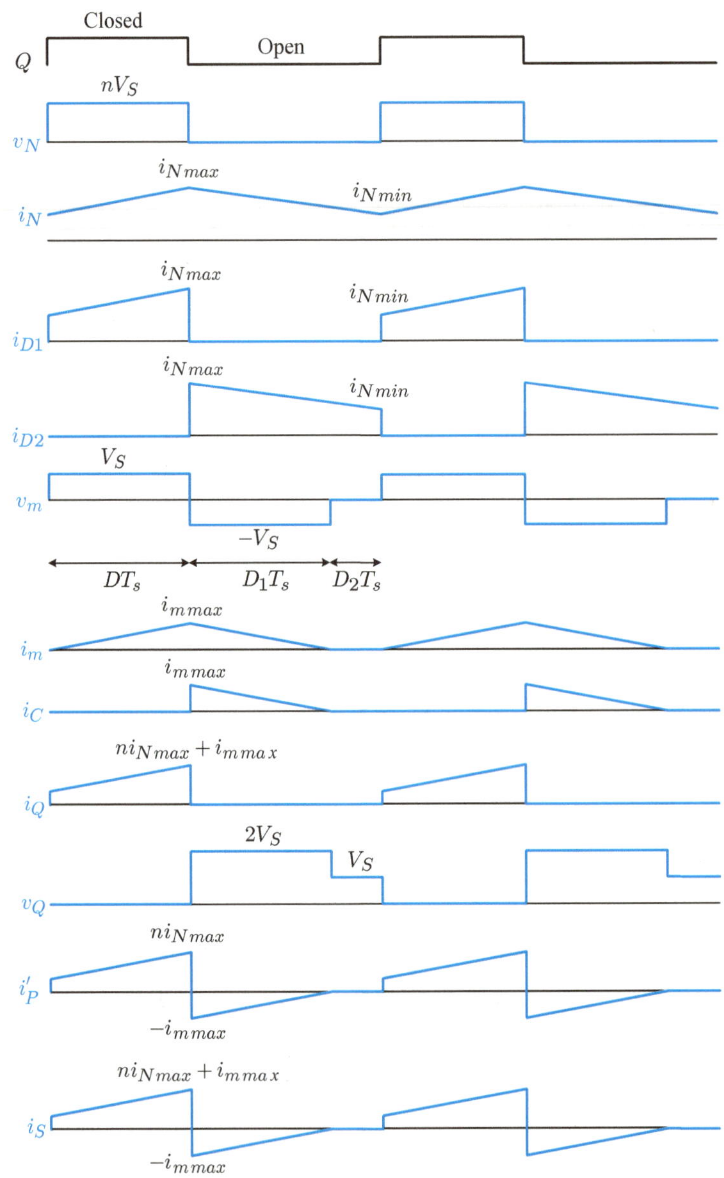

Figure 4.33 Waveforms of tertiary-winding reset forward converter.

Table 4.7 Expressions for Circuit Variables of Tertiary-Winding Reset Forward Converter

	On-time period	Reset period	Recess period
v_N	nV_S	0	0
v_m	V_S	$-V_S$	0
i_m	$\frac{V_S}{L_m}t$	$-\frac{V_S}{L_m}t$	0
i_C	0	$-\frac{V_S}{L_m}t$	0
i_N	$\frac{nV_S - V_O}{L}t$	$-\frac{V_O}{L}t$	$-\frac{V_O}{L}t$
i_{D1}	i_N	0	0
i_{D2}	0	i_N	i_N
i'_P	ni_N	$-i_m$	0
i_Q	$i_m + i'_P$	0	0
i_S	$i_m + i'_P$	$-i_C$	0
v_Q	0	$2V_S$	V_S

From these values, the peak and valley values of the other current waveforms are readily evaluated.

The tertiary-winding reset forward converter is one of most popular converter topologies. Advantages of this converter include the lossless reset and minimal component count. Due to the lossless reset, efficiency of the converter can be high, compared with other forward converters with a lossy reset scheme. This converter uses only one active switch, while handling the power level comparable to that of the half-bridge converter.

There are many other variations of the forward converter with lossless reset. One widely-used converter topology with lossless reset is the forward converter that

Table 4.8 Steady-State Analysis of Tertiary-Winding Reset Forward Converter

Circuit variable	Expression
V_O	$nDV_S = 0.5 \cdot 0.25 \cdot 64 = 8\ \text{V}$
v_{Nmax}	$nV_S = 0.5 \cdot 64 = 32\ \text{V}$
I_N	$\frac{V_O}{R} = \frac{8}{1} = 8\ \text{A}$
Δi_N	$\frac{nV_S - V_O}{L} DT_s = \frac{0.5 \cdot 64 - 8}{40 \times 10^{-6}} 0.25 \cdot 20 \times 10^{-6} = 3.0\ \text{A}$
i_{Nmax}	$I_N + \frac{1}{2}\Delta i_N = 8.0 + \frac{1}{2}3.0 = 9.5\ \text{A}$
i_{Nmin}	$I_N - \frac{1}{2}\Delta i_N = 8.0 - \frac{1}{2}3.0 = 6.5\ \text{A}$
v_{mmax}	$V_S = 64\ \text{V}$
v_{mmin}	$-V_S = -64\ \text{V}$
i_{mmax}	$\frac{V_S}{L_m} DT_s = \frac{64}{200 \times 10^{-6}} 0.25 \cdot 20 \times 10^{-6} = 1.6\ \text{A}$
v_{Qmax}	$2V_S = 2 \cdot 64 = 128\ \text{V}$

employs two active switches and one two-winding transformer, known as the two-switch forward converter. The operation of this converter is described in the next section.

EXAMPLE 4.6 Steady-State Operation of Forward Converter

This example illustrates the steady-state analysis and major circuit waveforms of a tertiary-winding reset forward converter. The operational conditions and circuit parameters of the forward converter are $V_S = 64\ \text{V}$, $L_m = 200\ \mu\text{H}$, $N_P : N_S : N_T = 1 : 0.5 : 1$, $L = 40\ \mu\text{H}$, $C = 400\ \mu\text{F}$, $R = 1\ \Omega$, $T_s = 20\ \mu\text{s}$, and $D = 0.25$. For the

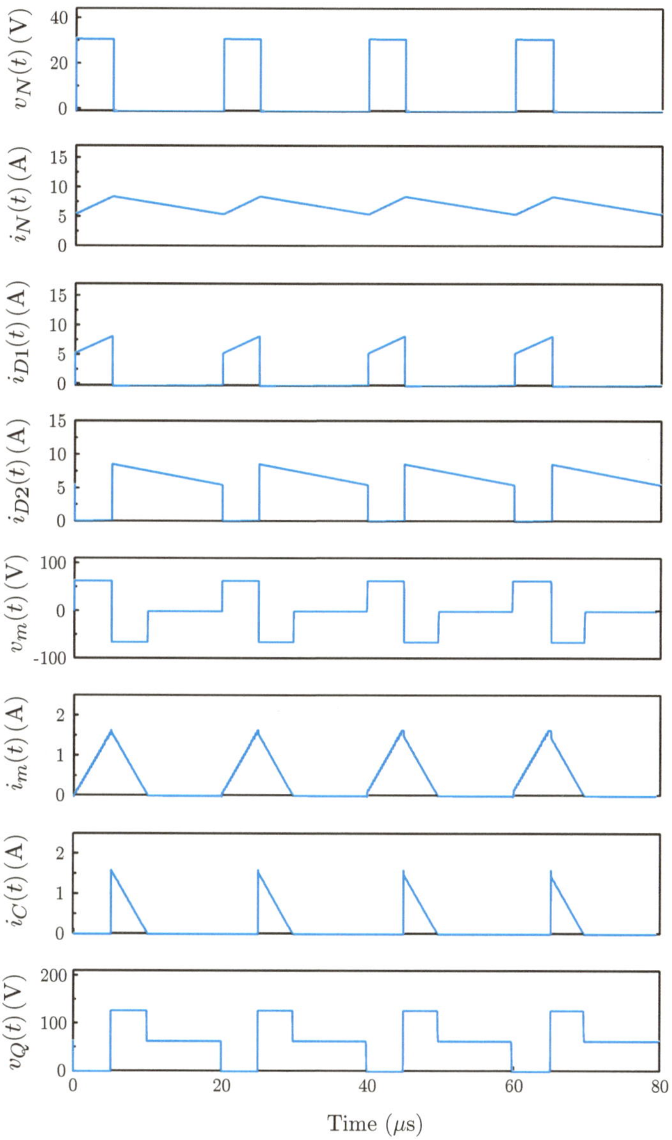

Figure 4.34 Circuit waveforms of tertiary-winding reset forward converter.

given power stage parameters and operating conditions, steady-state values of major circuit variables are evaluated using the previous analysis results. The steady-state values are listed in Table 4.8 and the simulated circuit waveforms are given in Fig. 4.34.

4.6.3 Two-Switch Forward Converter

As another variation of the forward converter, Fig. 4.35(a) shows the two-switch forward converter. The converter employs two synchronized active switches, Q_1 and Q_2, and two diodes, D_1 and D_2, in the primary side of the switch network. The remaining part of the switch network contains a two-winding transformer and two other rectification diodes. When the two synchronized active switches are turned-on, the converter presents the on-time subcircuit. The on-time subcircuit of this converter is functionally identical to that of the previous tertiary-winding reset forward converter.

When the two synchronized switches are turned-off, the magnetizing current i_m turns on the two diodes, D_1 and D_2, in the primary side of the circuit, and freewheels through the loop formed by L_m, D_1, D_2, and the voltage source V_S. When D_1 and D_2 conduct, the input voltage V_S is applied to the magnetizing inductance L_m in the negative polarity. Thus, the magnetizing current ramps down with a slope of $-V_S/L_m$. This period is referred to as the reset period, and the diodes D_1 and D_2 are called the reset diodes. In the reset period, the energy stored in the magnetizing inductance is recovered by the voltage source V_S, thereby offering a lossless reset.

When the condition $i_m = i_D = 0$ is reached, the reset diodes turn off and the recess period begins. Figure 4.35(b) shows the three subcircuits, each valid in the respective operational period of the converter. Figure 4.36 shows the circuit waveforms of the converter, constructed using the subcircuits in Fig. 4.35(b). The shapes and peak values of the circuit waveforms are nearly identical to those of the tertiary-winding reset forward converter. The only difference is the peak value of the switch voltage v_Q during the reset and recess periods, which is half the previous case.

4.7 CHAPTER SUMMARY

In this chapter, we studied an important family of dc-to-dc converters, focusing on their topological origins, operational principles, and circuit waveforms. We discussed the evolution of the buck converter into two other basic converters. The boost converter is viewed as an *inverse* buck converter, while the buck/boost converter is derived from the cascaded connection of the buck converter and boost converter.

Addition of a transformer to the three basic converters creates a number of new isolated dc-to-dc converters. The simplest among them is the flyback converter whose operation closely mimics that of the buck/boost converter while providing an input-output isolation.

A number of isolated dc-to-dc converters are derived from the buck converter by replacing the single-pole double-throw (SPDT) switch with a switch network.

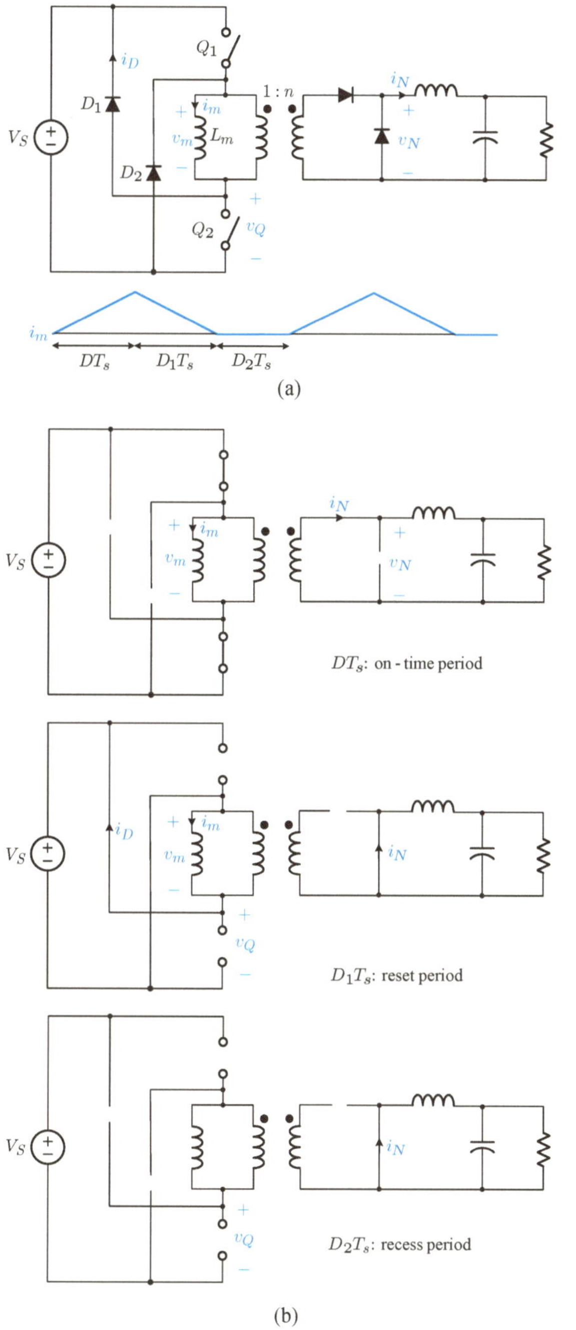

Figure 4.35 Two-switch forward converter. (a) Circuit diagram. (b) Three subcircuits.

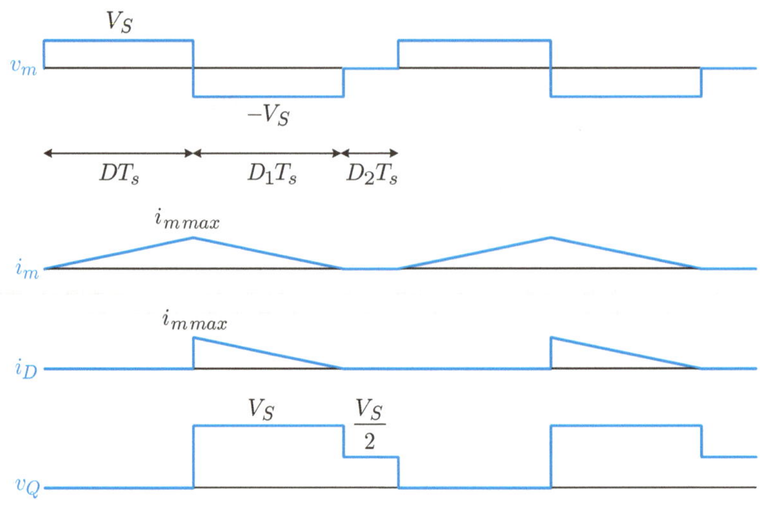

Figure 4.36 Circuit waveforms of two-switch forward converter.

Depending on the structure of the switch network, these converters are classified as the full-bridge, half-bridge, and push-pull converters. The operation of these bridge-type buck-derived converters has shown to be equivalent to that of the buck converter.

As another example of buck-derived isolated converters, the forward converter is studied, which employs a variety of auxiliary circuits to reset the transformer in the switch network. The two most widely-used forward converters, the tertiary-winding reset forward converter and two-switch forward converter, are studied in detail.

REFERENCES

1. R. W. Erickson and D. Maksimović, *Fundamentals of Power Electronics*, Kluwer Academic Publishers, 2001.
2. R. Severns and G. Bloom, *Modern Dc-to-Dc Switchmode Power Converter Circuits*, Van Nostrand Reinhold, 1985.
3. M. K. Kazimierczuk, *Pulse-width Modulated DC-DC Power Converters*, John Wiley & Sons, Hoboken, NJ, 2008.

PROBLEMS

4.1 Consider the circuits shown in Fig. P4.1 and answer the questions.

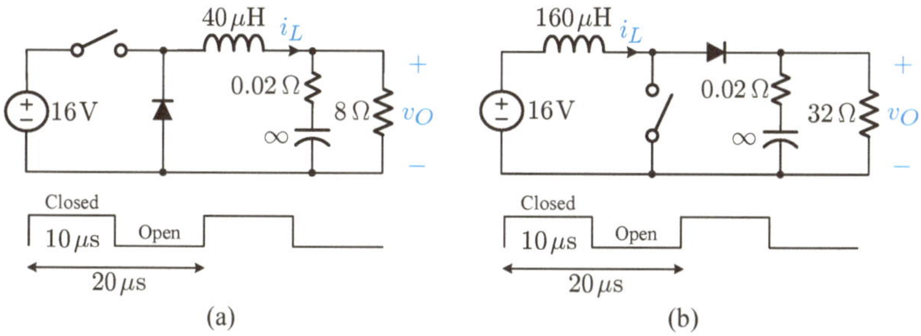

Fig. P4.1

a) For Circuit (a), sketch the steady-state waveforms of the inductor current, i_L, and output voltage ripple, $\tilde{v}_O = v_O - V_O$, for the two switching periods. Show the maximum and minimum values of the waveforms.

b) Repeat a) for Circuit (b).

4.2* In the two circuits shown in Fig. P4.2, Q_1 and Q_2 are ideal *bidirectional* switches which deliver the current in the forward and reverse directions. The switch drive signals shown in Fig. P4.2 are commonly applied to Circuit (a) and Circuit (b).

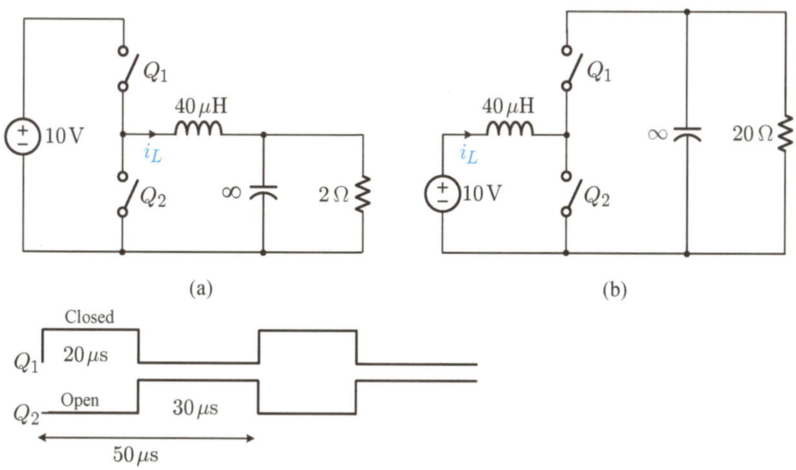

Fig. P4.2

a) For Circuit (a), sketch i_L for the two operational periods. Label the maximum and minimum values of the waveform.

b) Repeat a) for Circuit (b).

4.3* The maximum voltage that a switching device should block during the operation is called the voltage stress. Similarly, the maximum current that a switching device should carry is called the current stress. Referring to the boost converter shown in Fig. P4.3, answer the questions.

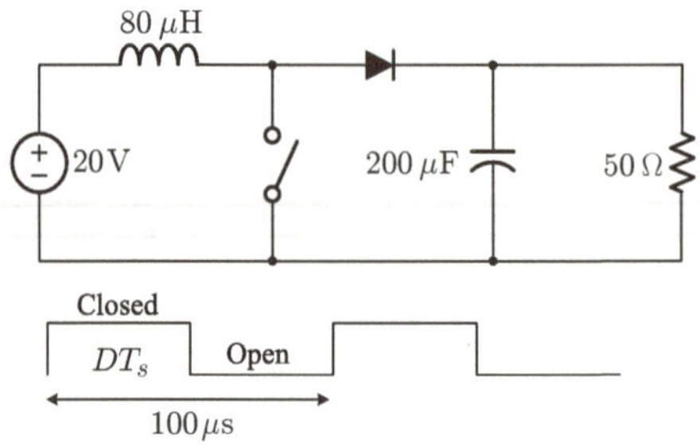

Fig. P4.3

a) Find the voltage stress and current stress of the MOSFET and diode when $D = 0.6$.

b) Repeat a) for $D = 0.9$.

4.4 Referring to Fig. P4.4, answer the following questions.

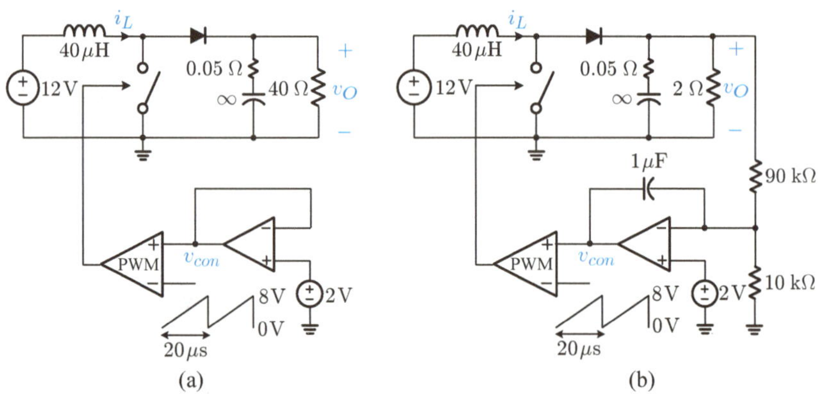

Fig. P4.4

a) For Converter (a), determine the following items:
   i) operational mode, CCM or DCM
   ii) average inductor current I_L
   iii) inductor current ripple Δi_L
   iv) output voltage ripple Δv_O
   v) average control voltage $\bar{v}_{con}$

b) Repeat a) for Converter (b).

4.5* Consider the two closed-loop controlled converters shown in Fig. P4.5 and answer the questions.

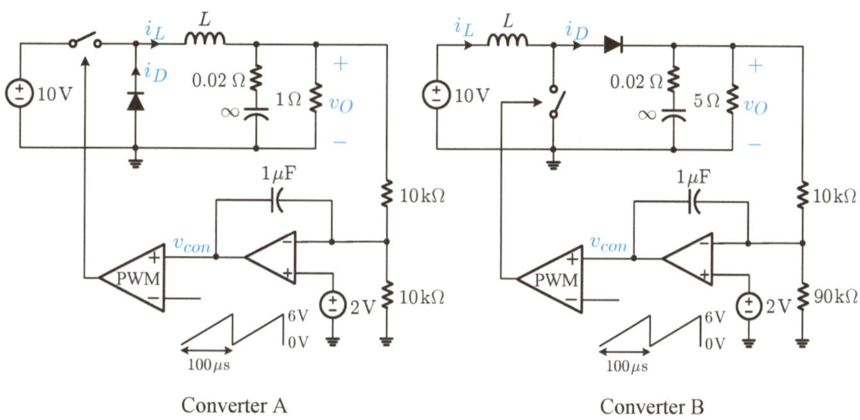

Fig. P4.5

a) For Converter A, determine the value of the inductance L so that the converter operates at the CCM/DCM borderline. You are not allowed to use the formula for the L_{crit} expression.
b) Assume that Converter A operates with the inductance L found in a). Under this condition, sketch the inductor current i_L, diode current i_D, control voltage v_{con}, and the ripple component of the output voltage, $\tilde{v}_O = v_O - V_O$, for the two operational periods. Show the maximum and minimum values.
c) Repeat a) and b) for Converter B.

4.6 Figure P4.6 shows a closed-loop controlled boost converter. Answer the questions.

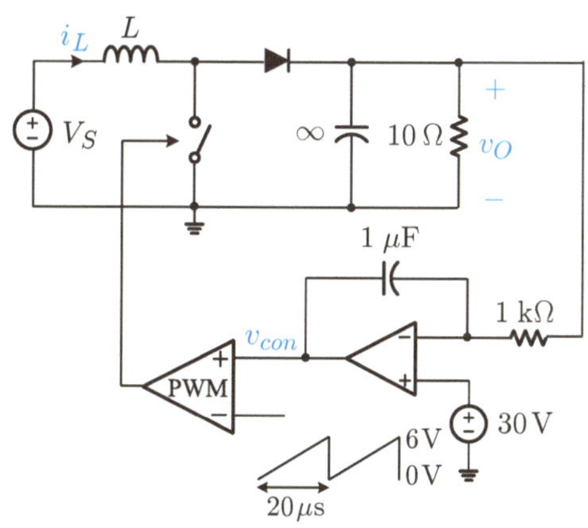

Fig. P4.6

a) Determine the value of the inductance L so that the converter operates at CCM/DCM borderline with $V_S = 12$ V.

b) Sketch the voltage gain V_O/V_S curve with respect to the duty ratio D when V_S varies between $6\text{ V} < V_S < 30\text{ V}$.

c) Now assume that $V_S = 6$ V. Determine the operational mode of the converter. Sketch the steady-state waveforms of i_L and v_{con}.

4.7 Assume all the circuit components are ideal in the converter circuits shown in Fig. P4.7. For each converter circuit, calculate the average values for the control voltage v_{con} and output voltage V_O in CCM operation.

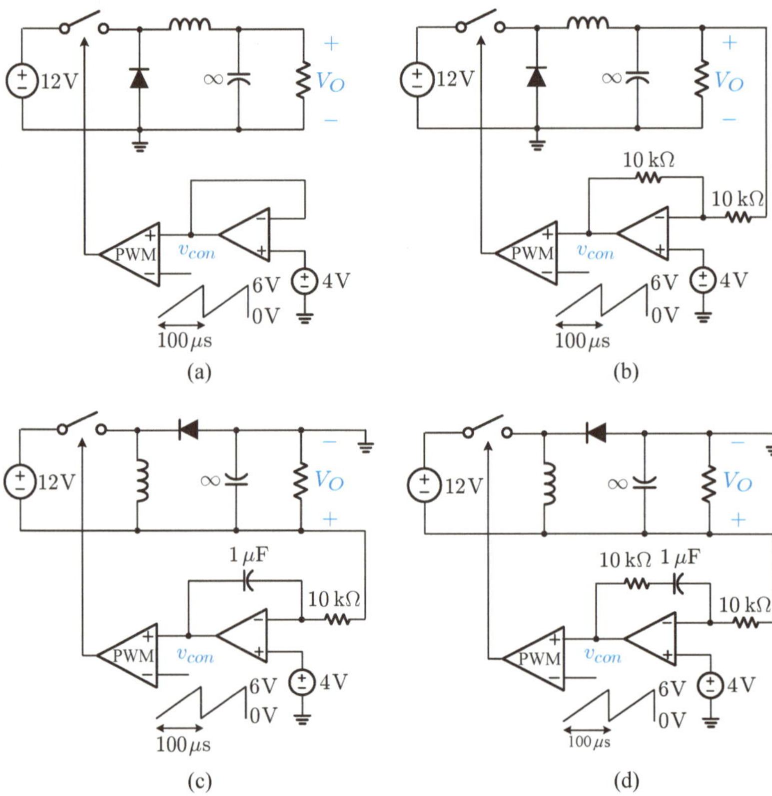

Fig. P4.7

4.8* Consider the four converter circuits shown in Fig. P4.8. For each converter circuit, calculate the reference voltage V_{ref} that is required to produce the output voltage $V_O = 20$ V in CCM operation.

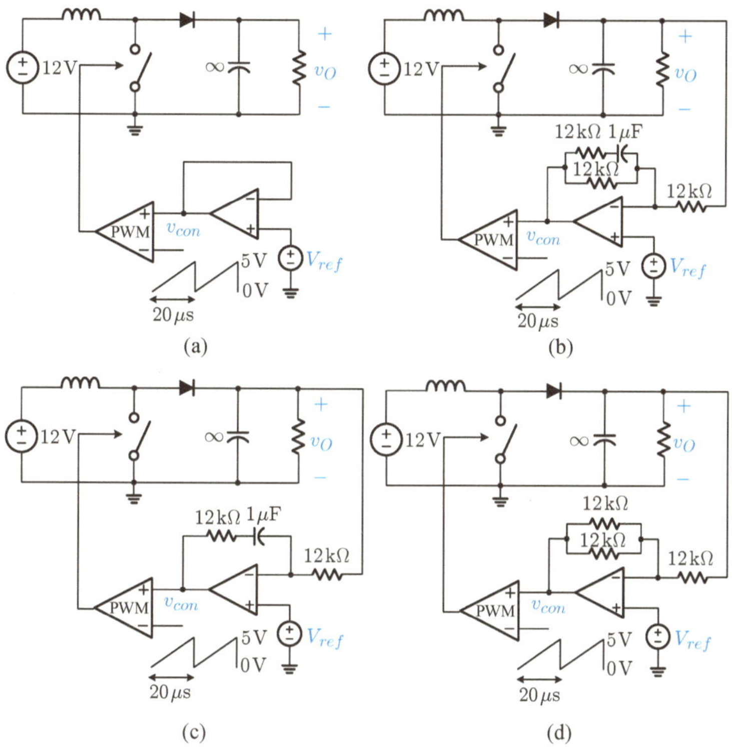

Fig. P4.8

4.9* Figure P4.9 shows three converter circuits. Assume the same switch drive signal for the three converters.

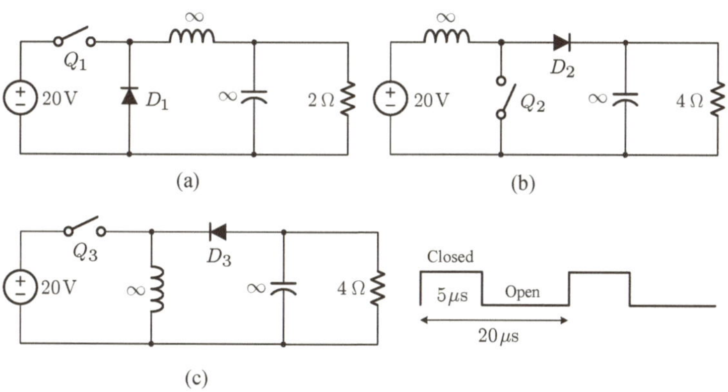

Fig. P4.9

a) Evaluate the average value of the three diode currents: $\bar{\imath}_{D1}$, $\bar{\imath}_{D2}$, and $\bar{\imath}_{D3}$.
b) Evaluate the average values of the three switch currents: $\bar{\imath}_{Q1}$, $\bar{\imath}_{Q2}$, and $\bar{\imath}_{Q3}$.
c) Find the voltage stress of the three switches: $v_{Q1\,max}$, $v_{Q2\,max}$, and $v_{Q3\,max}$.

4.10* Consider the four converter circuits shown in Fig. P4.10. For each converter circuit, find the average values of the control voltage $\bar{v}_{con}$ and output voltage $\bar{v}_O$.

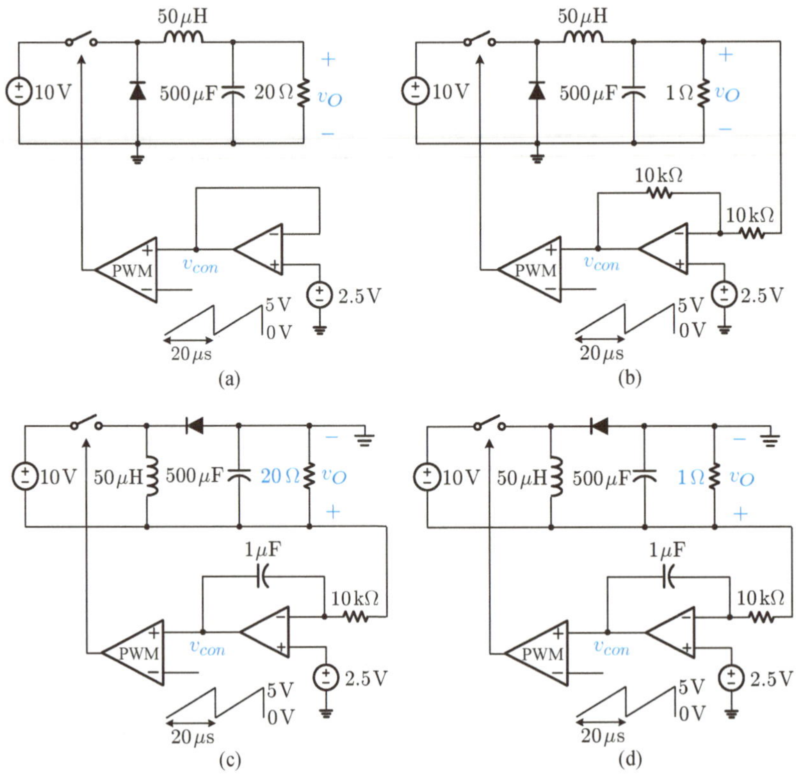

Fig. P4.10

4.11 Referring to the two converter circuit in Fig. P4.11, answer the following questions. The converters operate with $D = 0.6$ at the switching frequency $f_s = 50$ kHz.

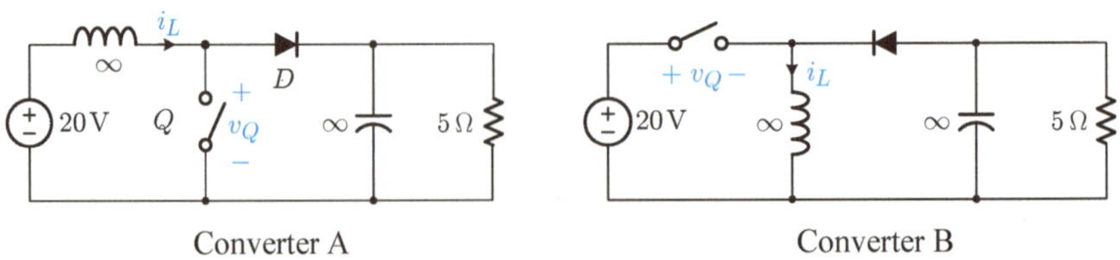

Fig. P4.11

a) Assume that all the circuit components are ideal. For each converter, evaluate the voltage stress of the active switch, Q_{max}, and the average value of the inductor current, $\bar{i}_L$.

b) For Converter A, assume the following practical devices for the semiconductor switches:
Q: MOSFET switch with the channel resistance of $R_{DS}(on) = 0.5\ \Omega$
D: Schottky diode with the turn-on voltage of $V_D(on) = 0.5$ V
The other circuit components are still assumed to be ideal. Evaluate the efficiency of Converter A.

4.12 The three converters in Fig. P4.12 generate the same inductor current waveform. For each converter, evaluate the output voltage V_O, load resistance R, and inductance L.

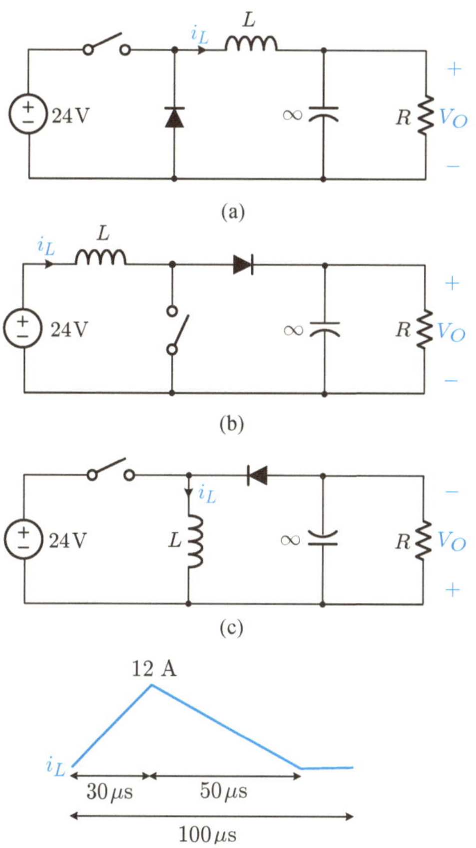

Fig. P4.12

4.13 Consider the two converter circuits shown in Fig. P4.13. The two converters generate the same inductor current waveform. For each converter, evaluate the output voltage V_O, load current I_O, inductance L, and load resistance R. You are not allowed to use the DCM gain formula in this problem.

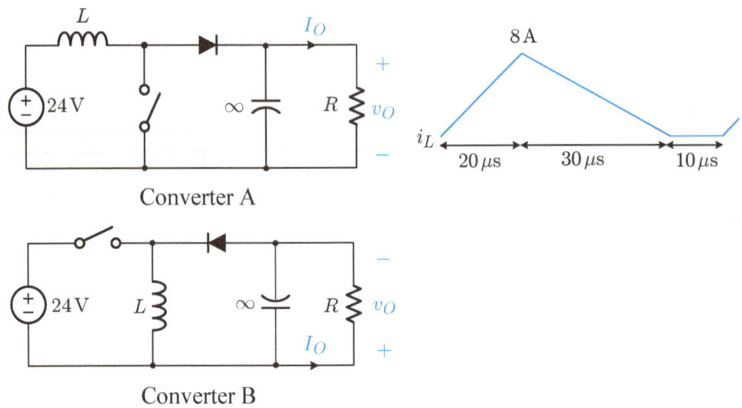

Fig. P4.13

4.14 In the dc-to-dc conversion system shown in Fig. P4.14, assume that all the active switches are driven from the same switch drive signal. Also, assume CCM operation for all three converters. Evaluate the average current running through each semiconductor switch: $\bar{\imath}_{Q1}$, $\bar{\imath}_{D1}$, $\bar{\imath}_{Q2}$, $\bar{\imath}_{D2}$, $\bar{\imath}_{Q3}$, and $\bar{\imath}_{D3}$.

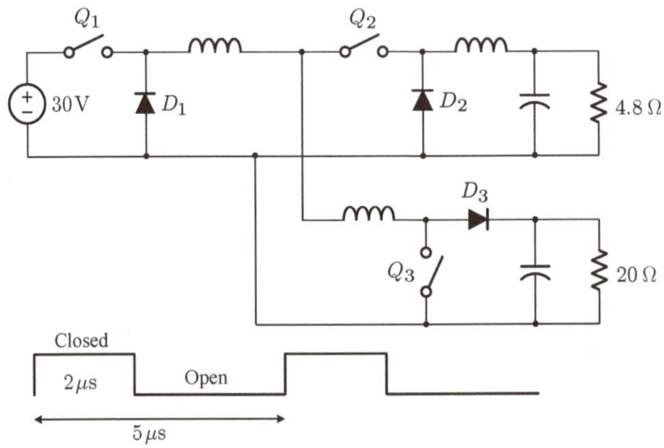

Fig. P4.14

4.15 Consider the dc-to-dc power conversion system shown in Fig. P4.15. All the converters operate in CCM with a duty ratio $D = 0.4$. Evaluate the average values of the inductor currents: $\bar{\imath}_{L1}$, $\bar{\imath}_{L2}$, and $\bar{\imath}_{L3}$.

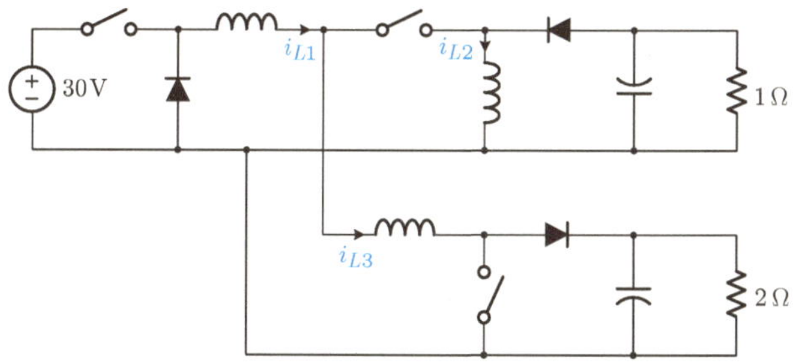

Fig. P4.15

4.16 Consider the power conversion system shown in Fig. P4.16. The three converters operate with $D = 0.5$ at the switching frequency $f_s = 50$ kHz. Assume that all the reactive components are so large that the converters are in CCM operation with negligible output voltage ripple.

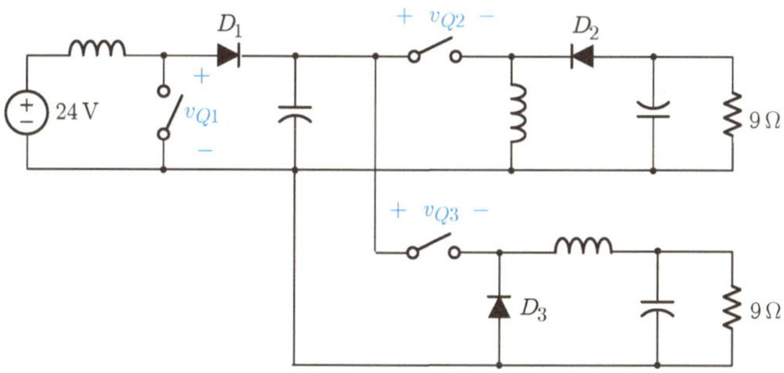

Fig. P4.16

a) Evaluate the voltage stress of the three

b) For the diodes in the circuit, assume the Schottky diodes with the turn-on voltage of $V_D(on) = 0.5$ V. Evaluate the average power consumptions in diodes, P_{D1}, P_{D2}, and P_{D3}.

4.17 For the three circuits shown in Fig. P4.17(a), assume that Q_1 and Q_2 are ideal *bidirectional* switches which deliver the current in both the forward and reverse directions.

a) Assume that the average value of v_2 is 14 V for all the three circuits. For the switch drive signals in Fig. P4.17(b), find the average value of v_1 in the three circuits in Fig. P4.17(a).

b) Repeat a) for the switch drive signals in Fig. P4.17(c).

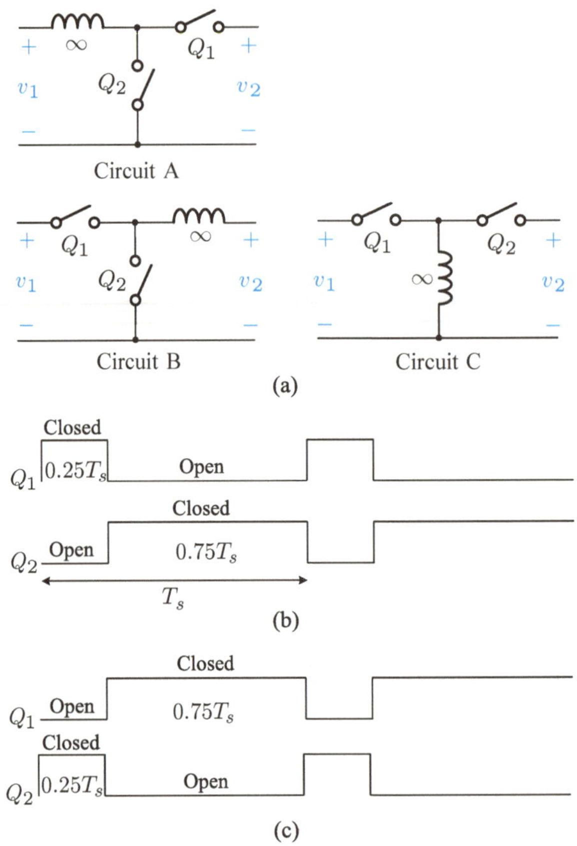

Fig. P4.17

4.18 Use the flux linkage balance condition and charge balance condition to derive the voltage gain expression

$$\frac{V_O}{V_S} = \frac{D}{1-D}\left(\frac{1}{1+\dfrac{1}{(1-D)^2}\dfrac{R_l}{R}}\right)$$

for the buck/boost converter with inductor winding resistance R_l.

4.19* Shown in Fig. P4.19 is a buck/boost converter built using the non-ideal switches and practical passive components.

a) During the normal operation, the load resistor is accidentally disconnected so that $R = \infty$. What do you think will happen to the circuit? If you claim a new steady-state operation, find the steady-state average values for the output voltage v_O and inductor current i_L. If you claim a catastrophic

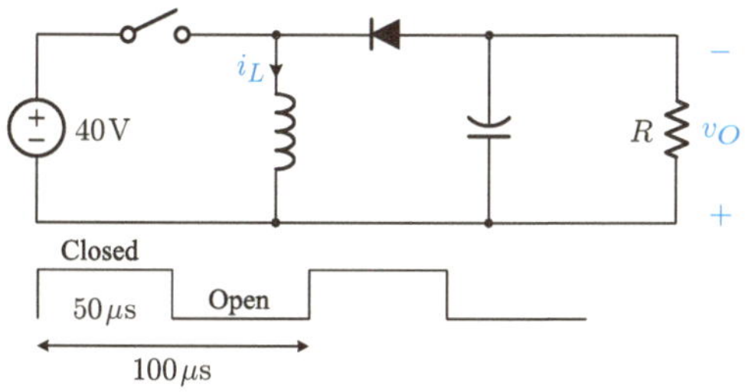

Fig. P4.19

failure of the circuit, identify the component that is responsible for the failure and state the origin/reason for the component breakdown.

b) Now assume that the load resistor is accidentally shorted, $R = 0$, during the normal operation of the converter. Repeat Problem a) under this assumption.

4.20* Figure P4.20 depicts a composite converter circuit and its switch drive signals.

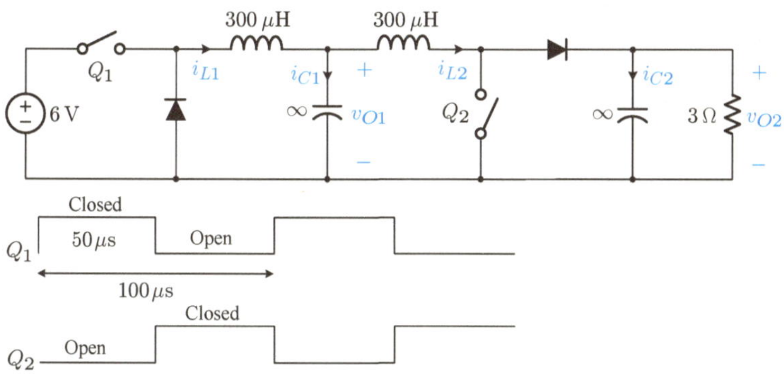

Fig. P4.20

a) Find the average values for v_{O1} and v_{O2}.

b) Sketch the steady-state waveforms of i_{L1} and i_{L2} for the two operational periods. Label the maximum and minimum values of each waveform.

c) Draw the steady-state waveforms of i_{C1} and i_{C2} for the two operational periods. Label the maximum and minimum values of each waveform.

4.21 Figure P4.21 depicts the circuit diagram of a composite converter, configured by cascading the boost converter and buck converter. Referring to the switch drive signals for Q_1 and Q_2, answer the questions.

a) Find the average values for v_{O1} and v_{O2}.

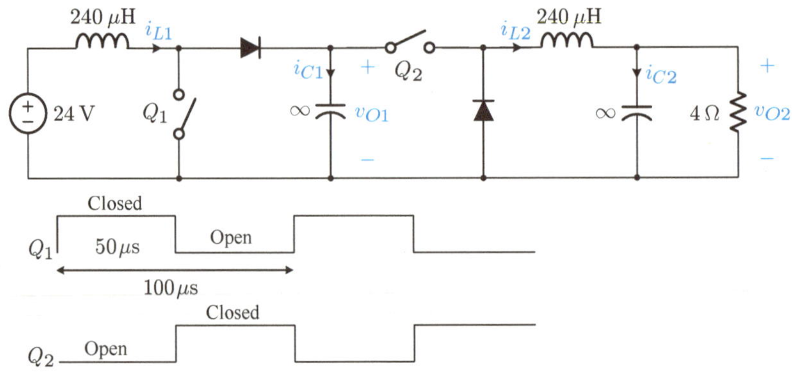

Fig. P4.21

b) Sketch the steady-state waveforms of i_{L1} and i_{L2} for the two switching periods. Label the maximum, minimum, and average values of each waveform.

c) Draw the steady-state waveforms of i_{C1} and i_{C2} for the two switching periods. Label the maximum, minimum, and average values of each waveform.

4.22* The forward converter shown in Fig. P4.22 is built using a practical transformer. The number of turns of the three-winding transformer is $N_P = N_S = N_T = 48$. The converter established steady-state operation with a duty ratio $D = 0.4$. Now assume that the following failure or change has occurred to the converter during its steady-state operation:

i) open-circuit failure of D_c ii) open-circuit failure of D_2
iii) open-circuit failure of C iv) decrease in the duty ratio to $D = 0.15$
v) increase in the duty ratio to $D = 0.65$

For each failure/change listed above, what do you think will happen to the converter? If you predict a catastrophic failure of the circuit, identify the component responsible to the failure and state the origin/reason of the component breakdown. If you claim a new steady-state operation, describe the major change in the circuit operation.

4.23* For the four isolated dc-to-dc converters shown in Fig. P4.23, assume that the transformer has a finite magnetizing inductance.

a) Answer the following questions.
   i) Which converter would result in the worst power conversion efficiency?
   ii) Which converter would offer the largest power handling capacity?
   iii) Which converter would minimize the manufacturing cost?

b) For each dc-to-dc converter, find the range of the duty ratio that guarantees proper operation of the converter. Express your answer in the form of $(\quad) < D < (\quad)$.

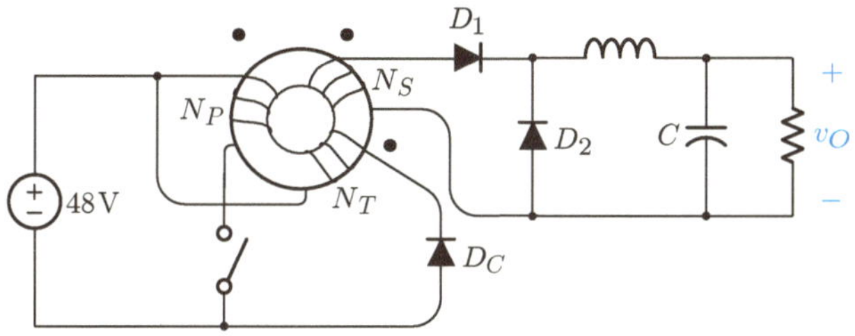

Fig. P4.22

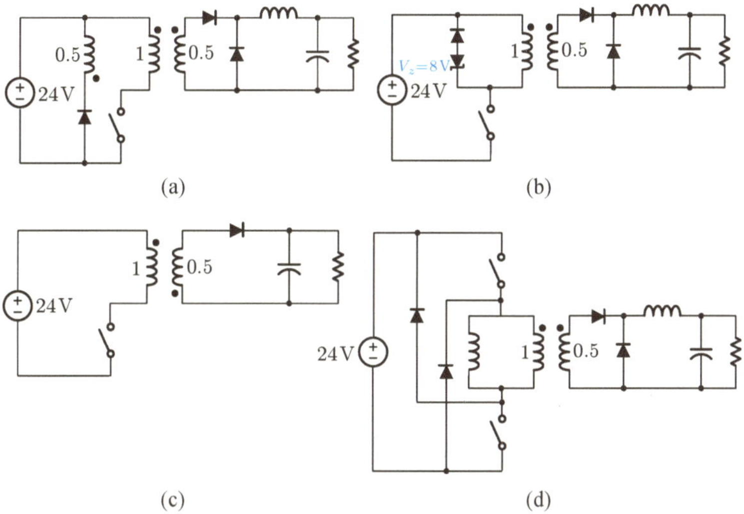

Fig. P4.23

c) Now assume that all the converters are closed-loop controlled to produce an output voltage of 6 V. Evaluate the voltage stress of the active switch employed in each circuit.

4.24* The following questions concern the topological structure, circuit components, and operational principles of various PWM converters. Answer the questions in a brief and precise manner.

a) The transformer employed in a flyback converter is typically built on the magnetic core to which an air gap is intentionally introduced. Explain the necessity of this practice.

b) While an appropriate reset circuit is required for the forward converter, the flyback converter and bridge-type converters do not require any reset circuit. Explain the reasons for this.

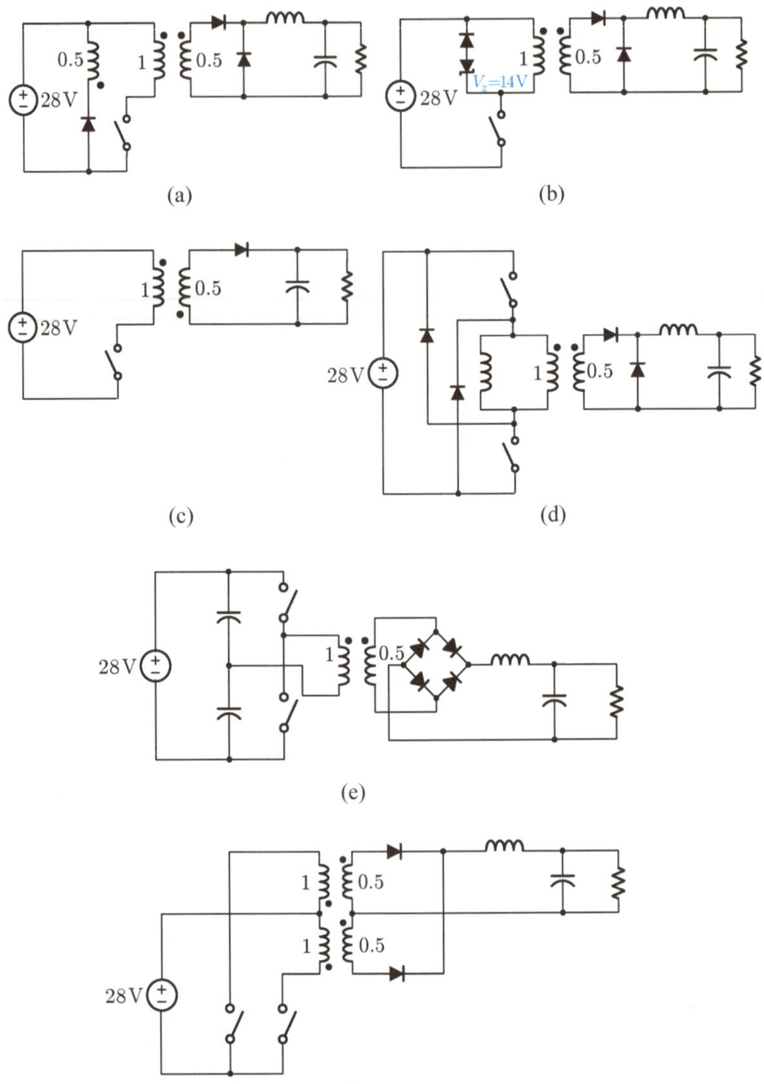

Fig. P4.24

c) Figure P4.24 shows six different converter topologies. For each converter, find the maximum possible value for the output voltage and the voltage stress of the active switch employed in the converter.

4.25* Shown in Fig. P4.25 are the five isolated PWM dc-to-dc converters. Assume all the components are ideal except for the practical transformer. For the above PWM converters, assume that
i) the magnetizing inductance of the practical transformer is $L_m = 200\ \mu\text{H}$

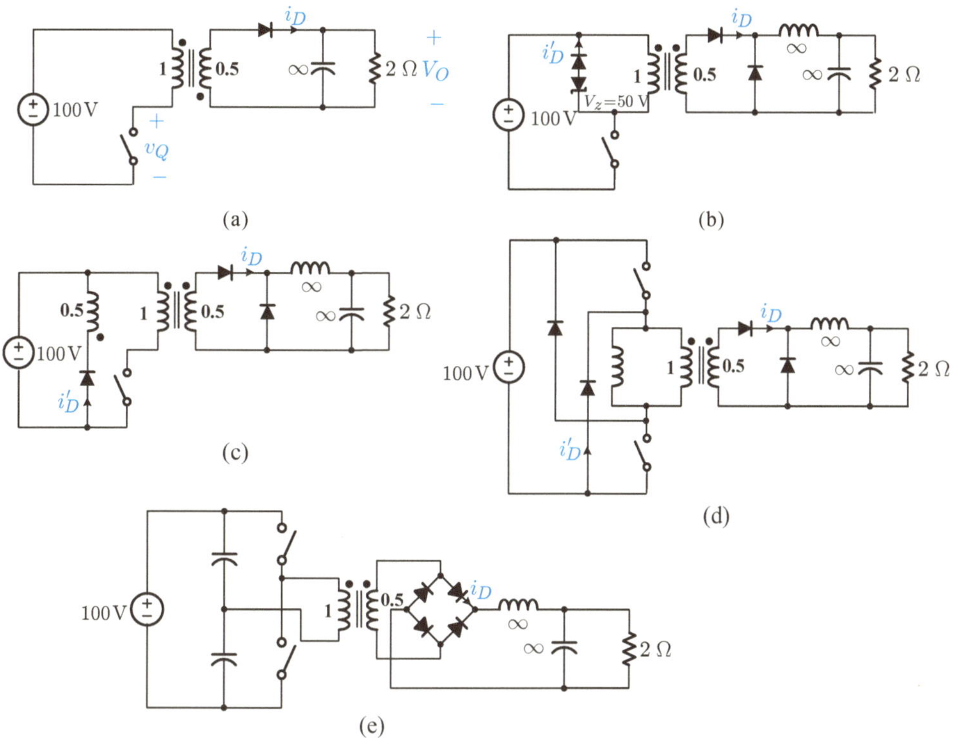

Fig. P4.25

ii) the (effective) duty ratio is $D = 0.2$ with respective to the switching period $T_s = 50\ \mu s$, thus yielding the turn-on time of $T_{on} = 10\ \mu s$. For each converter, evaluate the output voltage V_O, the peak value of the switch voltage $v_{Q\ peak}$, and the average values of the diode current, $\bar{i}_D$ and $\bar{i}'_D$, as appropriate.

PART II

MODELING AND DYNAMICS OF PWM CONVERTERS

PWM dc-to-dc converters are a nonlinear time-variant system to which conventional linear analysis techniques are not *directly* applicable for dynamic analyses. Thus, we first need to find an appropriate methodology to deal with PWM dc-to-dc converters. We circumvent the difficulties imposed by the nonlinear dynamics by developing an s-domain small-signal model which enables us to use conventional s-domain techniques to characterize nonlinear converter dynamics under certain assumptions. For most practical instances, we could perform dynamic analyses of nonlinear dc-to-dc converters in the same manner as we dealt with linear systems.

Chapter 5 is devoted to the s-domain modeling of PWM converters. Using several modeling techniques, we first develop small-signal models for power stage, PWM block, and feedback circuit of the three basic converters. Later, as the final outcome of this chapter, we establish a unified s-domain small-signal model for closed-loop controlled converters by integrating small-signal models of individual functional blocks of PWM converters. The unified s-domain model is applicable to all the closed-loop controlled PWM dc-to-dc converters in CCM operation or DCM operation.

In Chapter 6, we analyze transfer functions of converters' power stage. The transfer functions are derived from the s-domain small-signal model and later converted into Bode plots, in order to expose the frequency-domain dynamics. Results of this analysis are the basis for the control design and performance evaluation of closed-loop controlled converters. This chapter also introduces empirical approaches to small-signal analyses.

Chapter 7 covers stability and dynamic performance of PWM converters. This chapter first defines the performance metrics for PWM dc-to-dc converters as a voltage source. Backgrounds and implications of the stability analysis and performance evaluation using the small-signal models are addressed with practical examples. The validity and utility of the s-domain small-signal models are also demonstrated. The stability analysis of dc-to-dc converters using the classical analysis techniques, such as loop gain, Nyquist criterion, Bode plot, are underscored.

CHAPTER 5

MODELING PWM DC-TO-DC CONVERTERS

The first part of this book dealt with the power stage configuration and steady-state operation of PWM dc-to-dc converters. This static analysis constitutes one essential part of the dc-to-dc power conversion technology. As introduced in Chapter 1, the dynamic analysis is another critical area in studying the dc-to-dc power conversion. In the dynamic analysis, the dc-to-dc converter is viewed as a closed-loop controlled dynamic system and its characteristics are investigated using relevant dynamic models. As the prerequisite for the dynamic analysis, this chapter presents the modeling of PWM dc-to-dc converters.

PWM dc-to-dc converters are a time-variant system in the sense that their power stage structure constantly varies with time. Namely, dc-to-dc converters exhibit different power stage configurations depending on the status of the semiconductor switches. In addition, dc-to-dc converters employ the pulsewidth modulation (PWM) process to generate appropriate switch drive signals. The PWM process is a well-known nonlinear process in which the input and output variables are linked by a nonlinear relationship. The time variance of the power stage configuration and nonlinearity of the PWM process collectively classify dc-to-dc converters as a nonlinear time-variant system.

The dynamic analysis of a nonlinear time-variant system is known to be difficult. Conventional circuit analysis techniques are mainly intended for linear time-invariant systems and therefore cannot be directly applied to PWM dc-to-dc converters. Although certain analytical methods are available for nonlinear systems, the analysis becomes intractably complicated when adapted to PWM dc-to-dc converters. Accordingly, the analysis of the converter dynamics in the original form can be a very challenging task. The purpose of this chapter is to develop an analytical process, referred to as the modeling in this book, which enables us to overcome the obstacles presented by the time variance and nonlinearity of PWM dc-to-dc converters.

Generally, modeling implies the procedure of describing the dynamic characteristics of a given system in a desired format. Such desired format is determined by the object of the modeling. For example, to develop a *simplified model* for a complex system, the system behavior is expressed by the simple equations that only capture the major system dynamics while ignoring unimportant details. The object of the dc-to-dc converter modeling herein is to establish a systematic method of describing the dynamics of nonlinear time-variant dc-to-dc converters using the *language* and *format* that have been used for linear time-invariant systems. If we achieve this aim, we can analyze the dynamics of dc-to-dc converters in the exact same manner as we treat linear time-invariant systems.

5.1 OVERVIEW OF PWM CONVERTER MODELING

In this chapter, we develop *small-signal models* for PWM dc-to-dc converters using a series of modeling techniques. The small-signal model is a linear time-invariant circuit model to which all standard circuit analysis techniques can directly be applied.

Figure 5.1 shows a general structure of a closed-loop controlled PWM converter, powered by an ideal voltage source and loaded with a resistor. Discussions about the converter structure were given in Section 3.6. For the purpose of modeling, the closed-loop controlled PWM converter is divided into three functional blocks: power stage, PWM block, and voltage feedback circuit. First, each functional block is transformed

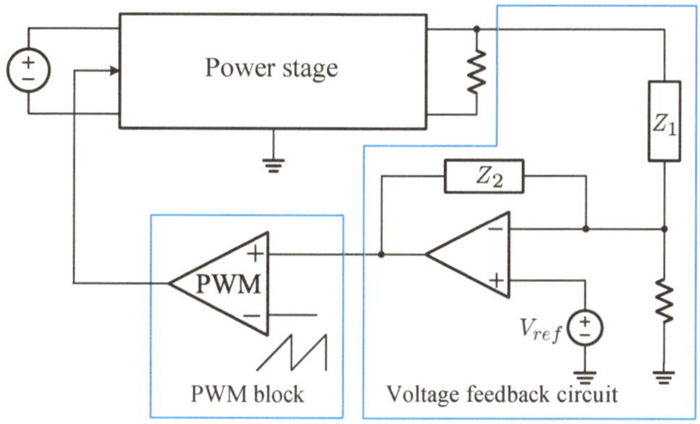

Figure 5.1 Structure of closed-loop controlled PWM converter.

into the respective small-signal model using various modeling techniques. The small-signal models of the three functional blocks are later merged together to yield a complete small-signal model for the closed-loop controlled PWM converter.

Power Stage Modeling

Figure 5.2 illustrates the procedures of power stage modeling. The method of averaging is first applied to the time-varying power stage dynamics. This method provides an average model in which the time variance is removed. In this process, however, certain nonlinear relationships among circuit variables are emerged and embedded into the average model. Thus, the average model is time-invariant but nonlinear. The average model describes the averaged time-domain dynamics of the nonlinear power stage.

As the second step, linearization is invoked to deal with the nonlinear relationships incurred during the averaging process. Linearization is a process of approximating the nonlinear relationships among the circuit variables to linear descriptions under *small-signal assumption*.† Linearization produces a linear time-invariant *small-signal model*. This model describes the averaged time-domain dynamics at the presence of small-signal excitations.

As the last step, the time-domain small-signal model is converted into a frequency-domain, or *s-domain*,‡ small-signal model, which provides transfer functions of power stage dynamics. The resulting transfer functions embrace all the standard s-domain analysis techniques and reveal the frequency-domain small-signal dynamics of power stage.

†The small-signal assumption implies that the system under consideration is excited only with sufficiently small input signals. The prefix *small-signal* is used to indicate the presence of the small excitation signals when the corresponding *item* is referred to. The formal definitions of the small-signal assumption, small-signal model, and other small-signal *entries* are given in Section 5.3.

‡The notation s in the s-domain refers to the complex frequency: $s = \sigma + j\omega$.

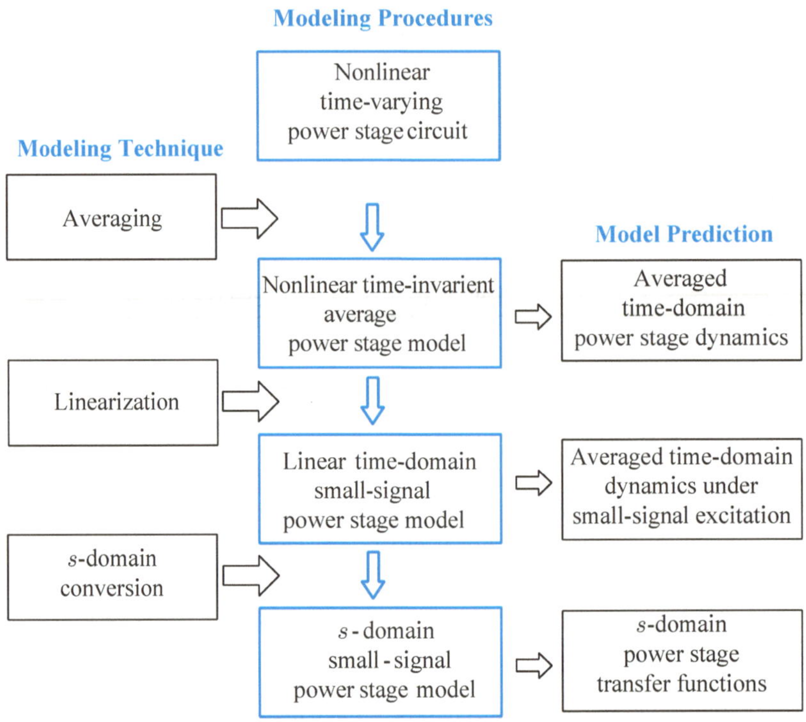

Figure 5.2 Steps of power stage modeling.

PWM Block Modeling

The PWM process employed in PWM dc-to-dc converters is a typical nonlinear operation. The PWM block receives the input signal from the voltage feedback circuit and produces the switch drive signal as its output. While the input of the PWM block is a continuous analog signal, the output is a periodic pulse waveform, whose cycle-by-cycle pulsewidth is modulated by the input signal. This highly nonlinear functionality can be approximated by a linear expression, assuming that the input signal only changes *narrowly* and *slowly* within the switching period. Because most PWM dc-to-dc converters meet this assumption, there exists a simple linear relationship between the input and output signals of the PWM block. This linear relationship is the small-signal gain of the PWM block, called the PWM gain or modulator gain.

Voltage Feedback Circuit and Small-Signal Model of PWM Converter

The voltage feedback circuit is a linear time-invariant circuitry which can readily be converted into the small-signal model using standard circuit analysis techniques. The small-signal model for the voltage feedback circuit will be described in Section 5.7.1.

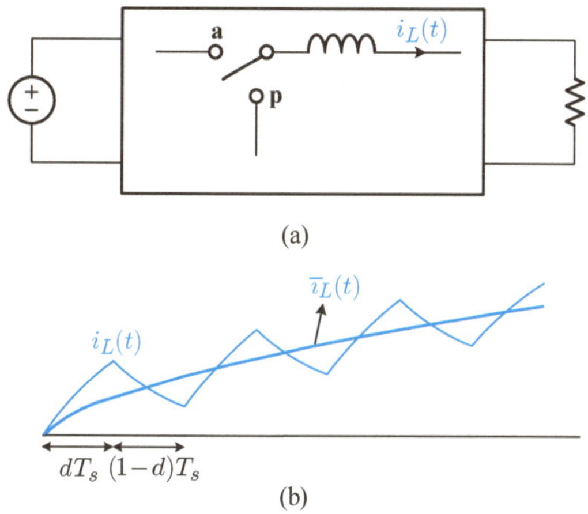

Figure 5.3 Averaging power stage dynamics. (a) Conceptual diagram of PWM dc-to-dc converter. (b) Concept of averaging.

The complete small-signal model for dc-to-dc converters can now be constructed by merging the small-signal models of the power stage, PWM block, and voltage feedback circuit. The resulting small-signal model allows us to perform the dynamic analysis of nonlinear time-variant dc-to-dc converters using the conventional s-domain analysis techniques.

5.2 AVERAGING POWER STAGE DYNAMICS

This section covers the averaging of power stage dynamics under continuous conduction mode (CCM) operation. The averaging process in discontinuous conduction mode (DCM) operation will be separately discussed in Section 5.5.2.

The method of power stage averaging is explained in Fig. 5.3. A conceptual diagram of a PWM dc-to-dc converter, containing one single-pole double-throw (SPDT) switch, is shown in Fig. 5.3(a). Figure 5.3(b) shows two different presentations of the inductor current i_L, which graphically illustrate the averaging process. Here, *the duty ratio of the SPDT switch is allowed to change with time.* This time-varying duty ratio is represented by ***d*** in Fig. 5.3(b).

Figure 5.3(b) illustrates the time variance of the power stage dynamics and the concept of averaging. When the SPDT switch is held at position **a** during dT_s, the power stage presents the on-time subcircuit. Likewise, with the SPDT switch at position **p** during $(1-d)T_s$, the power stage becomes the off-time subcircuit. During the on-time period dT_s, the inductor current $i_L(t)$ develops according to the circuit equations of the on-time subcircuit. During the off-time period $(1-d)T_s$, $i_L(t)$ evolves based on the circuit equations of the off-time subcircuit. Accordingly, the inductor current $i_L(t)$ changes its pattern at every switching instant.

The method of averaging is an analytical approach that attempts to produce a smooth waveform which follows the time-averaged trajectory of the inductor current, as illustrated with $\bar{i}_L(t)$ in Fig. 5.3(b). The method of averaging provides the circuit equations or circuit models that produce the time-averaged power stage waveforms of PWM dc-to-dc converters.

As the outcome of past research efforts, two useful averaging methods have been developed for PWM dc-to-dc converters. The first method manipulates the state-space description of the power stage so that the manipulated state-space description predicts the time-averaged dynamics of the power stage. This method is known as the *state-space averaging*.

The second method uses the time-averaged behavior of the circuit variables to develop an average model of the power stage. The average power stage model produces the continuous circuit waveforms that follow the averaged trajectories of the original waveforms. This method is referred to as the *circuit averaging* because it directly averages the circuit variables of the power stage.

This section presents both the state-space averaging method and circuit averaging technique, focusing on their theoretical basis and application to the PWM converter modeling. The relative merit of each method and equivalence between the two methods are also discussed.

5.2.1 State-Space Averaging Method

In the method of state-space averaging, an exact state-space description of the power stage is initially formulated, using the *switching function* which can have two different numerical values depending on time. The switching function becomes **1** during on-time periods and remains **0** during off-time periods. The resulting state-space description is referred to as the switched state-space model. Although the switched state-space model precisely describes the power stage dynamics, it is a time-variant model due to the presence of the switching function. The switched state-space model is then appropriately averaged to yield an averaged state-space model, which describes the time-averaged power stage dynamics.

Switched State-Space Model and Switching Function

As the initial step of deriving the switched state-space model, the power stage dynamics are described by separately formulating the state-space description of the on-time subcircuit and the state-space description of the off-time subcircuit. The power stage dynamics during an on-time period are expressed in the form of a state equation

$$\begin{aligned} \frac{\mathrm{d}\boldsymbol{x}(t)}{\mathrm{d}t} &= \boldsymbol{A}_{on}\boldsymbol{x}(t) + \boldsymbol{B}_{on}v_S(t) \\ v_O(t) &= \boldsymbol{C}_{on}\boldsymbol{x}(t) \end{aligned} \tag{5.1}$$

where $\boldsymbol{x}$ is the state vector, v_S is the input voltage, v_O is the output voltage, and $\{\boldsymbol{A}_{on}\ \boldsymbol{B}_{on}\ \boldsymbol{C}_{on}\}$ are the coefficient matrices of the on-time subcircuit. Similarly, another state equation is written for an off-time period

$$\begin{aligned} \frac{\mathrm{d}\boldsymbol{x}(t)}{\mathrm{d}t} &= \boldsymbol{A}_{off}\boldsymbol{x}(t) + \boldsymbol{B}_{off}v_S(t) \\ v_O(t) &= \boldsymbol{C}_{off}\boldsymbol{x}(t) \end{aligned} \tag{5.2}$$

where $\{\boldsymbol{A}_{off}\ \boldsymbol{B}_{off}\ \boldsymbol{C}_{off}\}$ are the coefficient matrices of the off-time subcircuit. The two state equations (5.1) and (5.2) are merged into one single state equation

$$\begin{aligned} \frac{\mathrm{d}\boldsymbol{x}(t)}{\mathrm{d}t} &= (q(t)\boldsymbol{A}_{on} + (1-q(t))\boldsymbol{A}_{off})\boldsymbol{x}(t) \\ &\quad + (q(t)\boldsymbol{B}_{on} + (1-q(t))\boldsymbol{B}_{off})v_S(t) \\ v_O(t) &= (q(t)\boldsymbol{C}_{on} + (1-q(t))\boldsymbol{C}_{off})\boldsymbol{x}(t) \end{aligned} \tag{5.3}$$

using the notation of the switching function $q(t)$, defined as

$$q(t) = \begin{cases} 1 & \text{for on-time period } dT_s \\ 0 & \text{for off-time period } (1-d)T_s \end{cases} \tag{5.4}$$

The equation (5.3) is called the *switched state-space model.* Detailed discussions about the switching function $q(t)$ will be given in the next section.

Switched State-Space Model of Ideal Buck Converter — EXAMPLE 5.1

This example illustrates the formation of the switched state-space model for the buck converter. The circuit diagram of an *ideal* buck converter without any parasitic resistances is shown in Fig. 5.4, along with its on-time and off-time subcircuits. When the state vector is defined as $\boldsymbol{x} = [i_L\ v_C]^T$, the coefficient matrices of the on-time and off-time subcircuits become

$$\boldsymbol{A}_{on} = \boldsymbol{A}_{off} = \begin{bmatrix} 0 & -\frac{1}{L} \\ \frac{1}{C} & -\frac{1}{CR} \end{bmatrix}$$

$$\boldsymbol{B}_{on} = \begin{bmatrix} \frac{1}{L} \\ 0 \end{bmatrix}$$

$$\boldsymbol{B}_{off} = \begin{bmatrix} 0 \\ 0 \end{bmatrix}$$

$$\boldsymbol{C}_{on} = \boldsymbol{C}_{off} = [\,0 \quad 1\,]$$

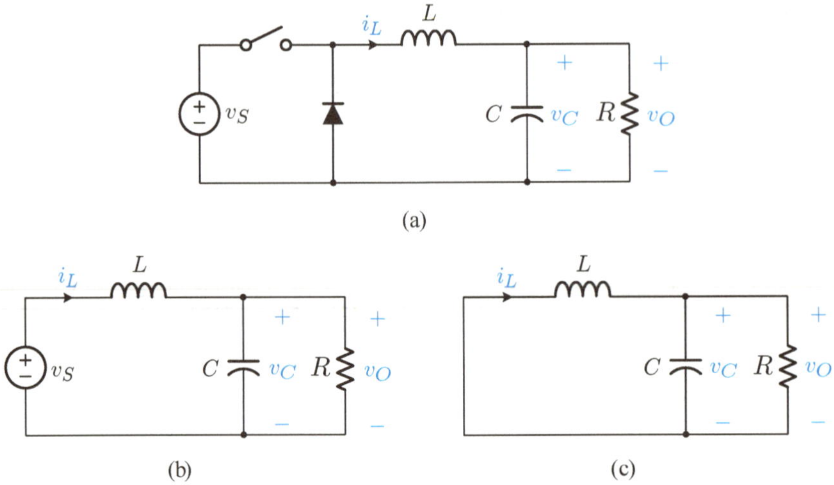

Figure 5.4 Ideal buck converter. (a) Circuit diagram. (b) On-time subcircuit. (c) Off-time subcircuit.

The switched state-space model is obtained by plugging the appropriate coefficient matrices into (5.3)

$$\frac{d\boldsymbol{x}(t)}{dt} = \left[q(t) \begin{bmatrix} 0 & -\frac{1}{L} \\ \frac{1}{C} & -\frac{1}{CR} \end{bmatrix} + (1-q(t)) \begin{bmatrix} 0 & -\frac{1}{L} \\ \frac{1}{C} & -\frac{1}{CR} \end{bmatrix} \right] \boldsymbol{x}(t)$$

$$+ \left[q(t) \begin{bmatrix} \frac{1}{L} \\ 0 \end{bmatrix} + (1-q(t)) \begin{bmatrix} 0 \\ 0 \end{bmatrix} \right] v_S(t)$$

$$v_O(t) = \left[q(t) \left[0 \quad 1 \right] + (1-q(t)) \left[0 \quad 1 \right] \right] \boldsymbol{x}(t)$$

EXAMPLE 5.2 Coefficient Matrices of Three Basic Converters

The coefficient matrices of the boost and buck/boost converters can readily be formulated from their circuit diagram. For the ideal boost converter and ideal buck/boost converter, the converter circuit is transformed into the on-time and off-time subcircuts. The state equation for each subcircuit is then written in the same manner as that of the previous buck converter case. From the state equations

Table 5.1 Coefficient Matrices of Three Basic Converters

	On-time subcircuit	Off-time subcircuit
Buck converter	$\boldsymbol{A}_{on} = \begin{bmatrix} 0 & -\frac{1}{L} \\ \frac{1}{C} & -\frac{1}{CR} \end{bmatrix}$	$\boldsymbol{A}_{off} = \begin{bmatrix} 0 & -\frac{1}{L} \\ \frac{1}{C} & -\frac{1}{CR} \end{bmatrix}$
	$\boldsymbol{B}_{on} = \begin{bmatrix} \frac{1}{L} & 0 \end{bmatrix}^T$	$\boldsymbol{B}_{off} = [0 \quad 0]^T$
	$\boldsymbol{C}_{on} = [0 \quad 1]$	$\boldsymbol{C}_{off} = [0 \quad 1]$
Boost converter	$\boldsymbol{A}_{on} = \begin{bmatrix} 0 & 0 \\ 0 & -\frac{1}{CR} \end{bmatrix}$	$\boldsymbol{A}_{off} = \begin{bmatrix} 0 & -\frac{1}{L} \\ \frac{1}{C} & -\frac{1}{CR} \end{bmatrix}$
	$\boldsymbol{B}_{on} = \begin{bmatrix} \frac{1}{L} & 0 \end{bmatrix}^T$	$\boldsymbol{B}_{off} = \begin{bmatrix} \frac{1}{L} & 0 \end{bmatrix}^T$
	$\boldsymbol{C}_{on} = [0 \quad 1]$	$\boldsymbol{C}_{off} = [0 \quad 1]$
Buck/boost converter	$\boldsymbol{A}_{on} = \begin{bmatrix} 0 & 0 \\ 0 & -\frac{1}{CR} \end{bmatrix}$	$\boldsymbol{A}_{off} = \begin{bmatrix} 0 & -\frac{1}{L} \\ \frac{1}{C} & -\frac{1}{CR} \end{bmatrix}$
	$\boldsymbol{B}_{on} = \begin{bmatrix} \frac{1}{L} & 0 \end{bmatrix}^T$	$\boldsymbol{B}_{off} = [0 \quad 0]^T$
	$\boldsymbol{C}_{on} = [0 \quad 1]$	$\boldsymbol{C}_{off} = [0 \quad 1]$

of the two subcircuits, the coefficient matrices are extracted. The coefficient matrices of the three basic dc-to-dc converters are summarized in Table 5.1. The coefficient matrices can be used to construct the switched state-space model of the respective converter.

Continuous Duty Ratio and Averaged State-Space Model

This section discusses the averaging process of the switched state-space model. As the onset of obtaining an averaged form of the switched state-space model, averaging operation is performed on the switching function $q(t)$, introduced in (5.4)

$$d(t) = \frac{1}{T_s} \int_{t-T_s}^{t} q(\tau)\, \mathrm{d}\tau \tag{5.5}$$

where T_s is the switching period. The $d(t)$ defined in (5.5) represents the averaged version of $q(t)$, evaluated while sliding the averaging period T_s in time. Accordingly, $d(t)$ is interpreted as the *moving average* or *local average* of $q(t)$. The $d(t)$ is called a *continuous duty ratio*. The implications of the continuous duty ratio $d(t)$ are summarized as follows.

1) It is recognized that $d(t)_{t=kT_s} = d(kT_s) \equiv d_k$ is the actual duty ratio in the kth switching period.

2) If $q(t)$ is periodic with a fixed on-time period, then it becomes that $d_k = d(t) = D$, where D is the steady-state duty ratio.

3) If $q(t)$ is non-periodic with a time-varying on-time period, then the actual duty ratio differs from the continuous duty ratio: $d_k \neq d(t)$. However, if the cycle-by-cycle variation in the on-time period is sufficiently small, the actual duty ratio can be approximated by the continuous duty ratio: $d_k \approx d(t)$.

Figure 5.5 illustrates the relationship among the switching function $q(t)$, actual cycle-by-cycle duty ratio d_k, and continuous duty ratio $d(t)$. The continuous duty ratio $d(t)$ is a continuous-time approximation of the discontinuous cycle-by-cycle duty ratio d_k. The error due to this approximation becomes negligible when the cycle-by-cycle variation in d_k is sufficiently small.

As the second step, averaging operation is performed on the switched state-space model of (5.3)

$$\begin{aligned} \frac{\mathrm{d}\bar{\boldsymbol{x}}(t)}{\mathrm{d}t} &= \overline{(q(t)\boldsymbol{A}_{on} + (1-q(t))\boldsymbol{A}_{off})\,\boldsymbol{x}(t)} \\ &\quad \overline{+\,(q(t)\boldsymbol{B}_{on} + (1-q(t))\boldsymbol{B}_{off})\,v_S(t)} \\ \bar{v}_O(t) &= \overline{(q(t)\,\boldsymbol{C}_{on} + (1-q(t))\boldsymbol{C}_{off})\,\boldsymbol{x}(t)} \end{aligned} \tag{5.6}$$

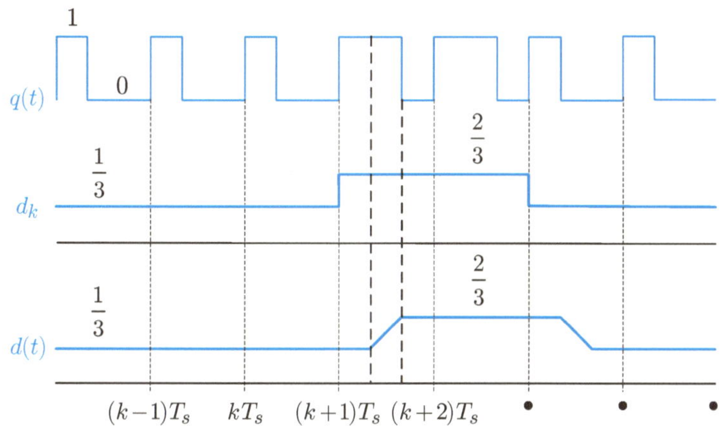

Figure 5.5 Relationships among switching function $q(t)$, cycle-by-cycle duty ratio d_k, and continuous duty ratio $d(t)$.

where the *overbar* denotes the moving average of the corresponding entry over the switching period T_s. For example, the *overbar* on $\boldsymbol{x}(t)$ indicates

$$\bar{\boldsymbol{x}}(t) = \frac{1}{T_s}\int_{t-T_s}^{t} \boldsymbol{x}(\tau)\,\mathrm{d}\tau \tag{5.7}$$

The expression (5.6) is approximated to

$$\begin{aligned} \frac{\mathrm{d}\bar{\boldsymbol{x}}(t)}{\mathrm{d}t} &= \left(\bar{q}(t)\boldsymbol{A}_{on} + (1-\bar{q}(t))\boldsymbol{A}_{off}\right)\bar{\boldsymbol{x}}(t) \\ &\quad + \left(\bar{q}(t)\boldsymbol{B}_{on} + (1-\bar{q}(t))\boldsymbol{B}_{off}\right)\bar{v}_S(t) \\ \bar{v}_O(t) &= \left(\bar{q}(t)\boldsymbol{C}_{on} + (1-\bar{q}(t))\boldsymbol{C}_{off}\right)\bar{\boldsymbol{x}}(t) \end{aligned} \tag{5.8}$$

based on the following facts and assumptions:

- The averaging is a linear operation.
- The coefficient matrices in (5.6) are constant matrices.
- If the state variables $\boldsymbol{x}(t)$ and input variable $v_S(t)$ vary only slowly, $\boldsymbol{x}(t)$ and $v_S(t)$ can be assumed that they remain approximately constant at $\bar{\boldsymbol{x}}(t)$ and $\bar{v}_S(t)$ over the averaging period.

Under above assumptions, it follows that

$$\begin{aligned} \overline{q(t)\boldsymbol{x}(t)} &\approx \overline{q(t)\bar{\boldsymbol{x}}(t)} = \bar{q}(t)\bar{\boldsymbol{x}}(t) \\ \overline{q(t)v_S(t)} &\approx \overline{q(t)\bar{v}_S(t)} = \bar{q}(t)\bar{v}_S(t) \end{aligned} \tag{5.9}$$

By noting that $\bar{q}(t)$ corresponds to the continuous duty ratio $d(t)$ defined in (5.5), (5.8) is rewritten as

$$\begin{aligned}\frac{\mathrm{d}\bar{\boldsymbol{x}}(t)}{\mathrm{d}t} &= \left(d(t)\boldsymbol{A}_{on} + (1-d(t))\boldsymbol{A}_{off}\right)\bar{\boldsymbol{x}}(t) \\ &\quad + \left(d(t)\boldsymbol{B}_{on} + (1-d(t))\boldsymbol{B}_{off}\right)\bar{v}_S(t) \\ \bar{v}_O(t) &= \left(d(t)\,\boldsymbol{C}_{on} + (1-d(t))\boldsymbol{C}_{off}\right)\bar{\boldsymbol{x}}(t)\end{aligned} \tag{5.10}$$

The description (5.10) is called the *averaged state-space model*. The averaged state-space model takes the continuous duty ratio $d(t)$ as an input variable and describes the time-averaged power stage dynamics. Namely, the averaged state-space model is a continuous-time approximation of the switched state-space model.

The continuous duty ratio $d(t)$ is used as an instrumental variable in transforming the switched state-space model into the averaged state-space model. In fact, the averaged state-space model is obtained by simply replacing the switching function $q(t)$ with the continuous duty ratio $d(t)$, and replacing the circuit variables with their average value. It should be emphasized that the averaged state-space model is a nonlinear model because the time-dependent variable $d(t)$ is multiplied with the coefficient matrices. The averaged state-space model is also called the *averaged state equation*.

The averaged state equation has been extensively used in the analysis of PWM dc-to-dc converters as well as other switching power converters. The steady-state equilibrium is found by solving the averaged state equation with the condition that all the state and input variables are frozen at constant: $\mathrm{d}\boldsymbol{x}(t)/\mathrm{d}t = 0$, $v_S(t) = V_S$, and $d(t) = D$.

More importantly, the averaged state equation has been used to simulate the converter waveforms, while focusing only on the time-averaged dynamics of the converter. Time-domain simulations on the averaged state equation provide practically the same information as the cycle-by-cycle simulations, yet with a greatly reduced simulation time.

Most significantly, the averaged state equation constitutes the foundation for developing the small-signal model for dc-to-dc converters. By linearizing the averaged state equation at a given operating point, a linear time-invariant small-signal model is obtained for dc-to-dc converters.

One disadvantage of the state-space averaging is the necessity of formulating and manipulating the state equation of the entire power stage, which can be time-consuming and tedious when the power stage contains a large number of reactive components. Another potential disadvantage is that the method only yields the final result in the format of the state equation, which is not directly programmable using standard circuit simulation software.

EXAMPLE 5.3 Averaged State-Space Model of Ideal Buck Converter

The averaged state-space model of an ideal buck converter can be obtained from the switched state-space model, given in Example 5.1, by simply replacing $q(t)$ with $d(t)$ and also replacing the circuit variables with their average value

$$
\begin{aligned}
\frac{\mathrm{d}\bar{\boldsymbol{x}}(t)}{\mathrm{d}t} &= \left[d(t) \begin{bmatrix} 0 & -\frac{1}{L} \\ \frac{1}{C} & -\frac{1}{CR} \end{bmatrix} + (1-d(t)) \begin{bmatrix} 0 & -\frac{1}{L} \\ \frac{1}{C} & -\frac{1}{CR} \end{bmatrix} \right] \bar{\boldsymbol{x}}(t) \\
&+ \left[d(t) \begin{bmatrix} \frac{1}{L} \\ 0 \end{bmatrix} + (1-d(t)) \begin{bmatrix} 0 \\ 0 \end{bmatrix} \right] \bar{v}_S(t) \\
\bar{v}_O(t) &= [\, d(t) [\, 0 \quad 1\,] + (1-d(t)) [\, 0 \quad 1\,]\,] \bar{\boldsymbol{x}}(t)
\end{aligned}
$$

with $\bar{\boldsymbol{x}} = [\bar{\imath}_L \; \bar{v}_C]^T$. The above expression is simplified to

$$
\begin{aligned}
\begin{bmatrix} \frac{\mathrm{d}\bar{\imath}_L(t)}{\mathrm{d}t} \\ \frac{\mathrm{d}\bar{v}_C(t)}{\mathrm{d}t} \end{bmatrix} &= \begin{bmatrix} 0 & -\frac{1}{L} \\ \frac{1}{C} & -\frac{1}{CR} \end{bmatrix} \begin{bmatrix} \bar{\imath}_L(t) \\ \bar{v}_C(t) \end{bmatrix} + \begin{bmatrix} \frac{d(t)}{L} \\ 0 \end{bmatrix} \bar{v}_S(t) \\
\bar{v}_O(t) &= \bar{v}_C(t)
\end{aligned}
$$

5.2.2 Circuit Averaging Technique

Another prevailing averaging technique is the circuit averaging technique. Unlike the previous state-space averaging technique dealing with the state-space equation, the circuit averaging method directly averages the circuit variables of the power stage. The circuit averaging technique yields an average power stage circuit model, which produces the averaged trajectories of the original power stage waveforms.

Averaging Switch Drive Signal

Most circuit variables of PWM converters are slowly-varying continuous waveforms. Averaging operation of such continuous waveforms can be easily conceived. However, the switch drive signal is a discontinuous signal, alternating between two discrete values to control the on/off status the switch. This section describes the averaging of the discontinuous switch drive signal in the perspective of the circuit averaging technique.

Figure 5.6 illustrates the averaging of the switch drive signal in a buck converter. The pulsewidth modulated switch drive signal in Fig. 5.6(a) is depicted as a pulse waveform, holding the value **1** for turning on and **0** for turning off the switch. The

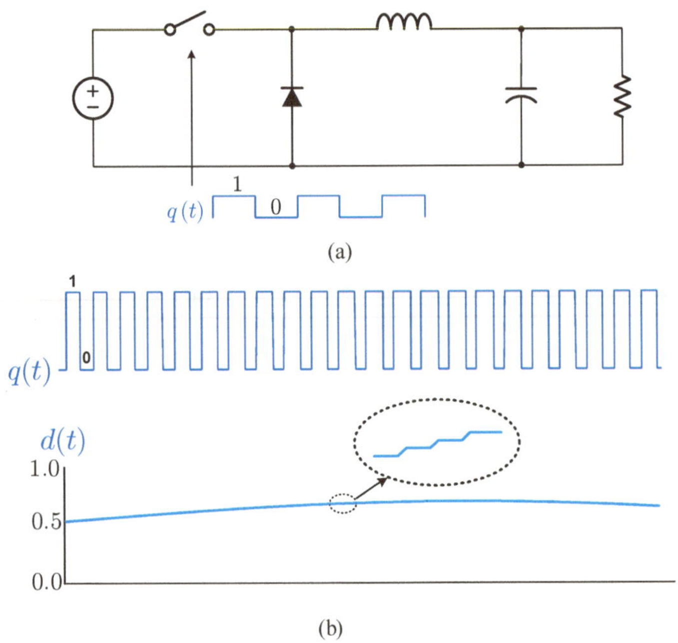

Figure 5.6 Switch drive signal, switching function, and continuous duty ratio of buck converter. (a) Switch drive signal of buck converter. (b) Conceptual illustration of switching function $q(t)$ and continuous duty ratio $d(t)$.

switch drive signal is then mathematically expressed as

$$q(t) = \begin{cases} 1 & \text{for on-time period } d\,T_s \\ 0 & \text{for off-time period } (1-d)T_s \end{cases}$$

which was previously referred to as the switching function in (5.4). This mathematical description enables us to perform the averaging of the switch drive signal.

Now, the averaging operation is performed on the switching function $q(t)$

$$d(t) = \frac{1}{T_s}\int_{t-T_s}^{t} q(\tau)\,\mathrm{d}\tau$$

resulting in the continuous duty ratio, formerly introduced in (5.5). Figure 5.6(b) is a conceptual illustration of the switching function $q(t)$ and continuous duty ratio $d(t)$. In this example, the continuous duty ratio varies around 0.5. The smooth-looking continuous duty ratio plot is actually the connection of many linear line segments, as highlighted in Fig. 5.6.

Procedure of Circuit Averaging

The method of the circuit averaging deals with the circuit variables associated with power stage circuit components. The continuous duty ratio $d(t)$ is naturally embedded into the averaging process. The circuit averaging technique might be best described by the steps of the modeling process.

1) Formulate an equation that describes the time-averaged behavior of the circuit variables associated with an individual component in the power stage. This equation contains the continuous duty ratio $d(t)$ and is referred to as the averaged circuit equation of the circuit component.
2) Synthesize a circuit model which satisfies the averaged circuit equation formulated in the previous step. This model is called the average model of the specific circuit component.
3) Put the resulting average model into the power stage in place of the original circuit component.
4) Repeat the above process for all the components in the power stage. When all the circuit components are replaced with their average model, the resulting circuit configuration becomes an *average circuit model* of the entire power stage. The average power stage model generates the circuit waveforms that track the moving averages of the original waveforms.

In circuit averaging, it is not always necessary to individually apply the averaging process to every single circuit component. In fact, several circuit components can be grouped together and treated as one composite circuit component. This practice will be explored in the next section.

PWM Switch

In principle, all the circuit components in the power stage should be replaced with their average model to yield an average model for the entire power stage. However, the linear circuit components, including the voltage source, inductor, and capacitor, remain invariant during the averaging process. In other words, linear circuit components can be used in their original form at their initial location when constructing an average model for the power stage. Accordingly, the circuit components that need to go through the formal averaging process are only the active and passive switches.

Figure 5.7 shows the circuit diagrams of the three basic PWM dc-to-dc converters, where the active-passive switch pair is identified as the composite circuit component that is subject to the averaging operation. The active-passive switch pair functions as a single-pole double-throw (SPDT) switch that periodically changes its structure based on the PWM principle. This active-passive switch pair is named as the *PWM switch*, in order to highlight its function as a three-terminal switching device operating under the principle of PWM.

Figure 5.8 shows the functional description, circuit symbol, and polarity/direction of the terminal circuit variables of the PWM switch. The node **a** stands for the

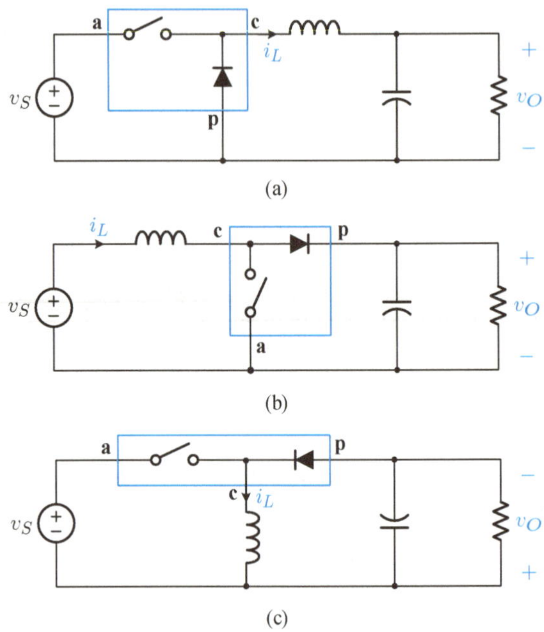

Figure 5.7 Circuit diagrams of three basic converters. (a) Buck converter. (b) Boost converter. (c) Buck/boost converter.

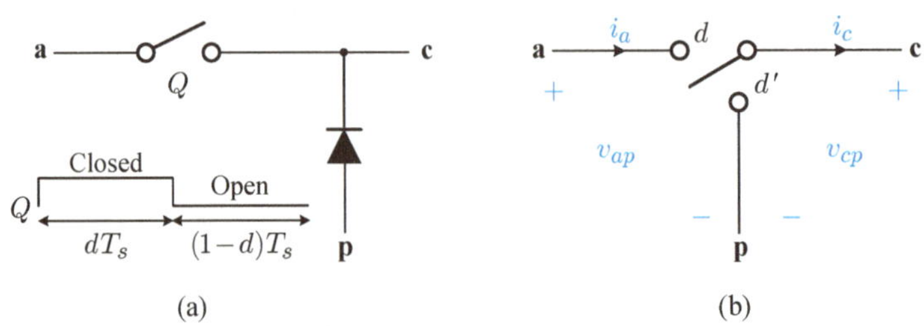

Figure 5.8 PWM switch. (a) Active-passive switch pair operating under PWM principle. (b) Circuit symbol for PWM switch.

active terminal directed to the active switch, **p** represents the passive terminal, and **c** denotes the common terminal. The circuit symbol also signifies that the common terminal is connected to the active terminal **a** for dT_s and to the passive terminal **p** for $d'T_s = (1-d)T_s$.

The PWM switch has a couple of characteristic features on its terminal waveforms, regardless of the converter topology in which the PWM switch is embedded. First, a dc voltage is applied across the active and passive terminals, as shown in Fig. 5.7.

The voltage between **a** and **p** terminals is

$$v_{ap}(t) = \begin{cases} v_S & \text{for buck converter} \\ -v_O & \text{for boost converter} \\ v_S + v_O & \text{for buck/boost converter} \end{cases} \tag{5.11}$$

which can be considered as dc value when the ripple component is ignored. The voltage across **c** and **p** terminals, v_{cp}, is a sampled replica of v_{ap}; v_{cp} equals to v_{ap} during dT_s and becomes zero during $d'T_s$.

Second, the common terminal **c** always carries the inductor current, which is a continuous triangular waveform,

$$i_c(t) = \begin{cases} i_L(t) & \text{for buck and buck/boost converters} \\ -i_L(t) & \text{for boost converter} \end{cases} \tag{5.12}$$

where i_c is the current coming out of the common terminal **c**. The active terminal current i_a is a sampled version of i_c, thus a pulsating discontinuous waveform. These facts will be used in the next section in formulating the circuit equations for the PWM switch.

Averaging PWM Switch

The PWM switch, identified as the three-terminal switching device embedded in all the three basic PWM converters, can be treated as an individual circuit component. Referring to Fig. 5.8(b), the circuit equations for the PWM switch are written given as

$$v_{cp}(t) = \begin{cases} v_{ap}(t) & \text{for } dT_s \\ 0 & \text{for } d'T_s \end{cases} \tag{5.13}$$

and

$$i_a(t) = \begin{cases} i_c(t) & \text{for } dT_s \\ 0 & \text{for } d'T_s \end{cases} \tag{5.14}$$

which can be written as

$$\boxed{\begin{aligned} v_{cp}(t) &= v_{ap}(t)q(t) \\ i_a(t) &= i_c(t)q(t) \end{aligned}} \tag{5.15}$$

using the switching function $q(t)$

$$\boxed{q(t) = \begin{cases} 1 & \text{for } dT_s \\ 0 & \text{for } d'T_s \end{cases}} \tag{5.16}$$

Equation (5.15) indicates that the continuous circuit variables v_{ap} and i_c are multiplied by the switching function $q(t)$, yielding the discontinuous circuit variables v_{cp} and

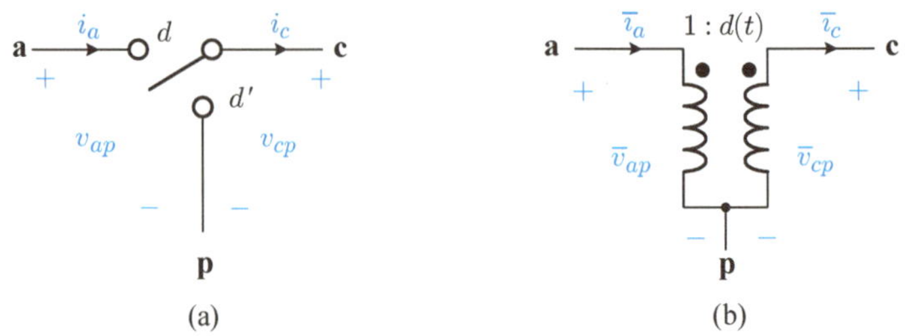

Figure 5.9 PWM switch and its average model. (a) PWM switch. (b) Average model.

i_a. This equation is a compact expression for the circuit variables of the PWM switch, formulated using the notation of the switching function $q(t)$.

By taking the local average of (5.15), it follows that

$$\begin{aligned} \bar{v}_{cp}(t) &= \overline{v_{ap}(t)q(t)} \\ \bar{i}_a(t) &= \overline{i_c(t)q(t)} \end{aligned} \tag{5.17}$$

If the continuous circuit variables $v_{ap}(t)$ and $i_c(t)$ change so slowly within the switching period that they remain nearly constant at their average, (5.17) can be approximated as

$$\begin{aligned} \bar{v}_{cp}(t) &= \overline{v_{ap}(t)q(t)} \approx \overline{\bar{v}_{ap}(t)q(t)} = \bar{v}_{ap}(t)\bar{q}(t) \\ \bar{i}_u(t) &- \overline{i_c(t)q(t)} \approx \overline{\bar{i}_c(t)q(t)} = \bar{i}_c(t)\bar{q}(t) \end{aligned} \tag{5.18}$$

Using the definition of the continuous duty ratio $d(t)$ in (5.5), the expressions (5.18) are rewritten as

$$\boxed{\begin{aligned} \bar{v}_{cp}(t) &= d(t)\bar{v}_{ap}(t) \\ \bar{i}_a(t) &= d(t)\bar{i}_c(t) \end{aligned}} \tag{5.19}$$

yielding the desired expressions for averaged circuit variables.

The averaged equations of the PWM switch in (5.19) are identical to the circuit equations of an ideal transformer, introduced in Section 2.2.3. Thus, the symbol of the ideal transformer can be used as an average model for the PWM switch. Figure 5.9 shows the PWM switch and its average model representation using the symbol of the ideal transformer. The ideal transformer in Fig. 5.9(b) has the following properties. First, the turns ratio is a time-dependent variable $d(t)$ rather than a constant. Second, the primary and secondary windings are tied together at the passive terminal. Therefore, the average model for the PWM switch is a time-variant three-terminal device which has the structure of the ideal transformer.

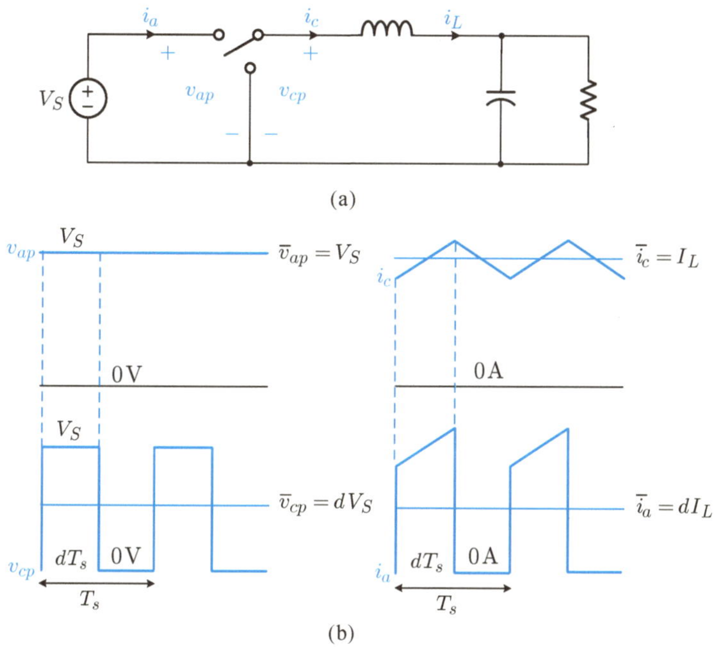

Figure 5.10 PWM switch in buck converter. (a) Circuit diagram. (b) Circuit waveforms.

Average Equations of PWM Switch in Buck Converter — EXAMPLE 5.4

This example illustrates the average equations for the PWM switch. Figure 5.10 shows the circuit waveforms of the PWM switch in a buck converter. Referring to the circuit waveforms, the average values of v_{ap} and i_c are written as

$$\begin{aligned}\bar{v}_{ap}(t) &= V_S \\ \bar{i}_c(t) &= I_L\end{aligned}$$

On the other hand, the average values of v_{cp} and i_a are expressed as

$$\begin{aligned}\bar{v}_{cp}(t) &= d(t)V_S \\ \bar{i}_a(t) &= d(t)I_L\end{aligned}$$

using the definition of the continuous duty ratio. It becomes apparent from the above expressions that

$$\begin{aligned}\bar{v}_{cp}(t) &= d(t)\bar{v}_{ap}(t) \\ \bar{i}_a(t) &= d(t)\bar{i}_c(t)\end{aligned}$$

which are the average equations for the PWM switch given in (5.19).

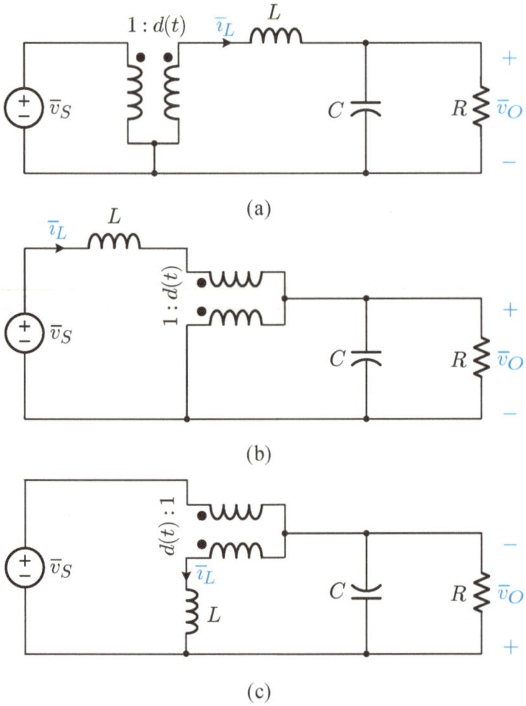

Figure 5.11 Average models for three basic converters. (a) Buck converter. (b) Boost converter. (c) Buck/boost converter.

Average Models for Three Basic PWM Converters

The average model for the PWM switch is derived while considering the PWM switch as a standalone individual device. Accordingly, the average model of the PWM switch can universally be used for all the three basic PWM converters, which commonly contain the PWM switch. The average models of the PWM converters are simply obtained with a pin-to-pin replacement of the PWM switch with its average model. Figure 5.11 shows the average models for the buck, boost, and buck/boost converters. Each average model describes the time-averaged circuit behavior of the respective PWM converter.

EXAMPLE 5.5 Responses of Switch Model and Average Model

This example illustrates the implication of the average model of dc-to-dc converters. Figure 5.12 compares the responses of the switch model and average model of a buck converter. Figure 5.12(a) is the simulation result using the switch model. For this simulation, the switch drive signal is appropriately modulated to produce the switching function $q(t)$ whose cycle-by-cycle duty ratio is varied in a sinusoidal fash-

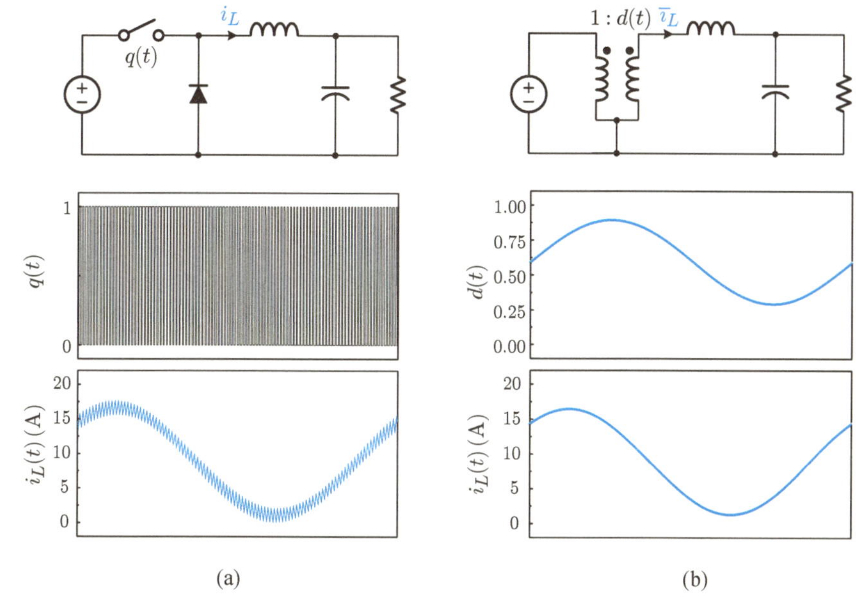

Figure 5.12 Responses of buck converter. (a) Switch model response. (b) Average model response.

ion. With this switching function, the inductor current $i_L(t)$ shows a low-frequency sinusoidal oscillation, as well as the high-frequency switching ripple.

Simulations with the average model are shown in Fig. 5.12(b). For this case, the switching function $q(t)$ is first converted into the continuous duty ratio $d(t)$ by performing the averaging operation

$$d(t) = \frac{1}{T_s} \int_{t-T_s}^{t} q(\tau) \, d\tau$$

The resulting continuous duty ratio $d(t)$, shown in Fig. 5.12(b), is then used as a time-varying turns ratio of the ideal transformer. The average model produces the continuous inductor current $\overline{i}_L(t)$ which accurately traces the moving average of the actual inductor current $i_L(t)$ of the switch model.

As shown in the previous example, the average model generates the continuous circuit waveforms that track the moving average of the actual responses. Accordingly, the average models are often used to simulate the time-averaged circuit waveforms of PWM dc-to-dc converters during transition periods.

The average model is also used to study the steady-state circuit operation. In the steady-state equilibrium, all the circuit variables denote dc quantities and the continuous duty ratio becomes the steady-state duty ratio D. In addition, as far as the dc circuit variables are concerned, the inductor behaves as a short circuit, while the capacitor functions as an open circuit. Under these conditions, the average model of the buck converter in Fig. 5.11(a) indicates that $V_O = DV_S$; the dc voltage gain of the converter equals to D. The use of the average model for the steady-state analysis is further illustrated in the following example.

EXAMPLE 5.6 Voltage Gain of Non-Ideal Boost Converter

In this example, we illustrate the use of the average model for the steady-state analysis of the boost converter with a parasitic resistance. The average model for the boost converter is shown in Fig. 5.13(a). In the average model, the resistance R_l represents the equivalent series resistance (esr) of the inductor. The inductor esr does not affect the averaging process. Thus, the esr is placed in the original location in the average model. The average model is transformed into a dc model, shown in Fig. 5.13(b), by shorting the inductor, opening the capacitor, and replacing the continuous duty ratio $d(t)$ with the steady-state duty ratio D. Figure 5.13(b) is further modified to Fig. 5.13(c) by replacing the ideal transformer with a pair of voltage source and current source.

The dc model can be used for the steady-state analysis at the presence of the esr of the inductor. For example, the voltage gain of the converter is evaluated as follows. Referring to Fig. 5.13(c), the average inductor current is given by

$$I_L = \frac{V_S + DV_O - V_O}{R_l} = \frac{V_S + (D-1)V_O}{R_l} \tag{5.20}$$

On the other hand, the output voltage is given by

$$V_O = (1-D)I_L R \tag{5.21}$$

From (5.20) and (5.21), it becomes

$$V_O = (1-D)\left(\frac{V_S + (D-1)V_O}{R_l}\right)R$$

which is rearranged as

$$\frac{V_O}{V_S} = \frac{1}{1-D}\frac{1}{1+\dfrac{1}{(1-D)^2}\dfrac{R_l}{R}}$$

yielding the voltage gain expression. This expression is the same as the gain formula derived in Section 4.1.4, using the flux linkage balance and charge balance conditions.

5.2.3 Circuit Averaging and State-Space Averaging

We studied the two important averaging techniques – the state-space averaging and circuit averaging. While the two techniques have their own merits, the circuit averaging has several advantages over the state-space averaging. First, the circuit averaging performs the averaging operation directly on the power stage circuit. Thus,

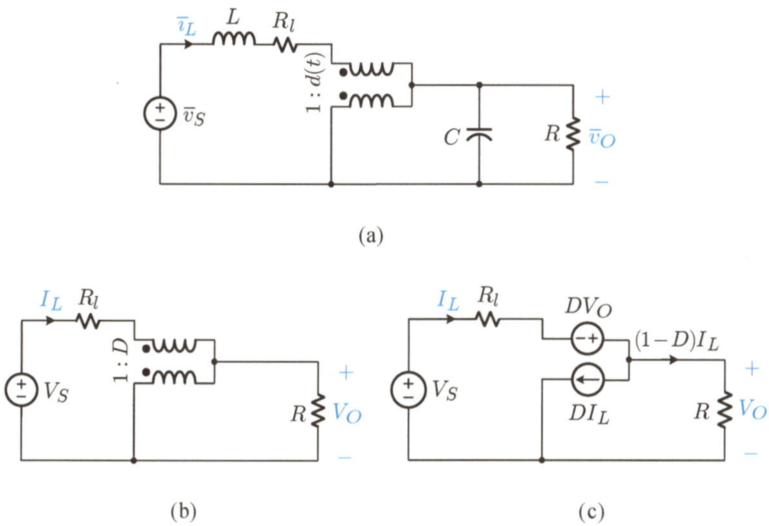

Figure 5.13 Average and dc models of boost converter. (a) Average model. (b) Dc model. (c) Modified model.

the manipulation of the state equation, which is needed for the state-space averaging, is not required for the circuit averaging technique. Second, the circuit averaging provides the average model in the structure that closely resembles the original power stage circuit diagram, which is not the case for the state-space averaging. Thus, the average model of the circuit averaging can readily be programmed with standard circuit simulation software.

In spite of the differences in the modeling approaches, the circuit averaging is functionally identical to the state-space averaging. A simple confirmation of the equivalence between these two techniques is given below.

The average model of the buck converter shown in Fig. 5.11(a) yields the circuit equation

$$\begin{aligned} L\frac{\mathrm{d}\bar{i}_L(t)}{\mathrm{d}t} &= d(t)\bar{v}_S(t) - \bar{v}_C(t) \\ C\frac{\mathrm{d}\bar{v}_C(t)}{\mathrm{d}t} &= \bar{i}_L(t) - \frac{\bar{v}_C(t)}{R} \end{aligned} \tag{5.22}$$

This equation is rearranged into a state equation

$$\begin{bmatrix} \frac{\mathrm{d}\bar{i}_L(t)}{\mathrm{d}t} \\ \frac{\mathrm{d}\bar{v}_C(t)}{\mathrm{d}t} \end{bmatrix} = \begin{bmatrix} 0 & -\frac{1}{L} \\ \frac{1}{C} & -\frac{1}{CR} \end{bmatrix} \begin{bmatrix} \bar{i}_L(t) \\ \bar{v}_C(t) \end{bmatrix} + \begin{bmatrix} \frac{d(t)}{L} \\ 0 \end{bmatrix} \bar{v}_S(t) \tag{5.23}$$

which is identical to the result of the state-space averaging given in Example 5.3.

The average models, derived either from the state-space averaging or from the circuit averaging, can be transformed into small-signal models through the linearization process. In this book, however, the average models derived from the circuit averaging are only considered for the linearization, primarily due to the simplicity in the linearization process. The derivation of average models using the state-pace averaging technique and linearization of the resulting average models can be found in [2].

5.3 LINEARIZING AVERAGED POWER STAGE DYNAMICS

While averaging eliminates the time variance from the power stage dynamics, the process of averaging brings certain nonlinearities into the average model of the power stage. In this section, we employ the linearization process to remove nonlinearities brought in during the averaging process. The linearized average model is the small-signal model of the power stage.

5.3.1 Linearization of Nonlinear Function and Small-Signal Model

Linearization is the process of approximating a nonlinear function to a linear relationship under certain assumptions. This section covers the linearization of both the single-variable and multiple-variable nonlinear functions.

Single-Variable Nonlinear Functions

For a single-variable nonlinear function $y = f(x)$, we assume that x deviates from a fixed value X by a small amount of $\hat{x}$; namely, $x = X + \hat{x}$ with $\hat{x} = |x - X| \ll 1$. The nonlinear function $y = f(x)$ is then expanded into Taylor series at $x = X$

$$y = f(X + \hat{x}) = f(X) + \left.\frac{\mathrm{d}f}{\mathrm{d}x}\right|_{x=X} \hat{x} + \frac{1}{2!}\left.\frac{\mathrm{d}^2 f}{\mathrm{d}x^2}\right|_{x=X} \hat{x}^2 + \cdots \tag{5.24}$$

We now retain only the constant and the first-order term of the series

$$\boxed{y = f(X + \hat{x}) \approx f(X) + \left.\frac{\mathrm{d}f}{\mathrm{d}x}\right|_{x=X} \hat{x}} \tag{5.25}$$

as an approximation of the nonlinear function near the quiescent point X.

In this book, the variable $\hat{x}$ denotes a *small sinusoidal* variation around X

$$\hat{x}(t) = x_s \sin \omega_s t \tag{5.26}$$

Thus, $\hat{x}$ is called the sinusoidal component or ac component, while X is referred to as the dc component. Figure 5.14 graphically illustrates the linearization process described by (5.25). It can be seen that the linearization is equivalent to replacing the nonlinear curve with its tangential line segment evaluated at $x = X$. In other words, the nonlinear relationship is *locally* linearized using the tangential line at

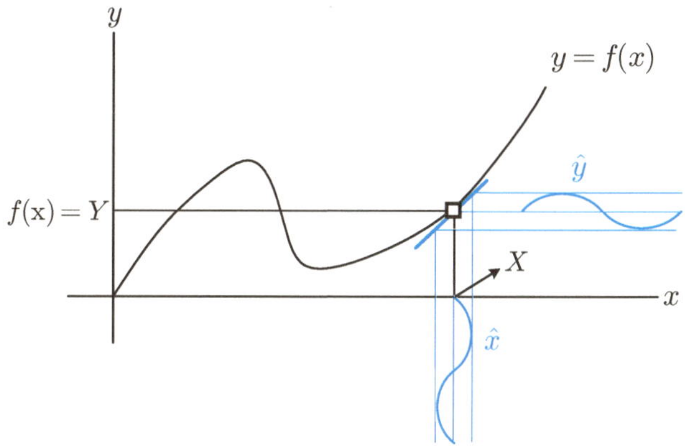

Figure 5.14 Linearization of nonlinear function.

the quiescent point. The assumption $\hat{x} = |x - X| \ll 1$ is necessary to ensure that the length of the tangential line segment is so short that the deviation between the nonlinear curve and the line segment is negligibly small.

With the input variable consisting of the dc and ac components, $x = X + \hat{x}$, the output variable y can be decomposed into a dc component Y and an ac component $\hat{y}$: $y = Y + \hat{y}$. From (5.25), the dc component of y is given by $Y = f(X)$, and the ac component of y then becomes

$$\hat{y} = \frac{\mathrm{d}f}{\mathrm{d}x}_{x=X} \hat{x} \tag{5.27}$$

The *small-signal gain* or *small-signal model* of the nonlinear equation is defined as the ratio of the ac components

$$\boxed{\frac{\hat{y}}{\hat{x}} = \frac{\mathrm{d}f}{\mathrm{d}x}_{x=X}} \tag{5.28}$$

The small-signal gain relates the ac components of input and output variables near the given dc operating point. The condition $\hat{x} = |x - X| \ll 1$ is called the *small-signal assumption* because it indicates that the ac component of the input variable is substantially smaller than the dc component. Correspondingly, $\hat{x}(t) = x_s \sin \omega t$ is referred to as the small signal, while X is called the large signal. The small-signal assumption is the necessary condition for the accuracy of small-signal model.

The process of finding a small-signal model is called the *small-signal modeling* of a nonlinear process. The previous analysis indicates that the small-signal modeling can be considered as the process of evaluating the slope or derivative of a nonlinear function at a given quiescent point.

The expression (5.28) can be written as

$$\boxed{\hat{y} = K\hat{x}} \tag{5.29}$$

where K is the linear gain between the first-order ac variables of the nonlinear equation. The constant K was determined as

$$K = \frac{\mathrm{d}f}{\mathrm{d}x}\bigg|_{x=X} \tag{5.30}$$

in the previous analysis.

The preceding argument hints at an alternative way of evaluating the small-signal gain of a nonlinear equation. For simple algebraic nonlinear equations, the small-signal model can be found by

1) evaluating the nonlinear equation with the variables consisting of dc and ac components, and

2) equating only the first-order ac components of input and output variables from the resulting expression.

The first step corresponds to Taylor series expansion of (5.24). The second step is the identification of the linear gain K, which is specified in (5.29) along with (5.30).

As a simple example, the small-signal gain of a nonlinear equation $y = x^2$ at $x = X = 2$ is found as follows. By evaluating the equation with the input and output variables consisting of both dc and ac components, it follows that

$$Y + \hat{y} = (X + \hat{x})^2 = X^2 + 2X\hat{x} + \hat{x}^2 \tag{5.31}$$

The first term in the right-hand side of (5.31), X^2, is a dc quantity and the third term $\hat{x}^2$ is a second-order expression of the ac variable. These two terms are irrelevant to the small-signal gain. By equating the first-order terms of the ac variables, the small-signal relationship is given by

$$\hat{y} = 2X\hat{x} \tag{5.32}$$

The evaluation of (5.32) at $X = 2$ leads to the desired small-signal gain

$$\frac{\hat{y}}{\hat{x}} = 2X\bigg|_{X=2} = 4 \tag{5.33}$$

On the other hand, a direct application of (5.28) to $y = x^2$ at $x = X = 2$ yields the small-signal gain

$$\frac{\hat{y}}{\hat{x}} = \frac{\mathrm{d}x^2}{\mathrm{d}x}\bigg|_{x=X=2} = 2x\bigg|_{x=X=2} = 4 \tag{5.34}$$

which is identical to (5.33).

Multiple-Variable Nonlinear Functions

The linearization process is now extended to multiple-variable nonlinear equations. A two-variable nonlinear equation $z = f(x, y)$ is expanded into Taylor series around a fixed operating point (X, Y) at the presence of the ac signals $\hat{x}$ and $\hat{y}$

$$\begin{aligned} z = Z + \hat{z} &= f(X + \hat{x}, Y + \hat{y}) \\ &= f(X, Y) + \left(\frac{\partial f}{\partial x}\hat{x} + \frac{\partial f}{\partial y}\hat{y}\right)_{x=X\ y=Y} \\ &+ \frac{1}{2!}\left(\frac{\partial^2 f}{\partial x^2}\hat{x}^2 + 2\frac{\partial^2 f}{\partial x \partial y}\hat{x}\hat{y} + \frac{\partial^2 f}{\partial y^2}\hat{y}^2\right)_{x=X\ y=Y} + \cdots \end{aligned} \tag{5.35}$$

When $\hat{x}$ and $\hat{y}$ are sufficiently small, (5.35) is approximated as

$$z = Z + \hat{z} = f(X, Y) + \left(\frac{\partial f}{\partial x}\hat{x} + \frac{\partial f}{\partial y}\hat{y}\right)_{x=X\ y=Y} \tag{5.36}$$

The small-signal relation of the ac variables becomes

$$\hat{z} = K_1\hat{x} + K_2\hat{y} \tag{5.37}$$

with

$$K_1 = \frac{\partial f}{\partial x}_{x=X\ y=Y} \quad \text{and} \quad K_2 = \frac{\partial f}{\partial y}_{x=X\ y=Y} \tag{5.38}$$

The equations (5.37) and (5.38) indicate that the small-signal model of the nonlinear equation $z = f(x, y)$ can be performed in two different ways, as is the case with the single-variable nonlinear equation. The equation (5.37) implies that the small signal gains, K_1 and K_2, are obtained by evaluating the nonlinear equation with the ac and dc variables and equating only the first-order ac variables. On the other hand, (5.38) infers that the small-signal gains are found by directly evaluating the partial derivatives of the nonlinear equation at the dc operating point. An example is given below to illustrate this fact.

Linearization of a Two-Variable Nonlinear Equation — EXAMPLE 5.7

This example shows the linearization of a nonlinear equation $z = x^2 y$ around (X, Y). First, the evaluation of the equation with ac and dc variables yields

$$\begin{aligned} Z + \hat{z} &= (X + \hat{x})^2(Y + \hat{y}) \\ &= (X^2 + 2X\hat{x} + \hat{x}^2)(Y + \hat{y}) \\ &= X^2Y + 2XY\hat{x} + Y\hat{x}^2 + X^2\hat{y} + 2X\hat{x}\hat{y} + \hat{x}^2\hat{y} \end{aligned}$$

By equating the first-order ac terms, the small-signal model is found

$$\hat{z} = 2XY\hat{x} + X^2\hat{y}$$

On the other hand, the evaluation of partial derivative

$$\frac{\partial(x^2y)}{\partial x}_{x=X\ y=Y} = 2XY \quad \text{and} \quad \frac{\partial(x^2y)}{\partial y}_{x=X\ y=Y} = X^2$$

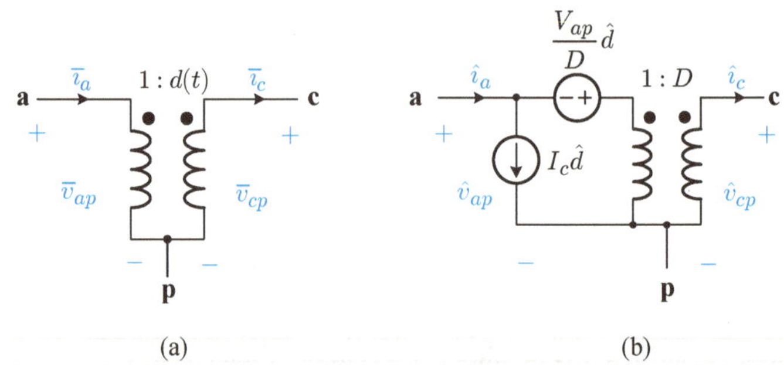

Figure 5.15 Dynamic models for PWM switch. (a) Average model of PWM switch. (b) Small-signal model of PWM switch.

results in the same small-signal model

$$
\begin{aligned}
\hat{z} &= \left.\frac{\partial(x^2y)}{\partial x}\right|_{x=X\,y=Y}\hat{x} + \left.\frac{\partial(x^2y)}{\partial y}\right|_{x=X\,y=Y}\hat{y} \\
&= 2XY\hat{x} + X^2\hat{y}
\end{aligned}
$$

5.3.2 Small-Signal Model of PWM Switch — The PWM Switch Model

The linearization is adapted to the average model of the PWM switch, shown in Fig. 5.15(a). The circuit equations for the average model of the PWM switch are repeated

$$
\boxed{
\begin{aligned}
\bar{v}_{cp}(t) &= d(t)\bar{v}_{ap}(t) \\
\bar{i}_a(t) &= d(t)\bar{i}_c(t)
\end{aligned}
}
\tag{5.39}
$$

Although the symbol *overbar* will be omitted in the following developments, all the circuit variables represent the averaged quantities.

Application of the linearization process to (5.39) yields

$$
\begin{aligned}
\underbrace{V_{cp}}_{\text{dc}} + \underbrace{\hat{v}_{cp}(t)}_{\text{ac}} &= (D + \hat{d}(t))(V_{ap} + \hat{v}_{ap}(t)) \\
&= \underbrace{DV_{ap}}_{\text{dc}} + \underbrace{\hat{d}(t)V_{ap} + D\,\hat{v}_{ap}(t)}_{\text{ac}} \\
&\quad + \underbrace{\hat{d}(t)\,\hat{v}_{ap}(t)}_{\text{second-order term}}
\end{aligned}
\tag{5.40}
$$

$$\begin{aligned}\underbrace{I_a}_{\text{dc}} + \underbrace{\hat{\imath}_a(t)}_{\text{ac}} &= (D + \hat{d}(t))(I_c + \hat{\imath}_c(t)) \\ &= \underbrace{DI_c}_{\text{dc}} + \underbrace{\hat{d}(t)\, I_c + D\,\hat{\imath}_c(t)}_{\text{ac}} \\ &+ \underbrace{\hat{d}(t)\,\hat{\imath}_c(t)}_{\text{second-order term}}\end{aligned} \tag{5.41}$$

where the capitalized variables are dc components and the variables with ^ are ac components. For example, D is the steady-state duty ratio, whereas the quantity $\hat{d}(t)$ denotes the small-signal ac component of the continuous duty ratio $d(t)$. Details about the ac component of $d(t)$ and other ac variables will be given in Section 5.4.1.

By equating the first-order ac components in (5.40) and (5.41), the small-signal representation of (5.39) is obtained

$$\boxed{\begin{aligned}\hat{v}_{cp}(t) &= V_{ap}\,\hat{d}(t) + D\,\hat{v}_{ap}(t) \\ \hat{\imath}_a(t) &= I_c\,\hat{d}(t) + D\,\hat{\imath}_c(t)\end{aligned}} \tag{5.42}$$

Equations (5.42) constitute the small-signal model or small-signal representation of the PWM switch. In (5.42), the left-hand side variables are expressed as a linear combination of the dc and ac components of other variables. In particular, V_{ap}, D, and I_c are the dc values of the corresponding circuit variables. Thus, the small-signal representation depends on the dc operating point of the PWM switch, as is always the case with the small-signal model of any nonlinear system.

The small-signal relationships (5.42) can be converted into an equivalent circuit model. The circuit representation is shown in Fig. 5.15(b). The circuit model was known as the *PWM switch model*, coined by Vatché Vorpérion who first proposed this model. The PWM switch model thus refers to the small-signal model of the PWM switch. As shown in Fig. 5.15(b), the PWM switch model contains two dependent signal sources, both controlled by the small-signal continuous duty ratio $\hat{d}$. It also includes an ideal transformer whose turns ratio is given by the steady-state duty ratio D of the converter. It is straightforward to show that the PWM switch model in Fig. 5.15(b) satisfies the circuit equation (5.42).

Table 5.2 compares the three different models for the PWM switch – the switch model, average model, and small-signal model. The switch model is a time-variant model in which the position of the SPDT switch represents the time dependency of the PWM switch. The switch model is converted into the time-invariant but nonlinear average model by adopting the continuous duty ratio $d(t)$, or the moving average of the switching function $q(t)$. The nonlinear average model is linearized under small-signal assumption, leading to the linear time-invariant small-signal model of the PWM switch – the PWM switch model. The circuit equations for the switch model, average model, and PWM switch model are shown in Table 5.2.

Table 5.2 Models for PWM Switch

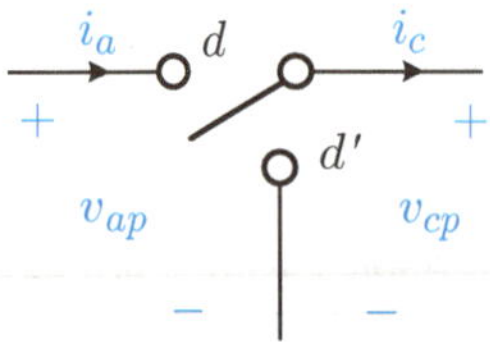

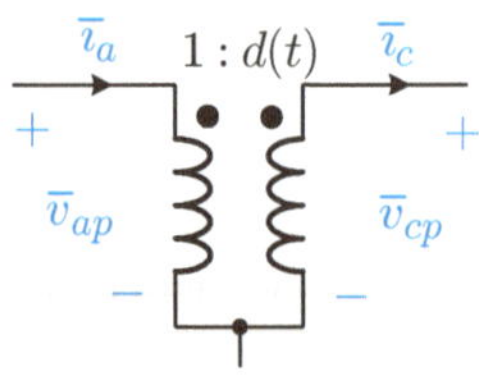

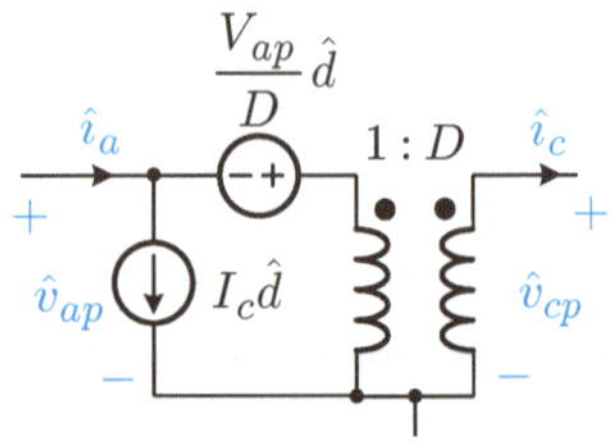

Switch model	Average model	Small-signal model
$v_{cp}(t) = q(t)v_{ap}(t)$	$\bar{v}_{cp}(t) = d(t)\bar{v}_{ap}(t)$	$\hat{v}_{cp}(t) = V_{ap}\hat{d}(t) + D\hat{v}_{ap}(t)$
$i_a(t) = q(t)i_c(t)$	$\bar{i}_a(t) = d(t)\bar{i}_c(t)$	$\hat{i}_a(t) = I_c\hat{d}(t) + D\hat{i}_c(t)$
$q(t) = \begin{cases} 1 \text{ for } dT_s \\ 0 \text{ for } (1-d)T_s \end{cases}$	$d(t) = \frac{1}{T_s}\int_{t-T_s}^{t} q(\tau)\mathrm{d}\tau$	

5.3.3 Small-Signal Model of Converter Power Stage

The linear circuit components are invariant to the linearization process and remain unaltered in constructing a small-signal model from the average model. Accordingly, the small-signal model for the converter power stage is obtained from the respective average model of the power stage, by simply replacing the average model of PWM switch with the small-signal model of the PWM switch and introducing appropriate small-signal sources as the input variables.

Figure 5.16(a) shows the small-signal model of the buck converter. In this model, $\hat{v}_s(t)$ is the small-signal input voltage, $\hat{d}(t)$ denotes the small-signal continuous duty ratio, and $\hat{i}_o(t)$ is the small-signal current source which represents the sinusoidal current deviation from the dc output current. The small-signal model shows that $V_{ap} = V_S$ and $I_c = I_L$ for the buck converter.

The small-signal power stage model in Fig. 5.16(a) is a time-domain model in which all the excitation sources and circuit variables are defined as time-dependent quantities. This time-domain small-signal model exhibits the transient response of the power stage in the presence of small-signal time-domain excitations. The small-signal model becomes far more useful when converted into the frequency-domain, or s-domain, model. The time-domain small-signal model is now transformed into the s-domain small-signal model, as shown in Fig. 5.16(b), by using s-domain expressions for the circuit variables and circuit components.

Figure 5.17 shows the s-domain small signal models for the three basic converters. The models are constructed from the average models by following the procedures explained above. The small-signal model of the boost converter incorporates that $V_{ap} = -V_O$ and $I_c = -I_L$, thus altering the polarity/direction of the dependent

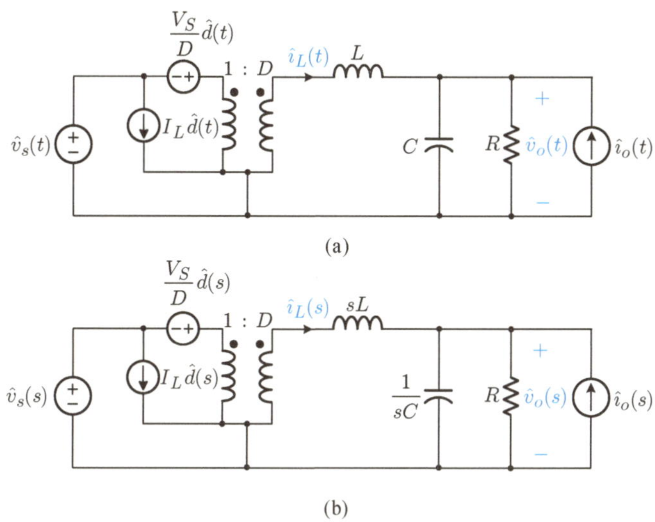

Figure 5.16 Small-signal models of buck converter. (a) Time-domain small-signal model. (b) s-domain small-signal model.

voltage/current source in the PWM switch model. The buck/boost converter model indicates that $V_{ap} = V_S + V_O$ and $I_c = I_L$. The s-domain small-signal models are a linear time-invariant model to which all classical circuit analysis techniques can be directly applied. In Chapter 6, the power stage dynamics of the three basic PWM dc-to-dc converters will be investigated using these models.

5.4 FREQUENCY RESPONSE OF CONVERTER POWER STAGE

The s-domain small-signal models can be used to study the frequency response of PWM converters. While the frequency response of linear time-invariant systems is well known, the frequency response of the PWM converter requires some explanation about its background.

The frequency response is evolved from the sinusoidal response. This section first discusses the sinusoidal response and later presents the frequency response of the converter power stage.

5.4.1 Sinusoidal Response of Power Stage

The sinusoidal response is originally established for linear time-invariant systems. When a sinusoidal input is applied to a linear time-invariant system, the output of the system is a sinusoid with the same frequency as that of the input sinusoid. However, the magnitude and phase of the output sinusoid are usually altered. The changes to the magnitude and phase of the input sinusoid are referred to as the sinusoidal

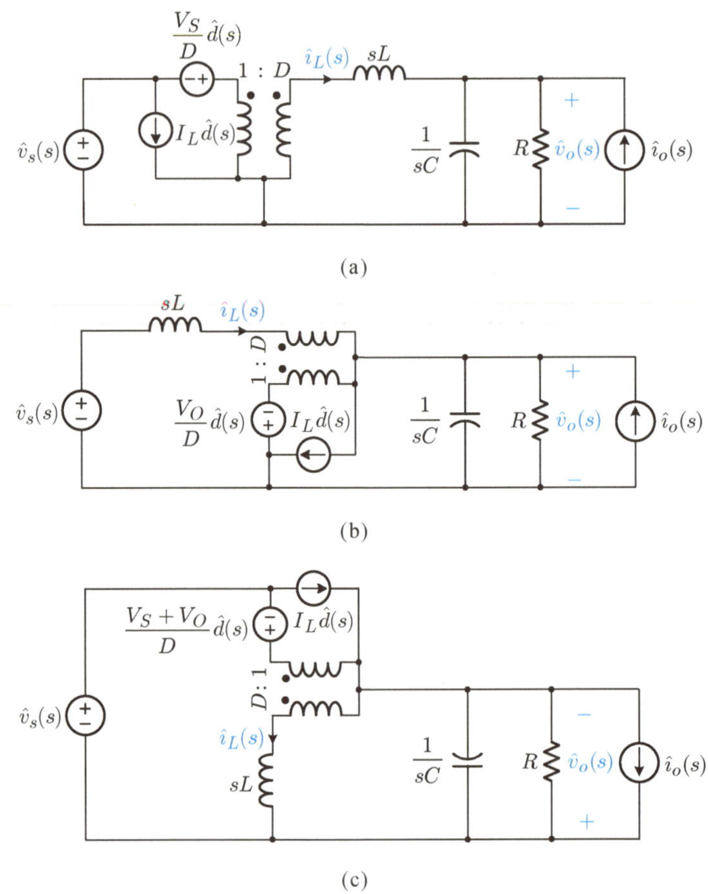

Figure 5.17 Small-signal models of three basic converters. (a) Buck converter. (b) Boost converter. (c) Buck/boost converter.

response of a linear time-invariant system. The concept of the sinusoidal response is now extended to PWM dc-to-dc converters.

First, it is presumed that the switch drive signal $q(t)$ is periodic with a fixed on-time period, resulting in a constant continuous duty ratio, $d(t) = D$. However, it is also assumed that the input voltage of the converter contains an ac component $\hat{v}_s$ on top of the dc component V_S

$$v_S(t) = V_S + \hat{v}_s(t) \tag{5.43}$$

The ac component is a sinusoid at frequency ω_s

$$\boxed{\hat{v}_s(t) = v_s \sin \omega_s t} \tag{5.44}$$

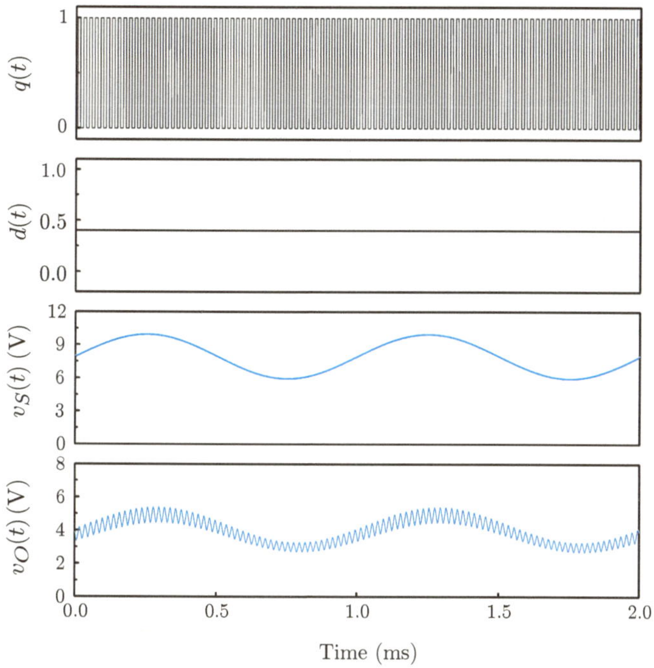

Figure 5.18 Output voltage response to sinusoidal change in input voltage.

When the small-signal assumption $v_s \ll V_S$ is imposed, the output voltage $v_O(t)$ will exhibit a sinusoidal excursion around a dc value, with the switching ripple superimposed on it. This situation is illustrated in Fig. 5.18, where the input voltage $v_S(t)$ contains a sinusoidal component, whereas the switch drive signal $q(t)$ is periodic and the continuous duty ratio $d(t)$ is a constant. When the switching ripple is ignored, the output voltage $v_O(t)$ is expressed as a combination of dc and ac components

$$v_O(t) = V_O + \hat{v}_o(t) \tag{5.45}$$

The ac component will be a sinusoid at frequency ω_s

$$\boxed{\hat{v}_o(t) = v_o \sin(\omega_s t + \theta_s)} \tag{5.46}$$

From (5.44) and (5.46), the input-to-output sinusoidal response at ω_s is defined as

- magnitude response at $\omega_s = \dfrac{|\hat{v}_o(t)|}{|\hat{v}_s(t)|} = \dfrac{v_o}{v_s}$
- phase response at $\omega_s = \angle\hat{v}_o(t) - \angle\hat{v}_s(t) = \theta_s$

Another example of the sinusoidal response is shown in Fig. 5.19. For this case, the input voltage is fixed at $v_S = V_S$ but the switch drive signal is modulated cycle-by-cycle, as illustrated by the switching function $q(t)$ in Fig. 5.19. The modulation

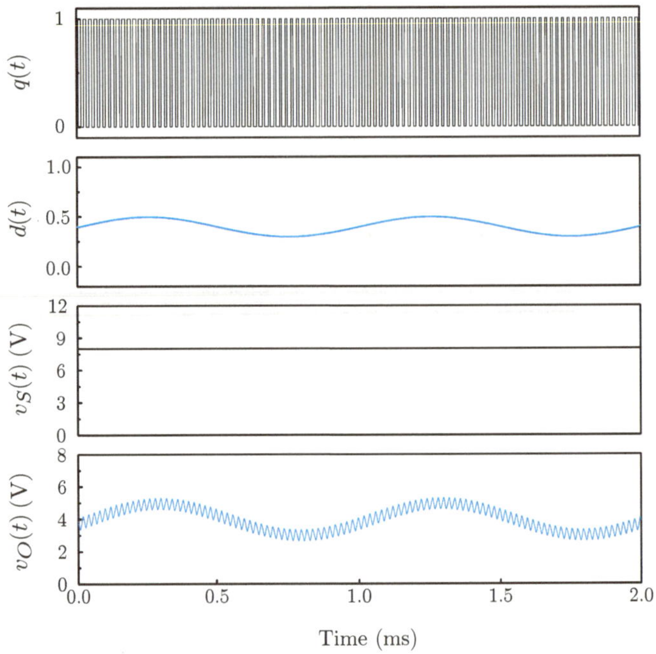

Figure 5.19 Output voltage response to sinusoidal change in continuous duty ratio.

in the switch drive signal has occurred in a sinusoidal fashion so that the continuous duty ratio $d(t)$ is expressed as

$$d(t) = D + \hat{d}(t) \tag{5.47}$$

with

$$\boxed{\hat{d}(t) = d \sin \omega_d t} \tag{5.48}$$

Under this circumstance, the output voltage shows an ac sinusoidal excursion from the dc value, along with the switching ripple. When the switching ripple is ignored, the output voltage can also be represented by the sum of dc and ac components

$$v_O(t) = V_O + \hat{v}_o(t) \tag{5.49}$$

with

$$\boxed{\hat{v}_o(t) = v_o \sin(\omega_d t + \theta_d)} \tag{5.50}$$

The sinusoidal response between the continuous duty ratio and output voltage is defined as

- magnitude response at $\omega_d = \dfrac{|\hat{v}_o(t)|}{|\hat{d}(t)|} = \dfrac{v_o}{d}$
- phase response at $\omega_d = \angle\hat{v}_o(t) - \angle\hat{d}(t) = \theta_d$

The sinusoidal response from the load current to the output voltage is defined in the same manner.

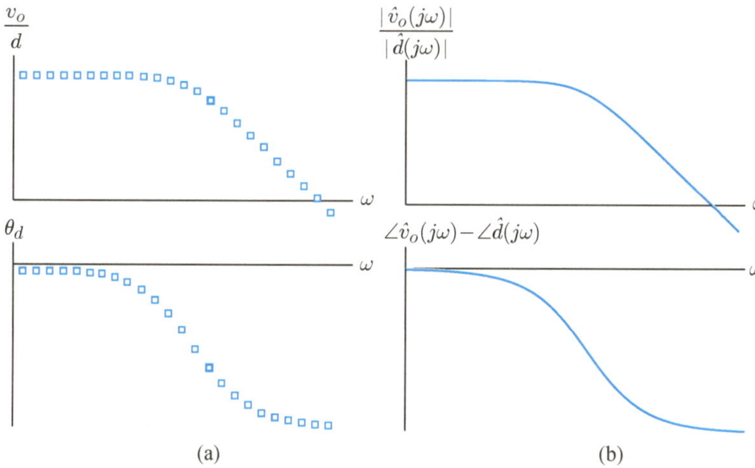

Figure 5.20 Sinusoidal response and frequency response. (a) Sinusoidal response. (b) Frequency response.

5.4.2 Frequency Response and s-Domain Small-Signal Model

The sinusoidal response of the power stage is now extended to the frequency response. Figure 5.20 exemplifies the duty ratio-to-output sinusoidal response and its extension to the frequency response. The magnitude and phase responses are evaluated at discrete frequencies and individually displayed on two separate plots, as shown in Fig. 5.20(a).

When the distance between the evaluation frequencies becomes vanishingly small, two continuous curves will be produced, as shown in Fig. 5.20(b). These two curves are referred to as the magnitude response curve and phase response curve. The magnitude and phase response curves collectively constitute the frequency response plot. The frequency response plot is usually displayed in the Bode plot format. Details about the Bode plot representation are given in Chapter 6.

The frequency response plot can be constructed analytically. The vital step in constructing the frequency response plot is the knowledge about the s-domain transfer functions of power stage. Once the s-domain transfer functions are known, the frequency response plots are created by evaluating the magnitude and phase of the transfer functions with the substitution of $s = j\omega$ and plotting the results in the Bode plot format.

The significant value of the s-domain small-signal model of the converter power stage lies in the the generation of the power stage transfer functions. The s-domain transfer functions are derived using conventional circuit analysis techniques. The resulting transfer functions are then converted into Bode plots to portray the frequency response of the power stage.

The frequency response carries the complete information about the power stage dynamics, as is the case with all linear time-invariant systems. The frequency response

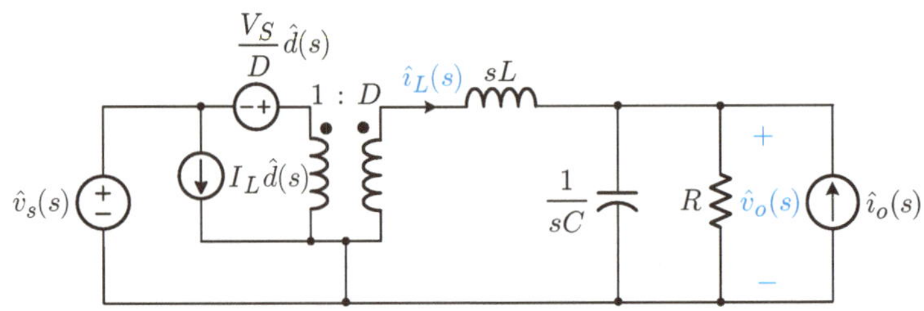

Figure 5.21 s-domain small-signal model of buck converter.

reveals the static and dynamic behaviors of the power stage and offers the foundation of the feedback controller design. Construction of the frequency response plots and analysis of the frequency response of PWM dc-to-dc converters will be covered in Chapter 6.

EXAMPLE 5.8 Power Stage Transfer Functions of Buck Converter

This example shows the evaluation of the frequency response of a converter power stage using its s-domain small-signal model. Figure 5.21 shows the s-domain small-signal model of an ideal buck converter. Considering the small-signal output voltage as the output variable, three power stage transfer functions are defined as follows:

- $G_{vs}(s) \equiv \left.\frac{\hat{v}_o(s)}{\hat{v}_s(s)}\right|_{\hat{d}(s)=\hat{\imath}_o(s)=0}$: input-to-output transfer function
- $G_{vd}(s) \equiv \left.\frac{\hat{v}_o(s)}{\hat{d}(s)}\right|_{\hat{v}_s(s)=\hat{\imath}_o(s)=0}$: duty ratio-to-output transfer function
- $Z_p(s) \equiv \left.\frac{\hat{v}_o(s)}{\hat{\imath}_o(s)}\right|_{\hat{v}_s(s)=\hat{d}(s)=0}$: load current-to-output transfer function

Expressions for these transfer functions are readily determined from the s-domain small-signal model in Fig. 5.21. The input-to-output transfer function is given by

$$G_{vs}(s) = \left.\frac{\hat{v}_o(s)}{\hat{v}_s(s)}\right|_{\hat{d}(s)=\hat{\imath}_o(s)=0} = D\frac{\frac{1}{sC} \parallel R}{sL + \frac{1}{sC} \parallel R} = \frac{D}{1 + \frac{s}{Q\omega_o} + \frac{s^2}{\omega_o^2}} \tag{5.51}$$

with

$$Q = R\sqrt{\frac{C}{L}} \tag{5.52}$$

and

$$\omega_o = \frac{1}{\sqrt{LC}} \tag{5.53}$$

The duty ratio-to-output transfer function, $G_{vd}(s)$, is found by evaluating $\hat{v}_o(s)/\hat{d}(s)$ with the condition $\hat{v}_s(s) = \hat{\imath}_o(s) = 0$

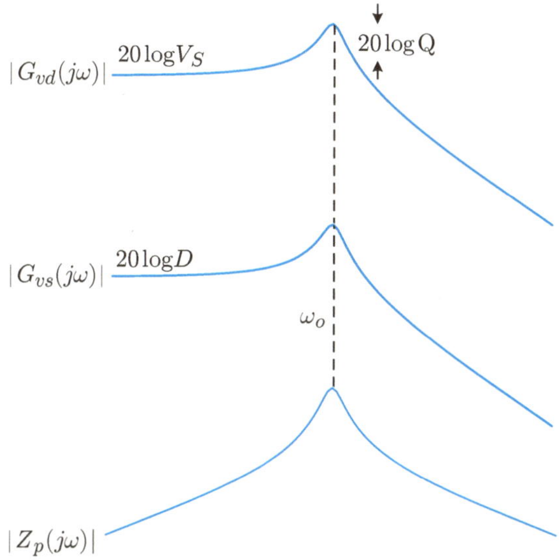

Figure 5.22 Bode plots of power stage transfer functions of buck converter.

$$G_{vd}(s) = \left.\frac{\hat{v}_o(s)}{\hat{d}(s)}\right|_{\hat{v}_s(s)=\hat{i}_o(s)=0} = \frac{V_S}{D} D \frac{\frac{1}{sC} \parallel R}{sL + \frac{1}{sC} \parallel R} = \frac{V_S}{1 + \frac{s}{Q\omega_o} + \frac{s^2}{\omega_o^2}} \tag{5.54}$$

The load current-to-output transfer function becomes

$$Z_p(s) = \left.\frac{\hat{v}_o(s)}{\hat{i}_o(s)}\right|_{\hat{v}_s(s)=\hat{d}(s)=0} = sL \parallel \frac{1}{sC} \parallel R = \frac{sL}{1 + \frac{s}{Q\omega_o} + \frac{s^2}{\omega_o^2}} \tag{5.55}$$

The transfer functions can be converted into Bode plots to reveal frequency response characteristics of the power stage. Figure 5.22 shows the magnitude response curve of power stage transfer functions of the ideal buck converter. Details about the Bode plot construction will be given in the next chapter.

5.5 GENERALIZATION OF POWER STAGE MODELING

In the earlier sections, the power stage modeling has been discussed only for the three basic non-isolated PWM converter operating in continuous conduction mode (CCM) with ideal circuit components. Relatively simple modeling procedures are applied to the buck, boost, and buck/boost converters operating in CCM. These modeling procedures need to be extended to include the following general cases:

- the cases where the reactive circuit components have parasitic resistances,
- the cases where dc-to-dc converters operate in discontinuous condition mode (DCM), and
- the cases where other isolated PWM converters are considered.

This section now starts with the impacts of the parasitic resistances on the power stage modeling. Considerations for the DCM operation and isolated PWM converters will be covered in succeeding sections.

5.5.1 Converter Modeling with Parasitic Resistances †

This section presents the power stage modeling at the presence of the parasitic resistances. Although the derivations are lengthy, the results are rather simple.

As mentioned earlier, linear circuit components remain invariant during both the averaging and linearization processes. Accordingly, the parasitic resistances of reactive components can directly be included in the average or small-signal model of the converter power stage. In Example 5.6, the equivalent series resistance (esr) of the inductor was added to the average power stage model of the boost converter and the resulting model was used to find the dc voltage gain of the converter.

Unlike the esr of the inductor, which can be added to the average or small-signal model without affecting other circuit components, the esr of the capacitor could alter the circuit waveforms of the PWM switch. Accordingly, the average and small-signal models of the PWM switch need to be modified at the existence of the esr of the capacitor, even though the esr itself can be added at its original location.

The effects of the esr of the capacitor will be covered in this section, yet it is now mentioned that the consequence of this modification is negligible for most situations. Correspondingly, it can broadly be considered that the esrs of the reactive components do not cause any practical changes to the models. Therefore, all the parasitic resistances can be included in the average or small-signal model at their original place without causing any notable error. As the justification of this statement, we proceed with the investigation on the impacts of esr.

Buck Converter with Ideal Voltage Source

The buck converter powered by an ideal voltage source is one special case where the parasitic resistances *do not* affect the modeling process. Figure 5.23(a) shows the circuit diagram of a buck converter where the inductor and capacitor both contain a parasitic resistance. Figure 5.23(b) illustrates the circuit waveforms of the PWM switch. It is evident from Fig. 5.23(b) that the equations for averaged circuit variables

† This section is included for the completeness of the power stage modeling. Readers could skip this section and move forwards Section 5.5.2 with no loss in continuity.

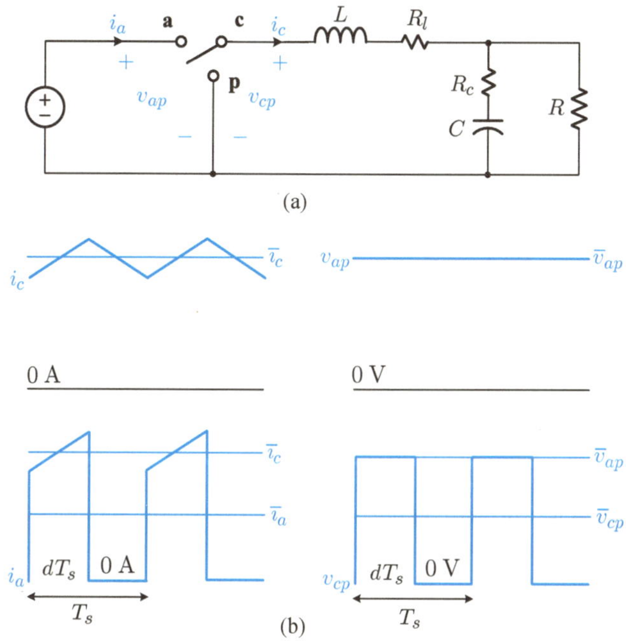

Figure 5.23 Buck converter with ideal voltage source. (a) Circuit diagram. (b) PWM switch waveforms.

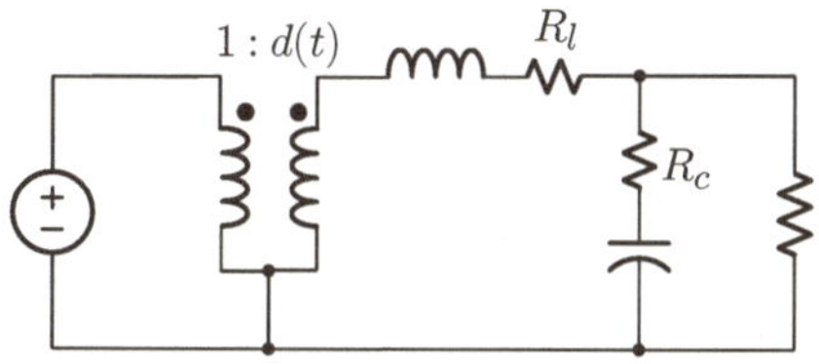

Figure 5.24 Average model of buck converter.

remain the same as those of Section 5.2.2

$$
\begin{aligned}
\bar{i}_a(t) &= d\,\bar{i}_c(t) \\
\bar{v}_{cp}(t) &= d\,\bar{v}_{ap}(t)
\end{aligned}
$$

regardless of the presence of the parasitic resistances. Accordingly, the parasitic resistances can be included as add-ons to the average model. Figure 5.24 shows the average model of the buck converter where the parasitic resistances of the inductor and capacitor are both included.

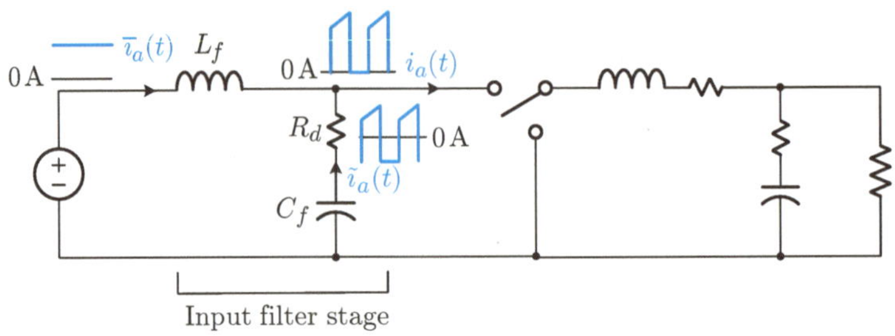

Figure 5.25 Buck converter with input filter.

Buck Converter with Input Filter

As shown in Fig. 5.23, the input current of the buck converter is a pulsating discontinuous current. If the pulsating input current is directly drawn from the voltage source, substantial harmonic current components circulate through the voltage source. The harmonic current components in turn produce excessive conducted electromagnetic interference (EMI), thereby failing to meet regulatory EMI standards. To avoid such a situation, a filter stage is placed between the voltage source and the converter's power stage. This filter stage is called the input filter. The input filter stage locally passes the ac component of the input current so that only the smoothly-filtered dc current flows out of the voltage source.

The operation of an input filter is illustrated in Fig. 5.25, where the reactive filter components, L_f and C_f, are sufficiently large. The voltage source delivers the dc component of the input current, $\bar{i}_a(t)$, while the R_d-C_f branch of the input filter carries the ac component of the input current, $\tilde{i}_a(t) = i_a(t)-\bar{i}_a(t)$. The resistance R_d provides appropriate damping for the filter stage.

The PWM switch waveforms with an input filter stage are shown in Fig. 5.26. As shown in Fig. 5.26(b), the input filter stage does not alter the average current equation of the PWM switch: $\bar{i}_a(t) = d\,\bar{i}_c(t)$. However, the voltage equation is changed by the input filter stage. The voltage across the the active-passive terminal, v_{ap}, is divided into two components: the voltage across the filter capacitor, v_{C_f}, and the voltage across the damping resistor (in the opposite polarity to v_{C_f}), v_{R_d}. The sufficiently large filter capacitor supports the average value or dc portion of v_{ap}

$$v_{C_f}(t) = \bar{v}_{ap}(t) \tag{5.56}$$

On the other hand, the damping resistor voltage, v_{R_d}, is an ac voltage developed by the ac portion of $i_a(t)$

$$v_{R_d}(t) = \big(i_a(t) - \bar{i}_a(t)\big)R_d = \tilde{i}_a(t)R_d \tag{5.57}$$

The voltage v_{R_d} is a scaled replica of $i_a(t)$ with zero average value. The voltage waveforms v_{C_f} and v_{R_d} are illustrated in Fig. 5.26(c), where the average voltage swing of v_{R_d}, Δv_{R_d}, is defined as

$$\Delta v_{R_d} = \bar{i}_c(t)R_d \tag{5.58}$$

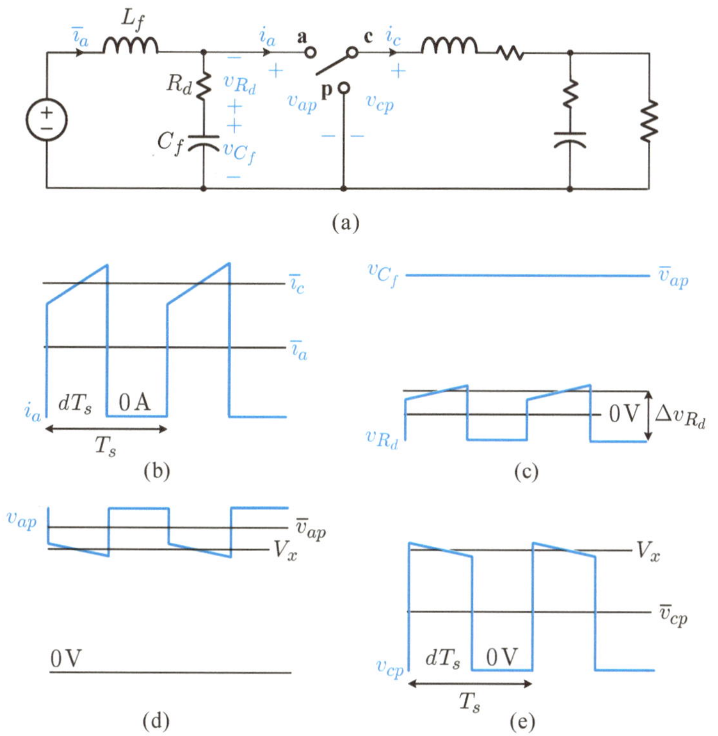

Figure 5.26 Buck converter with input filter stage. (a) Circuit diagram. (b) – (e) Circuit waveforms of PWM switch.

The active-passive terminal voltage v_{ap} is now expressed as

$$v_{ap}(t) = v_{C_f}(t) - v_{R_d}(t) = \bar{v}_{ap}(t) - \tilde{i}_a(t)R_d \tag{5.59}$$

This waveform is shown in Fig. 5.26(d), where another dc voltage level, V_x, is defined to facilitate the derivation of the voltage equation. Lastly, the voltage across the common and passive terminals, $v_{cp}(t)$, is illustrated in Fig. 5.26(e).

It is evident from Fig. 5.26(e) that the average value of v_{cp} is given by

$$\bar{v}_{cp}(t) = d\,V_x \tag{5.60}$$

The above equation differs from the previous voltage equation of $\bar{v}_{cp}(t) = d\,\bar{v}_{ap}(t)$ because $V_x \neq \bar{v}_{ap}(t)$. Now the remaining task is to find the expression for V_x, which is done using the detailed v_{ap} waveform in Fig. 5.27. From the graphical construction, it is recognized that the areas of the two shaded rectangles in Fig. 5.27 should be the same

$$(\bar{v}_{ap}(t) - V_x)d\,T_s = (V_x + \Delta v_{R_d} - \bar{v}_{ap}(t))d'T_s \tag{5.61}$$

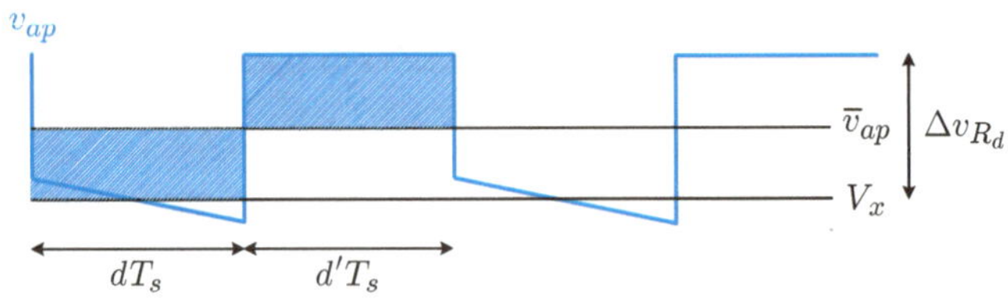

Figure 5.27 Active-passive terminal voltage waveform.

Thus, the expression for V_x is given by

$$V_x = \bar{v}_{ap}(t) - \Delta v_{R_d} d' = \bar{v}_{ap}(t) - R_d \bar{\imath}_c(t) d' \tag{5.62}$$

Using the notation of $R_d = r_{ap}$, the above expression is rewritten as

$$V_x = \bar{v}_{ap}(t) - r_{ap} \bar{\imath}_c(t) d' \tag{5.63}$$

The symbol r_{ap} signifies the ac resistance between the active and passive terminals, which can be found by opening L_f and shorting C_f. The final averaged voltage equation is then given by

$$\boxed{\bar{v}_{cp}(t) = dV_x = d\big(\bar{v}_{ap}(t) - r_{ap}\bar{\imath}_c(t)d'\big)} \tag{5.64}$$

Linearization of Averaged PWM Switch Equation

The results of the previous analysis are generalized for the three basic converters shown in Fig. 5.28. For all the three converters, the averaged circuit equations are given by

$$\boxed{\begin{aligned} \bar{\imath}_a(t) &= d\,\bar{\imath}_c(t) \\ \bar{v}_{cp}(t) &= d\left(\bar{v}_{ap}(t) - r_{ap}\bar{\imath}_c(t)d'\right) \end{aligned}} \tag{5.65}$$

where

$$\boxed{r_{ap} = \begin{cases} 0 & \text{for buck converter without input filter} \\ R_d & \text{for buck converter with input filter} \\ R_c \parallel R & \text{for boost and buck/boost converters} \end{cases}}$$

The ac resistance between the active-passive terminal, r_{ap}, is identified from Fig. 5.28. The r_{ap} carries the ac component of the active terminal current, $\tilde{\imath}_a(t) = i_a(t) - \bar{\imath}_a(t)$, as shown in Fig. 5.28. For the standalone buck converter, it is obvious that $r_{ap} = 0$.

Application of the linearization process to (5.65) yields

$$\begin{aligned} I_a + \hat{\imath}_a(t) &= (D + \hat{d}(t))(I_c + \hat{\imath}_c(t)) \\ V_{cp} + \hat{v}_{cp}(t) &= (D + \hat{d}(t)) \\ &\quad \Big((V_{ap} + \hat{v}_{ap}(t)) - r_{ap}(I_c + \hat{\imath}_c(t))(1 - (D + \hat{d}))\Big) \end{aligned} \tag{5.66}$$

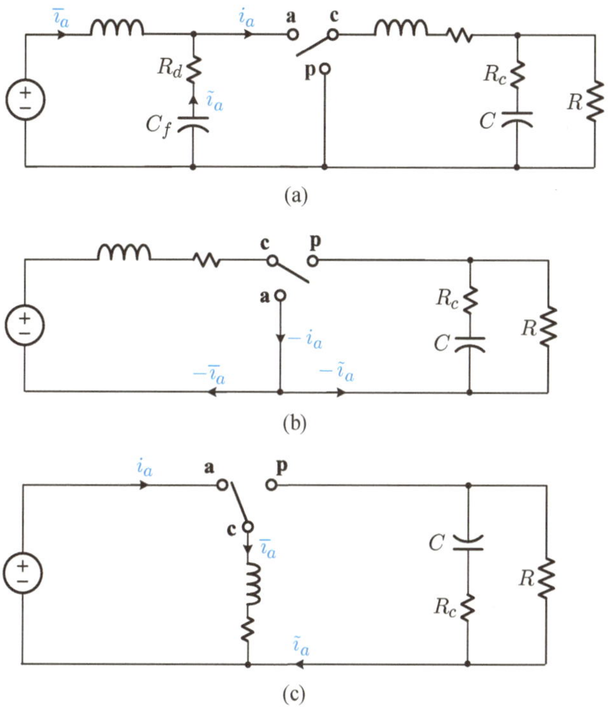

Figure 5.28 Circuit diagram of three basic converters. (a) Buck converter with input filter. (b) Boost converter. (c) Buck/boost converter.

By equating the first-order ac components from (5.66), the small-signal representation of (5.65) is obtained

$$\boxed{\begin{aligned} \hat{i}_a(t) &= D\,\hat{i}_c(t) + I_c\,\hat{d}(t) \\ \hat{v}_{cp}(t) &= D\,\hat{v}_{ap}(t) + V_D\,\hat{d}(t) - \hat{i}_c(t)\,DD'\,r_{ap} \end{aligned}} \tag{5.67}$$

with

$$\boxed{V_D = V_{ap} + I_c(D - D')r_{ap}} \tag{5.68}$$

The expressions for V_{ap} and I_c are given in Section 5.3.3. The circuit model for the above small-signal equations is given in Fig. 5.29. With $r_{ap} = 0$, the circuit model reduces to the previous small-signal model which does not incorporate the effects of parasitic resistances.

Predictions of Refined Small-Signal Model

As shown in (5.67) and (5.68), and Fig. 5.29, the ac resistance between the active-passive terminal, r_{ap}, alters the expression of the dependent voltage source and introduces an additional resistance to the small-signal model of the PWM switch.

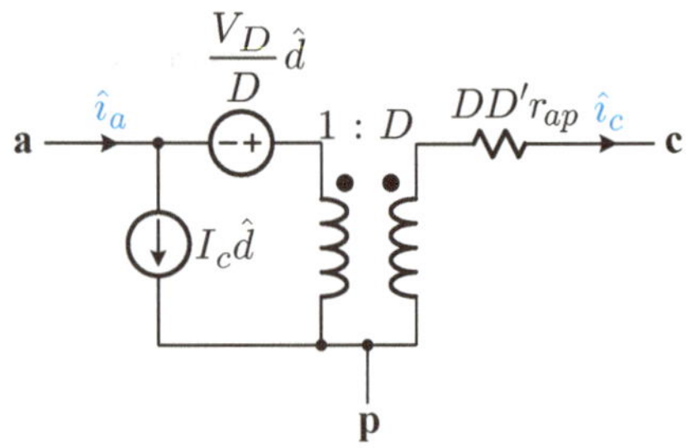

Figure 5.29 Refined small-signal model for PWM switch.

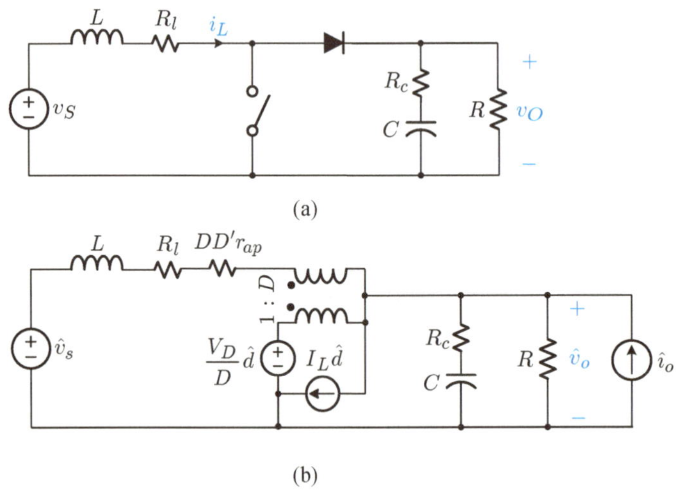

Figure 5.30 Boost converter. (a) Circuit diagram. (b) Small-signal model.

These changes only influence the dc gain and quality factor of transfer functions and their consequential effects will not be substantial unless the ac resistance r_{ap} is unusually large. Accordingly, the transfer functions of the refined model will largely be the same as those of the previous model which ignores the impact of r_{ap}.

EXAMPLE 5.9 Boost Converter Example

This example shows the prediction of the refined small-signal model. Figure 5.30 shows a boost converter and its small-signal model. In the small-signal model, the ac resistance r_{ap} is given by

$$r_{ap} = R_c \parallel R$$

and the dc voltage source V_D is expressed as

$$V_D = -V_O - I_L(D - D')r_{ap}$$

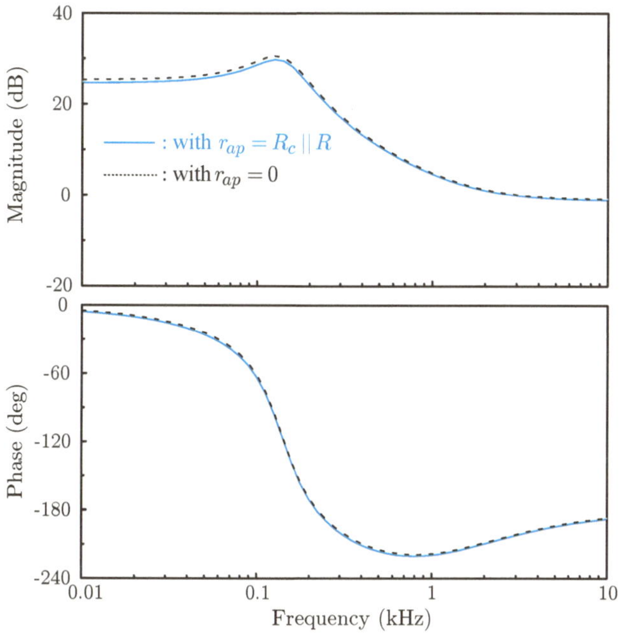

Figure 5.31 Duty ratio-to-output transfer function of boost converter.

The operational conditions and circuit parameters of the boost converter are $V_S = 15$ V, $L = 800$ μH, $R_l = 0.01$ Ω, $C = 1000$ μF, $R_c = 0.1$ Ω, $R = 2$ Ω, $f_s = 10$ kHz, and $D = 0.25$. Figure 5.31 compares the duty ratio-to-output transfer functions of the converter, obtained from PSpice®simulations using Fig. 5.30(b). One simulation is with $r_{ap} = R \parallel R_c$ and the other simulation is with $r_{ap} = 0$, that is, ignoring the consequence of r_{ap}. The difference between the transfer functions is nearly undetectable.

As demonstrated in the previous example, the impacts of r_{ap} are negligible for most cases. Thus, r_{ap} can be ignored in the PWM switch model without causing any notable error. In this book, the earlier small-signal model without r_{ap} will be used in theoretical discussions. Nonetheless, the refined model with r_{ap} in Fig. 5.29 will be employed to all forthcoming PSpice®simulations of transfer functions.

5.5.2 Modeling PWM Converters in DCM Operation

The operation of practical dc-to-dc converters usually spans both CCM and DCM regions. As a dc-to-dc converter departs from CCM operation and enters DCM operation, its small-signal equations will be altered, thereby requiring a new small-signal model. This section deals with the converter modeling under DCM operation.

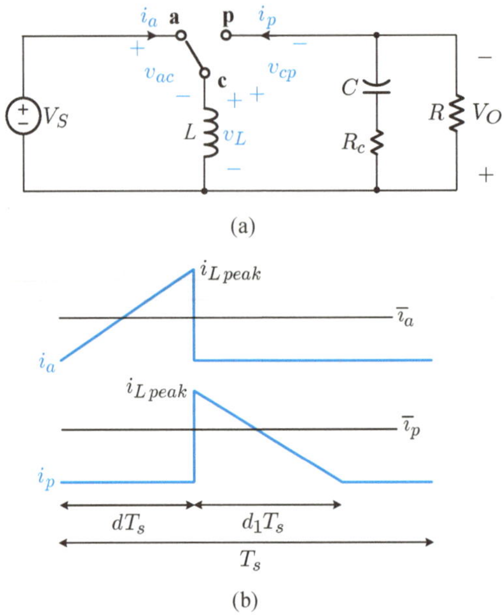

Figure 5.32 Buck/boost converter and DCM waveforms of PWM switch. (a) Buck/boost converter. (b) PWM switch waveforms in DCM.

Averaged Equations for PWM Switch in DCM

The buck/boost converter is selected to illustrate the modeling of DCM dynamics. However, the modeling results are invariant with the selected converter topology and thus can be extended to all the three basic dc-to-dc converters. The buck/boost converter circuit and its current waveforms in DCM operation are shown in Fig. 5.32. In Fig. 5.32(a), the following relationships hold

$$\begin{aligned} \bar{v}_{ac}(t) &= V_S \\ \bar{v}_{cp}(t) &= V_O \end{aligned} \tag{5.69}$$

because the average value of the inductor voltage is zero, $\bar{v}_L = 0$, owing to the volt-sec balance condition.

The current waveforms of the PWM switch in Fig. 5.32(b) indicate that

$$\begin{aligned} \bar{i}_a(t) &= \frac{1}{2}\frac{i_{Lpeak}\,dT_s}{T_s} = \frac{i_{Lpeak}}{2}d \\ \bar{i}_p(t) &= \frac{1}{2}\frac{i_{Lpeak}\,d_1T_s}{T_s} = \frac{i_{Lpeak}}{2}d_1 \end{aligned} \tag{5.70}$$

where i_{Lpeak} is the peak value of the inductor current, dT_s is the on-time period and d_1T_s is the portion of the off-time period in which the inductor current exists.

Expressions (5.70) yield the averaged current equation

$$\bar{\imath}_a(t) = \frac{d}{d_1}\bar{\imath}_p(t) \tag{5.71}$$

Two different expressions for i_{Lpeak} are formulated from Fig. 5.32

$$\begin{aligned} i_{Lpeak} &= \frac{V_S}{L}dT_s = \frac{\bar{v}_{ac}(t)}{L}dT_s \\ i_{Lpeak} &= \frac{V_O}{L}d_1T_s = \frac{\bar{v}_{cp}(t)}{L}d_1T_s \end{aligned} \tag{5.72}$$

to produce the averaged voltage equation for the PWM switch

$$\bar{v}_{ac}(t) = \frac{d_1}{d}\bar{v}_{cp}(t) \tag{5.73}$$

Using (5.70) and (5.72), the parameter d_1 is expressed in terms of the average circuit variables and operational conditions

$$d_1 = \frac{2Lf_s}{d}\frac{\bar{\imath}_a(t)}{\bar{v}_{cp}(t)} = \frac{2Lf_s}{d}\frac{\bar{\imath}_p(t)}{\bar{v}_{ac}(t)} \tag{5.74}$$

where $f_s = 1/T_s$ is the switching frequency. Expressions (5.71), (5.73), and (5.74) are combined together to establish a set of equations that describe the averaged dynamics of the PWM switch in DCM operation

$$\boxed{\begin{aligned} \bar{\imath}_a(t) &= \mu\,\bar{\imath}_p(t) \\ \bar{v}_{cp}(t) &= \mu\,\bar{v}_{ac}(t) \end{aligned}} \tag{5.75}$$

with

$$\boxed{\mu = \frac{d}{d_1} = \frac{d^2\,\bar{v}_{cp}(t)}{2Lf_s\bar{\imath}_a(t)} = \frac{d^2\,\bar{v}_{ac}(t)}{2Lf_s\bar{\imath}_p(t)}} \tag{5.76}$$

The above equations are to be linearized to yield the small-signal equations.

Linearization of Averaged Equation and Small-Signal Circuit Model

By combining (5.75) and (5.76) and linearizing the resulting equations, a set of small-signal equations are obtained [4]

$$\boxed{\begin{aligned} \hat{\imath}_a(t) &= \frac{1}{r_i}\hat{v}_{ac}(t) + k_i\hat{d}(t) \\ \hat{\imath}_p(t) &= g_f\,\hat{v}_{ac}(t) + k_o\hat{d}(t) + \frac{1}{r_o}\hat{v}_{pc}(t) \end{aligned}} \tag{5.77}$$

with

$$r_i = \frac{V_{ac}}{I_a} \quad k_i = \frac{2I_a}{D} \quad g_f = \frac{2I_p}{V_{ac}} \quad k_o = \frac{2I_p}{D} \quad \text{and} \quad r_o = \frac{V_{cp}}{I_p} \tag{5.78}$$

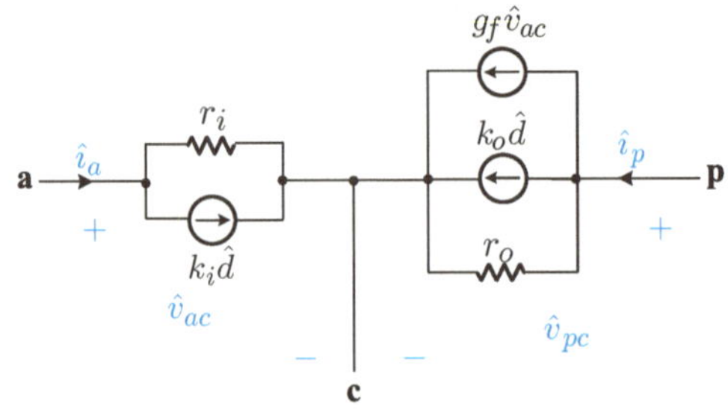

Figure 5.33 Small-signal model for PWM switch in DCM operation.

The above equations become the small-signal equation of the PWM switch in DCM operation. Derivations of (5.77) and (5.78) are discussed in Problem 5.13 at the end of this chapter.

A circuit representation of the small-signal equations (5.77) is shown in Fig. 5.33. This circuit model is referred to as the *DCM PWM switch model.* The DCM PWM switch model includes two resistive parameters, r_i and r_o. These resistive parameters dictate the small-signal dynamics of the model, as will be demonstrated in Section 6.3.1. Now, DCM small-signal models for the three basic converters are obtained by replacing the PWM switch with the DCM PWM switch model. As an example, Fig. 5.34 shows a buck/boost converter and its DCM small-signal model. Details about the DCM small-signal models and their use in DCM dynamics analysis will be given in Section 6.3.1.

5.5.3 Modeling Isolated PWM Converters

Small-signal models for the three basic non-isolated PWM converters were derived using the concept of the PWM switch. The modeling procedures for non-isolated converters converters are straightforward because each converter contains the PWM switch in its original structure. This section presents the extension of the modeling procedures to isolated dc-to-dc converters.

Isolated PWM converters contain a number of active and passive switches, which are separated by an isolation transformer. Thus, the PWM switch structure is not obvious in isolated converters. However, the operation of isolated converters reveals that the active and passive switches, even though located in distance with isolation, collectively execute the function of the PWM switch. Accordingly, the PWM switch can still be used as an effective tool in developing the dynamic models for isolated converters.

As demonstrated in Chapter 4, each isolated converter has a forerunning non-isolated converter from which that isolated converter is evolved. The modeling of an isolated converter is facilitated by using the existing model of the forerunning non-isolated converter. For example, the models of buck-derived dc-to-dc converters can readily be derived from the model of the buck converter. Similarly, the flyback converter can conveniently be modeled using the model of the buck/boost converter.

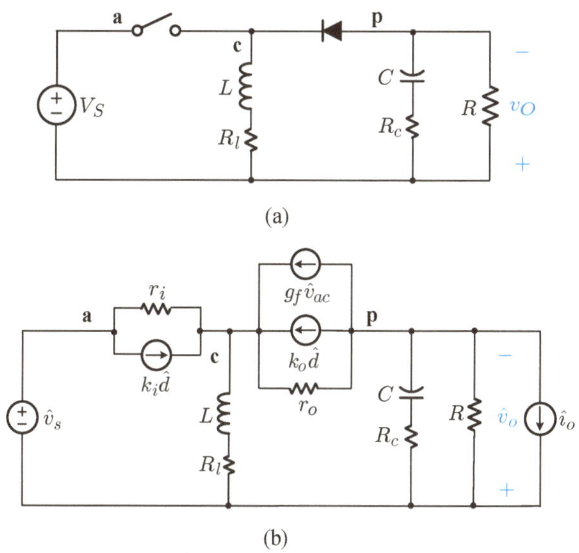

Figure 5.34 Buck/boost converter and DCM small-signal model. (a) Buck/boost converter. (b) DCM small-signal model.

Modeling Forward Converter and Bridge-Type Converters

The modeling of buck-derived isolated converters is illustrated using the forward converter. Figure 5.35 shows the modeling procedures of the tertiary-winding reset forward converter. The original circuit of the forward converter in Fig. 5.35(a) is modified to the functional model in Fig. 5.35(b) by recognizing the following two facts.

1) The reset circuit, only employed to reset the magnetizing inductance of the isolation transformer, does not interfere with the converter dynamics. Thus, the reset circuit, along with the magnetizing inductance itself, can be removed for the purpose of modeling.

2) As far as the power stage operation is concerned, the forward converter is a functional equivalent of the buck converter combined with an isolation transformer upstream. The turns ratio of the isolation transformer is $1:n$ with $n = N_S/N_P$.

The presence of the PWM switch is apparent in the functional model of Fig. 5.35(b). The average model of the forward converter is constructed as shown in Fig. 5.35(c) by replacing the PWM switch with its average model. The two ideal transformers in Fig. 5.35(c), the $1:n$ isolation transformer and the $1:d$ transformer in the PWM switch, are merged into a single $1:nd$ transformer in the simplified average model of Fig. 5.35(d). A set of averaged equations are written from Fig. 5.35(d)

$$\begin{aligned} \bar{i}_a(t) &= nd\,\bar{i}_c(t) \\ \bar{v}_{cp}(t) &= nd\,\bar{v}_{ap}(t) \end{aligned} \tag{5.79}$$

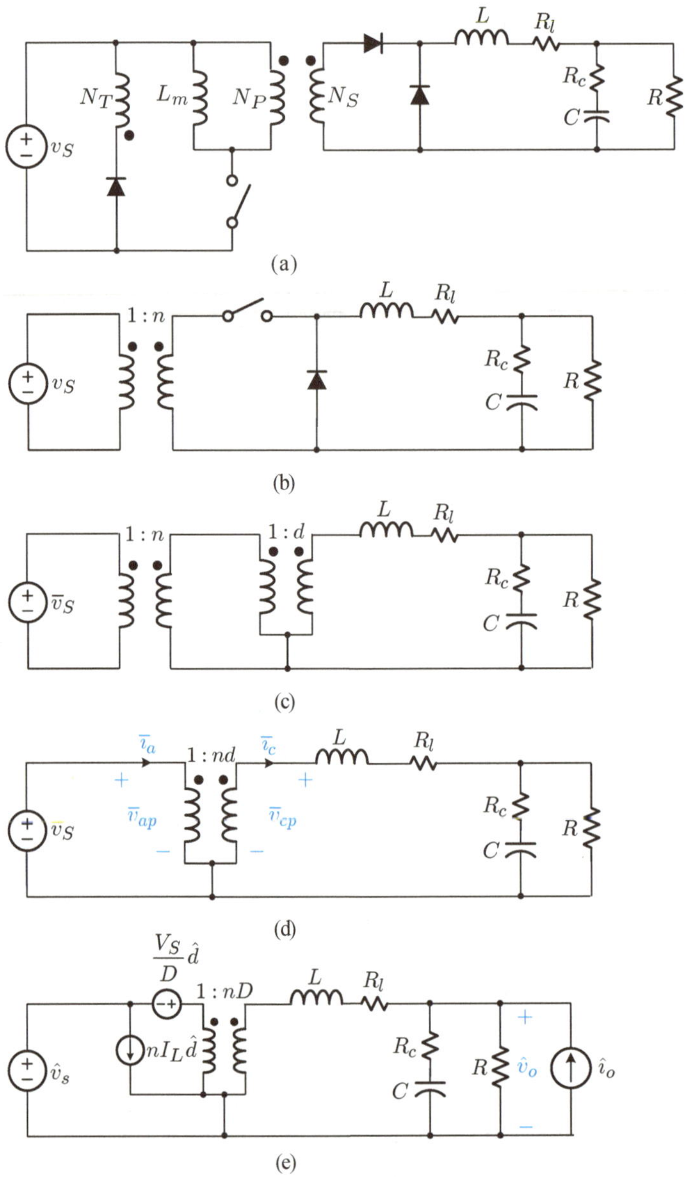

Figure 5.35 Modeling of tertiary-winding reset forward converter. (a) Original circuit. (b) Functional model. (c) Average model. (d) Simplified average model. (e) Small-signal model.

By linearizing (5.79), the small-signal expressions are obtained

$$\begin{array}{rcl} \hat{i}_a(t) & = & n\big(I_c\hat{d}(t) + D\,\hat{i}_c(t)\big) \\ \hat{v}_{cp}(t) & = & n\big(D\hat{v}_{ap}(t) + V_{ap}\hat{d}(t)\big) \end{array} \tag{5.80}$$

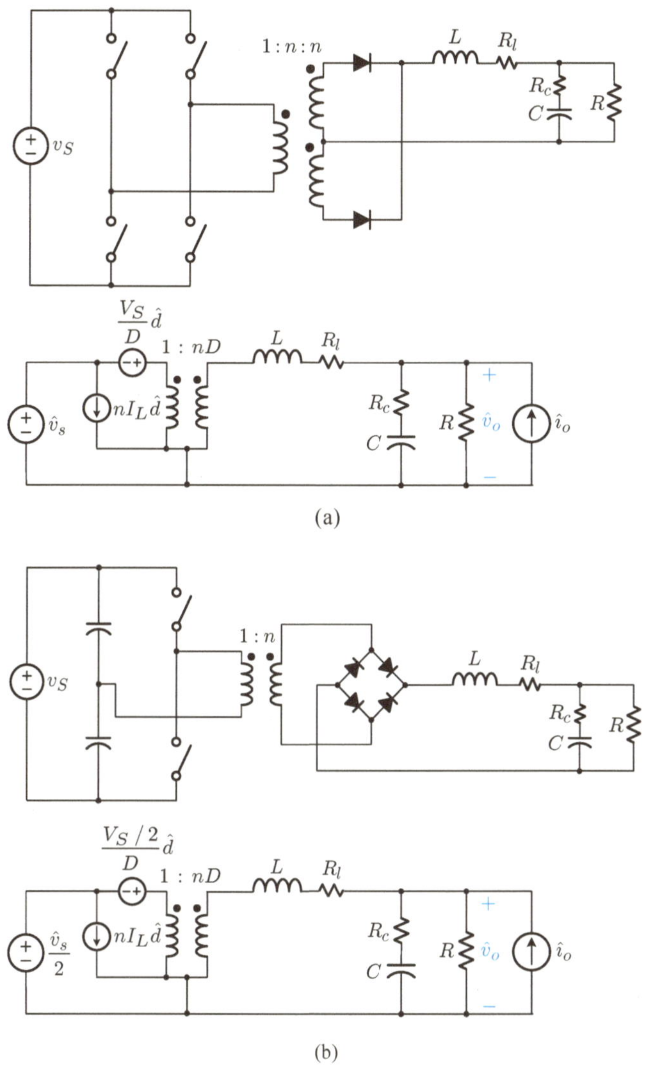

Figure 5.36 Modeling of bridge-type buck-derived converters. (a) Full-bridge converter. (b) Half-bridge converter.

With the incorporation of $I_c = I_L$ and $V_{ap} = V_S$, the final small-signal circuit model is derived as shown in Fig. 5.35(e).

The small-signal model of Fig. 5.35(e) is also valid for other types of forward converters including the two-switch forward converter. Furthermore, the preceding procedures are applicable to other buck-derived isolated converters. Figure 5.36(a) shows the full-bridge converter and its small-signal model, while Fig. 5.36(b) depicts those of the half-bridge converter.

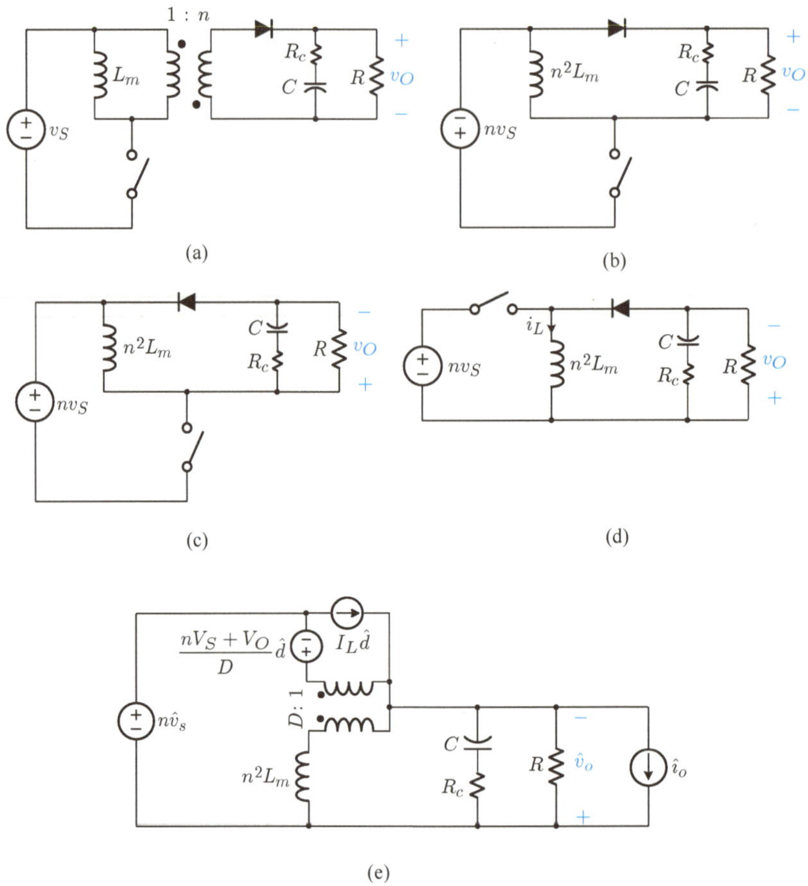

Figure 5.37 Modeling of flyback converter. (a) Original flyback converter. (b) Modified model. (c) Modified model. (d) Equivalent buck/boost converter. (e) Small-signal model.

Modeling Flyback Converter

This section deals with the modeling of the flyback converter. For the purpose of modeling, the flyback converter is successively modified as illustrated in Fig. 5.37.

The original circuit of the flyback converter is shown in Fig. 5.37(a). Here, the magnetizing inductance serves as the power stage inductor and thus should be included in the dynamic models. Figure 5.37(a) is altered to Fig. 5.37(b), where the ideal transformer is removed by reflecting the magnetizing inductance and the voltage source into the secondary side. Figure 5.37(b) is modified to Fig. 5.37(c) by changing the polarity of the input and output voltages and reversing the orientation of the diode. The power stage circuit is further modified into Fig. 5.37(d) by relocating the active switch without changing the circuit operation. Figure 5.37(d) is now recognized as a buck/boost converter whose input voltage and inductance are scaled by the turns ratio of the transformer in the flyback converter. Finally, the small-signal model of the flyback converter is obtained as shown in Fig. 5.37(e), by adapting the small-signal model of the buck/boost converter.

5.6 SMALL-SIGNAL GAIN OF PWM BLOCK

The small-signal model of closed-loop PWM converters can be obtained by combining the individual small-signal models of the power stage, PWM block, and voltage feedback compensation. The previous sections provided the small-signal models for the power stage of the three basic PWM converters. This section now presents the small-signal modeling of the PWM block. While the procedure and outcome of the modeling are relatively simple, the PWM block is an essential element of the small-signal model of closed-loop controlled PWM converters.

Figure 5.38 illustrates the operation of the PWM block in a closed-loop controlled PWM converter. The PWM block compares the control signal v_{con} against the ramp signal V_{ramp} to generate the pulsewidth modulated switch drive signal. The PWM output is represented by the switching function $q(t)$, while the cycle-by-cycle duty ratio of the PWM output is depicted as d_{k-1}, d_k, d_{k+1}, and d_{k+2}. From the PWM waveforms shown in thick lines in Fig. 5.38, the following relationship can be noted

$$\boxed{d_k T_s : T_s = v_{con}(t^*) : V_m} \tag{5.81}$$

where d_k denotes the duty ratio in the kth switching cycle, V_m is the height of the ramp signal, and t^* is the time instant at which the ramp signal V_{ramp} intersects the control voltage v_{con}. The relationship (5.81) is rearranged as

$$d_k = \frac{v_{con}(t^*)}{V_m} \tag{5.82}$$

When the control voltage v_{con} varies so *narrowly* that it remains nearly constant within the kth switching period, (5.82) can be approximated as

$$d_k = \frac{v_{con}(t)}{V_m} \tag{5.83}$$

If v_{con} is further restricted to change *slowly* over several switching periods, the cycle-by-cycle variation in the on-time period is small and d_k thus can be approximated by the continuous duty ratio $d(t)$: $d_k \approx d(t)$. The averaged functional description of the PWM block is now obtained from (5.83) by replacing d_k with $d(t)$ and taking the average of v_{con}

$$\boxed{d(t) = \frac{\bar{v}_{con}(t)}{V_m}} \tag{5.84}$$

Application of the linearization process to (5.84) yields

$$D + \hat{d}(t) = \frac{1}{V_m}(V_{con} + \hat{v}_{con}(t)) \tag{5.85}$$

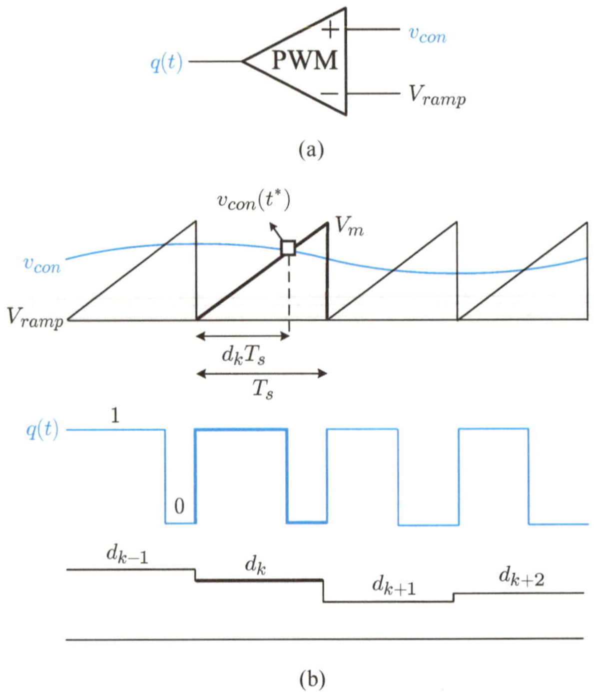

Figure 5.38 PWM block and its major waveforms. (a) PWM block. (b) PWM waveforms.

The small-signal equation of the PWM process is now given by

$$\hat{d}(t) = \frac{1}{V_m}\hat{v}_{con}(t) \tag{5.86}$$

leading to the small-signal gain of the PWM block

$$\boxed{\frac{\hat{d}(s)}{\hat{v}_{con}(s)} \equiv F_m = \frac{1}{V_m}} \tag{5.87}$$

The constant small-signal gain of the PWM block is termed as the *PWM gain* or *modulator gain*, F_m. The PWM gain is given by the inverse of the height of the ramp signal: $F_m = 1/V_m$. This simple result is the outcome of the fundamental assumption that the control voltage v_{con} does not vary widely within one switching period and only changes slowly over several switching periods.

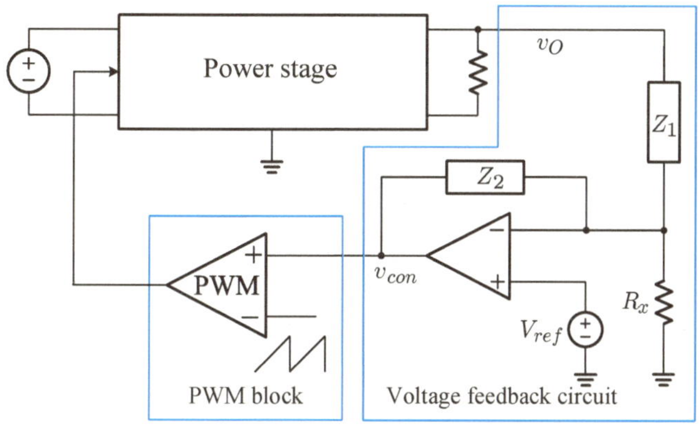

Figure 5.39 PWM converter and voltage feedback circuit.

5.7 UNIVERSAL SMALL-SIGNAL MODEL FOR PWM DC-TO-DC CONVERTERS

This section presents the complete small-signal model of closed-loop controlled PWM converters. Figure 5.39 shows the block diagram representation of a closed-loop controlled PWM converter. The small-signal models for the power stage and PWM block are already developed, as such, the modeling of the entire PWM converter will be completed by incorporating the small-signal characteristics of the voltage feedback circuit. This section first discusses the voltage feedback circuit and later presents a universal small-signal model for all isolated/non-isolated PWM dc-to-dc converters.

5.7.1 Voltage Feedback Circuit

The general structure of the voltage feedback circuit is shown in Fig. 5.39. While the voltage feedback circuit retains the same configuration as that of Section 3.6.1, it has an additional resistance R_x. When R_x is not present, the output voltage is regulated at the reference voltage, $V_O = V_{ref}$. The resistance R_x in the voltage feedback circuit provides a means of controlling the output voltage with a fixed reference voltage V_{ref}. This section first deals with the output voltage control and subsequently describes the small-signal transfer function of the voltage feedback circuit.

Output Voltage Control

Referring to Fig. 5.39, the node equation at the inverting terminal of the op amp is given by

$$\boxed{\frac{v_O - V_{ref}}{Z_1(s)} - \frac{V_{ref}}{R_x} = \frac{V_{ref} - v_{con}}{Z_2(s)}} \tag{5.88}$$

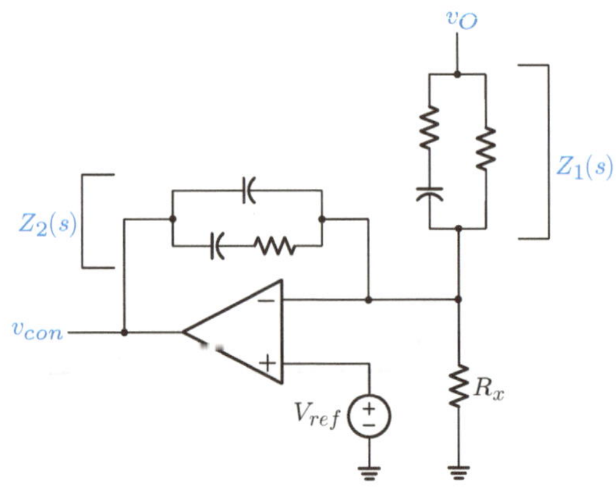

Figure 5.40 Voltage feedback circuit in which $|Z_2(j0)|$ is infinite while $|Z_1(j0)|$ is finite.

which is rearranged as

$$V_{ref} - v_{con} = \frac{Z_2(s)}{Z_1(s)}\left(v_O - V_{ref}\left(1 + \frac{Z_1(s)}{R_x}\right)\right) \tag{5.89}$$

The output voltage control is explained using the right-hand side expression of (5.89). If the magnitude of the impedance ratio $|Z_2(j0)/Z_1(j0)|$ is infinite, the term $V_O - V_{ref}(1+|Z_1(j0)|/R_x)$ should converge to zero, in order to make their product finite.† When $Z_1(s)$ and $Z_2(s)$ are selected such that $|Z_2(j0)| = \infty$, while $|Z_1(j0)|$ is finite, the steady-state output voltage is given by

$$V_O - V_{ref}\left(1 + \frac{|Z_1(j0)|}{R_x}\right) \doteq 0 \quad \Rightarrow \quad \boxed{V_O = V_{ref}\left(1 + \frac{|Z_1(j0)|}{R_x}\right)} \tag{5.90}$$

An example of the voltage feedback circuit, in which $|Z_2(j0)| = \infty$ and $|Z_1(j0)|$ is finite, is shown in Fig. 5.40. In fact, this circuit was introduced in Section 3.6.1 as the voltage feedback circuit for the buck converter.

Equation (5.90) indicates that the output voltage is controlled by varying R_x. While alternative schemes can be employed, the feedback circuit structure in Fig. 5.39 has an advantage over other methods. This advantage will become clear when its small-signal transfer function is analyzed in the next section.

Voltage Feedback Compensation

The transfer function of the voltage feedback circuit is readily found by noting that the inverting terminal of the op amp in Fig. 5.39 is an ac ground. For the small-signal ac analysis, the fixed dc voltage V_{ref} is replaced with a short circuit and the inverting

†Details about this statements were given in Section 3.6.1.

terminal thus becomes a virtual ground for ac signal. With this observation, a node equation is written at the inverting terminal of the op amp

$$\boxed{\frac{\hat{v}_o(s) - 0}{Z_1(s)} = \frac{0 - \hat{v}_{con}(s)}{Z_2(s)}} \tag{5.91}$$

which yields the transfer function of the voltage feedback circuit

$$\frac{\hat{v}_{con}(s)}{\hat{v}_o(s)} = -\frac{Z_2(s)}{Z_1(s)} \equiv -F_v(s) \tag{5.92}$$

with

$$\boxed{F_v(s) = \frac{Z_2(s)}{Z_1(s)}} \tag{5.93}$$

The transfer function $F_v(s)$ is called the *voltage feedback compensation*. Equation (5.93) indicates that the voltage feedback compensation is not affected by the output voltage-controlling resistor R_x. Accordingly, the voltage feedback circuit could control the output voltage without altering its transfer function. This desirable feature may not exist if other structures than Fig. 5.39 are selected for the voltage feedback circuit.

5.7.2 Universal Small-Signal Model for PWM Converters

The small-signal models for all the functional blocks are all prepared to compose a complete small-signal model for PWM dc-to-dc converters. Figure 5.41 shows a general functional diagram of closed-loop controlled PWM dc-to-dc converters. The PWM switch is combined with the inductor and the resulting circuitry is considered as a single composite circuit element. The outward terminal at the inductor side is denoted as **i**. Figure 5.41 could represent any of the three basic PWM converters by using different power stage connections. With the connections $\{\mathrm{a-X\ p-Y\ i-Z}\}$, Fig. 5.41 represents the buck converter. Likewise, the connections $\{\mathrm{i-X\ a-Y\ p-Z}\}$ lead to the boost converter, while the connections $\{\mathrm{a-X\ i-Y\ p-Z}\}$ yield the buck/boost converter.

The small-signal model for the PWM converters can be obtained from Fig. 5.41 by 1) replacing the PWM switch, PWM block, and voltage feedback circuit with their respective small-signal model, and 2) introducing appropriate small-signal excitations. When the continuous condition mode (CCM) operation is considered, the PWM switch is replaced with the PWM switch model to yield the CCM small-signal model of PWM converters. Figure 5.42 shows the CCM small-signal model. The small-signal model includes the parasitic resistances of the reactive components: R_l and R_c in Fig. 5.42. This figure can be arranged to represent the small-signal model of the buck, boost, or buck/boost converter. For example, Fig. 5.42 becomes the small-signal model of the buck converter when the connections $\{\mathrm{a-X\ p-Y\ i-Z}\}$ are assumed.

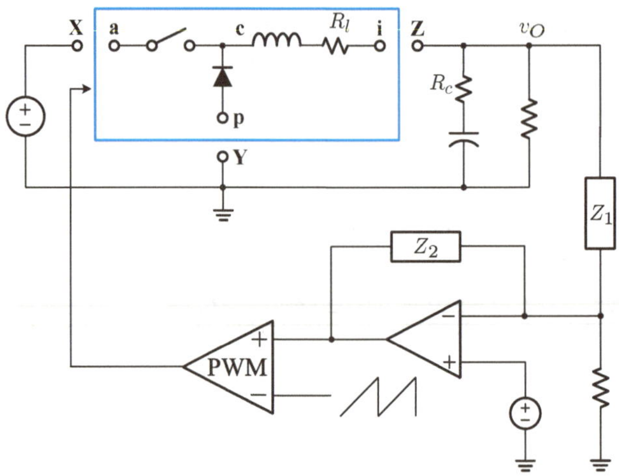

Figure 5.41 General functional diagram for three basic PWM converters.

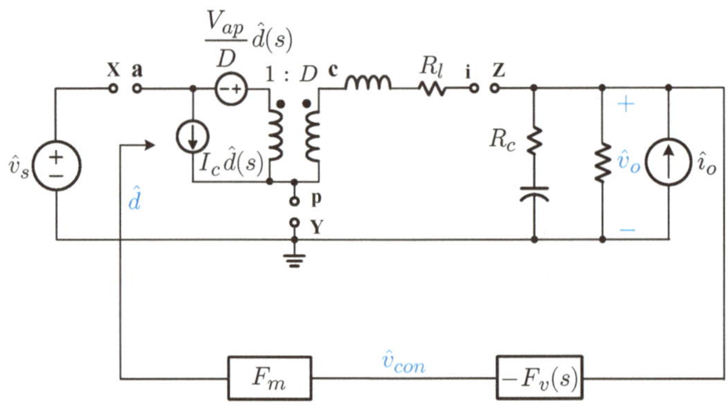

Figure 5.42 Universal CCM small-signal model for PWM converters.

As demonstrated in Section 5.5.3, isolated PWM converters have the same structure as that of the three basic PWM converters, as far as their power stage dynamic models are concerned. Thus, for example, the small-signal model of the tertiary-winding reset forward converter has the structure of Fig. 5.42 with the connections {a−X p−Y i−Z}. Accordingly, Fig. 5.42 serves as a universal CCM small-signal model for all isolated/non-isolated PWM converters

When the PWM switch model in Fig. 5.42 is replaced with the DCM PWM switch model, the DCM small-signal model for PWM converters is produced. Figure 5.43 shows the universal DCM small-signal model for isolated/non-isolated PWM converters.

The small-signal models in Figs. 5.42 and 5.43 are converted into a common block diagram representation shown in Fig. 5.44. In the block diagram, the small-

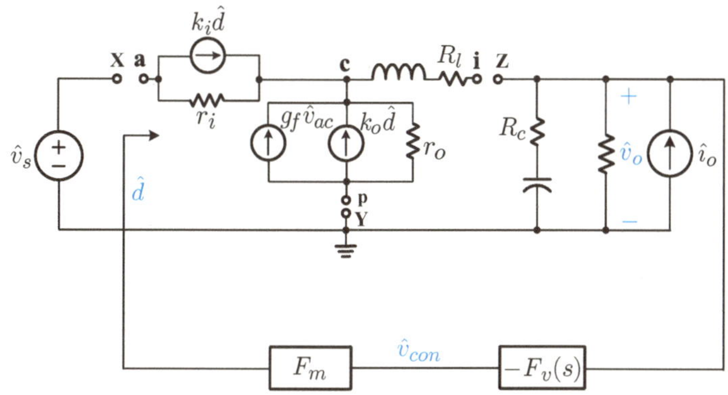

Figure 5.43 Universal DCM small-signal model for PWM converters.

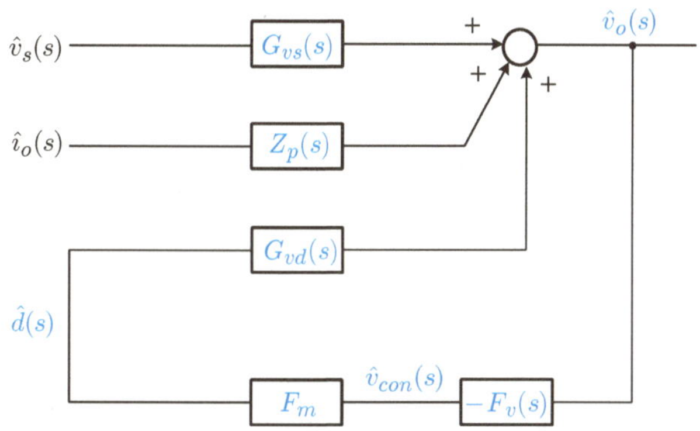

Figure 5.44 Block diagram representation of small-signal model.

signal sources $\hat{v}_s(s)$ and $\hat{i}_o(s)$ are the input variables, $\hat{v}_o(s)$ is the output variable, and $\hat{d}(s)$ is the control variable. Regarding the gain blocks, F_m is the PWM gain, $F_v(s)$ is the voltage feedback compensation, and the other gain blocks represent the power stage transfer functions: $G_{vs}(s) = \hat{v}_o(s)/\hat{v}_s(s)$, $G_{vd}(s) = \hat{v}_o(s)/\hat{d}(s)$, and $Z_p(s) = \hat{v}_o(s)/\hat{i}_o(s)$. The $-$ sign in front of the voltage feedback compensation $F_v(s)$ signifies the *negative* feedback control embedded in the control scheme.

The small-signal model of the closed-loop controlled converters or its block diagram representation is a linear time-invariant model to which all the conventional s-domain analysis techniques can directly be applied. The utility and versatility of the small-signal model will be demonstrated in the later chapters which deal with the small-signal analysis and control design of closed-loop controlled PWM dc-to-dc converters.

5.8 CHAPTER SUMMARY

In this chapter, we developed the s-domain small-signal models for all PWM dc-to-dc converters operating in CCM and DCM. As a linear time-invariant model, each small-signal model allows us to investigate the dynamics of PWM converters using standard s-domain analysis techniques. The small-signal model is derived in a general and unified way so that a single model could represent all the isolated/non-isolated PWM converters in CCM operation or DCM operation.

We employed several modeling techniques to derive the small-signal models for PWM converters. The averaging method is used to eliminate the time variance from the power stage configuration and the linearization process is employed to remove nonlinearities from the power stage dynamics and PWM process. Application of the circuit averaging to the PWM switch and ensuing linearization led to the PWM switch models, which are utilized as an instrumental tool in deriving the universal small-signal models for PWM converters.

The frequency response of PWM dc-to-dc converters is discussed in this chapter. The role of the small-signal models in the frequency response analysis is addressed. Further details about the frequency response of PWM convertes will be given in Chapter 6.

The small-signal model of the power stage is first developed for the three basic non-isolated converters operating in CCM with ideal circuit components. The modeling process is later extended to include all isolated/non-isolated PWM dc-to-dc converters functioning in both CCM and DCM with non-ideal circuit components.

The small-signal model for the PWM block was found to be a constant PWM gain, provided that the control signal only varies narrowly and slowly within the switching period of the converter: namely, under the small-signal assumption and low-frequency excitation. The PWM switch models and PWM gain are combined with the voltage feedback compensation, leading to the complete small-signal models of Fig. 5.42 and Fig. 5.43, which are applicable to all the closed-loop controlled isolated/non-isolated PWM dc-to-dc converters. With these small-signal models, we can perform the dynamic analysis of the nonlinear time-variant PWM converters in the manner that has been exercised for conventional linear time-invariant systems. The value of the small-signal models will be demonstrated in the later chapters dealing with the dynamic analysis and control design of PWM dc-to-dc converters.

REFERENCES

1. J. G. Kassakian, M. F. Schlecht, and G. C. Verghese, *Principles of Power Electronics*, Addison-Wesley Publishing Co, 1991.
2. R. D. Middlebrook and S. Ćuk, *Advances in Switch-Mode Power Conversion*, TESLAco, Pasadena, CA, 1983.
3. V. Vorpérian, *Fast Analytical Techniques for Electrical and Electronic Circuits*, Cambridge University Press, 2002.
4. V. Vorpérian, "Simplified analysis of PWM converters using the model of the PWM switch: Part I and II," *IEEE Trans. Aerosp. Electron. Syst.*, vol. AES-26, no. 3, pp. 490-505, May 1990.

PROBLEMS

5.1 Discuss or describe the following aspects of the small-signal modeling of PWM dc-to-dc converters.

a) Purpose of modeling: Describe the final goal of the small-signal modeling.

b) Two critical techniques: Illustrate two essential techniques to the modeling of PWM converters

c) Two underlying assumptions: Describe the two assumptions used behind the modeling process of the power stage and PWM block of dc-to-dc converters.

5.2 Consider the circuits shown in Fig. P5.2 and answer the questions. Assume CCM operation for the switching device.

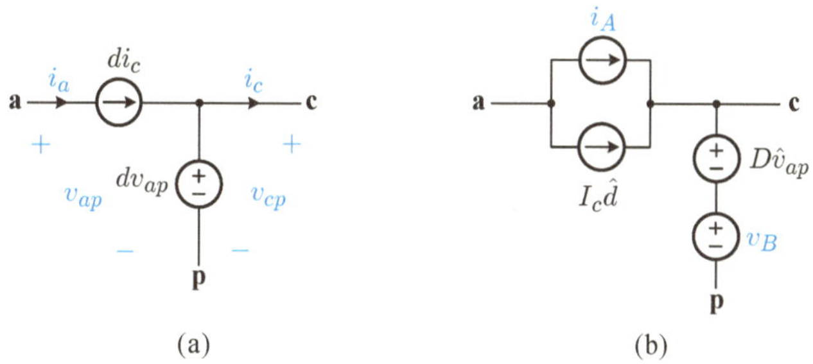

Fig. P5.2

a) Shown in Fig. P5.2(a) is an average model of a nonlinear three-terminal switching device. The variable d represents the continuous duty ratio of the active switch. Construct the nonlinear switching device using semiconductor switches.

b) Prove that the small-signal equations of the nonlinear switching device can be represented by the circuit model in Fig. P5.2(b). Find the expressions for the dependent current source i_A and voltage source v_B.

c) Construct a small-signal model of the buck converter using Fig. P5.2(b). Express the dependent sources in terms of the operating conditions and circuit variables of the buck converter.

d) Construct an average model of the boost converter using Fig. P5.2(a). Show all the model parameters.

e) Build a small-signal model of the boost converter using Fig. P5.2(b). Show all the model parameters.

5.3 Figure P5.3 shows the circuit diagrams of the three basic PWM converters with the parasitic resistances. Assume CCM operation for the converters.

a) Formulate the switched state-space model for the three basic converters.

b) Convert the switched state-space models you derived in a) into the averaged state-space models for the three basic converters.

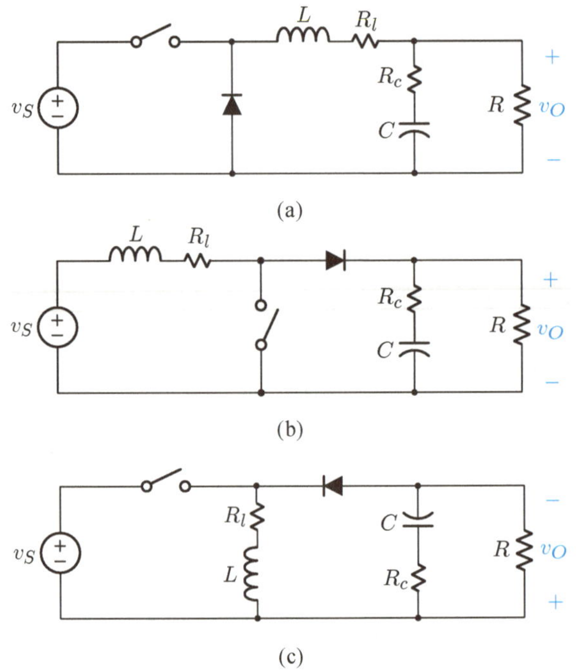

Fig. P5.3

5.4* Shown in Fig. P5.4 is a boost converter operating in CCM.

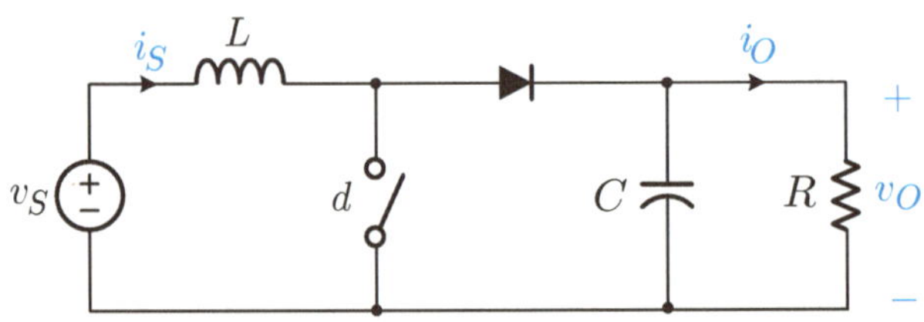

Fig. P5.4

a) Draw a nonlinear circuit model that predicts the time-averaged power stage dynamics of the converter.
b) Draw a dc model of the boost converter. Using the resulting model, find the dc voltage gain V_O/V_S and dc current gain I_O/I_S of the converter.
c) Draw the frequency-domain ac model of the converter when $v_S(t) = V_S + \hat{v}_s(t)$ and $d(t) = D + \hat{d}(t)$. Express all the model parameters using the given circuit parameters and the dc input voltage V_S.
d) Repeat c) when $v_S(t) = V_S$ and $d(t) = D + \hat{d}(t)$.
e) Repeat c) when $v_S(t) = V_S + \hat{v}_s(t)$ and $d(t) = D$.

5.5 Referring to the circuit in Fig. P5.5(a), answer the questions. Assume CCM operation for the converter.

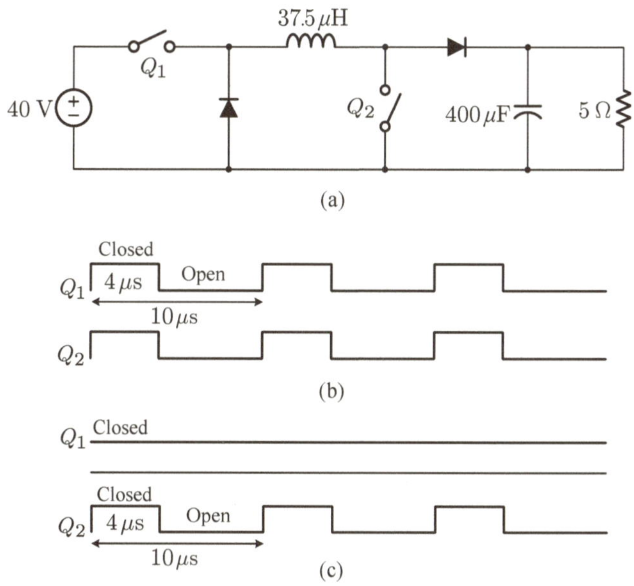

Fig. P5.5

a) The switch drive signals shown in Fig. P5.5(b) are applied to the circuit. Draw an average circuit model that predicts the time-averaged behavior of the converter. Show all the model parameters.

b) Repeat a) for the switch drive signal in Fig. P5.(c).

5.6* Figure P5.6 shows the circuit diagram of a buck/boost converter with the parasitic resistance of the inductor. Assume CCM operation for the converter.

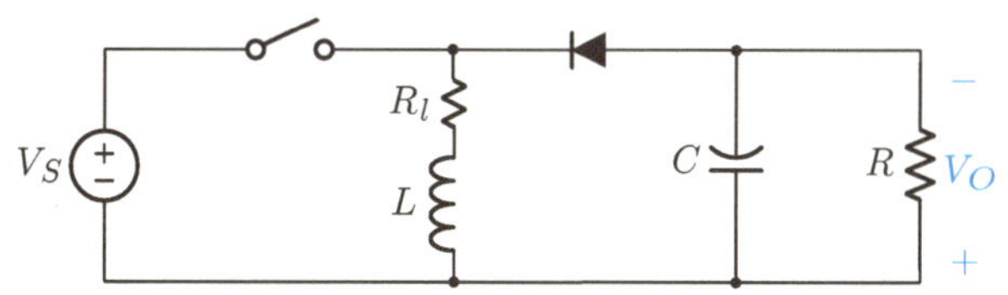

Fig. P5.6

a) Use the flux linkage balance and charge balance conditions to find the expression for the voltage gain V_O/V_S of the converter.

b) Use the averaged model of the power stage to derive the voltage gain expression of the converter. You need to convert the averaged model into the dc model.

5.7* Shown in Fig. P5.7 is the $v-i$ characteristics of a nonlinear device.

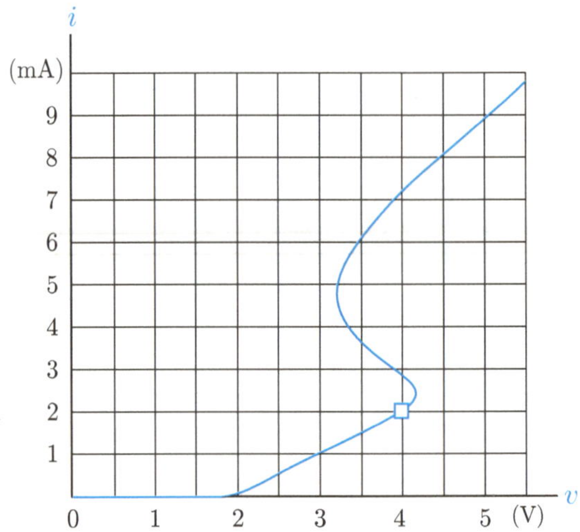

Fig. P5.7

a) Find the small-signal gain, $\hat{\imath}/\hat{v}$, at the operating point of $v = 4$ V and $i = 2$ mA.
b) Estimate the maximum magnitude of $\hat{v}(t)$ which validates the accuracy of the small-signal gain you found in a).
c) Assume $v(t) = 4 + 0.1 \sin 20t$ and find the expression for $i(t)$.
d) Now assume $v(t) = 4 + 0.5 \sin 20t$ and evaluate the range of $|i(t)|$: $(\quad) < |i(t)| < (\quad)$.

5.8 Assume that the functional relationship between the duty ratio d and the control voltage v_{con} of a certain PWM converter is given by

$$d(t) = 1 - \frac{0.2}{v_{con}(t)}$$

a) Find the expression for the small-signal modulator gain, $F_m = \hat{d}/\hat{v}_{con}$.
b) Find the numeric value for $F_m = \hat{d}/\hat{v}_{con}$ when $V_{con} = 2$ V.

5.9 The linearization of a multi-variable nonlinear function $z = f(x, y)$ at the operating point (X, Y) results in a linearized small-signal model. Derive the linearized small-signal model of the following two nonlinear functions in the form of $\hat{z} = K_1\hat{x} + K_2\hat{y}$ at the given operating point. Use the two different linrarization methods discussed in the text.

i) $z = x^2 y$ at (2,1) ii) $z = 2x/y$ at (2,1)

5.10* In Fig. P5.10, a family of cascade converters is configured by combining the buck converter, boost converter, and buck/boost converter. Assume that the composite converters are in CCM operation. For each converter, answer the following instructions or questions.

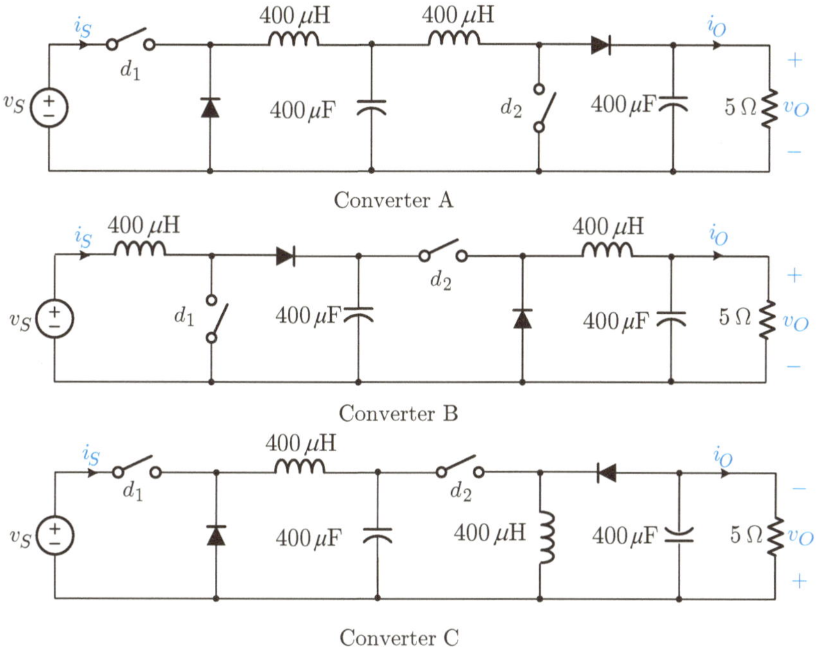

Fig. P5.10

a) Draw a nonlinear circuit model that predicts the time-averaged power stage dynamics of the converter.
b) Modify the circuit model found in a) into a dc model by replacing all the circuit variables/parameters with their dc quantities, opening the capacitors, and shorting the inductors.
c) Using the dc model found in b), find the dc voltage gain V_O/V_S and dc current gain I_O/I_S of the converter.
d) Now assume that $v_S = 40 + \hat{v}_s$, $d_1 = 0.25 + \hat{d}_1$, and $d_2 = 0.6 + \hat{d}_2$. Draw a linear circuit model that predicts the small-signal dynamics of the converter. Show the numeric values of the model parameters based on the operational conditions of the converters.

5.11 Shown in Fig. P5.11 is the CCM small-signal model of the buck converter with the parasitic resistances of reactive components. Find the accurate expressions for the three power stage transfer functions, $G_{vs}(s)$, $G_{vd}(s)$, and $Z_p(s)$.

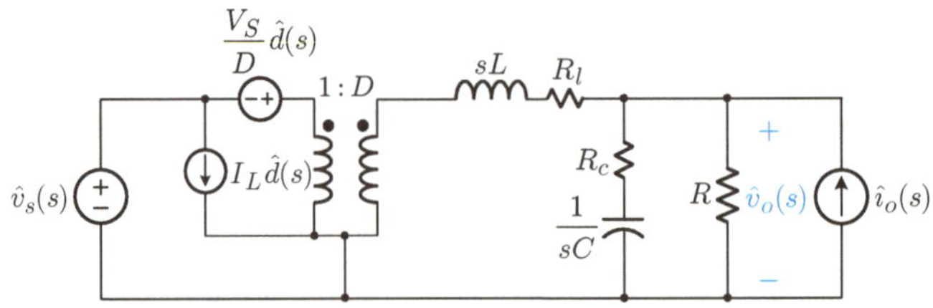

Fig. P5.11

5.12 Figure P5.12 is a PWM converter configured using a nonlinear switching network. The time-averaged expressions for the circuit variables associated with the nonlinear switching network are given by

$$\begin{aligned}\bar{v}_{ac}(t) &= (1 - d(t))\,\bar{v}_{bc}(t)\\ \bar{i}_b(t) &= (1 - d(t))\,\bar{i}_a(t)\end{aligned}$$

where d represents the continuous duty ratio of the active switch.

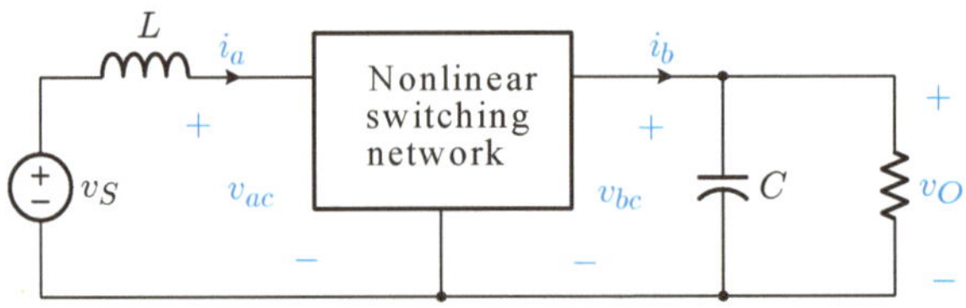

Fig. P5.12

a) Derive a set of equations that describe the small-signal dynamics of the nonlinear switching network.
b) Draw a small-signal circuit model for the nonlinear switching network.
c) Sketch the small-signal circuit model of the entire power converter.
d) Derive the input-to-output transfer function, $G_{vs}(s) = \hat{v}_o(s)/\hat{v}_s(s)$, in terms of the circuit parameters of the small-signal model.

5.13 Derive the small-signal equations of the DCM PWM switch model shown in (5.77) and (5.78) in the text. The derivation can be expedited by incorporating (5.76) into (5.75) and taking the partial derivative of the resulting equations.

5.14 Construct the DCM small-signal model for the buck converter and boost converter. Specify the five model parameters, r_i, k_i, g_f, k_o, and r_o, in terms of the operational conditions and steady-state circuit variables of each converter.

5.15* Consider the forward converter shown in Fig. P5.15.

a) Assume $L_m = 480\ \mu$H and answer the following questions.
   i) Sketch the steady-state waveforms of i_{D1}, i_{D2}, and i_C for the two operational periods. Label the maximum, minimum, and average values of each waveform.
   ii) Draw the small-signal model of the power stage. Show all the model parameters.

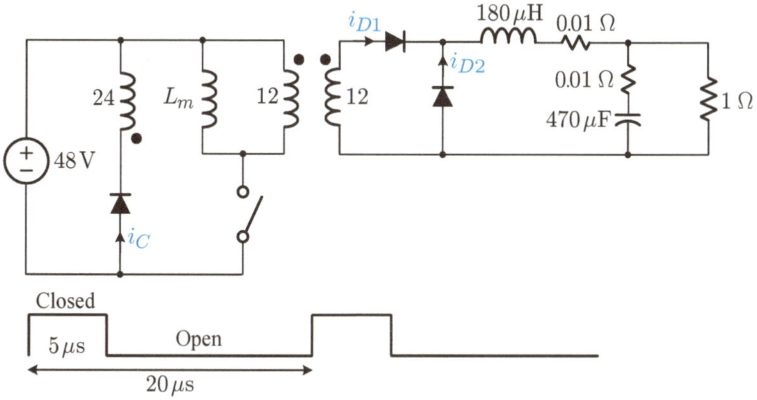

Fig. P5.15

b) Now assume $L_m = \infty$ and repeat a).

5.16 A two-switch forward converter is shown in Fig. P5.16. Answer the questions.

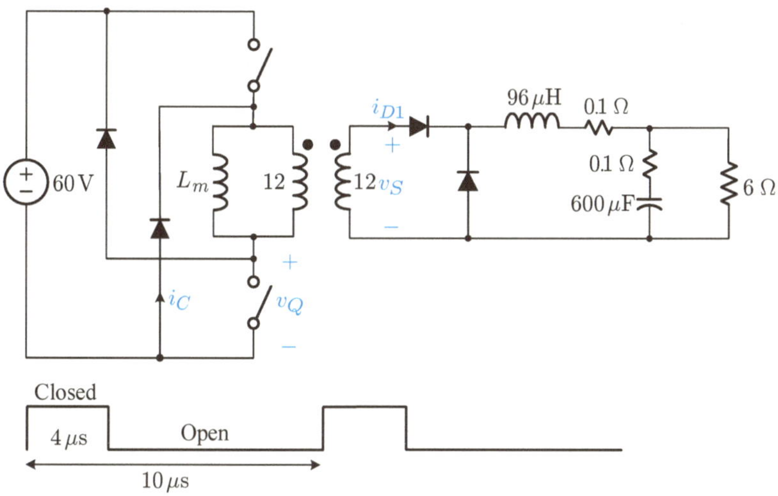

Fig. P5.16

a) Assume $L_m = 72$ μH for the magnetizing inductance of the transformer and sketch the steady-state waveforms of i_{D1}, i_C, v_S, and v_Q for the two operational periods. Label the maximum and minimum values of each waveform.

b) Draw the small-signal model of the power stage. Show all the model parameters.

5.17 Figure P5.17 shows a full-bridge converter. Assume CCM operation for the converter.

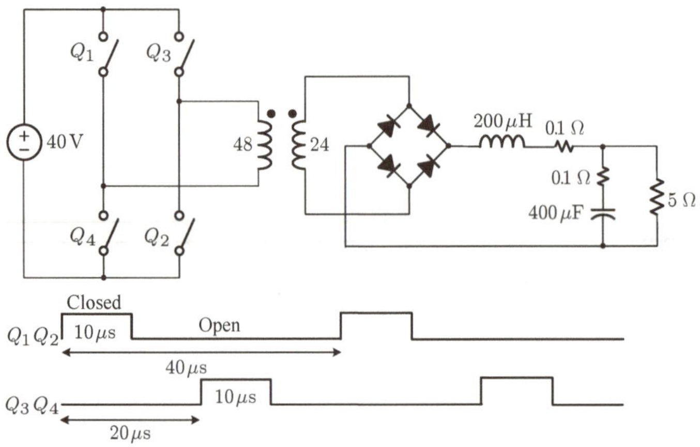

Fig. P5.17

a) Referring to the circuit diagram and the switch drive signals, draw an average model that predicts the time-averaged power stage dynamics. Show all the model parameters.

b) Construct the small-signal model of the power stage. Show all the model parameters.

5.18* Consider the flyback converter shown in Fig. P5.18 and answer questions.

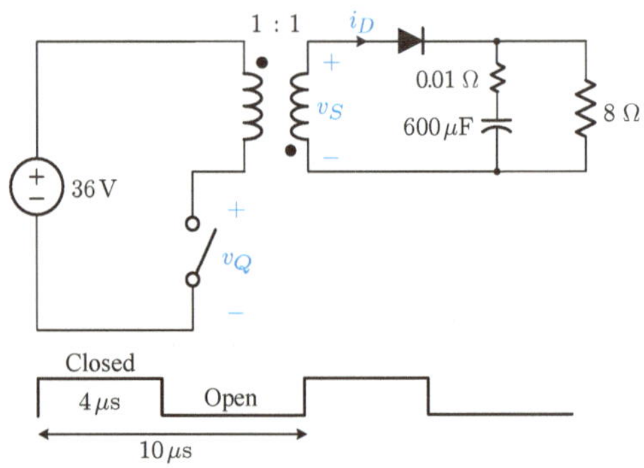

Fig. P5.18

a) Assume $L_m = 72\,\mu$H for the magnetizing inductance of the transformer and sketch the steady-state waveforms of i_D, v_S, and v_Q for the two operational periods. Label the maximum and minimum values of each waveform.

b) Draw a circuit model that predicts the time-averaged dynamics of the power stage.

c) Draw the small-signal model of the power stage. Show all the model parameters.

5.19* The PWM can be performed by comparing the control signal v_{con} against the carrier signal. Different carrier signals can be used to implement various PWM schemes. Three examples of such PWM schemes are shown in Fig. P5.19. For each PWM scheme, find the expression for the modulator gain F_m.

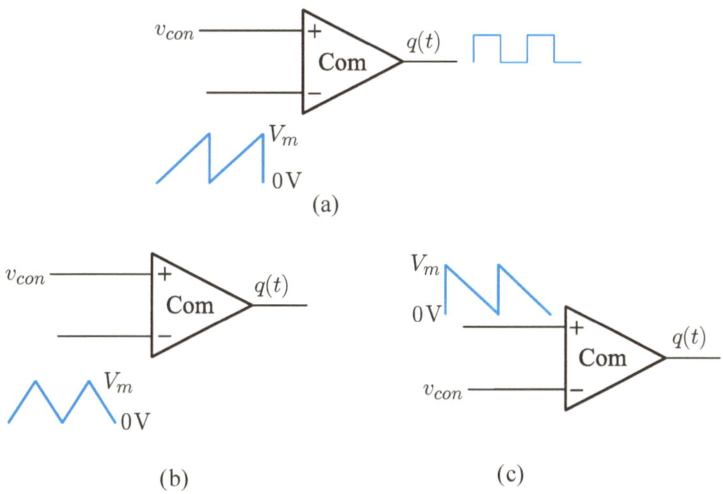

Fig. P5.19

CHAPTER 6

POWER STAGE TRANSFER FUNCTIONS

In the previous chapter, we developed s-domain small-signal models for PWM dc-to-dc converters. Using these models, we could perform the dynamic analysis and control design of nonlinear time-variant PWM dc-to-dc converters, in the same way as we do for conventional linear time-invariant systems.

In preparation for the analysis and design of PWM dc-to-dc converters, this chapter analyzes power stage transfer functions focusing on their frequency response characteristics. The first section covers the Bode plot representation of s-domain transfer functions. The later sections present analytical expressions, Bode plot portraits, and salient features of power stage transfer functions of isolated/non-isolated PWM converters. This chapter also discusses empirical approaches to the power stage dynamics analysis.

6.1 BODE PLOT FOR TRANSFER FUNCTIONS

The most useful outcome of the s-domain small-signal model is the transfer functions, which allow us to investigate the frequency response of the converter power stage. From the s-domain small-signal model, power stage transfer functions are derived via standard circuit analyses. The resulting s-domain transfer functions are then transformed into Bode plots to reveal the frequency response characteristics. This section deals with the construction of Bode plots for s-domain transfer functions, thereby providing the underlying basics for the forthcoming power stage analysis.

6.1.1 Basic Definitions

Discussions start with the fundamental definitions regarding the frequency response plots for s-domain transfer functions. These definitions include the transfer function, frequency response, and Bode plot representation.

Transfer Function

The transfer function is defined for linear time-invariant systems or circuits as the ratio of the Laplace transform of the output to the Laplace transform of the input, under the assumption that all initial conditions are zero

$$\boxed{T(s) \equiv \frac{v_o(s)}{v_s(s)}} \tag{6.1}$$

where $v_o(s)$ is the s-domain expression, or Laplace transform, of the output variable while $v_s(s)$ is the s-domain expression of the input variable. For general cases, the transfer function is obtained by performing the Laplace transform operation on the differential equation or state equation of the system. On the other hand, for linear time-invariant circuits, the transfer function can directly be derived from the circuit diagram by adopting s-domain representations for circuit variables and performing circuit analyses.

Although originally defined for linear time-invariant systems, the transfer function can be extended to nonlinear time-variant PWM converters to describe the frequency response of power stage circuit variables. For this case, the transfer functions are derived from the s-domain small-signal power stage model, as illustrated in Example 5.8. Readers may refer to Section 5.4 for the basics of the frequency response and transfer function for PWM dc-to-dc converters.

Frequency Response

The frequency response denotes the input-output relationship of linear or linearized systems, under the assumption that the system is excited by sinusoidal inputs. Accordingly, the frequency response is evaluated from the s-domain transfer function by replacing the complex frequency s with $j\omega$

$$T(j\omega) \equiv \frac{v_o(j\omega)}{v_s(j\omega)} \tag{6.2}$$

where ω is the frequency of sinusoidal excitations. The magnitude relationship of (6.2) becomes the magnitude response

$$|T(j\omega)| \equiv \frac{|v_o(j\omega)|}{|v_s(j\omega)|} \tag{6.3}$$

and the phase relationship of (6.2) is the phase response of the system

$$\angle T(j\omega) \equiv \angle v_o(j\omega) - \angle v_s(j\omega) \tag{6.4}$$

The magnitude and phase responses are collectively called the frequency response. Equations (6.3) and (6.4) indicate that the frequency response is found by evaluating the magnitude and phase of the transfer function with $s = j\omega$, while sweeping the frequency ω for the range of interest.

Polar Plot and Bode Plot Representations

The frequency response can be graphically displayed in the polar plot format or Bode plot format. Each of these graphical representations has its own value in s-domain analysis. Figure 6.1(a) illustrates the polar plot representation of the frequency response, where the frequency response is converted into the polar form and portrayed as a single curve in s-plane. Each point in the polar plot denotes the sinusoidal response in the polar coordinate, from which the magnitude and phase responses at a specific frequency ω_s are defined as illustrated in Fig. 6.1(a).

Figure 6.1(b) shows the Bode plot representation of the frequency response. The magnitude response is first calculated in dB scale, $20 \log |T(j\omega)|$, while the phase is expressed in degree °. The magnitude and phase responses are then individually displayed on the two separate plots. In the magnitude plot, the dB-scale magnitude is shown on the y-axis, while the frequency ω is displayed in log scale on the x-axis. In the phase plot, the linear-scale phase is shown against the log-scale frequency axis. Each point in the magnitude or phase plot represents the corresponding magnitude or phase response evaluated at a specific frequency ω_s, as shown in Fig. 6.1(b).

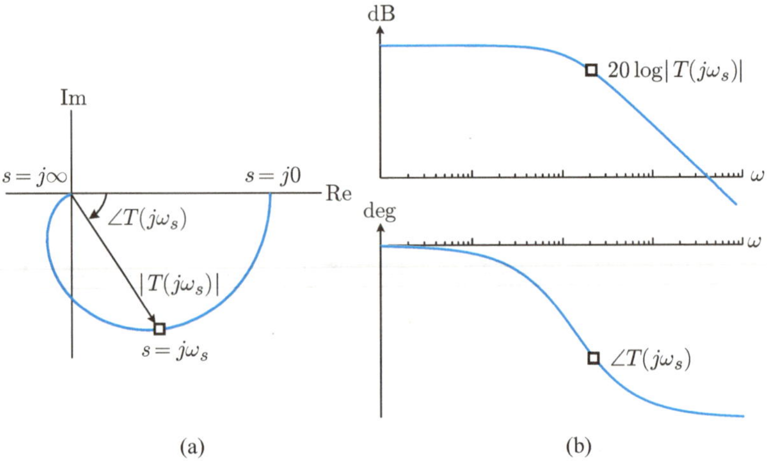

Figure 6.1 Graphical representation of frequency response. (a) Polar plot. (b) Bode plot.

6.1.2 Bode Plots for Multiplication Factors

It is assumed that the transfer function $T(s)$ for the Bode plot construction has been factorized into the following format:

$$T(s) = K\frac{1}{s^{\pm j}}\frac{1}{1+\dfrac{s}{\omega_p}}\left(1+\frac{s}{\omega_z}\right)\cdots\frac{1}{1+\dfrac{s}{Q_p\omega_o}+\dfrac{s^2}{\omega_o^2}}\left(1+\frac{s}{Q_z\omega_o}+\frac{s^2}{\omega_o^2}\right)\cdots \tag{6.5}$$

which is known as the time constant form. Each term in (6.5) is called a multiplication factor. The Bode plot for the transfer function $T_s(s)$ can be constructed by

1) finding the Bode plot of each individual multiplication factor, and

2) combining the resulting individual Bode plots.

This section discusses Bode plots of multiplication factors. The construction of the composite Bode plots for transfer functions will be treated in the next section.

Constant

Figure 6.2 shows the Bode plot for a constant K. The magnitude plot is a flat line of $20 \log K$ for both the positive K and negative K. However, the phase is 0° for $K > 0$ and −180° for $K < 0$.

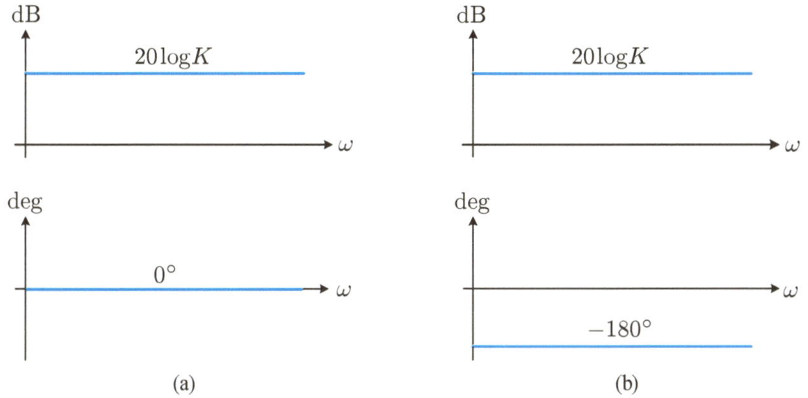

Figure 6.2 Bode plots for constant K. (a) $K > 0$. (b) $K < 0$.

Single and Double Integration Functions

The single integration function

$$F(s) = \frac{K_i}{s} \tag{6.6}$$

is evaluated with $s = j\omega$

$$F(j\omega) = \frac{K_i}{j\omega} \tag{6.7}$$

for the Bode plot construction. The above expression is split into the magnitude response

$$20 \log |F(j\omega)| = 20 \log \left(\frac{K_i}{\omega} \right) \tag{6.8}$$

and the phase response

$$\angle F(j\omega) = -90^\circ \tag{6.9}$$

Figure 6.3(a) shows the Bode plot based on (6.8) and (6.9). The magnitude linearly decays with a -20 dB/dec slope while the phase stays at $-90°$. The magnitude plot crosses the 0 dB line at $\omega = K_i$: $20 \log K_i/K_i = 0$ dB.

For notational simplicity, the following numeric symbols will be used to represent the slope of the magnitude curve

- -2 slope: -40 dB/dec slope — -1 slope: -20 dB/dec slope
- $+1$ slope: $+20$ dB/dec slope — $+2$ slope: $+40$ dB/dec slope

The double integration function

$$F(s) = \frac{K_i}{s^2} \tag{6.10}$$

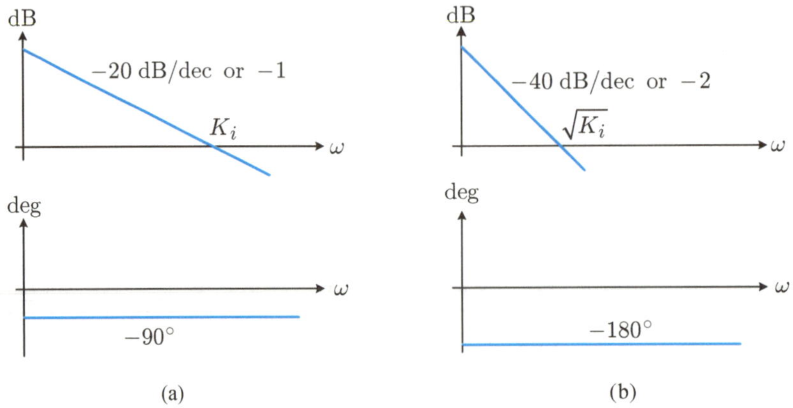

Figure 6.3 Bode plots for single and double integration functions. (a) $F(s) = K_i/s$. (b) $F(s) = K_i/s^2$.

is evaluated as

$$F(j\omega) = -\frac{K_i}{\omega^2} \tag{6.11}$$

leading to

$$20 \log |F(j\omega)| = 20 \log \left(\frac{K_i}{\omega^2} \right) \tag{6.12}$$

and

$$\angle F(j\omega) = -180° \tag{6.13}$$

The Bode plot is shown in Fig. 6.3(b). The descending slope of the magnitude plot is -40 dB/dec slope, or -2 slope, and the phase remains at $-180°$. The 0 dB crossover now occurs at $\omega = \sqrt{K_i} : 20 \log K_i/\sqrt{K_i}^2 = 0$ dB.

Single and Double Differentiation Functions

The single differentiation function $F(s) = K_d\, s$ is evaluated as $F(j\omega) = K_d\, j\omega$, yielding the magnitude response

$$20 \log |F(j\omega)| = 20 \log K_d\, \omega \tag{6.14}$$

and phase response

$$\angle F(j\omega) = 90° \tag{6.15}$$

The Bode plot for the single differentiation is shown in Fig. 6.4(a). The magnitude plot linearly increases with a 20 dB/dec slope, or $+1$ slope, and crosses the 0 dB line at $\omega = 1/K_d$, while the phase stays at 90°.

The double differentiation function $F(s) = K_d s^2$ is converted into $F(j\omega) = K_d(j\omega)^2$ for the Bode plot construction. The magnitude and phase responses become

$$20 \log |F(j\omega)| = 20 \log K_d\, \omega^2 \tag{6.16}$$

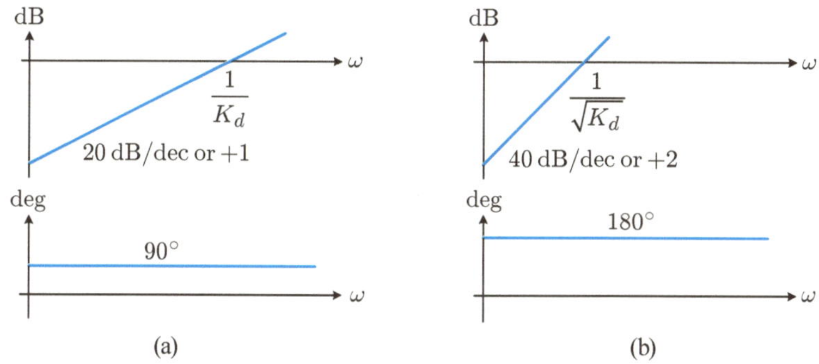

Figure 6.4 Bode plots for single and double differentiation functions. (a) $F(s) = K_d s$. (b) $F(s) = K_d s^2$.

and

$$\angle F(j\omega) = 180^\circ \tag{6.17}$$

leading to the Bode plot in Fig. 6.4(b). The magnitude plot increases with a 40dB/dec slope, or +2 slope, and the 0 dB crossover occurs at $\omega = 1/\sqrt{K_d}$. The phase remains at 180° for this case.

Single Pole and Single Zero Functions

The single pole function

$$\boxed{F(s) = \frac{1}{1 + \dfrac{s}{\omega_p}}} \tag{6.18}$$

is evaluated with $s = j\omega$

$$\boxed{F(j\omega) = \frac{1}{1 + \dfrac{j\omega}{\omega_p}}} \tag{6.19}$$

yielding the magnitude response

$$20 \log |F(j\omega)| = 20 \log \frac{1}{\sqrt{1 + \left(\dfrac{\omega}{\omega_p}\right)^2}} \tag{6.20}$$

and phase response

$$\angle F(j\omega) = -\tan^{-1} \frac{\omega}{\omega_p} \tag{6.21}$$

Although the Bode plot can accurately be drawn from (6.20) and (6.21), it is usually unnecessary to find the exact Bode plot. Instead, the asymptotic plot that describes the asymptotic behavior of the Bode plot is sufficient for most situations. The asymptotic plots can quickly be sketched from the transfer function and provide all the important information about the frequency response. Furthermore, an asymptotic plot can readily be refined so that the resulting plot closely resembles the actual frequency response.

For the asymptotic Bode plot construction, (6.19) is split into the three expressions

$$F(j\omega)=\frac{1}{1+\dfrac{j\omega}{\omega_p}}\approx\begin{cases}1 & \text{at frequencies where } \omega<\omega_p\\[2ex] \dfrac{1}{1+j} & \text{at frequency } \omega=\omega_p\\[2ex] \dfrac{\omega_p}{j\omega} & \text{at frequencies where } \omega>\omega_p\end{cases} \tag{6.22}$$

using the pole frequency ω_p as the borderline for approximations.

The top expression in the right-hand side of (6.22) describes the asymptotic behavior of $F(j\omega)$ at frequencies below ω_p, while the bottom expression approximates $F(j\omega)$ at frequencies beyond ω_p. In other words, the top expression provides the low-frequency asymptotes for the magnitude and phase plots: $|F|=0$ dB and $\angle F=0°$. Similarly, the bottom expression produces the high-frequency asymptotes: $20\log|F(j\omega)|=20\log(\omega_p/\omega)$ and $\angle F(j\omega)=-90°$. The high-frequency asymptotes are the same as those of a single integration function passing through the 0 dB line at ω_p. The middle equation is the exact frequency response evaluated at ω_p

$$20\log|F(j\omega_p)|=20\log\left|\frac{1}{1+j}\right|=20\log\frac{1}{\sqrt{2}}\approx-3\text{ dB} \tag{6.23}$$

$$\angle F(j\omega_p)=\angle\frac{1}{1+j}=-45° \tag{6.24}$$

Figure 6.5(a) shows the asymptotic plots of the single pole function, in comparison with the exact plots. The thick blue line is the asymptotic plots while the thin black line describes the exact plots. The asymptotic magnitude plot is formed by merging the low- and high-frequency asymptotes at ω_p. The asymptotic phase plot is constructed by bridging the low- and high-frequency asymptotes through a line segment that ramps down from $0.1\,\omega_p$ to $10\,\omega_p$ with $-45°/\text{dec}$ slope.

As shown in Fig. 6.5(a), the asymptotic magnitude plot shows small deviations from the exact plot in the frequency range $0.5\,\omega_p<\omega<2\,\omega_p$ with a maximum 3 dB error at ω_p; for this reason, the pole frequency ω_p is called the 3 dB frequency. The asymptotic phase plot also provides a good approximation for the exact plot. In particular, the asymptotic plot shows the exact $-45°$ phase at ω_p.

As illustrated in Fig. 6.5(a), an accurate Bode plot can be constructed, with negligible and predictable errors, by smoothly connecting the low- and high-frequency asymptotes so that the combined curves pass the exact values at ω_p: -3 dB for $|F(j\omega_p)|$ and $-45°$ for $\angle F(j\omega_p)$.

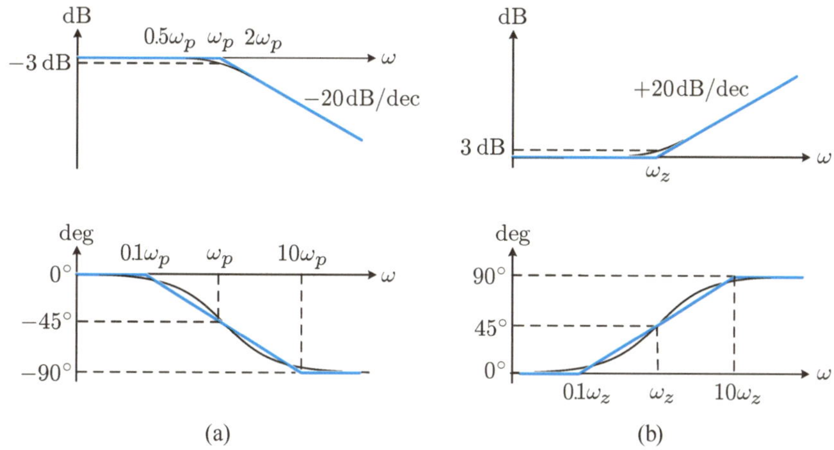

Figure 6.5 Bode plots for single pole and single zero functions. (a) Single pole function. (b) Single zero function.

The single zero function

$$\boxed{F(s) = 1 + \frac{s}{\omega_z}} \tag{6.25}$$

is approximated to

$$F(j\omega) = 1 + \frac{j\omega}{\omega_z} \approx \begin{cases} 1 & \text{at frequencies where } \omega < \omega_z \\ 1 + j & \text{at frequency } \omega = \omega_z \\ j\dfrac{\omega}{\omega_z} & \text{at frequencies where } \omega > \omega_z \end{cases} \tag{6.26}$$

for the asymptotic Bode plot construction. The bottom term in (6.26) constitutes the high-frequency asymptote, which is a single differentiation function crossing the 0 dB line at ω_z. The middle term is the exact frequency response at ω_z

$$\boxed{20 \log |F(j\omega_z)| = 20 \log |1 + j\,| \approx 3 \text{ dB}} \tag{6.27}$$

and

$$\boxed{\angle F(j\omega_z) = \angle(1 + j\,) = 45^\circ} \tag{6.28}$$

Figure 6.5(b) shows the asymptotic and exact Bode plots of the single zero function. The construction and interpretation of Fig. 6.5(b) are virtually identical to the cases of the previous single pole function.

Double Pole and Double Zero Functions

The double pole and double zero functions frequently appear in the upcoming power stage transfer functions. The double pole function is given by

$$F(s) = \frac{1}{1 + \dfrac{s}{Q_p \omega_o} + \dfrac{s^2}{\omega_o^2}} \tag{6.29}$$

where ω_o is the double pole frequency and Q_p is the quality factor. The double pole function with $Q_p > 0.5$ is approximated to

$$F(j\omega) = \frac{1}{1 + \dfrac{j\omega}{Q_p \omega_o} - \dfrac{\omega^2}{\omega_o^2}} \approx \begin{cases} 1 & \text{at frequencies where } \omega < \omega_o \\ \dfrac{Q_p}{j} & \text{at frequency } \omega = \omega_o \\ -\dfrac{\omega_o^2}{\omega^2} & \text{at frequencies where } \omega > \omega_o \end{cases} \tag{6.30}$$

The low-frequency asymptote is the 0 dB line, as is the case with the single pole function. On the other hand, the high-frequency asymptote is a double integration function passing through the 0 dB line at ω_o. The exact response at ω_o is

$$20 \log |F(j\omega_o)| = 20 \log \left| \frac{Q_p}{j} \right| = 20 \log Q_p \tag{6.31}$$

and

$$\angle F(j\omega_o) = \angle \frac{Q_p}{j} = -90° \tag{6.32}$$

Figure 6.6(a) shows the asymptotic plots of the double pole function, in parallel with the exact plots. The asymptotic magnitude plot is formed by merging low- and high-frequency asymptotes at ω_o. The phase plot is created by connecting the low- and high-frequency asymptotes with a line segment linearly declining at frequencies around ω_o. The declining slope of the phase asymptote is proportional to the quality factor Q_p; a lager Q_p produces a steeper slope.

The asymptotic magnitude plot shows a noticeable difference from the exact plot in the frequency range $0.3\,\omega_o < \omega < 3\,\omega_o$. In particular, the exact magnitude plot shows a peaking of $20 \log Q_p$ at ω_o, which is not accounted for in the asymptotic plot. On the other hand, the asymptotic phase plot passes the exact $-90°$ point at ω_o, while producing some error at neighboring frequencies.

The accurate magnitude and phase characteristics of the double pole function can be determined by evaluating (6.29) at frequencies around ω_o. Such an analysis reveals that the transition patterns of the magnitude and phase curves are strongly affected by the value of Q_p. Figure 6.7 shows the exact Bode plots of the double pole function

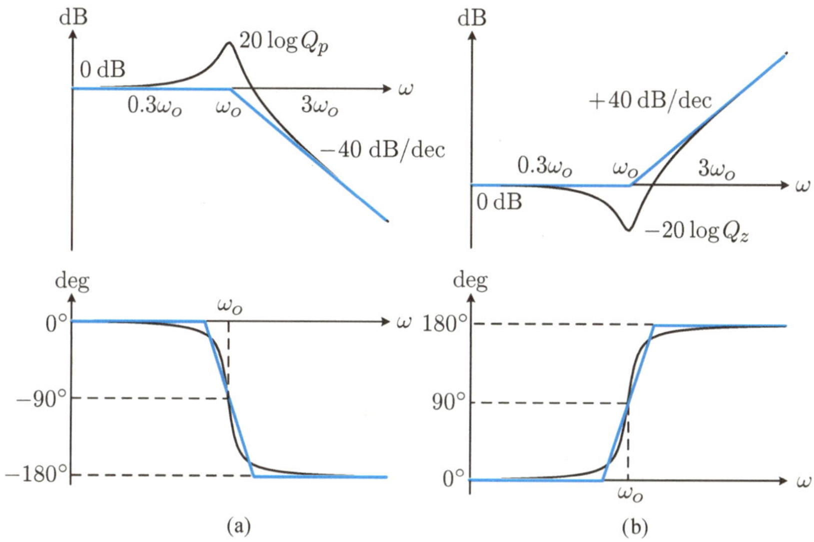

Figure 6.6 Bode plots for double pole and double zero functions. (a) Double pole function. (b) Double zero function.

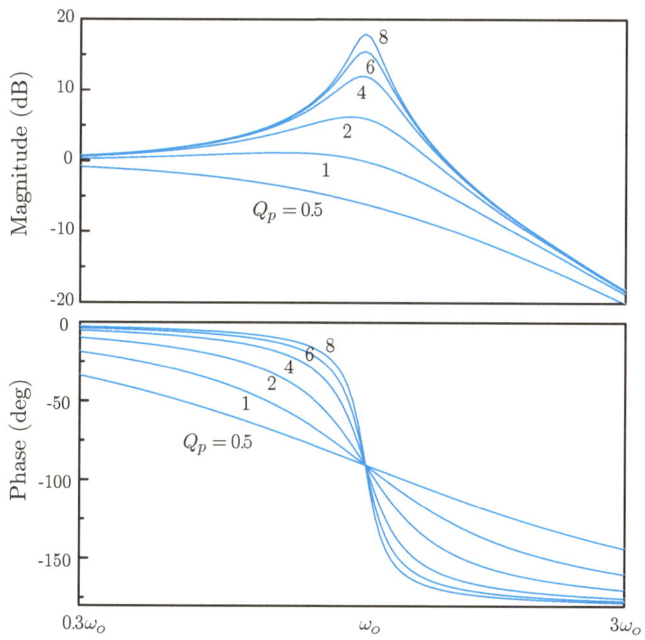

Figure 6.7 Exact Bode plots of double pole function with $0.5 < Q_p < 8$.

in the frequency range $0.3\,\omega_o < \omega < 3\,\omega_o$ with different Q_p values of $0.5 < Q_p < 8$. In the magnitude response, a larger Q_p generates a higher peaking at ω_o, $20\log Q_p$, resulting in a wider gap between the asymptote and exact plots. Regarding the phase response, a larger Q_p accelerates the decaying rate of the phase plot around ω_o, thereby exhibiting a more abrupt phase change in the narrower frequency range. These characteristics need to be incorporated when the actual frequency response is predicted from the asymptotic plots.

The double zero function

$$\boxed{F(s) = 1 + \frac{s}{Q_z\omega_o} + \frac{s^2}{\omega_o^2}} \tag{6.33}$$

with $Q_z > 0.5$ is approximated to

$$F(j\omega) = 1 + \frac{j\omega}{Q_z\omega_o} - \frac{\omega^2}{\omega_o^2} \approx \begin{cases} 1 & \text{at frequencies where } \omega < \omega_o \\ \dfrac{j}{Q_z} & \text{at frequency } \omega = \omega_o \\ -\dfrac{\omega^2}{\omega_o^2} & \text{at frequencies where } \omega > \omega_o \end{cases} \tag{6.34}$$

The bottom term in (6.34) indicates that the high-frequency asymptote is a double differentiation function crossing the 0 dB line at ω_o. Figure 6.6(b) shows the asymptotic and exact plots for the double zero function. The construction and accuracy of the asymptotic plots are very similar to those of the double pole case. The magnitude plot shows a dipping of $20\log Q_z$ at ω_o, which is unaccounted for in the asymptotic plot.

RHP Pole and RHP Zero Functions

The single pole function given by

$$\boxed{F(s) = \frac{1}{1 - \dfrac{s}{\omega_p}} = \frac{1}{1 + \dfrac{s}{-\omega_p}}} \tag{6.35}$$

is called the right-half plane (RHP) pole function in the sense that the pole $s = \omega_p$ is located in the right-half side of s-plane.

The RHP pole has unique characteristics and thus deserves special attentions. For the frequency response analysis, the transfer function is evaluated with $s = j\omega$

$$F(s) = \frac{1}{1 + \dfrac{j\omega}{-\omega_p}} \tag{6.36}$$

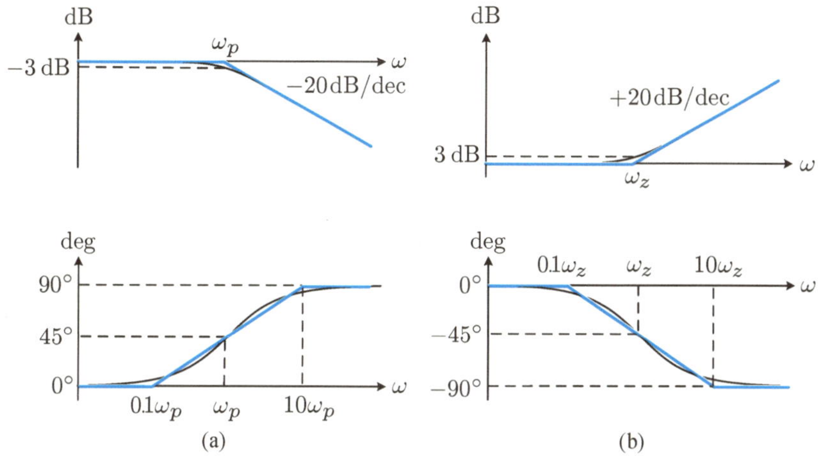

Figure 6.8 Bode plots for RHP pole and RHP zero functions. (a) RHP pole function. (b) RHP zero function.

yielding the magnitude response

$$20\log|F(j\omega)| = 20\log\frac{1}{\sqrt{1+\left(\dfrac{\omega}{-\omega_p}\right)^2}} = 20\log\frac{1}{\sqrt{1+\left(\dfrac{\omega}{\omega_p}\right)^2}} \tag{6.37}$$

and phase response

$$\angle F(j\omega) = -\tan^{-1}\frac{\omega}{-\omega_p} = \tan^{-1}\frac{\omega}{\omega_p} \tag{6.38}$$

Equation (6.37) indicates that the magnitude response of the RHP pole is the same as that of a regular pole function. However, (6.38) reveals that the phase response follows the pattern of a zero function, rather than a pole function. Figure 6.8(a) shows the asymptotic and exact plots of the RHP pole function. The flat magnitude asymptote breaks at ω_p and rolls down by -20dB/dec, or -1, slope thereafter, while the phase increases from 0° to 90° over the frequency range $0.1\,\omega_p < \omega < 10\,\omega_p$.

The frequency response of an RHP zero function

$$F(s) = 1 - \frac{s}{\omega_z} = 1 + \frac{s}{-\omega_z} \tag{6.39}$$

is evaluated as

$$|F(j\omega)| = 20\log\sqrt{1+\left(\frac{\omega}{-\omega_z}\right)^2} = 20\log\sqrt{1+\left(\frac{\omega}{\omega_z}\right)^2} \tag{6.40}$$

and

$$\angle F(j\omega) = \tan^{-1}\frac{\omega}{-\omega_z} = -\tan^{-1}\frac{\omega}{\omega_z} \tag{6.41}$$

Regarding the phase characteristics, the RHP zero follows the pattern of a single pole function, not a zero function. Figure 6.8(b) shows the asymptotic and exact plots for the RHP zero function. In the neighborhood of the RHP zero ω_z, the 0 dB magnitude plot starts to ramp up with a 20dB/dec, or +1, slope, while the phase drops linearly from 0° to −90°. The RHP zero often appears in the power stage transfer functions of PWM dc-to-dc converters, as will be discussed later in this chapter.

6.1.3 Bode Plot Construction for Transfer Functions

Bode plots for transfer functions are constructed by combining Bode plots of individual multiplication factors. Techniques for the Bode plot construction are illustrated with several examples.

Examples of Bode Plot Construction

The first example considers the Bode plot for the following transfer function

$$T(s) = 10s\frac{1}{1+\dfrac{s}{10^2}}\frac{1}{1+\dfrac{s}{10^5}} \tag{6.42}$$

The magnitude and phase responses are evaluated as

$$20\log|T(j\omega)| = 20\log|10j\omega| + 20\log\left|\frac{1}{1+\dfrac{j\omega}{10^2}}\right| + 20\log\left|\frac{1}{1+\dfrac{j\omega}{10^5}}\right| \tag{6.43}$$

and

$$\angle T(j\omega) = \angle 10j\omega + \angle\frac{1}{1+\dfrac{j\omega}{10^2}} + \angle\frac{1}{1+\dfrac{j\omega}{10^5}} \tag{6.44}$$

Figure 6.9 illustrates the procedure of building asymptotic plots for the transfer function. The magnitude plots for the multiplication factors are first drawn individually and then added together to construct the composite magnitude plot for the transfer function. The magnitude plot initially ramps up with a 20 dB/dec, or +1, slope due to the single differentiation function. After passing through the 0 dB line at $\omega = 0.1$, the magnitude plot becomes flat at the first pole $\omega = 10^2$. The flat mid-band magnitude is evaluated as $20\log|10j\omega|_{\omega=10^2} = 60$ dB. The 60 dB mid-band magnitude starts declining by −20 dB/dec, or −1, slope at the second pole $\omega = 10^5$, thereby crossing the 0 dB line at $\omega = 10^8$.

The phase plot is constructed by adding the phase plots of the individual multiplication factors. The phase starts with 90° due to the single differentiation function.

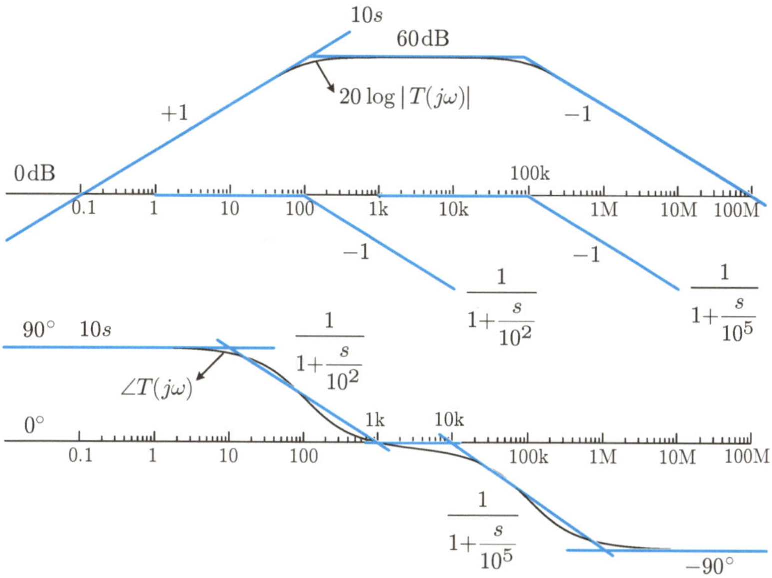

Figure 6.9 Example of Bode plot construction: $\frac{10s}{(1+s/10^2)(1+s/10^5)}$

The 90° low-frequency phase decreases to 0° over the first pole $\omega = 10^2$. The 0° mid-band phase reduces to the −90° final phase over the second pole $\omega = 10^5$. Figure 6.9 also shows the exact Bode plots of the transfer function.

From the previous example, the general transition patterns of Bode plots are described as follows. The Bode plot initially starts with a differentiation function, integration function, or constant. Then, the evolution of the Bode plot is changed by the poles or zeros of the transfer function. Here, we assume that the poles and zeros appear in the left-half plane of s-plane.†

1) A single pole ω_p decreases the magnitude slope by 20 dB/dec at the pole frequency, while causing a 90° phase lag over the frequency range $0.1\,\omega_p < \omega < 10\,\omega_p$.

2) A double pole ω_o reduces the magnitude slope by 40 dB/dec, while lagging the phase by 180° over the frequencies around ω_o.

3) A single zero ω_z boosts the magnitude slope by 20 dB/dec, while causing a 90° phase lead over the frequency range $0.1\,\omega_z < \omega < 10\,\omega_z$.

4) A double zero ω_o increases the magnitude slope by 40 dB/dec, while leading the phase by 180° over the frequencies around ω_o.

†For the right-half plane (RHP) poles and zeros, the phase patterns are opposite to those of the regular poles and zeros.

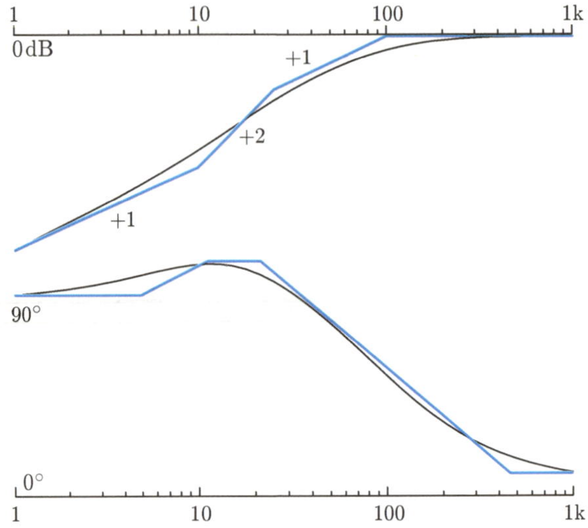

Figure 6.10 Example of Bode plot construction: $\frac{s(s+10)}{(s+25)(s+100)}$

The pole or zero frequency is called the corner frequency because the magnitude asymptote shapes a corner at pole and zero frequencies by changing the ascending or descending slopes. The aforementioned evolution patterns serve as the general rules in constructing Bode plots for given transfer functions.

The second example deals with the following transfer function

$$T(s) = \frac{s(s+10)}{(s+25)(s+100)} \tag{6.45}$$

The transfer function is first written in the time constant form

$$T(s) = \frac{s}{250} \frac{1+\dfrac{s}{10}}{\left(1+\dfrac{s}{25}\right)\left(1+\dfrac{s}{100}\right)} \tag{6.46}$$

The asymptotic plot is drawn in Fig. 6.10 by applying the Bode plot construction rules. The high-frequency magnitude is found from (6.45)

$$\lim_{\omega\to\infty} |T(j\omega)| = \left| \frac{j\omega(j\omega+10)}{(j\omega+25)(j\omega+100)} \right|_{\omega=\infty} = 1 \quad \Rightarrow \quad 0\,\text{dB} \tag{6.47}$$

while the high-frequency phase is evaluated as

$$\begin{aligned} \lim_{\omega\to\infty} \angle T(j\omega) &= \lim_{\omega\to\infty} \angle \frac{j\omega(j\omega+10)}{(j\omega+25)(j\omega+100)} \\ &= (90^\circ + 90^\circ) - (90^\circ + 90^\circ) = 0^\circ \end{aligned} \tag{6.48}$$

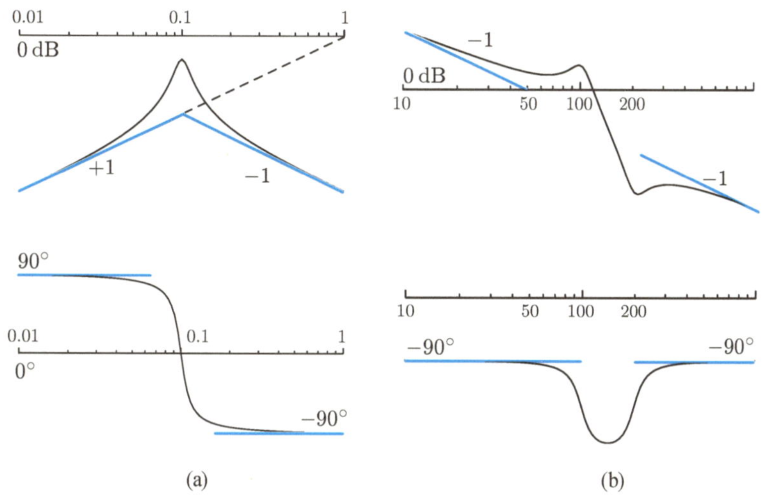

Figure 6.11 Examples of Bode plots. (a) $\frac{s}{1+s/0.5+s^2/0.1^2}$ (b) $\frac{50}{s}\frac{1+s/800+s^2/200^2}{1+s/400+s^2/100^2}$

By incorporating the actual frequency response characteristics near the corner frequencies, each asymptotic Bode plot is modified to a smooth curve, which closely resembles the exact Bode plot, as shown in Fig. 6.10.

For another example, the following transfer function is considered

$$T(s) = \frac{s}{1 + \dfrac{s}{0.5} + \dfrac{s^2}{0.1^2}} \tag{6.49}$$

whose denominator is the double pole function at $\omega_o = 0.1$ with $Q_p = 5$. Figure 6.11(a) shows the Bode plot, obtained by adding the $20 \log 5$ peaking at $\omega_o = 0.1$ to the asymptotic plot.

As an additional example, the Bode plot of the transfer function

$$T(s) = \frac{50}{s} \frac{1 + \dfrac{s}{800} + \dfrac{s^2}{200^2}}{1 + \dfrac{s}{400} + \dfrac{s^2}{100^2}} \tag{6.50}$$

can be constructed as shown in Fig. 6.11(b) by recognizing $Q_p = 4$ and $\omega_o = 100$ for the denominator, and $Q_z = 4$ and $\omega_o = 200$ for the numerator.

Non-Minimum Phase System

The dynamic system containing a right-half plane (RHP) zero or RHP pole in its transfer function is called the non-minimum phase system. To be specific, the following transfer function

$$T(s) = \frac{1 + \dfrac{s}{\omega_z}}{1 + \dfrac{s}{Q_p \omega_o} + \dfrac{s^2}{\omega_o^2}} \tag{6.51}$$

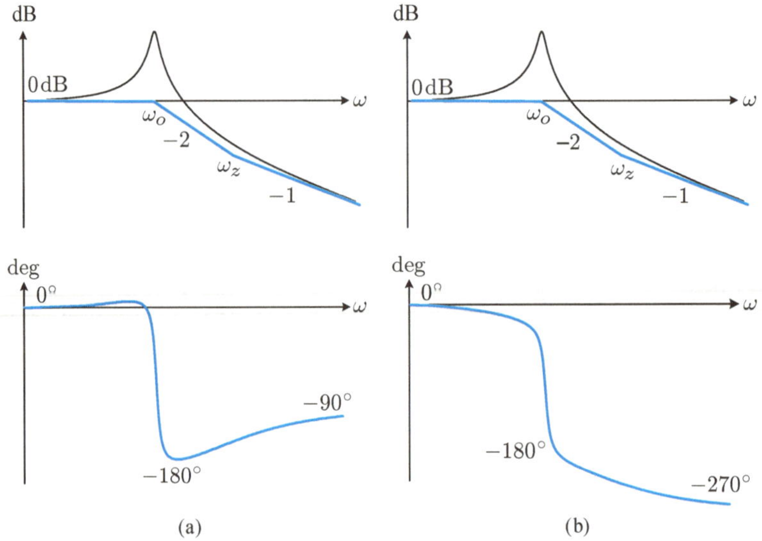

Figure 6.12 Bode plots for regular and non-minimum phase second-order systems. (a) Regular second-order system. (b) Non-minimum phase second-order system.

is a regular second-order system. In contrast, the transfer function given by

$$T(s) = \frac{1 - \dfrac{s}{\omega_z}}{1 + \dfrac{s}{Q_p\omega_o} + \dfrac{s^2}{\omega_o^2}} \tag{6.52}$$

is a non-minimum phase second-order system, due to the presence of an RHP zero function in the numerator.

Figure 6.12 compares the Bode plots of the regular second-order system and the non-minimum phase second-order system with the assumption $\omega_o < \omega_z$. The magnitude plots of the two systems are identical, but the phase responses differ from each other. Due to the presence of the RHP zero at ω_z, which causes a 90° phase delay, the phase of the non-minimum phase system decreases from −180° to −270° over ω_z, resulting in the overall phase variation between 0° and −270°. On the other hand, the phase of the regular system only changes between 0° and −180°. The non-minimum phase system is named so because the phase variation is wider than that of the regular system.

6.1.4 Identification of Transfer Function from Bode Plot

In many dynamic system analyses, the asymptotic plots of certain transfer functions are determined in advance and it then becomes necessary to extract analytical expressions of those transfer functions. This task is fulfilled by taking the inverse steps of the Bode plot construction.

From the given asymptotic plot, the structure of the transfer function is first determined. The transfer function is written in a time constant form which includes the leading coefficient in front; the general structure of the time constant form is given in (6.5). The leading coefficient is then calculated from the value of the low-frequency asymptote, high-frequency asymptote, 0 dB crossover frequency, or one of the corner frequencies. This technique will be repeatedly used in the later chapters which deal with the dynamic analysis and control design of dc-to-dc converters. Examples follow to show details about this technique.

Identification of Transfer Function — EXAMPLE 6.1

This example illustrates the procedures of identifying the transfer function from its asymptotic plot. The first illustration considers the asymptotic plot shown in Fig. 6.13(a), where the mid-frequency asymptote is given by $20\log K_m$. The asymptotic plot is converted into an analytic equation

$$T(s) = K_d s \frac{1+\dfrac{s}{\omega_z}}{\left(1+\dfrac{s}{\omega_{p1}}\right)\left(1+\dfrac{s}{\omega_{p2}}\right)} \tag{6.53}$$

by taking the inverse steps of the Bode plot construction. The mid-frequency magnitude of the asymptotic plot is determined as

$$|K_d s|_{s=j\omega_{p1}} = 20\log(K_d\,\omega_{p1}) = 20\log K_m \tag{6.54}$$

which indicates

$$K_d\,\omega_{p1} = K_m \tag{6.55}$$

The expression of the leading coefficient is now given by

$$K_d = \frac{K_m}{\omega_{p1}} \tag{6.56}$$

The second example deals with the asymptotic plot in Fig. 6.13(b). The expression for the transfer function is the same, but the leading coefficient K_d is different. The value for K_d is now found from the high-frequency asymptote of the transfer function. The high-frequency asymptote is evaluated from the $T(s)$ expression

$$\lim_{\omega\to\infty}|T(j\omega)| = 20\log\left(K_d\frac{\omega_{p1}\omega_{p2}}{\omega_z}\right) \tag{6.57}$$

On the other hand, the asymptotic plot indicates that

$$|T(j\infty)| = 20\log K_h \tag{6.58}$$

By equating (6.57) and (6.58), the leading coefficient is given by

$$K_d = K_h\frac{\omega_z}{\omega_{p1}\omega_{p2}} \tag{6.59}$$

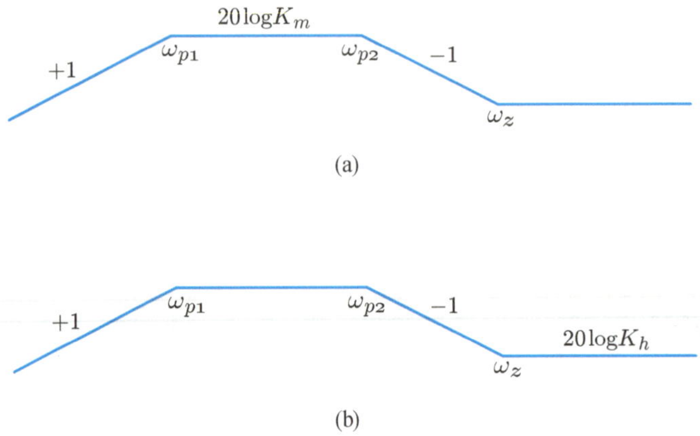

Figure 6.13 Example of asymptotic plots. (a) Case where mid-band gain is given. (b) Case where high-frequency asymptote is known.

6.2 POWER STAGE TRANSFER FUNCTIONS OF THREE BASIC CONVERTERS IN CCM OPERATION

The previous section dealt with the construction of Bode plots for s-domain transfer functions. The current section now investigates the frequency response characteristics of the power stage transfer functions of dc-to-dc converters. Power stage transfer functions of PWM dc-to-dc converters are derived from the s-domain small-signal model of the power stage. The Bode plot drawing techniques are then applied to the power stage transfer functions to reveal the frequency response characteristics.

This section investigates the power stage transfer functions of the three basic non-isolated dc-to-dc converters, buck converter, boost converter, and buck/boost converter, in continuous conduction mode (CCM) operation. Power stage transfer functions of isolated converters and transfer functions under discontinuous conduction mode (DCM) operation will be discussed in later sections.

6.2.1 Power Stage Transfer Functions of Buck Converter

Figure 6.14(a) shows the circuit diagram of the buck converter where the parasitic resistances of the inductor and capacitor are included. Figure 6.14(b) is the small-signal model of Fig. 6.14(a), obtained by replacing the PWM switch with its small-signal model and introducing appropriate small-signal sources. Readers may refer to Section 5.3.3 for the details about the small-signal model.

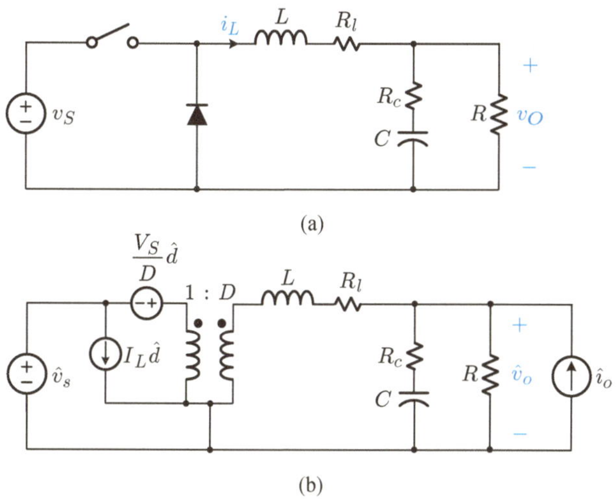

Figure 6.14 Small-signal model of buck converter. (a) Circuit diagram. (b) Small-signal model.

Input-to-Output Transfer Function

The standard circuit analysis of Fig. 6.14(b) with the condition $\hat{d}(s) = \hat{i}_o(s) = 0$ yields the input-to-output transfer function

$$G_{vs}(s) = \frac{\hat{v}_o(s)}{\hat{v}_s(s)} = K_{vs} \frac{1 + \dfrac{s}{\omega_{esr}}}{1 + \dfrac{s}{Q\omega_o} + \dfrac{s^2}{\omega_o^2}} \tag{6.60}$$

with

$$K_{vs} = \frac{D}{1 + \dfrac{R_l}{R}} \approx D \tag{6.61}$$

and

$$\omega_{esr} = \frac{1}{CR_c} \tag{6.62}$$

The ω_{esr}, appearing in the numerator due to the equivalent series resistance (esr) of the output capacitor, is named the *esr zero*. The power stage double pole ω_o and the quality factor Q are given by

$$\omega_o = \sqrt{\frac{1}{LC}\frac{R + R_l}{R + R_c}} \approx \frac{1}{\sqrt{LC}} \tag{6.63}$$

and

$$Q = \frac{1}{\omega_o} \frac{R + R_l}{L + C(R_l R_c + R_l R + R_c R)} \approx R\sqrt{\frac{C}{L}} \tag{6.64}$$

The approximations in (6.61), (6.63), and (6.64) become accurate with the conditions $R \gg R_l$ and $R \gg R_c$.

When the expression (6.60) is compared with the $G_{vs}(s)$ of an ideal buck converter, given by (5.51) in Example 5.8, the effects of the parasitic resistances can be seen. Most notably, the esr of the output capacitor introduces the esr zero. The effects of the esr zero on the frequency response will be described shortly.

While the parasitic resistances also alter the expressions for the pole frequency and quality factor, these changes are usually insignificant. When the parasitic resistances are sufficiently small, the dc gain, pole frequency and quality factor can be approximated to those of the ideal buck converter, as shown in (6.61), (6.63) and (6.64).

Figure 6.15 shows the asymptotic magnitude plot $|G_{vs}|$ and asymptotic phase plot $\angle G_{vs}$ of the transfer function, formed by following the Bode plot construction rules with assumptions $\omega_o \ll \omega_{esr}$ and $Q > 0.5$. The low-pass filter characteristics, originating from the circuit structure and operational principle of the buck converter, are apparent in Fig. 6.15. As a typical second-order low pass filter, the flat low-frequency asymptote of $|G_{vs}|$ falls off at the double pole frequency ω_o with a -40 dB/dec, or -2, slope. At the esr zero ω_{esr}, the -40 dB/dec slope is changed to the -20 dB/dec, or -1, slope. The low-frequency asymptote of $|G_{vs}|$ is the voltage gain of the buck converter, $|G_{vs}(j0)| = 20\log K_{vs} \approx 20\log D$.

The $\angle G_{vs}$ starts with 0° at low frequencies, drops down to $-180°$ in the mid-frequency band, and finally converges to $-90°$ at high frequencies.† The esr zero in effect provides a 90° phase boost for $\angle G_{vs}$.

The frequency response of the input-to-output transfer function should be interpreted based on the definition discussed in Section 5.4.1. The frequency response only describes the ac component of the circuit variables, as such, the dc component and switching ripple component should be separately considered.

EXAMPLE 6.2 Input-to-Output Transfer Function

This example shows the input-to-output transfer function of a buck converter and illustrates its relationship with time-domain circuit waveforms. Consider a buck converter operating with $V_S = 16$ V, $L = 40\ \mu$F, $R_l = 0.1\ \Omega$, $C = 470\ \mu$F, $R_c = 0.05\ \Omega$, $R = 1\ \Omega$, $f_s = 20$ kHz, and $D = 0.25$. The input-to-output transfer function of the converter is given by

$$G_{vs}(s) = K_{vs}\frac{1+\dfrac{s}{\omega_{esr}}}{1+\dfrac{s}{Q\omega_o}+\dfrac{s^2}{\omega_o^2}}$$

†The asymptotic plot in Fig. 6.15 is drawn with the assumption that the corner frequencies of the transfer function are *widely* separated so that the phase undergoes a complete 90° or 180° change over corner frequencies. However, the transfer function will *not* go through the *complete* phase change due to finite distances among corner frequencies. Thus, the phase does not reduce to $-180°$ after the power stage double pole ω_o but remains between $-90° < \angle G_{vs} < -180°$.

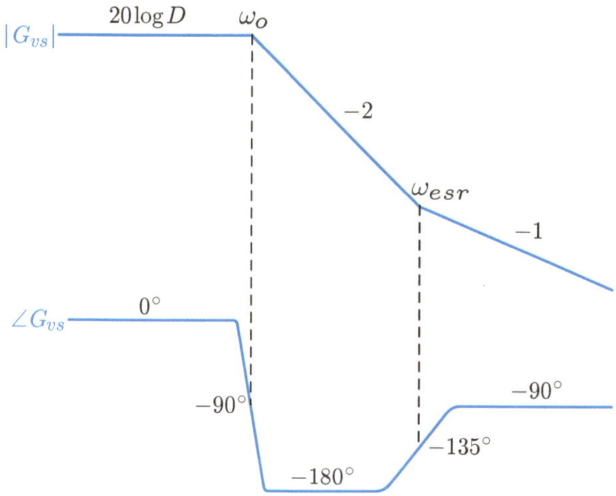

Figure 6.15 Asymptotic plot of input-to-output transfer function.

with

$$K_{vs} \approx D = 0.25 \quad \Rightarrow -12 \text{ dB}$$

$$Q \approx R\sqrt{\frac{C}{L}} = 1\sqrt{\frac{470 \times 10^{-6}}{40 \times 10^{-6}}} = 3.43$$

$$\omega_o \approx \frac{1}{\sqrt{LC}} = \frac{1}{\sqrt{40 \times 10^{-6}\, 470 \times 10^{-6}}} = 2\pi \cdot 1.16 \times 10^3 \text{ rad/s}$$

and

$$\omega_{esr} = \frac{1}{CR_c} = \frac{1}{470 \times 10^{-6}\, 0.05} = 2\pi \cdot 6.77 \times 10^3 \text{ rad/s}$$

Figure 6.16 shows the input-to-output transfer function obtained from PSpice® simulations using the s-domain small-signal model. The magnitude and phase responses at $f = 2$ kHz or $\omega = 2\pi \cdot 2 \times 10^3$ rad/s, highlighted with a rectangle in Fig. 6.16, are used to illustrate the implication of the frequency response. When a 2 kHz sinusoid is added to the dc input voltage, it will propagate to the output with about a 20 dB attenuation in magnitude and a 130° delay in phase. The output voltage is given by the sum of the dc value, switching ripple, and sinusoidal component originating from the input sinusoid. To verify this prediction, the time-domain simulation is performed with

$$v_S(t) = 16 + 0.5 \sin 2\pi \cdot 2 \times 10^3 t$$

Based on the converter operation and prediction of $G_{vs}(j2\pi \cdot 2 \times 10^3)$, the output voltage will be

$$\begin{aligned} v_O(t) &= 0.25 \cdot 16 + 0.1 \cdot 0.5 \sin(2\pi \cdot 2 \times 10^3 t - 130°) + \text{ switching ripple} \\ &= 4 + 0.05 \sin(2\pi \cdot 2 \times 10^3 t - 130°) + \text{switching ripple} \end{aligned}$$

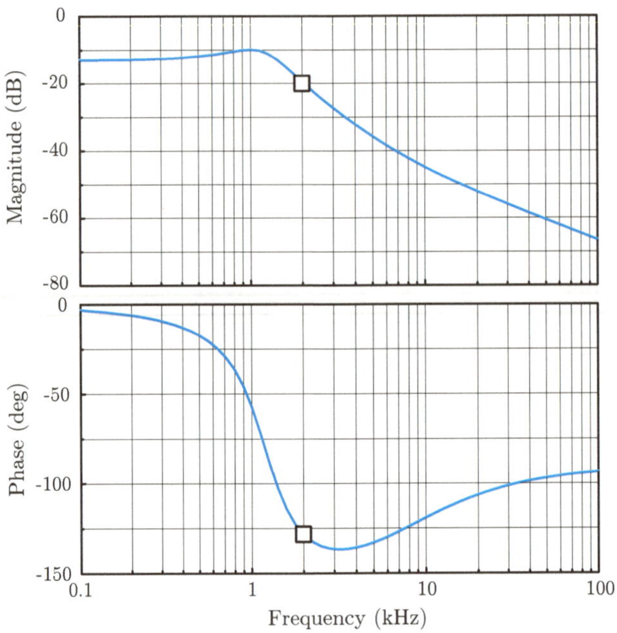

Figure 6.16 Input-to-output transfer function of buck converter.

Figure 6.17 shows the simulation results. When the switching ripple is ignored, the output voltage exhibits a 0.1 V sinusoidal swing as predicted from the input-to-output transfer function. The $-130°$ phase delay between the sinusoidal components of the input and output voltages is apparent.

Duty Ratio-to-Output Transfer Function

The duty ratio-to-output transfer function is evaluated from Fig. 6.14(b) with the condition $\hat{v}_s(s) = \hat{i}_o(s) = 0$

$$G_{vd}(s) = \frac{\hat{v}_o(s)}{\hat{d}(s)} = K_{vd}\frac{1+\dfrac{s}{\omega_{esr}}}{1+\dfrac{s}{Q\omega_o}+\dfrac{s^2}{\omega_o^2}} \tag{6.65}$$

with

$$K_{vd} = \frac{V_S}{1+\dfrac{R_l}{R}} \approx V_S \tag{6.66}$$

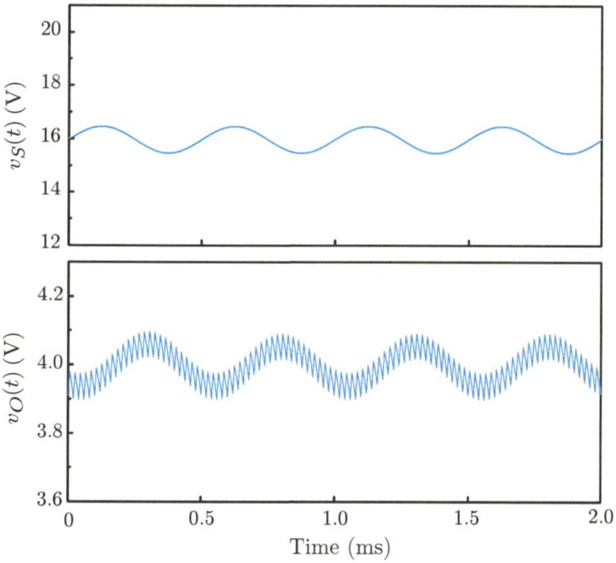

Figure 6.17 Time-domain response of input voltage $v_S(t)$ and output voltage $v_O(t)$ of buck converter.

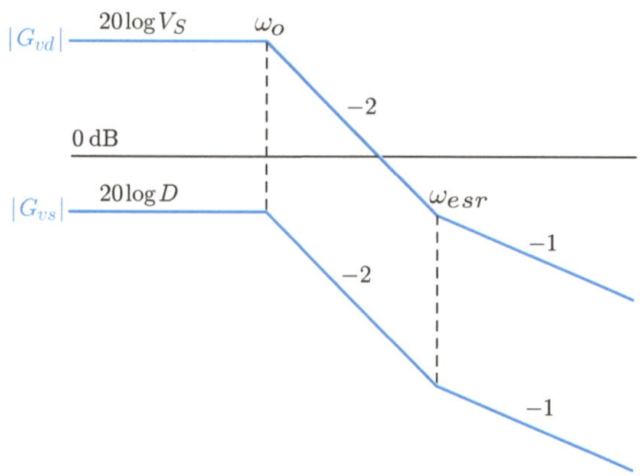

Figure 6.18 Asymptotic plot of input-to-output transfer function and duty ratio-to-output transfer function.

Figure 6.18 shows a typical $|G_{vd}|$ in comparison with $|G_{vs}|$. While the structure of $|G_{vd}|$ is identical to $|G_{vs}|$, the low-frequency asymptote of the transfer function, $|G_{vd}(j0)| \approx 20 \log V_S$, is usually much larger than $|G_{vs}(j0)| \approx 20 \log D$.

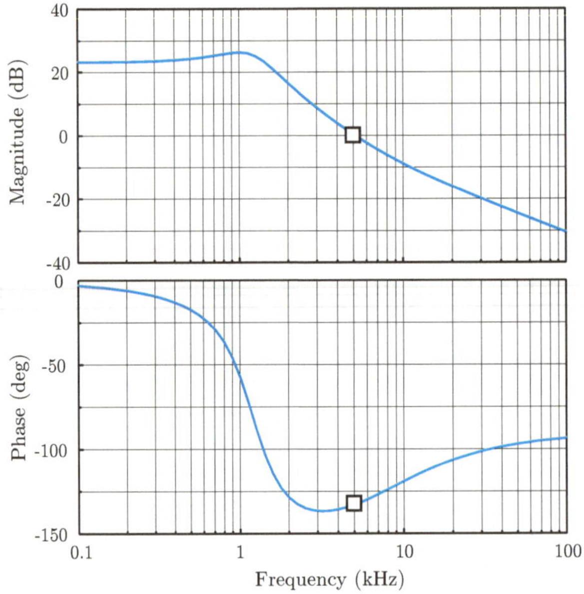

Figure 6.19 Duty ratio-to-output transfer function of buck converter.

EXAMPLE 6.3 Duty Ratio-to-Output Transfer Function

This example investigates the duty ratio-to-output transfer function of the buck converter introduced in Example 6.2. The duty ratio-to-output transfer function is given by

$$G_{vd}(s) = K_{vd}\frac{1+\dfrac{s}{\omega_{esr}}}{1+\dfrac{s}{Q\omega_o}+\dfrac{s^2}{\omega_o^2}}$$

with $K_{vd} \approx V_S = 16 \Rightarrow 24$ dB. Other parameters are the same as those of Example 6.2. Figure 6.19 shows the simulation of $G_{vd}(s)$. Similar to Example 6.2, the time-domain response between the duty ratio and output voltage can be predicted from Fig. 6.19. As highlighted with a rectangle in Fig. 6.19, the 5 kHz sinusoidal variation in the duty ratio will propagate to the output voltage with the same magnitude but with about 135° phase delay. When the switch drive signal is modulated to produce the following continuous duty ratio

$$d(t) = 0.25 + 0.05\sin 2\pi\cdot 5\times 10^3 t$$

the output voltage will be

$$\begin{aligned} v_O(t) &= 0.25\cdot 16 + 1\cdot 0.05\sin(2\pi\cdot 5\times 10^3 t - 135^\circ) + \text{switching ripple} \\ &= 4 + 0.05\sin(2\pi\cdot 5\times 10^3 t - 135^\circ) + \text{switching ripple} \end{aligned}$$

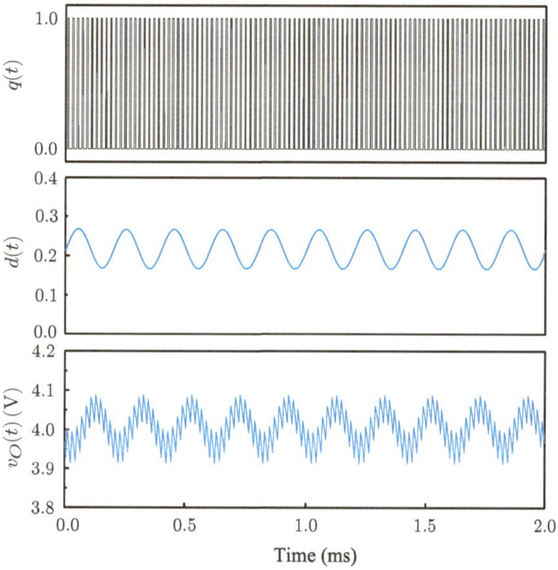

Figure 6.20 Time-domain response of switch drive signal $q(t)$, continuous duty ratio $d(t)$, and output voltage $v_O(t)$.

Figure 6.20 illustrates the time-domain response of the converter, simulated with the aforementioned operational conditions. The switch drive signal $q(t)$ is modulated to produce a 5 kHz sinusoidal variation in $d(t)$ around the steady-state duty ratio. When the switching ripple is neglected, the magnitude of the sinusoidal variation in the output voltage is approximately the same as that of the continuous duty ratio. On the other hand, the 135° phase difference between the continuous duty ratio and the output voltage is evident.

Load Current-to-Output Transfer Function

The load current-to-output voltage transfer function of the converter is derived from Fig. 6.14(b) with the condition $\hat{v}_s(s) = \hat{d}(s) = 0$

$$Z_p(s) = \frac{\hat{v}_o(s)}{\hat{i}_o(s)} = R \parallel R_l \frac{\left(1+\dfrac{s}{\omega_z}\right)\left(1+\dfrac{s}{\omega_{esr}}\right)}{1+\dfrac{s}{Q\omega_o}+\dfrac{s^2}{\omega_o^2}}$$

$$\approx \boxed{R_l \frac{\left(1+\dfrac{s}{\omega_z}\right)\left(1+\dfrac{s}{\omega_{esr}}\right)}{1+\dfrac{s}{Q\omega_o}+\dfrac{s^2}{\omega_o^2}}} \tag{6.67}$$

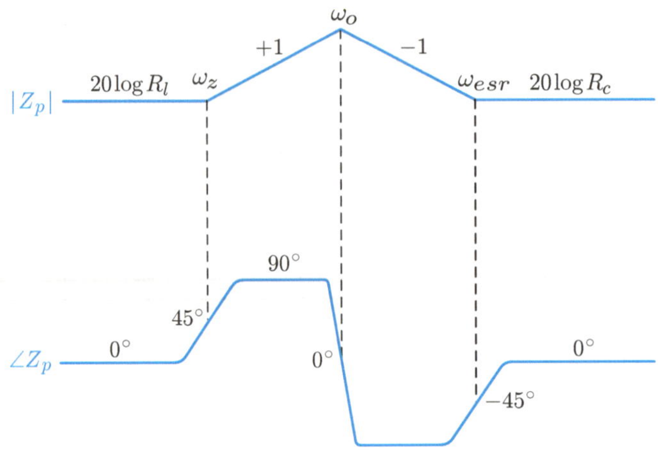

Figure 6.21 Asymptotic plot of load current-to-output transfer function.

with

$$\omega_z = \frac{R_l}{L} \tag{6.68}$$

In addition to ω_{esr}, the transfer function has another zero ω_z, created by the esr of the inductor. While ω_{esr} appears at high frequencies, ω_z is usually located at low frequencies. Accordingly, the low-frequency characteristics of the transfer function are mainly influenced by the esr of the inductor.

Figure 6.21 shows the asymptotic plots for $|Z_p|$ and $\angle Z_p$. The low-frequency asymptote of $|Z_p|$ is the parallel connection of the load resistor and esr of the inductor, which can be approximated to $|Z_p(j0)| = 20\log(R\,\|\,R_l) \approx 20\log R_l$ for most practical cases. The high-frequency asymptote can be found from the small-signal circuit model in Fig. 6.14(b). At high frequencies, the inductance behaves open-circuited and the capacitance presents very low impedance. Thus, the high-frequency asymptote is the parallel combination of the load resistor and esr of the output capacitor, which is approximated to $|Z_p(j\infty)| = 20\log(R\,\|\,R_c) \approx 20\log R_c$.

6.2.2 Power Stage Transfer Functions of Boost Converter

This section analyzes the power stage transfer functions of the boost converter. Figure 6.22(a) shows the power stage of the boost converter and Fig. 6.22(b) depicts its small-signal model. Details about the small-signal model were presented in Section 5.3.3. The analysis starts with the input-to-output transfer function.

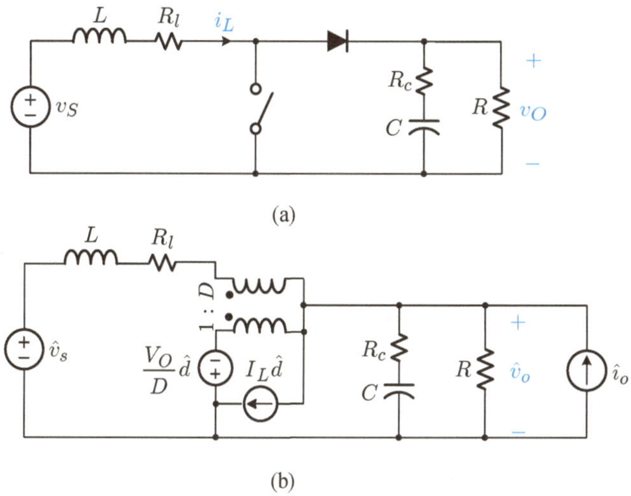

Figure 6.22 Small-signal model of boost converter. (a) Circuit diagram. (b) Small-signal model.

Input-to-Output Transfer Function

The input-to-output transfer function is derived from Fig. 6.22(b) with the condition $\hat{d}(s) = \hat{\imath}_o(s) = 0$

$$G_{vs}(s) = K_{vs} \frac{1 + \dfrac{s}{\omega_{esr}}}{1 + \dfrac{s}{Q\omega_o} + \dfrac{s^2}{\omega_o^2}} \tag{6.69}$$

where

$$K_{vs} = \frac{1}{D'\left(1 + \dfrac{R_l}{RD'^2}\right)} \approx \frac{1}{D'} \tag{6.70}$$

with

$$D' = 1 - D \tag{6.71}$$

and

$$\omega_{esr} = \frac{1}{CR_c} \tag{6.72}$$

$$\omega_o = \sqrt{\frac{1}{LC}\frac{RD'^2 + R_l}{R + R_c}} \approx \frac{D'}{\sqrt{LC}} = \frac{1}{\sqrt{L_e C}} \tag{6.73}$$

with

$$L_e = \frac{L}{D'^2} \tag{6.74}$$

and

$$Q = \frac{1}{\omega_o} \frac{RD'^2 + R_l}{L + C(R_l R_c + R_l R + R_c R D'^2)} \approx R\sqrt{\frac{C}{L_e}} \tag{6.75}$$

An interesting property of the boost converter is observed from the input-to-output transfer function. The transfer function shows the low-pass filter characteristics, as is the case with the buck converter. However, the inductive parameter appears as $L_e = L/D'^2$ in the double pole frequency ω_o and quality factor Q, instead of the original inductance L. This is attributed to the fact that the inductor is not directly connected to the output capacitor but separated by the PWM switch. The inductance located at one side of the PWM switch should be altered when the filter transfer function is evaluated at the other side. The transfer function expression also indicates that the pole frequency and quality factor depend on the duty ratio of the converter. Thus, the power stage dynamics will change when the duty ratio varies.

The low-frequency value of the transfer function is approximated to $|G_{vs}(j0)| = 20\log K_{vs} \approx 20\log(1/D')$, which corresponds to the voltage gain of an ideal boost converter. The asymptotic plots for $|G_{vs}|$ and $\angle G_{vs}$ are essentially the same as those of the buck converter.

Duty Ratio-to-Output Transfer Function and RHP Zero

With the condition $\hat{v}_s(s) = \hat{i}_o(s) = 0$, the duty ratio-to-output transfer function is derived as

$$\boxed{G_{vd}(s) = K_{vd}\frac{\left(1 - \dfrac{s}{\omega_{rhp}}\right)\left(1 + \dfrac{s}{\omega_{esr}}\right)}{1 + \dfrac{s}{Q\omega_o} + \dfrac{s^2}{\omega_o^2}}} \tag{6.76}$$

with

$$K_{vd} = \frac{V_S}{D'^2}\frac{1 - \dfrac{R_l}{RD'^2}}{1 + \dfrac{R_l}{RD'^2}} \approx \frac{V_S}{D'^2} \tag{6.77}$$

and

$$\omega_{rhp} = \frac{D'^2 R}{L}\left(1 - \frac{R_l}{RD'^2}\right) \approx \frac{D'^2 R}{L} = \frac{R}{L_e} \tag{6.78}$$

with $L_e = L/D'^2$.

The most distinctive feature of the transfer function is the presence of a right-half plane (RHP) zero ω_{rhp} in the numerator of (6.76). The zero, ω_{rhp}, is located in the right-hand side of s-plane and the subscript *rhp* is used to signify this fact. The expression for the RHP zero, $\omega_{rhp} = R/L_e = (1 - D)^2 R/L$, indicates that the zero frequency is affected by the duty ratio. Accordingly, the zero frequency will move around when the duty ratio is changed; for this reason, ω_{rhp} is called the *moving* RHP zero.

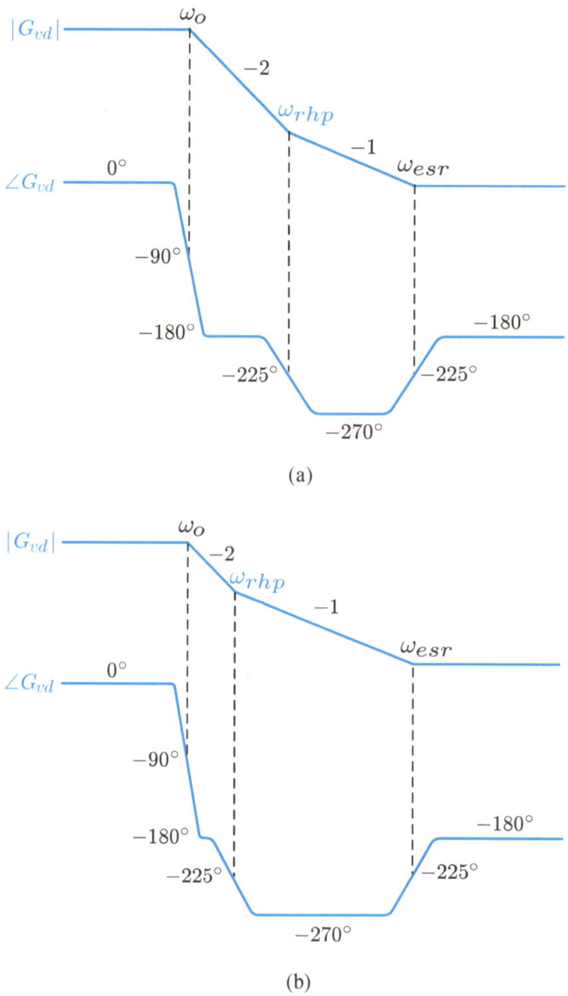

Figure 6.23 Effects of ω_{rhp} on duty ratio-to-output transfer function. (a) Case with nominal operational condition. (b) Case with increased duty ratio or reduced load resistance.

As discussed in Section 6.1.3, an RHP zero increases the magnitude slope by 20 dB/dec but it decreases the phase by 90°. The impacts of the moving RHP zero ω_{rhp} on the duty ratio-to-output transfer function are explained in Fig. 6.23. As far as the magnitude of the transfer function is concerned, ω_{rhp} has the same effect as a regular zero. However, ω_{rhp} causes a 90° phase delay to $\angle G_{vd}$, in contrast to the 90° phase boost in the case of a regular zero. Thus, ω_{rhp} brings down $\angle G_{vd}$ by 90°, while increasing the slope of $|G_{vd}|$ by 20 dB/dec.

Typical asymptotic plots for $|G_{vd}|$ and $\angle G_{vd}$ are shown in Fig. 6.23(a), with the assumption $\omega_o \ll \omega_{rhp} \ll \omega_{esr}$. At frequencies beyond ω_o, $\angle G_{vd}$ mainly stays

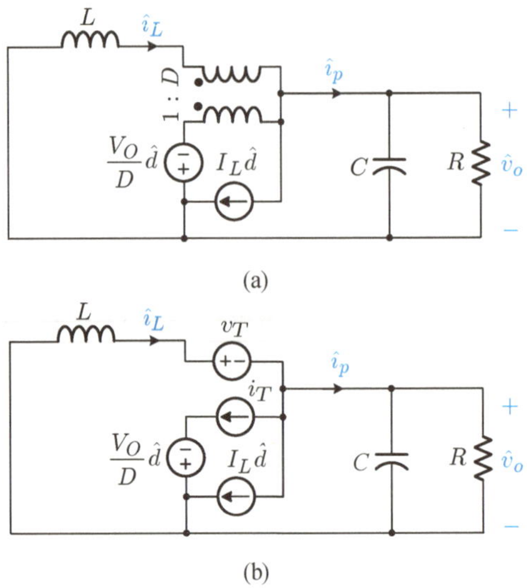

Figure 6.24 Small-signal model of ideal boost converter. (a) Original model. (b) Modified model.

below $-180°$, with a $-270°$ basin in the frequency range of $\omega_{rhp} < \omega < \omega_{esr}$. When the duty ratio is increased from the previous value or the load resistance becomes smaller, ω_{rhp} shifts towards lower frequencies, resulting in the asymptotic plots shown in Fig. 6.23(b). For this case, $\angle G_{vd}$ would stay at the $-270°$ basin for a wider frequency range. This trend intensifies as ω_{rhp} shifts further lower frequencies and the extreme happens when the input voltage is minimum and the load current is maximum within a given operational range. As will be detailed in Section 8.4.7, these phase characteristics present considerable difficulties to the design of the voltage feedback circuit. Accordingly, the control design of the boost converter is more challenging than that of the buck converter. The impact of ω_{rhp} on the control design and converter performance will be covered in later chapters.

EXAMPLE 6.4 RHP Zero in Duty Ratio-to-Output Transfer Function

To confirm the existence of the RHP zero, this example derives the duty ratio-to-output transfer function of the boost converter. Figure 6.24(a) shows the small-signal model of an ideal boost converter. This model is derived from Fig. 6.22(b) for the evaluation of $G_{vd}(s)$ with the conditions $\hat{v}_s(s) = \hat{i}_o(s) = 0$ and $R_l = R_c = 0$. Figure 6.24(a) is modified to Fig. 6.24(b) by replacing the ideal transformer with a pair of voltage source v_T and current source i_T

$$v_T = -\left(\hat{v}_o + \frac{V_O}{D}\hat{d}\right) D = -(D\hat{v}_o + V_O\hat{d}) \tag{6.79}$$

$$i_T = D\hat{\imath}_L \tag{6.80}$$

The inductor current $\hat{\imath}_L$ is then given by

$$\hat{\imath}_L = -\left(\frac{\hat{v}_o + v_T}{sL}\right) = \frac{-(1-D)\hat{v}_o + V_O\hat{d}}{sL} \tag{6.81}$$

The current coming out of the passive terminal of the PWM switch, $\hat{\imath}_p$ in Fig. 6.24(b), is given by

$$\hat{\imath}_p = \hat{\imath}_L - i_T - I_L\hat{d} = \hat{\imath}_L - D\hat{\imath}_L - I_L\hat{d} = (1-D)\hat{\imath}_L - I_L\hat{d} \tag{6.82}$$

The output voltage $\hat{v}_o$ is now determined as

$$\hat{v}_o = \hat{\imath}_p\left(\frac{1}{sC} \parallel R\right) \tag{6.83}$$

By substituting (6.81) and (6.82) into (6.83), it follows that

$$\hat{v}_o = \left((1-D)\frac{-(1-D)\hat{v}_o + V_O\hat{d}}{sL} - I_L\hat{d}\right)\frac{R}{1+sCR} \tag{6.84}$$

which can be arranged in a transfer function form

$$G_{vd}(s) = \frac{\hat{v}_o(s)}{\hat{d}(s)} = \frac{V_O}{1-D}\frac{1-\dfrac{sL}{(1-D)^2R}}{1+\dfrac{sL}{(1-D)^2R}+\dfrac{s^2LC}{(1-D)^2}} \tag{6.85}$$

The presence of the RHP zero, $\omega_{rhp} = (1-D)^2R/L = R/L_e$, is explicit in (6.85). While the derivation becomes somewhat complicated when the parasitic resistances are included, the same procedure leads to the equation (6.76).

Duty Ratio-to-Output Transfer Function of Boost Converter EXAMPLE 6.5

This example shows the duty ratio-to-output transfer function of a boost converter. The operational conditions of the boost converter are $V_S = 12$ V, $L = 800$ μH, $R_l = 0.01$ Ω, $C = 1000$ μF, $R_c = 0.05$ Ω, $R = 2$ Ω, $f_s = 10$ kHz, and $D = 0.25$. The duty ratio-to-output transfer function is given by

$$G_{vd}(s) = K_{vd}\frac{\left(1-\dfrac{s}{\omega_{rhp}}\right)\left(1+\dfrac{s}{\omega_{esr}}\right)}{1+\dfrac{s}{Q\omega_o}+\dfrac{s^2}{\omega_o^2}} \tag{6.86}$$

with

$$K_{vd} \approx \frac{V_S}{(1-D)^2} = \frac{12}{(1-0.25)^2} = 21.3 \quad \Rightarrow 26.6 \text{ dB}$$

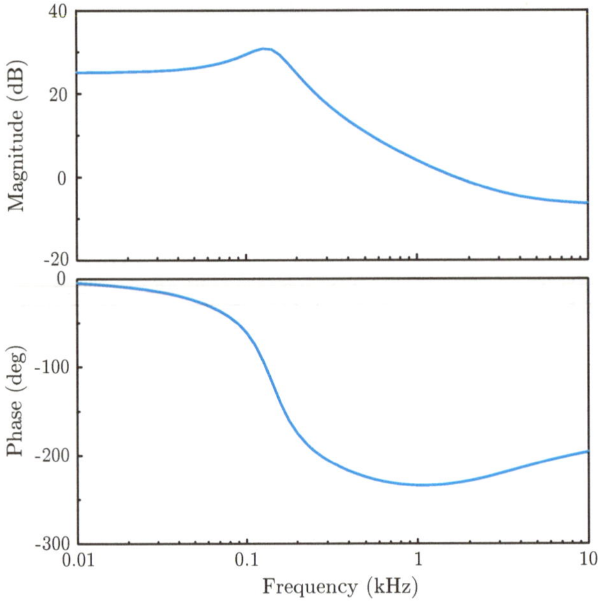

Figure 6.25 Duty ratio-to-output transfer function of boost converter.

$$\begin{aligned}\omega_o &\approx \sqrt{\frac{(1-D)^2}{LC}} = \sqrt{\frac{(1-0.25)^2}{800\times10^{-6}\,1000\times10^{-6}}} \\ &= 2\pi\cdot 133 \text{ rad/s}\end{aligned}$$

$$Q \approx R\sqrt{\frac{(1-D)^2C}{L}} = 2\sqrt{\frac{(1-0.25)^2 1000\times10^{-6}}{800\times10^{-6}}} = 1.68$$

$$\begin{aligned}\omega_{rhp} &\approx \frac{(1-D)^2}{L}R = \frac{(1-0.25)^2}{800\times10^{-6}}2 \\ &= 2\pi\cdot 224 \text{ rad/s}\end{aligned}$$

and

$$\omega_{esr} = \frac{1}{CR_c} = \frac{1}{1000\times10^{-6}\,0.05} = 2\pi\cdot 3.18\times10^3 \text{ rad/s}$$

Figure 6.25 shows the duty ratio-to-output transfer function obtained from PSpice® simulations using Fig. 6.22(b). The impact of the RHP zero at $\omega_{rhp} = 2\pi\cdot 224$ rad/s is clear in the Bode plot. In particular, the phase remains lower than $-200°$ after the power stage double pole, $\omega_o = 2\pi\cdot 133$ rad/s. As will be demonstrated in Section 8.4.7, these phase characteristics complicate the control design and hinder the boost converter from having acceptable closed-loop performance.

Load Current-to-Output Transfer Function

The load current-to-output transfer function is evaluated as

$$\boxed{Z_p(s) = K_p \frac{\left(1+\frac{s}{\omega_z}\right)\left(1+\frac{s}{\omega_{esr}}\right)}{1+\frac{s}{Q\omega_o}+\frac{s^2}{\omega_o^2}}} \tag{6.87}$$

with

$$K_p = R \parallel \frac{R_l}{(1-D)^2} \approx \frac{R_l}{(1-D)^2} \tag{6.88}$$

and

$$\omega_z = \frac{R_l}{L} \tag{6.89}$$

The asymptotic plots for $|Z_p|$ and $\angle Z_p$ are the same as those of the buck converter, shown in Fig. 6.21, except for the low-frequency magnitude of Z_p: $|Z_p(0)| = 20\log(R_l/(1-D)^2)$.

Physical Origin of RHP Zero

As the most salient feature, the boost converter has an RHP zero in its duty ratio-to-output transfer function. While the direct evaluation of the transfer function led to the expression of the RHP zero, the existence of the RHP zero can also be confirmed from the operational principle of the boost converter. For this purpose, we need to investigate the time-domain response of the system with an RHP zero in its transfer function; this type of system was previously referred to as the non-minimum phase system. To be specific, assume the following second-order equation as the transfer function of a non-minimum phase system

$$\frac{v_o(s)}{d(s)} = T(s) = \frac{1-\frac{s}{\omega_z}}{1+\frac{s}{Q\omega_o}+\frac{s^2}{\omega_o^2}} \tag{6.90}$$

The unit step response of the system is given by

$$v_O(t) = \mathcal{L}^{-1}(d(s)T(s)) = \mathcal{L}^{-1}\left(\frac{1}{s}\frac{1-\frac{s}{\omega_z}}{1+\frac{s}{Q\omega_o}+\frac{s^2}{\omega_o^2}}\right) \tag{6.91}$$

using the fact $d(s) = 1/s$ for the unit step change in the continuous duty ratio. The above equation is split into two parts

$$\begin{aligned} v_O(t) &= \mathcal{L}^{-1}\left(\frac{1}{s}\frac{1}{1+\dfrac{s}{Q\omega_o}+\dfrac{s^2}{\omega_o^2}}\right) \\ &- \frac{1}{\omega_z}\mathcal{L}^{-1}\left(s\left(\frac{1}{s}\frac{1}{1+\dfrac{s}{Q\omega_o}+\dfrac{s^2}{\omega_o^2}}\right)\right) \end{aligned} \tag{6.92}$$

where $\mathcal{L}^{-1}$ denotes the inverse Laplace transform operation. By noting that the multiplication with s in s-domain corresponds to the derivative operation in time-domain, the expression (6.92) is written as

$$v_O(t) = \tilde{v}_O(t) - \frac{1}{\omega_z}\frac{d\tilde{v}_O(t)}{dt} \tag{6.93}$$

with

$$\tilde{v}_O(t) = \mathcal{L}^{-1}\left(\frac{1}{s}\frac{1}{1+\dfrac{s}{Q\omega_o}+\dfrac{s^2}{\omega_o^2}}\right) \tag{6.94}$$

The transient response is expressed as the sum of two terms. The first term is a transient waveform given by (6.94). The second term is generated by taking the derivative of the first term and multiplying it by the negative inverse of the RHP zero frequency. Figure 6.26 illustrates the construction of v_O based on (6.93). As shown in this figure, v_O dips into the negative value before it proceeds towards the positive value. As the RHP zero frequency ω_z moves closer to the origin, the dip becomes deeper. This initial dip is the distinctive transitional behavior of a non-minimum phase system and thus can be a criterion to judge the existence of the RHP zero.

The presence of an RHP zero in the duty ratio-to-output transfer function is deduced from the transient waveforms of the boost converter. Figure 6.27 illustrates the circuit waveforms of a boost converter experiencing a step increase in its duty ratio at $t = t^*$. Although the duty ratio is instantly increased, the inductor current slowly rises towards the final value based on the power stage dynamics. The diode current i_D, which corresponds to the off-time inductor current, gradually increases in its peak value but abruptly decreases in width. Accordingly, the moving average of the diode current, $\bar{i}_D$, initially droops until the inductor current becomes sufficiently large. This situation is illustrated in Fig. 6.27.

The output voltage is proportional to the moving average of the diode current $\bar{i}_D$. Thus, the initial droop in $\bar{i}_D$ causes the transitional undershoot in the output voltage at the beginning of the transition period. This validates the presence of an RHP zero

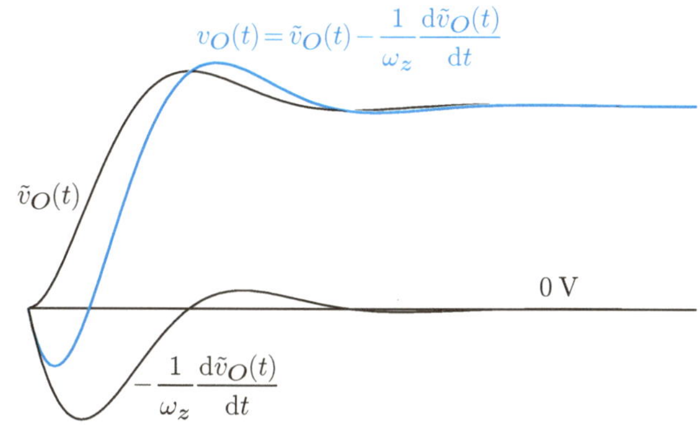

Figure 6.26 Unit step response of non-minimum phase system.

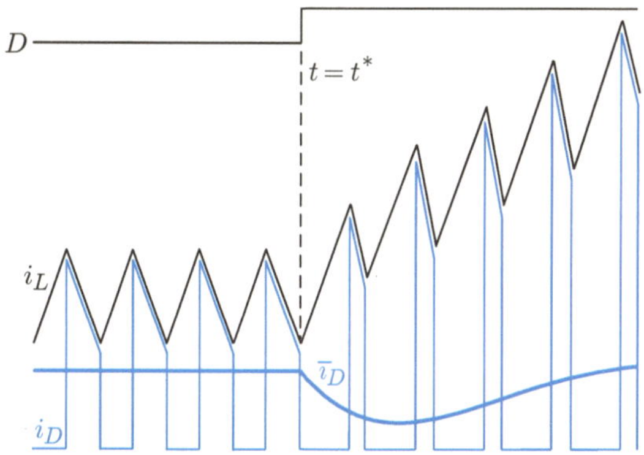

Figure 6.27 Transient response of boost converter with step increase in duty ratio.

in the duty ratio-to-output transfer function. This phenomenon commonly occurs in all the PWM converters in which the load current is supported by the diode current, such as the boost converter, buck/boost converter, and all isolated converters derived from these two converters.

Transient Response of Boost Converter — EXAMPLE 6.6

The transitional behavior of a boost converter, illustrated in Fig. 6.27, is verified by time-domain simulations. Figure 6.28 shows the transient response of the boost converter introduced in Example 6.5. The boost converter has an RHP zero at $\omega_{rhp} = 2\pi \cdot 224$ rad/s. The duty ratio of the boost converter undergoes a step increase from $D = 0.25$ to 0.5 at $t = 5$ ms. As shown in Fig. 6.28, the output voltage v_O reveals an initial droop before it proceeds towards a new steady-state

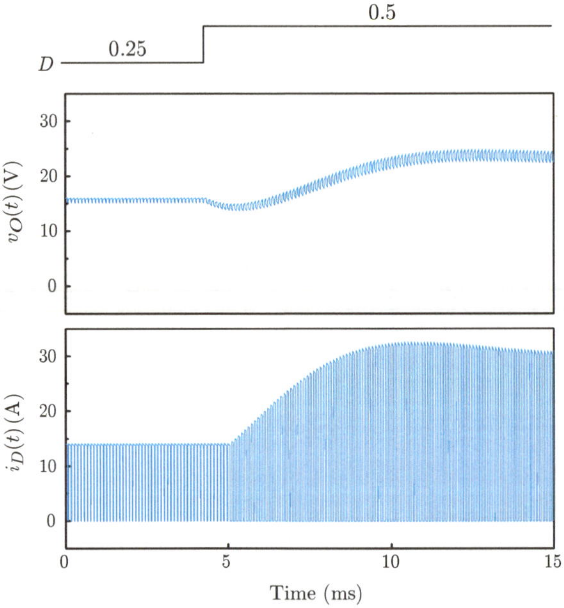

Figure 6.28 Transient waveform of output voltage v_O and diode current i_D of boost converter.

value, thereby verifying the existence of the RHP zero. Figure 6.28 also shows the diode current i_D during the transient period. The abrupt decrease in the width of the diode current is responsible for the initial droop in the output voltage.

6.2.3 Power Stage Transfer Functions of Buck/Boost Converter

This section deals with the power stage transfer functions of the buck/boost converter. Figure 6.29(a) depicts the power stage of the buck/boost converter and Fig. 6.29(b) is the small-signal model obtained by adopting the PWM switch model. Section 5.3.3 presented the details about the small-signal model of the buck/boost converter.

With the condition $\hat{d}(s) = \hat{\imath}_o(s) = 0$, the input-to-output transfer function is derived as

$$G_{vs}(s) = K_{vs}\frac{1+\dfrac{s}{\omega_{esr}}}{1+\dfrac{s}{Q\omega_o}+\dfrac{s^2}{\omega_o^2}} \tag{6.95}$$

where

$$K_{vs} = \frac{D}{D'\left(1+\dfrac{R_l}{RD'^2}\right)} \approx \frac{D}{D'} \tag{6.96}$$

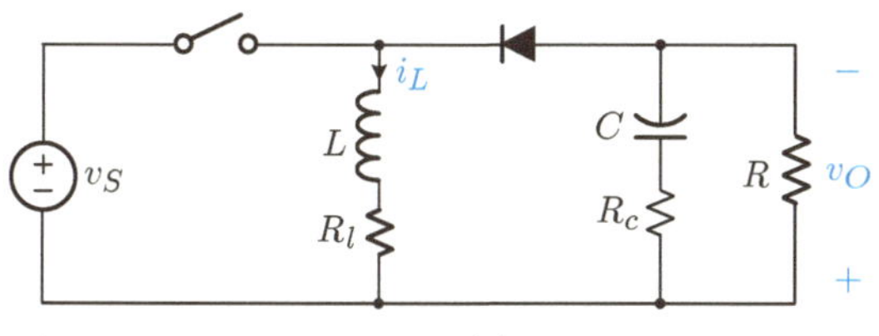

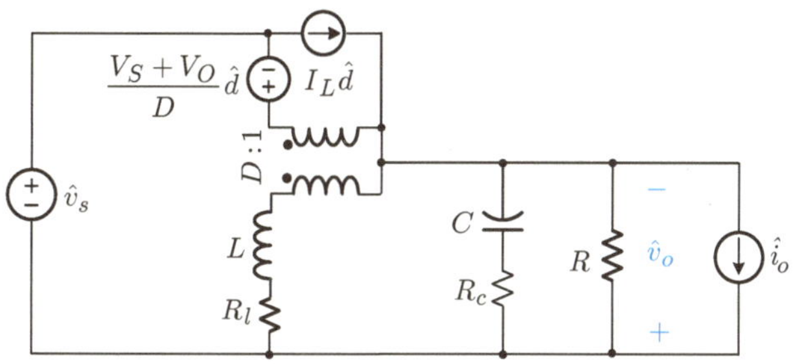

Figure 6.29 Small-signal model of buck/boost converter. (a) Circuit diagram. (b) Small-signal.

$$\omega_{esr} = \frac{1}{CR_c} \tag{6.97}$$

$$\omega_o = \sqrt{\frac{1}{LC}\frac{RD'^2 + R_l}{R + R_c}} \approx \frac{D'}{\sqrt{LC}} = \frac{1}{\sqrt{L_e C}} \tag{6.98}$$

$$Q = \frac{1}{\omega_o}\frac{RD'^2 + R_l}{L + C(R_l R_c + R_l R + R_c R D'^2)} \approx R\sqrt{\frac{C}{L_e}} \tag{6.99}$$

with

$$L_e = \frac{L}{D'^2} \tag{6.100}$$

The low-frequency value of the transfer function is the voltage gain of the buck/boost converter: $20 \log K_{vs} \approx 20 \log(D/D')$.

The duty ratio-to-output transfer function is given by

$$G_{vd}(s) = K_{vd}\frac{\left(1 - \frac{s}{\omega_{rhp}}\right)\left(1 + \frac{s}{\omega_{esr}}\right)}{1 + \frac{s}{Q\omega_o} + \frac{s^2}{\omega_o^2}} \tag{6.101}$$

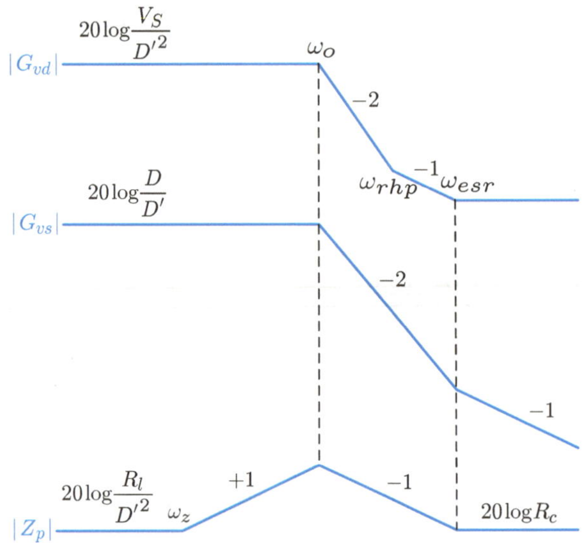

Figure 6.30 Asymptotic plots for transfer functions of buck/boost converter.

where

$$K_{vd} = \frac{V_S}{D'^2}\frac{1-\dfrac{R_l D}{RD'^2}}{1+\dfrac{R_l}{RD'^2}} \approx \frac{V_S}{D'^2} \tag{6.102}$$

and

$$\omega_{rhp} = \frac{D'^2 R}{DL}\left(1-\frac{R_l D}{RD'^2}\right) \approx \frac{D'^2 R}{DL} = \frac{R}{DL_e} \tag{6.103}$$

with $L_e = L/D'^2$.

As expected from the circuit structure and operational principle of the converter, the transfer function has a moving RHP zero, ω_{rhp}, as is the case with the boost converter. Finally, the load current-to-output transfer function is derived as

$$Z_p(s) = K_p\frac{\left(1+\dfrac{s}{\omega_z}\right)\left(1+\dfrac{s}{\omega_{esr}}\right)}{1+\dfrac{s}{Q\omega_o}+\dfrac{s^2}{\omega_o^2}} \tag{6.104}$$

with

$$K_p = R \parallel \frac{R_l}{D'^2} \approx \frac{R_l}{D'^2} \tag{6.105}$$

and

$$\omega_z = \frac{R_l}{L} \tag{6.106}$$

Asymptotic plots for the power stage transfer functions of the buck/boost converter are shown in Fig. 6.30. The structure and attributes of the asymptotic plots are very similar to those of the boost converter.

6.3 POWER STAGE TRANSFER FUNCTIONS IN DCM OPERATION

The previous section analyzed the power stage transfer functions of the three basic converters under continuous conduction mode (CCM) operation. This section now extends the transfer function analysis to the discontinuous conduction mode (DCM) operation. The distinctive features of the DCM transfer functions are addressed, focusing on the duty ratio-to-output transfer function of the three basic converters.

6.3.1 Evaluation of DCM Transfer Functions

When dc-to-dc converters enter DCM operation, their power stage waveforms go through substantial changes from those of the CCM counterpart. Based on the DCM power stage waveforms, the DCM small-signal model for the PWM switch was developed in Section 5.5.2. The resulting *DCM PWM switch model* can be used to analyze DCM transfer functions of converter power stages.

As an illustration, Fig. 6.31 shows the circuit diagram of a buck/boost converter and its DCM small-signal model, obtained by replacing the PWM switch with the DCM PWM switch model. Figure 6.32 shows the duty ratio-to-output transfer function of the buck/boost converter, simulated using the DCM small-signal model of Fig. 6.31(b).

The operational conditions of the buck/boost converter are $V_S = 18\,\text{V}, L = 40\,\mu\text{H}$, $R_l = 0.01\ \Omega$, $C = 470\ \mu\text{F}$, $R_c = 0.03\ \Omega$, $R = 20\ \Omega$, $f_s = 50$ kHz, and $D = 0.25$. The critical resistance that places the converter on the CCM/DCM boundary is

$$R_{crit} = \frac{2L}{(1-D)^2 T_s} = \frac{2 \cdot 40 \times 10^{-6}}{(1-0.25)^2\, 20 \times 10^{-6}} = 7.1\ \Omega$$

As shown in Fig. 6.31(b), the DCM PWM switch model contains the five parameters, whose expressions were given in (5.78) in Section 5.5.2. Referring to (5.78), the parameters of DCM PWM switch model are evaluated as follows. First, the DCM voltage gain M and output voltage V_O are evaluated using the fact $V_O/V_S = I_S/I_O$

$$M = \frac{V_O}{V_S} = D\sqrt{\frac{RT_s}{2L}} = 0.25\sqrt{\frac{20 \cdot 20 \times 10^{-6}}{2 \cdot 40 \times 10^{-6}}} = 0.56$$

$$V_O = MV_S = 0.56 \cdot 18 = 10.08\ \text{V}$$

The inverse current gain μ and input current I_S are determined as

$$\mu = \frac{I_S}{I_O} = M = 0.56 \qquad I_S = \mu\frac{V_O}{R} = 0.56\frac{10.08}{20} = 0.28\ \text{A}$$

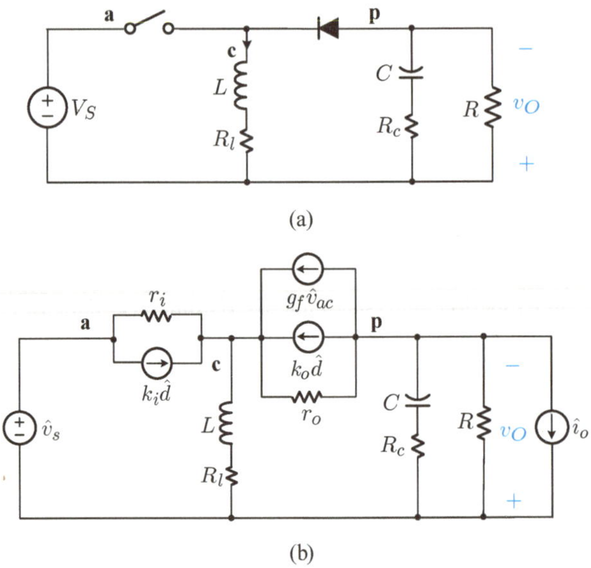

Figure 6.31 Buck/boost converter and DCM small-signal model. (a) Buck/boost converter. (b) DCM small-signal model.

The five parameters for the DCM PWM switch model are now given by

$$r_i = \frac{V_O}{MI_S} = \frac{10.08}{0.56 \cdot 0.28} = 64.3 \qquad k_i = \frac{2I_S}{D} = \frac{2 \cdot 0.28}{0.25} = 2.24$$

$$g_f = 2\frac{I_S M}{\mu V_O} = \frac{2 \cdot 0.28}{10.08} = 5.56 \times 10^{-2} \qquad k_o = \frac{2I_S}{\mu D} = \frac{2 \cdot 0.28}{0.5 \cdot 0.25} = 4$$

$$r_o = \frac{\mu V_O}{I_S} = \frac{0.56 \cdot 10.08}{0.28} = 20$$

The above parameters are incorporated into Fig. 6.31(b) to produce the Bode plot in Fig. 6.32. The Bode plot demonstrates the distinctive features of the DCM transfer function, which will be analyzed in the next section.

6.3.2 Analysis of DCM Duty Ratio-to-Output Transfer Function

Although the DCM small-signal model of the power stage is suited for frequency-domain simulations, further analyses of the small-signal model are necessary, in order to understand the distinct DCM dynamics demonstrated in Fig. 6.32. Such analyses were performed in [4] and the equations for the duty ratio-to-output transfer functions of the three basic converters were provided. Table 6.1 shows the outcome of these

Table 6.1 Expressions for $G_{vd}(s)$ for Three Basic Converters in DCM

$$G_{vd}(s) \approx K_d \frac{\left(1 - \frac{s}{\omega_{rhp}}\right)\left(1 + \frac{s}{\omega_{esr}}\right)}{\left(1 + \frac{s}{\omega_{p1}}\right)\left(1 + \frac{s}{\omega_{p2}}\right)}$$

Buck converter

$$K_d = \frac{2V_O}{D}\frac{1-M}{2-M} \quad \text{with} \quad M = \frac{2D}{D + \sqrt{D^2 + \frac{8L}{RT_s}}}$$

$$\omega_{p1} = \frac{1}{CR}\frac{2-M}{1-M} \qquad \omega_{p2} = 2f_s\left(\frac{M}{D}\right)^2$$

$$\omega_{rhp} = \infty \qquad \omega_{esr} = \frac{1}{CR_c}$$

Boost converter

$$K_d = \frac{2V_O}{D}\frac{M-1}{2M-1} \quad \text{with} \quad M = \frac{1}{2}\left(1 + \sqrt{1 + \frac{2D^2RT_s}{L}}\right)$$

$$\omega_{p1} = \frac{1}{CR}\frac{2M-1}{M-1} \qquad \omega_{p2} = 2f_s\left(\frac{1-1/M}{D}\right)^2$$

$$\omega_{rhp} = \frac{R}{M^2L} \qquad \omega_{esr} = \frac{1}{CR_c}$$

Buck/boost converter

$$K_d = \frac{V_O}{D} \quad \text{with} \quad M = D\sqrt{\frac{RT_s}{2L}}$$

$$\omega_{p1} = \frac{2}{CR} \qquad \omega_{p2} = 2f_s\left(\frac{1/D}{1+1/M}\right)^2$$

$$\omega_{rhp} = \frac{R}{M(1+M)L} \qquad \omega_{esr} = \frac{1}{CR_c}$$

For the buck converter, ω_{rhp} does not exist so $\omega_{rhp} = \infty$ is used.

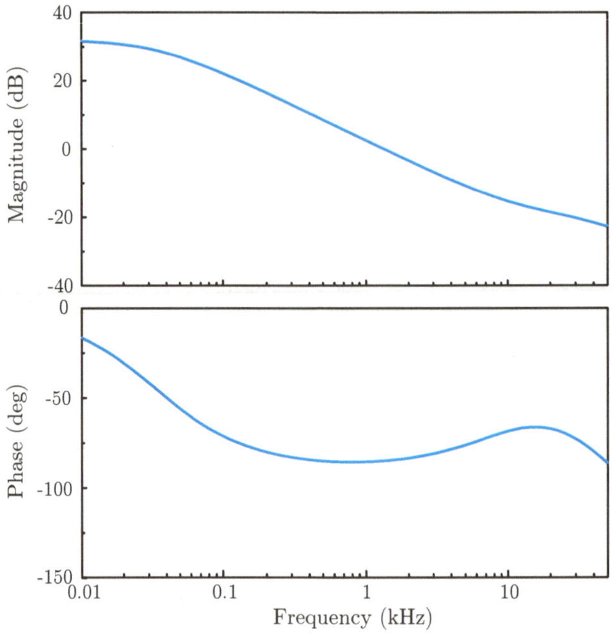

Figure 6.32 Duty ratio-to-output transfer function of buck/boost converter in DCM.

analyses. As shown in the table, the duty ratio-to-output transfer function, $G_{vd}(s)$, of all the three basic PWM converters has the common structure

$$G_{vd}(s) \approx K_d \frac{\left(1 - \dfrac{s}{\omega_{rhp}}\right)\left(1 + \dfrac{s}{\omega_{esr}}\right)}{\left(1 + \dfrac{s}{\omega_{p1}}\right)\left(1 + \dfrac{s}{\omega_{p2}}\right)} \tag{6.107}$$

The denominator of the DCM $G_{vd}(s)$ consists of two real poles, in contrast to a complex double pole in the CCM case. The two real poles produce the specific DCM transfer function characteristics, which were observed in Fig. 6.32 and will be further discussed in the following example.

EXAMPLE 6.7 Analysis of DCM Duty Ratio-to-Output Transfer Function

This example presents qualitative discussions about the DCM duty ratio-to-output transfer function of the buck/boost converter. Figure 6.33 shows a family of the duty ratio-to-output transfer functions of the buck/boost converter used in Fig. 6.32. The DCM transfer functions, evaluated with $R = 7.1\ \Omega$, $10\ \Omega$, and $20\ \Omega$, are shown in comparison with the CCM transfer functions with $R = 0.2\ \Omega\,(Q \approx D'R\sqrt{C/L} = 0.5)$, $0.4\ \Omega\,(Q = 1)$, and $1\ \Omega\,(Q = 2.5)$. The critical resistance of the converter was $R_{crit} = 7.1\ \Omega$.

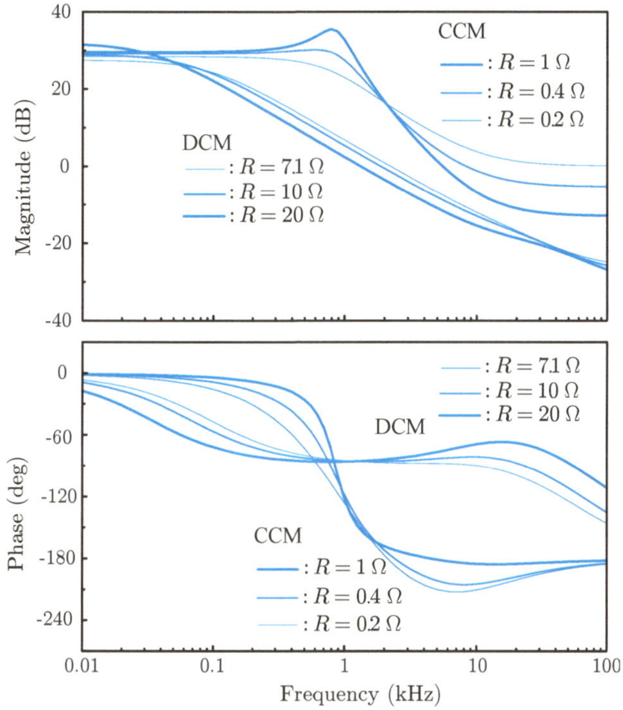

Figure 6.33 Duty ratio-to-output transfer functions of buck/boost converter in CCM and DCM operation.

The CCM transfer functions demonstrate the presence of the double pole at $\omega_o \approx D'\sqrt{1/(LC)} = 2\pi \cdot 871$ rad/s with the respective peaking of $20 \log Q$. As the converter moves into DCM operation, the transfer functions exhibit notable changes.

1) The double pole is heavily damped and split into two real poles, ω_{p1} and ω_{p2} in (6.107). The cause of this damping is the two resistive parameters, r_i and r_o, in the DCM PWM switch model. The damping will intensify as the load resistance R becomes larger. Referring to the $G_{vd}(s)$ expression of the buck/boost converter in Table 6.1, the behavior of the two real poles are described below.

   - The first real pole, $\omega_{p1} = 2/(CR)$, appears at low frequencies and approaches the origin as R increases towards infinity.
   - The second pole ω_{p2} usually occurs at higher frequencies than ω_s/π.

2) In the frequencies below ω_s/π, the transfer function is dominated by the first pole ω_{p1} and exhibits the first-order system dynamics. The second pole ω_{p2} only becomes influential at very high frequencies and virtually disappears from the converter dynamics at the frequencies of practical importance. The first-order system behavior is apparent in both magnitude and phase characteristics.

3) Due to the presence of the first pole ω_{p1} at low frequencies, the magnitude of the transfer function is substantially reduced. The reduction in the magnitude will adversely affect the closed-loop performance in DCM, as will be shown later in Chapter 8.

As demonstrated in the preceding buck/boost converter example, the DCM operation transforms the second-order system into the first-order system. By examining the $G_{vd}(s)$ expressions in Table 6.1, the same conclusion is extended to buck and boost converters. This apparent change in the power stage transfer function should be considered for dc-to-dc converters encountering both CCM and DCM operations. Chapter 8 covers the details about this topic.

6.4 POWER STAGE TRANSFER FUNCTIONS OF ISOLATED CONVERTERS

This section analyzes power stage transfer functions of isolated converters. As discussed in Section 5.5.3, the small-signal model of an isolated converter was derived from the small-signal of the non-isolated converter from which that converter is evolved. Once the small-signal model is found, the power stage transfer functions of the isolated converter can be analyzed in the exact same manner as the case of the previous non-isolated converters. This section presents examples of such analyses using a tertiary-winding reset forward converter and flyback converter.

Tertiary-Winding Reset Forward Converter

As the first example, Fig. 6.34 shows a tertiary-winding reset forward converter and its small-signal model. Details about the small-signal model were given in Section 5.5.3. The transfer functions of the tertiary-winding reset forward converter are easily inferred from those of the buck converter. By comparing Fig. 6.34(b) with Fig. 6.14(b), the duty ratio-to-output transfer function is determined as

$$G_{vd}(s) = \frac{\hat{v}_o(s)}{\hat{d}(s)} = K_{vd}\frac{1+\dfrac{s}{\omega_{esr}}}{1+\dfrac{s}{Q\omega_o}+\dfrac{s^2}{\omega_o^2}} \tag{6.108}$$

with

$$K_{vd} = \frac{nV_S}{1+\dfrac{R_l}{R}} \approx nV_S \tag{6.109}$$

$$\omega_{esr} = \frac{1}{CR_c} \tag{6.110}$$

$$\omega_o = \sqrt{\frac{1}{LC}\frac{R+R_l}{R+R_c}} \approx \frac{1}{\sqrt{LC}} \tag{6.111}$$

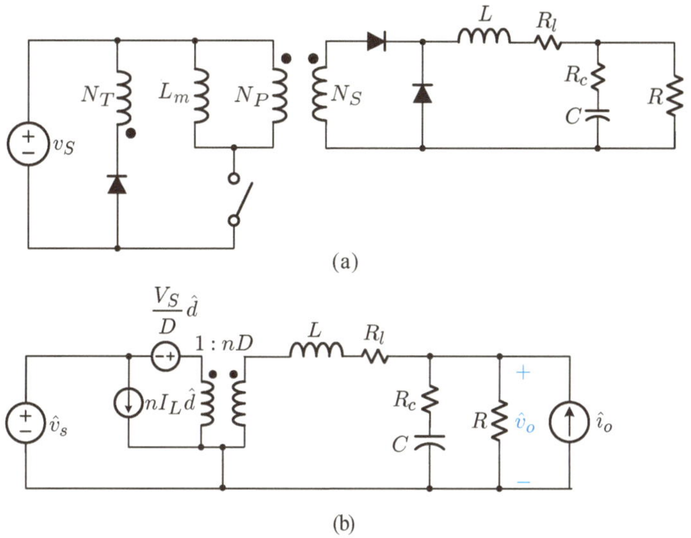

Figure 6.34 Tertiary-winding reset forward converter. (a) Original model. (b) Small-signal model.

and

$$Q = \frac{1}{\omega_o} \frac{R + R_l}{L + C(R_l R_c + R_l R + R_c R)} \approx R\sqrt{\frac{C}{L}} \tag{6.112}$$

Figure 6.35 shows the duty ratio-to-output transfer function of a tertiary-winding reset forward converter. The operational conditions and circuit parameters are $V_S = 64\,\text{V}$, $L = 40\,\mu\text{H}$, $R_l = 0.01\,\Omega$, $C = 400\,\mu\text{F}$, $R_c = 0.02\,\Omega$, $R = 1\,\Omega$, $f_s = 50\,\text{kHz}$, and $D = 0.25$. The turns ratio of the ideal three-winding transformer is $N_P : N_S : N_T = 36{:}18{:}36$ and the magnetizing inductance is $L_m = 200\,\mu\text{H}$. The parameters of the duty ratio-to-output transfer function are evaluated as

$$K_{vd} \approx nV_S = 0.5 \cdot 64 = 32 \quad \Rightarrow \quad 30\text{ dB}$$

$$\omega_{esr} = \frac{1}{CR_c} = \frac{1}{400 \times 10^{-6} 0.02} = 2\pi \cdot 1.99 \times 10^4\text{ rad/s}$$

$$\omega_o \approx \frac{1}{\sqrt{LC}} = \frac{1}{\sqrt{40 \times 10^{-6} 400 \times 10^{-6}}} = 2\pi \cdot 1.26 \times 10^3\text{ rad/s}$$

and

$$Q \approx R\sqrt{\frac{C}{L}} = 1\sqrt{\frac{400 \times 10^{-6}}{40 \times 10^{-6}}} = 3.16$$

The transfer function reveals the same characteristics as those of the buck converter shown in Example 6.3.

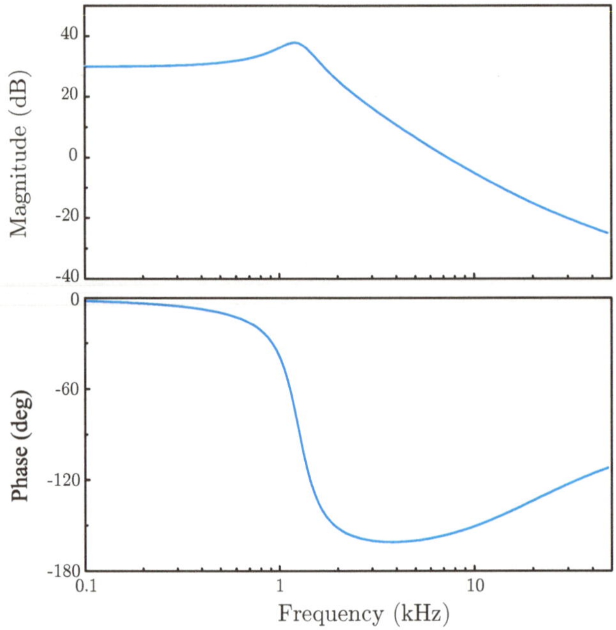

Figure 6.35 Duty ratio-to-output transfer function of tertiary-winding reset forward converter.

Flyback Converter

This section analyzes the duty ratio-to-output transfer function of the flyback converter. Figure 6.36 shows the original circuit diagram, equivalent buck/boost converter representation, and small-signal model of a flyback converter, all discussed in detail in Section 5.5.3. Similar to the previous case, the duty ratio-to-output transfer function of the flyback converter is evaluated as

$$G_{vd}(s) = K_{vd} \frac{\left(1 - \frac{s}{\omega_{rhp}}\right)\left(1 + \frac{s}{\omega_{esr}}\right)}{1 + \frac{s}{Q\omega_o} + \frac{s^2}{\omega_o^2}} \tag{6.113}$$

with

$$K_{vd} \approx \frac{nV_S}{D'^2}$$

$$\omega_{rhp} \approx \frac{D'^2 R}{Dn^2 L_m}$$

$$\omega_{esr} = \frac{1}{CR_c}$$

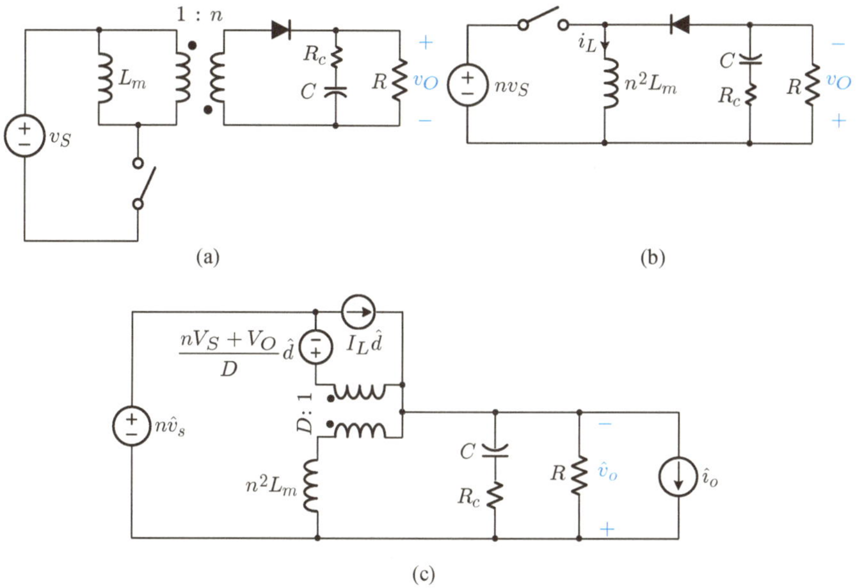

Figure 6.36 Flyback converter. (a) Original flyback converter. (b) Equivalent buck/boost converter. (c) Small-signal model.

$$\omega_o = \sqrt{\frac{1}{n^2L_mC}\frac{RD'^2}{R+R_c}} \approx \frac{D'}{\sqrt{n^2L_mC}}$$

and

$$Q = \frac{1}{\omega_o}\frac{RD'^2}{L+CR_cRD'^2)} \approx R\sqrt{\frac{CD'^2}{n^2L_m}}$$

Figure 6.37 shows the duty ratio-to-output transfer function of a flyback converter operating with $V_S = 180$ V, $L_m = 4.0$ mH, $C = 470\ \mu$F, $R_c = 0.03\ \Omega$, $R = 1\ \Omega$, $f_s = 50$ kHz, and $D = 0.25$. The turns ratio of the flyback transformer is 60:6. The parameters of the duty ratio-to-output transfer function are determined as

$$K_d = \frac{nV_S}{D'^2} = \frac{0.1 \cdot 180}{0.75^2} = 32 \quad \Rightarrow \quad 30\text{ dB}$$

$$\omega_{rhp} = \frac{D'^2R}{Dn^2L_m} = \frac{0.75^21}{0.25 \cdot 0.1^24 \times 10^{-3}} = 2\pi \cdot 8.96 \times 10^3\text{ rad/s}$$

$$\omega_{esr} = \frac{1}{CR_c} = \frac{1}{470 \times 10^{-6}0.03} = 2\pi \cdot 1.13 \times 10^4\text{ rad/s}$$

$$\omega_o = \frac{D'}{\sqrt{n^2L_mC}} = \frac{0.75}{\sqrt{0.1^24.0 \times 10^{-3}470 \times 10^{-6}}} = 2\pi \cdot 871\text{ rad/s}$$

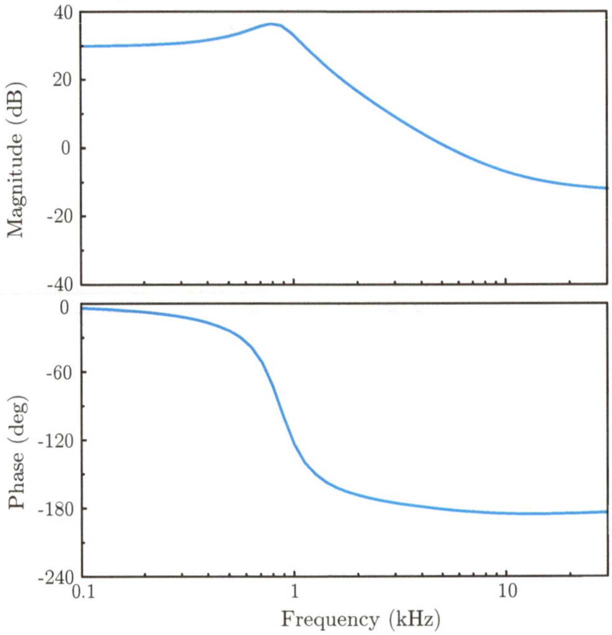

Figure 6.37 Duty ratio-to-output transfer function of flyback converter.

and

$$Q = R\sqrt{\frac{CD'^2}{n^2 L_m}} = 1\sqrt{\frac{0.75^2 470 \times 10^{-6}}{0.1^2 4 \times 10^{-3}}} = 2.57$$

6.5 EMPIRICAL METHODS FOR SMALL-SIGNAL ANALYSIS

In addition to the analytical method discussed in the previous sections, empirical approaches have been employed to investigate the frequency response of dc-to-dc converters. These empirical methods can be used as a means of verifying theoretical predictions or as a substitute for analytical analyses.

Figure 6.38 shows an empirical construction to investigate the duty ratio-to-output transfer function, $\hat{v}_o(s)/\hat{d}(s)$. When the small-signal source $\hat{v}_p$ is *not* excited, a fixed dc voltage V_{dc} is applied as the control voltage v_{con} to the PWM block. For this case, the switch drive signal is periodic and the converter establishes a steady-state equilibrium with a constant duty ratio D.

When the small-signal source $\hat{v}_p$ is activated, the control signal is perturbed

$$v_{con}(t) = V_{dc} + \hat{v}_p(t) \tag{6.114}$$

thereby introducing a sinusoidal ac component in the duty ratio

$$d(t) = D + \hat{d}(t) \tag{6.115}$$

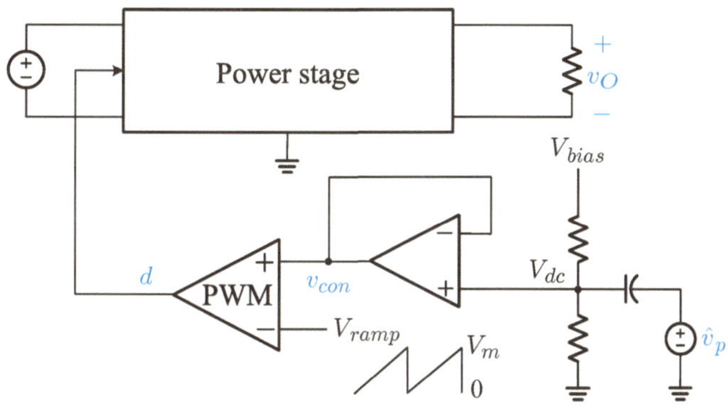

Figure 6.38 Experimental set-up for duty ratio-to-output transfer function measurement.

The agitated duty ratio in turn generates a perturbed output voltage, consisting of the dc and small-signal components

$$v_O(t) = V_O + \hat{v}_o(t) \tag{6.116}$$

By evaluating the ratio $\hat{v}_o(s)$ to $\hat{v}_p(s)$

$$\frac{\hat{v}_o(s)}{\hat{v}_p(s)} = \frac{\hat{d}(s)}{\hat{v}_p(s)} \frac{\hat{v}_o(s)}{\hat{d}(s)} \tag{6.117}$$

and incorporating the small-signal gain of the PWM block with the condition $\hat{v}_{con}(s) = \hat{v}_p(s)$

$$\frac{\hat{d}(s)}{\hat{v}_{con}(s)} = \frac{\hat{d}(s)}{\hat{v}_p(s)} = F_m \tag{6.118}$$

the desired duty ratio-to-output transfer function is now obtained

$$\frac{\hat{v}_o(s)}{\hat{d}(s)} = \frac{1}{F_m} \frac{\hat{v}_o(s)}{\hat{v}_p(s)} \tag{6.119}$$

The evaluation of the $\hat{v}_o(s)/\hat{v}_p(s)$ ratio can be performed with two different empirical methods; one is an experimental method and the other is a computational method. The first experimental method is to measure the $\hat{v}_o(s)/\hat{v}_p(s)$ ratio from the operational converter using an *impedance analyzer*. The impedance analyzer injects the input sinusoid while sweeping the perturbation frequency, extracts the output sinusoid in order to compare it with the input sinusoid, and finally generates the magnitude and phase plots for the $\hat{v}_o(s)/\hat{v}_p(s)$ ratio.

The second *computational method* is to calculate the $\hat{v}_o(s)/\hat{v}_p(s)$ ratio based on time-domain simulations using circuit simulation software. In this method, circuit simulation software is employed as functional equivalent to the impedance analyzer. The simulation software performs a series of time-domain simulations while sweeping the perturbation frequency for the range of interest. The software processes the results

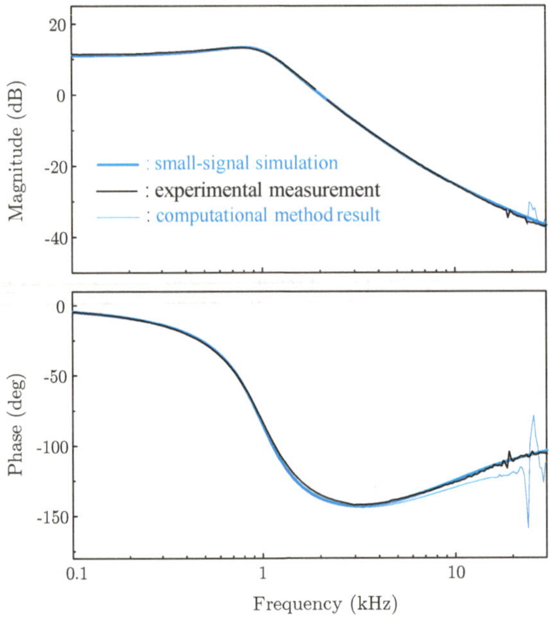

Figure 6.39 Duty ratio-to-output transfer function of buck converter.

of the simulation to yield the magnitude and phase plots for the $\hat{v}_o(s)/\hat{v}_p(s)$ ratio. Some commercial circuit simulation software offers an automated execution of this type of computational procedures. In this book, the computational method will be used as a means of verifying the analytical results of the small-signal analysis. Whenever appropriate and informative, theoretical results of small-signal analysis will be compared and correlated with the results of the computational method.

EXAMPLE 6.8 Comparison of Duty Ratio-to-Output Transfer Functions

This example compares the prediction of the s-domain small-signal model with the outcomes of the two empirical methods. Figure 6.39 shows the duty ratio-to-output transfer function of the buck converter, operating with $V_S = 15$ V, $L = 68$ μH, $R_l = 0.16$ Ω, $C = 430$ μF, $R_c = 0.05$ Ω, $R = 1$ Ω, $f_s = 50$ kHz, and $D = 0.33$. The outcome of the small-signal simulation is compared with both the experimental measurement using an impedance analyzer and the results of the computational method using circuit simulation software.

At low and mid frequencies, the small-signal simulation shows a close match with both the experimental and computational results. On the other hand, there are some discrepancies at high frequencies. The mismatch with the experimental measurement is due to the high-frequency switching noises and parasitic circuit components, which are ignored in the small-signal analysis but affect the experimental data. The disparity with the result of the computational method is caused by the sensitivity of the numerical algorithm of the circuit simulator. These discrepancies will be observed, yet only at high frequencies, in all the forthcoming transfer functions. Example 8.5 in Chapter 8 provides further discussions about the results of the computational method.

6.6 CHAPTER SUMMARY

This chapter investigated the power stage transfer functions of PWM dc-to-dc converters. The input-to-output transfer function, duty ratio-to-output transfer function, and load current-to-output transfer function are analyzed, focusing on their frequency response characteristics.

The three basic PWM converters commonly revealed the low-pass filter characteristics. For the buck converter, the power stage inductor L directly combines with the output capacitor to form a low pass filter. For the boost and buck/boost converters, an effective inductance $L_e = L/(1-D)^2$ appears in the power stage transfer functions as the inductive parameter.

The boost and buck/boost converters contain a right-half plane (RHP) zero in their duty ratio-to-output transfer function. In addition to the analytical evaluation of the transfer function, some physical explanations are given to support the existence of the RHP zero. The impact of the RHP zero is analyzed concentrating on the phase characteristics of the duty ratio-to-output transfer function. As will be demonstrated in Section 8.4.7, the RHP zero incurs considerable difficulties to the feedback compensation design and prevents the converter from securing stable operation with good performance.

Analytical expressions and asymptotic plots for power stage transfer functions in continuous conduction mode (CCM) operation are given for the buck, boost, and buck/boost converters. These results will be used in later chapters for the feedback compensation design and closed-loop analysis. Table 6.2 summarizes the expressions for power stage transfer functions of the three basic PWM converters in CCM operation.

The discontinuous conduction mode (DCM) operation causes notable changes in transfer functions of PWM converters. In DCM operation, the power stage double pole splits into two real poles. The first pole appears at low frequencies and the second pole occurs at significantly high frequencies. Therefore, the transfer functions exhibit the first-order system dynamics for the frequency range of practical importance. The changes in the transfer functions are demonstrated in Example 6.7 using the duty ratio-to-output transfer function of a buck/boost converter.

The transfer functions of isolated converters have the same structure as that of the three basic non-isolated converters. The results of the transfer function analysis for the three basic converters extend to the isolated converters and Table 6.2 is thus applicable to all isolated/non-isolated PWM converters with appropriate scalings in the power stage parameters.

Table 6.2 Transfer Functions of Three Basic Converters in CCM Operation

Buck converter

$$G_{vs} = D\frac{1+\dfrac{s}{\omega_{esr}}}{1+\dfrac{s}{Q\omega_o}+\dfrac{s^2}{\omega_o^2}} \qquad G_{vd} = V_S\frac{1+\dfrac{s}{\omega_{esr}}}{1+\dfrac{s}{Q\omega_o}+\dfrac{s^2}{\omega_o^2}} \qquad Z_p = R_l\frac{\left(1+\dfrac{s}{\omega_z}\right)\left(1+\dfrac{s}{\omega_{esr}}\right)}{1+\dfrac{s}{Q\omega_o}+\dfrac{s^2}{\omega_o^2}}$$

$$Q = R\sqrt{\frac{C}{L}} \qquad \omega_o = \frac{1}{\sqrt{LC}}$$

$$\omega_{esr} = \frac{1}{CR_c} \qquad \omega_z = \frac{R_l}{L}$$

Boost converter

$$G_{vs} = \frac{1}{D'}\frac{1+\dfrac{s}{\omega_{esr}}}{1+\dfrac{s}{Q\omega_o}+\dfrac{s^2}{\omega_o^2}} \qquad G_{vd} = \frac{V_S}{D'^2}\frac{\left(1-\dfrac{s}{\omega_{rhp}}\right)\left(1+\dfrac{s}{\omega_{esr}}\right)}{1+\dfrac{s}{Q\omega_o}+\dfrac{s^2}{\omega_o^2}} \qquad Z_p = \frac{R_l}{D'^2}\frac{\left(1+\dfrac{s}{\omega_z}\right)\left(1+\dfrac{s}{\omega_{esr}}\right)}{1+\dfrac{s}{Q\omega_o}+\dfrac{s^2}{\omega_o^2}}$$

$$D' = 1-D \qquad Q = D'R\sqrt{\frac{C}{L}} \qquad \omega_o = D'\frac{1}{\sqrt{LC}}$$

$$\omega_{esr} = \frac{1}{CR_c} \qquad \omega_z = \frac{R_l}{L} \qquad \omega_{rhp} = D'^2\frac{R}{L}$$

Boost/boost converter

$$G_{vs} = \frac{D}{D'}\frac{1+\dfrac{s}{\omega_{esr}}}{1+\dfrac{s}{Q\omega_o}+\dfrac{s^2}{\omega_o^2}} \qquad G_{vd} = \frac{V_S}{D'^2}\frac{\left(1-\dfrac{s}{\omega_{rhp}}\right)\left(1+\dfrac{s}{\omega_{esr}}\right)}{1+\dfrac{s}{Q\omega_o}+\dfrac{s^2}{\omega_o^2}} \qquad Z_p = \frac{R_l}{D'^2}\frac{\left(1+\dfrac{s}{\omega_z}\right)\left(1+\dfrac{s}{\omega_{esr}}\right)}{1+\dfrac{s}{Q\omega_o}+\dfrac{s^2}{\omega_o^2}}$$

$$D' = 1-D \qquad Q = D'R\sqrt{\frac{C}{L}} \qquad \omega_o = D'\frac{1}{\sqrt{LC}}$$

$$\omega_{esr} = \frac{1}{CR_c} \qquad \omega_z = \frac{R_l}{L} \qquad \omega_{rhp} = \frac{D'^2}{D}\frac{R}{L}$$

The transfer functions are approximations whose accuracy improves with the conditions $R \gg R_c$ and $R \gg R_l$. The expressions are also applicable to isolated converters with *scaled* power stage parameters.

This chapter presented the construction of Bode plots for general transfer functions, which was used to illustrate the frequency response characteristics of the dc-to-dc converters. This chapter also introduced experimental and computational approaches to the power stage dynamic analysis. The computational method will be used in later chapters, as a method of verifying predictions of the small-signal analysis and control design.

REFERENCES

1. R. W. Erickson and D. Maksimović, *Fundamentals of Power Electronics*, Kluwer Academic Publishers, 2001.
2. V. Vorpérian, *Fast Analytical Techniques for Electrical and Electronic Circuits*, Cambridge University Press, 2002.
3. K. Ogata, *Modern Control Engineering*, Prentice Hall, 4th ed., 2002.
4. V. Vorpérian, "Simplified analysis of PWM converters using the model of the PWM switch: Part I and II," *IEEE Trans. Aerosp. Electron. Syst.,* vol. AES-26, no. 3, pp. 490-505, May 1990.

PROBLEMS

6.1 For each asymptotic magnitude curve in Fig. P6.1, derive the expression for the corresponding transfer function in the time constant form.

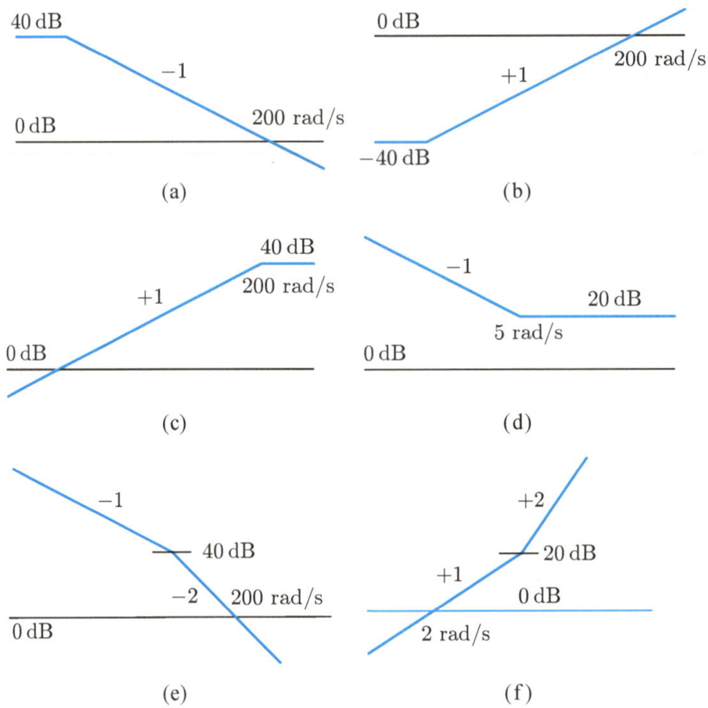

Fig. P6.1

6.2 The input-to-output transfer function of a two-port network is given by

$$\frac{v_o(s)}{v_i(s)} = \frac{\left(1+\dfrac{s}{\omega_z}\right)\left(1+\dfrac{s^2}{\omega_{o1}^2}\right)}{\left(1+\dfrac{s}{\omega_p}\right)\left(1+\dfrac{s^2}{\omega_{o2}^2}\right)}$$

For each of the following different input signals, find the expression for amplitude of $v_O(t)$.

i) $v_I(t) = V_m \sin \omega_z t$ ii) $v_I(t) = V_m \sin \omega_{o1} t$

iii) $v_I(t) = V_m \sin \omega_p t$ iv) $v_I(t) = V_m \sin \omega_{o2} t$

6.3 For asymptotic magnitude curves in Fig. P6.3, derive the expressions for the corresponding transfer functions in the time constant form.

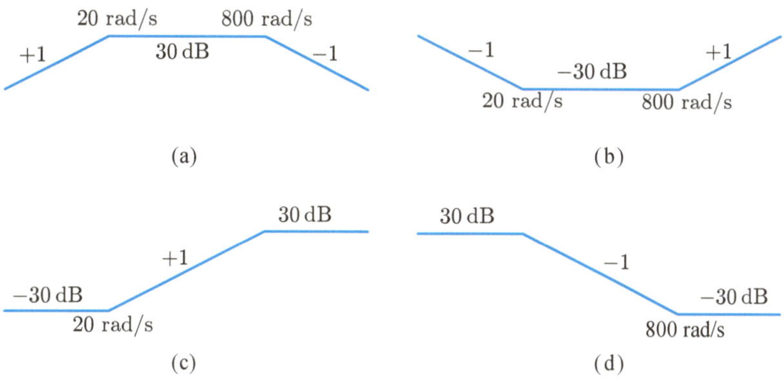

Fig. P6.3

6.4 Consider the following transfer function

$$T(s) = K_t \frac{(s+\omega_{z1})(s+\omega_{z2})}{(s+\omega_{p1})(s+\omega_{p2})}$$

a) Assume $K_t > 1$, $\omega_{p1} < \omega_{z1} < \omega_{z2} < \omega_{p2}$, and $|\omega_{z1} - \omega_{p1}| < |\omega_{p2} - \omega_{z2}|$. Sketch the asymptotic plot of $|T|$. Find the expressions for the maximum and minimum values of $|T|$.

b) Assume $K_t < 1$, $\omega_{z1} < \omega_{p1} < \omega_{z2} < \omega_{p2}$, and $|\omega_{p1} - \omega_{z1}| > |\omega_{p2} - \omega_{z2}|$. Repeat a).

6.5 Sketch the asymptotic plot for $|T|$ of the following transfer functions. Find the expression of the maximum value of $|T|$. Show the corner frequencies of $|T|$.

a) $T(s) = K_t s \dfrac{1}{1 + \dfrac{s}{\omega_p}}$

b) $T(s) = K_t s \dfrac{1}{\left(1 + \dfrac{s}{\omega_{p1}}\right)\left(1 + \dfrac{s}{\omega_{p2}}\right)}$ with $\omega_{p1} \ll \omega_{p2}$

c) $T(s) = K_t s \dfrac{1 + \dfrac{s}{\omega_z}}{\left(1 + \dfrac{s}{\omega_{p1}}\right)\left(1 + \dfrac{s}{\omega_{p2}}\right)}$ with $\omega_{p1} \ll \omega_z \ll \omega_{p2}$

d) $T(s) = K_t s \dfrac{1 + \dfrac{s}{\omega_z}}{\left(1 + \dfrac{s}{\omega_{p1}}\right)\left(1 + \dfrac{s}{\omega_{p2}}\right)}$ with $\omega_{p1} \ll \omega_{p2} \ll \omega_z$

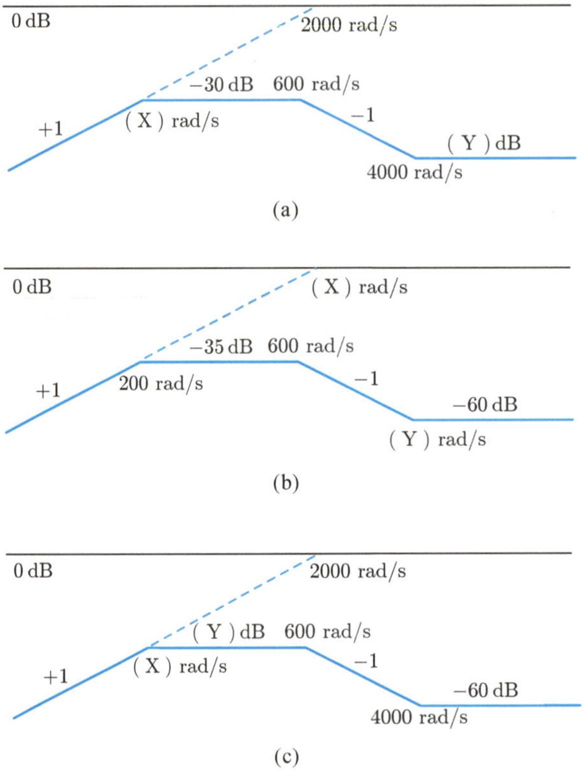

Fig. P6.6

6.6* For the asymptotic plots shown in Fig. P6.6, find the numerical values for X and Y specified in the blanks.

6.7 Sketch the asymptotic magnitude plot $|T|$ of the following transfer function for the three cases listed below.

$$T(s) = \frac{K_t}{s} \frac{\left(1 + \dfrac{s}{\omega_{z1}}\right)\left(1 + \dfrac{s}{\omega_{z2}}\right)}{\left(1 + \dfrac{s}{Q\omega_o} + \dfrac{s^2}{\omega_o^2}\right)\left(1 + \dfrac{s}{\omega_p}\right)}$$

i) $\omega_{z1} < \omega_o < \omega_{z2} < \omega_p$

ii) $\omega_{z1} < \omega_{z2} < \omega_o < \omega_p$

iii) $\omega_o < \omega_{z1} < \omega_{z2} < \omega_p$

6.8 Referring to the asymptotic magnitude plots shown in Fig. P6.8, find the numerical values for X and Y specified in the blanks.

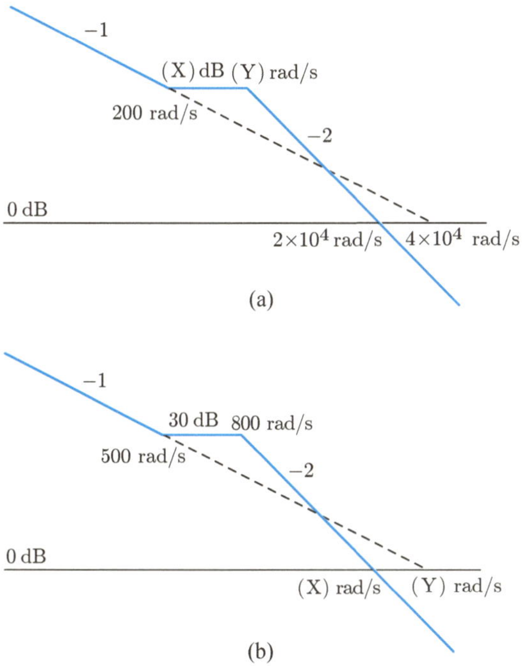

Fig. P6.8

6.9 Sketch the asymptotic magnitude plot for the following transfer functions. Show the corner frequency, 0 dB crossover frequency, and slopes of asymptotes.

i) $T(s) = \dfrac{\omega_o^2}{s^2}$ ii) $T(s) = \dfrac{s^2}{\omega_o^2}$

iii) $T(s) = 1 + \dfrac{\omega_o}{s}$ iv) $T(s) = \dfrac{1}{1 + \dfrac{\omega_o}{s}}$

6.10 Construct the magnitude and phase asymptotic plots for the following transfer functions. Show the corner frequency, 0 dB crossover frequency, and slopes of magnitude asymptotes, as appropriate.

i) $T(s) = \dfrac{1-s}{1+s}$ ii) $T(s) = \dfrac{1-\dfrac{s}{2}}{1+s}$

iii) $T(s) = \dfrac{1-s+s^2}{1+s+s^2}$ iv) $T(s) = \dfrac{1-\dfrac{s}{2}+\dfrac{s^2}{4}}{1+s+s^2}$

6.11* Figure P6.11 shows the circuit diagram of a buck converter and its duty ratio-to-output transfer function, $|G_{vd}|$, and load current-to-output transfer function, $|Z_p|$. Based on the information given in the Bode plots, estimate the values of the circuit components L, R_l, C, R_c, R, and the input voltage V_S of the buck converter. Assume $R \gg R_l$ and $R \gg R_c$.

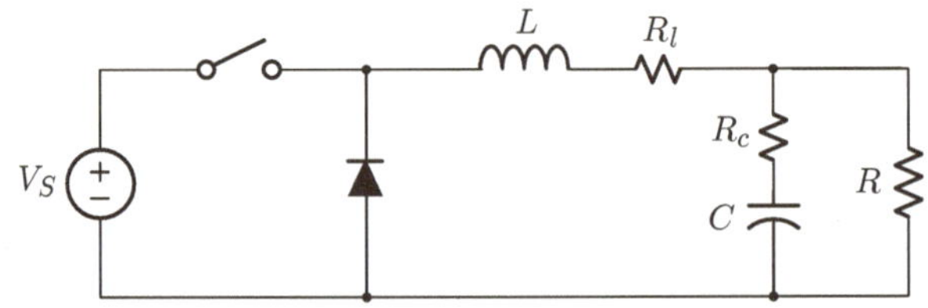

(a)

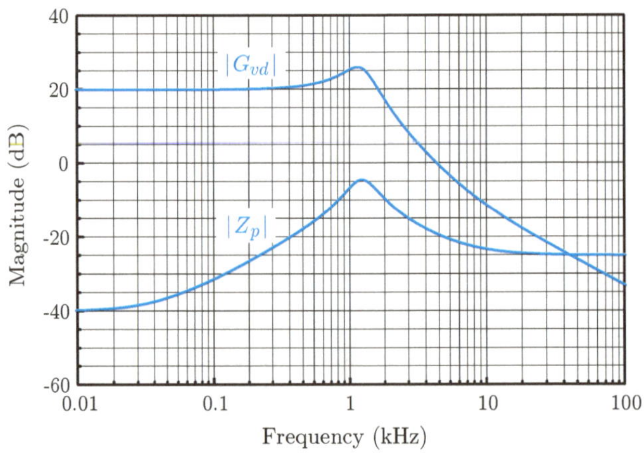

(b)

Fig. P6.11

6.12 Derive the duty ratio-to-output transfer function, $G_{vd}(s)$, of the boost converter and buck/boost converter with the parasitic resistances of the reactive components: $R_l \neq 0$ and $R_c \neq 0$. Assume CCM operation for the converters.

6.13 Figure P6.13 shows the circuit diagrams of the three basic dc-to-dc converters. For each converter, sketch the asymptotic plots for the magnitude and phase responses of the three transfer functions of $G_{vs}(s)$, $G_{vd}(s)$, and $Z_p(s)$. Note that $R_l = 0$, yet $R_c \neq 0$. Express the corner frequencies and asymptotic values of the magnitude plot in terms of the circuit parameters and operating conditions. Assume $Q > 0.5$ and $\omega_o \ll \omega_{esr}$ for the buck converter, and $\omega_o \ll \omega_{rhp} \ll \omega_{esr}$ for boost and buck/boost converters.

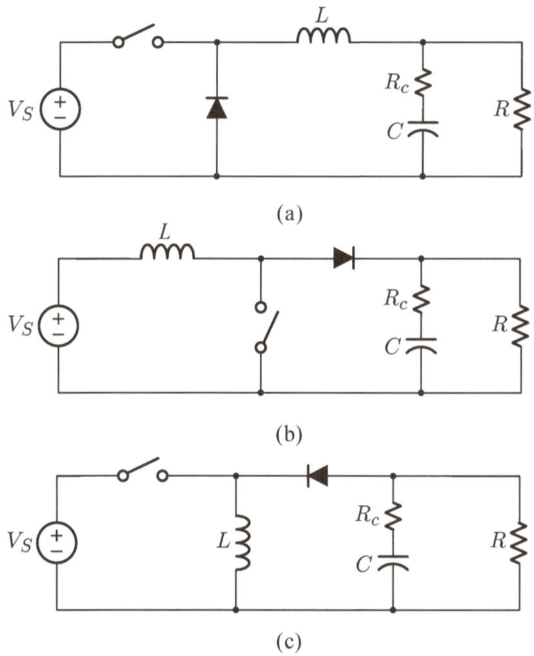

Fig. P6.13

6.14 This problem deals with the identification of the transfer function from the given Bode plot.

a) Assume that the Bode plot in Fig. P6.14(a) is generated from the transfer function

$$T(s) = \frac{K_t}{s} \frac{1 + \dfrac{s}{\omega_z}}{1 + \dfrac{s}{Q\omega_o} + \dfrac{s^2}{\omega_o^2}}$$

Find the values for the parameters, K_t, ω_z, Q, and ω_o.

b) Figure P6.14(b) is the Bode plot of the transfer function

$$T(s) = K_t s \frac{1}{\left(1 + \dfrac{s}{\omega_{p1}}\right)\left(1 + \dfrac{s}{\omega_{p2}}\right)}$$

Determine the values for the parameters, K_t, ω_{p1}, and ω_{p2}.

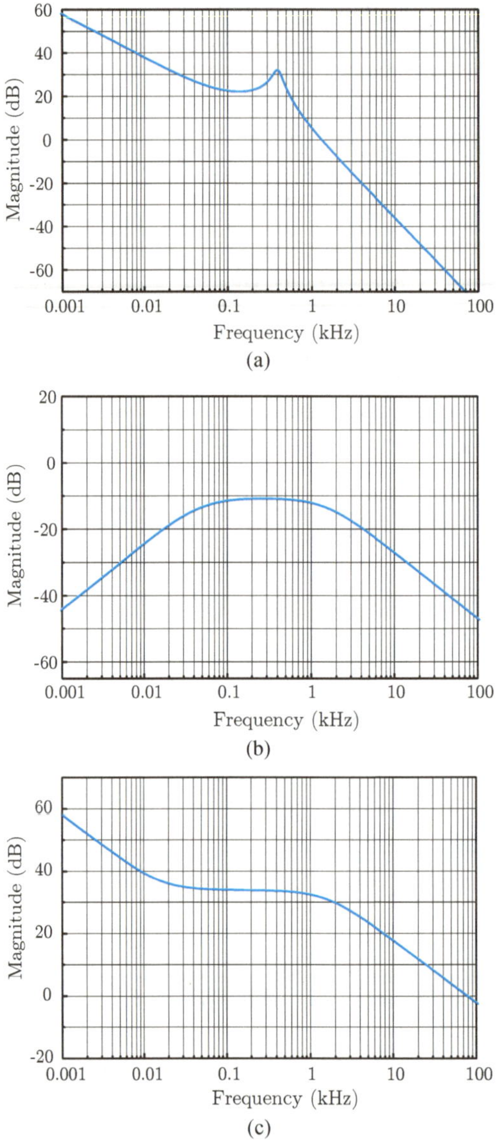

Fig. P6.14

c) Figure P6.14(c) is generated from the transfer function

$$T(s) = \frac{K_t}{s} \frac{1 + \dfrac{s}{\omega_z}}{1 + \dfrac{s}{\omega_p}}$$

Find the values for the parameters, K_t, ω_z, and ω_p.

6.15* Shown in Fig. P6.15 are the the circuit diagram of a buck converter and its input-to-output transfer function, $G_{vs}(s) = v_o(s)/v_s(s)$, and load current-to-output transfer function, $Z_p(s) = v_o(s)/i_o(s)$.

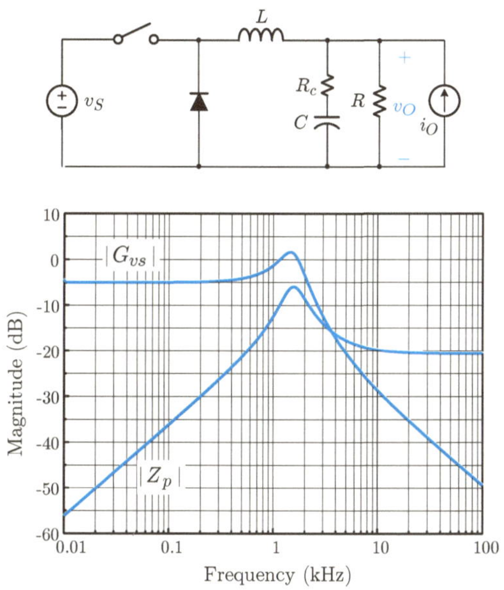

Fig. P6.15

a) Referring to the Bode plot of the power stage transfer functions, determine the values for the circuit components, L, C, R, and R_c, within reasonable accuracy.

b) Assume the converter is excited with $v_S = 40 + \sin 2\pi \cdot 200t$ and $i_O = 0$. Find the expression v_O and sketch v_O waveform for the two operational periods. Ignore the ripple component in v_O and show the maximum/minimum values.

c) Now assume the buck converter is excited with $v_S = 40$ and $i_O = \sin 2\pi \cdot 200\,t$. Under this condition, repeat b).

6.16* An experimental circuit is built to extract the frequency response of a buck converter power stage. Using the experimental set-up in Fig. P6.16(a), the frequency response of $\hat{v}_o(s)/\hat{v}_p(s)$ is measured as shown in Fig. P6.15(b).

a) Find an analytical expression for the duty ratio-to-output transfer function, $G_{vd}(s) = \hat{v}_o(s)/\hat{d}(s)$, of the buck power stage.

b) Find an expression for the input-to-output transfer function, $G_{vs}(s) = \hat{v}_o(s)/\hat{v}_s(s)$, of the buck converter converter power stage.

c) Find the average value of the output voltage of the buck converter.

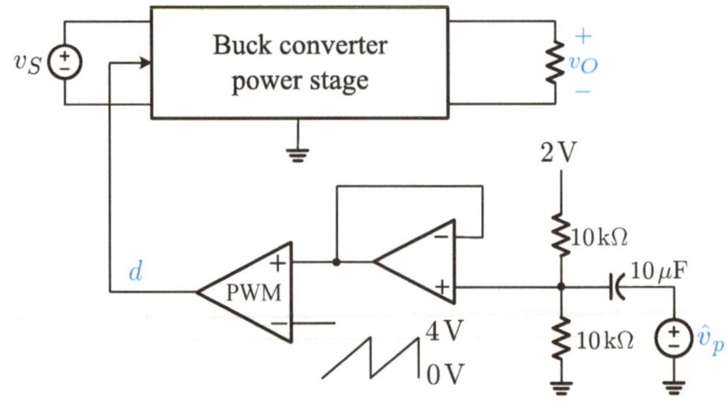

Buck converter
power stage
v_S
v_O
2 V
10 kΩ
10 μF
10 kΩ
$\hat{v}_p$
d
PWM
4 V
0 V

(a)

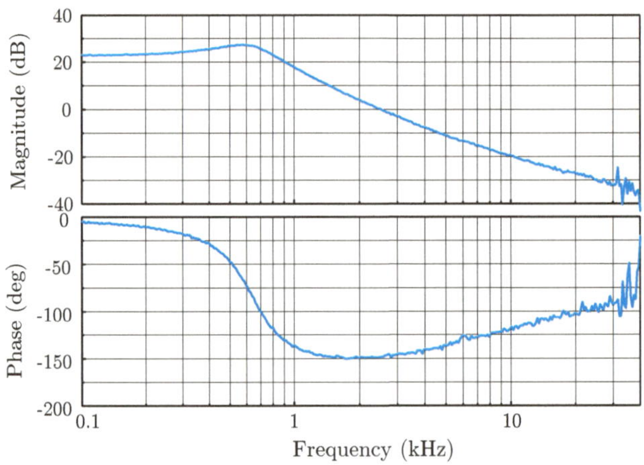

Magnitude (dB)
40
20
0
-20
-40
Phase (deg)
0
-50
-100
-150
-200
0.1
1
10
Frequency (kHz)

(b)

Fig. P6.16

CHAPTER 7

DYNAMIC PERFORMANCE OF PWM DC-TO-DC CONVERTERS

Dc-to-dc converters are intended to function as an effective and reliable voltage source. Accordingly, there are certain metrics to assess the performance of dc-to-dc converters as a voltage source. These performance metrics are typically classified into two categories. The first category is the static or dc metrics which characterize the converters' performance in steady state. The static performance metrics include the power handing capacity, current and voltage stresses of switches, and output voltage ripple. These metrics are determined by power stage parameters and are irrelevant to the feedback controller.

The second category is the dynamic or ac performance metrics which describe the converters' ability to withstand the external and internal disturbances or transient behavior at the presence of certain changes in operational conditions. The dynamic performance metrics include stability, frequency-domain transfer functions, and time-domain transient responses. In contrast to the static performance, the dynamic performance is solely determined by the feedback controller. Thus, two dc-to-dc converters with identical power stage parameters could show utterly different dynamic performance, depending on their feedback controller.

There are several requirements for engineers to build dc-to-dc converters for satisfactory dynamic performance. The first requirement is the comprehension about the small-signal dynamics of the power stage. Chapters 5 and 6, dealt with the small-signal modeling and analysis of the converter power stage, serve this purpose. The second requirement is to understand the implication of the dynamic performance of closed-loop controlled dc-to-dc converters. The final requirement is the implementation of the feedback controller which could provide desired dynamic performance. The purpose of this chapter is to cover the second requirement, namely, investigating the dynamic performance of dc-to-dc converters. The design and implementation of the feedback controller are studied in Chapters 8 through 10.

7.1 STABILITY

In previous chapters, it is implicitly conceived that dc-to-dc converters always establish periodic operation in steady state. However, this is only the case when dc-to-dc converters operate in an open-loop fashion with a fixed duty ratio. When closed-loop feedback control is employed to regulate the output voltage, dc-to-dc converters only reach periodic steady-state operation, given that their feedback controller is properly designed to meet the stability criterion.

Figure 7.1(a) shows the circuit diagram of a closed-loop controlled buck converter. This converter will be used throughout this chapter to illustrate the stability and other dynamic performance of closed-loop controlled dc-to-dc converters. Figure 7.1(b) shows the output voltage v_O and inductor current i_L of the buck converter, which goes through a transition from stable operation to unstable operation. Initially, the feedback controller is duly designed for stable operation. In the middle of stable operation, the circuit parameters of the feedback compensation are altered to new values that violate the stability criterion. With the new unstable compensation parameters, the circuit waveforms start to show a growing oscillation.

Stability analysis can be performed using the familiar classical control theory because we have already developed the s-domain small-signal model for dc-to-dc converters. As will be explained in Section 7.5.3, stability is judged by investigating the loop gain characteristics of the converter. First, we evaluate the loop gain from the small-signal model of the converter. Then, the loop gain is portrayed either in the Bode plot format to assess stability margins or in the polar plot format to apply the

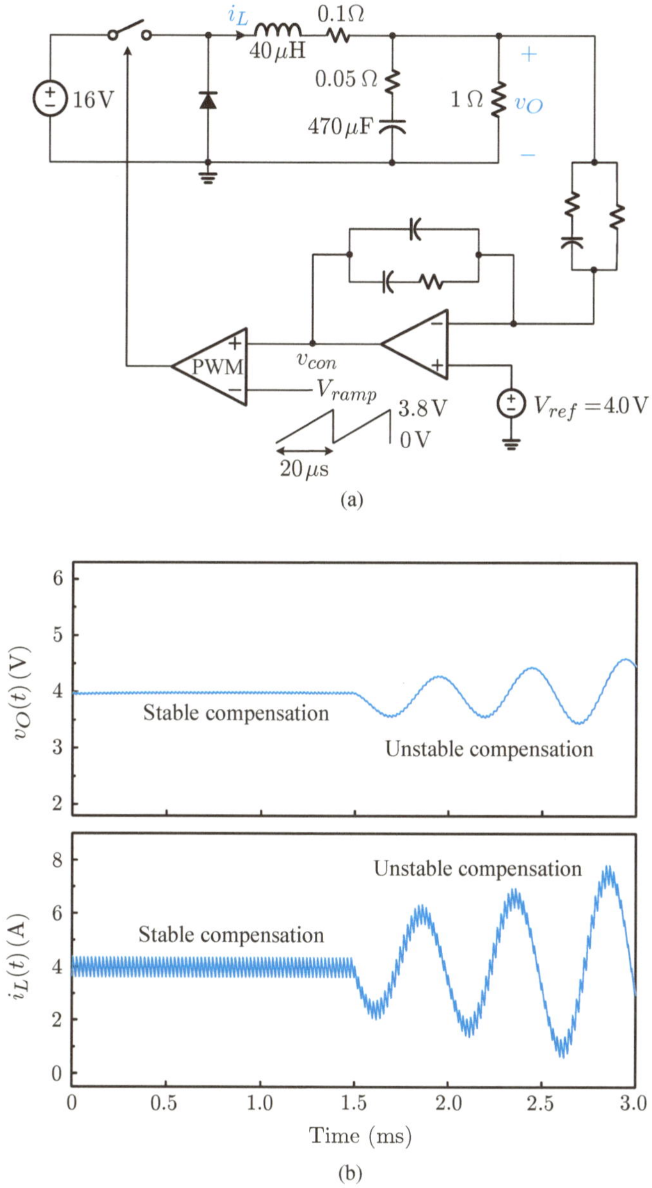

Figure 7.1 Closed-loop controlled buck converter. (a) Circuit diagram. (b) Circuit waveforms showing transition from stable operation to unstable operation.

Nyquist stability criterion. While the detailed stability analysis using the s-domain small-signal model will be covered later in this chapter, the results of such an analysis are shown in the following example, in order to highlight the utility of the s-domain small-signal model.

EXAMPLE 7.1 Stability Analysis of Buck Converter

This example provides an introductory overview of the upcoming stability analysis based on the classical control theory. The simplicity and accuracy of the stability analysis are demonstrated using the buck converter shown in Fig. 7.1. The loop gain of the converter, obtained from the s-domain small-signal model, is employed as the basis for the stability analysis. Figure 7.2(a) shows the loop gain of the buck converter in the Bode plot format, while Fig. 7.2(b) depicts the same loop gain in the polar plot form. The loop gain is evaluated with two different feedback compensations: the original stable compensation and the modified unstable compensation.

With the original compensation, the Bode plot indicates that the converter is stable with a sufficient phase margin and the polar plot well satisfies the Nyquist stability criterion. With the modified compensation, in contrast, the phase margin vanishes and the polar plot encircles the $(-1, 0)$ point, thereby both showing that the converter is unstable. The predictions of Fig. 7.2 are surely consistent with the circuit waveforms in Fig. 7.1.

Although the stability analysis is rather straightforward and standard, readers should be reminded that this simple analysis is possible due to the existence of the s-domain small-signal model. If the small-signal model were not available, the stability analysis would be a very challenging task.

7.2 FREQUENCY-DOMAIN PERFORMANCE CRITERIA

There are three frequency-domain transfer functions that are meaningful and useful as the dynamic performance for a closed-loop controlled converter. They are the loop gain, audio-susceptibility, and output impedance. This section describes the definition and implication of these performance metrics.

7.2.1 Loop Gain

As demonstrated in Example 7.1, the loop gain carries the whole information about the converter stability. Figure 7.3 is the small-signal block diagram of a closed-loop controlled dc-to-dc converter. The gain block F_m represents the small-signal gain of the PWM block, $F_v(s)$ is the voltage feedback compensation, and the other gain blocks denote the small-signal transfer functions of the power stage. Power stage transfer functions for PWM converters were studied in Chapter 6.

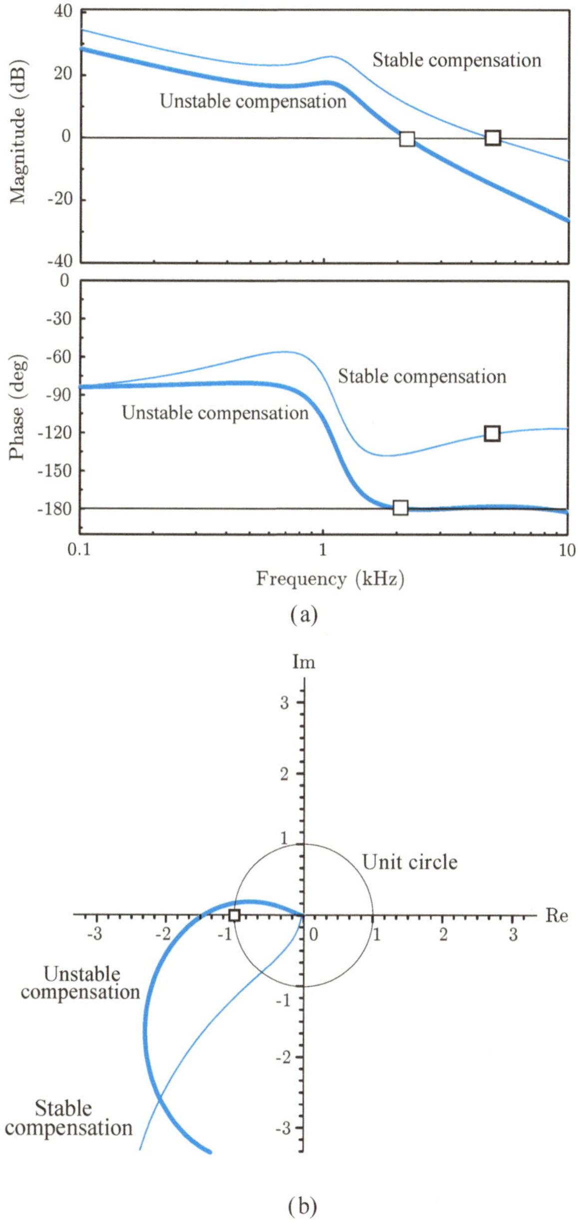

Figure 7.2 Stability analysis of buck converter. (a) Bode plot of loop gain. (b) Polar plot of loop gain.

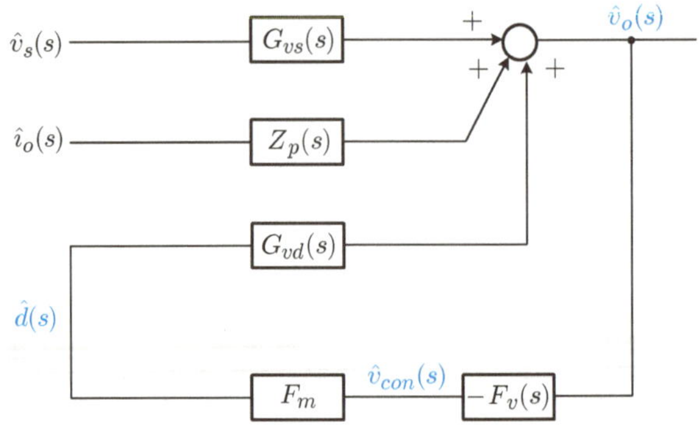

Figure 7.3 Small-signal block diagram of closed-loop controlled dc-to-dc converter.

The loop gain is defined as the *negative* product of all the gain blocks located along the feedback path. From Fig. 7.3, the loop gain is determined as

$$\begin{aligned} T_m(s) \equiv (-)\frac{\hat{v}_o(s)}{\hat{d}(s)}\frac{\hat{v}_{con}(s)}{\hat{v}_o(s)}\frac{\hat{d}(s)}{\hat{v}_{con}(s)} &= (-)G_{vd}(s)(-)F_v(s)F_m \\ &= G_{vd}(s)F_v(s)F_m \end{aligned} \tag{7.1}$$

where $G_{vd}(s)$ is the duty ratio-to-output transfer function. Once evaluated from the individual gain blocks, the loop gain can be converted into the Bode plot or polar plot for stability analysis. Examples of the loop gain plots were shown in Example 7.1.

In addition to stability, the loop gain is also closely related with other frequency-domain transfer functions – the audio-susceptibility and output impedance. Thus, the loop gain lies at the center of the small-signal analysis and control design, and governs all dynamic performance of the converter. For the given duty ratio-to-output transfer function and PWM gain, the voltage feedback compensation $F_v(s)$ is the only gain block that can be freely designed for good loop gain characteristics. Therefore, the dynamic analysis and control design are eventually directed by the design of $F_v(s)$. This topic will be treated in Chapter 8.

7.2.2 Audio-Susceptibility

Dc-to-dc converters usually receive the input from a non-ideal voltage source, rather than a pure dc source. For example, the input of an off-line dc-to-dc converter† is the rectified utility line, which usually contains line-frequency ripple component. Also, in dc power distribution systems where several dc-to-dc converters are employed together, the input of one dc-to-dc converter may pick up various noises generated by the other converters in the system.

† An off-line converter receives the input power from the utility line and employs a rectification circuit in front.

A dc-to-dc converter is required to generate a constant dc voltage as its output, regardless of the ripple or noise component in the input voltage. Thus, it becomes necessary to define the input-to-output noise rejection as one of the performance metrics. The *audio-susceptibility* refers to the input-to-output voltage transfer function evaluated under the closed-loop condition, which in fact represents the input-to-output noise rejection capacity of dc-to-dc converters.

In the past, dc-to-dc converters operated with the switching frequency that falls into the audible frequency range, for example $f_s = 10$ kHz. When excited by the 10 kHz switching frequency, the magnetic components in the converter generate audible noises that would propagate to the converter output. The term audio-susceptibility was originally coined to assess the immunity of the converter against audible frequency noises. Although the switching frequency of modern dc-to-dc converters now extends far beyond the audible frequency range, the audio-susceptibility is still used to describe the noise transmission characteristics of dc-to-dc converters.

By applying Mason's gain rule to Fig. 7.3, the audio-susceptibility is evaluated as

$$A_u(s) \equiv \left.\frac{\hat{v}_o(s)}{\hat{v}_s(s)}\right|_{\text{closed}} = \frac{G_{vs}(s)}{1+T_m(s)} \tag{7.2}$$

where $G_{vs}(s)$ is the open-loop input-to-output transfer function of the power stage and $T_m(s)$ is the loop gain defined in (7.1). With a given $G_{vs}(s)$, the audio-susceptibility can be altered by changing the loop gain. One objective of the loop gain design would be minimizing the audio-susceptibility for all frequencies, thereby providing a constant output voltage with minimal noise contamination.

Audio-Susceptibility of Buck Converter **EXAMPLE 7.2**

Figure 7.4 shows the audio-susceptibility of the buck converter introduced in Example 7.1. The prediction of the small-signal model is shown in comparison with the numerical result, obtained from the computational method discussed in Section 6.5. The use and accuracy of the computational method for the dynamic analysis of dc-to-dc converters will be described in detail in Example 8.6 in Chapter 8. The audio-susceptibility curve indicates that the converter provides sufficient noise rejection at low and high frequencies yet it passes mid-frequency noise components with about 33 dB attenuation.

7.2.3 Output Impedance

Dc-to-dc converters are required to maintain the output voltage constant at the presence of variations or fluctuations in the load current. One way to characterize this feature is to investigate the closed-loop transfer function from the load current to output voltage. This transfer function is referred to as the *output impedance.*

Same as the audio-susceptibility case, the output impedance should be minimized for all frequencies. As the output impedance becomes smaller, the dc-to-dc converter

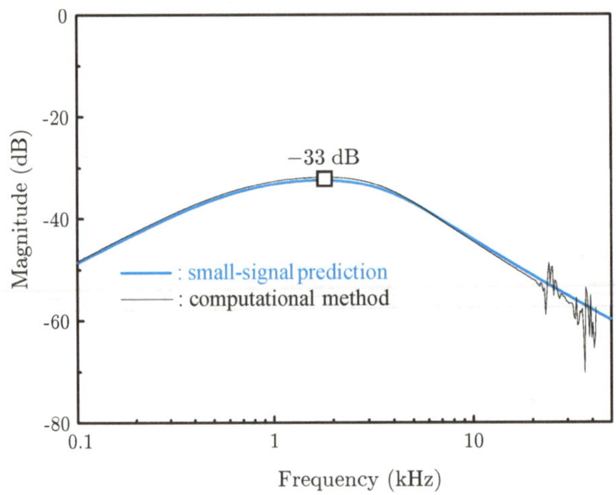

Figure 7.4 Audio-susceptibility of buck converter.

more closely resembles an ideal voltage source, which has *zero* output impedance. Thus, minimizing the output impedance is another important objective of the controller design.

From the small-signal block diagram of Fig. 7.3, the output impedance is evaluated as

$$Z_o(s) \equiv \left.\frac{\hat{v}_o(s)}{\hat{\imath}_o(s)}\right|_{\text{closed}} = \frac{Z_p(s)}{1+T_m(s)} \tag{7.3}$$

where $Z_p(s)$ is the (open-loop) load current-to-output transfer function. The loop gain affects the output impedance in the same manner as that of the audio-susceptibility case.

EXAMPLE 7.3 Output Impedance of Buck Converter

Figure 7.5 shows the theoretical and computational output impedances of the buck converter used in Example 7.1. The output impedance starts with a very small magnitude at low frequencies and gradually increases until it approaches a constant value at high frequencies. As will be shown in Section 8.2.2, the high-frequency asymptote is determined by the equivalent series resistance (esr) of the output capacitor: $|Z_o(j\infty)| = 20\log R_c$. For the power stage parameters shown in Fig. 7.1(a), the high-frequency asymptote is given by $20\log 0.05 = -26$ dB.

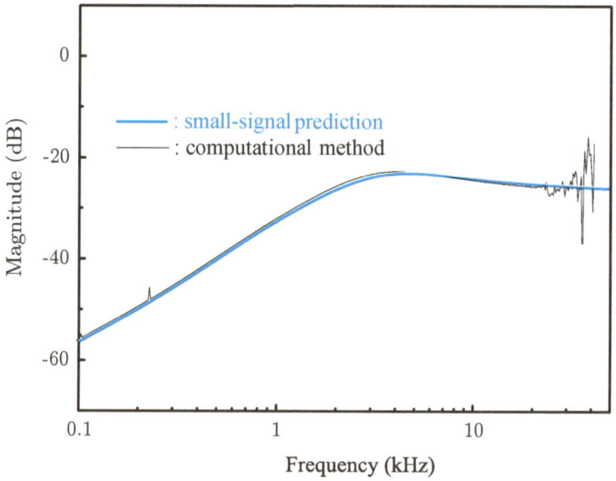

Figure 7.5 Output impedance of buck converter.

7.3 TIME-DOMAIN PERFORMANCE CRITERIA

The time-domain performance is also important when evaluating dc-to-dc converters as a voltage source. The time-domain performance includes the transient response of the output voltage due to a sudden change in the load current or input voltage.

Transient responses of a buck converter have already been presented in Section 3.6.2, in order to illustrate the principles of the closed-loop control using the PWM scheme. In this section, the transient responses are investigated as the dynamic performance of dc-to-dc converters.

7.3.1 Step Load Response

The step load response refers to the transient response of the output voltage due to a step change in the load current. The step load response has been of concern for two reasons. First, dc-to-dc converters in many applications encounter step changes in the load current. For example, dc-to-dc converters powering digital equipment downstream frequently experience sudden changes in the load current. Second, the step load response is usually used as a means of assessing the transient performance in general. The converter producing a good step load response will also offer good transient responses upon other changes in operational conditions.

The time-domain transient response can be analyzed using the *appropriate* s-domain transfer function. Precisely, the step load response is investigated through the output impedance, which is the closed-loop load current-to-output transfer function. First, the s-domain expression of the output voltage is obtained by multiplying the s-domain expression of the step load change by the output impedance

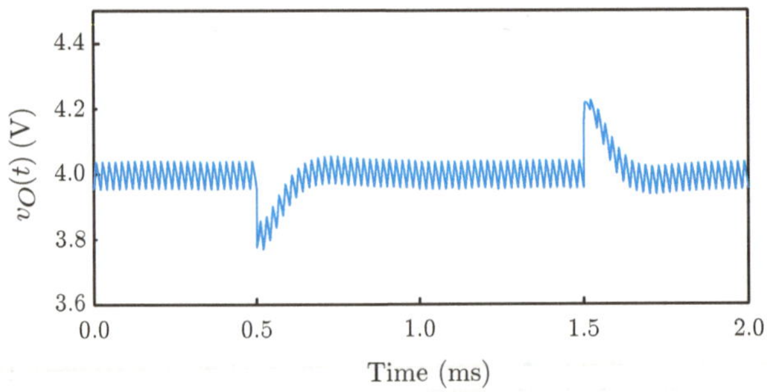

Figure 7.6 Step load response of buck converter.

$$\hat{v}_o(s) = \hat{\imath}_o(s) \cdot \frac{\hat{v}_o(s)}{\hat{\imath}_o(s)} = \frac{I_{step}}{s} Z_o(s) \tag{7.4}$$

where I_{step} is the magnitude of the step change in the load current. Now, the inverse Laplace transform is performed on (7.4) to yield the time-domain expression for the output voltage

$$v_O(t) = \mathcal{L}^{-1}\left(\frac{I_{step}}{s} Z_o(s)\right) \tag{7.5}$$

where $\mathcal{L}^{-1}$ represents the inverse Laplace transform. This relationship will be used in later chapters to investigate the transient behavior of the output voltage.

EXAMPLE 7.4 Step Load Response of Buck Converter

This example shows the step load response of the buck converter used in the previous examples. Figure 7.6 illustrates the output voltage of the converter going through a series of step changes in the load current: $I_O = 4\text{ A} \Rightarrow 8\text{ A} \Rightarrow 4\text{ A}$. Qualitative explanations about the output voltage response were given in Section 3.6.2.

The step load response is mainly characterized by two parameters: the peak overshoot/undershoot and settling time. The peak overshoot/undershoot is the maximum transitional deviation of the output voltage from the steady-state waveform. The settling time is the time interval until the output voltage settles within $\pm 5\%$ of the final value. Figure 7.6 indicates that the peak overshoot/undershoot is limited by ± 0.2 V and the settling time is less than 0.2 ms.

7.3.2 Step Input Response

The step input response is the transient behavior of the output voltage due to a step change in the input voltage. Examples of the step input response were presented in Section 3.6.2. Similar to the step load response case, the step input response is

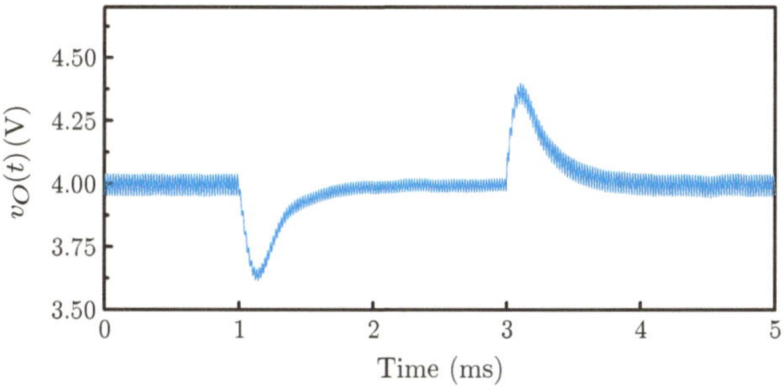

Figure 7.7 Step input response of buck converter.

analyzed using the audio-susceptibility

$$v_o(t) = \mathcal{L}^{-1}\left(\frac{V_{step}}{s} A_u(s)\right) \tag{7.6}$$

where V_{step} is the magnitude of the step change in the input voltage.

Step Input Response of Buck Converter **EXAMPLE 7.5**

This example illustrates the step input response of the previous buck converter. Figure 7.7 is the output voltage of the buck converter experiencing a series of step changes in the input voltage: V_S = 16 V ⇒ 8 V ⇒ 16 V. Same as the step load response case, the peak overshoot/undershoot and settling time are the important parameters for the step input response. Figure 7.7 shows a ±0.4 V peak overshoot/undershoot which returns to the steady-state value within a 0.7 ms settling time.

7.4 STABILITY OF DC-TO-DC CONVERTERS

Section 7.1 introduced the concept of stability and demonstrated the behavior of an unstable dc-to-dc converter. In this section, we now study the stability theory adapted to dc-to-dc converters. The classical stability theory for linear time-invariant (LTI) systems is first reviewed. Then, we discuss how the classical stability theory is extended to nonlinear time-variant PWM dc-to-dc converters.

7.4.1 Stability of Linear Time-Invariant Systems

The stability theory for LTI systems is closely related with the transfer function and characteristic equation. The transfer function of an LTI system is defined as the ratio of the Laplace transform of the output to the Laplace transform of the input under the assumption that all initial conditions are zero. The transfer function is given by a ratio of s-domain polynomials

$$F(s) = \frac{b_o + b_1 s + b_2 s^2 + \cdots\cdots + b_{m-1} s^{m-1} + b_m s^m}{a_o + a_1 s + a_2 s^2 + \cdots\cdots + a_{n-1} s^{n-1} + a_n s^n} \tag{7.7}$$

The characteristic equation is then defined as

$$a_o + a_1 s + a_2 s^2 + \cdots\cdots + a_{n-1} s^{n-1} + a_n s^n = 0 \tag{7.8}$$

from the denominator of the transfer function. The characteristic equation contains all the important information about the dynamic properties of the system. Most significantly, the solutions of the characteristic equation are called the poles or natural modes of the system. The location of the poles in s-plane determines stability of the system.

LTI systems are defined *unstable* if any of their poles is located in the right-hand plane (RHP) of s-plane or on the imaginary axis. An unstable system would produce exponentially or sinusoidally growing responses if it has any poles in the RHP of s-plane. Also, the system would exhibit a sustained oscillation if its poles are located on the imaginary axis.

Conversely, an LTI system is defined *stable* if its poles are all located in the left-hand plane (LHP) of s-plane. A stable LTI system does not contain any growing or oscillating terms and all transient responses will eventually settle down to their steady-state value. Accordingly, a stable LTI system establishes a stationary operating point in steady state. Furthermore, if the system is perturbed by certain external or internal disturbances, a stable system always returns to the original operating point.

7.4.2 Small-Signal Stability of Dc-to-Dc Converters

The classical stability theory is now extended to nonlinear time-variant PWM dc-to-dc converters. We consider a dc-to-dc converter stable if the converter produces periodic circuit waveforms in steady state. The inductor current of a stable converter settles into a periodic triangular waveform and the output voltage of the converter periodically *ripples* on top of the desired dc component. If the averaging operation discussed in Section 5.2.2 is executed on stable periodic waveforms, the resulting waveforms will not show any behavior of an unstable LTI system. In other words, the averaged circuit waveforms of a stable dc-to-dc converter do not resemble the circuit waveforms produced by an unstable LTI system. Conversely, the averaged circuit waveforms of an unstable dc-to-dc converter will exhibit the circuit behavior of an unstable LTI system.

As elaborated in Section 5.2.2, the average model of dc-to-dc converters is developed to accurately predict the averaged circuit waveforms of converters. Accordingly, we could adapt the stability theory to the average model of a dc-to-dc converter to test whether the converter establishes a periodic steady-state operation, or equivalently, to determine stability of the converter. However, the classical stability theory cannot be directly applied to the average model of the converter, because the average model is still nonlinear although it is time invariant. As a functional alternative, the s-domain small-signal model obtained by linearizing the average model is used for stability analysis.

The small-signal model is a linear approximation of the nonlinear average model. The approximation is based on the assumption that the converter remains in close proximity to the initial operating point. Therefore, the validity of the small-signal model is limited to the neighborhood of the given operating point. This implies that the result of the stability analysis using the small-signal model can only be used for the *local* stability. In other words, the stability theory only provides the information about the converter's behavior near the given operating point and is not suited to the converters that experience large excursions in operational conditions. In this sense, the stability defined on the small-signal model is also called the *small-signal* stability. Despite the aforementioned limitations, the small-signal stability still is a very useful and viable method for analyzing the stability of the nonlinear time-variant dc-to-dc converters.

Based on the preceding discussions, we apply the classical stability theory to small-signal models of PWM dc-to-dc converters. If the small-signal model is found to be stable, we assure that the converter would attain a periodic steady-state operation. In addition, the converter will withstand external and internal disturbances and return to the original state if the magnitude of the disturbance is not too large.

7.5 NYQUIST CRITERION

The Nyquist criterion plays the central role in the stability analysis. The Nyquist criterion allows us to judge stability of closed-loop controlled PWM converters only using the Bode plot of their loop gain. This section first reviews the theoretical foundation of the Nyquist criterion and later demonstrates its use in the stability analysis of PWM dc-to-dc converters.

7.5.1 Theoretical Foundation of Nyquist Criterion

The Nyquist criterion is based on Cauchy's theorem regarding the contour mapping in s-plane. As the theoretical basis for the Nyquist criterion, the s-plane contour mapping and Cauchy's theorem are discussed in this section.

Contour Mapping from s-Plane to $F(s)$-Plane

The contour mapping is a mathematical operation of projecting a contour in s-plane into another plane by a function $F(s)$. Because the variable s is a complex variable, $s = \sigma + j\omega$, the function $F(s)$ is also a complex expression, which can be described as $F(s) = u + jv$. Thus, $F(s)$ can be portrayed on the complex **$F(s)$-plane** with coordinates u and v. As an example, a square contour in s-plane, shown in Fig. 7.8(a), is mapped into another plane by a function $F(s) = 2s + 1$. The mapping of

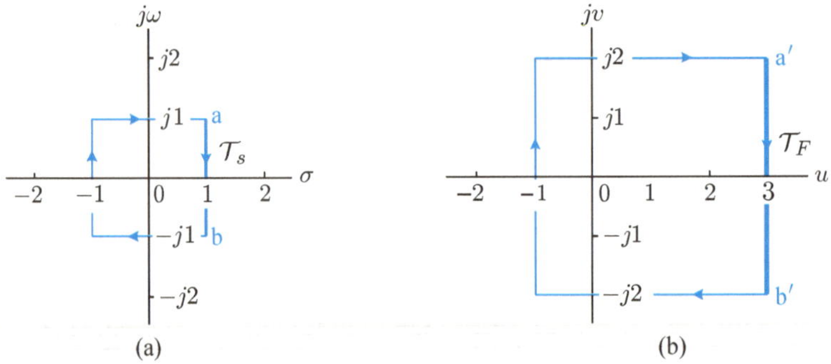

Figure 7.8 Mapping square contour by function of $F(s) = 2s + 1$. (a) s-plane contour $\mathcal{T}_s$. (b) $F(s)$-plane contour $\mathcal{T}_F$.

the s-plane square contour into the $F(s)$-plane is accomplished based on the equation

$$F(s) = 2s + 1 = 2(\sigma + j\omega) + 1 = u + jv \quad \Rightarrow \quad u = 2\sigma + 1 \quad v = 2\omega$$

The square contour $\mathcal{T}_s$ in s-plane is mapped into another square contour $\mathcal{T}_F$ in the $F(s)$-plane, as shown in Fig. 7.8(b). To illustrate the mapping process, the line segment a − b, defined by $\sigma = 1$ and $-j1 < j\omega < j1$ in s-plane, is projected into another line segment a′ − b′, defined by $u = 2\sigma + 1 = 3$ and $-j2 < jv = j2\omega < j2$ in the $F(s)$-plane.

Cauchy's Theorem

In contour mapping, there exists an intrinsic connection between the pattern of the s-plane contour $\mathcal{T}_s$ and the pattern of the $F(s)$-plane contour $\mathcal{T}_F$. This connection is determined by both the property of the mapping function $F(s)$ and the pattern of $\mathcal{T}_s$. The mapping function $F(s)$ in general is an s-domain rational function which has a finite number of poles and zeros. The s-plane contour $\mathcal{T}_s$ may encircle some of the poles and zeros of $F(s)$. The pattern of the $F(s)$-plain contour $\mathcal{T}_F$, created by the mapping $\mathcal{T}_s$ through $F(s)$, is governed by the numbers of the poles and zeros of $F(s)$ that are encircled by $\mathcal{T}_s$. The impact of the numbers of the encircled poles and zeros is described by the following Cauchy's theorem.

Cauchy's Theorem: Assume that $\mathcal{T}_s$ traverses in the clockwise direction in s-plane and encircles some poles and zeros of $F(s)$, while not passing through any other poles and zeros of $F(s)$. If $\mathcal{T}_s$ encircles $\mathcal{Z}$ zeros and $\mathcal{P}$ poles of $F(s)$, $\mathcal{T}_F$ encircles the origin of the $F(s)$-plane $\mathcal{N} = \mathcal{Z} - \mathcal{P}$ times in the clockwise direction.

The proof of Cauchy's theorem, which is given in most textbooks on the complex variable or feedback control, is not repeated here. Instead, two illustrative examples are given in Fig. 7.9. Let's consider that $F(s) = (s+j3)(s-j3)/((s+3)(s-1)(s-2))$

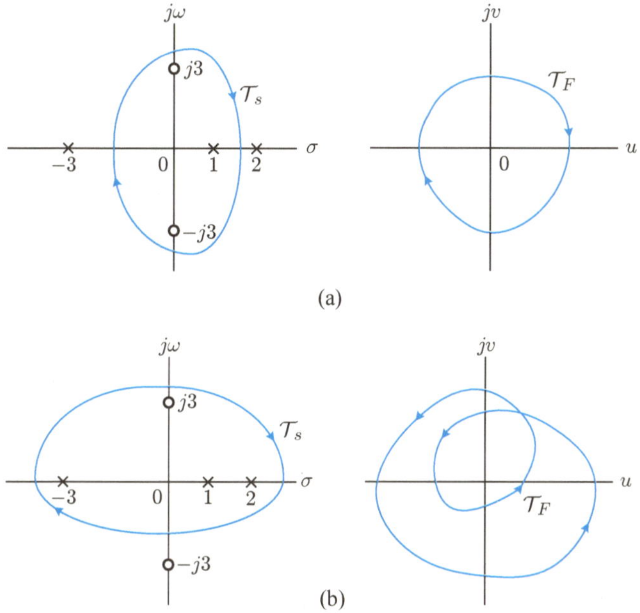

Figure 7.9 Example of Cauchy's theorem. (a) $\mathcal{N} = \mathcal{Z} - \mathcal{P} = 2 - 1 = 1$. (b) $\mathcal{N} = \mathcal{Z} - \mathcal{P} = 1 - 3 = -2$

is the mapping function. Figure 7.9(a) shows the first example of the s-plane contour $\mathcal{T}_s$ and the corresponding $F(s)$-plane contour $\mathcal{T}_F$. For this case, $\mathcal{T}_s$ encircles two zeros and one pole of $F(s)$. Thus, $\mathcal{T}_F$ should encircle the origin once in the clockwise direction: $\mathcal{N} = \mathcal{Z} - \mathcal{P} = 2 - 1 = 1$. For the second example of Fig. 7.9(b) where $\mathcal{T}_s$ encircles one zero and three poles, $\mathcal{T}_F$ should encircle the origin twice in the counter clockwise direction: $\mathcal{N} = \mathcal{Z} - \mathcal{P} = 1 - 3 = -2$.

Cauchy's Theorem Applied to $F(s) = 1 + T(s)$

Now assume that we desire to know whether the following equation has any RHP roots

$$\boxed{1 + T(s) = 0} \tag{7.9}$$

To answer this question, we employ Cauchy's theorem through the mapping equation of $F(s) = 1 + T(s)$ with the $\mathcal{T}_s$ encompassing the entire right-hand plane (RHP) of s-plane; this RHP-spanning contour is referred to as the *Nyquist contour* or *Nyquist path*. Figure 7.10(a) shows the Nyquist path that traverses the whole RHP of s-plane in the clockwise direction. Figure 7.10(b) is an example of $F(s)$-plane contour $\mathcal{T}_F$, traced by projecting the Nyquist path through the relation $F(s) = 1 + T(s)$. In this example, $\mathcal{T}_F$ encircles the origin in the $F(s)$-plane twice in the clockwise direction: thus, $\mathcal{N} = 2$ for this case.

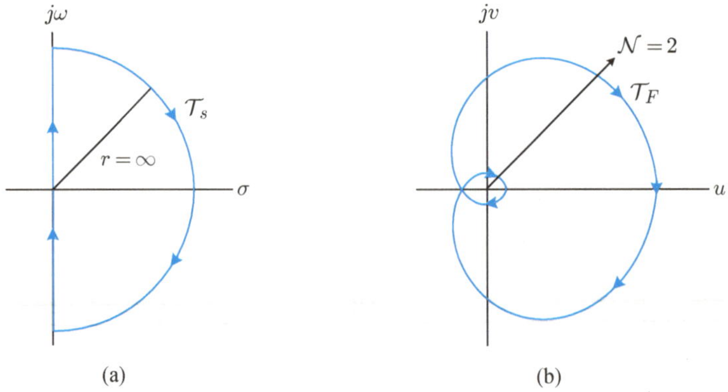

Figure 7.10 Application of Cauchy's theorem to determine presence of any RHP roots. (a) Nyquist path $\mathcal{T}_s$ encompassing whole RHP plane of s-plane. (b) Example of $F(s)$-plane contour $\mathcal{T}_F$.

Cauchy's theorem asserts that

$$\boxed{\mathcal{N} = \mathcal{Z} - \mathcal{P}} \tag{7.10}$$

where

- $\mathcal{N}$ denotes the number of the clockwise encirclements of the origin made by the contour $\mathcal{T}_F$.
- $\mathcal{Z}$ indicates the number of the zeros in $1 + T(s)$ encircled by the Nyquist path. *The number $\mathcal{Z}$ does correspond to the number of RHP roots in the equation of* $1 + T(s) = 0$.
- $\mathcal{P}$ represents the number of the poles in $1 + T(s)$ encircled by the Nyquist path. It is easily inferred that *the number $\mathcal{P}$ also indicates the number of the RHP poles in $T(s)$.*

Once the values for $\mathcal{N}$ and $\mathcal{P}$ are determined, the number of the RHP roots in $1 + T(s) = 0$ is evaluated from the equation

$$\boxed{\mathcal{Z} = \mathcal{N} + \mathcal{P}} \tag{7.11}$$

Thus, the number of RHP roots in $1+T(s) = 0$ is determined without actually solving the equation. In other words, the number of RHP roots is indirectly calculated by assessing $\mathcal{N}$ and $\mathcal{P}$, which is a much simpler task than the actual evaluation of the roots.

The preceding argument is reformulated into a more convenient and useful form by modifying the mapping relation into

$$F'(s) = F(s) - 1 = 1 + T(s) - 1 = T(s)$$

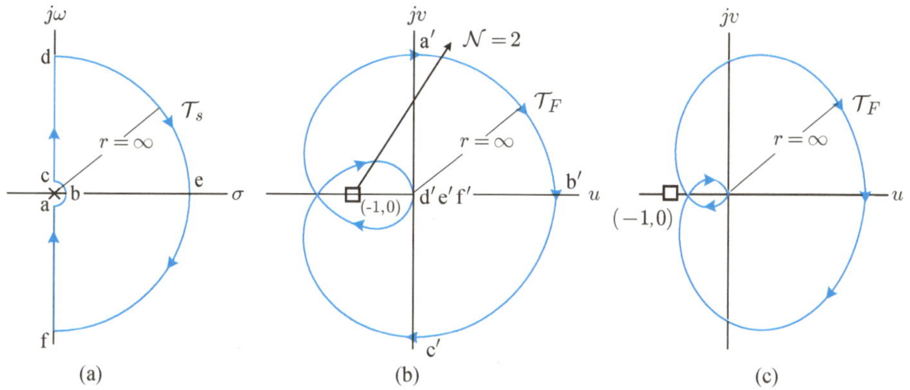

Figure 7.11 Mapping Nyquist contour through $T(s) = K/(s(1+s/\omega_{p1})(1+s/\omega_{p2}))$. (a) s-plane contour. (b) $T(s)$-plane contour with $\mathcal{N} = 2$. (c) $T(s)$-plane contour with $\mathcal{N} = 0$.

With this modification, the Nyquest path is now mapped through the relation $T(s)$ rather than $1 + T(s)$. Accordingly, the number $\mathcal{N}$, which initially corresponded to the number of the clockwise encirclements of the origin of the $1 + T(s)$-plane, now becomes the number of the clockwise encirclements of the $(-1, 0)$ point in the $T(s)$-plane. In other words, when Cauchy's theorem is adapted to evaluate the number of RHP roots in the equation $1 + T(s) = 0$, the Nyquist path is mapped through $T(s)$ and the number of the clockwise encirclements of the $(-1, 0)$ point is evaluated. The result is then used to determine the number of RHP roots in $1 + T(s) = 0$ using the relationship (7.11). This specific application of Cauchy's theorem is most conveniently and frequently used for stability analysis; this achievement is called the Nyquist criterion.

Mapping Nyquist path with $T(s) = K/(s(1+s/\omega_{p1})(1+s/\omega_{p2}))$ EXAMPLE 7.6

This example illustrates the mapping of the Nyquist path into the $T(s)$-plane using

$$T(s) = \frac{K}{s\left(1 + \dfrac{s}{\omega_{p1}}\right)\left(1 + \dfrac{s}{\omega_{p2}}\right)}$$

The mapping function $T(s)$ has a single pole at the origin of s-plane. Figure 7.11(a) shows the Nyquist contour $\mathcal{T}_s$ spanning the entire RHP. A tiny semicircle near the origin is necessary to prevent the Nyquist contour from passing the $T(s)$ pole at the origin. Figure 7.11(b) shows an example of $T(s)$-plane contour. The construction of Fig. 7.11(b) is explained as follows:

1) The a − b − c path in s-plane: the small semicircular detour around the $T(s)$ pole at the origin is mapped into the whole RHP of the $T(s)$-plane, as explained below. With the notation of $s = \epsilon e^{j\phi}$, the a − b − c path in s-plane is described as $-90^\circ < \phi < 90^\circ$ with $\epsilon \to 0$. The mapping the a − b − c path through $T(s)$

is translated as

$$\lim_{\epsilon \to 0} |T(\epsilon e^{j\phi})| \to \infty$$

and

$$-90^\circ < \phi < 90^\circ \quad \Rightarrow \quad 90^\circ > \angle T(\epsilon e^{j\phi}) > -90^\circ$$

The resulting $T(s)$-plane contour becomes the semicircle a′ − b′ − c′ of an infinite radius.

2) The c − d path: the portion of the Nyquist contour $\mathcal{T}_s$ from $\omega = +0$ to $\omega = +\infty$ is mapped as the real-frequency polar plot of $T(j\omega)$. In other words, mapping the c − d path through $T(s)$ is the same as drawing the polar plot of $T(j\omega)$ over $+0 < \omega < +\infty$. From $T(s)$ expression, it follows that

$$+0 < \omega < +\infty \quad \Rightarrow \quad \infty > |T(j\omega)| > 0 \quad \text{and} \quad -90^\circ > \angle T(j\omega) > -270^\circ$$

The c′ − d′ path in Fig. 7.11(b) represents the resulting $T(s)$-plane contour.

3) The d − e − f path: the portion of the $\mathcal{T}_s$ contour from $\omega = +\infty$ to $\omega = -\infty$ is projected into the origin in the $T(s)$-plane by noting that

$$\lim_{\epsilon \to \infty} |T(\epsilon e^{j\phi})| \quad \to \quad 0$$

4) The f − a path: the portion of $\mathcal{T}_s$ from $\omega = -\infty$ to $\omega = -0$ is mapped as $T(-j\omega)$, which is the complex conjugate of $T(j\omega)$. Therefore, the resulting $T(s)$-plane contour is the mirror image of the polar plot of $T(j\omega)$, reflected by the real axis in the $T(s)$-plane. This contour corresponds to the d′ − a′ path in Fig. 7.11(b).

By connecting the preceding contour segments, the whole $T(s)$-plane contour is completed as shown in Fig. 7.11(b). For this case, the $T(s)$-plane contour encircles the $(-1, 0)$ point twice in the clockwise direction: $\mathcal{N} = 2$. Figure 7.11(c) depicts another example of $T(s)$-plane contour which does not encircle the $(-1, 0)$ point: $\mathcal{N} = 0$.

The preceding example demonstrated the full details of the $\mathcal{T}_F$ formation. However, it is usually unnecessary to go through the entire steps of the $\mathcal{T}_F$ construction. The construction of the $T(s)$-plane contour can be simplified if the mapping function $T(s)$ meets the following conditions.

- $T(s)$ contains a single pole at the origin.
- The order of the denominator is lager than the order of the numerator.

These requirements ensure that the real-frequency polar plot of $T(j\omega)$ starts from the infinite on the negative imaginary axis and terminates at the origin, as shown in Fig. 7.11.

The mapping function $T(s)$, associated with the stability analysis of feedback controlled systems, usually meets the aforementioned conditions and the corresponding $T(s)$-plane contour $\mathcal{T}_F$ thus follows the pattern of Fig. 7.11. For such cases, the $T(s)$-plane contour can be envisaged by sketching the polar plot of $T(j\omega)$ over $+0 < \omega < +\infty$ and conceptually completing the remaining contour by adopting the pattern of Fig. 7.11. Accordingly, for most usual $T(s)$ cases, the number of the

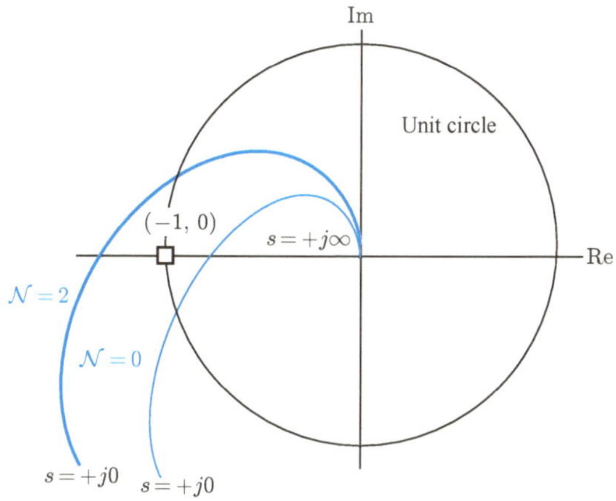

Figure 7.12 Polar plot of $T(j\omega)$.

encirclements of the $(-1, 0)$ point can be determined by only investigating the real frequency polar plot of $T(j\omega)$.

7.5.2 Nyquist Stability Criterion

The results of the previous section are recapitulated as the Nyquist criterion or Nyquist stability criterion. The Nyquist criterion provides a graphical means of determining the number of RHP roots in the equation of the following specific form

$$\boxed{1 + T(s) = 0} \tag{7.12}$$

where $T(s)$ is an s-domain rational function. The Nyquist criterion is based on the following relationship,

$$\boxed{\mathcal{Z} = \mathcal{N} + \mathcal{P}} \tag{7.13}$$

where $\mathcal{Z}$ denotes the number of RHP roots in $1 + T(s) = 0$, $\mathcal{N}$ represents the number of clockwise encirclements of the $(-1, 0)$ point made by the $T(s)$-plane contour, and $\mathcal{P}$ is the number of RHP poles in $T(s)$ itself.

Figure 7.12 shows the polar plots of $T(j\omega)$, drawn under the assumption that $T(s)$ meets the two conditions discussed in the previous section. The polar plot displayed with thin line does not encircle the $(-1, 0)$ point and consequently $\mathcal{N} = 0$ for this case. On the other hand, the polar plot depicted with thick line encircles the $(-1, 0)$ point. When the entire Nyquist path is considered, the second $T(s)$-plane contour encircles the $(-1, 0)$ point twice in the clockwise direction, as illustrated in Fig. 7.11(b). Accordingly, it becomes that $\mathcal{N} = 2$ for this case. Now, if the number of RHP poles in $T(s)$, $\mathcal{P}$, is known, the number of RHP roots in $1 + T(s) = 0$, $\mathcal{Z}$, is determined as $\mathcal{Z} = \mathcal{N} + \mathcal{P}$.

7.5.3 Application of Nyquist Stability Criterion to Dc-to-Dc Converters

The Nyquist criterion is now adapted to the stability analysis of dc-to-dc converters. The audio-susceptibility of the closed-loop controlled converter is evaluated as

$$A_u(s) = \left.\frac{\hat{v}_o(s)}{\hat{v}_s(s)}\right|_{\text{closed}} = \frac{G_{vs}(s)}{1 + G_{vd}(s)F_v(s)F_m} \tag{7.14}$$

by applying Mason's gain rule to the small-signal model of the converter in Fig. 7.3. When the input-to-output transfer function $G_{vs}(s)$ is denoted as

$$G_{vs}(s) = \frac{\mathcal{N}(s)}{\mathcal{D}(s)} \tag{7.15}$$

the audio-susceptibility becomes

$$\boxed{A_u(s) = \frac{\mathcal{N}(s)}{\mathcal{D}(s)\big(1 + G_{vd}(s)F_v(s)F_m\big)}} \tag{7.16}$$

The characteristic equation is then given by

$$\boxed{\mathcal{D}(s)\big(1 + G_{vd}(s)F_v(s)F_m\big) = 0} \tag{7.17}$$

If the equation $\mathcal{D}(s) = 0$ does not contain any RHP roots, stability is determined by investigating the existence of any RHP root in $1 + G_{vd}(s)F_v(s)F_m = 0$, that is, by applying the Nyquist criterion to $1 + T_m(s) = 0$ where $T_m(s) = G_{vd}(s)F_v(s)F_m$ is the loop gain defined in the previous section.

There are several remarks regarding the preceding discussions. First, while the audio-susceptibility is used in the previous analysis, any closed-loop transfer function can be used for the stability analysis, because all the transfer functions have the same denominator. Second, the power stage transfer functions of PWM converters do not have any RHP pole; more specifically, the equation $\mathcal{D}(s) = 0$ does not have any RHP roots as presumed in the previous discussions. Finally, the loop gains, $T_m(s) = G_{vd}(s)F_v(s)F_m$, of PWM converters do not contain any RHP poles: $\mathcal{P} = 0$. Accordingly, the Nyquist criterion is simplified as follows. The number of the RHP roots in the characteristic equation is the same as the number of the $(-1, 0)$ point encirclements made by the polar plot of the loop gain

$$\boxed{\mathcal{Z} = \mathcal{N}} \tag{7.18}$$

In this case, stability is simply tested using the polar plot drawn only for real frequencies; **for stability, the polar plot should not encircle the $(-1, 0)$ point.**

The preceding discussions about the Nyquist criterion provide a graphical method for the stability analysis using the converter loop gain. Figure 7.13 shows the four different cases of the polar plot of the converter loop gain, $T_m(s) = G_{vd}(s)F_v(s)F_m$.

Figure 7.13(a) shows the stable case in which the polar plot of the converter loop gain does not encircle the $(-1, 0)$ point. Figure 7.13(b) is the unstable polar plot,

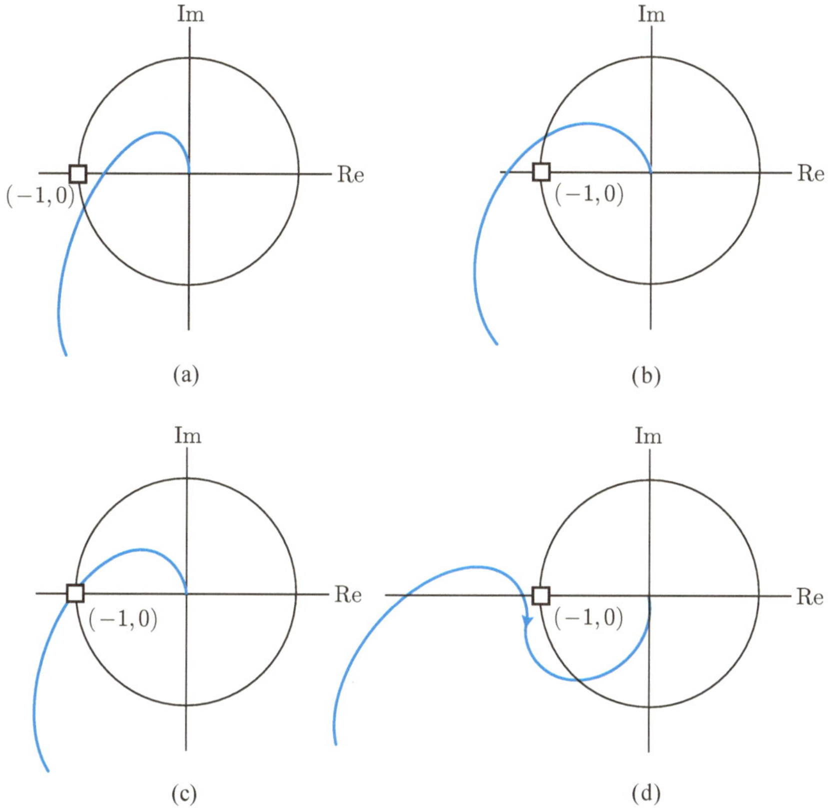

Figure 7.13 Polar plot of converter loop gain and stability. (a) Stable system. (b) Unstable system. (c) Marginally stable system. (d) Conditionally stable system.

encircling the $(-1, 0)$ point. Figure 7.13(c) is the marginally stable case where the polar plot just passes the $(-1, 0)$ point. The marginally stable case is the borderline between stability and instability. For this case, the system has a pair of poles on the imaginary axis and develops sustained oscillations in time-domain responses. The behavior of a marginally stable dc-to-dc converter will be analyzed in Example 7.7.

Figure 7.13(d) shows a special case where the polar plot does not encircle the $(-1, 0)$ point but exhibits a distinctive pattern. The polar plot initially crosses the − real axis outside the unit circle. Later, the polar plot passes the − real axis again before intersecting the unit circle, and finally approaches the origin without encircling the $(-1, 0)$ point. The system with these loop gain characteristics is called a *conditionally* stable system. Although stable, the conditionally stable system is problematic in real applications due to the distinguishing polar plot pattern. The feature of the conditionally stable system, along with its potential problem in practical applications, will be analyzed in Example 7.8.

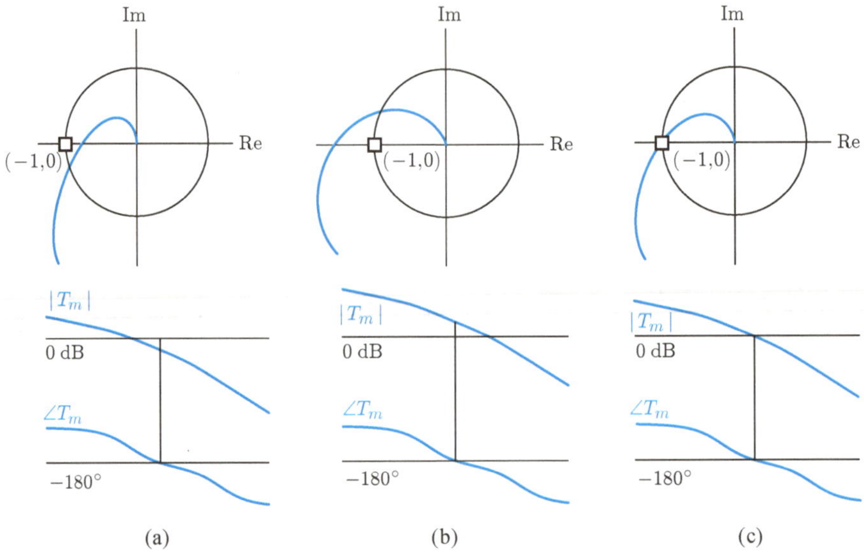

Figure 7.14 Stability analysis using polar plot and Bode plot of converter loop gain. (a) Stable case. (b) Unstable case. (c) Marginally stable case.

The Nyquist criterion can also be applied to the converter loop gain portrayed in the Bode plot format. To illustrate this, the first three cases in Fig. 7.13 are redrawn in Fig. 7.14 in both the Bode plot and polar plot formats.

1) Stable case in Fig. 7.14(a): The polar plot passes the – real axis through a point located between the $(-1, 0)$ point and the origin. In other words, the polar plot crosses the – real axis with its magnitude smaller than unity: $|T_m| < 0$ dB at the frequency where $\angle T_m$ becomes $-180°$. This situation is illustrated in the Bode plot in Fig. 7.14(a).

2) Unstable case in Fig. 7.14(b): The polar plot crosses the – real axis with its magnitude larger than unity: $|T_m| > 0$ dB at the frequency where $\angle T_m$ is $-180°$. This condition is depicted in the Bode plot in Fig. 7.14(b).

3) Marginally stable case in Fig. 7.14(c): The polar plot passes the – real axis with its magnitude exactly being unity. In other words, the polar plot hits the $(-1, 0)$ point. This circumstance corresponds to the case where $|T_m|$ crosses the 0 dB line with $\angle T_m = -180°$.

As illustrated in Fig. 7.14, stability is judged directly from the Bode plot of the loop gain. For stability, $|T_m|$ should be smaller than 0 dB when $\angle T_m$ falls to $-180°$, or equivalently $\angle T_m$ should be larger than $-180°$ (or less negative than $-180°$) when $|T_m|$ reduces to 0 dB. This condition is identical to the requirement of *not* encircling the $(-1, 0)$ point in the polar plot. Example 7.1 illustrated the application of this stability theory to a practical dc-to-dc converter.

Marginally Stable Buck Converter — EXAMPLE 7.7

The purpose of this example is to demonstrate the loop gain characteristics and time-domain response of a marginally stable buck converter. The feedback compensation parameters used in the previous buck converter examples are modified to yield a marginally stable case. Figure 7.15 shows the loop gain curves of the buck converter. The Bode plot in Fig. 7.15(a) indicates that $|T_m| = 0$ dB and $\angle T_m = -180°$ at the frequency $f_c = 2$ kHz. Figure 7.15(b) shows that the polar plot traverses the critical $(-1, 0)$ point. Thus, the buck converter is marginally stable according to the definition of stability. Furthermore, the loop gain indicates that

$$T_m(j2\pi f_c) = T_m(j2\pi \cdot 2 \times 10^3) = 1\angle - 180° = -1 \tag{7.19}$$

which is rearranged as

$$1 + T_m(j2\pi \cdot 2 \times 10^3) = 1 - 1 = 0 \tag{7.20}$$

Equation (7.20) implies that $s = j2\pi \cdot 2 \times 10^3$ is one root of the characteristic equation. Then, $s = -j2\pi \cdot 2 \times 10^3$ becomes another root of the characteristic equation. With a pair of poles at $s = \pm j2\pi \cdot 2 \times 10^3$, the buck converter would produce a sustained oscillation at the frequency $\omega_c = 2\pi \cdot 2 \times 10^3$ rad/s.

Figure 7.16 is the output voltage of the converter, which indeed oscillates at the expected frequency $\omega_c = 2\pi \cdot 2 \times 10^3$ rad/s, or equivalently with the period $t_{os} = 2\pi/\omega_c = 0.5$ ms. This example again confirms that the classical stability theory, when applied to the s-domain small-signal model, accurately predicts the dynamics of the nonlinear time-variant dc-to-dc converters.

Conditionally Stable System — EXAMPLE 7.8

This example illustrates the potential problem of a conditionally stable system. A conditionally stable system stays stable at the given operating point. However, when the operating point is altered, the conditionally stable system could become unstable. This *destabilizing effect* could take place in real applications and threaten the system operation.

Figure 7.17(a) shows a family of the loop gain plots for a conditionally stable system under the assumption that the magnitude of the loop gain is successively decreased, while its phase characteristics remain unchanged. The decrease in the loop gain magnitude results in a contraction in the polar plot. The contraction in turn incurs the encirclement of the $(-1, 0)$ point. Thus, a conditionally stable system becomes unstable when the loop gain magnitude is unduly lessened.

Most closed-loop controlled systems employ an amplifier in their feedback path. For this case, the loop gain magnitude can inadvertently be decreased when the operating point shifts from the initial position. Figure 7.17(b) shows the input-output transfer characteristics of a practical amplifier which encounters saturations when the operating point moves largely away from the origin. The slope of the input-output transfer curve is the incremental gain of the amplifier. With the initial operating point lying in Region A, the incremental gain of the amplifier is large.

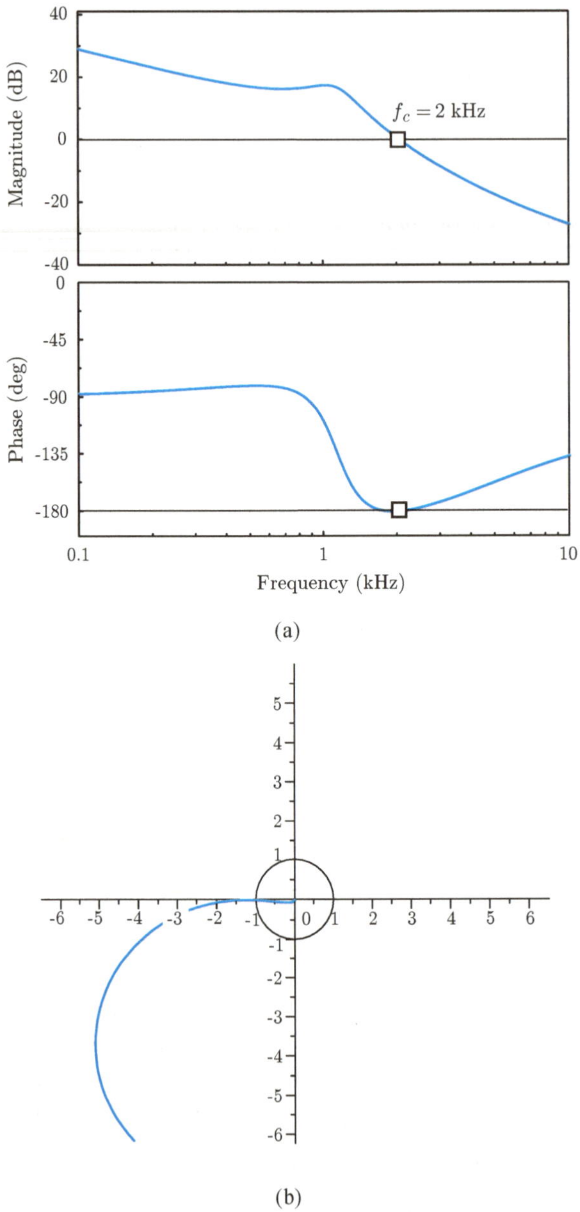

Figure 7.15 Loop gain of marginally stable buck converter. (a) Bode plot of loop gain. (b) Polar plot of loop gain.

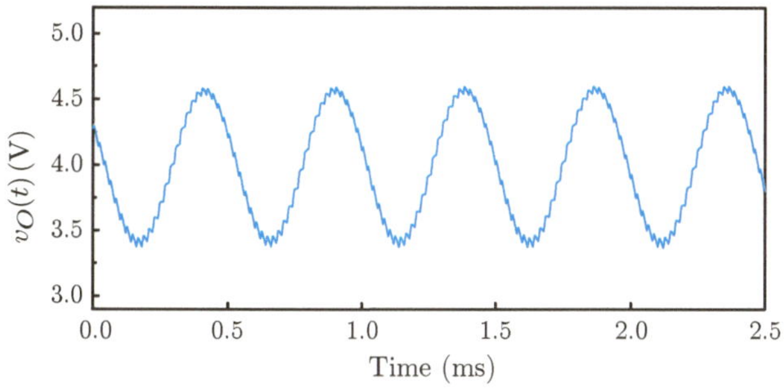

Figure 7.16 Output voltage of marginally stable buck converter.

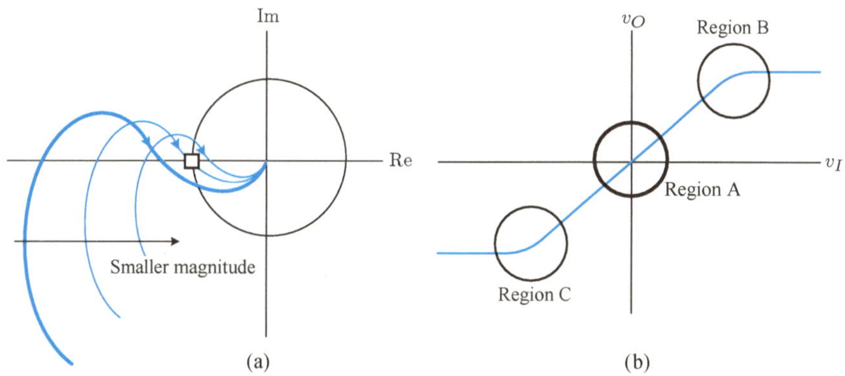

Figure 7.17 Problem of conditionally stable system. (a) Loop gain of conditionally stable system. (b) Input-to-output characteristics of practical amplifier.

However, when the operating point shifts towards Region B or Region C, the amplifier starts being saturated. Consequently, the slope of the input-output curve will be decreased, resulting in considerable reduction in the loop gain magnitude. The reduction in gain could destabilize the system as explained earlier. These unstable shifts in the operating point could readily occur when the operational conditions undergo large fluctuations during the start-up procedures, protective operations, or large-scale load/input changes.

Most stable systems also become unstable when the magnitude of the loop gain is sufficiently increased so that the polar plot stretches out to encircle the $(-1, 0)$ point. This would occur when the amplifier gain is overly increased. However, such a large gain increase will not take place in practice because the maximum allowable gain of the amplifier is mostly limited by the amplifier design. Thus, unlike the case for conditionally stable systems, the destabilizing effect of the gain increase in stable systems will not be a real problem.

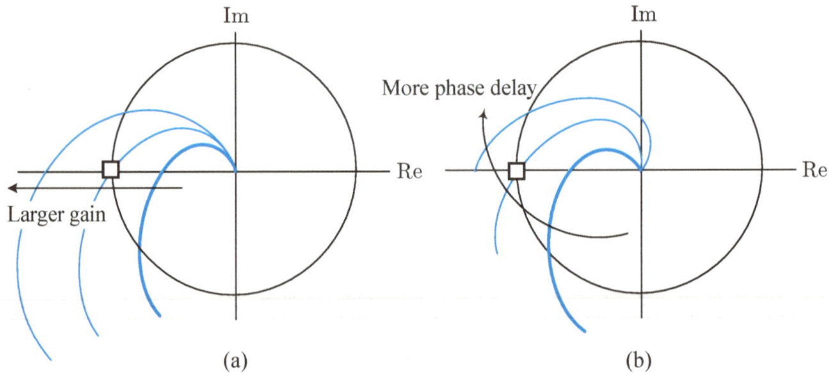

Figure 7.18 Destabilizing effect of gain increase and phase delay. (a) Increase in gain. (b) Increase in phase delay.

7.6 RELATIVE STABILITY: GAIN MARGIN AND PHASE MARGIN

Stability is the most important issue for any closed-loop controlled dynamic systems. Accordingly, we are firstly concerned with the absolute stability; we primarily want to know whether the system is currently stable or not. In addition to the absolute stability, it is also important to know the *firmness* of stability. We may want to know 1) how easily a currently stable system could become unstable or 2) whether one system is more or less stable than the other system.

The absolute stability is determined by applying the Nyquist criterion to the loop gain. The polar plot of a stable loop gain does not encircle the $(-1, 0)$ point. However, the increase either in the loop gain magnitude or in the phase delay would cause the polar plot to violate the Nyquist stability criterion. Figure 7.18 shows the effects of gain increase or phase delay on the polar plot.

1) An increase in gain with a fixed phase delay results in a proportional expansion of the polar plot. As shown in Fig. 7.18(a), a continuous gain increase eventually ends up with the encirclement of the $(-1, 0)$ point, thus making the system unstable.

2) An increase in phase delay with a fixed gain causes a clockwise rotation of the polar plot. An excessive phase delay also brings in the encirclement of the $(-1, 0)$ point, as shown in Fig. 7.18(b).

Because the increase in the loop gain magnitude or in the phase delay could destabilize the system, the relative stability needs to be specified by two separate margins: the gain margin and the phase margin.

Gain Margin: The gain margin is the amount of gain increase that can be added to $|T_m|$ before a stable system becomes marginally stable or unstable, with the assumption that phase characteristics remain the same. Figure 7.19 is an illustration of the gain margin. Figure 7.19(a) indicates that the polar plot currently

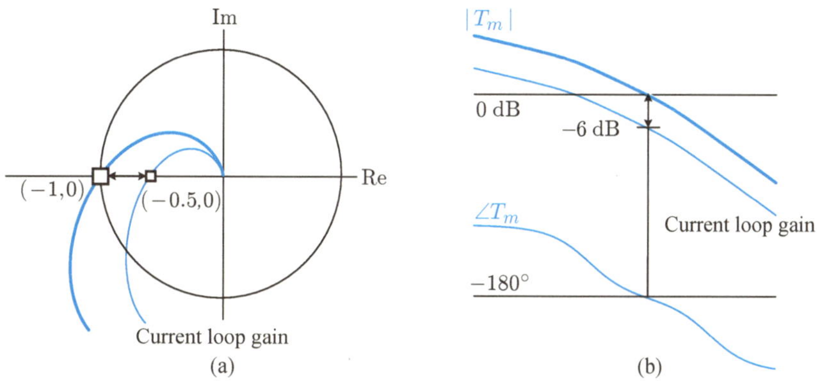

Figure 7.19 Gain margin representation. (a) Polar plot. (b) Bode plot.

crosses the $(-0.5, 0)$ point. The polar plot will pass the $(-1, 0)$ point, as shown with the thick line, when the loop gain magnitude is doubled. For this case, the gain margin is given by $20 \log 2 \approx 6$ dB. Figure 7.19(b) is the Bode plot of the same loop gain. It can be inferred that $|T_m|$ should be -6 dB at the frequency where $\angle T_m$ falls to $-180°$ so that the Bode plot also predicts a 6 dB gain margin. In other words, when $|T_m|$ is raised by 6 dB as shown with the thick line, the system encounters the condition $|T_m| = 0$ dB with $\angle T_m = -180°$, which implies the polar plot passing the $(-1, 0)$ point.

Phase Margin: The phase margin denotes the amount of phase delay that can be added to $\angle T_m$ with fixed $|T_m|$, while retaining stability. Figure 7.20 shows the loop gain plots with a 45° phase margin. Figure 7.20(a) indicates that the polar plot crosses the unit circle with a $-135°$ phase angle. When the polar plot is rotated by 45° in the clockwise direction, the polar plot passes the $(-1, 0)$ point and the system loses stability. Figure 7.20(b) shows that $\angle T_m$ becomes $-135°$ when $|T_m|$ curve crosses the 0 dB line. If $\angle T_m$ is lowered by 45°, the loop gain again encounters the condition $|T_m| = 0$ dB with $\angle T_m = -180°$ and the system becomes marginally stable. This indicates the phase margin is 45°.

The gain margin and phase margin, individually shown in Figs. 7.19 and 7.20, are simultaneously displayed on the loop gain plots in Fig. 7.21. Figure 7.21(a) depicts the gain margin and phase margin on the polar plot.

1) Gain margin, $\text{GM} = 20 \log(1/k)$: The gain margin specifies to the distance from the $(-1, 0)$ point to the point at which the polar plot passes the – real axis. When the polar plot crosses the $(-k, 0)$ point, the gain margin is given by $20 \log(1/k)$.

2) Phase margin, PM: The phase margin denotes the angle measured between the $(-1, 0)$ point and the point at which the polar plot intersects the unit circle.

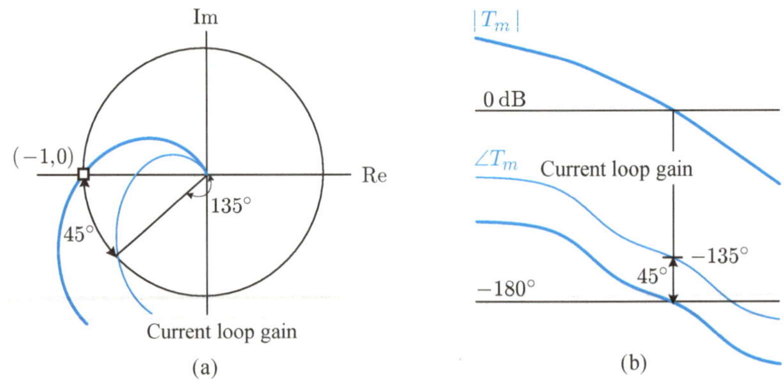

Figure 7.20 Phase margin representation. (a) Polar plot. (b) Bode plot.

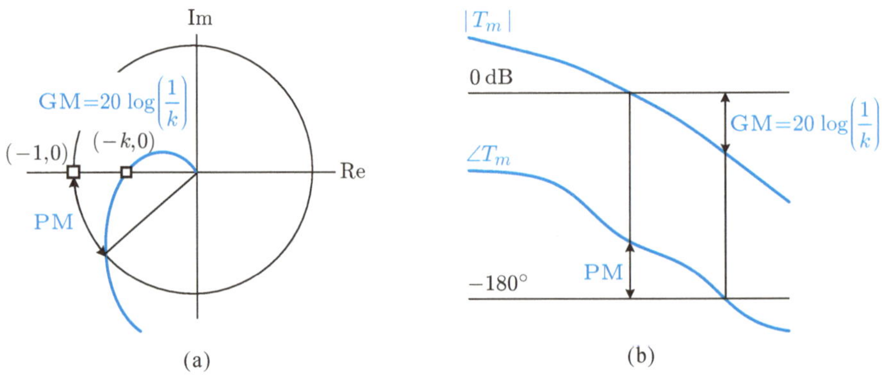

Figure 7.21 Gain margin and phase margin representation. (a) Polar plot. (b) Bode plot.

Figure 7.21(b) illustrates the gain margin and phase margin on the Bode plot.

1) Gain margin GM: The gain margin is the difference between 0 dB and $|T_m|$ evaluated at the frequency where $\angle T_m$ falls to $-180°$.

2) Phase margin PM: The phase margin corresponds to the difference between $-180°$ and $\angle T_m$ evaluated at the frequency where $|T_m|$ reduces to 0 dB.

It is easily inferred from Fig. 7.21 that both the gain margin and phase margin reduce to zero for marginally stable systems and become negative for unstable systems.

Although originally intended to quantify the relative stability, the stability margins also serve as metrics for the frequency- and time-domain performance of closed-loop controlled systems. The relationship between the stability margins and dynamic performance is based on the following arguments. From the definition of the relative stability, it is evident that the stability margins account for the nearness of the polar plot to the $(-1, 0)$ point. When the polar plot passes the $(-1, 0)$ point, the system is

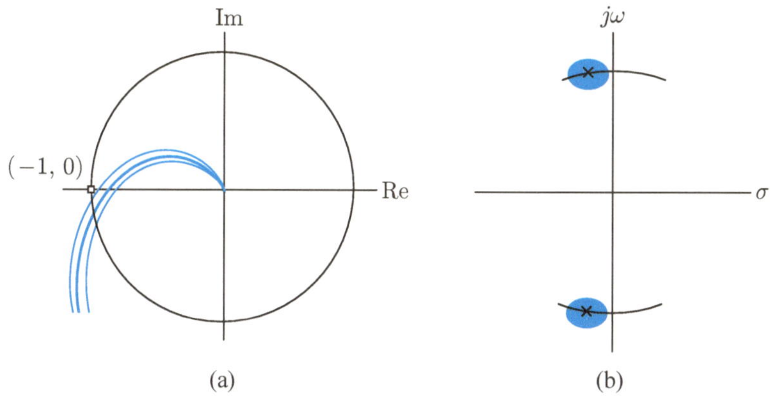

Figure 7.22 System with small stability margins. (a) Polar plot with small stability margins. (b) Pole locations with small stability margins.

marginally stable and the stability margins reduce to zero. For this case, the system has a pair of poles on the imaginary axis, as demonstrated in Example 7.7. When the polar plot bypasses the $(-1, 0)$ point, in close proximity to but not encircling the $(-1, 0)$ point, the system is barely stable with small stability margins. For this case, the system poles locate near the imaginary axis although they are in the LHP of s-plane. This instance is illustrated in Fig. 7.22.

The position of the system poles influences both the frequency- and time-domain performance. First, the poles located nearby $\pm j\omega_c$ points induce a peaking in s-domain transfer functions at the frequencies near ω_c. The nearer the poles to $\pm j\omega_c$ points, the larger the peaking in transfer functions. When the system poles are located at $\pm\ j\omega_c$ points, the peaking becomes infinite. Proof of the preceding statements can be found in most textbooks on control theory or circuit analysis. Second, the pole location affects the transient behavior of time-domain responses. As demonstrated in Example 7.7, the poles located at $\pm j\omega_c$ points cause a sustained oscillation at the frequency of ω_c. The period of this oscillation is given by $t_{os} = 2\pi/\omega_c$. When the poles are located in the LHP neighborhood of $\pm j\omega_c$ points, the system exhibits a decaying sinusoidal oscillation at ω_c.

The peaking in transfer functions or oscillation in transient responses worsen the closed-loop performance of the system. These detrimental effects become more pronounced as the poles move closer to the imaginary axis, or equivalently the stability margins become smaller. Accordingly, the stability margins can be considered as a gauge for the frequency- and time-domain performance. For this reason, the lower limits of the stability margins are recommended for acceptable closed-loop performance. As a rule of thumb, *a* $45°$ *limit is generally accepted for the phase margin, while a* 12 dB *threshold is adopted for the gain margin.*

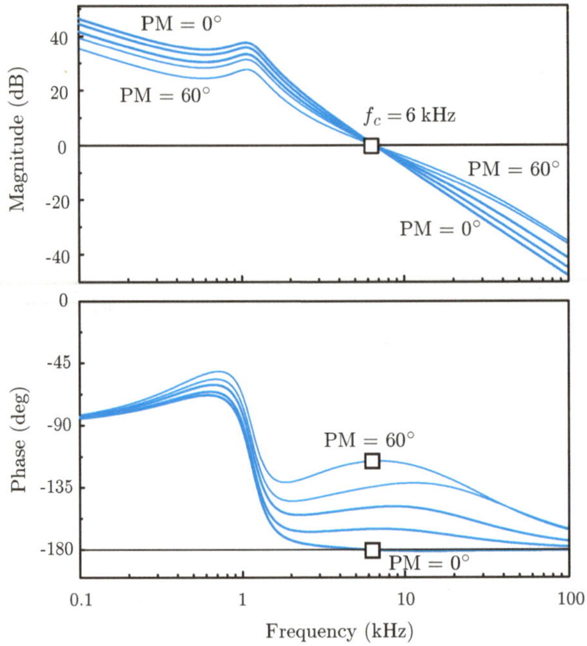

(a)

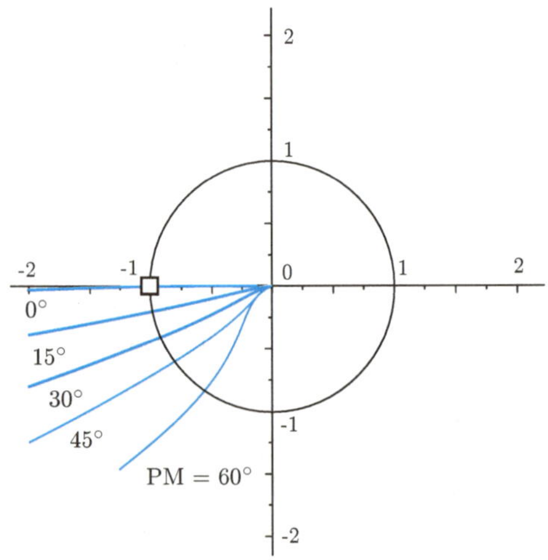

(b)

Figure 7.23 Loop gains with different phase margins. (a) Bode plot. (b) Polar plot.

Phase Margin and Closed-Loop Performance — EXAMPLE 7.9

This example illustrates the consequences of a small phase margin on the dynamic performance of the buck converter used in the previous examples. Figure 7.23(a) shows the Bode plots of the converter loop gain. For this example, the feedback compensation is selected to produce five different phase margins, decreasing successively from PM $= 60°$ to $0°$ by a $15°$ step, while maintaining the 0 dB crossover frequency at $f_c = 6$ kHz. Figure 7.23(b) shows the loop gains displayed in the polar plot format. The decrease in the phase margin is evident in the polar plot.

Figure 7.24 shows the output impedances of the converter with the five different phase margins. When the phase margin is lower than 60°, the output impedance produces a peaking due to the nearness of the system poles to the imaginary axis. The peaking occurs at the loop gain crossover frequency, $f_c = 6$ kHz.

The magnitude of the peaking is inversely proportional to the phase margin; the smaller the phase margin, the nearer the system poles to the imaginary axis and the larger the peaking. In fact, there is an exact relationship between the magnitude of the peaking and the phase margin, PM

$$|\text{peaking}| = 20\log\left(\frac{1}{\sqrt{2-2\cos \text{PM}}}\right) \tag{7.21}$$

This expression will be discussed in detail in Section 8.4.4.

Finally, Fig. 7.25 shows the transient responses of the output voltage v_O due to a step change in the load current: $I_O = 4\text{ A} \Rightarrow 2\text{ A}$. As the phase margin is lessened, the transient response becomes increasingly oscillatory until it develops a full oscillation with PM $= 0°$. The frequency of the transitional or sustained oscillation coincides with the loop gain crossover frequency: $\omega_{os} = 2\pi \cdot 6 \times 10^3$ rad/s $\Rightarrow t_{os} = 1/6 \times 10^3 = 0.17$ ms.

7.7 CHAPTER SUMMARY

Dc-to-dc converters are intended to function as a voltage source that resembles the ideal voltage source as closely as possible. Accordingly, a dc-to-dc converter is required to possess certain properties as its dynamic performance.

1) A dc-to-dc converter should establish a periodic steady-state operation, thereby producing a predetermined output voltage. More importantly, when perturbed by certain external or internal disturbances, the converter should return to the original operating point.

2) A dc-to-dc converter should maintain the output voltage as cleanly as possible in the presence of the high-frequency noises or sinusoidal variations either in the input voltage or load current.

3) A dc-to-dc converter should minimize the fluctuation in the output voltage in response to sudden changes in the load current or input voltage.

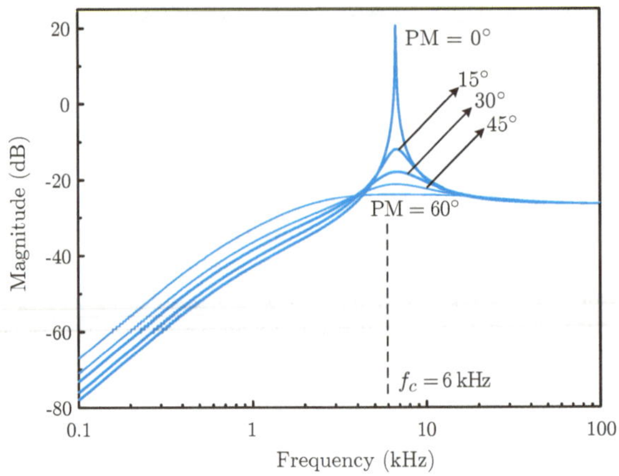

Figure 7.24 Output impedance with different phase margins.

The first property among these three items is defined as the stability of dc-to-dc converters. The second requirement is quantified by the audio-susceptibility and output impedance. Finally, the third aspect is specified by the step load response and step input response. This chapter presented practical examples to illustrate the implication and importance of these performance metrics.

We demonstrated that all the aforementioned metrics can be analyzed using the conventional control theory or circuit analysis techniques. The s-domain small-signal model of dc-to-dc converters is the instrumental vehicle that links the conventional techniques, originally developed only for linear systems, to the dynamic analysis of nonlinear time-variant dc-to-dc converters.

While the dynamic analysis of dc-to-dc converters can mostly be performed in the exact same manner as that of linear time-invariant systems, the results should be interpreted based on the assumptions and limitations of the small-signal analysis; that is, the analysis results are only valid when dc-to-dc converters do not significantly depart from the initial operating point.

This chapter provided selective summaries of the classical control theory, in preparation for upcoming details about the converter dynamics analysis. The current chapter reviewed the Nyquist stability criterion and presented the loop gain analysis using both the polar plot and Bode plot. The implication and significance of stability margins are addressed. The role of the stability margins as the performance index of dc-to-dc converters is emphasized. While this chapter highlighted theoretical essentials, further details about the classical control theory can be found in most standard textbooks.

REFERENCES

1. K. Ogata, *Modern Control Engineering*, Prentice Hall, 4th Ed., 2002.
2. B. H. Cho and F. C. Lee, *VPEC Power Electronics Professional Seminar*, Course I–Control Design, VPEC, 1991.

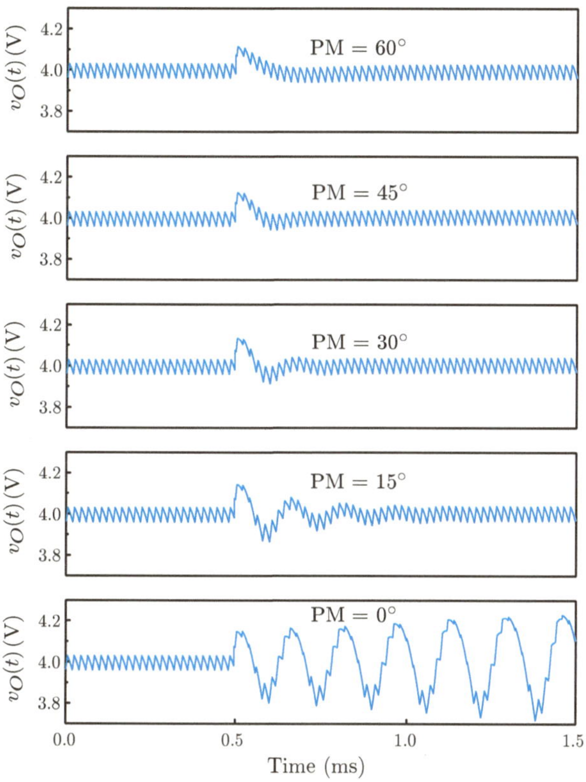

Figure 7.25 Step load response of output voltage v_O with different phase margins.

3. R. W. Erickson and D. Maksimović, *Fundamentals of Power Electronics*, Kluwer Academic Publishers, 2001.

4. R. D. Middlebrook and S. Ćuk, *Advances in Switch-Mode Power Conversion*, TESLAco, Pasadena, CA, 1983.

PROBLEMS

7.1 Shown in Fig. P7.1 is the circuit diagram of a closed-loop controlled buck converter.

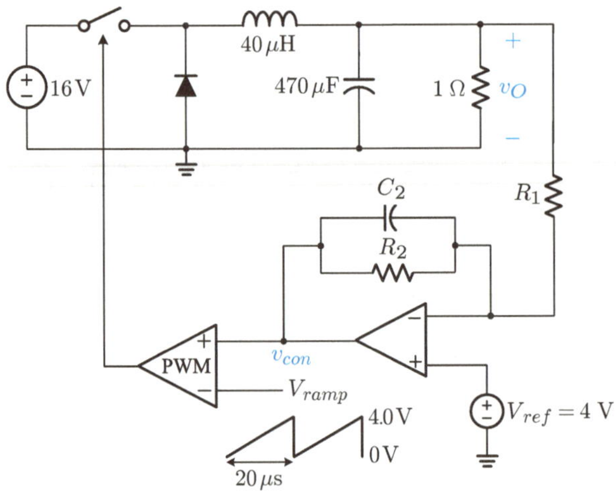

Fig. P7.1

a) Assume $R_1 = 10\,\text{k}\Omega$, $R_2 = 100\,\text{k}\Omega$, and $C_2 = 0$ and answer the questions.
   i) Find the expression for the loop gain $T_m(s)$.
   ii) Sketch the asymptotic plots for $|T_m|$ and $\angle T_m$.
   iii) Find the average values for v_O and v_{con}.

b) Repeat a) for $R_1 = 10\text{ k}\Omega$, $R_2 = \infty$, and $C_2 = 32$ nF.

c) Repeat a) for $R_1 = 10\text{ k}\Omega$, $R_2 = 100\text{ k}\Omega$, and $C_2 = 32$ nF.

7.2* Figure P7.2 is the Bode plot of the audio-susceptibility of a closed-loop controlled buck converter. The buck converter is operating at $f_s = 50$ kHz with the duty ratio $D = 0.25$. Assume that the audio-susceptibility is expressed as

$$A_u(s) = K_a s \frac{1}{\left(1 + \dfrac{s}{\omega_{p1}}\right)\left(1 + \dfrac{s}{\omega_{p2}}\right)} \quad \text{with} \quad \omega_{p1} \ll \omega_{p2}$$

a) Use the relationship (7.6) to derive an analytical expression for the output voltage v_O when a step increase of V_{step} is occurred in the input voltage. Sketch the general shape of v_O. Show all the important information on your sketch.

b) Based on the information given in Fig. P7.2, estimate the parameters, K_a, ω_{p1}, and ω_{p2}, which appeared in the audio-susceptibility expression.

c) Use the results of a) and b) to find the numeric expression for v_O with $V_{step} = 2$ V.

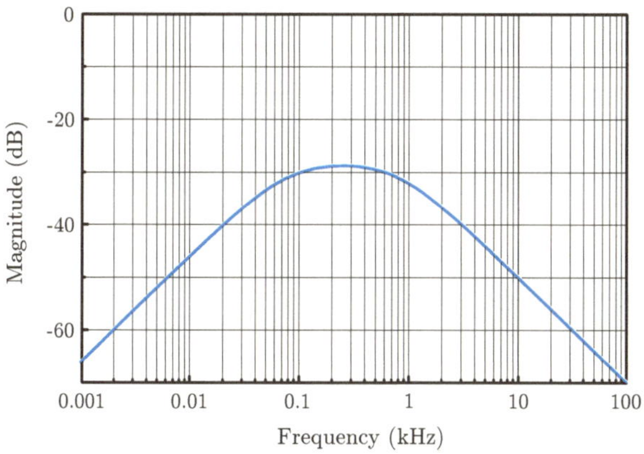

Fig. P7.2

7.3* Consider the audio-susceptibility curve shown in Fig. P7.2. The audio-susceptibility curve is generated from a buck converter with the duty ratio of $D = 0.5$. Now assume that $v_S(t) = 12 + \sin 2\pi 20t$ is applied to the input of the converter. Find the expression for the steady-state output voltage while ignoring the switching ripple component.

7.4 Shown in Fig. P7.4 is the Bode plot of the output impedance of a closed-loop controlled buck converter. Assume that the output impedance is expressed by the equation

$$Z_o(s) = K_z s \frac{1}{1 + \dfrac{s}{\omega_p}}$$

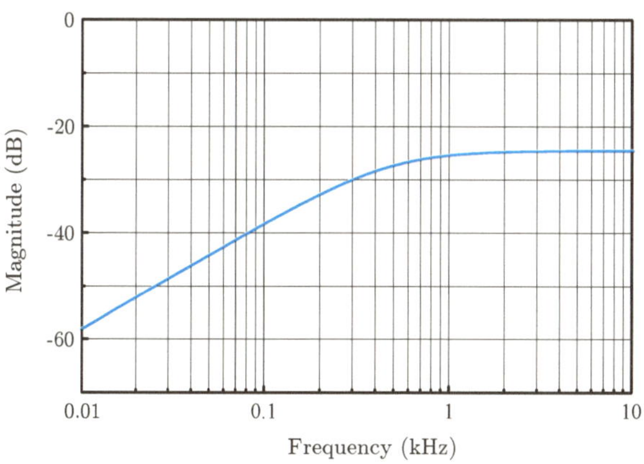

Fig. P7.4

a) Using the relationship (7.5), derive an analytic equation for the output voltage v_O when a step decrease of I_{step} occurred in the load current. Sketch the general shape of v_O and show all the important information on your sketch.

b) Refer to the information given in the Bode plot to determine the parameters K_z and ω_p, used in the output impedance expression.

c) Using the results of a) and b), find the numeric expression for v_O with $I_{step} = 2$ A.

7.5 Consider the following expressions as the output impedance of a closed-loop controlled buck converter.

i) $Z_o(s) = K_z s \dfrac{1}{1+\dfrac{s}{\omega_p}}$

ii) $Z_o(s) = K_z s \dfrac{1+\dfrac{s}{\omega_z}}{\left(1+\dfrac{s}{\omega_{p1}}\right)\left(1+\dfrac{s}{\omega_{p2}}\right)}$ with $\omega_{p1} \ll \omega_z \ll \omega_{p2}$

iii) $Z_o(s) = K_z s \dfrac{1+\dfrac{s}{\omega_z}}{\left(1+\dfrac{s}{\omega_{p1}}\right)\left(1+\dfrac{s}{\omega_{p2}}\right)}$ with $\omega_{p1} \ll \omega_{p2} \ll \omega_z$

Now assume that a step decrease of I_{step} is occurred in the load current. For each of the above expressions, derive an analytic equation for the transient response of the output voltage v_O using the relationship (7.5). Also, sketch the general shape of the transitional output voltage.

7.6 Figure P7.6 is the magnitude plot of the loop gain $|T_m|$ and the load current-to-output transfer function $|Z_p|$ of a closed-loop controlled buck converter. As discussed in Section 7.2.3, the output impedance is expressed as

$$Z_o(s) = \frac{Z_p(s)}{1+T_m(s)}$$

This relationship can be split into the approximations

$$Z_o(s) = \frac{Z_p(s)}{1+T_m(s)} \approx \begin{cases} \dfrac{Z_p(s)}{T_m(s)} & \text{at frequencies where } |T_m| > 1 \\ Z_p(s) & \text{at frequencies where } |T_m| < 1 \end{cases}$$

to construct the asymptotic plot for $|Z_o|$.

a) Sketch the asymptotic plot of the output impedance $|Z_o|$. Label the corner frequency, slope, and peak value of $|Z_o|$.

b) Use the outcome of a) to find the expression for $Z_o(s)$.

c) Derive the equation for the output voltage v_O due to a 5 A step decrease in the load current. Sketch v_O and show the time constant and peak undershoot of v_O.

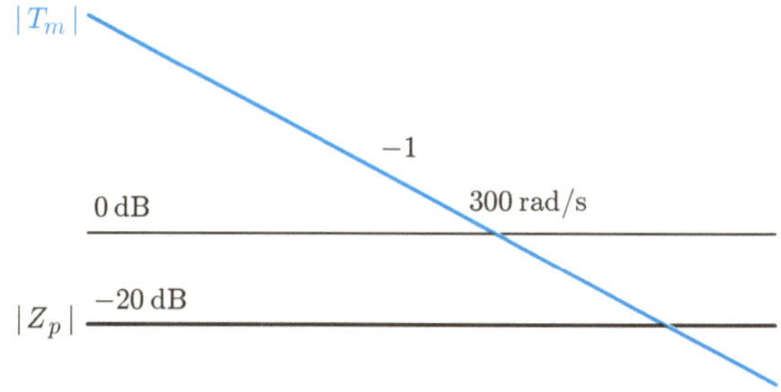

Fig. P7.6

7.7* A conditionally stable converter shows some characteristic features. *Graphically* illustrate the notable features in the following closed-loop performance of a conditionally stable converter. Assume that the final phase of the loop gain is $-270°$: $\angle T_m(j\infty) = -270°$.
i) Polar plot of the loop gain
ii) Bode plot of the loop gain
iii) Output voltage of the converter with a sufficient increase in the feedback gain
iv) Output voltage of the converter with a sufficient decrease in the feedback gain. Show the difference in the voltage waveform in comparison with iii)
v) A practical operational scenario where the conditionally stable converter becomes problematic.

7.8* A composite system shown in Fig. P7.8(a) is configured using a voltage source and two two-port systems.

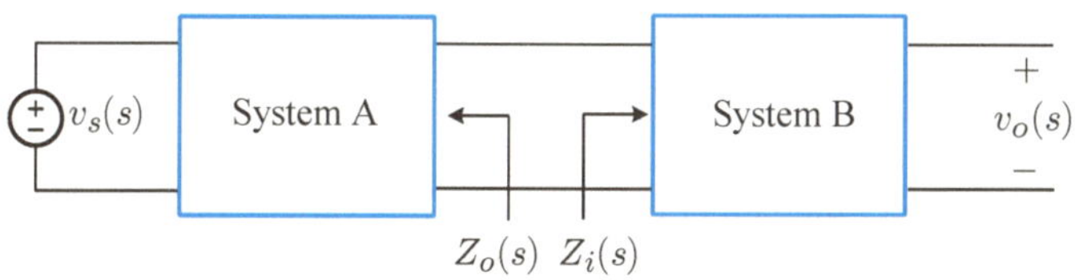

Fig. P7.8(a)

a) Find the expression for the input-to-output transfer function of the composite system, $v_o(s)/v_s(s)$, in terms of the following transfer functions defined individually for System A and System B:
$F_A(s)$: the input-to-output transfer function of System A
$F_B(s)$: the input-to-output transfer function of System B
$Z_o(s)$: the output impedance of System A
$Z_i(s)$: the input impedance of System B

b) Assume that System A and System B are individually stable. Now argue that the stability of the composite system can be assessed by applying the Nyquist criterion to the impedance ratio $Z_o(s)/Z_i(s)$.

c) Derive a sufficient, but not necessary, condition to ensure the stability of the composite system.

d) Now assume that $|T_m(j0)| = |T_m(j\infty)| = 0$. Sketch a typical example of the polar plot for the $T_m(j\omega) = Z_o(j\omega)/Z_i(j\omega)$ that satisfies the sufficient condition found in c).

e) Assume $|T_m(j0)| = |T_m(\infty)| = 0$ and sketch a typical polar plot of the $T_m(j\omega) = Z_o(j\omega)/Z_i(j\omega)$ for each of the following conditions:
i) the sufficient condition is violated and the system is unstable
ii) the sufficient condition is violated and the system is marginally stable
iii) the sufficient condition is violated but the system still remains stable.

f) Shown in Fig. P7.8(b) are two different cases of the polar plot for the impedance ratio of $Z_o(j\omega)/Z_i(j\omega)$ of the composite system in Fig. P7.8(a). Referring the polar plots, answer the questions.

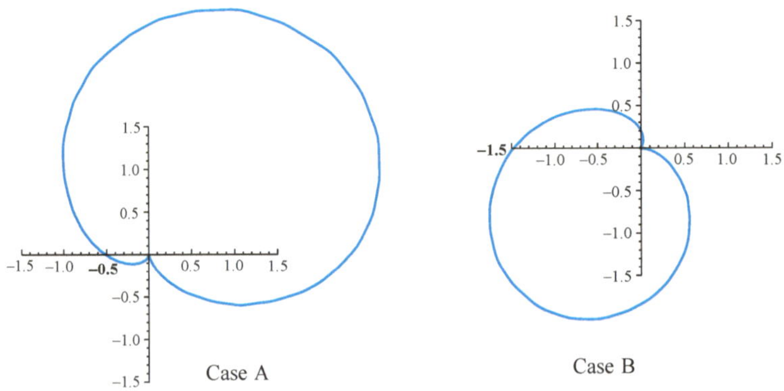

Fig. P7.8(b)

i) For Case A, determine the stability of the composite system. Now assume the magnitude of the output impedance $|Z_s|$ is varied while its phase characteristics remain the same. The input impedance $Z_i(s)$ remains unchanged in both its magnitude and phase. State the condition that makes the system marginally stable.
ii) Repeat i) for Case B.

7.9* A marginally stable system shows some characteristic features in the closed-loop performance. Describe or illustrate the characteristic feature(s) in the following closed-loop performance of a marginally stable system. For the definiteness of your argument, assume that 0 dB crossover frequency of the loop gain occurs at $\omega = \omega_c$.

i) Polar plot of loop gain
ii) Bode plot of loop gain
iii) Pole location in s-plane
iv) Time-domain waveforms
v) Closed-loop transfer functions

7.10* Shown in Fig. P7.10 are the polar plot for loop gain and output voltages of a closed-loop controlled dc-to-dc converter.

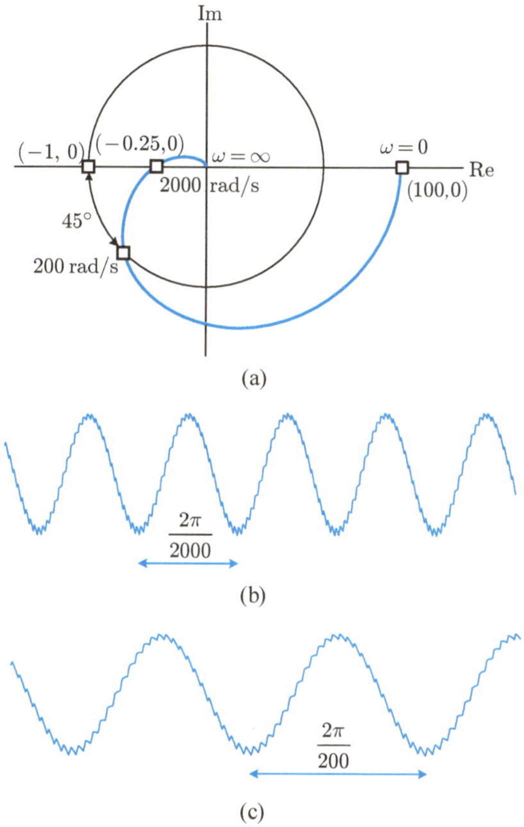

Fig. P7.10

a) Referring to Fig. P7.10(a), determine stability of the converter. Evaluate the gain margin, phase margin, and 0 dB crossover frequency of the loop gain.
b) Convert the polar plot into the Bode plot for $|T_m|$ and $\angle T_m$. Show all the prominent features on the Bode plot.
c) Assume that the converter produces the output voltage waveform shown in Fig. P7.10(b). What do you think happened to the voltage feedback compensation of the converter?
d) Now assume that the converter output generates the waveform in Fig. P7.10(c). What do you think occurred in the voltage feedback compensation?

7.11 The Bode plot shown in Fig. P7.11 is the loop gain of a closed-loop controlled dc-to-dc converter.

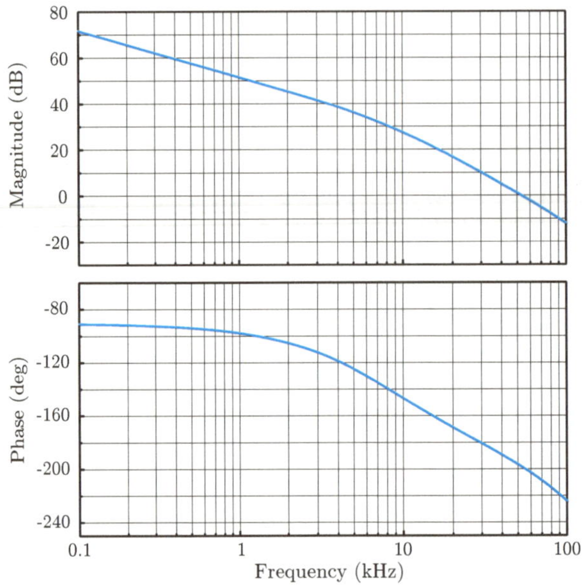

Fig. P7.11

a) Determine stability of the converter. Evaluate the gain margin (positive or negative), phase margin (positive or negative), and 0 dB crossover frequency of the loop gain.

b) You are only allowed to change the magnitude of the voltage feedback compensation to meet the following design objectives:
i) stability with the phase margin in the neighborhood of 20°,
ii) stability with the phase margin in the neighborhood of 60°, and
iii) stability with the gain margin in the neighborhood of 20 dB
What would you do to achieve each specification described above?

7.12* Figure P7.12 is the polar plot of the loop gain of a PWM dc-to-dc converter.

a) Determine stability of the converter.

b) The gain of the voltage feedback compensation is currently $K_v = 100$. Now assume that the gain is varied between $1 < K_v < 500$ while other compensation parameters remain the same. Under this condition, find the range(s) of K_v in which the converter remains stable.

c) What would happen to the converter output when the magnitude of the current feedback gain, $K_v = 100$, is changed to 10, 20, and 400?

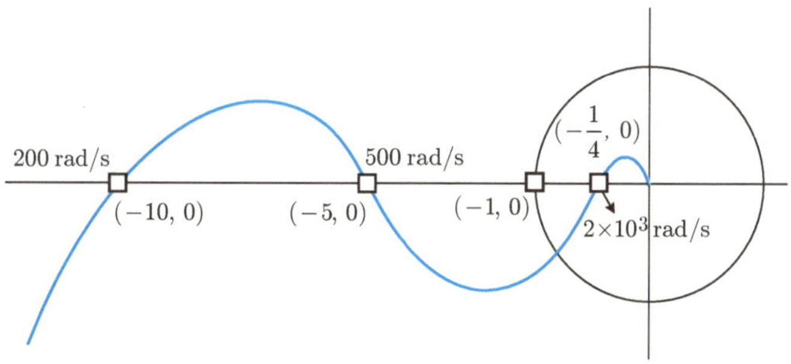

Fig. P7.12

7.13 Figure P7.13 is the Bode plot of the loop gain of a closed-loop controlled buck converter.

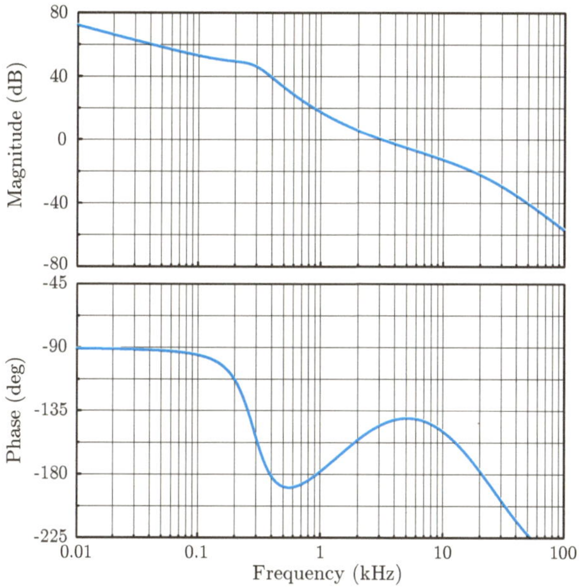

Fig. P7.13

a) Determine the stability of the converter.
b) Convert the Bode plot into the polar plot. Show all the prominent features of the loop gain on the polar plot.
c) Assume that the gain of the voltage feedback compensation is $K_v = 10^3$ for the current design. Now assume that the gain is varied between $1 < K_v < 10^5$. Find the range(s) of K_v which secure(s) the stability of the converter.

PART III

CONTROL SCHEMES AND CONVERTER PERFORMANCE

The small-signal modeling and s-domain analysis of PWM converters are aimed at achieving systematic design of feedback controller. As the heart of this book, Part III covers the control design for stability and performance of PWM dc-to-dc converters. The underlying fundamentals and engineering practices of the control design are described along with several examples. The two prevalent control schemes, voltage mode control and current mode control, are treated in great detail.

Chapter 8 deals with voltage mode control, which employs the output voltage as the sole feedback signal. The design and assessment of voltage mode control are discussed in consideration of the frequency- and time-domain performance of a buck converter. Relations between the feedback parameters and closed-loop performance are analyzed, leading to step-by-step design procedures for voltage mode control. Limitation of voltage mode control are discussed, while addressing the need for current mode control for boost and buck/boost converters.

Chapters 9 and 10 cover current mode control, which uses an additional feedback from the inductor current on top of the previous output voltage feedback. Chapter

9 first discusses the evolution of current mode control and benefits of the current feedback. The control design and performance analysis, studied in Chapter 8 for voltage mode control, are reinforced and extended to cover current model control.

In current mode control, the inductor current is sampled and held constant for every single switching period and used as the control variable in the next switching period. Due to this feature, current mode control implants certain attributes of the sampled-data system into the converter dynamics. This facet of the control scheme is referred to as the sampling effects of current mode control. Chapter 10 is dedicated to the sampling effects of current mode control. Consequences of the sampling effects on the converter dynamics and closed-loop performance are investigated. Alternative design procedures, based on the analysis of the sampling effect, are developed for current mode control. The new control design procedures are compared and contrasted with the previous design procedures covered in Chapter 9, providing more informative insight onto current mode control. As an application example, the final section of Chapter 10 provides the design and evaluation of current mode control, adapted to an off-line flyback converter with an optocoupler isolated feedback control.

CHAPTER 8

FEEDBACK COMPENSATION AND CLOSED-LOOP PERFORMANCE

The dynamic performance of closed-loop controlled dc-to-dc converters is solely determined by the feedback controller. The ultimate goal of the small-signal modeling and dynamic analysis is to develop systematic procedures that allow us to design the feedback controller for optimal dynamic performance.

The transfer function of the feedback controller is denoted as the voltage feedback compensation. Thus, the dynamic performance of the converter is governed by the voltage feedback compensation. This chapter covers the dynamic analysis and control design, and provides design guidelines of the voltage feedback compensation for optimal closed-loop performance of PWM dc-to-dc converters.

The dynamic performance includes the loop gain, frequency-domain transfer functions, and time-domain transient responses. These performance metrics are closely related to each other and, as such, they must be collectively considered when designing the voltage feedback compensation. The first step towards a successful controller design is to understand the relationships among various performance metrics and to analyze the impacts of the voltage feedback compensation on these metrics. Once such issues have been resolved, it then becomes possible to develop design principles for the voltage feedback compensation.

This chapter covers the dynamic analysis and feedback controller design of closed-loop controlled PWM dc-to-dc converters. The first two sections analyze the dynamic performance focusing on the relationship between the loop gain and frequency-domain performance. The following sections deal with the design of voltage feedback compensation. The desired structure for the voltage feedback circuit is identified and step-by-step guidelines for selecting compensation parameters are presented. Later sections illustrate the impacts of the voltage feedback compensation on the frequency- and time-domain performance. The continuous conduction mode (CCM) is initially treated in the dynamic analysis and control design. Considerations for the discontinuous conduction mode (DCM) will be later addressed in a separate section.

The graphical *asymptotic analysis method* is used as the principal means for the closed-loop analysis and feedback compensation design. The concept and examples of the asymptotic analysis are introduced in the first section. The asymptotic analysis method is then used in the forthcoming sections to illustrate the outcomes of the closed-loop analysis, design steps of voltage feedback compensation, and impacts of feedback compensation on converter performance.

8.1 ASYMPTOTIC ANALYSIS METHOD

This section presents the concept and examples of the asymptotic analysis method. The motivation of the asymptotic analysis is first introduced. Practical details about the proposed method are then illustrated with examples.

8.1.1 Concept of Asymptotic Analysis Method

As discussed in Chapter 7, the frequency-domain performance includes the loop gain, audio-susceptibility, and output impedance. This section presents a systematic approach to analyzing such performance metrics. The proposed approach is referred

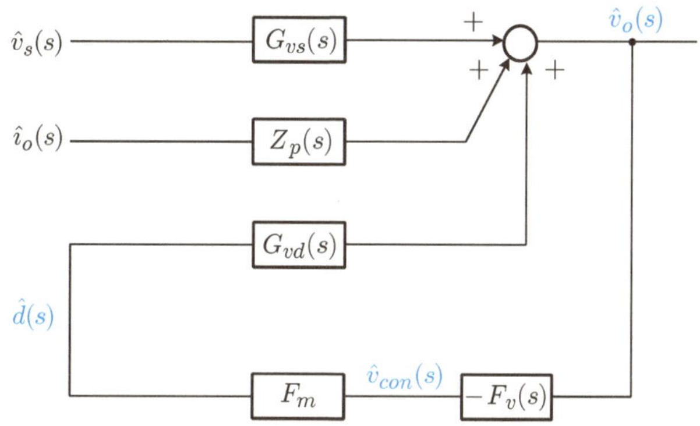

Figure 8.1 Small-signal block diagram of PWM converter.

to as the asymptotic analysis method because it uses the asymptotic expressions of the transfer functions as the basis for the analysis.

Figure 8.1 shows the small-signal block diagram of a closed-loop controlled dc-to-dc converter. The gain block F_m is the PWM gain, $F_v(s)$ is the voltage feedback compensation, and the other small-signal gain blocks denote the power stage transfer functions. From Fig. 8.1, the loop gain is determined as

$$T_m(s) = -\frac{\hat{v}_o(s)}{\hat{d}(s)}\frac{\hat{v}_{con}(s)}{\hat{v}_o(s)}\frac{\hat{d}(s)}{\hat{v}_{con}(s)} = (-)G_{vd}(s)(-)F_v(s)F_m = G_{vd}(s)F_v(s)F_m \tag{8.1}$$

The application of Mason's gain rule to Fig. 8.1 yields the expressions for the frequency-domain performance metrics. The audio-susceptibility is given by

$$A_u(s) = \left.\frac{\hat{v}_o(s)}{\hat{v}_s(s)}\right|_{\text{closed}} = \frac{G_{vs}(s)}{1+T_m(s)} \tag{8.2}$$

If the small-signal gain blocks are all known, the expression (8.2) can be used to find an equation for the audio-susceptibility. However, this *direct method* would not yield any useful results for the feedback compensation design.

As an alternative to the direct method, an asymptotic analysis method is used to simplify the analysis and, more importantly, to obtain design information. The expression (8.2) is split into two asymptotic approximations

$$A_u(s) = \frac{G_{vs}(s)}{1+T_m(s)} \approx \begin{cases} \dfrac{G_{vs}(s)}{T_m(s)} & \text{at frequencies where } |T_m| > 1 \\ G_{vs}(s) & \text{at frequencies where } |T_m| < 1 \end{cases} \tag{8.3}$$

For practical converters, $|T_m|$ is very large at low frequencies, crosses the 0 dB line at mid-frequencies, and continuously decreases at high frequencies. Therefore, the 0 dB crossover frequency of the loop gain serves as the natural borderline for the asymptotic approximation: $A_u(s) \approx G_{vs}(s)/T_m(s)$ at frequencies before the 0 dB crossover frequency and $A_u(s) \approx G_{vs}(s)$ thereafter.

The asymptotic approximation of (8.3) is combined with the graphical Bode plot analysis technique. The resulting analysis method will be referred to as the *asymptotic analysis method*. The asymptotic analysis method explicitly shows the impacts of the voltage feedback compensation on the audio-susceptibility $A_u(s)$, thereby providing easy and clear design information. This method does not require any complex analytical treatments and quickly yields the expression for $A_u(s)$ in a factorized form. For most cases, a factorized expression for $A_u(s)$ can immediately be written by inspection.

The expression for the output impedance is found from Fig. 8.1 and put into the approximation

$$\begin{aligned} Z_o(s) &= \left.\frac{\hat{v}_o(s)}{\hat{i}_o(s)}\right|_{\text{closed}} \\ &= \frac{Z_p(s)}{1+T_m(s)} \approx \begin{cases} \dfrac{Z_p(s)}{T_m(s)} & \text{at frequencies where } |T_m| > 1 \\ Z_p(s) & \text{at frequencies where } |T_m| < 1 \end{cases} \end{aligned} \tag{8.4}$$

for the asymptotic analysis. Because the expressions (8.3) and (8.4) have the same structure, the asymptotic analysis method can consistently be applied to both the audio-susceptibility and output impedance.

8.1.2 Examples of Asymptotic Analysis Method

Before embarking on the converter performance analysis, it is first necessary to familiarize with the proposed asymptotic analysis method. Details about the asymptotic analysis are illustrated using the following equation

$$F(s) = \frac{G(s)}{1+T(s)} \approx \begin{cases} \dfrac{G(s)}{T(s)} & \text{at frequencies where } |T| > 1 \\ G(s) & \text{at frequencies where } |T| < 1 \end{cases} \tag{8.5}$$

with simple expressions for $G(s)$ and $T(s)$. Three different cases are considered to provide practical insights on the asymptotic analysis method.

Case A: $G(s) = K_g < 1$ and $T(s) = \dfrac{\omega_c}{s}$

As the simplest example, a constant is used for $G(s)$ while a single integration function is assumed for $T(s)$: $G(s) = K_g < 1$ and $T(s) = \omega_c/s$. Figure 8.2 displays

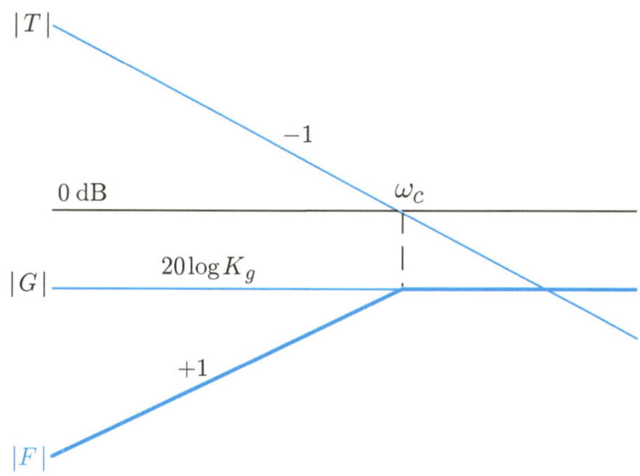

Figure 8.2 Asymptotic analysis for Case A.

the construction of the asymptotic plot for $|F|$ with the given $G(s)$ and $T(s)$. As previously discussed, the 0 dB crossover frequency of $|T|$, denoted as ω_c in Fig. 8.2, serves as the boundary for the approximations. For frequencies beyond ω_c, $|F|$ follows $|G|$. For frequencies below ω_c, $|F|$ is formed by subtracting $|T|$ from $|G|$, because the division in the linear scale corresponds to the subtraction in the log scale. As shown in Fig. 8.2, the ascending slope and corner frequency of $|F|$ are readily determined from the graphical construction.

The asymptotic plot of $|F|$ is transformed into an analytic equation for $F(s)$

$$F(s)_{\mathrm{asym}} = K_f s \frac{1}{1+\dfrac{s}{\omega_c}} \tag{8.6}$$

by applying the inverse process of the Bode plot construction, as discussed in Section 6.1.4. The leading coefficient K_f is found by equating $|F_{\mathrm{asym}}|$ at ω_c to $|G|$

$$\begin{aligned} |F(s)_{\mathrm{asym}}|_{s=j\omega_c} &= |K_f s|_{s=j\omega_c} = |K_f\omega_c| = |G| \\ &\Rightarrow 20\log(K_f\omega_c) = 20\log K_g \end{aligned} \tag{8.7}$$

to yield the relationship

$$K_f\,\omega_c = K_g \tag{8.8}$$

The desired expression for K_f is obtained from (8.8)

$$K_f = \frac{K_g}{\omega_c} \tag{8.9}$$

resulting in the expression

$$F(s)_{\mathrm{asym}} = \frac{K_g}{\omega_c} s \frac{1}{1+\dfrac{s}{\omega_c}} \tag{8.10}$$

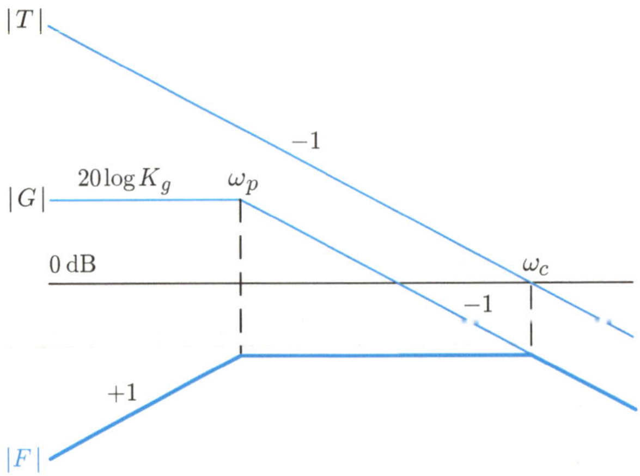

Figure 8.3 Asymptotic analysis for Case B.

On the other hand, the direct evaluation of the original equation $F(s) = G(s)/(1+T(s))$ with $G(s) = K_g$ and $T(s) = \omega_c/s$ yields the exact equation for $F(s)$

$$F(s)_{\text{exac}} = \frac{K_g}{1+\dfrac{\omega_c}{s}} = \frac{K_g}{\omega_c} s \frac{1}{1+\dfrac{s}{\omega_c}} \tag{8.11}$$

For this particular case, the asymptotic analysis is the same as the exact analysis: $F(s)_{\text{asym}} = F(s)_{\text{exac}}$. However, the sameness is due to the simple structure of $T(s)$ and is not always true for general cases.

Case B: $G(s) = \dfrac{K_g}{1+s/\omega_p}$ with $K_g > 1$ and $T(s) = \dfrac{\omega_c}{s}$

In this case, a low-pass filter function is considered for $G(s)$ while the single integration function is used for $T(s)$: $G(s) = K_g/(1+s/\omega_p)$ and $T(s) = \omega_c/s$. Figure 8.3 shows the construction of the asymptotic plot for $|F|$. The $|F|$ plot is obtained by subtracting $|T|$ from $|G|$ at frequencies prior to ω_c, and tracing $|G|$ thereafter. The asymptotic plot for $|F|$ in Fig. 8.3 is transformed into a factorized equation for $F(s)$

$$F(s)_{\text{asym}} = K_f s \frac{1}{\left(1+\dfrac{s}{\omega_p}\right)\left(1+\dfrac{s}{\omega_c}\right)} \tag{8.12}$$

The value for K_f is determined by evaluating the magnitude of $F(s)_{\text{asym}}$ at ω_p

$$\begin{aligned} |F(s)_{\text{asym}}|_{s=j\omega_p} &= |K_f s|_{s=j\omega_p} \\ &= |G(s)|_{s=j\omega_p} - |T(s)|_{s=j\omega_p} \end{aligned} \tag{8.13}$$

which is transformed to

$$K_f\,\omega_p = \frac{K_g}{\dfrac{\omega_c}{\omega_p}} \tag{8.14}$$

leading to the expression

$$K_f = \frac{K_g}{\omega_c} \tag{8.15}$$

On the other hand, the direct evaluation of the equation $F(s) = G(s)/(1+T(s))$ with the given transfer functions produces the exact expression for the transfer function for $F(s)$

$$F(s)_{\text{exac}} = \frac{\dfrac{K_g}{1+\dfrac{s}{\omega_p}}}{1+\dfrac{\omega_c}{s}} = \frac{K_g}{\omega_c}s\frac{1}{\left(1+\dfrac{s}{\omega_p}\right)\left(1+\dfrac{s}{\omega_c}\right)} \tag{8.16}$$

which is identical to the result of the preceding asymptotic analysis.

Figures 8.2 and 8.3 provide insights about the asymptotic analysis, leading to the general procedures for this graphical approach. The procedures and rules for the asymptotic analysis are summarized in Tables 8.1 and 8.2 at the end of this section.

Case C: $G(s) = K_g < 1$ and $T(s) = \dfrac{\omega_c}{s}\dfrac{1+s/\omega_z}{1+s/\omega_p}$ with $\omega_p < \omega_z$

As the last example, it is assumed that $G(s) = K_g < 1$ and $T(s) = (\omega_c/s)(1 + s/\omega_z)/(1 + s/\omega_p)$ with $\omega_p < \omega_z$. Figure 8.4 illustrates the construction of the asymptotic plot for $|F|$. In Fig. 8.4, ω_c' corresponds to the 0 dB crossover frequency of $|T|$ and serves as the borderline for the asymptotic construction. The asymptotic plot is transformed into a factorized equation for $F(s)$

$$F(s)_{\text{asym}} = \frac{K_g}{\omega_c}s\frac{1+\dfrac{s}{\omega_p}}{\left(1+\dfrac{s}{\omega_z}\right)\left(1+\dfrac{s}{\omega_c'}\right)} \tag{8.17}$$

based on the procedures in Table 8.2. The leading coefficient of (8.17) is determined using the same procedure as that of Case B.

The direct evaluation of the expression $F(s) = G(s)/(1 + T(s))$ with the given $G(s)$ and $T(s)$ expressions yields

$$F(s)_{\text{exac}} = \frac{K_g}{1+\dfrac{\omega_c}{s}\dfrac{1+\dfrac{s}{\omega_z}}{1+\dfrac{s}{\omega_p}}} = \frac{K_g}{\omega_c}s\frac{1+\dfrac{s}{\omega_p}}{1+\dfrac{s}{\omega_z}+\dfrac{s}{\omega_c}+\dfrac{s^2}{\omega_p\omega_c}} \tag{8.18}$$

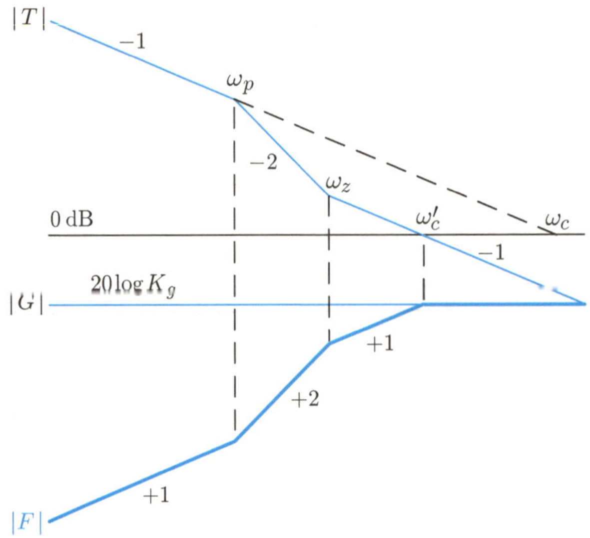

Figure 8.4 Asymptotic analysis for Case C.

Comparing (8.17) and (8.18), it becomes apparent that the asymptotic analysis is an approximation to the exact evaluation. Even so, this approximation is accurate for the following reasons. First, the two equations have the same leading coefficient and numerator. Because these terms determine the low-frequency characteristics, the two transfer functions will show the same behavior at low frequencies. Second, the high-frequency asymptote of the approximated equation $F(s)_{\text{asym}}$ is given by

$$|F(j\infty)_{\text{asym}}| = 20\log\left(\lim_{s\to\infty} F(s)_{\text{asym}}\right) = 20\log\left(\frac{K_g\,\omega_z\,\omega_c'}{\omega_c\,\omega_p}\right) \tag{8.19}$$

On the other hand, it can be shown that

$$\frac{\omega_z}{\omega_p} = \frac{\omega_c}{\omega_c'} \tag{8.20}$$

from the asymptotic plot in Fig. 8.4. By incorporating (8.20) into (8.19), the high-frequency asymptote of $F(s)_{\text{asym}}$ is given by

$$|F(j\infty)_{\text{asym}}| = 20\log K_g \tag{8.21}$$

which is identical to the high-frequency asymptote of the exact equation

$$|F(j\infty)_{\text{exac}}| = 20\log\left(\frac{K_g\omega_p\,\omega_c}{\omega_c\,\omega_p}\right) = 20\log K_g \tag{8.22}$$

Thus, the high-frequency asymptotes of the two equations are also identical.

The previous analysis concludes that the asymptotic approximation duplicates the exact equation at low and high frequencies, yet with some deviation in the mid-frequency range. This mid-frequency error is negligibly small for most cases.

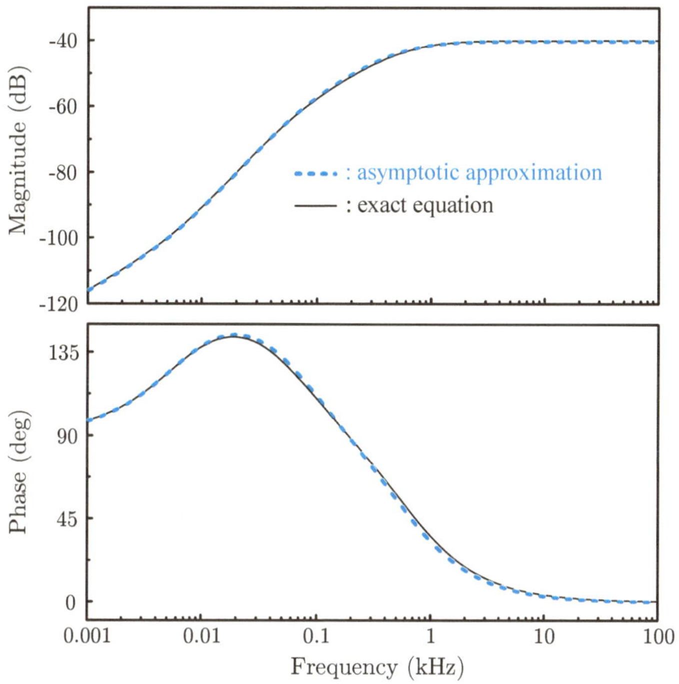

Figure 8.5 Accuracy of asymptotic approximation.

Accuracy of Asymptotic Approximation — EXAMPLE 8.1

This example substantiates the accuracy of the asymptotic approximation. As an illustration, Case C in the previous three cases is selected with $G(s) = 0.01$ and

$$T(s) = \frac{4 \times 10^4}{s} \frac{1 + \dfrac{s}{400}}{1 + \dfrac{s}{40}}$$

Figure 8.5 shows the Bode plot of the asymptotic approximation of (8.17) with $K_g = 0.01$, $\omega_c = 4 \times 10^4$ rad/s, $\omega_c' = 4 \times 10^3$ rad/s, $\omega_z = 400$ rad/s, and $\omega_p = 40$ rad/s, in comparison with the exact equation of (8.18). As predicted, the asymptotic approximation ideally matches with the exact equation at low and high frequencies, while showing nearly-undetectable error in the mid-band. This confirms that the asymptotic analysis offers good accuracy and can generally be used for the closed-loop dynamic analysis of dc-to-dc converters.

Table 8.1 Procedures for $|F(j\omega)|$ Sketch

Basic equation
$$F(s) = \frac{G(s)}{1+T(s)} \approx \begin{cases} \dfrac{G(s)}{T(s)} & \text{at frequencies where } \lvert T \rvert > 1 \\ G(s) & \text{at frequencies where } \lvert T \rvert < 1 \end{cases}$$
Sketch rules for asymptotic plot of $\lvert F \rvert$

1) Sketch $|G|$ and $|T|$ for given equations.
2) Start sketching $|F|$ from high frequencies.
3) Sketch $|F|$ based on the following guidelines.
   a) Frequencies beyond the 0 dB crossover frequency of $|T|$: $|F|$ duplicates $|G|$.
   b) At the 0 dB crossover frequency of $|T|$: $|F|$ intersects $|G|$.
   c) Frequencies below the 0 dB crossover frequency of $|T|$: $|F|$ is constructed by subtracting $|T|$ from $|G|$. There is a simple relationship among the ascending and descending slopes of the transfer functions

   $$\text{slope of } |F| = \text{slope of } |G| - \text{slope of } |T|$$

   For example, when $|G|$ decays with -1 slope and $|T|$ also declines with -1 slope, $|F|$ stays at a flat value with *zero* slope: $-1-(-1)=0$. Similarly, when $|G|$ ascends with $+1$ slope and $|T|$ descends with -1 slope, $|F|$ increases with $+2$ slope: $+1-(-1)=2$.

Procedures for Asymptotic Analysis

As illustrated in the previous analyses, the graphical asymptotic analysis of $F(s) = G(s)/(1+T(s))$ involves the following two steps.

1) The first step is to sketch the asymptotic plot for $|F(j\omega)| = |G(j\omega)|/|(1+T(j\omega))|$ for the known $G(s)$ and $T(s)$ expressions.

2) The second step is to construct the $F(s)$ expression from the $|F(j\omega)|$ sketch obtained in the first step.

Tables 8.1 and 8.2 present the general rules for the asymptotic analysis. Table 8.1 illustrates the steps of sketching the asymptotic plot for $|F(j\omega)|$, while Table 8.2

Table 8.2 Procedures for $F(s)$ Construction

Basic equation
$F(s) = K_f \frac{1}{s^{\pm j}} \frac{1}{1+\frac{s}{\omega_p}} \cdots \left(1+\frac{s}{\omega_z}\right) \cdots \frac{1}{1+\frac{s}{Q_p \omega_o}+\frac{s^2}{\omega_o^2}} \cdots \left(1+\frac{s}{Q_z \omega_o}+\frac{s^2}{\omega_o^2}\right) \cdots$
Construction rules for factorized equation of $F(s)$
1) Construct $F(s)$ in a time constant form.
2) Start writing $F(s)$ expression from low-frequency terms.
3) Determine the poles and zeros for $F(s)$ from $\lvert F(j\omega)\rvert$ sketch. The poles and zeros appear at the frequencies where $\lvert F\rvert$ changes its slope. When the slope is changed by *one*, a single pole or single zero occurs. Similarly, when the slope is changed by *two*, a double pole or double zero appears.
4) Determine the value for the leading coefficient K_f. The K_f value can be determined from the low-frequency asymptote, high-frequency asymptote, 0 dB crossover frequency, or at one of corner frequencies of $\lvert F\rvert$.

summarizes the rules of writing $F(s)$ expression from $|F(j\omega)|$ sketch. Readers are urged to review Tables 8.1 and 8.2 in preparation for upcoming details about the dynamic analysis and control design based on the asymptotic analysis.

8.2 ANALYSIS OF FREQUENCY-DOMAIN PERFORMANCE IN CCM

This section investigates the frequency-domain performance of dc-to-dc converters using the asymptotic analysis method. Relationships between the loop gain and closed-loop transfer functions are graphically illustrated. Factorized s-domain expressions for the transfer functions are provided. A closed-loop controlled buck converter is used as an illustrative example, nonetheless, the results can be extended to all PWM dc-to-dc converters. This section starts with the audio-susceptibility analysis for a buck converter in CCM operation.

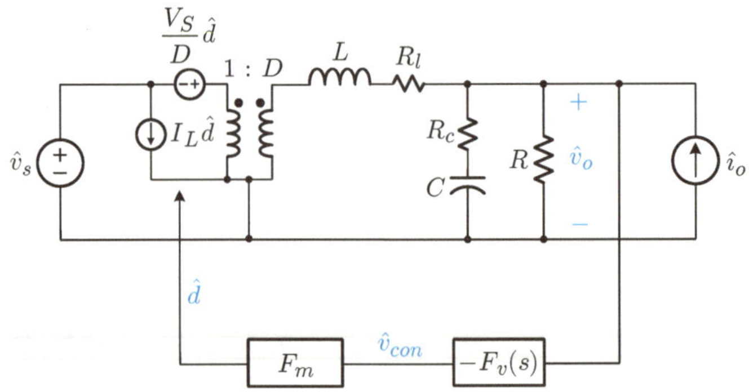

Figure 8.6 Small-signal model of buck converter.

8.2.1 Audio-Susceptibility Analysis

Figure 8.6 shows the small-signal model of a closed-loop controlled buck converter in CCM operation. The input-to-output transfer function is found from Fig. 8.6

$$G_{vs}(s) \equiv \left.\frac{\hat{v}_o(s)}{\hat{v}_s(s)}\right|_{\text{open}} \approx D\frac{1+\dfrac{s}{\omega_{esr}}}{1+\dfrac{s}{Q\omega_o}+\dfrac{s^2}{\omega_o^2}} \tag{8.23}$$

where

$$\omega_{esr} = \frac{1}{CR_c} \tag{8.24}$$

$$\omega_o = \sqrt{\frac{1}{LC}} \tag{8.25}$$

and

$$Q = R\sqrt{\frac{C}{L}} \tag{8.26}$$

with assumptions $R \gg R_l$ and $R \gg R_c$.

Figure 8.7 shows the construction of the asymptotic plot for $|A_u|$ based on the rules described Table 8.1. While $|G_{vs}|$ is formed using (8.23), $|T_m|$ is drawn assuming a single integration function

$$T_m(s) = \frac{\omega_c}{s} \tag{8.27}$$

The ascending/descending slopes and corner frequencies of $|A_u|$ can readily be determined from the asymptote sketch rules given in Table 8.1.

The effects of the loop gain on the audio-susceptibility are clearly seen in Fig. 8.7. The loop gain provides attenuation up to the 0 dB crossover frequency. The amount

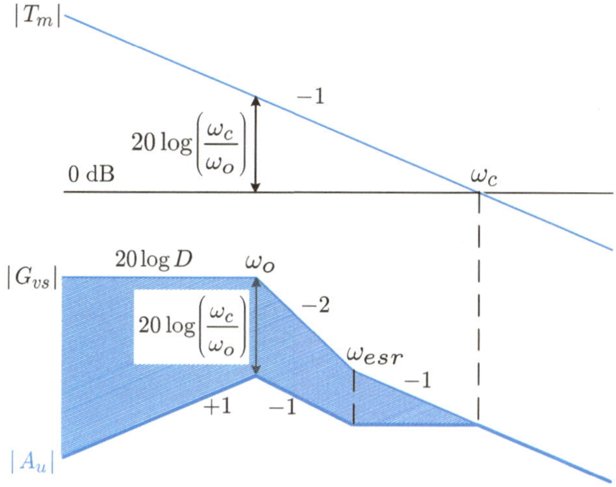

Figure 8.7 Construction of asymptotic plot for audio-susceptibility.

of attenuation, highlighted by the shaded region in Fig. 8.7, is equal to the area of the triangle formed by the $|T_m|$ curve and 0 dB line. The range and extent of the attenuation will be increased by pushing the loop gain crossover frequency towards higher frequencies, or equivalently by increasing the integrator gain ω_c.

The asymptotic plot of $|A_u|$ is converted into a factorized expression for $A_u(s)$ based on the rules presented in Table 8.2

$$A_u(s) = K_a s \frac{1 + \dfrac{s}{\omega_{esr}}}{\left(1 + \dfrac{s}{Q\omega_o} + \dfrac{s^2}{\omega_o^2}\right)\left(1 + \dfrac{s}{\omega_c}\right)} \tag{8.28}$$

The leading coefficient K_a is found by evaluating the magnitude of the transfer function at the power stage double pole $s = j\omega_o$

$$|A_u(j\omega_o)| = |K_a s|_{s=j\omega_o} = |G_{vs}(j\omega_o)| - |T_m(j\omega_o)| \tag{8.29}$$

which is transformed to

$$20\log(K_a\omega_o) = 20\log D - 20\log\left(\frac{\omega_c}{\omega_o}\right) \tag{8.30}$$

This relationship is illustrated in Fig. 8.7. Equation (8.30) is converted to the linear relationship

$$K_a\omega_o = \frac{D}{\dfrac{\omega_c}{\omega_o}} \tag{8.31}$$

yielding the expression for K_a

$$K_a = \frac{D}{\omega_c} \tag{8.32}$$

8.2.2 Output Impedance Analysis

The load current-to-output transfer function is derived from Fig. 8.6

$$Z_p(s) \equiv \left.\frac{\hat{v}_o(s)}{\hat{i}_s(s)}\right|_{\text{open}} \approx R_l \frac{\left(1+\dfrac{s}{\omega_z}\right)\left(1+\dfrac{s}{\omega_{esr}}\right)}{1+\dfrac{s}{Q\omega_o}+\dfrac{s^2}{\omega_o^2}} \tag{8.33}$$

where

$$\omega_z = \frac{R_l}{L} \tag{8.34}$$

Figure 8.8 shows the construction of the asymptotic plot for the output impedance $|Z_o|$ using the expression (8.33) and $T_m(s) = \omega_c/s$. The asymptotic plot is converted into a factorized equation

$$Z_o(s) = K_z s \frac{\left(1+\dfrac{s}{\omega_z}\right)\left(1+\dfrac{s}{\omega_{esr}}\right)}{\left(1+\dfrac{s}{Q\omega_o}+\dfrac{s^2}{\omega_o^2}\right)\left(1+\dfrac{s}{\omega_c}\right)} \tag{8.35}$$

The expression for K_z is found using the high-frequency asymptote of the output impedance, $|Z_o(j\infty)|$. As shown in Fig. 8.8, $|Z_o(j\infty)|$ converges to the high-frequency asymptote of $|Z_p|$ because $|T_m| \ll 1$ at high frequencies: $|Z_o(j\infty)| = |Z_p(j\infty)|$. Now, it is deduced from the small-signal model in Fig. 8.6 that $|Z_p(j\infty)|$ is the parallel connection of the load resistor and esr of the output capacitor, because the inductor L behaves as an open circuit while the capacitor C becomes a short circuit at high frequencies. Accordingly, it follows that

$$|Z_o(j\infty)| = |Z_p(j\infty)| = 20\log(R \parallel R_c) \approx 20\log R_c \tag{8.36}$$

On the other hand, $|Z_o(j\infty)|$ is also found from (8.35)

$$|Z_o(j\infty)| = 20\log\left(\lim_{s\to\infty} Z_o(s)\right) = 20\log\left(\frac{K_z\,\omega_o^2\,\omega_c}{\omega_z\,\omega_{esr}}\right) \tag{8.37}$$

By equating (8.36) and (8.37), K_z is given by

$$K_z = R_c \frac{\omega_z\,\omega_{esr}}{\omega_o^2\,\omega_c} \tag{8.38}$$

The influence of the loop gain on the output impedance could be understood in the same manner as the audio-susceptibility case.

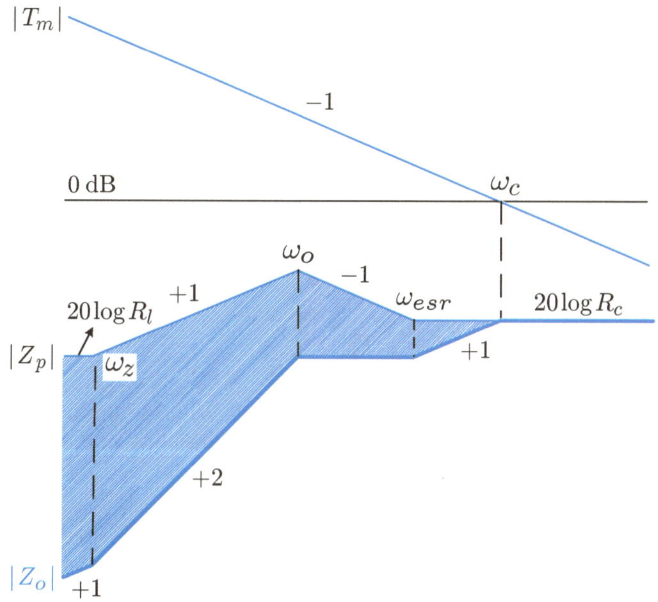

Figure 8.8 Construction of asymptotic plot for output impedance.

8.3 VOLTAGE FEEDBACK COMPENSATION AND LOOP GAIN IN CCM

In the previous section, we investigated the relationship between the loop gain and closed-loop transfer functions. To focus on the effects of the loop gain characteristics, a single integration function was assumed for the loop gain. However, the loop gain is the very quantity that must be properly designed in order to obtain good closed-loop transfer functions. Thus, it is first necessary to investigate the loop gain itself.

As described in (8.1) along with Fig. 8.1, the loop gain is given by the product of the three factors — the duty ratio-to-output transfer function $G_{vd}(s)$, PWM gain F_m, and voltage feedback compensation $F_v(s)$. Among these three factors, the voltage feedback compensation is the only element that can be altered by design. For the given $G_{vd}(s)$ and F_m, the voltage feedback compensation $F_v(s)$ should be designed for good loop gain characteristics.

The objective of the voltage feedback compensation is to shape the loop gain into the desired structure. The primary consideration in this *loop-gain shaping* is stability of the converter. In addition, the loop gain should provide good audio-susceptibility and output impedance characteristics. This section presents detailed steps of determining the structure and parameters of the voltage feedback compensation, in order to obtain the desired loop gain characteristics.

From Fig. 8.6, the duty ratio-to-output transfer function of the buck converter in CCM is given by

$$G_{vd}(s) \equiv \left.\frac{\hat{v}_o(s)}{\hat{d}(s)}\right|_{\text{open}} \approx V_S \frac{1+\dfrac{s}{\omega_{esr}}}{1+\dfrac{s}{Q\omega_o}+\dfrac{s^2}{\omega_o^2}} \tag{8.39}$$

In Section 5.6, the PWM gain was found as

$$F_m = \frac{1}{V_m} \tag{8.40}$$

where V_m is the height of the PWM ramp signal. With the knowledge of the duty ratio-to-output transfer function and PWM gain, the structure of the voltage feedback compensation $F_v(s)$ is now investigated.

8.3.1 Problems of Single Integrator

As discussed in Section 3.6.1, the voltage feedback compensation should have an infinite dc gain, $|F_v(j0)| = \infty$, in order to regulate the output voltage at the desired value. The simplest circuit with such a property is the single integrator. As an initial attempt, a single integrator is considered for the voltage feedback compensation

$$F_v(s) = \frac{K_v}{s} \tag{8.41}$$

The expression for the loop gain then becomes

$$\begin{aligned} T_m(s) = G_{vd}(s)\,F_v(s)\,F_m \quad &= \underbrace{V_S \frac{1+\dfrac{s}{\omega_{esr}}}{1+\dfrac{s}{Q\omega_o}+\dfrac{s^2}{\omega_o^2}}}_{G_{vd}(s)} \underbrace{\frac{K_v}{s}}_{F_v(s)} \underbrace{\frac{1}{V_m}}_{F_m} \\ &= \frac{K_t}{s} \frac{1+\dfrac{s}{\omega_{esr}}}{1+\dfrac{s}{Q\omega_o}+\dfrac{s^2}{\omega_o^2}} \end{aligned} \tag{8.42}$$

with

$$K_t = \frac{V_S K_v}{V_m} \tag{8.43}$$

Figure 8.9 shows the asymptotic plots for $|T_m|$ and $\angle T_m$ in comparison with $|G_{vd}|$. The integrator causes $|T_m|$ to decrease with -1 slope at low frequencies. The initial -1 slope is altered to -3 slope at the double pole frequency ω_o. The -3

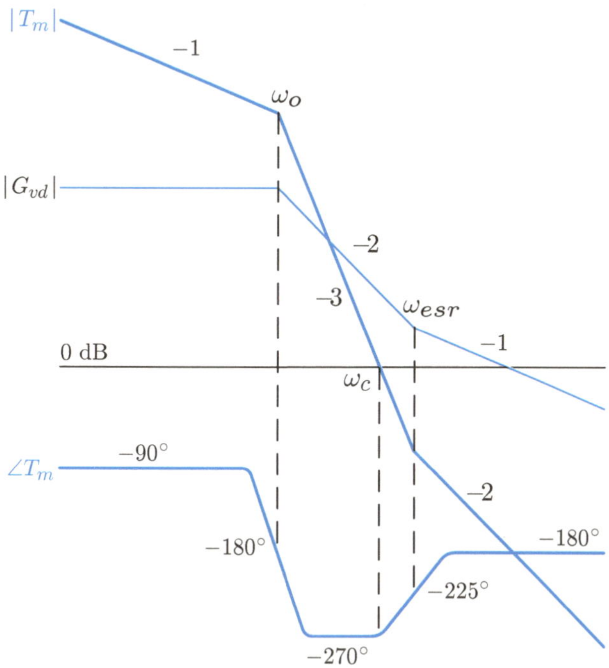

Figure 8.9 Unstable loop gain with $F_v(s) = K_v/s$.

mid-frequency slope is finally changed to the -2 high-frequency slope at the esr zero ω_{esr}. Figure 8.9 assumes that the crossover frequency of the loop gain, ω_c, occurs at the frequency between the power stage double pole and esr zero: $\omega_o < \omega_c < \omega_{esr}$.

The phase of the loop gain starts with $-90°$ due to the integrator. The initial $-90°$ phase dips down to $-270°$ over ω_o. At high frequencies, $\angle T_m$ settles into the $-180°$ final phase, owing to the $90°$ phase boost at the esr zero.

The problem of the voltage feedback compensation is clear in Fig. 8.9. Because $\angle T_m$ falls well below $-180°$ at the loop gain crossover frequency, the loop gain violates the Nyquist stability criterion and the converter becomes unstable. The only way to secure stability is to significantly reduce the integrator gain K_v so that $|T_m|$ crosses the 0 dB line at the frequencies before the power stage double pole: $\omega_c < \omega_o$.

The loop gain with this modification is shown in Fig. 8.10. Even though the converter is stable, both the phase and gain margins will be very small. Furthermore, $|T_m|$ is significantly reduced with a small integrator gain. The reduced $|T_m|$ only provides limited attenuation for closed-loop transfer functions, as highlighted by the shaded triangle in Fig. 8.10. For this case, the magnitude of the closed-loop transfer functions would be large. The small stability margins and large closed-loop transfer function magnitude both make the voltage feedback compensation unacceptable.

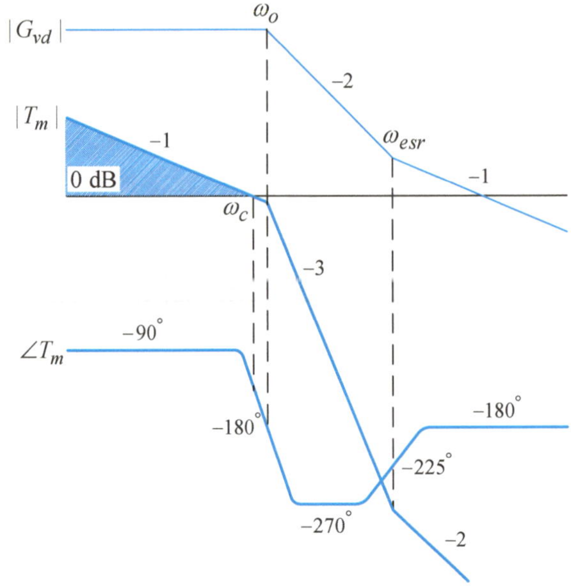

Figure 8.10 Stable loop gain with $F_v(s) = K_v/s$.

8.3.2 Voltage Feedback Compensation

The origin of the problem in the single integrator compensation is the phase delay incurred by the integrator and power stage double pole. The integrator forces the loop gain to start with the $-90°$ initial phase. The power stage double pole brings in an additional $-180°$ phase delay, causing $\angle T_m$ to decline to $-270°$. This makes the converter highly prone to instability. To secure stability under this situation, the loop gain crossover should occur before the power stage double pole at the expense of the poor closed-loop performance.

The voltage feedback compensation should always start with an integrator structure to achieve the output voltage regulation with the condition $|F_v(j0)| = \infty$. After the initial integrator, the voltage feedback compensation could incorporate two zeros to boost the phase of the loop gain. When placed in the neighborhood of the power stage double pole, the two zeros compensate for the $-180°$ phase delay incurred by the power stage double pole. This allows the 0 dB crossover frequency of the loop gain to be placed at the frequencies beyond the power stage double pole, while maintaining good phase characteristics.

Besides the two zeros, two poles should be added to the voltage feedback compensation for the following reason. The voltage feedback circuit receives its input signal from the output of the converter, which abounds with high-frequency switching noises. The high-frequency noises will be transmitted to the PWM block through the voltage feedback compensation. To prevent malfunctions triggered by the high-frequency noises at the PWM block, the voltage feedback compensation must provide

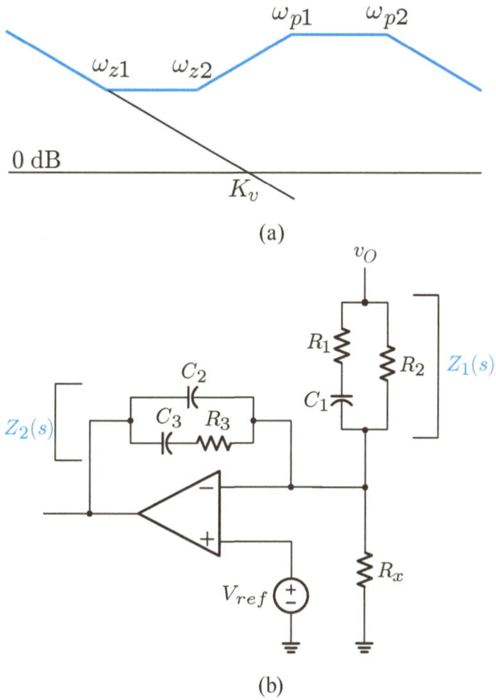

Figure 8.11 Three-pole two-zero feedback compensation. (a) Structure of voltage feedback compensation. (b) Circuit implementation.

a fair amount of attenuation at high frequencies. To provide this high-frequency noise attenuation, the magnitude of the voltage feedback compensation should have a descending high-frequency asymptote. At the presence of the two zeros, two poles are necessary to provide a -20 dB/dec roll-off at high frequencies.

From the preceding discussions, the desired structure of the voltage feedback compensation is identified as

$$F_v(s) = \frac{K_v}{s} \frac{\left(1+\dfrac{s}{\omega_{z1}}\right)\left(1+\dfrac{s}{\omega_{z2}}\right)}{\left(1+\dfrac{s}{\omega_{p1}}\right)\left(1+\dfrac{s}{\omega_{p2}}\right)} \tag{8.44}$$

Figure 8.11(a) shows the asymptotic plot of the voltage feedback compensation. The exact locations of the zeros and poles, along with the integrator gain K_v, must be determined in consideration of their impacts on the stability margins and closed-loop transfer functions. This issue will be treated in Section 8.4. The voltage feedback compensation in Fig. 8.11(a) can be constructed using an op amp and passive circuit components. Figure 8.11(b) shows such implementation, in which the impedance ratio $Z_2(s)/Z_1(s)$ constitutes the voltage feedback compensation. The resistance R_x in Fig. 8.11(b) is added to control the magnitude of the output voltage: $V_O = V_{ref}(1 + R_2/R_x)$. As discussed in Section 5.7.1, this resistance is irrelevant to the voltage feedback compensation. By directly evaluating $F_v(s) = Z_2(s)/Z_1(s)$, the compensation parameters are expressed in terms of the circuit components

$$F_v(s) = \frac{K_v}{s}\frac{\left(1+\frac{s}{\omega_{z1}}\right)\left(1+\frac{s}{\omega_{z2}}\right)}{\left(1+\frac{s}{\omega_{p1}}\right)\left(1+\frac{s}{\omega_{p2}}\right)} \tag{8.45}$$

with

$$K_v = \frac{1}{R_2(C_2+C_3)}$$

$$\omega_{z1} = \frac{1}{R_3C_3} \qquad \omega_{z2} = \frac{1}{(R_1+R_2)C_1}$$

$$\omega_{p1} = \frac{1}{R_1C_1} \qquad \omega_{p2} = \frac{1}{R_3\left(\frac{C_2C_3}{C_2+C_3}\right)}$$

Once the integrator gain and corner frequencies are selected, the circuit components in Fig. 8.11(b) are determined using the above equations. Among the six circuit components, one can be arbitrarily chosen and the other components are found using the equations in (8.45). This compensation circuit, named the three-pole two-zero compensation after its transfer function, was previously introduced in Section 3.6.1 and Section 5.7.1.

EXAMPLE 8.2 Alternative Three-pole Two-zero Compensation

The three-pole two-zero compensation in Fig. 8.11(b) sometimes become inappropriate for dc-to-dc converters for high voltage applications. When the output voltage is large, an excessive dc current could flow through the resistors R_2 and R_x, thereby reducing the efficiency of the converter — R_2 and R_x are determined only considering the voltage sensing ratio and transfer function of the voltage feedback circuit, as such, the resistances R_2 and R_x can be small and could carry a large current.

As an alternative, the three-pole two-zero compensation shown in Fig. 8.12 can be employed. Here, the resistances R_s and R_x will be selected for a small sensing current. The transfer function of the feedback circuit is given by

$$F_v(s) = \frac{K_v}{s}\frac{\left(1+\frac{s}{\omega_{z1}}\right)\left(1+\frac{s}{\omega_{z2}}\right)}{\left(1+\frac{s}{\omega_{p1}}\right)\left(1+\frac{s}{\omega_{p2}}\right)}$$

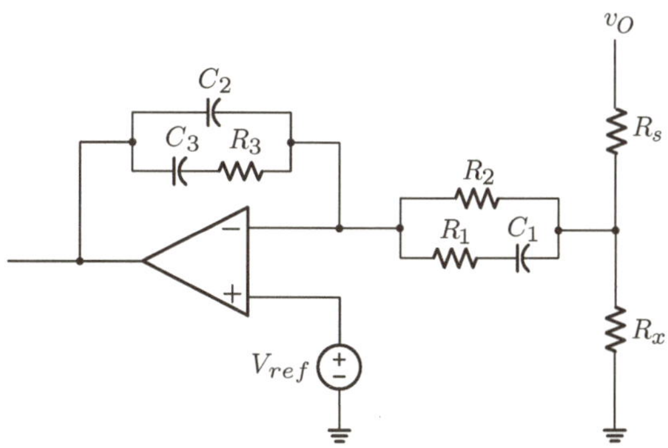

Figure 8.12 Alternative three-pole two-zero compensation for high voltage applications.

with

$$K_v = \frac{R_x}{(R_s + R_x)(R_2 + R_s \parallel R_x)(C_2 + C_3)}$$

$$\omega_{z1} = \frac{1}{R_3 C_3} \qquad \omega_{z2} = \frac{1}{(R_1 + R_2)C_1}$$

$$\omega_{p1} = \frac{1}{(R_1 + R_2 \parallel R_s \parallel R_x)C_1} \qquad \omega_{p2} = \frac{1}{R_3\left(\dfrac{C_2 C_3}{C_2 + C_3}\right)}$$

As one demerit of the feedback circuit, the two compensation parameters, K_v and ω_{p1}, will be altered when R_x is varied to adjust the output voltage of the converter.

8.4 COMPENSATION DESIGN AND CLOSED-LOOP PERFORMANCE IN CCM

The design of the voltage feedback compensation involves the selection of the five compensation parameters, $\omega_{z1}, \omega_{z2}, \omega_{p1}, \omega_{p2}$, and K_v, while simultaneously meeting the following design objectives:

- stability with adequate phase and gain margins,
- small audio-susceptibility, and
- small output impedance.

This section provides design principles to properly select the five compensation parameters. The current section mainly deals with the buck converter in CCM operation. Section 8.4.7 addresses the issues of the control design for the boost and buck/boost converters. In addition, considerations for DCM operation are given in Section 8.5.

8.4.1 Voltage Feedback Compensation and Loop Gain

The selection of the compensation parameters is investigated in this section, focusing on loop gain characteristics of a buck converter in CCM operation. With the three-pole two-zero compensation, the loop gain is expressed as

$$\begin{aligned} T_m(s) &= G_{vd}(s)F_v(s)F_m \\ &= \underbrace{V_S\frac{1+\dfrac{s}{\omega_{esr}}}{1+\dfrac{s}{Q\omega_o}+\dfrac{s^2}{\omega_o^2}}}_{G_{vd}(s)}\underbrace{\frac{K_v}{s}\frac{\left(1+\dfrac{s}{\omega_{z1}}\right)\left(1+\dfrac{s}{\omega_{z2}}\right)}{\left(1+\dfrac{s}{\omega_{p1}}\right)\left(1+\dfrac{s}{\omega_{p2}}\right)}}_{F_v(s)}\underbrace{\frac{1}{V_m}}_{F_m} \end{aligned} \tag{8.46}$$

which is arranged as

$$T_m(s)=\frac{K_t}{s}\frac{\left(1+\dfrac{s}{\omega_{esr}}\right)\left(1+\dfrac{s}{\omega_{z1}}\right)\left(1+\dfrac{s}{\omega_{z2}}\right)}{\left(1+\dfrac{s}{Q\omega_o}+\dfrac{s^2}{\omega_o^2}\right)\left(1+\dfrac{s}{\omega_{p1}}\right)\left(1+\dfrac{s}{\omega_{p2}}\right)} \tag{8.47}$$

with

$$K_t=\frac{V_SK_v}{V_m} \tag{8.48}$$

For the given power stage parameters and operational conditions, the zeros and poles of the voltage feedback compensation should be selected to provide desirable loop gain characteristics, which in turn offer good audio-susceptibility and output impedance characteristics.

Figure 8.13 shows the asymptotic plots for $|T_m|$, $|G_{vd}|$, and $\angle T_m$. The asymptotic plots are constructed with the assumption $\omega_{z1} < \omega_o < \omega_{z2} < \omega_c < \omega_{p1} = \omega_{esr} < \omega_{p2}$,† where ω_c denotes the 0 dB crossover frequency of the loop gain. Justifications for the selection of the compensation parameters are given as follows:

1) The first compensation zero ω_{z1} should be placed prior to the power stage double pole ω_o: $\omega_{z1} < \omega_o$. This step is necessary to prevent the converter from being a conditionally stable system. If the compensation zero is not present until the power stage double pole appears, the initial $-90°$ loop gain phase will dip down well below $-180°$ over the power stage double pole ω_o, which appears before the 0 dB frequency of the loop gain ω_c. In this case, the two compensation zeros should be placed between ω_o and the loop gain crossover frequency ω_c, in order

†The asymptotic plot in Fig. 8.13 is drawn with the assumption that the corner frequencies of the loop gain are *widely* separated so that the phase undergoes a complete 90° or 180° change over corner frequencies. However, the loop gain will not go through the *complete* phase change due to the finite distance among the corner frequencies. Thus, the phase does not reduce to $-180°$ after the power stage double pole ω_o and the phase at the crossover frequency ω_c does not increase to $-90°$.

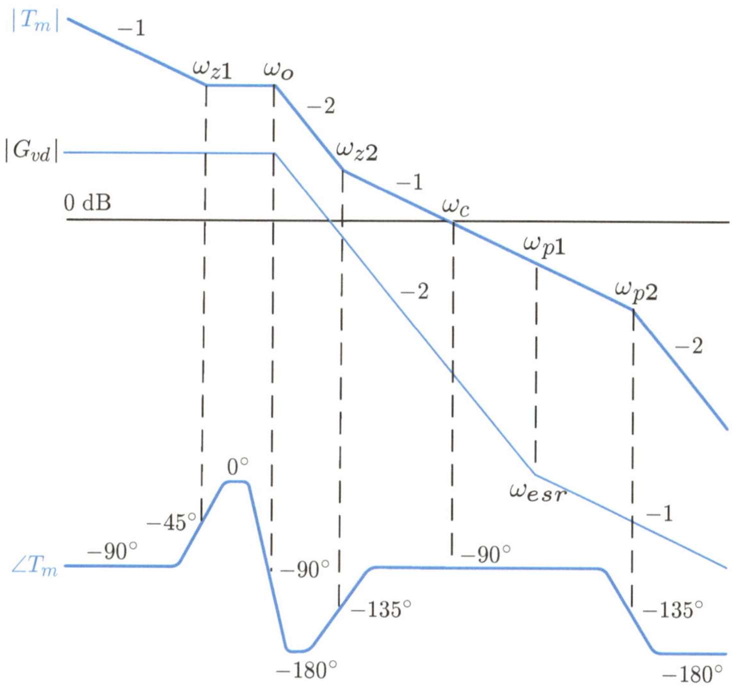

Figure 8.13 Construction of loop gain asymptotic plot.

to provide the phase boost needed for a positive phase margin.

Although the converter could be stable in this case, there surely is the frequency range in which $\angle T_m$ is less than $-180°$ while $|T_m|$ is larger than unity. Figure 8.14(a) is the polar plot of the loop gain for such a case. The polar plot indicates that the converter is a conditionally stable system which is problematic in real operations, as illustrated in Example 7.8.

Conversely, if the first compensation zero precedes the power stage double pole, $\angle T_m$ stays above $-180°$ due to the early phase boost and the converter thus secures stability without being conditionally stable. Figure 8.14(b) is the loop gain plot for this case.

2) The second compensation zero ω_{z2} needs to be placed after the power stage double pole ω_o but before the loop gain crossover frequency ω_c, $\omega_o < \omega_{z2} < \omega_c$, in order to provide adequate phase boost at the crossover frequency. If ω_{z2} is placed after ω_c, it only makes limited contribution to the phase margin and its phase boosting effect is mostly wasted. The loop gain magnitude $|T_m|$ retains the initial -20 dB/dec slope after ω_{z2}, as shown in Fig. 8.13.

3) The first compensation pole ω_{p1} is placed at the esr zero, $\omega_{p1} = \omega_{esr}$, to maintain the -20 dB/dec slope by canceling the esr zero with the first compensation pole.

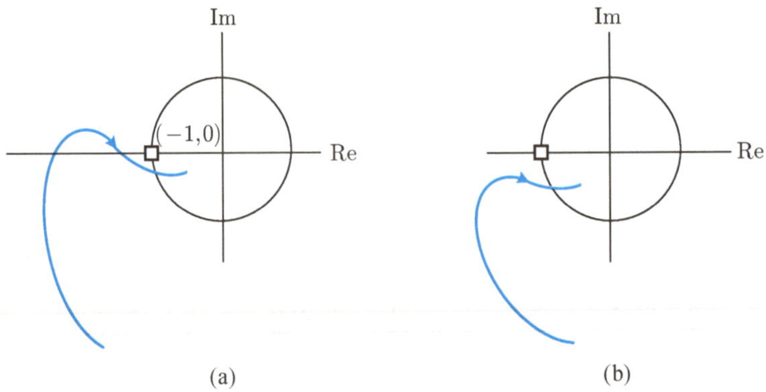

Figure 8.14 Polar plot of loop gain. (a) Conditionally stable loop gain with $\omega_o < \omega_{z1}$. (b) Stable loop gain with $\omega_{z1} < \omega_o$.

The ideal structure of the loop gain is in fact the single integration function that has a fixed -20 dB/dec magnitude slope for all frequencies, as illustrated in Section 8.2. Although this ideal loop gain structure is not feasible due to the power stage dynamics, it is always desirable to increase the frequency range in which $|T_m|$ maintains a -20 dB/dec slope. This design objective is achieved by placing ω_{p1} at ω_{esr}.

4) Finally, the second compensation pole ω_{p2} is placed at high frequencies to provide the high-frequency noise attenuation.

With the compensation parameters described above, the loop gain crossover frequency can be placed at higher frequencies, while securing sufficient phase margin. Although Fig. 8.13 hints that the phase margin could reach 90°, this theoretical maximum cannot be obtained due to the finite distances among the corner frequencies of the loop gain. The actual phase margin is determined by the relative locations of the compensation parameters and usually falls in the range of $45° - 70°$.

8.4.2 Feedback Compensation Design Guidelines

Based on the previous discussions, step-by-step compensation design guidelines are established as follows.

1) Place the first compensation pole ω_{p1} at the esr zero, $\omega_{p1} = \omega_{esr}$, to nullify the effects of the esr zero. This provides a -20 dB/dec roll-off for a wider frequency range.

2) Locate the first compensation zero ω_{z1} before the power stage double pole ω_o to provide phase boost without yielding a conditionally stable system. As will be shown in Section 8.4.5, the location of this compensation zero determines the speed of the step input response. The response becomes faster as the zero is

positioned at higher frequencies. Thus, ω_{z1} should be placed as high as possible, yet still should not exceed the power stage double pole, in order *not* to render the converter conditionally stable. As a rule of thumb, it is recommended that $\omega_{z1} = (0.6 - 0.8)\,\omega_o$.

3) Place the second compensation zero ω_{z2} after the power stage double pole but before the loop gain crossover frequency, to obtain an adequate phase boost at the crossover frequency. As will be discussed in Section 8.4.6, the position of this zero determines the speed of the step load response. For faster responses, the zero should be placed at higher frequencies. However, the phase boosting effect at the crossover frequency will be diminished as ω_{z2} is pushed towards higher frequencies. It is generally recommended that $\omega_{z2} = (1.5 - 3.0)\,\omega_o$ to trade off the phase boosting effect and speed of the step load response.

4) Locate the second compensation pole ω_{p2} at high frequencies to provide switching noise attenuation. In general, ω_{p2} can be placed around $50\% - 80\%$ of the switching frequency ω_s for acceptable switching ripple attenuation: $\omega_{p2} = (0.5 - 0.8)\,\omega_s$. If ω_{p2} comes earlier nearing the loop gain crossover frequency, the phase margin will be reduced.

5) Select the desired frequency for the loop gain crossover, ω_c. It is good practice to place the crossover frequency around 10%–30% of the switching frequency ω_s: $\omega_c = (0.1 - 0.3)\,\omega_s$. Discussions about the selection of ω_c will be given in the next section. From the asymptotic plot of $|T_m|$ in Fig. 8.13 and the expression (8.48), the following relationship is derived

$$20\log\left(\frac{V_S K_v}{V_m \omega_{z1}}\right) - 40\log\left(\frac{\omega_{z2}}{\omega_o}\right) - 20\log\left(\frac{\omega_c}{\omega_{z2}}\right) = 0\text{ dB} \tag{8.49}$$

where ω_c is the desired location of the crossover frequency. The preceding equation is converted into the design equation

$$\boxed{\frac{V_S K_v}{V_m \omega_{z1}}\left(\frac{\omega_o}{\omega_{z2}}\right)^2\left(\frac{\omega_{z2}}{\omega_c}\right) = 1 \quad\Rightarrow\quad K_v = \frac{V_m}{V_S}\frac{\omega_{z1}\,\omega_{z2}\,\omega_c}{\omega_o^2}} \tag{8.50}$$

This equation is used to find K_v for a preselected ω_c.

6) Check the phase margin and adjust the integrator gain K_v, if necessary, to secure a $45° - 70°$ phase margin.

The above design procedures are summarized in Table 8.3 for reference. An application of this design approach will be illustrated in Example 8.3.

8.4.3 Voltage Feedback Compensation and Closed-Loop Performance

The voltage feedback compensation governs the loop gain characteristics, which in turn determine the closed-loop performance of the converter. This section investigates the impacts of the voltage feedback compensation on the closed-loop performance of

Table 8.3 Compensation Design Procedures for Buck Converter

Loop gain expression

$$T_m(s) = G_{vd}(s)F_v(s)F_m$$

$$= \underbrace{V_S \frac{1+\dfrac{s}{\omega_{esr}}}{1+\dfrac{s}{Q\omega_o}+\dfrac{s^2}{\omega_o^2}}}_{G_{vd}(s)} \underbrace{\frac{K_v}{s}\frac{\left(1+\dfrac{s}{\omega_{z1}}\right)\left(1+\dfrac{s}{\omega_{z2}}\right)}{\left(1+\dfrac{s}{\omega_{p1}}\right)\left(1+\dfrac{s}{\omega_{p2}}\right)}}_{F_v(s)} \underbrace{\frac{1}{V_m}}_{F_m}$$

$$= \frac{V_S K_v/V_m}{s} \frac{\left(1+\dfrac{s}{\omega_{esr}}\right)\left(1+\dfrac{s}{\omega_{z1}}\right)\left(1+\dfrac{s}{\omega_{z2}}\right)}{\left(1+\dfrac{s}{Q\omega_o}+\dfrac{s^2}{\omega_o^2}\right)\left(1+\dfrac{s}{\omega_{p1}}\right)\left(1+\dfrac{s}{\omega_{p2}}\right)}$$

Compensation design guidelines

1) Set the first compensation pole: $\omega_{p1} = \omega_{esr}$.
2) Choose the first compensation zero: $\omega_{z1} = (0.6-0.8)\,\omega_o$.
3) Select the second compensation zero: $\omega_{z2} = (1.5-3.0)\,\omega_o$.
4) Choose the second compensation pole: $\omega_{p2} = (0.5-0.8)\,\omega_s$.
5) Select the loop gain crossover frequency: $\omega_c = (0.1-0.3)\,\omega_s$.
6) Evaluate the integrator gain: $K_v = \dfrac{V_m}{V_S}\dfrac{\omega_{z1}\,\omega_{z2}\,\omega_c}{\omega_o^2}$.
7) Check the phase margin and adjust K_v to secure a 45°–70° phase margin.
8) Evaluate the circuit components for the voltage feedback compensation using (8.45).

the buck converter in CCM operation. Figure 8.15 illustrates the relationship among the loop gain, open-loop transfer functions, and closed-loop transfer functions. The asymptotic plots are constructed using the rules given in Table 8.1. The impacts of the loop gain characteristics, or equivalently the effects of the voltage feedback

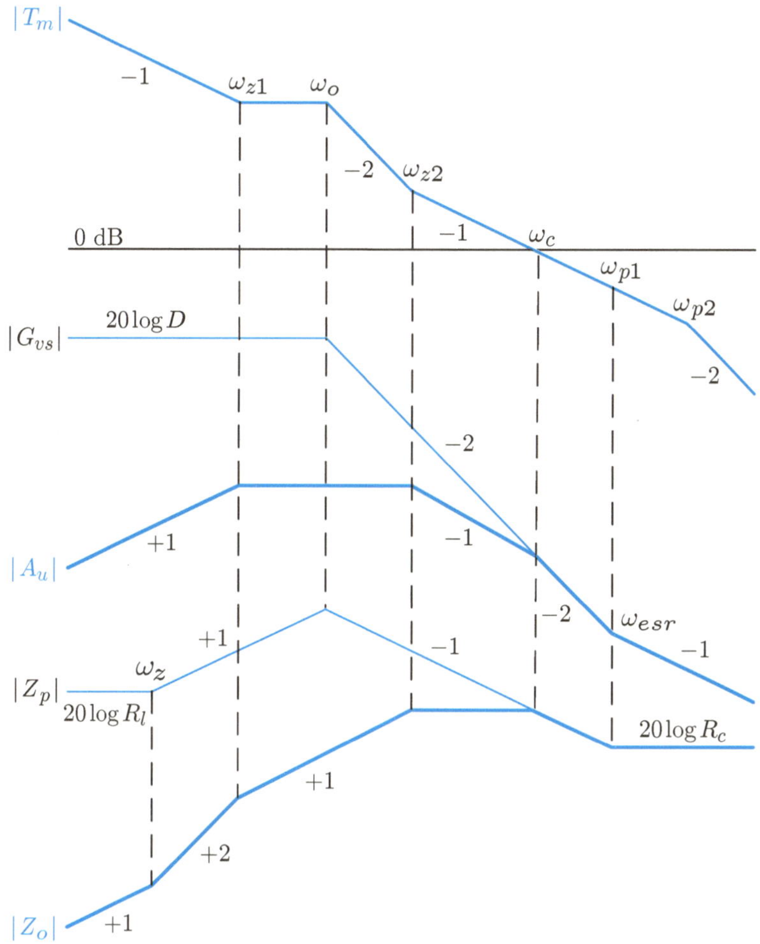

Figure 8.15 Voltage feedback compensation and closed-loop transfer functions.

compensation, are clear in Fig. 8.15. The open-loop transfer functions are attenuated up to the loop gain crossover frequency, ω_c. The power stage double pole is not apparent in the closed-loop transfer functions because the second-order feature of the open-loop transfer functions is canceled by the loop gain curve, which itself carries the same second-order feature — $A_u(s) \approx G_{vs}(s)/T_m(s)$ and $Z_o(s) \approx Z_p(s)/T_m(s)$ for the frequencies where $|T_m| \gg 1$.

The asymptotic plot reveals the structure of the closed-loop transfer functions and also provides the information to judge the influence of each compensation parameter on the converter performance. For example, when the integrator gain K_v is increased within a limited range while other parameters remain unchanged, the following changes will occur:

- increase in the loop gain crossover frequency,
- decrease in the peak value of the audio-susceptibility $|A_u|_{peak}$, and
- decrease in the peak value of the output impedance $|Z_o|_{peak}$.

On the other hand, an excessive increase in K_v lessens the phase margin. The converter could lose stability when the crossover frequency is pushed beyond ω_{p2} with a very large K_v.

The asymptotic plot of $|A_u|$ in Fig. 8.15 is converted into an analytical expression, based on the rules in Table 8.2

$$A_u(s) = K_a s \frac{1+\dfrac{s}{\omega_{esr}}}{\left(1+\dfrac{s}{\omega_{z1}}\right)\left(1+\dfrac{s}{\omega_{z2}}\right)\left(1+\dfrac{s}{\omega_c}\right)} \tag{8.51}$$

The expression for K_a is found from the magnitude relationship among the transfer functions at ω_{z1}

$$|A_u(j\omega_{z1})| = |K_a s|_{s=j\omega_{z1}} = |G_{vs}(j\omega_{z1})| - |T_m(j\omega_{z1})| \tag{8.52}$$

Referring to (8.23) and (8.47), the above equation is translated as

$$K_a \omega_{z1} = \frac{\dfrac{D}{K_t}}{\omega_{z1}} \tag{8.53}$$

yielding the expression for K_a

$$K_a = \frac{D}{K_t} \tag{8.54}$$

where $K_t = (V_S K_v)/V_m$, as given by (8.48). The peak value of $|A_u|$ is approximated from Fig. 8.15

$$|A_u|_{peak} = |A_u|_{s=j\omega_{z1}} = |K_a s|_{s=j\omega_{z1}} = 20 \log\left(\frac{D}{K_t}\omega_{z1}\right) \tag{8.55}$$

From Fig. 8.15, the closed-loop output impedance is expressed as

$$Z_o(s) = K_z s \frac{\left(1+\dfrac{s}{\omega_z}\right)\left(1+\dfrac{s}{\omega_{esr}}\right)}{\left(1+\dfrac{s}{\omega_{z1}}\right)\left(1+\dfrac{s}{\omega_{z2}}\right)\left(1+\dfrac{s}{\omega_c}\right)} \tag{8.56}$$

The value for K_z is found from the high-frequency asymptote of the output impedance. The high-frequency asymptote of $|Z_o|$ is given by

$$|Z_o(j\infty)| = 20 \log\left(\lim_{s\to\infty} Z_o(s)\right) = 20 \log\left(\frac{K_z\, \omega_{z1}\, \omega_{z2}\, \omega_c}{\omega_z\, \omega_{esr}}\right) \tag{8.57}$$

As discussed in Section 8.2.2, the high-frequency asymptote of $|Z_o|$ approaches the esr of the output capacitor

$$|Z_o(j\infty)| = 20\log(R \parallel R_c) \approx 20\log R_c \tag{8.58}$$

By equating (8.57) and (8.58), K_z is given by

$$K_z = R_c \frac{\omega_z\,\omega_{esr}}{\omega_{z1}\,\omega_{z2}\,\omega_c} \tag{8.59}$$

The peak value of the output impedance is found form Fig. 8.15

$$|Z_o|_{peak} - 20\log\left(\frac{\omega_{esr}}{\omega_c}\right) = 20\log R_c$$

$$\Rightarrow |Z_o|_{peak} = 20\log R_c + 20\log\left(\frac{\omega_{esr}}{\omega_c}\right) = 20\log\left(R_c\frac{\omega_{esr}}{\omega_c}\right) \tag{8.60}$$

Compensation Design Example — EXAMPLE 8.3

This example demonstrates the procedures of the compensation design. Figure 8.16 shows the circuit diagram of a closed-loop controlled buck converter. The converter regulates the output voltage at 4 V using the three-pole two-zero compensation. From the circuit parameters, the power stage double pole and esr zero are determined as

$$\omega_o \approx \frac{1}{\sqrt{LC}} = \frac{1}{\sqrt{40\times10^{-6}\,470\times10^{-6}}} = 2\pi\cdot 1.16\times10^3 \text{ rad/s}$$

and

$$\omega_{esr} = \frac{1}{CR_c} = \frac{1}{470\times10^{-6}\,0.05} = 2\pi\cdot 6.77\times10^3 \text{ rad/s}$$

The switching frequency of the converter is set at $\omega_s = 2\pi\cdot 5.0\times10^4$ rad/s and the height of the ramp signal is $V_m = 3.8$ V. Based on the compensation design guidelines, the corner frequencies of the three-pole two-zero compensation are selected

- $\omega_{z1} = 0.8\,\omega_o = 2\pi\cdot 928$ rad/s
- $\omega_{z2} = 1.5\,\omega_o = 2\pi\cdot 1.74\times10^3$ rad/s
- $\omega_{p1} = \omega_{esr} = 2\pi\cdot 6.77\times10^3$ rad/s
- $\omega_{p2} = 0.8\,\omega_s = 2\pi\cdot 4\times10^4$ rad/s

The 0 dB crossover frequency is chosen at $\omega_c = 0.116\,\omega_s = 2\pi\cdot 5.8\times10^3$ rad/s. Based on the design equation, the integrator gain K_v is determined as

$$\begin{aligned} K_v &= \frac{V_m}{V_S}\frac{\omega_{z1}\,\omega_{z2}\,\omega_c}{\omega_o^2} \\ &= \frac{3.8}{16}\frac{(2\pi\cdot 928)(2\pi\cdot 1.74\times10^3)(2\pi\cdot 5.8\times10^3)}{(2\pi\cdot 1.16\times10^3)^2} \\ &= 1.04\times10^4 \end{aligned}$$

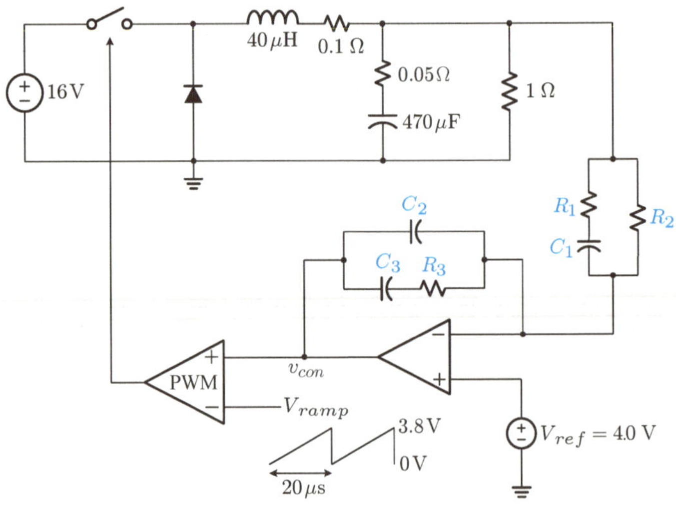

Figure 8.16 Closed-loop controlled buck converter: $R_1 = 2.2$ kΩ, $C_1 = 11$ nF, $R_2 = 6.4$ kΩ, $C_2 = 365$ pF, $R_3 = 11$ kΩ, and $C_3 = 15$ nF.

The loop gain characteristics with the selected parameters are shown in Fig. 8.17. The loop gain crossover occurs at the exact target frequency of 5.8 kHz with 65°phase margin. The loop gain plot, as well as other upcoming closed-loop transfer functions, is obtained from PSpice®simulations using the small-signal model of the converter.

From the loop gain in Fig. 8.17, it can be noticed that the integrator gain can be further increased to place the crossover frequency around 8 kHz where the largest phase margin is expected. However, this extreme design is not desirable in practice and could end up with inferior performance, compared with the current design of $K_v = 1.04 \times 10^4$. Justifications for *not* increasing K_v beyond the current value will be given in Example 8.5, which covers the issues related with the fidelity and limitation of the small-signal model.

The integrator gain $K_v = 1.04 \times 10^4$ is now selected as the final design, while other compensation parameters were chosen earlier. The circuit components for the voltage feedback circuit are determined from the selected compensation parameters. One circuit component is arbitrarily chosen as $R_1 = 2.2$ kΩ and the remaining circuit components are determined as $C_1 = 11$ nF, $R_2 = 6.4$ kΩ, $C_2 = 365$ pF, $R_3 = 11$ kΩ, and $C_3 = 15$ nF, based on the equation (8.45).

Figure 8.18 illustrates the closed-loop performance of the converter. Figure 8.18(a) shows the audio-susceptibility $|A_u|$ along with the input-to-output transfer function $|G_{vs}|$ and loop gain $|T_m|$. Figure 8.18(b) displays the output impedance $|Z_o|$, load current-to-output transfer function $|Z_p|$, and loop gain $|T_m|$. The validity and usefulness of the asymptotic analysis are confirmed when the Bode plots in Fig. 8.18 are compared with the asymptotic plots in Fig. 8.15.

The compensation design is also evaluated by time-domain simulations. Figure 8.19 shows the transient response of the inductor current and output voltage due to a series of step changes in the load current: $I_O = 4$ A $\Rightarrow$ 8 A $\Rightarrow$ 4 A. The transient

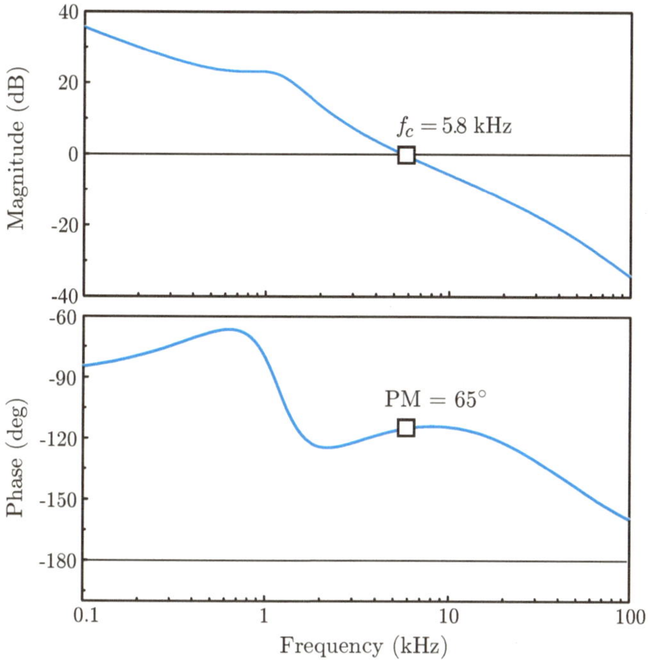

Figure 8.17 Loop gain characteristics with $K_v = 1.04 \times 10^4$.

waveforms exhibit very stable and well-controlled behavior. The detailed analysis of the step load response will be given in Section 8.4.6. The compensation design presented in this example was used in Section 3.6.2 to illustrate the dc regulation and transient behavior of a closed-loop controlled buck converter.

Accuracy of Asymptotic Approximation — EXAMPLE 8.4

The closed-loop transfer functions are expressed in a factorized analytical form based on the asymptotic analysis. In this example, the accuracy of this asymptotic approximation is assessed using the buck converter in the previous example. The audiosusceptibility was given by

$$A_u(s) = K_a s \frac{1 + \dfrac{s}{\omega_{esr}}}{\left(1 + \dfrac{s}{\omega_{z1}}\right)\left(1 + \dfrac{s}{\omega_{z2}}\right)\left(1 + \dfrac{s}{\omega_c}\right)} \tag{8.61}$$

The leading coefficient K_a is determined as

$$K_a = \frac{V_m}{V_S}\frac{D}{K_v} \tag{8.62}$$

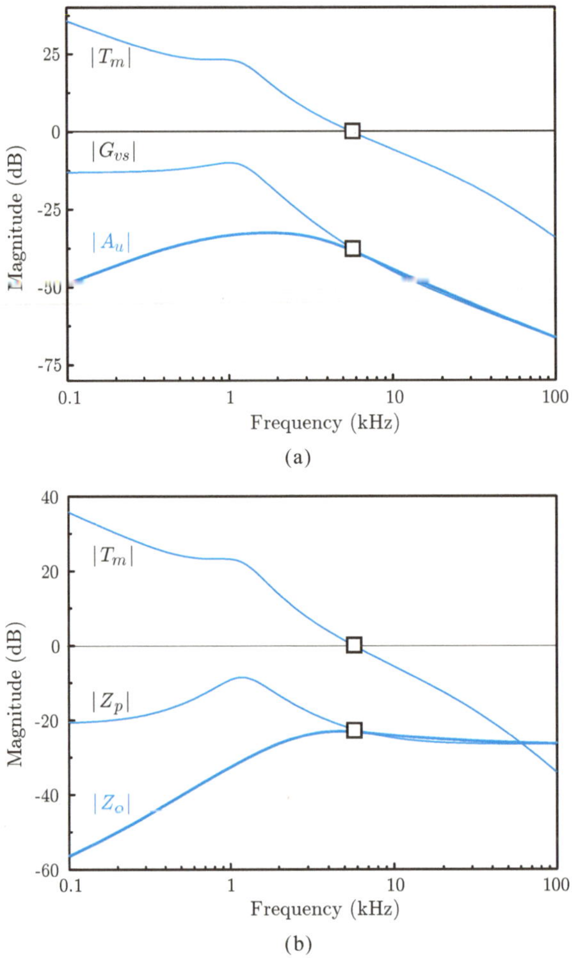

Figure 8.18 Closed-loop performance. (a) Audio-susceptibility. (b) Output impedance.

from (8.48) and (8.54). Based on the information shown in Fig. 8.16 and the integrator gain selected in Example 8.3, K_a is found as

$$K_a = \frac{V_m}{V_S}\frac{D}{K_v} = \frac{3.8}{16}\frac{0.25}{1.04 \times 10^4} = 5.71 \times 10^{-6}$$

The corner frequencies of (8.61) are given in Example 8.3 — $\omega_{z1} = 2\pi \cdot 928$ rad/s, $\omega_{z2} = 2\pi \cdot 1.74 \times 10^3$ rad/s, $\omega_{esr} = 2\pi \cdot 6.77 \times 10^3$ rad/s, and $\omega_c = 2\pi \cdot 5.8 \times 10^3$ rad/s.

Figure 8.20 displays the Bode plot of (8.61) in comparison with the simulated audio-susceptibility. The simulated audio-susceptibility refers to the PSpice® simulation obtained from the small-signal model of the converter. The asymptotic transfer function shows a good correlation with the actual transfer function, except for some

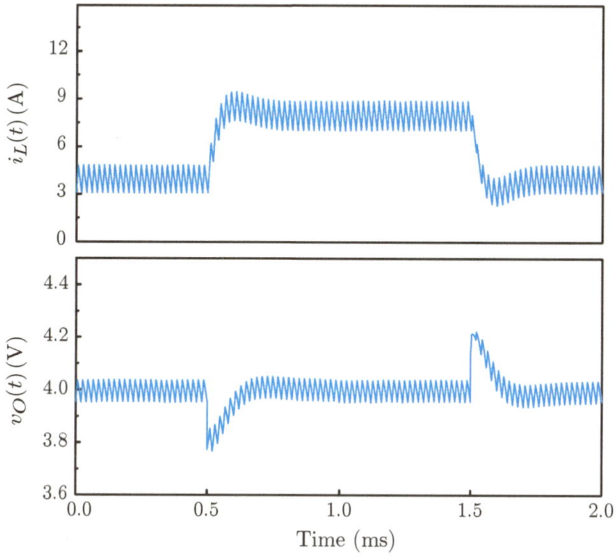

Figure 8.19 Step load transient response of inductor current i_L and output voltage v_O.

deviation in the frequency range of 1 kHz – 20 kHz. The asymptotic analysis assumes that $|T_m| \gg 1$ for all the frequencies before the loop gain crossover, and $|T_m| \ll 1$ for all the frequencies thereafter. However, this assumption is not quite satisfied in the frequency range centered around the loop gain crossover frequency, because the condition $|T_m| \approx 1$ prevails in these frequencies. For the given converter, this range falls into the frequency region of 1 kHz – 20 kHz and the asymptotic approximation produces a 3 dB – 4 dB error in this region. The asymptotic analysis predicts the peak value of $|A_u|$

$$\begin{aligned} |A_u|_{peak} &= 20\log(K_a\omega_{z1}) \\ &= 20\log(5.71 \times 10^{-6}\ 2\pi \cdot 928) = -29.6\ \text{dB} \end{aligned}$$

The asymptotic approximation for the output impedance was given by

$$Z_o(s) = K_z s \frac{\left(1+\dfrac{s}{\omega_z}\right)\left(1+\dfrac{s}{\omega_{esr}}\right)}{\left(1+\dfrac{s}{\omega_{z1}}\right)\left(1+\dfrac{s}{\omega_{z2}}\right)\left(1+\dfrac{s}{\omega_c}\right)} \tag{8.63}$$

where

$$\omega_z = \frac{R_l}{L} = \frac{0.1}{40 \times 10^{-6}} = 2\pi \cdot 398\ \text{rad/s}$$

$$\begin{aligned} K_z = R_c\frac{\omega_z\,\omega_{esr}}{\omega_{z1}\omega_{z2}\,\omega_c} &= 0.05\frac{(2\pi \cdot 398)(2\pi \cdot 6.77 \times 10^3)}{(2\pi \cdot 928)(2\pi \cdot 1.74 \times 10^3)(2\pi \cdot 5.8 \times 10^3)} \\ &= 2.29 \times 10^{-6} \end{aligned}$$

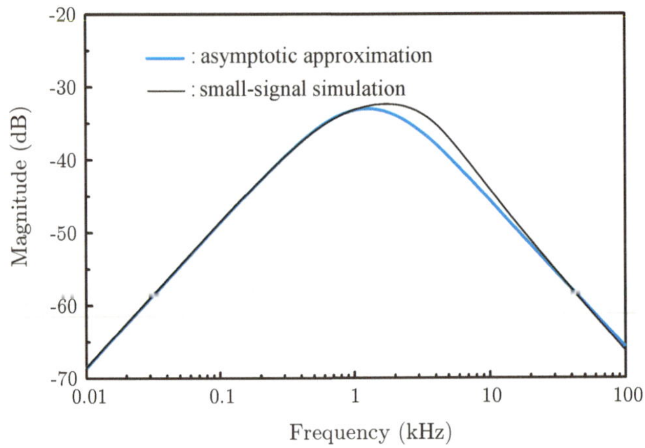

Figure 8.20 Audio-susceptibility comparison.

The peak value of the output impedance is predicted as

$$
\begin{aligned}
|Z_o|_{peak} &= 20\log\left(R_c \frac{\omega_{esr}}{\omega_c}\right) \\
&= 20\log\left(0.05\,\frac{2\pi\cdot 6.77\times 10^3}{2\pi\cdot 5.8\times 10^3}\right) = -24.7\text{ dB}
\end{aligned}
$$

Figure 8.21 compares the Bode plot of (8.63) with the simulated output impedance. Same as the audio-susceptibility case, the asymptotic approximation exhibits good accuracy except for the frequency range of 1 kHz–20 kHz, where either the condition $|T_m| \gg 1$ or $|T_m| \ll 1$ is not quite satisfied.

EXAMPLE 8.5 Accuracy of Small-Signal Model

The accuracy of the asymptotic analysis is verified in the previous example. This example investigates the accuracy of the small-signal model itself. The buck converter in Example 8.3 was built and its loop gain is measured using an impedance analyzer. The measured loop gain characteristics are then compared with the results of the small-signal simulation. The comparison is shown in Fig. 8.22. While affected by some measurement noises, the experimental data clearly exhibit the loop gain characteristics of the converter.

The small-signal simulation shows a close correlation with the experimental data. However, the simulations deviate from the measurements in two different frequency ranges. The first deviation is observed in the neighborhood of the power stage double pole at $f_o = \omega_o/2\pi = 1.16$ kHz and the second disagreement is found at high frequencies.

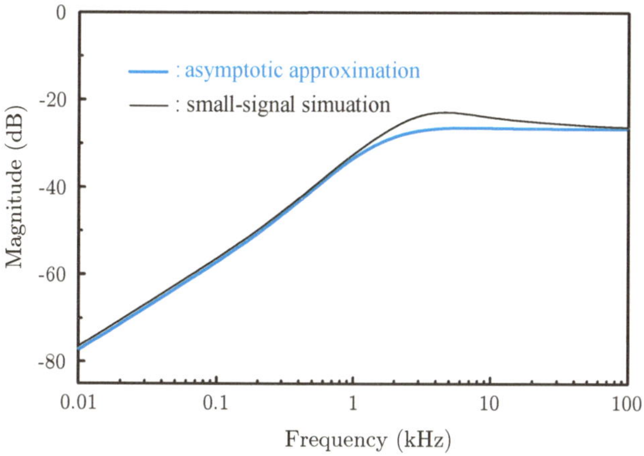

Figure 8.21 Output impedance comparison.

The first deviation is due to the difference in the power stage damping. In the small-signal model, the esrs of the output capacitor and inductor are only considered as the parasitic resistances that contribute to the power stage damping. In the experimental converter, there are many other parasitic resistances which are not accounted for in the small-signal model. These unaccounted parasitic resistances offer additional contribution to the damping and the experimental data show less peaking in the magnitude and more gradual change in the phase. The accuracy of the small-signal simulation will improve if the actual parasitic resistances are properly included in the small-signal model.

The second disagreement at the high frequencies stems from the limitation of the small-signal model of the PWM block. The constant modulator gain F_m, defined as the small-signal model of the PWM process in Section 5.6, is an approximation based on the presumption that the input signal to the PWM block does not significantly vary within one switching period. This presumption only holds true when the frequency response is evaluated at significantly lower frequencies than the switching frequency. As the evaluation frequency is increased, this assumption becomes less accurate and so does the PWM gain.

When the evaluation frequency increases so high that the PWM input signal varies substantially within one switching period, the nonlinear time-varying effect of the PWM process starts being pronounced. In this frequency range, the frequency response is strongly influenced by the sideband components of the PWM process; more specifically, the frequency response contains the sideband offshoots of the PWM process, as well as the response to the original small-signal excitation. This frequency-dependent nonlinear time-varying effect of the PWM process is not included in the small-signal PWM gain.

As shown in Fig. 8.22, the error in the small-signal prediction becomes larger as the evaluation frequency is increased. In particular, the measured phase curve drops noticeably as the evaluation frequency approaches half the switching frequency, $0.5f_s = 25$ kHz. In addition, the measured magnitude curve shows a dip at the integer multiples of the switching frequency, 50 kHz and 100 kHz. In fact,

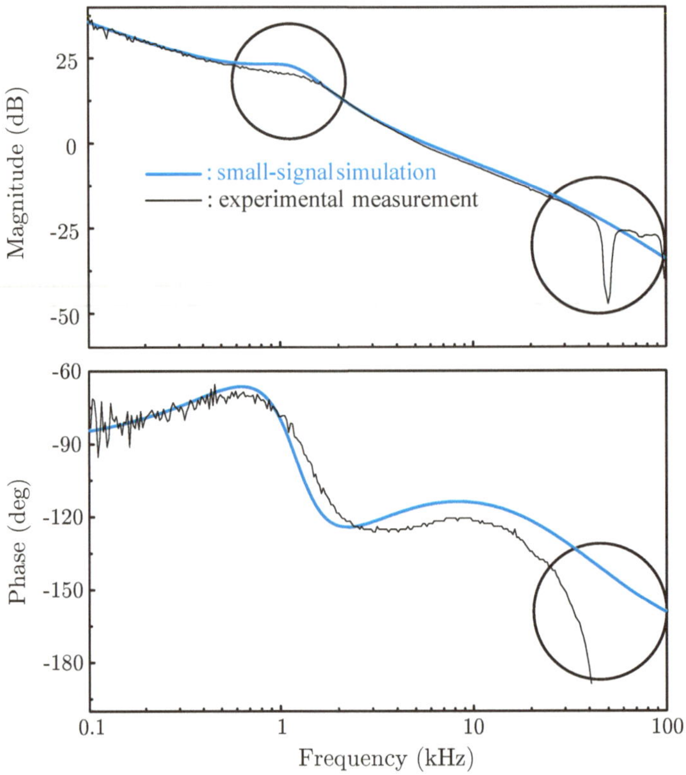

Figure 8.22 Loop gain comparison.

these are the singular points where the nonlinear effect of the PWM process is most pronounced and the small-signal model produces the largest error.

Figure 8.22 illustrates the danger of overly increasing the loop gain crossover frequency. The phase characteristics indicate that the phase margin of the actual converter is smaller than the estimation based on the small-signal model. The error becomes increasingly larger as the crossover frequency is pushed towards higher frequencies. The crossover frequency is recommended to be placed around 10%–30% of the switching frequency, in order to avoid the risk of overestimating the phase margin and to secure the fidelity of the analysis. In addition, it is good engineering practice to consider the additional phase drop at high frequencies when designing the loop gain. For example, in anticipation of the additional phase drop which is not accounted for in the small-signal model, the crossover frequency can be placed *before* the frequency at which the small-signal model predicts the largest phase margin, as previously exercised in Example 8.3.

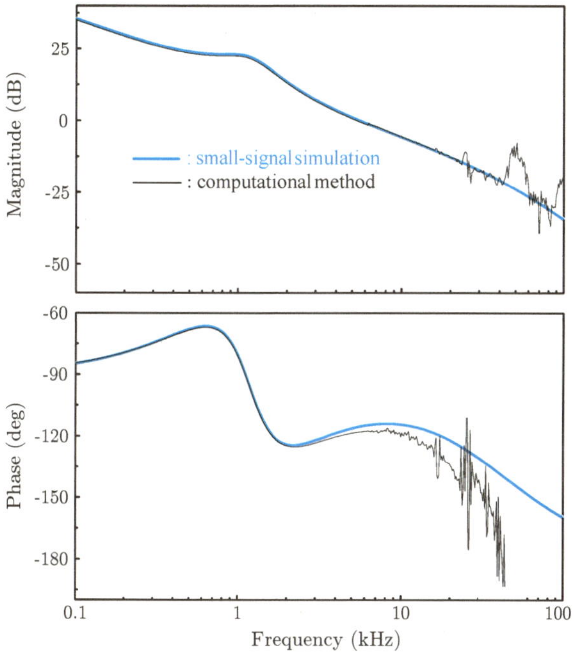

Figure 8.23 Computational method for loop gain evaluation.

Computational Method for Frequency-Domain Analysis EXAMPLE 8.6

As introduced in Section 6.5, the computational method can be used as a means of verifying predictions of the small-signal model. In the computational method, the time-domain simulation is performed with a perturbed input signal, and the output at the perturbation frequency is extracted and compared with the input signal. By sweeping the perturbation frequency for the range of interest and recording the ratio between the input and output signals, the Bode plot of the corresponding frequency response is obtained. *This computational method is a functional duplication of the experimental measurement using an impedance analyzer.*

Some commercial circuit simulation software offers an automated process for the computational method. In this book, small-signal model predictions are compared with the results of the computational method using PSIM® from Powersim Inc [2].

Figure 8.23 shows the loop gain of the buck converter used in Example 8.3. The predictions of the small-signal model are compared with the outcome of the computational method. At low frequencies, the disparity between the small-signal simulation and the computational method is undetectable. On the other hand, some discrepancies are noticeable at high frequencies. The computational data show irregular spikes in both magnitude and phase due to the sensitivity of the numerical algorithms. The discrepancies usually occur at high frequencies, well above the loop gain crossover frequency, and do not interfere with the evaluation of the loop gain performance. Even so, the computational data exhibit discernible patterns. The magnitude envelope reasonably follows the result of the small-signal model. Very interestingly, the high-frequency phase envelope deviates from the small-signal model prediction but tracks the measured phase response demonstrated in Fig. 8.22. This indicates that *the computational method duly captures the frequency-dependent nonlinear time-varying effects of the PWM process and provides accurate high-frequency dynamics of the converter.*

8.4.4 Phase Margin and Closed-Loop Performance

As shown in the audio-susceptibility and output impedance analysis, the closed-loop transfer function $G(s)_{\text{closed}}$ is related with the open-loop transfer function $G(s)_{\text{open}}$ through the following equation

$$\boxed{G(s)_{\text{closed}} = \frac{G(s)_{\text{open}}}{1+T_m(s)}} \tag{8.64}$$

where $T_m(s)$ is the loop gain of the converter. This equation is divided into the two approximations

$$\begin{aligned} G(s)_{\text{closed}} &= \frac{G(s)_{\text{open}}}{1+T_m(s)} \\ &\approx \begin{cases} \dfrac{G(s)_{\text{open}}}{T_m(s)} & \text{at frequencies where } |T_m| > 1 \\ \\ G(s)_{\text{open}} & \text{at frequencies where } |T_m| < 1 \end{cases} \end{aligned} \tag{8.65}$$

for the asymptotic analysis. This asymptotic analysis is accurate in the frequency range where either the condition $|T_m| \gg 1$ or $|T_m| \ll 1$ is met. However, neither of these assumptions is satisfied in the neighborhood of the loop gain crossover frequency where $|T_m| \approx 1$. The accuracy of the asymptotic approximation near the loop gain crossover was broadly discussed in Example 8.4. A more precise analysis around the crossover frequency is now given.

The exact behavior of the closed-loop transfer function can be investigated by evaluating the magnitude of the closed-loop transfer function

$$|G_{\text{closed}}| = \frac{|G_{\text{open}}|}{|1+T_m|} \tag{8.66}$$

at the loop gain crossover frequency. Figure 8.24 illustrates the evaluation of the denominator of (8.66) based on the following facts:

- The denominator of (8.66) is the magnitude of the composite vector $|\vec{1}+\vec{T_m}|$.
- The magnitude of the loop gain is unity, $|T_m| = 1$, at the 0 dB crossover frequency.

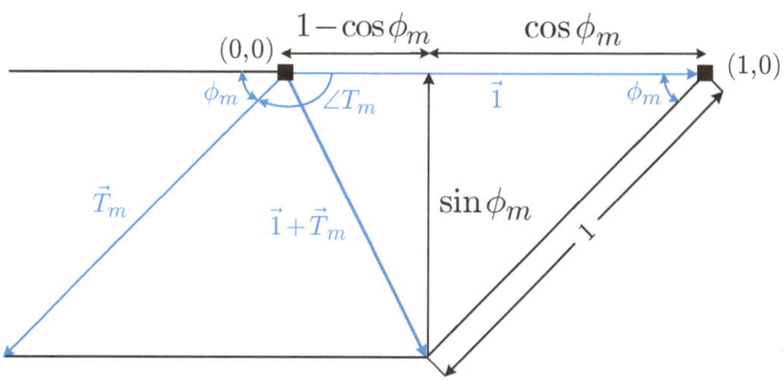

Figure 8.24 Evaluation of $|\vec{1}+\vec{T}_m|$ at loop gain crossover frequency.

- The angle ϕ_m in Fig. 8.24, which corresponds to the difference between $-180°$ and $\angle T_m$ at the crossover frequency, is the phase margin of the loop gain.

From the vector addition rule and trigonometric relationships, the denominator of (8.66) becomes

$$|\vec{1}+\vec{T}_m| = \sqrt{\sin^2\phi_m + (1-\cos\phi_m)^2} = \sqrt{2-2\cos\phi_m} \tag{8.67}$$

Accordingly, the relationship (8.66) becomes

$$|G_{\text{closed}}| = \frac{|G_{\text{open}}|}{\sqrt{2-2\cos\phi_m}} \tag{8.68}$$

If the phase margin ϕ_m is smaller than 60°, the denominator of (8.68) is less than unity and $|G_{\text{closed}}|$ will surge from $|G_{\text{open}}|$ by the amount of

$$\boxed{|G|_{\text{peaking}} = 20\log\left(\frac{1}{\sqrt{2-2\cos\phi_m}}\right)} \tag{8.69}$$

The magnitude of the peaking is inversely proportional to the phase margin; the smaller the phase margin, the larger the peaking. If the phase margin reduces to zero, the peaking becomes infinite.

When the phase margin ϕ_m is *not* significantly lower than 60°, the $|G_{\text{closed}}|$ curve is obtained by smoothly connecting the asymptotes of $|G_{\text{open}}|/|T_m|$ and $|G_{\text{open}}|$ at the $|T_m|$ crossover frequency, as suggested in the asymptotic analysis rules in Table 8.1. If ϕ_m is noticeably lower than 60°, the peaking at the crossover frequency of $|T_m|$ should be considered when predicting the $|G_{\text{closed}}|$ curve.

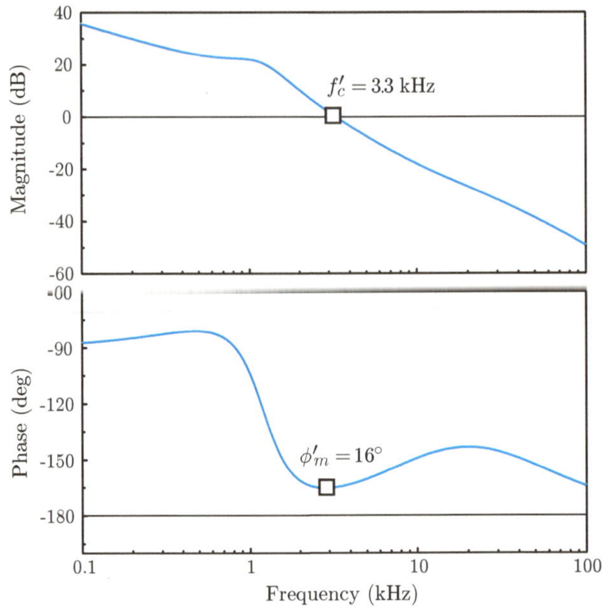

Figure 8.25 Loop gain characteristics with modified compensation design.

EXAMPLE 8.7 Effects of Small Phase Margin

This example demonstrates the detrimental effects of a small phase margin. The small phase margin was shown to cause a peaking in the closed-loop transfer functions. It will also be argued that the peaking in turn results in an oscillatory behavior in the transient responses.

The voltage feedback compensation, employed to the buck converter in the previous examples, is now modified to

$$F_v(s) = \frac{1.04 \times 10^4}{s} \frac{\left(1 + \frac{s}{2\pi \cdot 928}\right)\left(1 + \frac{s}{2\pi \cdot 9.9 \times 10^3}\right)}{\left(1 + \frac{s}{2\pi \cdot 6.77 \times 10^3}\right)\left(1 + \frac{s}{2\pi \cdot 4 \times 10^4}\right)}$$

to produce the loop gain with a small phase margin. Compared to the original design, only the second compensation zero is increased from $\omega_{z2} = 2\pi \cdot 1.74 \times 10^3$ rad/s to $\omega'_{z2} = 2\pi \cdot 9.9 \times 10^3$ rad/s. The loop gain characteristics with this modification are shown in Fig. 8.25. The 0 dB crossover now occurs at $f'_c = 3.3$ kHz and the phase margin is reduced to $\phi'_m = 16°$. The magnitude of the peaking is calculated as

$$|G|_{\text{peaking}} = 20 \log\left(\frac{1}{\sqrt{2 - 2\cos 16°}}\right) \approx 11 \text{ dB}$$

This peaking is expected to occur at the loop gain crossover frequency, $f'_c = 3.3$ kHz, of the new loop gain, T'_m. Figure 8.26 shows the output impedance $|Z'_o|$ with the

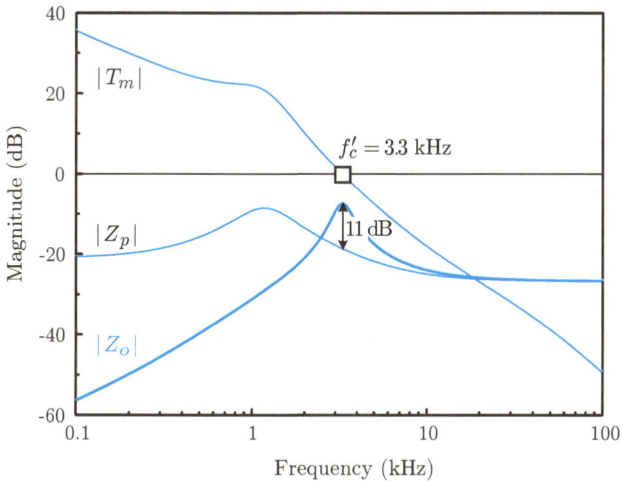

Figure 8.26 Output impedance characteristics with modified compensation design.

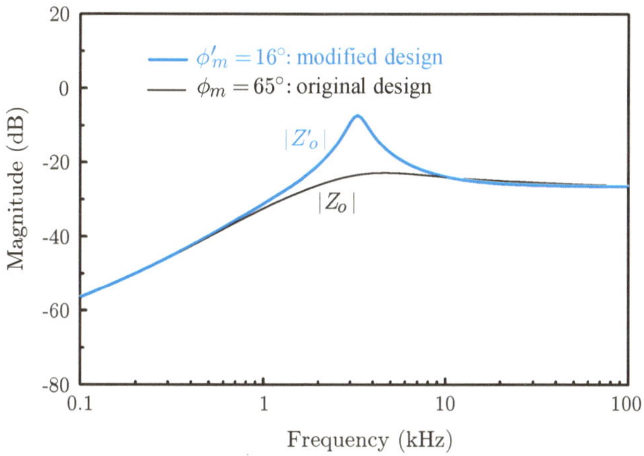

Figure 8.27 Phase margin and output impedance.

modified compensation design, along with the load current-to-output transfer function $|Z_p|$ and loop gain $|T'_m|$. The output impedance $|Z'_o|$ follows the curve formed by $|Z_p| - |T'_m|$ at low frequencies and it converges to $|Z_p|$ at high frequencies. However, in the neighborhood of the loop gain crossover frequency, the output impedance deviates from the respective approximations, showing the expected 11 dB upsurge from $|Z_p|$ at the crossover frequency $f'_c = 3.3$ kHz. The effect of the small phase margin is clear in Fig. 8.27, which compares the output impedance of the modified design, $|Z'_o|$, with that of the original design, $|Z_o|$. The phase margin of the original design was $\phi_m = 65°$, while that of the modified design is $\phi'_m = 16°$.

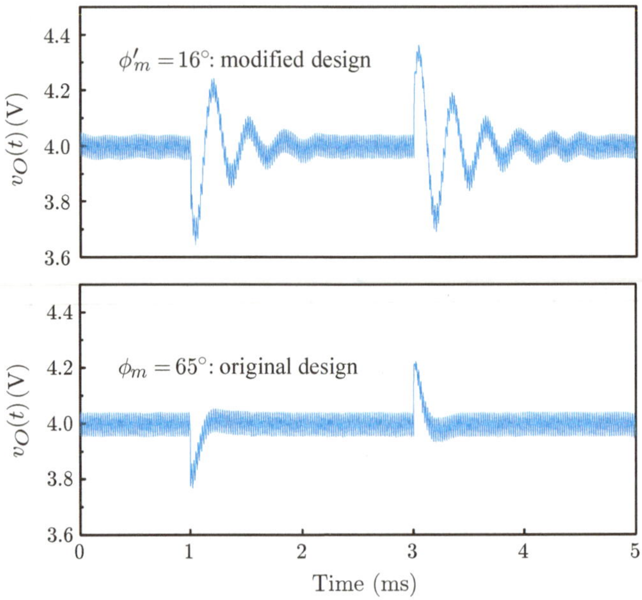

Figure 8.28 Phase margin and step load response of output voltage.

As discussed in Section 7.3.1, the transient response of the output voltage due to the step change in the load current, I_{step}, is given by[†]

$$v_O(t) = \mathcal{L}^{-1}\left(\frac{I_{step}}{s} Z_o(s)\right) \tag{8.70}$$

where $\mathcal{L}^{-1}$ is the inverse Laplace transform operator. The impact of the output impedance peaking is deduced from this relationship. The peaking in $|Z_o|$ indicates the existence of an underdamped second-order term in the output impedance. When the inverse Laplace transform is taken, the underdamped second-order term produces a decaying sinusoidal component in the time-domain equation. Accordingly, the peaking in $|Z_o|$ signals an oscillatory behavior in the output voltage. The frequency of the oscillation coincides with the frequency of the output impedance peaking, which is the loop gain crossover frequency.

Figure 8.28 shows the step load response of the modified design, in comparison with that of the original design. A series of step changes in the load current, $I_O = 4\ \text{A} \Rightarrow 8\ \text{A} \Rightarrow 4\ \text{A}$, is assumed in Fig. 8.28. The output voltage of the modified design shows a decaying oscillation, while the original design does not show any

[†] Theoretically speaking, the relationship (8.70) is only valid when I_{step} is sufficiently small because the output impedance is derived under the small-signal assumption. Even so, it was observed that the predictions of the output impedance exhibit markedly good correlations to transient responses with relatively large changes: for example, up to 50% change in the load current or in the input voltage.

oscillatory behavior. The period of the oscillation is predicted as

$$t'_{osc} = \frac{1}{f'_c} = \frac{1}{3.3 \times 10^3} = 0.3 \text{ ms}$$

where $f'_c = 3.3$ kHz is the loop gain crossover frequency of the modified design.

As demonstrated in the previous example, there is a close relationship among the phase margin, output impedance peaking, and output voltage oscillation. A small phase margin causes a peaking in the output impedance, which in turn invokes an oscillation in the output voltage. The smaller phase margin intensifies the peaking and the intensified peaking makes the oscillation more persistent. When the phase margin is reduced to zero, the output impedance peaks to infinity and the output voltage shows a sustained oscillation as the sign of instability.

The effects of a small phase margin were previously addressed in Section 7.6. However, the perspective of Section 7.6 differs from that of the current section. Section 7.6 dealt with the problem in a qualitative manner, in conjunction with the location of the system poles, while this section provided an explicit equation of (8.69). It is informative to compare the analytical results of this section with the qualitative discussions of Section 7.6. Readers are urged to review Example 7.9 which substantiates and reinforces the results of Example 8.7.

8.4.5 Compensation Zeros and Speed of Transient Responses

In the previous section, the three-pole two-zero circuit

$$F_v(s) = \frac{K_v}{s} \frac{\left(1 + \frac{s}{\omega_{z1}}\right)\left(1 + \frac{s}{\omega_{z2}}\right)}{\left(1 + \frac{s}{\omega_{p1}}\right)\left(1 + \frac{s}{\omega_{p2}}\right)} \tag{8.71}$$

was identified as the optimal structure for the voltage feedback compensation. The selection of the compensation parameters is discussed focusing on the frequency-domain performance metrics.

As will be shown in this section, the compensation parameters also affect the time-domain performance. In particular, the compensation zeros, ω_{z1} and ω_{z2}, determine the speed of transient responses. The first compensation zero ω_{z1} governs the speed of step input responses, while the second compensation zero ω_{z2} dictates the speed of step load responses. Accordingly, the compensation zeros should be selected considering their impacts on the transient responses. This section analyzes the impact of the first compensation zero ω_{z1} and the next section deals with the second compensation zero ω_{z2}.

The impact of the first compensation zero ω_{z1} is explained using the audio-susceptibility and its relationship to the step input response. Figure 8.29 shows the asymptotic plot of the audio-susceptibility $|A_u|$, along with the asymptotic plots of

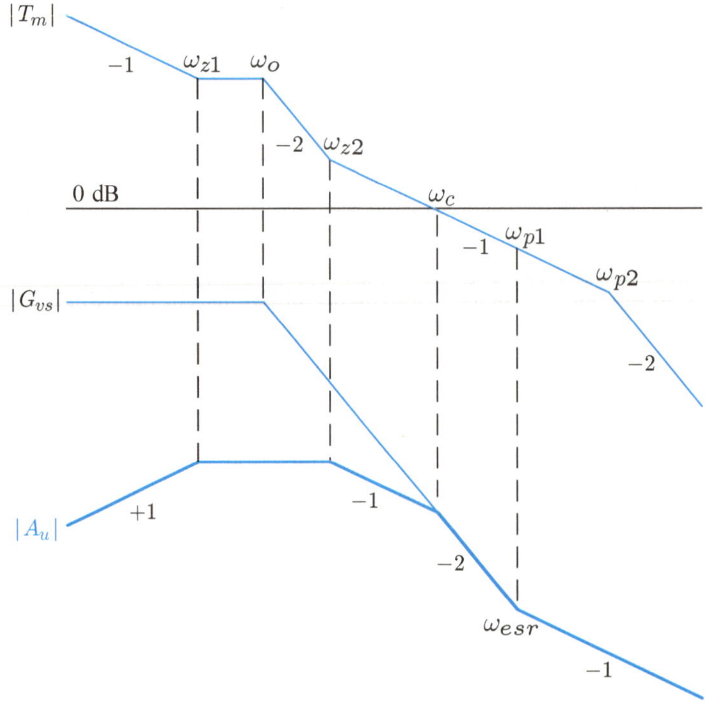

Figure 8.29 Construction of asymptotic plot for audio-susceptibility.

$|T_m|$ and $|G_{vs}|$. The audio-susceptibility is written as

$$A_u(s) = K_a s \frac{1 + \dfrac{s}{\omega_{esr}}}{\left(1 + \dfrac{s}{\omega_{z1}}\right)\left(1 + \dfrac{s}{\omega_{z2}}\right)\left(1 + \dfrac{s}{\omega_c}\right)} \tag{8.72}$$

with the condition $\omega_{z1} \ll \omega_{z2} \ll \omega_c \ll \omega_{esr}$.

The first compensation zero ω_{z1} becomes a pole in the $A_u(s)$ expression in (8.72). The step input response of the output voltage is given by

$$\boxed{v_O(t) = \mathcal{L}^{-1}\left(\frac{V_{step}}{s} A_u(s)\right)} \tag{8.73}$$

where V_{step} denotes the magnitude of the step input change. From (8.72) and (8.73), the output voltage is expressed as

$$v_O(t) = \mathcal{L}^{-1}\left(V_{step} \frac{K_a \left(1 + \dfrac{s}{\omega_{esr}}\right)}{\left(1 + \dfrac{s}{\omega_{z1}}\right)\left(1 + \dfrac{s}{\omega_{z2}}\right)\left(1 + \dfrac{s}{\omega_c}\right)} \right) \tag{8.74}$$

When the inverse Laplace transform is performed, the output voltage expression would contain three different exponential terms

$$\begin{aligned} v_O(t) = & \text{ the term associated with } e^{-\omega_{z1}t} \\ & + \text{the term associated with } e^{-\omega_{z2}t} \\ & + \text{the term associated with } e^{-\omega_{c}t} \end{aligned} \tag{8.75}$$

The first term in (8.75) is the slowest mode due to the prevailing condition of $\omega_{z1} \ll \omega_{z2} \ll \omega_c$; in other words, the first compensation zero ω_{z1} becomes the dominant pole. Accordingly, the time constant of the transient response is $\tau = 1/\omega_{z1}$. The settling time of the transient response is determined as

$$\boxed{t_s = 3\tau = 3\frac{1}{\omega_{z1}}} \tag{8.76}$$

For faster response, ω_{z1} should be selected at higher frequencies. However, if ω_{z1} is placed after the power stage double pole ω_o, $\angle T_m$ could temporarily drop below $-180°$ making the converter conditionally stable, as discussed in Section 8.4.1. Thus, the location of the first compensation zero ω_{z1}, whilst the higher frequencies are preferred, should be limited below the power stage double pole ω_o. It was recommended that $\omega_{z1} = (0.6-0.8)\,\omega_o$ in the previous section.

Compensation Zero and Step Input Response — EXAMPLE 8.8

This example substantiates the previous discussions about the impacts of the first compensation zero. The voltage feedback compensation of the buck converter is now modified to

$$F_v(s) = \frac{3.9 \times 10^3}{s} \frac{\left(1+\dfrac{s}{2\pi \cdot 312}\right)\left(1+\dfrac{s}{2\pi \cdot 1.74 \times 10^3}\right)}{\left(1+\dfrac{s}{2\pi \cdot 6.77 \times 10^3}\right)\left(1+\dfrac{s}{2\pi \cdot 4.0 \times 10^4}\right)}$$

Compared with the original design, the first compensation zero is reduced from $\omega_{z1} = 2\pi \cdot 928$ rad/s to $\omega'_{z1} = 2\pi \cdot 312$ rad/s and the integrator gain is decreased from $K_v = 1.04 \times 10^4$ to $K'_v = 3.9 \times 10^3$, while the other parameters are unchanged. Figure 8.30 shows the audio-susceptibility of the modified design in comparison with the original transfer function.

Figure 8.31 compares the output voltage of the buck converter in response to the step changes in the input voltage: $V_S = 16$ V $\Rightarrow 8$ V $\Rightarrow 16$ V. As expected, the modified design produces a slower response. The settling time of the modified design is predicted as

$$t'_s = 3\tau' = 3\frac{1}{\omega'_{z1}} = 3\frac{1}{2\pi \cdot 312} = 1.53 \text{ ms}$$

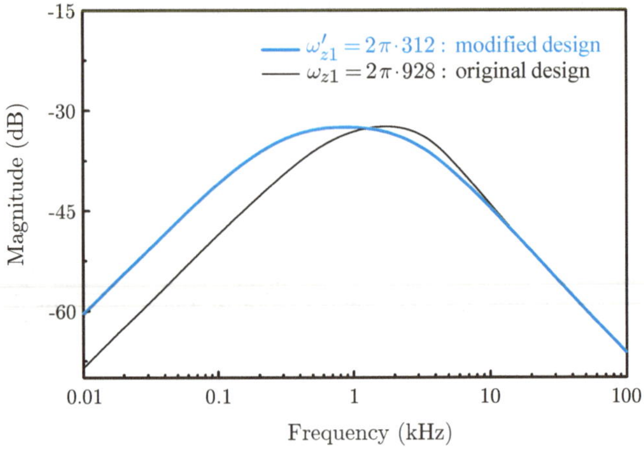

Figure 8.30 Compensation zero and audio-susceptibility.

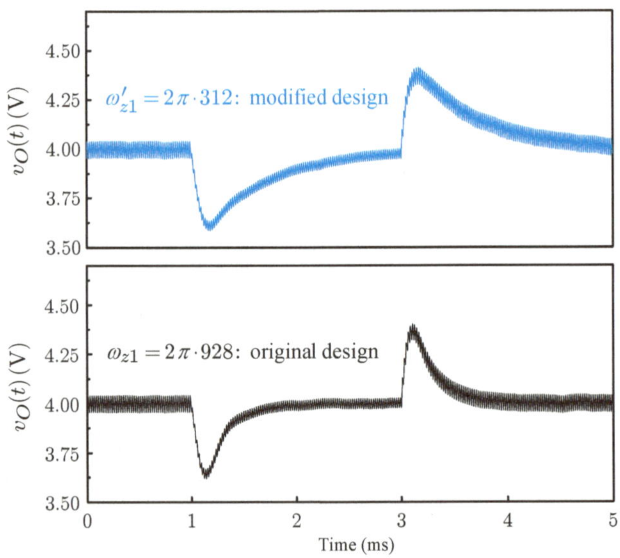

Figure 8.31 Compensation zero and step input response.

while that of the original design is estimated as

$$t_s = 3\tau = 3\frac{1}{\omega_{z1}} = 3\frac{1}{2\pi \cdot 928} = 0.51 \text{ ms}$$

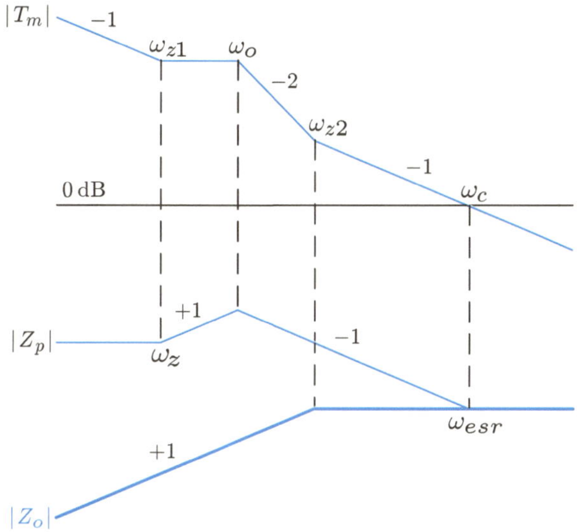

Figure 8.32 Construction of asymptotic plot for output impedance.

8.4.6 Step Load Response

The step load response is a very important performance metric for many dc-to-dc converters. As discussed in Section 7.3.1, the transient response of the output voltage can be analyzed using the output impedance. This section illustrates such an analysis with some simplifying assumptions on the output impedance characteristics. We will show that the speed of the step load response is determined by the position of the second compensation zero, ω_{z2} in (8.71).

The asymptotic plot for the output impedance, formerly introduced in Fig. 8.15, is modified in Fig. 8.32 with the following assumptions.

1) The first compensation zero ω_{z1} is placed at the low-frequency zero ω_z of the load current-to-output transfer function: $\omega_{z1} = \omega_z$.

2) The loop gain crossover frequency ω_c is placed at the esr zero ω_{esr}: $\omega_c = \omega_{esr}$.

As will be shown in Example 8.9, these assumptions are broadly satisfied in many practical dc-to-dc converters. The asymptotic plot for $|Z_o|$ is redrawn in Fig. 8.33(a). The first compensation zero ω_{z1}, which determines the speed of the step input response, does not appear in the output impedance. From Fig. 8.33(a), the output impedance is written as

$$Z_o(s) = \frac{s}{\omega_m} \frac{1}{1 + \dfrac{s}{\omega_{z2}}} \tag{8.77}$$

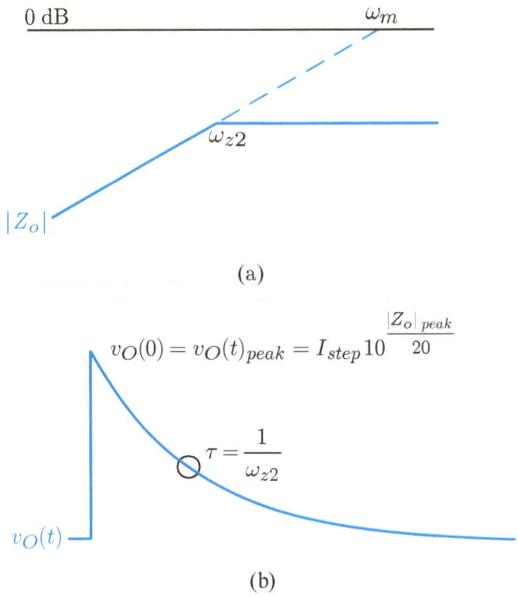

Figure 8.33 Output impedance and step load response. (a) Output impedance. (b) Step load response.

where ω_m represents the frequency at which the initial line segment of $|Z_o|$ projects to the 0 dB line. The peak magnitude of the output impedance is given by

$$|Z_o(j\omega)|_{peak} = |Z_o(j\infty)| = 20 \log \left(\frac{\omega_{z2}}{\omega_m} \right) \tag{8.78}$$

The transient response of the output voltage due to the step load change I_{step} is expressed as

$$v_O(t) = \mathcal{L}^{-1} \left(\frac{I_{step}}{s} \frac{s}{\omega_m} \frac{1}{1 + \frac{s}{\omega_{z2}}} \right) = I_{step} \frac{\omega_{z2}}{\omega_m} e^{-\omega_{z2} t} \tag{8.79}$$

Figure 8.33(b) depicts the waveform of the output voltage. The peak overshoot of the output voltage is given by

$$\Delta v_O(t)_{peak} = v_O(0) = I_{step} \frac{\omega_{z2}}{\omega_m} = I_{step} 10^{\frac{|Z_o|_{peak}}{20}} \tag{8.80}$$

based on (8.78) and (8.79). The output voltage decays from the peak with a time constant of $\tau = 1/\omega_{z2}$. Thus, the settling time becomes

$$\boxed{t_s = 3\tau = 3\frac{1}{\omega_{z2}}} \tag{8.81}$$

As shown in the preceding analysis, there are simple, yet very practical and useful, relationships among the output impedance characteristics and step load transient response. The settling time of the output voltage is dictated by the second compensation zero, $t_s = 3/\omega_{z2}$. The peak value of the output voltage is determined by the product of the size of step load change and the peak magnitude of the output impedance: $\Delta v_O(t)_{peak} = I_{step}\, 10^{|Z_o|_{peak}/20}$.

Output Impedance and Step Load Response **EXAMPLE 8.9**

The accuracy of the preceding step load response analysis is confirmed in this example. For the buck converter in Example 8.3, the corner frequencies are given as $\omega_{z1} = 2\pi \cdot 928$ rad/s, $\omega_z = 2\pi \cdot 398$ rad/s, $\omega_c = 2\pi \cdot 5.80 \times 10^3$ rad/s, and $\omega_{esr} = 2\pi \cdot 6.67 \times 10^3$ rad/s, thereby broadly satisfying the conditions $\omega_{z1} \approx \omega_z$ and $\omega_c \approx \omega_{esr}$.

Figure 8.34 shows the output impedance $|Z_o|$, load current-to-output transfer function $|Z_p|$, and loop gain $|T_m|$ of the converter. The simulated output impedance shows good correlation with the theoretical prediction of Fig. 8.32.

Figure 8.35 illustrates the transient response of the output voltage with $I_{step} =$ 4 A. The peak value of the output voltage is predicted as

$$\begin{aligned} v_O(t)_{peak} &= 4.0 + \Delta v_O(t)_{peak} = 4.0 + I_{step}\, 10^{\frac{|Z_o|_{peak}}{20}} \\ &= 4.0 + 4 \cdot 10^{-26/20} = 4.2 \text{ V} \end{aligned}$$

and the settling time is estimated as

$$t_s = 3\tau = 3\frac{1}{\omega_{z2}} = 3\frac{1}{2\pi \cdot 1.74 \times 10^3} = 0.27 \text{ ms}$$

8.4.7 Non-Minimum Phase System Case: Boost and Buck/Boost Converters

The earlier sections investigated the control design and performance evaluation of the closed-loop controlled buck converter. The current section now deals with the boost and buck/boost converters.

As discussed in Section 6.2, the boost and buck/boost converters contain a right-half plane (RHP) zero in the duty ratio-to-output transfer function $G_{vd}(s)$. The RHP

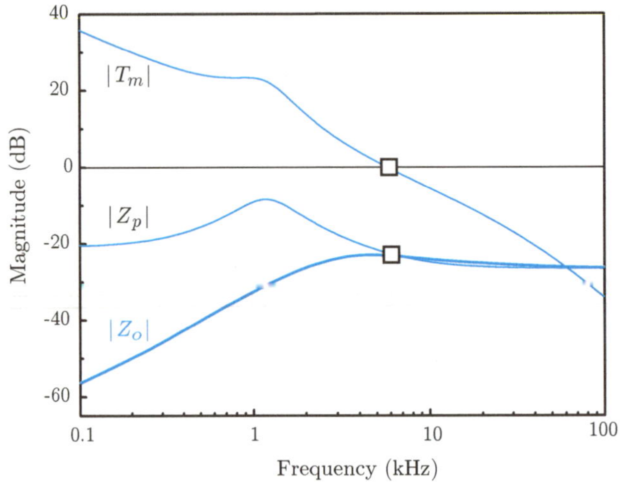

Figure 8.34 Output impedance of buck converter.

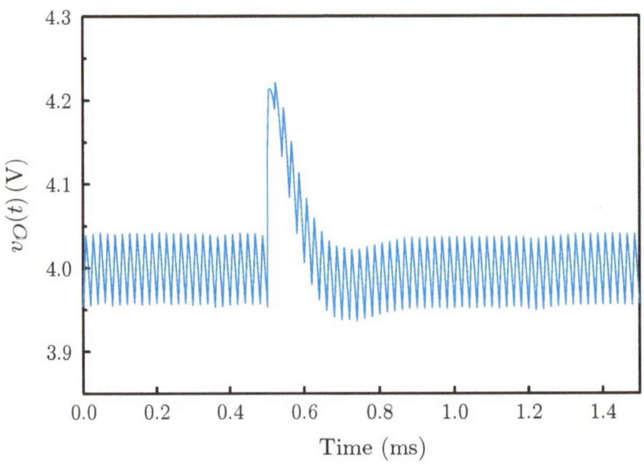

Figure 8.35 Step load response of buck converter.

zero induces a 90° phase delay to $\angle G_{vd}$ while increasing the slope of $|G_{vd}|$ by 20 dB/dec. The dynamic system with an RHP zero is called the non-minimum phase system in the sense that the total phase variation in the transfer function is larger than that of the system with a regular left-half plane zero. Thus, the boost and buck/boost converters are a typical example of the non-minimum phase system.

The non-minimum phase system presents considerable difficulties to the feedback compensation design. The boost and buck/boost converters indeed do not render themselves to the application of the feedback control scheme discussed in this chapter. In this section, we investigate the problem of the boost and buck/boost converters

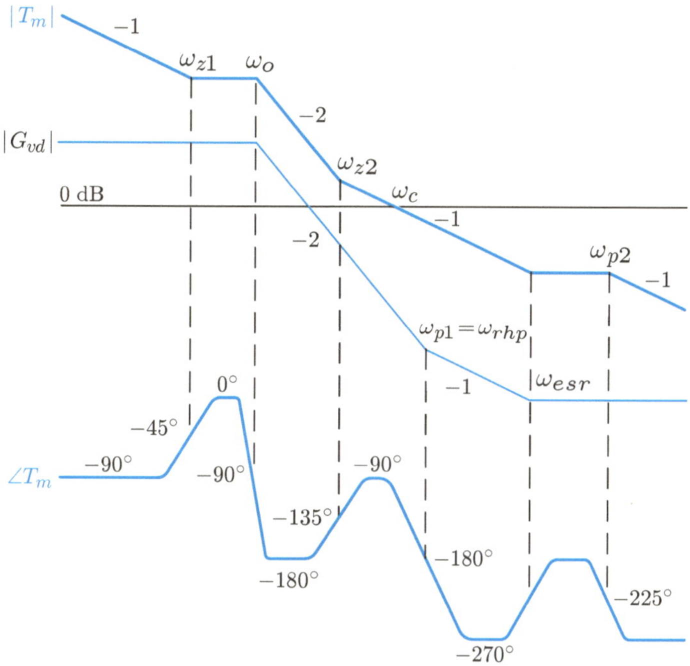

Figure 8.36 Loop gain characteristics of boost or buck/boost converter with three-pole two-zero compensation.

in regard to their control design. This section also briefly introduces a new control scheme which could be adapted to the boost and buck/boost converters to mitigate the problem of the RHP zero.

Boost and Buck/Boost Converters

The RHP zero imposes demanding constraints on the compensation design and hinders the boost or buck/boost converter from having acceptable closed-loop performance. The problem of the RHP zero can readily be seen in Fig. 8.36, which shows the asymptotic plots for the $|G_{vd}|$ and loop gain of a boost or buck/boost converter. The previous three-pole two-zero circuit is assumed for the voltage feedback compensation.

The compensation parameters are selected based on the design strategy established in Section 8.4.2. In particular, the first compensation pole is placed at the RHP zero, $\omega_{p1} = \omega_{rhp}$, in an attempt to extend the frequency range in which $|T_m|$ maintains the desirable -20 dB/dec slope.

The phase characteristics of the loop gain uncover the problem of the RHP zero. As shown in $\angle T_m$ in Fig. 8.36, the loop gain phase drops by $180°$ over the RHP zero, thereby worsening the high-frequency phase characteristics. The $90°$ phase

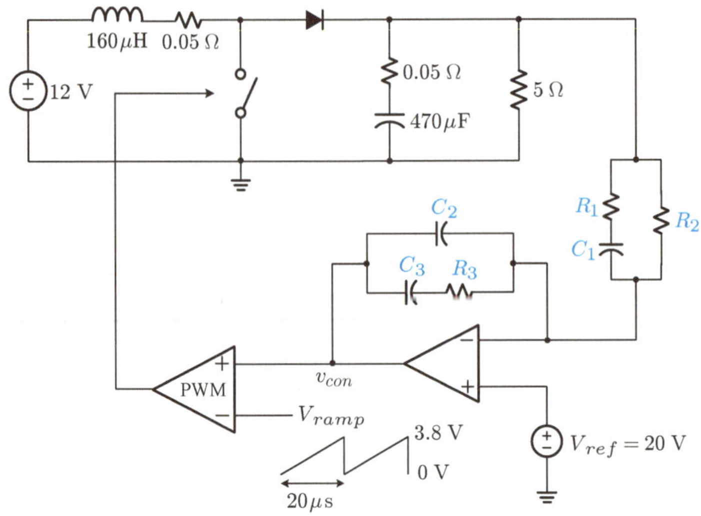

Figure 8.37 Closed-loop controlled boost converter: $R_1 = 2.2$ kΩ, $C_1 = 40.4$ nF, $R_2 = 7.2$ kΩ, $C_2 = 1.7$ nF, $R_3 = 2.4$ kΩ, and $C_3 = 0.28$ μF.

drop caused by ω_{rhp} is augmented by the additional 90° phase delay brought in by ω_{p1}, resulting in 180° phase decay while $|T_m|$ stays in -20 dB/dec slope. Under this situation, $\angle T_m$ mainly remains near or below $-180°$ from the mid-frequencies to high frequencies. Due to these unfavorable phase characteristics, it is very difficult to achieve satisfactory closed-loop performance while securing sufficient phase margin.

EXAMPLE 8.10 Boost Converter Example

This example demonstrates the difficulty in designing feedback compensation for a practical boost converter. Figure 8.37 shows the circuit diagram of a closed-loop controlled boost converter. The converter produces a 20 V output from a 12 V input source using the three-pole two-zero voltage feedback circuit. The steady-state duty ratio of the converter is determined as

$$\frac{V_O}{V_S} = \frac{1}{1-D} \quad \Rightarrow \quad \frac{20}{12} = \frac{1}{1-D} \quad \Rightarrow \quad D = 0.4$$

From the duty ratio and power stage circuit parameters, the corner frequencies of the duty ratio-to-output transfer function are determined as

$$\omega_o \approx \frac{1-D}{\sqrt{LC}} = \frac{1-0.4}{\sqrt{160\times 10^{-6} 470\times 10^{-6}}} = 2\pi \cdot 348 \text{ rad/s}$$

$$\omega_{rhp} \approx \frac{(1-D)^2 R}{L} = \frac{(1-0.4)^2 5}{160\times 10^{-6}} = 2\pi \cdot 1.79 \times 10^3 \text{ rad/s}$$

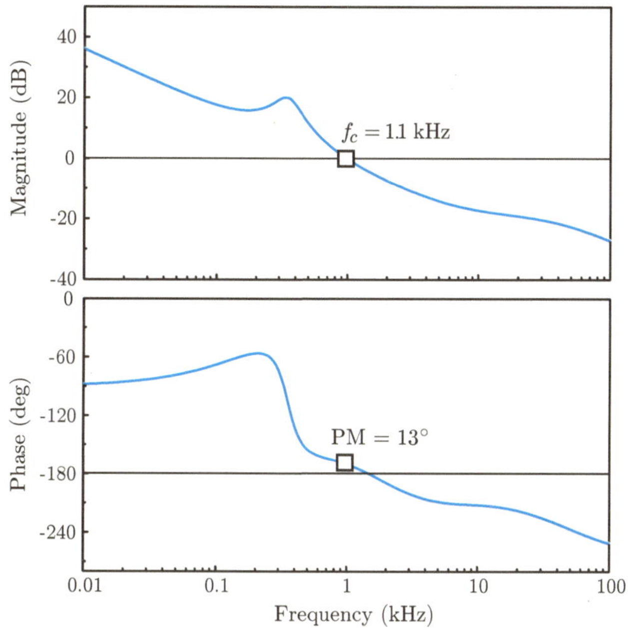

Figure 8.38 Loop gain characteristics of boost converter.

and

$$\omega_{esr} = \frac{1}{CR_c} = \frac{1}{470 \times 10^{-6}\, 0.05} = 2\pi \cdot 6.77 \times 10^3 \text{ rad/s}$$

Based on the design guidelines for the three-pole two-zero compensation, the compensation parameters are selected as

- $\omega_{z1} = 0.7\omega_o = 2\pi \cdot 244$ rad/s
- $\omega_{z2} = 1.2\,\omega_o = 2\pi \cdot 418$ rad/s
- $\omega_{p1} = \omega_{rhp} = 2\pi \cdot 1.79 \times 10^3$ rad/s
- $\omega_{p2} = 0.8\,\omega_s = 2\pi \cdot 4.0 \times 10^4$ rad/s
- $K_v = 500$

The loop gain characteristics are shown in Fig. 8.38. When compared with the asymptotic plots in Fig. 8.36, $\angle T_m$ curve shows a rather gradual decay at the frequencies beyond the power stage double pole. This is because the corner frequencies of the voltage feedback compensation are not as widely separated as presumed in the asymptotic analysis.

The loop gain crossover frequency is located at $\omega_c = 2\pi \cdot 1.1 \times 10^3$ rad/s with a 13° phase margin. This design would not be acceptable for most applications due to the insufficient phase margin. A possible design change for an adequate phase margin is to place the loop gain crossover frequency near or before the power stage double pole. However, such a design would end up with poor closed-loop performance with overly large output impedance and audio-susceptibility.

Alternative Control Scheme: Current Mode Control

The difficulty in the compensation design in Example 8.10 originates not from the structure of the feedback circuit but from the existence of the RHP zero in the boost converter. While the previous example underscored the problem of the three-pole two-zero compensation circuit, it is readily inferred that other compensation structures would worsen the situation.

The feedback control scheme covered in this chapter employs the output voltage as the only feedback signal. In this sense, this control scheme is referred to as *voltage mode control*. Example 8.10 proved that voltage mode control is not suitable for the converters that have an RHP zero, such as the boost converter, the buck/boost converter, and all other isolated PWM converters derived from these two converters.

For the dc-to-dc converters with the RHP zero, an alternative control method is adapted to resolve the aforementioned problem. The alternative control scheme employs the feedback signal not only from the output voltage but also from the inductor current. This control scheme is called *current mode control* because it uses an additional feedback from the inductor current. Current mode control could provide good closed-loop performance for the boost and buck/boost converters in the presence of the RHP zero. Current mode control deserves rigorous treatments and thus is the topic of the next two chapters of this book.

8.5 CONSIDERATION OF DCM OPERATION

Earlier sections only considered the CCM operation in investigating the control design and dynamic performance of dc-to-dc converters. However, even if intended for CCM operation at the nominal conditions, all dc-to-dc converters invariably enter DCM operation when their load current is sufficiently reduced. Accordingly, the DCM operation should be considered in the control design of any practical dc-to-dc converters. In this section, we review the converter dynamics in DCM operations and discuss the control design strategy for both CCM operation and DCM operation.

8.5.1 Review of DCM Converter Dynamics

In Section 6.3, the converter transfer functions in DCM operation have been investigated. As uncovered in the prior analysis, the transfer functions of the three basic PWM converters undergo substantial changes in DCM operation. Most significantly, the power stage double pole splits into two real poles and the duty ratio-to-output

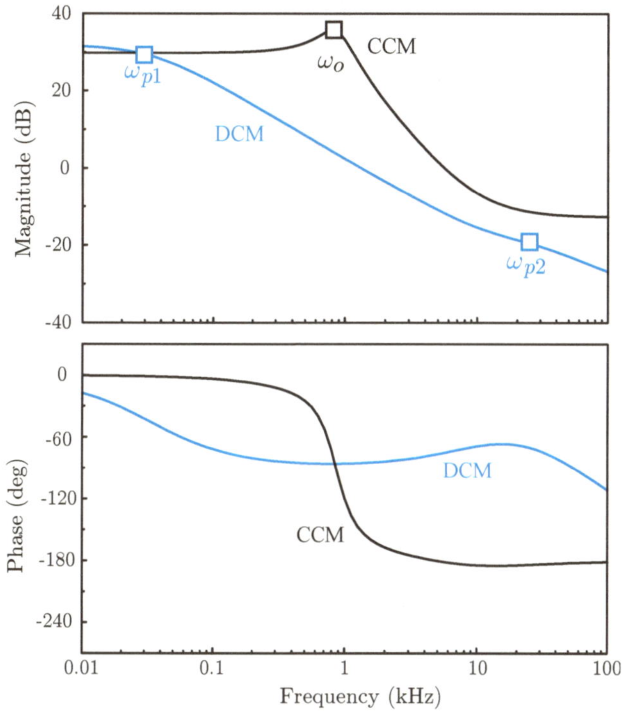

Figure 8.39 Example of $G_{vd}(s)$ characteristics of buck/boost converter in CCM and DCM operations.

transfer function $G_{vd}(s)$ is expressed as

$$G_{vd}(s) = K_d \frac{\left(1 - \frac{s}{\omega_{rhp}}\right)\left(1 + \frac{s}{\omega_{esr}}\right)}{\left(1 + \frac{s}{\omega_{p1}}\right)\left(1 + \frac{s}{\omega_{p2}}\right)} \tag{8.82}$$

for all three basic PWM converters in DCM operation.

Figure 8.39 shows the Bode plots of the $G_{vd}(s)$ characteristics of a buck/boost converter in both CCM operation and DCM operation. Details about the buck/boost converter used for Fig. 8.39 were given in Example 6.7. The $G_{vd}(s)$ curves in CCM elucidate the presence of the double pole at ω_o. On the other hand, the DCM transfer function plots visualize the separation of the double pole into the two real poles. The first real pole ω_{p1} appears at low frequencies and governs the low-frequency dynamics of the transfer function. In contrast, the second real pole ω_{p2} emerges at high frequencies and does not affect the transfer function up to mid frequencies. Accordingly, the DCM operation effectively changes the second-order system into the first-order system. The first-order system behavior is apparent in both magnitude and phase characteristics in Fig. 8.39. In addition, due to the appearance of the first pole ω_{p1} at low frequencies, the mid-frequency magnitude of the duty ratio-to-output transfer function, $|G_{vd}|$, is notably reduced.

As demonstrated above, the major changes in DCM dynamics include the presence of a low-frequency pole and ensuing magnitude reduction in $|G_{vd}|$ at mid-frequencies. These changes should be considered in determining the control design strategy.

8.5.2 Control Design Strategy and Converter Performance

Referring to Fig. 8.39, the $G_{vd}(s)$ in CCM presents worse characteristics than the $G_{vd}(s)$ in DCM. The double pole in the CCM transfer function causes a $180°$ phase drop to $\angle G_{vd}$. This abrupt and large phase delay was the main concern in designing the voltage feedback compensation. In contrast, the DCM transfer function only exhibits a gradual $90°$ phase delay.

The $G_{vd}(s)$ in CCM operation should be used as the basis for the control design for the following reasons. If the control is designed for DCM operation, the converter becomes unstable when it enters CCM operation, due to the excessive phase delay in $\angle G_{vd}$. On the contrary, the control design tailored for the worst case CCM operation will offer stable operation for both CCM and DCM. Thus, the rational choice is to design the controller for CCM operation and to predict the closed-loop performance in DCM operation.

The control can be optimally designed based on the CCM dynamics of the converter. When the converter enters DCM operation, $G_{vd}(s)$ will be altered as illustrated in Fig. 8.39. This change will affect the converter dynamics in various ways. In particular, the reduction in $|G_{vd}|$ will directly propagate to the loop gain characteristics and consequently alter the other closed-loop transfer functions. Even so, the change in $G_{vd}(s)$ does not jeopardize stability of the converter, as will be demonstrated in the following example.

EXAMPLE 8.11 Performance of Buck Converter in DCM Operation

This example investigates the performance of the buck converter in both CCM operation and DCM operation, whose control is designed for one particular CCM operation. For this purpose, the buck converter used in Example 8.3 is revisited and its performance is evaluated over a wide operational range to include both CCM and DCM. As demonstrated in Example 8.3, the control design is optimized for the CCM operation with $R = 1\ \Omega$. The critical resistance of the buck converter is determined as $R_{crit} = 2L/(D'T_s) = 5.33\ \Omega$. Thus, the converter crosses the CCM/DCM boundary at $R = R_{crit} = 5.33\ \Omega$.

Figure 8.40 shows the loop gain characteristics of the buck converter evaluated for both CCM operation and DCM operation. For CCM operation, $R = 0.5\ \Omega$, $R = 1\ \Omega$, and $R = 2\ \Omega$ are used. On the other hand, $R = R_{crit} = 5.33\ \Omega$, $R = 10\ \Omega$, and $R = 20\ \Omega$ are selected for DCM operation. The other conditions are the same as those of Example 8.3. The mid-band gain reduction in $|G_{vd}|$ flows into the DCM loop gain characteristics, producing much narrower crossover frequencies. Nonetheless,

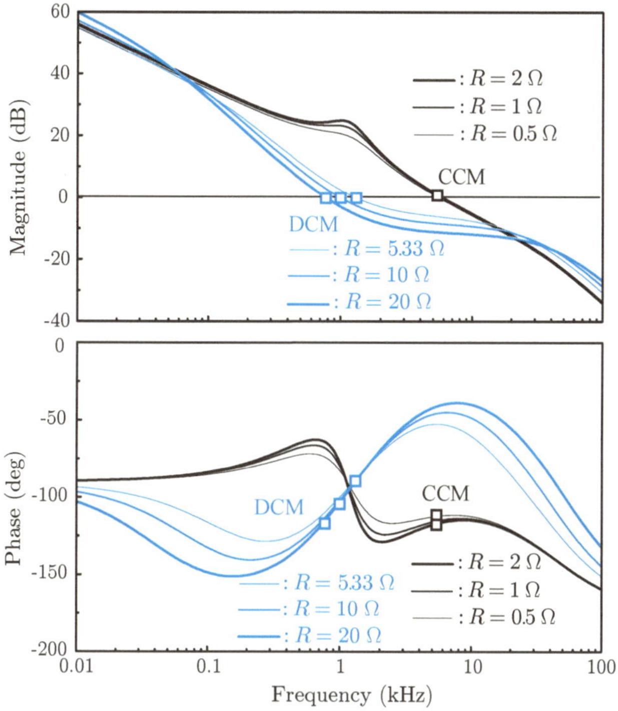

Figure 8.40 Loop gain characteristics of buck converter.

the loop gains demonstrate that the converter remains stable, even with a larger phase margin, in DCM operation. This justifies the validity of the control design — *the control design offering stable CCM operation also guarantees stability in DCM operation.*

The reduced mid-band gain and narrower crossover frequency only provide small attenuation for closed-loop transfer functions. Figure 8.41 compares the output impedances evaluated with the same conditions. As predicted, the DCM output impedances show inferior characteristics with much larger $|Z_o|_{peak}$. In addition, the first pole of the DCM output impedance, labeled as ω_{pz} in Fig. 8.41, occurs at much lower frequencies.

The degradation in the output impedance in DCM operation is the predicted consequential effect of using the voltage feedback compensation designed for CCM operation. The control might be initially optimized for the DCM operation for the improved output impedance characteristics. However, for that case, the converter becomes unstable when it enters CCM operation.

As analyzed in Section 8.4.6, there are direct correlations between the output impedance and step load response. The inferior output impedance in DCM would deteriorate the step load response. The larger $|Z_o|_{peak}$ increases the transitional deviation of the output voltage, while the lower ω_{pz} slows down the transient response. Figure 8.42 compares the output voltage of the converter in response to step

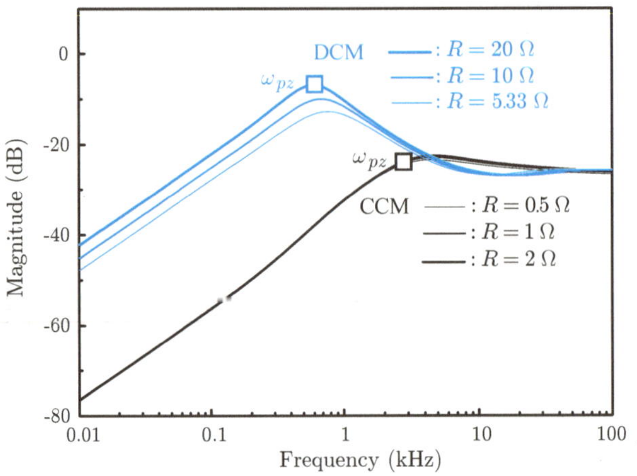

Figure 8.41 Output impedance characteristics of buck converter.

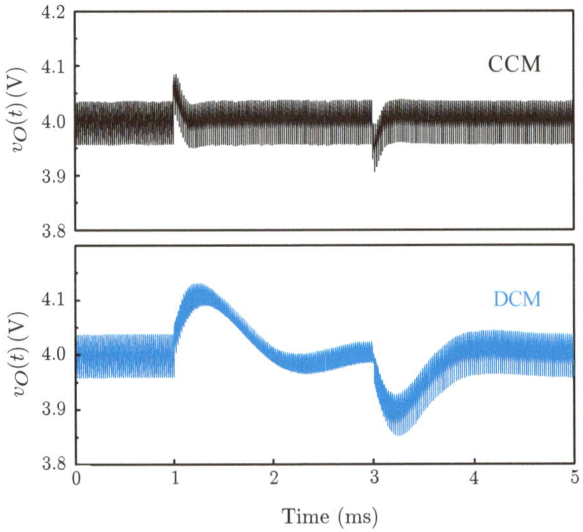

Figure 8.42 Step load responses of buck converter.

load changes in CCM operation and DCM operation. For CCM operation, the step changes of $R = 0.8\ \Omega \Rightarrow 1\ \Omega \Rightarrow 0.8\ \Omega$ are introduced, yielding $\Delta I_{step} = 1$ A. On the other hand, the step changes of $R = 5.3\ \Omega \Rightarrow 20\ \Omega \Rightarrow 5.3\ \Omega$ are exercised in DCM operation, thus producing $\Delta I_{step} = 0.55$ A. As anticipated from the output impedance characteristics, the DCM operation shows a sluggish response with larger overshoot and undershoot.

Although originated from the fluctuation of the power stage dynamics, the performance change over the CCM/DCM boundary is also related with the control scheme of the converter. The performance degradation in DCM operation is an innate limitation of voltage mode control. For voltage mode control, the fluctuation of the power stage dynamics directly propagates into the closed-loop performance. The converter remains stable, however, other closed-loop performance will be degraded in DCM operation.

As introduced in the previous section, there exists an alternative control scheme for PWM converters, current model control, which employs an additional feedback from the inductor current. It has been demonstrated that current mode control effectively alleviates the sensitivity of the converter dynamics and could offer near uniform loop gain characteristics for both CCM operation and DCM operation. The next chapter addresses this feature of current mode control, along with other details about the control scheme.

8.6 CHAPTER SUMMARY

This chapter presented the dynamic analysis and feedback design of closed-loop controlled PWM dc-to-dc converters. The graphical asymptotic analysis method is used to illustrate the results of the closed-loop analysis, principles of the feedback compensation design, and impacts of the compensation parameters. The asymptotic analysis has emerged as a systematic and viable means for the dynamic analysis and control design. This method provides a graphical process for the compensation design and closed-loop analysis, and also yields factorized equations for closed-loop transfer functions.

The three-pole two-zero circuit is identified as the optimal structure for the voltage feedback compensation for the buck converter. The relationships between the compensation parameters and performance metrics are investigated using the asymptotic analysis method. Straightforward design procedures for the voltage feedback compensation are established.

The frequency-domain analysis based on the small-signal model is an approximation and the results should be interpreted with care. In particular, the small-signal model does not account for the high-frequency phase lag, incurred by the nonlinear time-variant dynamics of the PWM process. The small-signal model thus underestimates the phase delay and the actual phase characteristics could be worse than the predictions of the small-signal model. The error steadily grows as the evaluation frequency increases towards the switching frequency. This fact should be incorporated in the design and analysis of the converter loop gain, in order *not* to misinterpret the phase characteristics of the loop gain. Especially, the loop gain crossover frequency should be located at the frequencies sufficiently lower than the switching frequency, typically $10\% - 30\%$ of the switching frequency, for the accuracy of the stability margins. Furthermore, the additional phase delay, which is not accounted for in the small-signal model, should be considered in assessing the phase margin.

The asymptotic analysis only uses the magnitude asymptotes while ignoring phase characteristics. This method is acceptable for most cases because the phase characteristics in general do not significantly affect the asymptotic analysis. However, there is one exception for this general trend. At the frequencies near the loop gain crossover, where $|T_m| \approx 1$, the underlying assumption of $|T_m| \gg 1$ or $|T_m| \ll 1$ is infringed and the closed-loop transfer function is strongly affected by the phase margin. In particular, insufficient phase margins induce a peaking in transfer functions, which the asymptotic analysis cannot predict. To avoid an excessive peaking, the phase margin of $45° - 70°$ is recommended as general guideline.

It was demonstrated that the results of the frequency-domain analysis could be used to predict and optimize time-domain transient responses. This chapter presented the three examples of such analyses.

1) The peaking in transfer functions, which appears at the loop gain crossover frequency when the phase margin is insufficient, is translated to a damped oscillation in transient responses. The frequency of the oscillation is the same as the frequency of the peaking.

2) The location of the first compensation zero dictates the speed of the step input response.

3) Finally, the location of the second compensation zero governs the speed of the step load response, while the peak value of the output impedance determines the magnitude of the output voltage excursion.

Although the above conclusions are drawn from the analyses using the small-signal model derived under the small-signal assumption, the results show good correlation with the transient responses produced by relatively large changes in operational conditions: for example, up to the 50% change in the load current or input voltage.

Impacts of DCM operation on the control design and converter performance are investigated. It was verified that the control design intended for CCM operation guarantees stability in DCM operation and therefore suited for both CCM operation and DCM operation. In contrast, the control design tailored for DCM operation usually becomes unstable when the converter encounters CCM operation. Accordingly, the control design should be based on CCM operation even if the converter occasionally or mainly remains in DCM operation.

The feedback control scheme covered in this chapter is not suitable for the boost and buck/boost converters which have an RHP zero in their power stage transfer functions. For boost and buck/boost converters, an advanced feedback control technique, called current mode control, needs to be employed. Current mode control utilizes an additional feedback from the inductor current, as well as the output voltage feedback. Current mode control is now widely adapted to PWM dc-to-dc converters due to its inherent advantages.

In contrast to current mode control, the control scheme covered in this chapter is called voltage mode control because it only utilizes the output voltage as the single feedback signal. Although voltage mode control has currently limited applications to buck or buck-derived isolated converters, it has profound significance in the realm of the PWM converter modeling and analysis. Before the advent of current mode control, voltage mode control was virtually the only available control scheme for PWM converters. Due to this long-time prevalence, the modeling and analysis of PWM dc-to-dc converters have advanced centering around voltage mode control. As the outcomes of decades of research efforts, many important and valuable results are established for voltage mode control. Most of these achievements can be directly, or with minor modifications, adapted to current mode control. Current mode control will be treated in great detail in the following two chapters.

REFERENCES

1. R. W. Erickson and D. Maksimović, *Fundamentals of Power Electronics*, Kluwer Academic Publishers, 2001.
2. *PSIM User's Guide*, Powersim Inc., Version 7.0, Mar. 2006.
3. B. H. Cho and F. C. Lee, *VPEC Power Electronics Professional Seminar*, Course I–Control Design, VPEC, 1991.

PROBLEMS

8.1 Consider the asymptotic plots for $|T|$ and $|G|$ shown in Fig. P8.1.

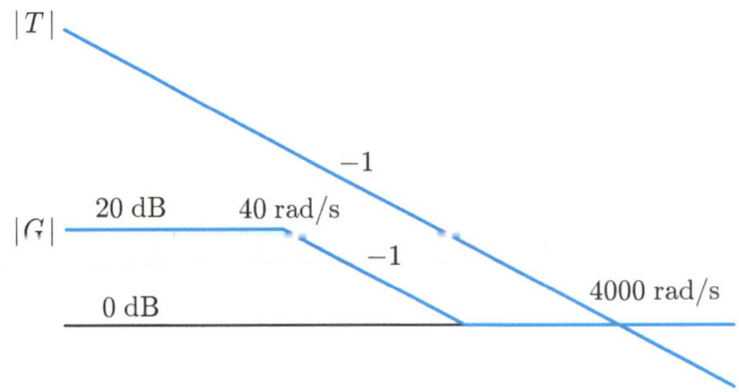

Fig. P8.1

a) Use the asymptotic analysis method to find an expression for $F(s) = G(s)/(1+T(s))$.
b) Directly evaluate $F(s) = G(s)/(1+T(s))$ to find the expression for $F(s)$.

8.2* Figure P8.2 illustrates the asymptotic analysis of the relationship $F(s) = G(s)/(1+T(s))$ with $G(s) = K_g < 1$ and $T(s) = \omega_c(1+s/\omega_z)/(s(1+s/\omega_p))$ with $\omega_p < \omega_z$. Based on the asymptotic analysis, $F(s)$ is approximated as

$$F(s) = K_f s \frac{1+s/\omega_p}{(1+s/\omega_z)(1+s/\omega_c')}$$

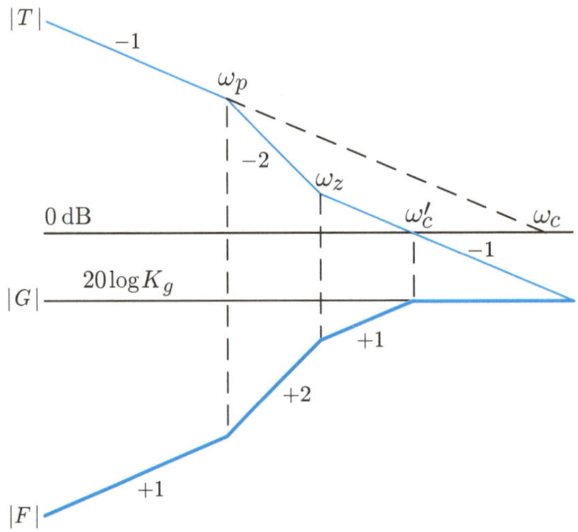

Fig. P8.2

a) Derive an expression for the differentiation gain K_f based on $|F(j\omega_p)|$.
b) Derive another expression for K_f using $|F(j\infty)|$.
c) Referring to $|F(j\omega)|$ plot, verify that the results of a) and b) are identical.

8.3* Figure P8.3 shows the circuit diagram of a closed-loop controlled buck converter.

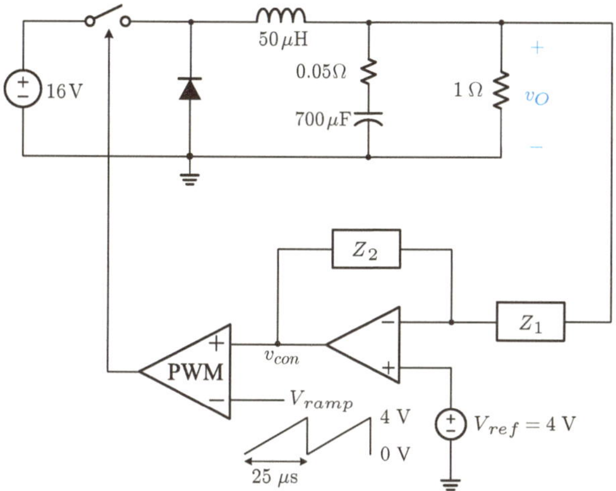

Fig. P8.3

a) Sketch the asymptotic plots for $|G_{vs}|$, $|G_{vd}|$, and $|Z_p|$. Show the values for the low- and high-frequency asymptotes and corner frequencies on your plots.

b) Assume that a single integration function, $F_v(s) = Z_2(s)/Z_1(s) = K_v/s$, is employed as the voltage feedback compensation. Argue that the converter becomes marginally stable when the 0 dB crossover frequency of the loop gain occurs at the power stage double pole. Based on your argument, find the value for K_v that makes the converter marginally stable.

c) Assume that the following three-pole two-zero circuit is now employed as the voltage feedback compensation

$$F_v(s) = \frac{K_v}{s} \frac{\left(1 + \dfrac{s}{\omega_{z1}}\right)\left(1 + \dfrac{s}{\omega_{z2}}\right)}{\left(1 + \dfrac{s}{\omega_{p1}}\right)\left(1 + \dfrac{s}{\omega_{p2}}\right)}$$

i) Select the numerical values for ω_{z1}, ω_{z2}, ω_{p1}, and ω_{p2}, based on the compensation design procedures discussed in Section 8.4.2.

ii) Determine the integrator gain K_v in order to place the loop gain crossover frequency at the esr zero, while using the feedback compensation parameters determined in i).

iii) Construct asymptotic plots for $|T_m|$, $|G_{vs}|$, and $|A_u|$, based on the results of i) and ii). Find the expression for the audio-susceptibility in a factorized form. Specify the corner frequencies and leading coefficient of the audio-susceptibility. Evaluate the peak value of $|A_u|$.

8.4* The loop gain of a closed-loop controlled buck converter is shown in Fig. P8.4. Now assume that the three-pole two-zero compensation is used for the voltage feedback compensation. The feedback compensation parameters are selected as outlined in Section 8.4.2.

$$F_v(s) = \frac{K_v}{s} \frac{\left(1+\frac{s}{\omega_{z1}}\right)\left(1+\frac{s}{\omega_{z2}}\right)}{\left(1+\frac{s}{\omega_{p1}}\right)\left(1+\frac{s}{\omega_{p2}}\right)} \quad \text{with} \quad K_v = 2500$$

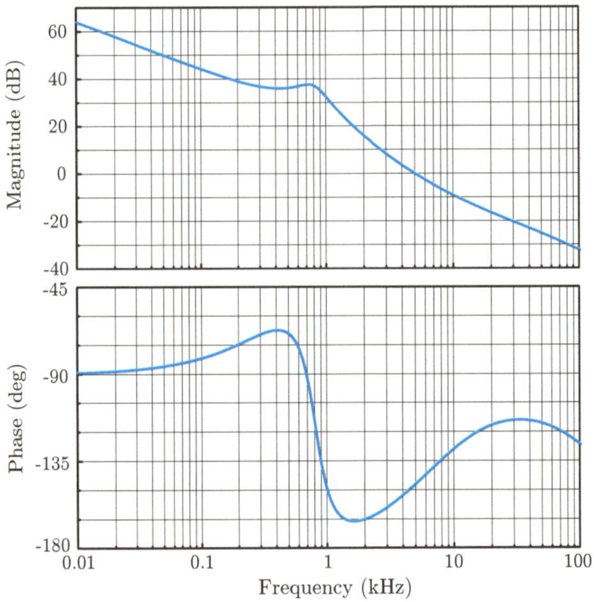

Fig. P8.4

a) Estimate the location of the power stage double pole.
b) Estimate the value for the first compensation zero ω_{z1}.
c) Find the 0 dB crossover frequency and phase margin of the loop gain.
d) Sketch the polar plot of the loop gain. Show the 0 dB crossover frequency and phase margin on your plot.
e) Assume that the input voltage of the converter is $V_S = 12$ V. Find the modulator gain F_m of the PWM block. The current integrator gain of the voltage feedback compensation is $K_v = 2500$.
f) Now assume that the integrator gain is varied between $250 < K_v < 25000$, while the other compensation parameters remain the same. The current integrator gain is $K_v = 2500$. The change in the integrator gain will alter the loop gain characteristics. Find the range in the 0 dB crossover frequency

f_c and phase margin ϕ_m of the loop gain. Arrange your answer in the form of ( ) $< f_c <$ ( ) and ( ) $< \phi_m <$ ( ).

8.5 Shown in Fig. P8.5 is the circuit diagram of a closed-loop controlled buck converter.

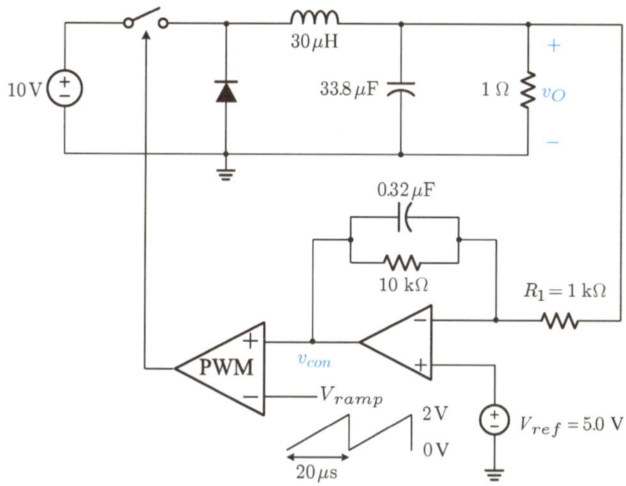

Fig. P8.5

a) Sketch the asymptotic plot for the loop gain magnitude. Show the values for the low-frequency asymptote, crossover frequency, and corner frequencies of the loop gain.
b) Find the average value for v_{con} and v_O. Specify your answer up to the three significant digits.
c) Sketch the asymptotic plot for the magnitude of the audio-susceptibility. Show the low-frequency asymptote value, peak value, and corner frequencies of the audio-susceptibility. Use the results of a) and b).
d) Find the new value for R_1 that makes the converter marginally stable.

8.6* Figure P8.6 is the asymptotic magnitude plot for the loop gain of a buck converter. The buck converter employs the three-pole two-zero compensation, thus yielding the loop gain expression of

$$T_m(s) = \frac{K_t}{s} \frac{\left(1 + \frac{s}{\omega_{esr}}\right)\left(1 + \frac{s}{\omega_{z1}}\right)\left(1 + \frac{s}{\omega_{z2}}\right)}{\left(1 + \frac{s}{Q\omega_o} + \frac{s^2}{\omega_o^2}\right)\left(1 + \frac{s}{\omega_{p1}}\right)\left(1 + \frac{s}{\omega_{p2}}\right)}$$

with $K_t = V_S F_m K_v$

a) Explain the reason for selecting $\omega_{z1} < \omega_o$.
b) Describe the considerations for choosing the position of ω_{z1}.

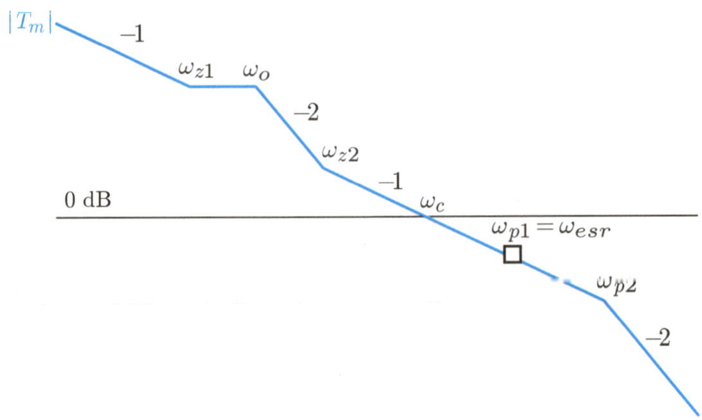

Fig. P8.6

Table P8.6

	Benefit of moderate change	Problem of excessive change
Increase in ω_{z1}	A faster step input response	( )
Increase in ω_{z2}	( )	( )
Decrease in ω_{p2}	( )	( )
Increase in K_v	( )	( )

c) Describe the considerations for selecting the location of ω_{z2}.

d) Explain the reason for choosing $\omega_{p1} = \omega_{esr}$.

e) Why the step load response is always faster than the step input response?

f) Selection of the compensation parameters affect the performance of the converter in various ways. Table P8.6 summarizes the impacts of changes in the value of the specific compensation parameters. Fill in the blanks in the table.

8.7 Figure P8.7 shows the loop gain of a closed-loop controlled buck converter.

a) Determine stability of the converter.

b) Sketch the polar plot of the loop gain. Show all the prominent features of the loop gain.

c) Assume that the integrator gain K_v of the voltage feedback compensation is increased by 10 times. The other compensation parameters remain the same. What response is expected in the converter output voltage?

d) Assume that the integrator gain K_v of the voltage feedback compensation is decreased by 10 times. The other compensation parameters remain the same. What response is expected in the converter output voltage?

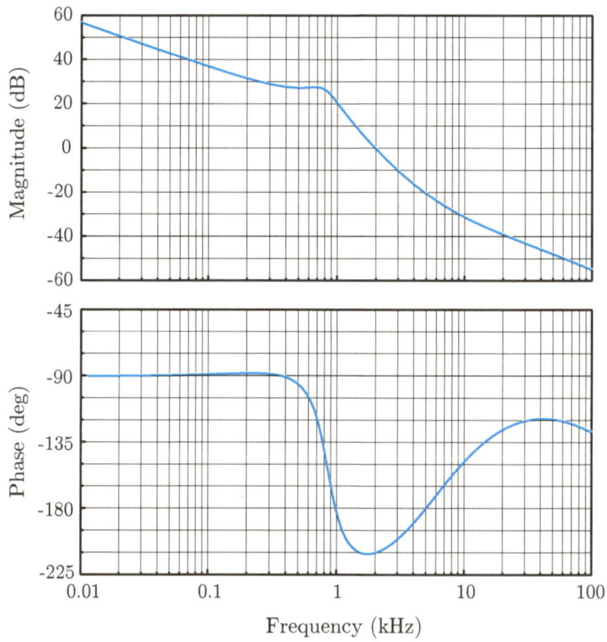

Fig. P8.7

e) There are two methods to stabilize the converter by only changing the magnitude of the integrator gain K_v. Describe the two methods. Do you think any of these method is desirable? Justify your answer.

8.8* Shown in Fig. P8.8 is the circuit diagram of a closed-loop controlled buck converter. Answer the questions.

a) Sketch the asymptotic plots for $|G_{vs}|$, $|G_{vd}|$, and $|Z_p|$. Show the values for the low- and high-frequency asymptotes. Also show the corner frequencies and slopes of the asymptotic plots.

Assume the following three-pole two-zero circuit for the voltage feedback compensation

$$F_v(s) = \frac{1.2 \times 10^4 \left(1 + \dfrac{s}{5.7 \times 10^3}\right)\left(1 + \dfrac{s}{1.1 \times 10^4}\right)}{s\left(1 + \dfrac{s}{4.0 \times 10^4}\right)\left(1 + \dfrac{s}{2.5 \times 10^5}\right)}$$

The integrator gain is selected to place the 0 dB crossover frequency at the esr zero.

b) Sketch the asymptotic plots for $|T_m|$, $|G_{vs}|$, and $|A_u|$. Show the corner frequencies and slopes of the asymptotic plots.

c) Sketch the asymptotic plots for $|T_m|$, $|Z_p|$, and $|Z_o|$. Show the corner frequencies and slopes of the asymptotic plots.

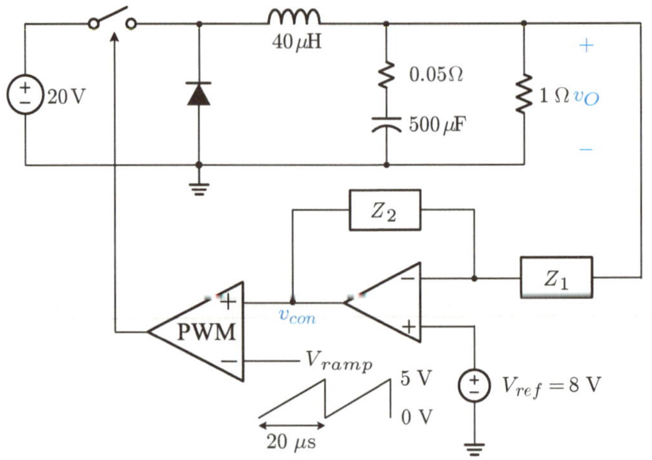

Fig. P8.8

8.9 The loop gain of a closed-loop controlled buck converter is shown in Fig. P8.9.

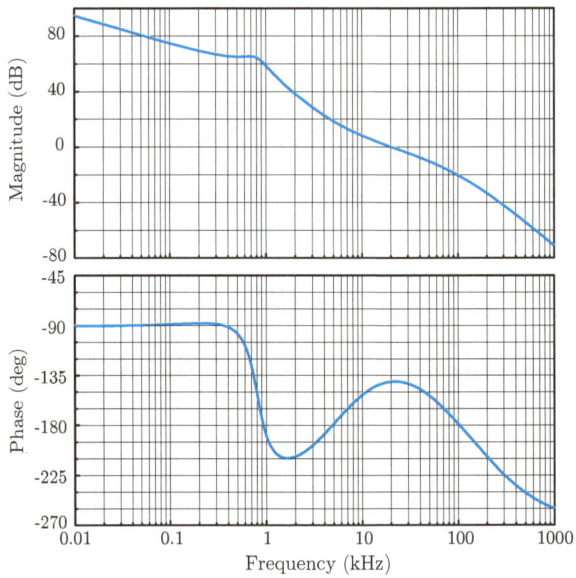

Fig. P8.9

a) Determine the stability of the converter.

b) Sketch the polar plot of the loop gain. Show all the prominent features of the polar plot.

Now assume the three-pole two-zero circuit is used for the voltage feedback compensation

$$F_v(s) = \frac{K_v}{s}\frac{\left(1+\dfrac{s}{\omega_{z1}}\right)\left(1+\dfrac{s}{\omega_{z2}}\right)}{\left(1+\dfrac{s}{\omega_{p1}}\right)\left(1+\dfrac{s}{\omega_{p2}}\right)}$$

The current integrator gain is $K_v = 2500$. However, the integrator gain K_v will be varied in the following problems.

c) Find the ranges for K_v that destabilize the converter.

d) Find the value for K_v that produces a sustained oscillation at $\omega = 2\pi \cdot 10^5$ rad/s.

8.10* The audio-susceptibility $A_u(s)$ of a closed-loop controlled converter is given by $A_u(s) = G_{vs}(s)/(1+T_m(s))$ where $G_{vs}(s)$ is the input-to-output transfer function and $T_m(s)$ is the loop gain. Shown in Fig. P8.10 are asymptotic plots for $|G_{vs}|$ and $|T_m|$ of a closed-loop controlled converter.

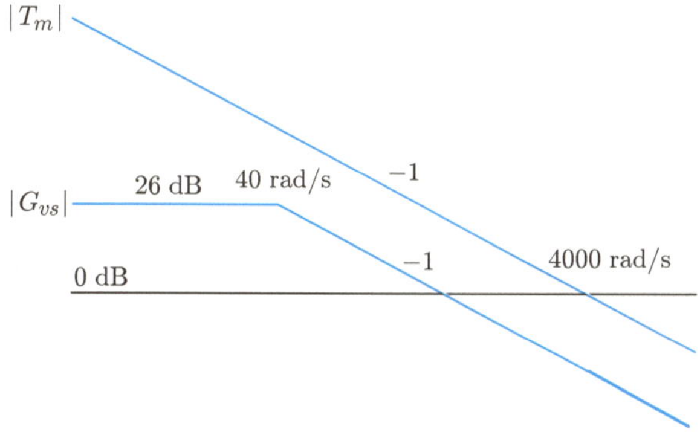

Fig. P8.10

a) Use the asymptotic analysis method to find an expression for $A_u(s)$.

b) Directly evaluate the relationship $A_u(s) = G_{vs}(s)/(1+T_m(s))$ to find the expression for $A_u(s)$.

c) Now assume that a step increase of $V_{step} = 10$ V is occurred in the input voltage. Find the expression for the transitional waveform of the output voltage v_O. Draw the general shape of v_O and show all the important features of v_O.

8.11* Consider Fig. P8.11 as the asymptotic plots for the load current-to-output transfer function $|Z_p|$ and the loop gain $|T_m|$ of a closed-loop controlled converter.

a) Sketch the asymptotic plot of the closed-loop output impedance $|Z_o|$. Show the peak value and corner frequencies of the asymptotic plot.

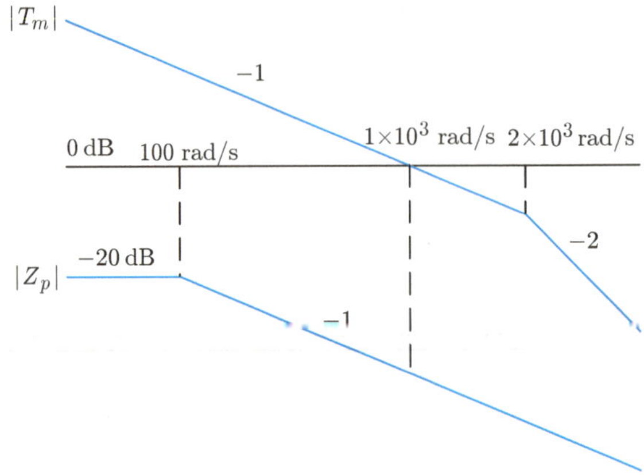

Fig. P8.11

b) Derive an expression for $Z_o(s)$.

c) Find an expression for the output voltage in response to a 0.1 A step decrease in the load current.

8.12* Shown in Fig. P8.12 is the circuit diagram of a closed-loop controlled buck converter. Assume that the following three-pole two-zero circuit is employed as the voltage feedback compensation

$$F_v(s) = \frac{Z_2(s)}{Z_1(s)} = \frac{K_v}{s} \frac{(1 + s/\omega_{z1})(1 + s/\omega_{z2})}{(1 + s/\omega_{p1})(1 + s/\omega_{p2})}$$

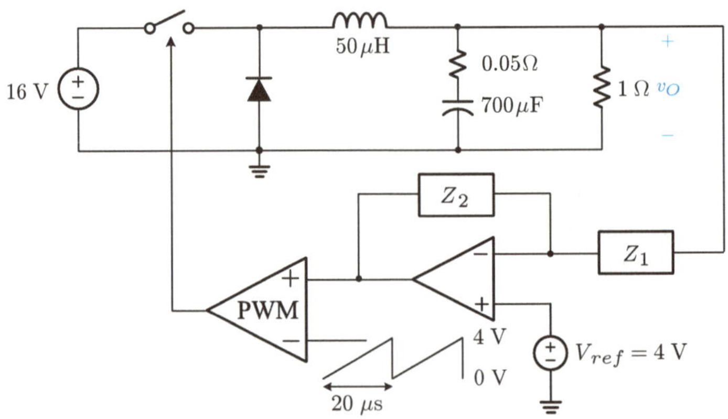

Fig. P8.12

a) Determine the parameters ω_{z1}, ω_{z2}, ω_{p1}, ω_{p2}, and K_v of the voltage feedback compensation $F_v(s)$ to meet the following design specifications

i) Settling time of the step input response: 0.7 ms

ii) Settling time of the step load response: 0.3 ms
iii) 0 dB crossover frequency of the loop gain $T_m(s)$ at the esr zero of the power stage
iv) 40 dB attenuation of the loop gain magnitude at the switching frequency: $|T_m(j\omega_s)| = -40$ dB

b) Based on your compensation design in a) and power stage parameters, sketch the asymptotic plots for $|T_m|$, $|G_{vs}|$, and $|A_u|$. Show the corner frequencies and slopes of the transfer functions. Show the peak value of the audio-susceptibility: $|A_u|_{peak}$. Find the expression for the audio-susceptibility, $A_u(s)$.

8.13* Consider the asymptotic plots of $|T_v|$ and $|T_i|$ shown in Fig. P8.13.

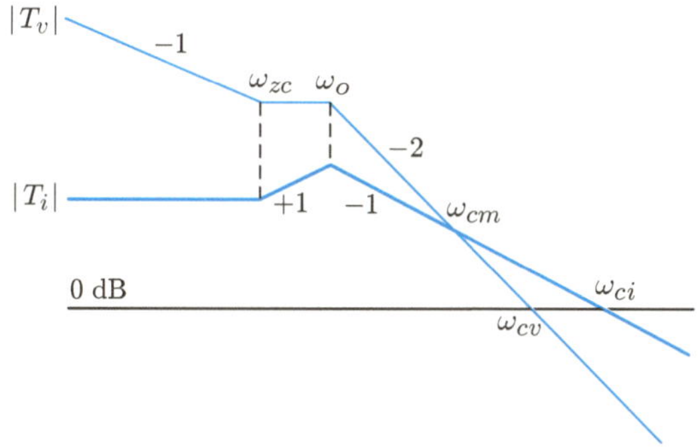

Fig. P8.13

Three different loop gains are defined using $T_v(s)$ and $T_i(s)$ shown above.

$$T_1(s) = \frac{T_v(s)}{1+T_i(s)} \qquad T_2(s) = \frac{T_i(s)}{1+T_v(s)} \qquad T_3(s) = T_v(s) + T_i(s)$$

a) Construct the asymptotic plot of $|T_1|$ and find the analytical expression of $T_1(s)$. Find the 0 dB crossover frequency and phase margin of $T_1(s)$.

b) Construct the asymptotic plot of $|T_2|$ and find the analytical expression of $T_2(s)$.

c) Prove that the absolute stability of the two converters, one having $T_1(s)$ as its loop gain and the other having $T_2(s)$ as its loop gain, is the same.

d) Sketch the polar plots of $T_1(s)$ and $T_2(s)$. Judge stability of the two converters, one having $T_1(s)$ as its loop gain and the other having $T_2(s)$ as its loop gain, based on your polar plot. Justify your answer.

e) Sketch the asymptotic plot of $|T_3|$. Evaluate the 0 dB crossover frequency, phase margin PM, and gain margin GM of $T_3(s)$. Also, find the analytical expression for $T_3(s)$.

8.14 Consider the relation of

$$Z_o(s) = \frac{Z_p(s)}{1 + T_m(s)}$$

with $T_m(j\omega) = |T_m|\angle(-180° + \theta)$.

a) Find the exact expression for $|Z_o| = |Z_p|/|1 + T_m|$ as a function of $|Z_p|$, $|T_m|$, and θ.

b) Use the result of a) to evaluate $|Z_o|$ for the following two cases:
i) $|T_m| = 1$ ii) $\theta = 0°$

c) Describe the use of b) in the small-signal analysis of PWM converters.

8.15* Figure P8.15 shows the circuit diagram of a buck converter and the Bode plots of two different load current-to-output transfer functions of the converter, each with a distinct operational condition. The load current-to-output transfer function $v_o(s)/i_o(s)_{\mathrm{A}}$ is obtained with the switch SW at A Position, while $v_o(s)/i_o(s)_{\mathrm{B}}$ is obtained with SW at B position.

a) Find the values for R_l, L, R_c, and C.

b) Assume the switch SW is at A position. Find the expression for v_O when $i_O = \sin 2\pi \cdot 0.1 \times 10^3 t + \sin 2\pi \cdot 4 \times 10^3 t$. Ignore the switching ripple component in v_O.

c) Evaluate the 0 dB crossover frequency and phase margin of the loop gain of the converter.

d) Find the expression for $v_o(s)/i_o(s)_{\mathrm{B}}$ as accurately as possible.

e) Assume the switch SW is at B position. Find the expression for v_O when $i_O = \sin 2\pi \cdot 100t + \sin 2\pi \cdot 4 \times 10^3 t$. Ignore the switching ripple component.

f) Assume the switch SW is at B position. Sketch v_O when a 5 A step increase is occurred in $i_O(t)$: $i_O = 5u(t)$.

g) Sketch the asymptotic plot of the loop gain magnitude between 100 Hz $< f < 4 \times 10^3$ Hz. Show the slopes and corner frequencies. Mark the loop gain magnitude at 100 Hz: $|T_m(2\pi \cdot 100)|$.

8.16 The loop gain T_m of a closed-loop controlled dc-to-dc converter is shown in Fig. P8.16.

a) Find the gain margin and phase margin of the converter.

b) As discussed in Section 8.4.4, the closed-loop transfer function exhibits a peaking at $\omega = \omega_c$ where $|T_m|$ crosses the 0 dB line. For the given system, evaluate the size of peaking in dB scale.

c) It can be shown that the closed-loop transfer function also shows a peaking at $\omega = \omega_c'$ where $\angle T_m$ becomes $-180°$. Derive an equation that describes the magnitude of the closed-loop transfer function at ω_c' and evaluate the size of the peaking in dB scale.

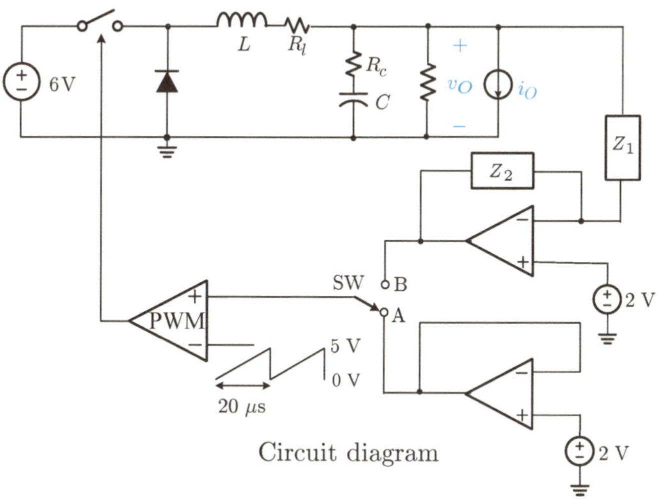

Circuit diagram

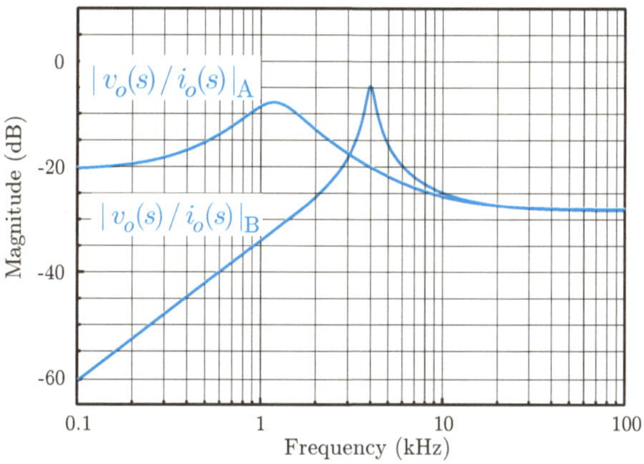

Bode plot

Fig. P8.15

8.17* Shown in Fig. P8.17 are the small-signal block diagram of a closed-loop controlled buck converter and the asymptotic plot for its loop gain $|T_m|$. Assume the following transfer functions for the gain blocks in the block diagram:

$$G_{vd}(s) = \frac{50\left(1+\dfrac{s}{2\times 10^4}\right)}{\left(1+\dfrac{s}{400}\right)\left(1+\dfrac{s}{4\times 10^5}\right)} \qquad Z_p(s) = \frac{10\left(1+\dfrac{s}{8\times 10^4}\right)}{1+\dfrac{s}{400}}$$

and $F_m = 0.2$

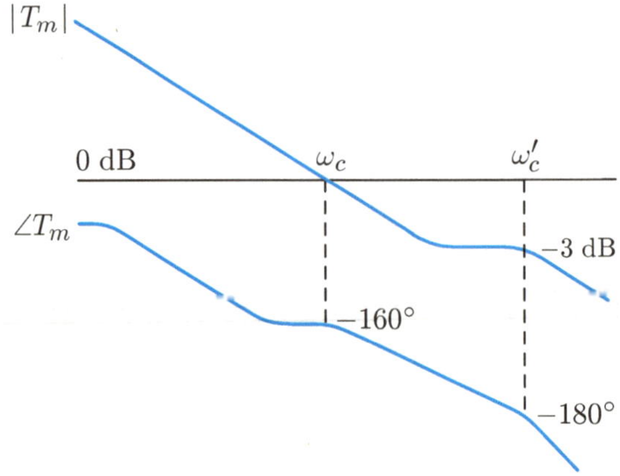

Fig. P8.16

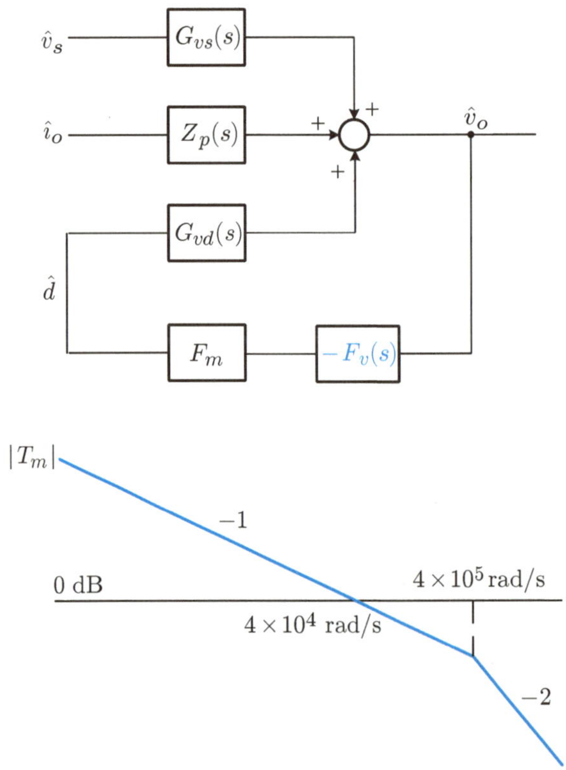

Fig. P8.17

a) Estimate the phase margin and gain margin of the converter.
b) Find an expression for the voltage feedback compensation $F_v(s)$.
c) Sketch the asymptotic plot for the output impedance $|Z_o| = |Z_p|/|1+T_m|$ and find the equation for $Z_o(s)$.

8.18 Figure P8.18 shows the loop gain characteristics of a closed-loop controlled buck converter. Assume the standard three-pole two-zero circuit for the voltage feedback compensation.

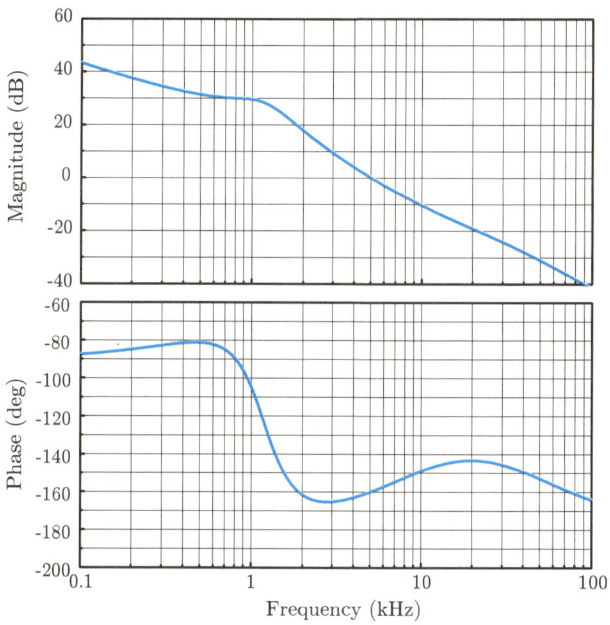

Fig. P8.18

$$F_v(s) = \frac{K_v}{s} \frac{\left(1+\dfrac{s}{\omega_{z1}}\right)\left(1+\dfrac{s}{\omega_{z2}}\right)}{\left(1+\dfrac{s}{\omega_{p1}}\right)\left(1+\dfrac{s}{\omega_{p2}}\right)} \quad \text{with} \quad K_v = 1500$$

a) Evaluate the crossover frequency and phase margin of the loop gain.
b) Describe the characteristic features of the output impedance with the current design.
c) Explain the characteristic features of the step load response with the current design.
d) Find the new value for K_v that offers a $60°$ phase margin. The current value of K_v is 1500. Assume the other compensation parameters remain unchanged. Although the new integrator gain provides the desired phase

margin, the resulting design is considered unacceptable. State the reason for this.

e) Find the new value for K_v that offers the largest possible phase margin while locating the crossover frequency beyond the current position. Assume the other compensation parameters remain the same.

f) It is not feasible to practically obtain the largest phase margin evaluated in e). Explain the reason for your answer and state a preferred design strategy.

8.19* Figure P8.19 shows the loop gain plots of a closed-loop controlled buck converter. Assume the operational conditions of $V_S = 16\,\text{V}$, $V_O = 4\,\text{V}$, $I_O = 4\,\text{A}$, and $f_s = 100$ kHz.

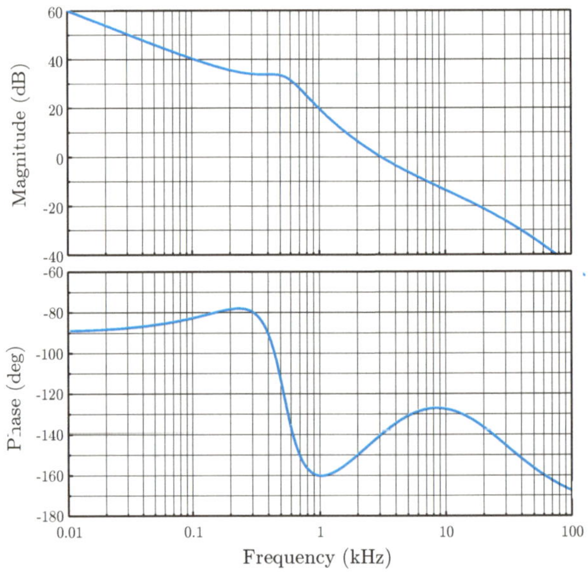

Fig. P8.19

a) Find the 0 dB crossover frequency ω_c and phase margin PM of the converter.

b) Convert the Bode plot into the polar plot. Show the PM and ω_c on the polar plot.

c) Find the magnitude of the audio-susceptibility at $\omega = 2\pi \cdot 100$, $|A_u(j2\pi \cdot 100)|$.

The converter employs a three-pole two-zero compensation with the integrator gain of $K_v = 2 \times 10^4$.

d) Find the height of the PWM ramp signal.

e) Now assume that the integrator gain is varied between $2 \times 10^4 \cdot 10^{-20/20} < K_v < 2 \times 10^4 \cdot 10^{10/20}$ while the other compensation parameters remain the same. Find the range in the 0 dB crossover frequency ω_c and the phase margin PM.

f) You are asked to destabilize the converter by changing one compensation parameter. What would you do to make the converter unstable?

8.20* The Bode plot of the input-to-output transfer function $|G_{vs}|$ and audio-susceptibility $|A_u|$ of a closed-loop controlled buck converter are shown in Fig. P8.20.

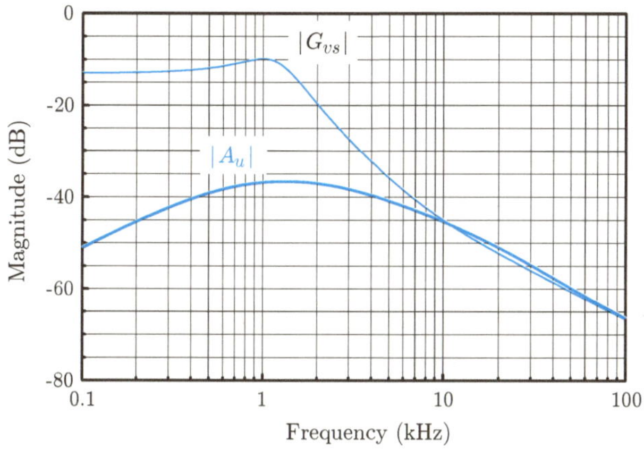

Fig. P8.20

a) Estimate the loop gain crossover frequency of the converter.
b) Estimate the duty ratio of the converter.
c) Assume that $v_S(t) = 10 + 2\sin 2\pi \cdot 400t$ is applied to the input of the converter. Find an expression for the output voltage. Ignore the switching ripple component.
d) Estimate the settling time of the output voltage due to a step input change.

8.21 The output impedance $Z_o(s)$ of a closed-loop controlled converter is given by $Z_o(s) = Z_p(s)/(1+T_m(s))$ where $Z_p(s)$ is the load current-to-output transfer function and $T_m(s)$ is the loop gain. Shown in Fig. P8.21 are the asymptotic plots for $|Z_p|$ and $|T_m|$ of the converter.

a) Find the expressions for $Z_p(s)$ and $T_m(s)$.
b) Use the asymptotic analysis method to find the expression for $Z_o(s)$.
c) Now assume that a step decrease of $I_{step} = 10$ A is occurred in the load current. Find the expression for the transitional waveform of the output voltage v_O. Sketch v_O to show all important features of the waveform. Ignore the switching ripple component.

8.22* Shown in Fig. P8.22 are the load current-to-output transfer function, $Z_p(s)$, and output impedances, $Z_o(s)$, of a closed-loop controlled buck converter. The two different output impedances are generated with the different compensation designs: Design A and Design B.

a) Consider Design A and answer the following questions.
   i) Evaluate the phase margin of the loop gain.

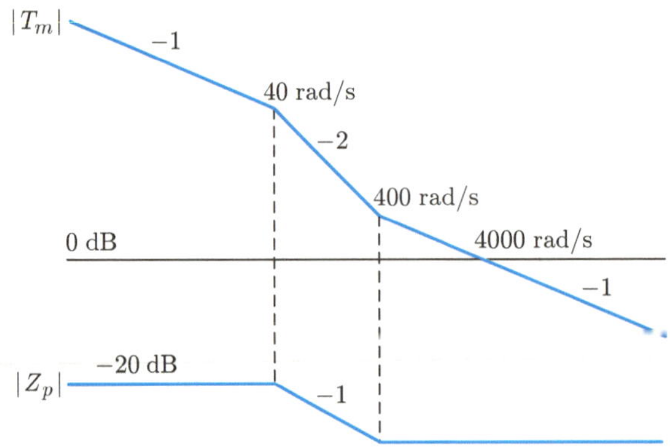

Fig. P8.21

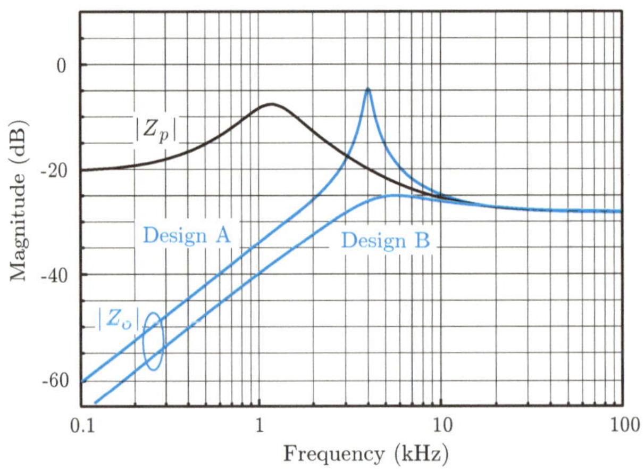

Fig. P8.22

ii) Determine the 0 dB crossover frequency of the loop gain.

iii) Sketch the transient response of the output voltage when a step decrease of $I_{step} = 3\,\text{A}$ is occurred in the load current. Show all the prominent features of the output voltage.

b) For Design B, answer the following questions.

i) Estimate the 0 dB crossover frequency of the loop gain.

ii) Evaluate the magnitude of the loop gain, $|T_m(s)|$, evaluated at $f = 1\,\text{kHz}$.

iii) Sketch the transient response of the output voltage when a step increase of $I_{step} = 3\,\text{A}$ is occurred in the load current. Specify the peak deviation and settling time of the output voltage.

8.23* Shown in Fig. P8.23 are the Bode plots of the input-to-output transfer function $|G_{vs}|$ and audio-susceptibility $|A_u|$ of another closed-loop controlled buck converter. Assume that the converter employs the same three-pole two-zero circuit discussed in the previous problem. Evaluate or illustrate the following items based on the information given in the Bode plot:

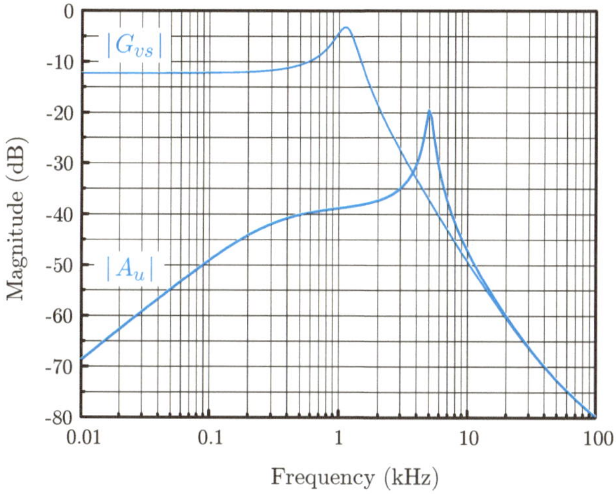

Fig. P8.23

a) the loop gain crossover frequency,
b) the phase margin of the loop gain,
c) the location of the first compensation zero ω_{z1},
d) the duty ratio of the converter D,
e) the settling time of the output voltage due to a 10 V step increase in the input voltage, and
f) the general shape of the output voltage due to a 10 V step increase in the input voltage.

CHAPTER 9

CURRENT MODE CONTROL (I) – FUNCTIONAL BASICS AND CLASSICAL ANALYSIS

The control scheme studied in Chapters 3 and 8 employs the output voltage as the only feedback signal in the process of generating the pulsewidth modulated (PWM) switch drive signal. This control scheme is called voltage mode control in the sense that the output voltage alone is involved in the PWM process. As introduced in Section 8.4.7, there exists an alternative control scheme for PWM dc-to-dc converters, called current mode control.

Current mode control refers to a class of control schemes that employs the inductor current as an additional component in the PWM process. Thus, current mode control uses both the output voltage and inductor current for the closed-loop PWM control. Current mode control is implemented in several different forms by changing the method of current sensing or the way of utilizing the sensed current. Among various current mode control schemes, the most popular is the *peak current mode control* which employs the peak value of the inductor current as a control variable.

In the previous chapters, we studied voltage mode control. All discussions about the dynamic analysis and control design were presented with regard to voltage-mode controlled PWM converters. However, modern PWM dc-to-dc converters extensively adapt current mode control rather than voltage mode control.

Though the theoretical basics acquired from voltage mode control are equally applicable to current mode control, additional knowledge is necessary for current mode control. As will be seen shortly, the principle of current mode control is rather simple and straightforward. However, its dynamic characteristics are complex and somewhat intriguing. In fact, the dynamic analysis of current mode control had been one challenging and active research topic in the late 1980s through the early 1990s.

This chapter deals with the peak current mode control, covering both functional basics and dynamic characteristics. Motivation, evolution, and implementation of the peak current mode control are first discussed. Then, the dynamic analysis and control design are covered. This chapter also investigates the closed-loop performance of peak current-mode controlled PWM converters. In particular, the design and performance of the peak current mode control for the boost and buck/boost converters, having the right-half plane (RHP) zero in their duty ratio-to-output transfer function, are studied in detail.

9.1 CURRENT MODE CONTROL BASICS

The current section presents the evolution, benefits, and issues of current mode control. While most contents are directed towards the peak current mode control, other types of current mode control are also treated in brief.

9.1.1 Evolution to Peak Current Mode Control

The concept of current mode control is illustrated in Fig. 9.1 which compares current mode control and voltage mode control, both adapted to a buck converter. Figure 9.1(a) shows voltage mode control, where the PWM is performed using the ramp signal V_{ramp} and the control voltage v_{con} derived from the voltage feedback circuit. At the onset of each operational period, the switch is turned-on and later turned-off at the instant V_{ramp} intersects v_{con}. The output voltage is regulated at the reference voltage, $V_O = V_{ref}$, by the condition $|Z_2(j0)|/|Z_1(j0)| = \infty$ in the voltage feedback circuit. The value of v_{con} is automatically adjusted to yield the required duty ratio for the output voltage regulation, $V_O = DV_S = V_{ref}$ for CCM operation.

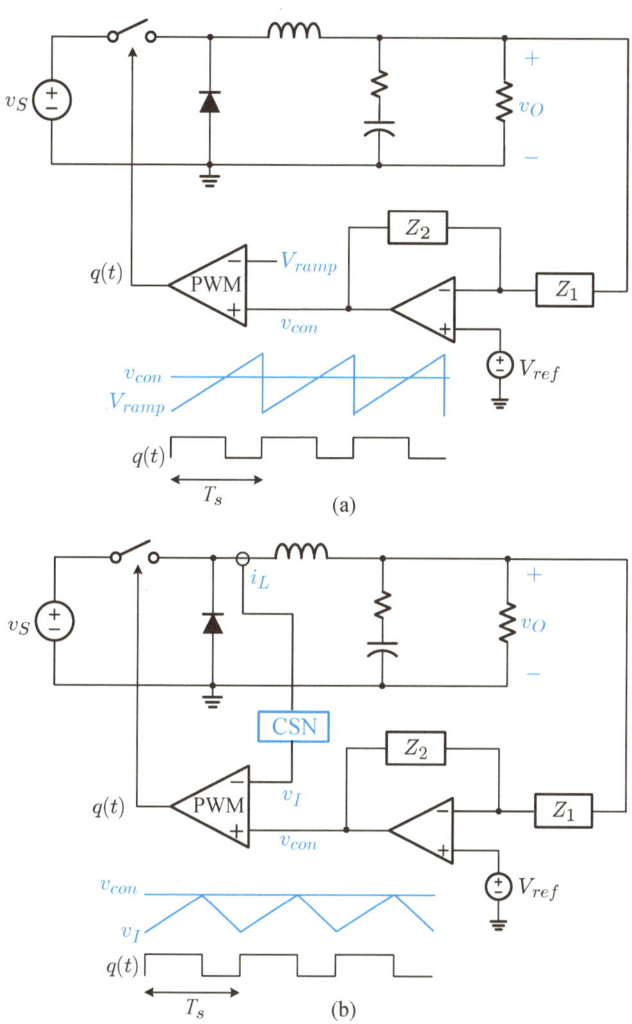

Figure 9.1 Control schemes for PWM converters. (a) Voltage mode control. (b) Current mode control.

A time-varying piecewise linear waveform is necessary for the PWM operation. In the case of voltage mode control, a ramp signal is generated inside the controller and used for the time-varying waveform. However, for the purpose of PWM, other linear waveforms can also be used. In particular, the triangular waveform can be employed for the PWM process. More so, the required triangular waveform is already available in the power stage waveforms. In all PWM converters, the inductor current linearly increases during the on-time period and decreases during the off-time period, thus becoming a triangular waveform.

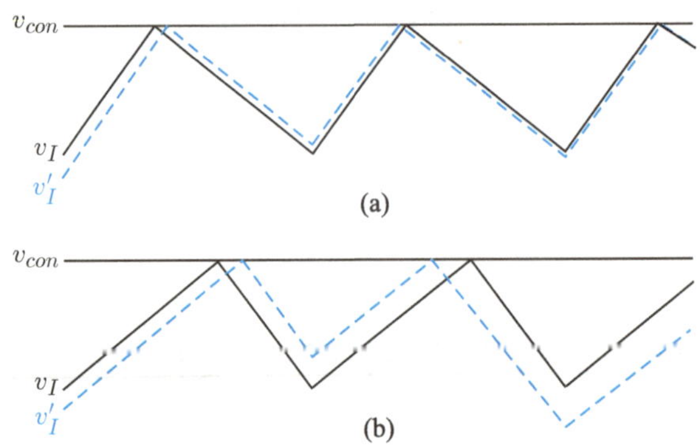

Figure 9.2 Propagation of current feedback signal disturbance. (a) Stable operation with $D < 0.5$. (b) Unstable operation with $D > 0.5$.

Figure 9.1(b) is an example of current mode control. The switch is turned-on at the beginning of a fixed switching period T_s. Once the switch is turned-on, the inductor current linearly increases. The inductor current is sensed via the current sensing network (CSN) and converted into the voltage signal, v_I in Fig. 9.1(b). The sensed voltage signal v_I is then compared against the control signal v_{con}, in order to determine the switch turn-off instant. When v_I hits v_{con}, the switch is turned-off and v_I linearly decreases. At the end of the switching period T_s, the switch is turned-on again to repeat the periodic operation.

Current mode control uses the triangular inductor current as a functional replacement of the ramp signal in voltage mode control. The structure and role of the voltage feedback circuit remain the same. The output voltage is regulated at $V_O = V_{ref}$ as far as the condition $|Z_2(j0)|/|Z_1(j0)| = \infty$ is met.

Compensation Ramp

The original motivation of current mode control was to exploit the *free* inductor current waveform in place of the *costly* ramp signal; in the past, the creation of the ramp signal was not as easy as is the case nowadays. However, it was soon found that current mode control has one critical problem. This problem is illustrated in Fig. 9.2, which shows the propagation of the disturbed current feedback signal for two different cases. Figure 9.2(a) is the case where the duty ratio is less than 0.5. The solid line v_I is the original current feedback signal, while the dashed line v'_I represents the perturbed current feedback signal. As time elapses, the distance between the two feedback signals shrinks and the initial disturbance eventually disappears; in short, the converter is stable. The case with the duty ratio larger than 0.5 is shown in Fig. 9.2(b). In contrast to the previous case, the initial disturbance successively grows, shortly developing into an erratic behavior. This unstable operation is called the sub-harmonic oscillation, referring to the nonlinear oscillation occurring at half the switching frequency. Thus, when the duty ratio D exceeds 0.5, current mode control becomes unstable and ends up with the sub-harmonic oscillation.

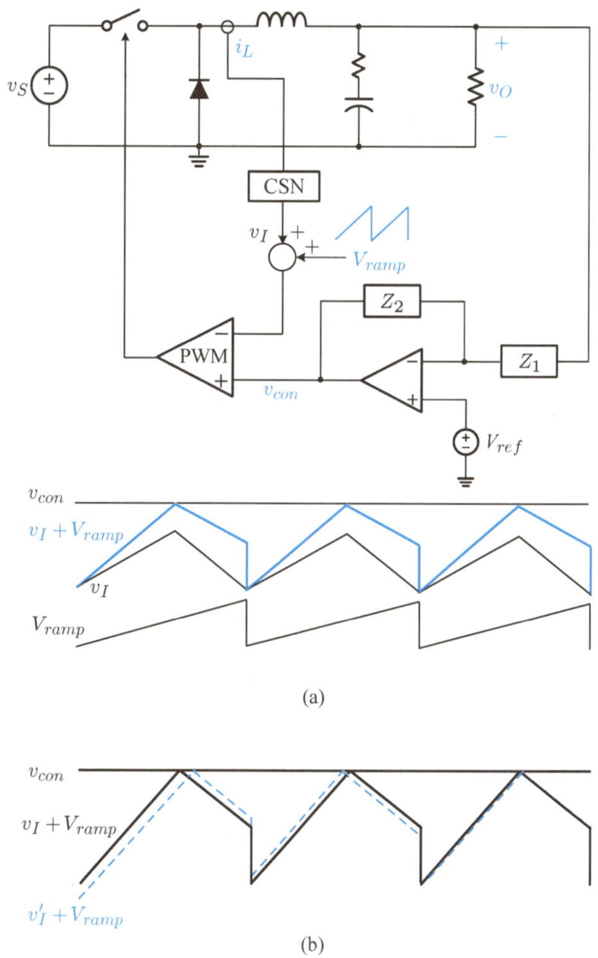

Figure 9.3 Addition of compensation ramp. (a) Control scheme. (b) Propagation of current feedback signal disturbance.

The remedy for the sub-harmonic oscillation is simple. The solution is to reintroduce the ramp signal. This situation is shown in Fig. 9.3(a), where the current feedback signal v_I is summed with V_{ramp} and the resulting signal is compared against v_{con}. Figure 9.3(b) illustrates the propagation of the disturbance in the current feedback signal. Unlike the previous case, the initial disturbance, brought in under the condition $D > 0.5$, gradually lessens until it vanishes. The ramp signal stabilizes the PWM process and the converter thus operates properly for the entire duty ratio range, $0 < D < 1$. The ramp signal employed for this purpose is called the *compensation ramp*.

Although current mode control was first conceived to remove the ramp signal in the PWM process, the compensation ramp is still necessary to avoid the sub-harmonic

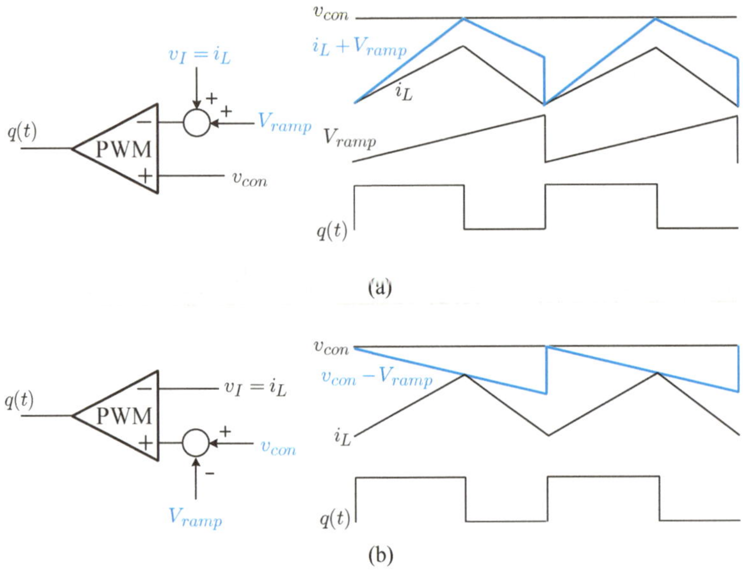

Figure 9.4 Structure and waveforms of PWM block for current mode control. (a) PWM with unity inductor current sensing. (b) Equivalent representation.

oscillation. Thus, the initial objective is not achieved. Instead, it was soon revealed that current mode control offers considerable advantages even though it still requires the compensation ramp. The merits of current mode control will be described in the next section.

The stabilizing effect of the compensation ramp is explained using the current feedback signal illustrated in Fig. 9.4 and Fig. 9.5. Figure 9.4(a) shows the structure and waveforms of the PWM for current mode control, where the current-to-voltage conversion gain of the current sensing network (CSN) is assumed unity for simplicity; thus, $v_I = i_L$ for this case. Figure 9.4(a) is rearranged into the equivalent form in Fig. 9.4(b), obtained by inverting the ramp signal V_{ramp} and moving the inverted ramp signal to the non-inverting terminal of the PWM comparator.

The PWM waveforms in Fig. 9.4(b) are closely analyzed in Fig. 9.5, which shows the propagation of the perturbed inductor current. In the enlarged illustration in Fig. 9.5, S_n is the slope of the on-time inductor current and S_f is the slope of the off-time inductor current, while S_e is the slope of the compensation ramp. The ΔdT_s denotes the deviation in the on-time period due to the inductor current perturbation. From the graphical construction, the initial distance between the original current i_L and the perturbed current i'_L is given by

$$|i_L(k) - i'_L(k)| = \Delta i_L(k) = S_n \Delta dT_s + S_e \Delta dT_s \tag{9.1}$$

The distance between the two inductor currents after one operational period is given by

$$|i_L(k+1) - i'_L(k+1)| = \Delta i_L(k+1) = S_f \Delta dT_s - S_e \Delta dT_s \tag{9.2}$$

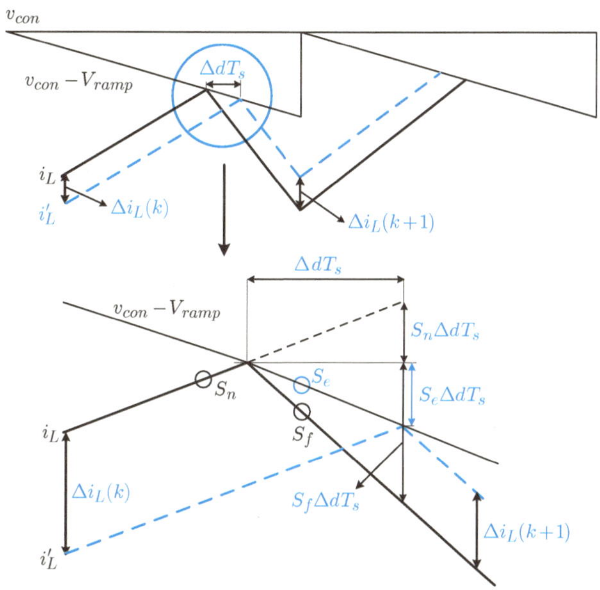

Figure 9.5 PWM waveforms.

For a successive decrease in the distance between i_L and i'_L in the ensuing operational periods, the condition

$$\boxed{\frac{\Delta i_L(k+1)}{\Delta i_L(k)} = \frac{S_f - S_e}{S_n + S_e} < 1} \tag{9.3}$$

is required, leading to the condition for the compensation ramp slope

$$\boxed{S_e > \frac{S_f - S_n}{2}} \tag{9.4}$$

for the stabilizing effect. The exact value of the compensation ramp slope should be determined in consideration of the closed-loop performance of the converter. In fact, the selection of the compensation ramp slope is the most important issue in the design of current mode control. The expression (9.3) also indicates that, when $S_e = 0$ with no compensation ramp, stability is only maintained under the condition $S_n > S_f$, which is true with the duty ratio $D < 0.5$. In other words, the converter is only stable for the duty ratio less than 0.5 at the absence of the compensation ramp.

Peak Current Mode Control

Current mode control can be implemented in various forms. The most popular among them is the peak current mode control, illustrated in Fig. 9.6. In this control scheme, the switch current is utilized in place of the inductor current. The switch current, which

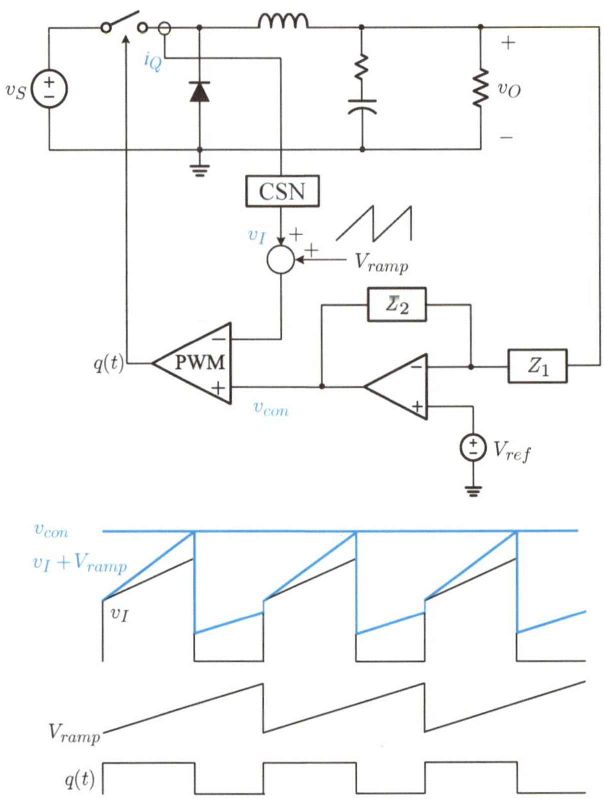

Figure 9.6 Peak current mode control.

corresponds to the on-time inductor current, is sensed through CSN and blended with the compensation ramp. The peak value of the switch current, which is the same as the peak value of the inductor current, is used to determine the instant to turn off the switch; thus, the control scheme is called the *peak current mode control*.† It should be noted that the switch current sensing is functionally identical to the inductor current sensing because the peak value of the inductor current is employed as the criterion to turn off the switch.

There are several advantages in sensing the switch current rather than the inductor current. The first is the simplicity in CSN. An implementation of the CSN for the peak current mode control is illustrated in a later example. As the second advantage, the sensed switch current can be used for the over-current protection of semiconductor switches. Due to these advantages, the peak current mode control is widely adapted to modern PWM dc-to-dc converters.

†The inductor current starts declining when the PWM block turns off the active switch. Thus, the control scheme ***does not use*** but ***determines*** the peak value of the inductor current. However, the operation of the peak current mode control is broadly interpreted as described in the text.

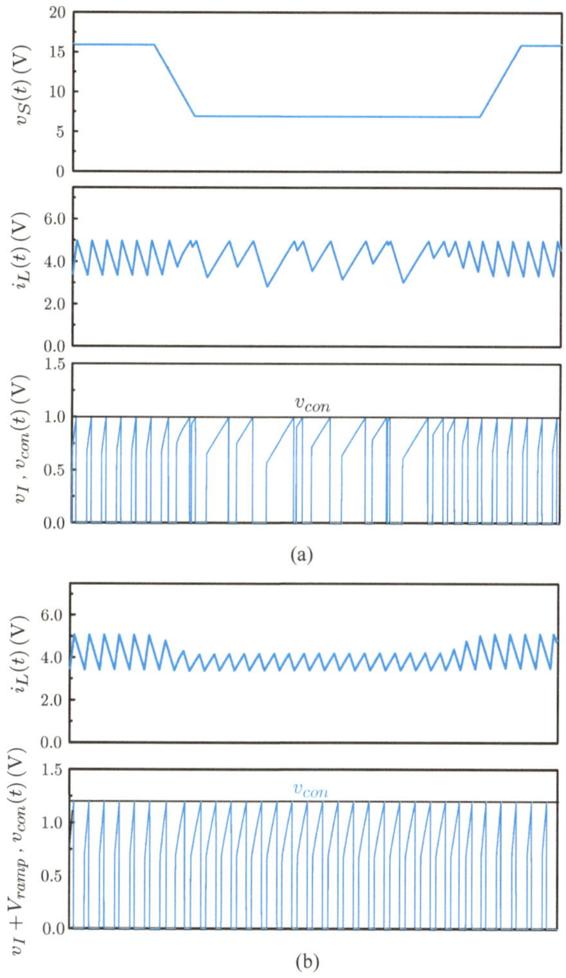

Figure 9.7 Sub-harmonic oscillation and effects of compensation ramp. (a) Sub-harmonic oscillation. (b) Effects of compensation ramp.

Sub-Harmonic Oscillation and Compensation Ramp — EXAMPLE 9.1

This example demonstrates the sub-harmonic oscillation and effects of the compensation ramp. A buck converter employing the peak current mode control is used in this example. The circuit parameters and operational conditions of the buck converter are $L = 40\ \mu$H, $C = 400\ \mu$F, $R_c = 0.01\ \Omega$, $R = 1\ \Omega$, and $f_s = 50$ kHz. The output of the converter is regulated at $V_O = 4$ V, while the input voltage is varied linearly from $V_S = 16$ V to 7 V, and later from $V_S = 7$ V to 16 V. Figure 9.7(a) shows the circuit waveforms where the CSN output alone is used for PWM without the compensation ramp.

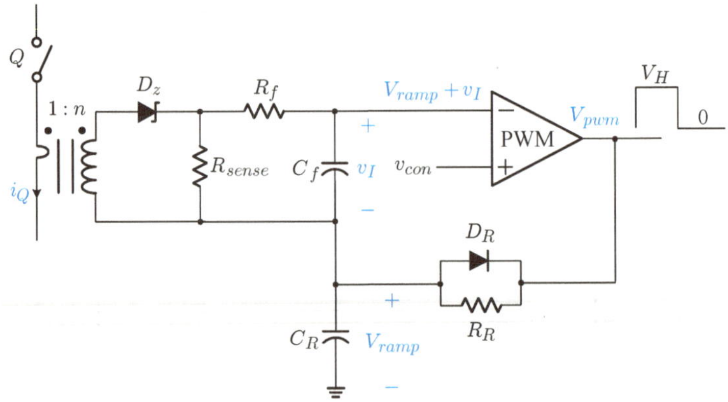

Figure 9.8 Current sensing network with compensation ramp.

When the input voltage is $V_S = 16$ V, thereby yielding the duty ratio $D = 4/16 = 0.25$, the inductor current and PWM waveforms confirm stable operation. When the input voltage starts declining from $V_S = 16$ V to 7 V, the waveforms depart from the stable pattern and finally develop the sub-harmonic oscillation when the duty ratio is increased to $D = 4/7 = 0.57$. Figure 9.7(b) illustrates the waveforms when an appropriate compensation ramp is added to the PWM block. The converter shows a stable operation regardless of the change in the input voltage, demonstrating the stabilizing effects of the compensation ramp.

EXAMPLE 9.2 Current Sensing Network with Compensation Ramp

This example presents a circuit implementation of CSN with the compensation ramp. The circuit is shown in Fig. 9.8. The switch current i_Q is sensed by a $1:n$ current transformer and converted into the voltage signal at the sensing resistor R_{sense}. The Zener diode D_z provides a path for the sensed current when the switch Q is closed, and resets the current transformer when Q is open. A first-order low pass filter is configured using R_f and C_f to remove the switching noise and produce an uncorrupted voltage signal v_I. The ratio of the sensed voltage signal v_I to the switch current i_Q constitutes the current-to-voltage conversion gain of CSN, denoted as R_i

$$R_i = \frac{v_I}{i_Q} = \frac{1}{n} R_{sense} \tag{9.5}$$

The compensation ramp is generated using R_R, C_R, and D_R, along with the output of the PWM block, V_{pwm}. At the onset of each operational period, the switch-on signal is issued and V_{pwm} is set at V_H. Then, C_R is charged towards V_H through R_R. When V_{pwm} is reset to zero to turn off the power switch Q, C_R is instantly discharged through D_R.

During the charging period, the voltage across C_R is given by

$$V_{ramp}(t) = V_H \left(1 - e^{-\frac{1}{R_R C_R}t}\right) \tag{9.6}$$

which is approximated as a linear ramp signal within the switching period T_s

$$\begin{aligned} V_{ramp}(t) &\approx V_H \left(1 - \left(1 - \frac{1}{R_R C_R}t\right)\right) \\ &= \frac{V_H}{R_R C_R}t \end{aligned}$$

with the assumption $T_s \ll R_R C_R$. The slope of the ramp signal is given by

$$S_e = \frac{V_H}{R_R C_R} \tag{9.7}$$

The ramp voltage V_{ramp} is summed with the sensed voltage signal v_I. The resulting signal is fed to the inverting terminal of the PWM comparator whose non-inverting terminal is connected to v_{con}.

9.1.2 Benefits and Issues of Peak Current Mode Control

Peak current mode control requires the compensation ramp to prevent the sub-harmonic oscillation. Thus, the initial attempt to remove the ramp signal from the PWM process is not fulfilled. Even so, peak current mode control is widely accepted because it offers significant advantages over conventional voltage mode control.

Benefits of Peak Current Mode Control

Advantages of the peak current mode control are mainly recognized in the dynamic performance.

1) Improved dynamic performance: Current mode control improves the dynamic performance of PWM dc-to-dc converters. These benefits are most pronounced in the boost and buck/boost converters that have the right-half plane (RHP) zero in their power stage transfer functions. In fact, current mode control is indispensable to these converters for stability and performance.

2) Reduced sensitivity of converter dynamics: Current mode control reduces the sensitivity of the converter performance to operational conditions. As an example, current-mode controlled converters undergo less changes in the dynamic performance at the switchover between CCM operation and DCM operation [1], when compared with voltage-mode controlled converters.

3) Simple compensation design: Current mode control simplifies the structure and design of the voltage feedback compensation. In contrast to voltage mode control requiring the three-pole two-zero compensation, current mode control employs

a simpler two-pole one-zero circuit. Furthermore, the voltage compensation can be standardized so that a single design procedures applies to all PWM dc-to-dc converters.

Issues of Peak Current Mode Control

While current mode control offers the aforementioned advantages, it also complicates the converter dynamics and presents challenges in the small-signal analysis and control design.

1) Dynamic modeling and analysis: Current mode control utilizes an additional feedback from the inductor current on top of the output voltage feedback. Thus, in control terminology, current-mode controlled PWM converters are a multi-loop controlled system in which multiple feedback loops are present. The previous analysis techniques, established in Chapters 7 and 8 for the single-loop voltage mode control, need to be reinforced and extended to deal with the multi-loop controlled system.

2) Sampling effects of current mode control: In the peak current mode control, the control action is periodically exerted at the instant the inductor current attains its peak value. In other words, the control action is executed by sampling the peak value of the fast-varying inductor current waveform. Due to this feature, the system exhibits the characteristics of sampled-data discrete-time systems.† This has been referred to as the *sampling effects* of current mode control. The sampling effects deserve special attention and require pertinent analysis.

9.1.3 Average Current Mode Control and Charge Control

In addition to peak current mode control, two other useful variations of current mode control are the average current mode control and charge control. This section briefly discusses the basics of these control schemes.

Average Current Mode Control

Average current mode control is illustrated in Fig. 9.9. Referring to the block diagram in Fig. 9.9(a) and the control waveforms in Fig. 9.9(b), the operation of the average current mode control is explained as follows. First, the triangular inductor current is sensed through the CSN. The CSN is usually a single resistor and converts the triangular inductor current into a triangular voltage signal, as shown in Fig. 9.9(b). The CSN output signal v_I is then processed through the current feedback circuit,

†In the exact sense, voltage mode control is also a discrete-time control because the duty ratio is updated only once in every single operational period. However, the control is performed by a slow-varying continuous output signal of the voltage feedback circuit. Thus, the discrete-time features of voltage mode control are much weaker than those of current mode control and are practically negligible.

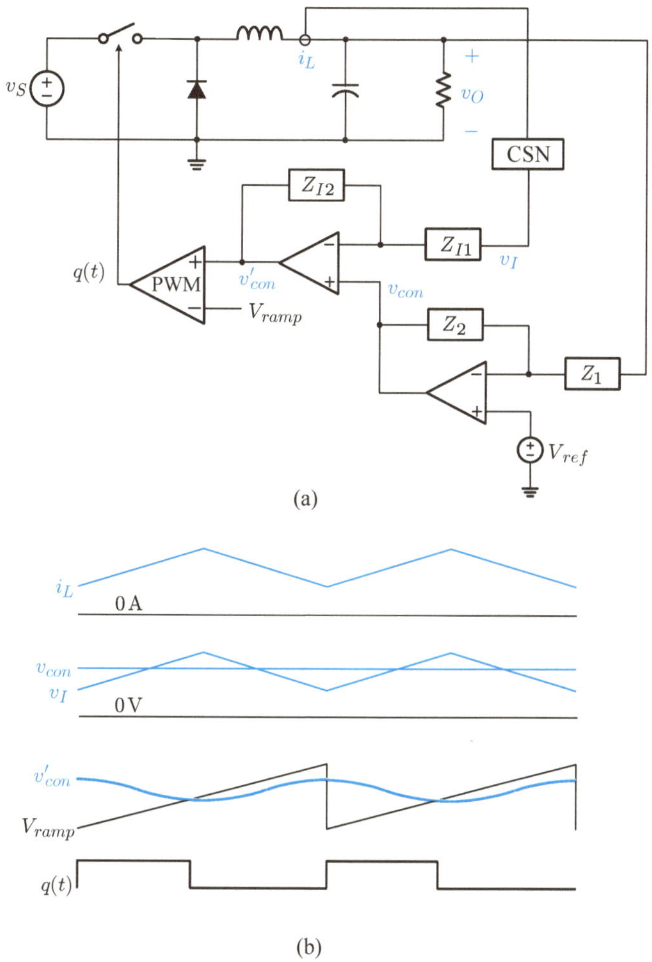

Figure 9.9 Average current mode control. (a) Functional block diagram. (b) Control waveforms.

which is composed of an op amp along with the current feedback compensation, Z_{I1} and Z_{I2}. If the current feedback compensation meets the condition

$$\frac{|Z_{I2}(j0)|}{|Z_{I1}(j0)|} = \infty \tag{9.8}$$

the sensed voltage signal v_I is forced to follow the control voltage v_{con} as closely as possible.

The sensed voltage signal v_I, which is a scaled replica of the triangular inductor current, is a triangular waveform. If v_{con} is a dc or slowly varying waveform, v_I cannot exactly follow the control voltage v_{con}. For this case, v_I only tracks v_{con} in the average sense; in other words, the moving average of the scaled inductor current will follow the control voltage v_{con}, as illustrated in the second plot of Fig. 9.9(b).

On the other hand, the output voltage is regulated at $V_O = V_{ref}$ as far as the voltage feedback circuit satisfies the condition $|Z_2(j0)|/|Z_1(j0)| = \infty$. The output of the current feedback compensation is given by

$$v'_{con}(t) = -\frac{Z_{I2}}{Z_{I1}} v_I(t) + \left(1 + \frac{Z_{I2}}{Z_{I1}}\right) v_{con}(t) \tag{9.9}$$

The composite control signal v'_{con} is compared with the ramp signal to produce the required PWM signal for the output voltage regulation.

The average current mode control forces the moving average of the scaled inductor current to follow the control voltage v_{con}. The potential merit of this feature is not fully revealed in dc-to-dc converters where v_{con} is usually a dc waveform. However, there are other applications where the control voltage v_{con} is programmed into a specific waveform. A standout example is power factor corrected (PFC) ac-to-dc converters, where v_{con} is given by a low-frequency sinusoidal waveform.

Figure 9.10 is an illustration of a PFC ac-to-dc converter operating with the average current mode control. The ac-to-dc converter is configured with a bridge rectifier and boost converter. The ac-to-dc converter receives an ac voltage from the utility line. The bridge rectifier performs the full-wave rectification and the resulting voltage waveform is supplied to the boost converter downstream. Accordingly, the input voltage of the boost converter is a rectified line-frequency sinusoid rather than a dc voltage, as illustrated by $v_S(t)$ in Fig. 9.10(b).

Assume that the boost converter is now required to draw its inductor current $i_L(t)$ in a special manner so that the moving average of $i_L(t)$ becomes the same sinusoid as the converter's input voltage $v_S(t)$. To achieve this goal, the control signal $v_{con}(t)$ is derived from the rectified line voltage $v_S(t)$, as shown in Fig. 9.10(a), and the average current mode control is adapted to the boost converter. The $v_{con}(t)$ is the line-frequency sinusoid and the inductor current $i_L(t)$ is forced to follow $v_{con}(t)$ in the averaged sense. The resulting inductor current waveform $i_L(t)$ is shown in Fig. 9.10(b). Figure 9.10(c) is an expanded comparison of the $i_L(t)$ and $v_{con}(t)$ waveforms. The moving average of $i_L(t)$ tracks $v_{con}(t)$, namely the scaled $v_S(t)$, as directed by the control law of the average current mode control.

While the full discussions about the PFC ac-to-dc converter are beyond the topic of this chapter, it is now seen that the average current mode control is a viable control scheme for PFC ac-to-dc converters. Operations and applications of average current-mode controlled PFC ac-to-dc converters are covered in [2, 3].

Charge Control

Another useful variation of current mode control is the charge control, shown in Fig. 9.11(a). The switch current i_Q is sensed through the CSN and the sensed current is immediately used to charge the capacitor C_I, resulting in the voltage signal v_I shown in Fig. 9.11(b). When v_I intersects the control voltage v_{con}, the switch-off signal is

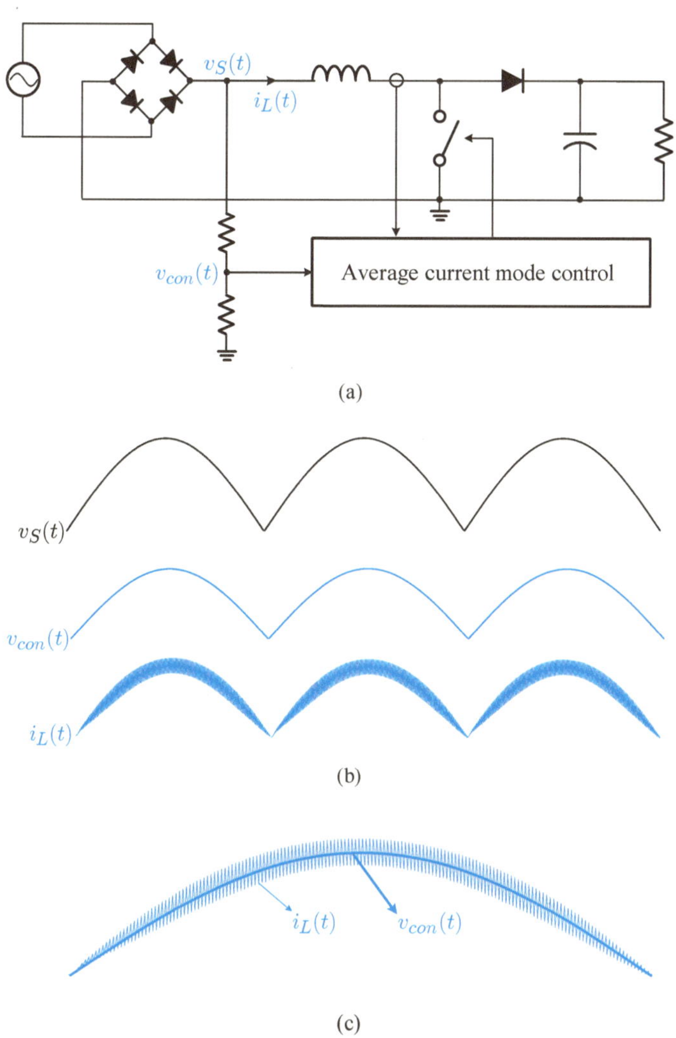

Figure 9.10 PFC ac-to-dc converter and control waveforms. (a) PFC ac-to-dc converter. (b) Major waveforms. (c) Expanded view of $i_L(t)$ and $v_{con}(t)$.

issued and, at the same time, the capacitor C_I is instantly discharged by closing the parallel switch Q_I.

One obvious advantage of the charge control is the enhanced noise immunity. This feature is shown in Fig. 9.11(c). The switch current i_Q is usually corrupted with the high-frequency ringings and spikes, which could cause false triggering when the peak current mode control is used. In charge control, the noisy switch current is effectively integrated by charging the capacitor C_I, resulting in the v_I waveform shown in Fig. 9.11(c). This eliminates the risk of the false triggering even in the presence of

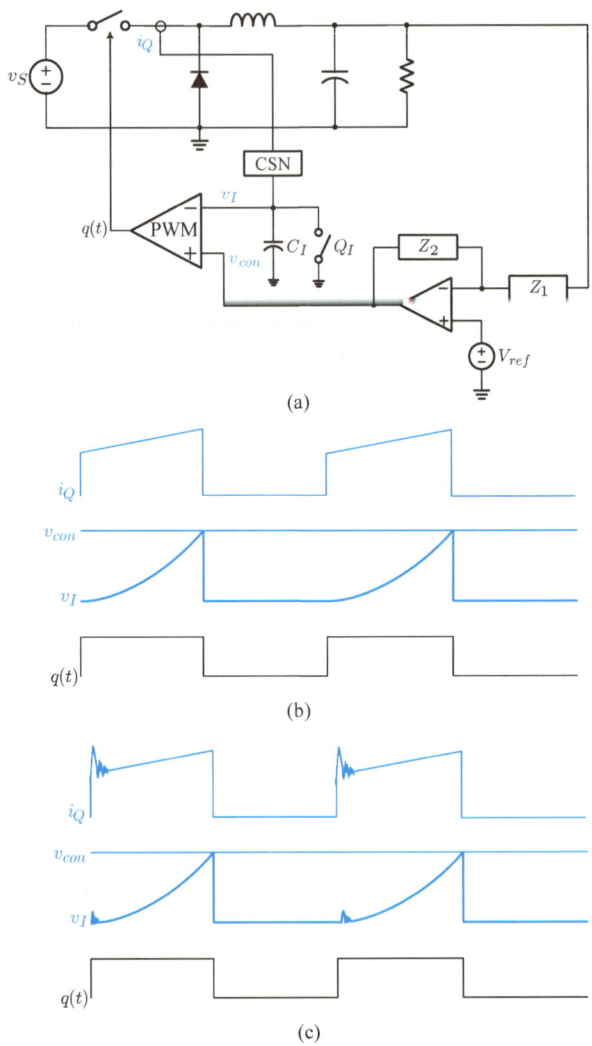

Figure 9.11 Charge control. (a) Functional block diagram. (b) Ideal control waveforms. (c) Practical control waveforms.

substantial high-frequency noises. Further details about the charge control are given in [4].

9.2 CLASSICAL ANALYSIS AND CONTROL DESIGN PROCEDURES

This section covers the analysis and control of the peak current mode control. The peak current mode control in Fig. 9.6 exerts the control action by periodically sampling the peak value of the inductor current. This control mechanism implants the discrete-time sampling effects into the converter dynamics. In principle, the sampling effects should be incorporated into the analysis and design of peak current mode control. This would require discrete-time analyses using z-domain techniques.

Conventionally, peak current mode control has been analyzed and designed based on the continuous-time s-domain techniques, implicitly assuming that the sampling effects only cause negligible consequences on the converter dynamics. This simplified analysis is referred to as the *classical analysis* in this book. The classical analysis, although it does not include the sampling effects, accurately describes the major dynamics of peak current mode control and provides correct design methodology for most cases.

In this chapter, the peak current mode control is investigated based on the classical s-domain analysis, while postponing the study of the sampling effects to the next chapter. The classical analysis will reveal its own value and advantages that would support its long-time prevalence. In particular, the classical analysis provides step-by-step design procedures which offer stability and performance for all PWM converters with peak current mode control.

9.2.1 Small-Signal Model for Peak Current Mode Control

As the onset of the classical analysis, this section discusses the small-signal model of peak current mode control. Figure 9.12 shows a general circuit diagram of PWM converters employing the peak current mode control. With the connections $\{\mathrm{a{-}X\ \ p{-}Y\ \ i{-}Z}\}$, Fig. 9.12 represents the buck converter. Similarly, the connections $\{\mathrm{i{-}X\ \ a{-}Y\ \ p{-}Z}\}$ lead to the boost converter, while the connections $\{\mathrm{a{-}X\ \ i{-}Y\ \ p{-}Z}\}$ yield the buck/boost converter.

Figure 9.13 is the small-signal model of the converter, obtained from Fig. 9.12 by replacing the PWM switch, PWM block, and voltage feedback circuit with their respective small-signal model, and by introducing appropriate small-signal excitations. In Fig. 9.13, *a feedback path is created from the inductor current $\hat{i}_L$, even though the feedback current is actually sensed from the switch.* The reason for this practice is as follows. The control action is executed at the moment the switch current reaches its peak value. At the instant of the control action, the switch current is the same as the inductor current. Thus, the switch current sensing is functionally identical to the inductor current sensing. In other words, there is no difference between the switch current sensing and inductor current sensing, as far as the control mechanism is concerned. The feedback path from the inductor current $\hat{i}_L$ accords with this fact.

The gain block $F_v(s)$ is the voltage feedback compensation

$$\boxed{F_v(s) = \frac{Z_2(s)}{Z_1(s)}} \tag{9.10}$$

and R_i is the CSN gain, given by

$$\boxed{R_i = \frac{1}{n} R_{sense}} \tag{9.11}$$

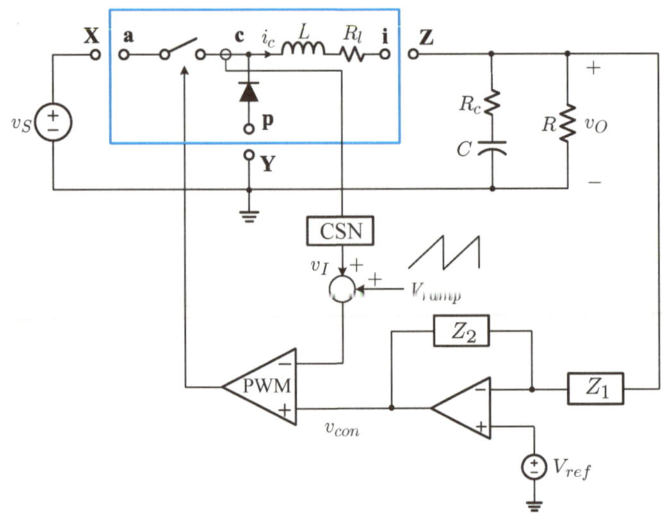

Figure 9.12 General circuit diagram for current-mode controlled PWM converters.

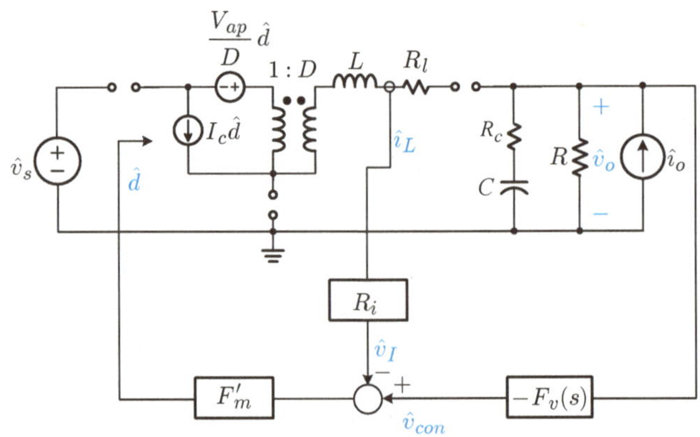

Figure 9.13 Small-signal model for current-mode controlled PWM converters.

for the circuit discussed in Example 9.2. The gain block F'_m represents the small-signal gain of the PWM block – the modulator gain of the peak current mode control. The modulator gain of the peak current mode control differs from that of voltage mode control and the notation F'_m is used to emphasize the difference.

Several different expressions for the modulator gain F'_m were proposed in the past. While there are subtle differences in the existing F'_m expressions, they do not cause significant consequences on the analysis and design of peak current mode control. This book adopts the modulator gain proposed by F. C. Lee [5, 6]

Table 9.1 Expressions for Slopes of PWM Waveforms

	Buck converter	Boost converter	Buck/boost converter
S_n	$\frac{V_S - V_O}{L} R_i$	$\frac{V_S}{L} R_i$	$\frac{V_S}{L} R_i$
S_f	$\frac{V_O}{L} R_i$	$\frac{\lvert V_S - V_O \rvert}{L} R_i$	$\frac{\lvert V_O \rvert}{L} R_i$

$$\boxed{F'_m = \frac{2}{(S_n - S_f + 2S_e)T_s}} \tag{9.12}$$

where S_n is the on-time slope and S_f is the off-time slope of the sensed current feedback signal v_I, while S_e is the slope of the compensation ramp. Derivations of the modulator gain are given in Example 9.3. The expressions for S_n and S_f for the three basic converters are summarized in Table 9.1.

Interestingly, the modulator gain in (9.12) correctly predicts the sub-harmonic oscillation, which occurs when the duty ratio D exceeds 0.5 in the absence of the compensation ramp. It can be seen that $S_n > S_f$ with $D < 0.5$, $S_n = S_f$ with $D = 0.5$, and $S_n < S_f$ with $D > 0.5$. When the compensation ramp is not present, so $S_e = 0$, F'_m approaches infinity with $D = 0.5$ and becomes negative with $D > 0.5$. These facts support the sub-harmonic oscillation occurring with $D \geq 0.5$ and $S_e = 0$.

Modulator Gain for Peak Current Mode Control F'_m — EXAMPLE 9.3

This example presents the derivation of the modulator gain given by (9.12). Figure 9.14 shows the modulator waveforms of the peak current mode control, in which the inductor current is assumed to increase cycle-by-cycle so that $i_L(k+1) > i_L(k)$ where $i_L(k)$ is the initial inductor current at the kth switching period. In Fig. 9.14, the current level $\bar{\imath}_{on}$ is the average value of the inductor current during the on-time period dT_s and $\bar{\imath}_{off}$ is that of the off-time period $(1-d)T_s$. The following relationships are derived from Fig. 9.14

$$\begin{aligned} \bar{\imath}_{on}(t) &= v_{con} - S_e d T_s - \frac{1}{2} S_n d T_s \\ \bar{\imath}_{off}(t) &= v_{con} - S_e d T_s - \frac{1}{2} S_f (1-d) T_s \end{aligned} \tag{9.13}$$

When the inductor current is assumed only to change slowly, $i_L(k) \approx i_L(k+1)$, it becomes $\bar{\imath}_{on} \approx \bar{\imath}_{off}$ so that either $\bar{\imath}_{on}$ or $\bar{\imath}_{off}$ can be considered as the average inductor current, $\bar{\imath}_L(t)$. Furthermore, the assumption $i_L(k) \approx i_L(k+1)$ validates the approximation

$$S_n d T_s = S_f (1-d) T_s \tag{9.14}$$

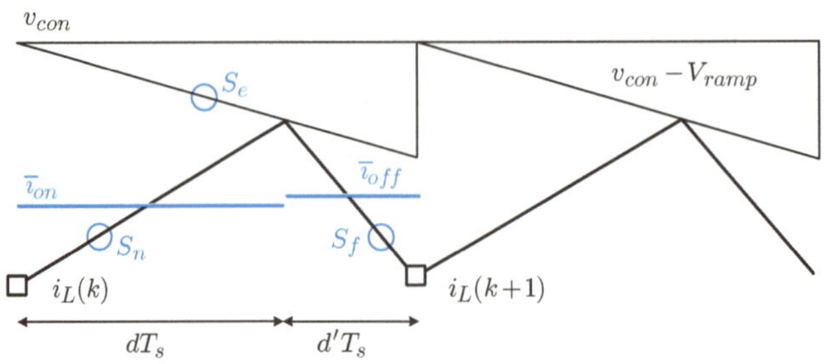

Figure 9.14 Peak current mode control waveforms.

which is rearranged as

$$(S_n + S_f)d = S_f \tag{9.15}$$

By considering $\bar{\imath}_{off}$ in (9.13) as the approximation of the averaged inductor current $\bar{\imath}_L(t)$ and using (9.15), the expression (9.13) is modified as

$$\bar{\imath}_{off}(t) = \bar{\imath}_L(t) = v_{con} - S_e d T_s - \frac{1}{2}(S_n + S_f)d(1-d)T_s \tag{9.16}$$

Linearization of (9.16) yields

$$\begin{aligned} I_L + \hat{\imath}_L \quad &= \quad (V_{con} + \hat{v}_{con}) - S_e(D + \hat{d})T_s \\ &\quad -\frac{1}{2}(S_n + S_f)(D + \hat{d})\big(1 - (D + \hat{d})\big)T_s \end{aligned} \tag{9.17}$$

By equating the ac terms in (9.17), it becomes

$$\hat{v}_{con} - \hat{\imath}_L = \left(S_e T_s + \frac{1}{2}(S_n + S_f)T_s(1 - 2D) \right) \hat{d} \tag{9.18}$$

Using the steady-state expression of (9.15), $(S_n + S_f)D = S_f$, the expression (9.18) is rearranged as

$$F'_m = \frac{\hat{d}}{\hat{v}_{con} - \hat{\imath}_L} = \frac{2}{(S_n - S_f + 2S_e)T_s} \tag{9.19}$$

which can be fitted into Fig. 9.13 with the assumption $R_i = 1$.

In this derivation, it was also assumed that the slopes of the inductor current remain unchanged. The cases where this assumption is discarded will be treated in the next chapter, which covers the sampling effects of current mode control.

The small-signal model in Fig. 9.13 is transformed into a block diagram. Figure 9.15 shows the block diagram representation of the small-signal model. The expressions of the power stage gain blocks can be obtained from the circuit analysis of Fig. 9.13. The block diagram clearly shows the two individual feedback loops, one originating from the output voltage $\hat{v}_o$ and the other stemming from the inductor current $\hat{\imath}_L$. The system is thus called a two-loop or multi-loop controlled system.

Table 9.2 Power Stage Transfer Functions of Three Basic Converters

Transfer functions

$$G_{vs}=K_{vs}\frac{1+\frac{s}{\omega_{esr}}}{1+\frac{s}{Q\omega_o}+\frac{s^2}{\omega_o^2}} \qquad G_{vd}=K_{vd}\frac{\left(1-\frac{s}{\omega_{rhp}}\right)\left(1+\frac{s}{\omega_{esr}}\right)}{1+\frac{s}{Q\omega_o}+\frac{s^2}{\omega_o^2}} \qquad G_{is}=K_{is}\frac{1+\frac{s}{\omega_{is}}}{1+\frac{s}{Q\omega_o}+\frac{s^2}{\omega_o^2}}$$

$$G_{id}=K_{id}\frac{1+\frac{s}{\omega_{id}}}{1+\frac{s}{Q\omega_o}+\frac{s^2}{\omega_o^2}} \qquad Z_p=K_p\frac{\left(1+\frac{s}{\omega_z}\right)\left(1+\frac{s}{\omega_{esr}}\right)}{1+\frac{s}{Q\omega_o}+\frac{s^2}{\omega_o^2}} \qquad Z_q=K_q\frac{1+\frac{s}{\omega_{esr}}}{1+\frac{s}{Q\omega_o}+\frac{s^2}{\omega_o^2}}$$

Expressions for dc gain and corner frequencies			
	Buck converter	Boost converter	Buck/boost converter
K_{vs}	D	$1/(1-D)$	$D/(1-D)$
K_{vd}	V_S	$V_S/(1-D)^2$	$V_S/(1-D)^2$
K_{is}	D/R	$1/((1-D)^2R)$	$D/((1-D)^2R)$
K_{id}	V_S/R	$2V_S/((1-D)^3R)$	$V_S(1+D)/((1-D)^3R)$
K_p	R_l	$R_l/(1-D)^2$	$R_l/(1-D)^2$
K_q	-1	$-1/(1-D)$	$-1/(1-D)$
ω_{esr}	$1/(CR_c)$	$1/(CR_c)$	$1/(CR_c)$
ω_{rhp}	∞	$(1-D)^2R/L$	$(1-D)^2R/(DL)$
ω_{is}	$1/(CR)$	$1/(CR)$	$1/(CR)$
ω_{id}	$1/(CR)$	$2/(CR)$	$(1+D)/(CR)$
ω_z	R_l/L	R_l/L	R_l/L
ω_o	$1/\sqrt{LC}$	$(1-D)/\sqrt{LC}$	$(1-D)/\sqrt{LC}$
Q	$R\sqrt{C/L}$	$(1-D)R\sqrt{C/L}$	$(1-D)R\sqrt{C/L}$

The expressions are approximations whose accuracy improves with the conditions $R \gg R_l$ and $R \gg R_c$.

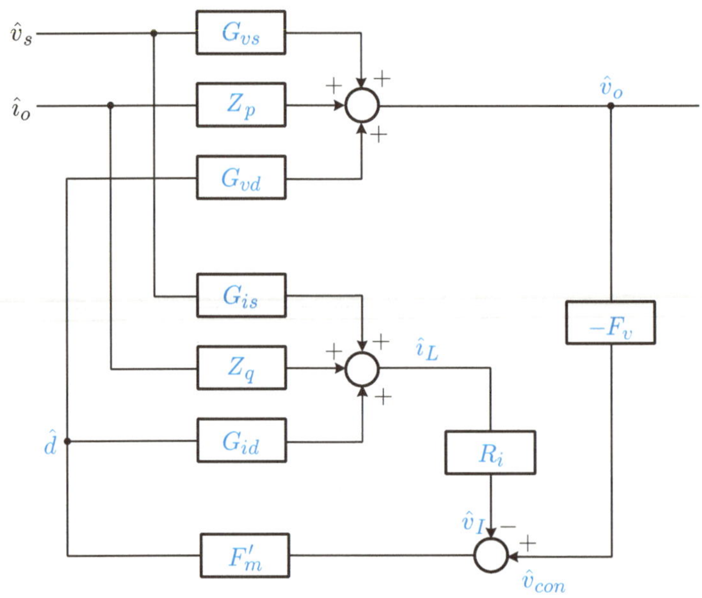

Figure 9.15 Small-signal block diagram representation of current-mode controlled PWM converters.

In addition to the three power stage gain blocks, $G_{vs}(s)$, $Z_p(s)$, and $G_{vd}(s)$, associated with the output voltage feedback, three additional gain blocks, $G_{is}(s)$, $Z_q(s)$, and $G_{id}(s)$, are necessary due to the presence of the inductor current feedback. The expressions for the six gain blocks of the three basic PWM converters are listed in Table 9.2. Now, the small-signal dynamics of current-mode controlled PWM converters are investigated using Fig. 9.15 and Table 9.2.

9.2.2 Loop Gain Analysis

In current-mode controlled PWM converters, there exists two individual feedback loops — one associated with the output voltage feedback and the other involved with the inductor current feedback. Furthermore, several different system loop gains can be defined by breaking the signal path at different locations in the system. These individual feedback loops and system loop gains should be first analyzed, in order to characterize the small-signal dynamics of the two-loop controlled systems.

Individual Feedback Loops

From the small-signal block diagram in Fig. 9.15, two individual feedback loops are identified. The two individual feedback loops, referred to as the current loop and voltage loop, are analyzed in this section.

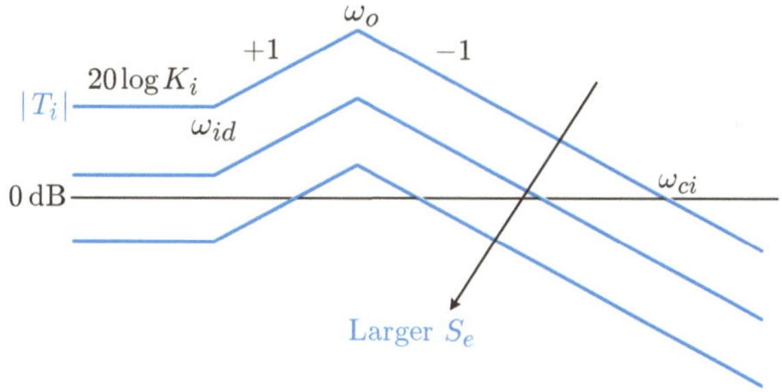

Figure 9.16 Asymptotic plot for $|T_i|$ and effects of compensation ramp slope.

Current loop $T_i(s)$: The current loop $T_i(s)$ is the *negative* gain product of the signal path created by the inductor current feedback

$$T_i(s) = -\frac{\hat{\imath}_L(s)}{\hat{d}(s)}\frac{\hat{v}_I(s)}{\hat{\imath}_L(s)}(-)\frac{\hat{d}(s)}{\hat{v}_I(s)} = G_{id}(s)\,R_i\,F'_m \tag{9.20}$$

Using the expressions in Table 9.2, the current loop $T_i(s)$ is determined as

$$\begin{aligned} T_i(s) &= \underbrace{K_{id}\frac{1+\dfrac{s}{\omega_{id}}}{1+\dfrac{s}{Q\omega_o}+\dfrac{s^2}{\omega_o^2}}}_{G_{id}(s)}\,R_i\,F'_m \\ &= K_i\frac{1+\dfrac{s}{\omega_{id}}}{1+\dfrac{s}{Q\omega_o}+\dfrac{s^2}{\omega_o^2}} \end{aligned} \tag{9.21}$$

where

$$K_i = K_{id}\,R_i\,F'_m = K_{id}\,R_i\,\frac{2}{(S_n - S_f + 2S_e)T_s} \tag{9.22}$$

Figure 9.16 is the asymptotic plot for $|T_i|$. The structure and corner frequencies of $|T_i|$ are fixed by the power stage parameters. Only the dc gain K_i can be altered by the CSN gain R_i and compensation ramp slope S_e. Equation (9.22) indicates that the dc gain K_i is inversely proportional to S_e. When S_e becomes larger, the magnitude of the current loop $|T_i|$ will be decreased, as illustrated in Fig. 9.16. With an excessively reduced $|T_i|$, the inductor current feedback becomes ineffective and only the output voltage feedback dictates the control mechanism. *This implies that, when the slope of the compensation ramp is overly increased, the control scheme practically alters to voltage mode control, even though it has the structure of the peak current mode control.*

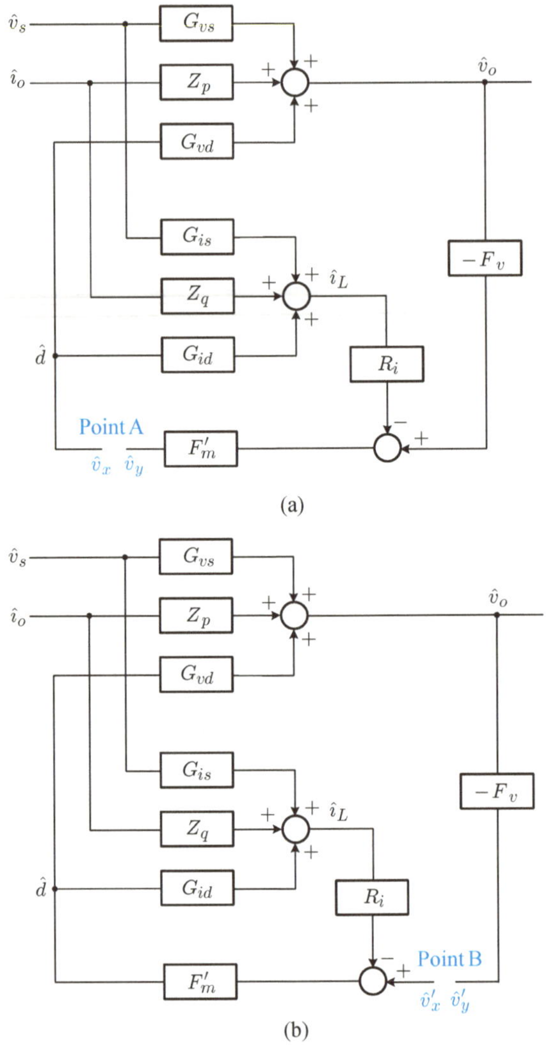

Figure 9.17 System loop gains. (a) Overall loop gain. (b) Outer loop gain.

Voltage loop $T_v(s)$: The negative gain product of the signal path associated with the output voltage feedback is referred to as the voltage loop $T_v(s)$

$$T_v(s) = -\frac{\hat{v}_o(s)}{\hat{d}(s)}\frac{\hat{v}_{con}(s)}{\hat{v}_o(s)}\frac{\hat{d}(s)}{\hat{v}_{con}(s)} = -G_{vd}(s)\,F_v(s)\,(-)\,F'_m = G_{vd}(s)\,F_v(s)\,F'_m \tag{9.23}$$

The voltage loop is directly affected by the voltage feedback compensation $F_v(s)$. Thus, the structure and parameters of $F_v(s)$ should be adequately determined in order to obtain desirable $T_v(s)$ characteristics.

Overall Loop Gain and Outer Loop Gain

For single-loop controlled systems, only one system loop gain exists in the system, as is the case of voltage mode control. In contrast, for multi-loop controlled systems, several system loop gains can be identified. Each system loop gain has its own implication and role in the dynamic analysis and control design. For current-mode controlled PWM converters, two particular system loop gains are important and useful. Figure 9.17 illustrates these two system loop gains, which are called the overall loop gain and outer loop gain.

Overall Loop Gain $T_1(s)$: The first system loop gain is defined by breaking the signal path at Point A in Fig. 9.17(a). By applying Mason's gain rule to Fig. 9.17(a), this loop gain is expressed as

$$T_1(s) = -\frac{\hat{v}_y(s)}{\hat{v}_x(s)} = T_i(s) + T_v(s) \tag{9.24}$$

where $T_i(s)$ and $T_v(s)$ are defined in (9.20) and (9.23). This loop gain is called the overall loop gain because it is defined by breaking the signal path inside the current loop and voltage loop.

Outer Loop Gain $T_2(s)$: The other system loop gain is defined at Point B in Fig. 9.17(b). Application of Mason's gain rule to Fig. 9.17(b) yields

$$T_2(s) = -\frac{\hat{v}'_y(s)}{\hat{v}'_x(s)} = \frac{T_v(s)}{1 + T_i(s)} \tag{9.25}$$

This loop gain is denoted as the outer loop gain because the loop gain is defined at the outer voltage feedback path.

Mason's Gain Rule and System Loop Gains — EXAMPLE 9.4

This example shows the derivation of the system loop gain expressions using Mason's gain rule. Mason's gain rule is a general formula to find the expression of a specific transfer function at the presence of multiple feedback loops

$$H(s) = \frac{1}{\Delta}\sum_{k=1}^{n} M_k \Delta_k$$

where

$H(s)$: transfer function of interest

Δ : 1 – (sum of the gains of all individual loops)

+ (sum of the gain products of all two non-touching loops) – ···

M_k : gain of the kth forward path

Δ_k : 1 – (sum of the gains of all individual loops not touched by the kth forward path) + (sum of the gain products of all two non-touching loops, not touched by the kth forward path) – ···

To evaluate the overall loop gain $T_1(s)$, Mason's rule is applied to Fig. 9.17(a), yielding $\Delta = 1$, $M_1 = G_{id}(s)R_iF'_m = T_i(s)$, Δ_1=1, $M_2 = G_{vd}(s)F_v(s)F'_m = T_v(s)$, and $\Delta_2 = 1$. The overall loop gain is now given by

$$T_1 = \frac{1}{\Delta}(M_1\Delta_1 + M_2\Delta_2) = T_i(s) + T_v(s)$$

Similarly, by applying Mason's rule to Fig. 9.17(b), it becomes that $\Delta = 1 + G_{id}(s)R_iF'_m = 1 + T_i(s)$, $M_1 = G_{vd}(s)F_v(s)F'_m = T_v(s)$, and $\Delta_1 = 1$. Thus, the outer loop gain is resulted as

$$T_2 = \frac{1}{\Delta}M_1\Delta_1 = \frac{T_v(s)}{1 + T_i(s)}$$

9.2.3 Stability Analysis

For voltage mode control, stability analysis was performed using the single loop gain in the system. In contrast, for current mode control, the overall loop gain and outer loop gain are both necessary and useful in determining the absolute stability and relative stability of the converter.

Absolute Stability

As discussed in Section 7.5, the absolute stability can be assessed using any closed-loop transfer function because all transfer functions have the same denominator. The closed-loop input-to-output transfer function, or audio-susceptibility, is derived by applying Mason's gain rule to the small-signal block diagram of Fig. 9.15.

$$A_u(s) = \frac{\hat{v}_o(s)}{\hat{v}_s(s)} = \frac{G_{vs}(1 + G_{id}R_iF'_m) - G_{is}R_iF'_mG_{vd}}{1 + G_{id}R_iF'_m + G_{vd}F_vF'_m} = \frac{F_p(s)}{1 + T_i(s) + T_v(s)} \tag{9.26}$$

with $F_p(s) = G_{vs}(1 + G_{id}R_iF'_m) - G_{is}R_iF'_mG_{vd}$.

It can be shown that the numerator of the above equation, $F_p(s)$, does not contain any right-half plane (RHP) poles. Thus, the absolute stability is judged by the existence of any RHP roots in the following equation

$$\boxed{1 + T_i(s) + T_v(s) = 0} \tag{9.27}$$

which is recognized as

$$\boxed{1 + T_1(s) = 0} \tag{9.28}$$

Equation (9.27) is written in an alternative form

$$1 + \frac{T_v(s)}{1 + T_i(s)} = 0 \tag{9.29}$$

which becomes

$$\boxed{1 + T_2(s) = 0} \tag{9.30}$$

This analysis implies that the Nyquist criterion can be applied to either $T_1(s)$ or $T_2(s)$, in order to check the existence of any RHP roots in $1 + T_i(s) + T_v(s) = 0$. The result, of course, should be the same regardless of the choice of the loop gain — if one loop gain predicts stability or instability, the other loop gain should indicate the same.

Absolute Stability and Polar Plots of $T_1(s)$ and $T_2(s)$ — EXAMPLE 9.5

This example demonstrates that the overall loop gain and outer loop gain carry the same information about the absolute stability of the converter. Figure 9.18 shows three cases of the polar plots of $T_1(s)$ and $T_2(s)$ of a current-mode controlled boost converter. Each case of Fig. 9.18 is evaluated at a stable, marginally stable, or unstable operating point. At the stable operating point, the polar plots of the two loop gains both do not encircle the $(-1, 0)$ point to confirm stability of the converter, as shown in Fig. 9.18(a). Now, the operating point is altered so that the converter becomes marginally stable. As shown in Fig. 9.18(b), the two polar plots identically traverse the $(-1, 0)$ point, thus indicating that the converter is on the verge of instability. Figure 9.18(c) is the polar plots when the operating point is further changed to encounter instability. For this case, the two loop gains encircle the $(-1, 0)$ point as the token of instability. Although the shape and transition pattern are utterly different, the two loop gains carry the same information about the absolute stability of the converter.

Relative Stability

Although the overall loop gain and outer loop gain provide the same information about the absolute stability, they are two different s-domain transfer functions. Therefore, the frequency responses of the two loop gains are quite distinct in both the shape and evolution pattern, thereby producing different phase and gain margins. The stability margins thus should be interpreted based on the original definitions. The frequency response of the overall loop gain, $T_1(s) = T_i(s) + T_v(s)$, is expressed as

$$T_1(j\omega) = |T_i(j\omega) + T_v(j\omega)|\angle\big(T_1(j\omega) + T_v(j\omega)\big) \tag{9.31}$$

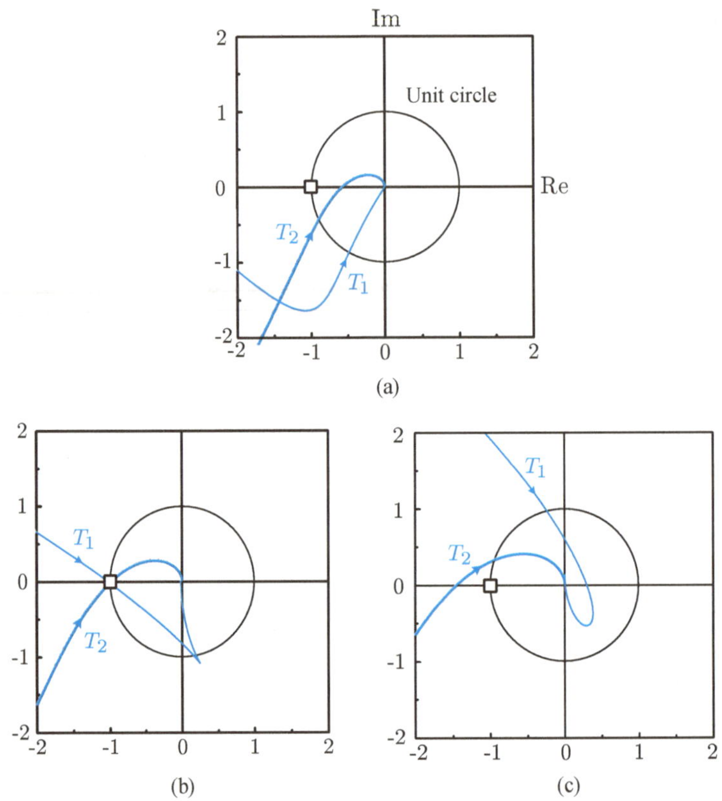

Figure 9.18 Polar plots of overall loop gain T_1 and outer loop gain T_2. (a) Stable case. (b) Marginally stable case. (c) Unstable case.

The gain margin is the extra gain that can be added to $|T_i + T_v|$ before the system becomes unstable. For example, if $T_1(s)$ has a 6 dB gain margin, the system remains stable until $|T_v + T_i|$ is doubled: $20 \log 2 \approx 6$ dB. Similarly, the phase margin is the additional delay that can be put in $\angle(T_i + T_v)$ while maintaining stability.

The frequency response of the outer loop gain $T_2(s) = T_v(s)/(1+T_i(s))$ is given by

$$T_2(j\omega) = \left|\frac{T_v(j\omega)}{1+T_i(j\omega)}\right| \angle\left(\frac{T_v(j\omega)}{1+T_i(j\omega)}\right) \tag{9.32}$$

The gain margin and phase margin of $T_2(s)$ can be interpreted in the same manner. For example, the gain margin is the additional gain increase that can be introduced to $|T_v/(1+T_i)|$. However, the stability margins become far more informative and useful when $T_i(s)$ is fixed prior to analyzing $T_2(s)$. For such a case, the stability margins of $T_2(s) = T_v(s)/(1 + T_i(s))$ are in fact those of $T_v(s)$. As will be shown in upcoming control design procedures, $T_i(s)$ is indeed prefixed and $T_v(s)$ is later designed for optimal $T_2(s)$ characteristics. Accordingly, the phase margin and gain margin of $T_2(s)$ can be used to assist and evaluate the design of the voltage loop $T_v(s)$. Further discussions about this topic are given later in Section 9.4.2.

9.2.4 Voltage Feedback Compensation

As shown in the previous section, the structure of the current loop $T_i(s)$ is fixed by the power stage transfer function. On the other hand, the structure of the voltage loop gain $T_v(s)$ can freely be chosen by altering the voltage feedback compensation $F_v(s)$. Accordingly, $F_v(s)$ needs to be tailored for the desired $T_v(s)$ structure.

This section investigates the design of $F_v(s)$ that would offer stability and good closed-loop performance. For simplicity, a buck converter is considered in this section. Conclusions of this section will be extended to boost and buck/boost converters in the next section.

Instability with Single Integrator

In order to obtain insight on the $F_v(s)$ design, it is first necessary to investigate the current loop $T_i(s)$ and voltage loop $T_v(s)$. For the buck converter, the expression for the current loop $T_i(s)$ is given by

$$\begin{aligned} T_i(s) &= G_{id}(s)\, R_i\, F_m' \\ &= \underbrace{\frac{V_S}{R}\frac{1+\dfrac{s}{\omega_{id}}}{1+\dfrac{s}{Q\omega_o}+\dfrac{s^2}{\omega_o^2}}}_{G_{id}(s)} R_i F_m' \end{aligned} \tag{9.33}$$

with $\omega_{id} = 1/(CR)$.

The expression for the voltage loop $T_v(s)$ is directly affected by the selection of the voltage feedback compensation $F_v(s)$. As the first candidate, a single integrator is considered for $F_v(s)$. With $F_v(s) = K_v/s$, the voltage loop $T_v(s)$ becomes

$$\begin{aligned} T_v(s) &= G_{vd}(s) F_v(s) F_m' \\ &= \underbrace{V_S\frac{1+\dfrac{s}{\omega_{esr}}}{1+\dfrac{s}{Q\omega_o}+\dfrac{s^2}{\omega_o^2}}}_{G_{vd}(s)} \underbrace{\frac{K_v}{s}}_{F_v(s)} F_m' \end{aligned} \tag{9.34}$$

The correctness of the $F_v(s)$ structure is judged by investigating the overall loop gain $T_1(s) = T_i(s)+T_v(s)$. The magnitude plot of the overall loop gain $|T_1| = |T_i+T_v|$ is depicted in Fig. 9.19, along with the asymptotic plots of $|T_i|$ and $|T_v|$. The asymptotic

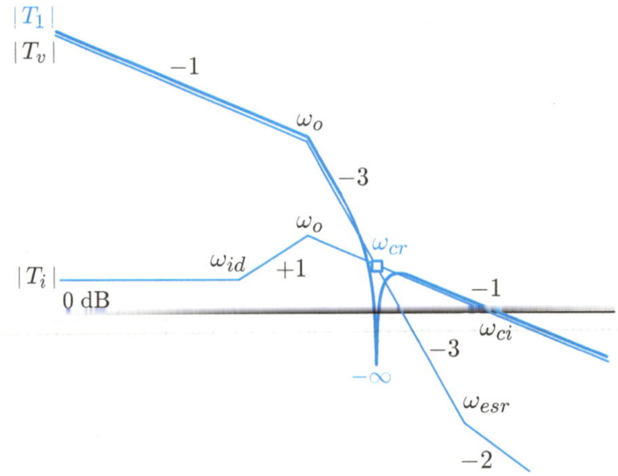

Figure 9.19 Individual feedback loops and overall loop gain with single integrator.

plot for $|T_1|$ is determined as explained below. Based on the asymptotic analysis, it follows that

$$\begin{aligned} T_1(s) &= T_i(s) + T_v(s) \\ &\approx \begin{cases} T_i(s) & \text{at frequencies where } |T_i| > |T_v| \\ T_v(s) & \text{at frequencies where } |T_i| < |T_v| \end{cases} \end{aligned}$$

Thus, $|T_1|$ follows either $|T_i|$ or $|T_v|$, whichever is larger in magnitude at the given frequencies. However, there is one singular point which could utterly depart from this general trend. At the frequency where $|T_i| = |T_v|$, denoted as ω_{cr} in Fig. 9.19, the overall loop gain T_1 is given by the sum of two equal-length vectors, $\vec{T_1} = \vec{T_i} + \vec{T_v}$. In this case, the magnitude of T_1 is strongly affected by the phase characteristics of $\vec{T_i}$ and $\vec{T_v}$. The slopes of $|T_i|$ and $|T_v|$ indicate that $\angle T_i \approx -90°$ and $\angle T_v \approx -270° = 90°$ at ω_{cr}. Now, the magnitude of T_1 at ω_{cr} becomes infinitely small, because $\vec{T_i}$ and $\vec{T_v}$ are 180° apart with an equal magnitude and therefore cancel each other. The sudden collapse in magnitude in turn implies an abrupt decrease in phase. Thus, $|T_1|$ prematurely crosses the 0 dB line with the phase far less than $-180°$; in short, the converter is unstable.

EXAMPLE 9.6 Instability with Single Integrator

This example demonstrates the instability of a single integrator. Figure 9.20 shows the Bode plots of $T_i(s)$, $T_v(s)$, and $T_1(s)$ of a buck converter which employs a single integrator. The overall loop gain exhibits the dip in $|T_1|$ and sudden drop in $\angle T_1$ at

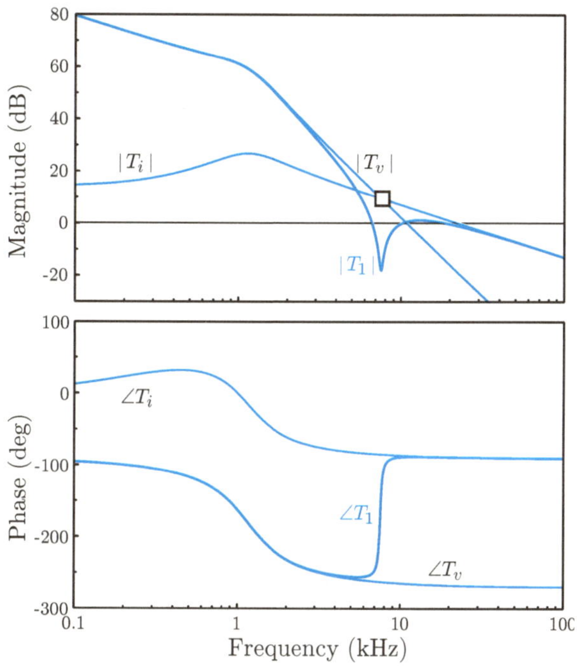

Figure 9.20 Examples of individual feedback loops and overall loop gain.

the frequencies where $|T_i| \approx |T_v|$. The converter is surely unstable with these loop gain characteristics.

Two-Pole One-Zero Compensation

As the second candidate for the voltage feedback compensation, a two-pole one-zero circuit

$$F_v(s) = \frac{K_v\left(1 + \dfrac{s}{\omega_{zc}}\right)}{s\left(1 + \dfrac{s}{\omega_{pc}}\right)} \tag{9.35}$$

is considered. For this case, the voltage loop $T_v(s)$ becomes

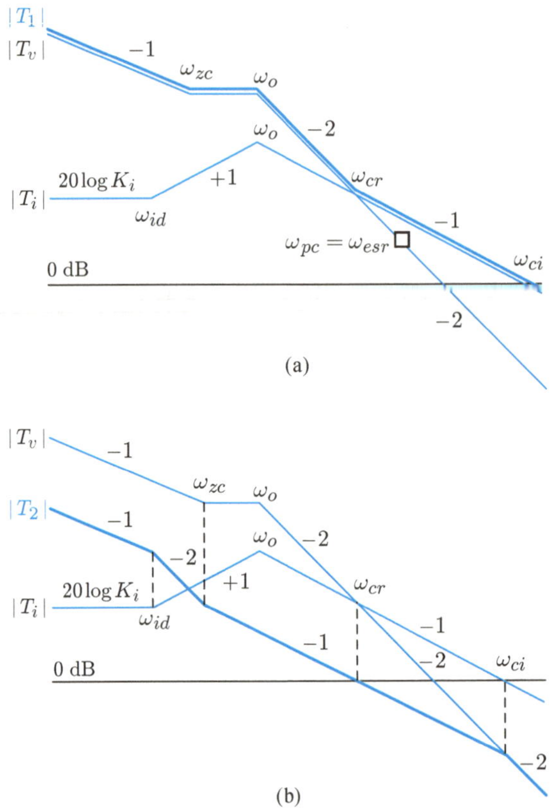

Figure 9.21 System loop gains. (a) Overall loop gain T_1. (b) Outer loop gain T_2.

$$
\begin{aligned}
T_v(s) &= G_{vd}(s)F_v(s)F_m' \\
&= \underbrace{V_S \frac{1+\dfrac{s}{\omega_{esr}}}{1+\dfrac{s}{Q\omega_o}+\dfrac{s^2}{\omega_o^2}}}_{G_{vd}(s)} \underbrace{\frac{K_v\left(1+\dfrac{s}{\omega_{zc}}\right)}{s\left(1+\dfrac{s}{\omega_{pc}}\right)}}_{F_v(s)} F_m'
\end{aligned} \tag{9.36}
$$

Figure 9.21(a) shows the asymptotic plots for $|T_i|$, $|T_v|$, and $|T_1|$. The asymptotic plot of $|T_v|$ indicates that the compensation pole ω_{pc} is placed at the esr zero, $\omega_{pc} = \omega_{esr}$, and the compensation zero ω_{zc} is located before the power stage double pole ω_o: $\omega_{zc} < \omega_o$. Justifications for these selections will be given shortly. Due to a $90°$ phase boost by ω_{zc}, $\angle T_v$ approaches $-180°$ at the frequency where $|T_i| = |T_v|$. The phase difference between $\angle T_i$ and $\angle T_v$ is $90°$, and the magnitude of the overall loop gain now becomes $|T_1| = \sqrt{2}\,|T_i| = \sqrt{2}\,|T_v|$ without showing any eccentric behavior. Thus, the two-pole one-zero circuit is well suited for the voltage feedback compensation. As will be confirmed in forthcoming discussions, the two-pole one-zero compensation is indeed the optimal structure for $T_v(s)$, which can be adopted to all the three basic PWM converters.

The asymptotic plot of the overall loop gain $|T_1|$ in Fig. 9.21(a) reveals the design strategy for the peak current mode control. *At low frequencies, the voltage loop should be large in magnitude for tight output regulation and good closed-loop performance, as is the case with voltage mode control.* The compensation zero ω_{zc} should be placed before the power stage double pole ω_o. As demonstrated in Section 8.4.2, the system becomes conditionally stable if ω_{zc} appears after ω_o. For this case, the converter could encounter the stability problem during transition periods in which the output of the feedback controller remains in saturation. To avoid this situation, ω_{zc} should appear before ω_o.

At high frequencies, the current loop should prevail over the voltage loop, $|T_i| \gg |T_v|$, so that $T_1(s) = T_i(s) + T_v(s) \approx T_i(s)$. The current loop $T_i(s)$ is inherently stable with a -20 dB/dec high frequency asymptote and $-90°$ final phase. Thus, the dominance of $T_i(s)$ at high frequencies ensures stability for $T_1(s)$. In fact, the crossover frequency and phase margin of $T_i(s)$ become those of the overall loop gain $T_1(s)$ if the $|T_i| \gg |T_v|$ condition is met at high frequencies.

Outer Loop Gain

The outer loop gain defined in (9.25) is now analyzed using the asymptotic method introduced in Section 8.1

$$
\begin{aligned}
T_2(s) &= \frac{T_v(s)}{1+T_i(s)} \\
&\approx \begin{cases} \dfrac{T_v(s)}{T_i(s)} & \text{at frequencies where } |T_i| > 1 \\[2ex] T_v(s) & \text{at frequencies where } |T_i| < 1 \end{cases}
\end{aligned}
\tag{9.37}
$$

Figure 9.21(b) shows the asymptotic plot for $|T_i|$, $|T_v|$, and $|T_2|$. The asymptotic plot for $|T_2|$ is constructed based on (9.37). The borderline of the approximation is the 0 dB crossover frequency of $|T_i|$, ω_{ci} in 9.21(b). The $|T_2|$ tracks $|T_v|$ at the frequencies beyond ω_{ci}. At frequencies lower than ω_{ci}, $|T_2|$ follows the line segments that are created from the asymptote sketch rules given in Table 8.1 in Chapter 8: $|T_2| = |T_v| - |T_i|$.

The outer loop gain $|T_2|$ is very different from the overall loop gain $|T_1|$ in structure, magnitude, and crossover frequency. Features of $|T_2|$ are summarized below.

1) The second-order power stage dynamics do not appear in $|T_2|$. The second-order dynamics, commonly appearing in both $T_i(s)$ and $T_v(s)$, are canceled in $T_2(s) \approx T_v(s)/T_i(s)$ and $|T_2|$ shows the -20 dB/dec slope for a wide frequency range.

2) The magnitude of $|T_2|$ is substantially smaller than the magnitude of $|T_1|$; that is $|T_2| \approx |T_v/T_i| \ll |T_1| = |T_i + T_v|$.

3) The crossover frequency of $|T_2|$ is much lower than that of $|T_1|$. In fact, $|T_2|$ crossover occurs at the frequency where $|T_v| = |T_i|$; namely, the condition $|T_v| = |T_i|$ indicates $|T_2| = |T_v|/|T_i| = 1 = 0$ dB. This frequency was previously denoted as ω_{cr}.

4) The outer loop gain has a high-frequency pole at the $|T_i|$ crossover frequency, denoted as ω_{ci} in Fig. 9.21(b). For sufficient phase margin, the $|T_2|$ crossover frequency ω_{cr} should appear well before this high-frequency pole ω_{ci}: $\omega_{cr} \ll \omega_{ci}$.

Loop Gain Characteristics

For current mode control, two system loop gains are defined at the different locations in the system. These two loop gains both provide useful information about the internal structure of the feedback controller, however, the informational contents of the two loop gains are very different. Thus, the roles of the two loop gains in the dynamic analysis and control design are distinct and unique. This situation is in sharp contrast to the case of voltage mode control, where only one system loop gain exists. For voltage mode control, the connections between the loop gain and closed-loop performance are direct and explicit. Thus, the loop gain analysis is straightforward.

For current mode control, the loop gain analysis is rather involved because the connections between the loop gains and closed-loop performance are indirect and implicit. Furthermore, the contributions of the two loop gains to the control design are very different. Discussions about the dynamic analysis and control design using the two loop gains are given in the next section.

Readers may be tempted to compare the loop gain characteristics of current mode control with those of voltage mode control. In that case, the comparison must be made with care, because a direct comparison between the loop gains of voltage mode control and current mode control does not provide consistent information about the performance of the converter or correctness of the control design. Indeed, the loop gain of voltage mode control usually *seems* superior to the outer loop gain of current mode control. However, this does not imply that voltage mode control outperforms current mode control, but does indicate that the outward message of the two loop gains should be interpreted differently and carefully in consideration of their connections to the closed-loop performance of the respective converter system.

9.2.5 Control Design Procedures

Based on the previous loop gain analysis, step-by-step design procedures for the peak current mode control are established. The design procedures are discussed in a general manner so that the results can be applied for all the three basic PWM converters.

As the first step, the current loop $T_i(s)$ is designed to offer good high-frequency characteristics for the overall loop gain $T_1(s)$. The second step is to create the voltage loop $T_v(s)$ for desirable properties of the outer loop gain $T_2(s)$.

Current Loop Design

The general expression for the current loop is given by

$$\begin{aligned} T_i(s) &= G_{id}(s)R_iF_m' \\ &= \underbrace{K_{id}\frac{1+\dfrac{s}{\omega_{id}}}{1+\dfrac{s}{Q\omega_o}+\dfrac{s^2}{\omega_o^2}}}_{G_{id}(s)} R_iF_m' = K_i\frac{1+\dfrac{s}{\omega_{id}}}{1+\dfrac{s}{Q\omega_o}+\dfrac{s^2}{\omega_o^2}} \end{aligned} \tag{9.38}$$

where

$$K_i = K_{id}R_iF_m' \tag{9.39}$$

The current loop $T_i(s)$ should dictate the overall loop $T_1(s)$ at high frequencies. This design objective is simply achieved by placing the 0 dB crossover frequency of $T_i(s)$ at higher frequencies. *The crossover frequency of $T_i(s)$ can be increased up to the frequencies where the validity and accuracy of the small-signal analysis are not impaired, typically $10\%-30\%$ of the switching frequency.* Readers may refer to Example 8.5 in Chapter 8 for the background of this statement. As will be demonstrated in the next chapter, this design strategy also prevents or minimizes the detrimental effects originated from the sampling effects of current mode control.

Once the $T_i(s)$ crossover frequency, denoted as ω_{ci} in Fig. 9.21, is selected, the current loop design proceeds based on the $T_i(s)$ expression in (9.38).

1) Determine the dc gain of the current loop so that the $T_i(s)$ crossover frequency occurs at the desired frequency. It is recommended to place the crossover frequency at $10\%-30\%$ of the switching frequency. From the asymptotic plot of $|T_i|$ in Fig. 9.21, the following relationship can be seen

$$20\log K_i + 20\log\left(\frac{\omega_o}{\omega_{id}}\right) - 20\log\left(\frac{\omega_{ci}}{\omega_o}\right) = 0\text{ dB} \tag{9.40}$$

where K_i is the dc gain of the current loop and ω_{ci} is the desired $T_i(s)$ crossover frequency. The preceding equation is converted into the design equation

$$K_i\left(\frac{\omega_o}{\omega_{id}}\right)\left(\frac{\omega_o}{\omega_{ci}}\right) = 1 \quad\Rightarrow\quad K_i = \frac{\omega_{id}\,\omega_{ci}}{\omega_o^2} \tag{9.41}$$

to determine the required K_i for the preselected ω_{ci}.

2) Determine the CSN gain R_i considering hardware constraints. The product of the peak inductor current $i_{L\,peak}$ and R_i should fall in the allowable voltage

range for the PWM block: $i_{L\,peak}R_i < V_{max}$ where V_{max} is the maximum allowable input voltage for the PWM block.

3) Determine the modulator gain F'_m using the relationship

$$K_i = K_{id}R_iF'_m \quad \Rightarrow \quad F'_m = \frac{K_i}{K_{id}R_i} \tag{9.42}$$

The K_{id} expressions for the three basic converters are given in Table 9.2. Once the modulator gain is fixed, the slope of the compensation ramp S_e is determined from

$$\boxed{F'_m = \frac{2}{(S_n - S_f + 2S_e)T_s} \quad \Rightarrow \quad S_e = \frac{1}{T_sF'_m} + \frac{S_f - S_n}{2}} \tag{9.43}$$

This design procedure places the $T_i(s)$ crossover frequency at the desired frequency. The $T_i(s)$ crossover frequency is the same as the crossover frequency of the overall loop gain $T_1(s) = T_i(s) + T_v(s)$, due to the condition $|T_i| \gg |T_v|$ at high frequencies.

Voltage Loop Design

The general expression for the voltage loop is given by

$$\begin{aligned} T_v(s) &= G_{vd}(s)F_v(s)F'_m \\ &= \underbrace{K_{vd}\frac{\left(1-\dfrac{s}{\omega_{rhp}}\right)\left(1+\dfrac{s}{\omega_{esr}}\right)}{1+\dfrac{s}{Q\omega_o}+\dfrac{s^2}{\omega_o^2}}}_{G_{vd}(s)} F_v(s)F'_m \end{aligned} \tag{9.44}$$

For buck converters, the RHP zero does not exist so $\omega_{rhp} = \infty$ in (9.44). The voltage loop should prevail over the current loop at low frequencies. For this purpose, the two-pole one-zero circuit is employed for the voltage feedback compensation

$$\boxed{F_v(s) = \frac{K_v\left(1+\dfrac{s}{\omega_{zc}}\right)}{s\left(1+\dfrac{s}{\omega_{pc}}\right)}} \tag{9.45}$$

The voltage loop $T_v(s)$ then becomes

$$T_v(s) = \underbrace{K_{vd}\frac{\left(1-\dfrac{s}{\omega_{rhp}}\right)\left(1+\dfrac{s}{\omega_{esr}}\right)}{1+\dfrac{s}{Q\omega_o}+\dfrac{s^2}{\omega_o^2}}}_{G_{vd}(s)} \underbrace{\frac{K_v\left(1+\dfrac{s}{\omega_{zc}}\right)}{s\left(1+\dfrac{s}{\omega_{pc}}\right)}}_{F_v(s)} F'_m \tag{9.46}$$

Referring to the asymptotic plots of the system loop gains in Fig. 9.21, the selection of the compensation parameters is explained below.

1) Place the compensation pole ω_{pc} at the lowest frequency among the RHP zero, esr zero, and half the switching frequency: $\omega_{pc} = \min\{\omega_{rhp}\ \omega_{esr}\ 0.5\,\omega_s\}$. The compensation pole ω_{pc} cancels ω_{rhp} or ω_{esr}, whichever comes first, and $T_v(s)$ thus maintains -40 dB/dec roll-off at high frequencies. This design strategy is necessary to ensure the dominance of the current loop at high frequencies.

2) Place the compensation zero ω_{zc} before the power stage double pole ω_o, in order to provide a 90° phase boost without rendering a conditionally stable system. As will be shown later, the position of ω_{zc} determines the speed of transient responses. For faster response, ω_{zc} should be placed at higher frequencies, yet still should not exceed the power stage double pole. As a rule of thumb, it is recommended $\omega_{zc} = (0.6-0.8)\,\omega_o$.

3) Adjust the integrator gain K_v for design trade-off. At this stage, the current loop $T_i(s)$ is fixed and the locations of the compensation pole and zero in $F_v(s)$ are determined. Now, the integrator gain K_v is the only undetermined design parameter. By changing K_v, the magnitude of the outer loop gain, $|T_2| = |T_v|/|1+T_i|$, is raised or lowered, thereby controlling the position of the $T_2(s)$ crossover frequency ω_{cr}. For design purposes, the position of ω_{cr} is first selected and the integrator gain K_v is determined later.

   - For buck converters, ω_{cr} can be positioned at high frequencies, approaching the esr zero: $\omega_{cr} = (0.3-1.0)\,\omega_{esr}$.
   - For boost and buck/boost converters, ω_{cr} should be selected at sufficiently lower frequencies than the RHP zero: $\omega_{cr} = (0.1-0.3)\,\omega_{rhp}$.

   Discussions about the selection of ω_{cr} will be given in Section 9.3.3 and Section 9.4.1. Once the $T_2(s)$ crossover frequency ω_{cr} is chosen, the integrator gain K_v is determined from the design equation

$$\frac{K_{vd}K_v}{K_{id}R_i\,\omega_{id}}\left(\frac{\omega_{id}}{\omega_{zc}}\right)^2\frac{\omega_{zc}}{\omega_{cr}} = 1 \quad\Rightarrow\quad K_v = \frac{K_{id}\,\omega_{cr}\,R_i\,\omega_{zc}}{\omega_{id}\,K_{vd}} \tag{9.47}$$

   Derivation of (9.47) is given in Example 9.7.

4) Check the phase margin of $T_2(s)$ and tune the integrator gain K_v to secure a $45° - 70°$ phase margin.

EXAMPLE 9.7 Design Equation for $T_2(s)$

This example shows the derivation of the design equation of $T_2(s)$ given in (9.47). Referring to Fig. 9.21(b), the magnitude of $T_2(s)$ at ω_{id} is given by

$$
\begin{aligned}
|T_2(j\omega_{id})| &\approx \left|\frac{T_v(j\omega_{id})}{T_i(j\omega_{id})}\right| \\
&= 20\log\left(\frac{K_{vd}\dfrac{K_v}{\omega_{id}}F'_m}{K_{id}R_iF'_m}\right) = 20\log\left(\frac{K_{vd}K_v}{K_{id}R_i\,\omega_{id}}\right)
\end{aligned}
$$

Figure 9.21(b) also indicates that

$$
20\log\left(\frac{K_{vd}K_v}{K_{id}R_i\,\omega_{id}}\right) - 40\log\left(\frac{\omega_{zc}}{\omega_{id}}\right) - 20\log\left(\frac{\omega_{cr}}{\omega_{zc}}\right) = 0\ \text{dB}
$$

which transforms to (9.47) in the linear scale.

Circuit for Two-Pole One-Zero Compensation

This section discusses the implementation of the voltage feedback compensation of (9.45). Figure 9.22 shows the circuit for the two-pole one-zero compensation. The resistance R_x is used to control the magnitude of the output voltage: $V_O = V_{ref}(1 + R_1/R_x)$. If R_x is not used, the output voltage is regulated at the reference voltage: $V_O = V_{ref}$ with $R_x = \infty$. This resistance is irrelevant to the voltage feedback compensation. Direct circuit analysis shows that

$$
F_v(s) = \frac{Z_2(s)}{Z_1(s)} = \frac{K_v\left(1+\dfrac{s}{\omega_{zc}}\right)}{s\left(1+\dfrac{s}{\omega_{pc}}\right)} \tag{9.48}
$$

with

$$
\begin{aligned}
K_v &= \frac{1}{R_1(C_2+C_3)} \\
\omega_{zc} &= \frac{1}{R_2C_2} \\
\omega_{pc} &= \frac{1}{R_2\left(\dfrac{C_2C_3}{C_2+C_3}\right)}
\end{aligned}
$$

Once the compensation parameters are selected, the circuit components in Fig. 9.22 are determined using the above equations. Among the four circuit components, one can arbitrarily be chosen and the other components are found using the equation (9.48).

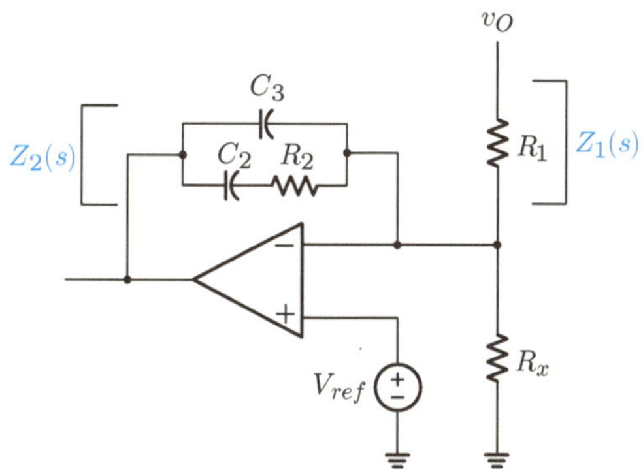

Figure 9.22 Circuit implementation of two-pole one-zero compensation.

The design procedures developed in this section are summarized in Table 9.3 for easy reference.

Buck Converter Design and Performance Evaluation — EXAMPLE 9.8

This example illustrates the control design and closed-loop performance of the peak current mode control adapted to a buck converter. Figure 9.23 shows the circuit diagram of the buck converter. The converter regulates the output voltage at 4 V using the two-pole one-zero compensation. From the power stage circuit parameters, the corner frequencies and dc gains of power stage transfer functions are determined as

$$\omega_o = \frac{1}{\sqrt{LC}} = \frac{1}{\sqrt{40 \times 10^{-6}\, 470 \times 10^{-6}}} = 2\pi \cdot 1.16 \times 10^3 \text{ rad/s}$$

$$\omega_{esr} = \frac{1}{CR_c} = \frac{1}{470 \times 10^{-6}\, 0.1} = 2\pi \cdot 3.39 \times 10^3 \text{ rad/s}$$

$$K_{vd} = V_s = 16$$

$$\omega_{id} = \frac{1}{CR} = \frac{1}{470 \times 10^{-6}\, 1} = 2\pi \cdot 339 \text{ rad/s}$$

$$K_{id} = \frac{V_S}{R} = \frac{16}{1} = 16$$

The switching frequency is $\omega_s = 2\pi \cdot 50 \times 10^3$ rad/s and the duty ratio is $D = 4/16 = 0.25$. The maximum input voltage for the PWM block is assumed as $V_{max} = 5.0$ V and the peak value of the inductor current is calculated as

$$\begin{aligned} i_{L\,peak} &= \frac{V_O}{R} + \frac{1}{2}\frac{V_S - V_O}{L} DT_s \\ &= \frac{4}{1} + \frac{1}{2}\frac{16-4}{40 \times 10^{-6}} 0.25 \cdot 20 \times 10^{-6} = 4.75 \text{ A} \end{aligned}$$

Table 9.3 Design Procedures for Peak Current Mode Control

Current loop design

$$T_i(s) = K_{id}\underbrace{\frac{1+\dfrac{s}{\omega_{id}}}{1+\dfrac{s}{Q\omega_o}+\dfrac{s^2}{\omega_o^2}}}_{G_{id}(s)} R_i F'_m = K_i \frac{1+\dfrac{s}{\omega_{id}}}{1+\dfrac{s}{Q\omega_o}+\dfrac{s^2}{\omega_o^2}} \quad \text{with } K_i = K_{id}\,R_i\,F'_m$$

1) Select $T_i(s)$ crossover frequency: $\omega_{ci} = (0.1 - 0.3)\,\omega_s$.
2) Evaluate the dc gain: $K_i = (\omega_{id}\,\omega_{ci})/\omega_o^2$.
3) Select the CSN gain R_i such that $i_{L\,peak} R_i < V_{max}$ where V_{max} is the maximum allowable input voltage for the PWM block.
4) Evaluate the modulator gain: $F'_m = K_i/(K_{id}R_i)$.
5) Evaluate the compensation ramp slope: $S_e = 1/(T_s F'_m) + (S_f - S_n)/2$.

Voltage loop design

$$T_v(s) = K_{vd}\underbrace{\frac{\left(1-\dfrac{s}{\omega_{rhp}}\right)\left(1+\dfrac{s}{\omega_{esr}}\right)}{1+\dfrac{s}{Q\omega_o}+\dfrac{s^2}{\omega_o^2}}}_{G_{vd}(s)}\underbrace{\frac{K_v\left(1+\dfrac{s}{\omega_{zc}}\right)}{s\left(1+\dfrac{s}{\omega_{pc}}\right)}}_{F_v(s)} F'_m$$

1) Set the compensation pole: $\omega_{pc} = \min\{\omega_{rhp}\;\; \omega_{esr}\;\; 0.5\,\omega_s\}$.
2) Select the compensation zero: $\omega_{zc} = (0.6 - 0.8)\,\omega_o$.
3) Set $T_2(s)$ crossover frequency:

   $\omega_{cr} = (0.3 - 1.0)\,\omega_{esr}$ for buck converters, and

   $\omega_{cr} = (0.1 - 0.3)\,\omega_{rhp}$ for boost and buck/boost converters.
4) Evaluate the integrator gain: $K_v = (K_{id}\,\omega_{cr}\,R_i\,\omega_{zc})/(\omega_{id}\,K_{vd})$.
5) Check $T_2(s)$ phase margin and adjust K_v for a $45° - 70°$ phase margin.
6) Evaluate the circuit components for voltage feedback compensation using (9.48).

For buck converters, ω_{rhp} does not exist so $\omega_{rhp} = \infty$.

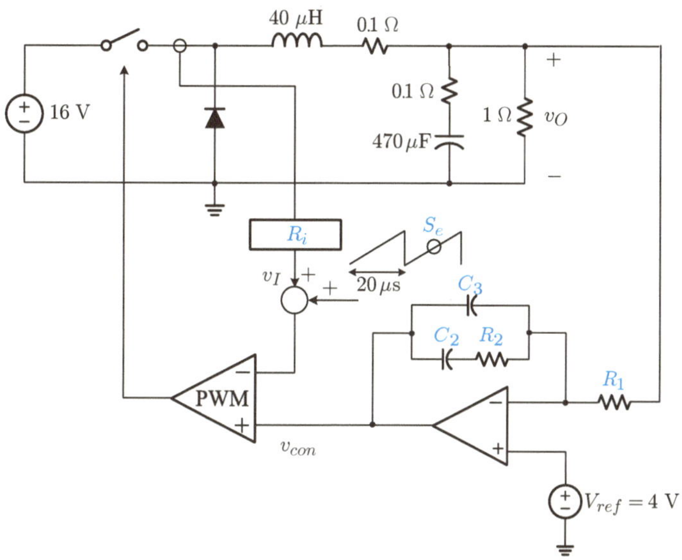

Figure 9.23 Current-mode controlled buck converter: $R_i = 0.67$, $S_e = 1.46 \times 10^5$ V/s, $R_1 = 10$ kΩ, $R_2 = 92.3$ $k\Omega$, $C_2 = 1.86$ nF, and $C_3 = 0.70$ nF.

Based on the proposed design procedures, the control design is performed as follows.

Current Loop Design

1) $T_i(s)$ crossover frequency: $\omega_{ci} = 0.2\,\omega_s = 0.2\,(2\pi \cdot 50 \times 10^3) = 2\pi \cdot 10 \times 10^3$ rad/s
2) Dc gain of $T_i(s)$:

$$K_i = \frac{\omega_{id}\,\omega_{ci}}{\omega_o^2} = \frac{(2\pi \cdot 339)(2\pi \cdot 10 \times 10^3)}{(2\pi \cdot 1.16 \times 10^3)^2} = 2.52$$

3) CSN gain:

$$R_i < \frac{V_{max}}{i_{L\,peak}} = \frac{5.0}{4.75} = 1.05 \quad \Rightarrow R_i = 0.67$$

4) Modulator gain:

$$F_m' = \frac{K_i}{K_{id} R_i} = \frac{2.52}{16 \cdot 0.67} = 0.235$$

5) Compensation ramp:

$$\begin{aligned} S_e &= \frac{1}{T_s F_m'} + \frac{S_f - S_n}{2} \\ &= \frac{1}{20 \times 10^{-6}\, 0.235} + \frac{\dfrac{4}{40 \times 10^{-6}} 0.67 - \dfrac{16-4}{40 \times 10^{-6}} 0.67}{2} \\ &= 1.46 \times 10^5 \text{ V/s} \end{aligned}$$

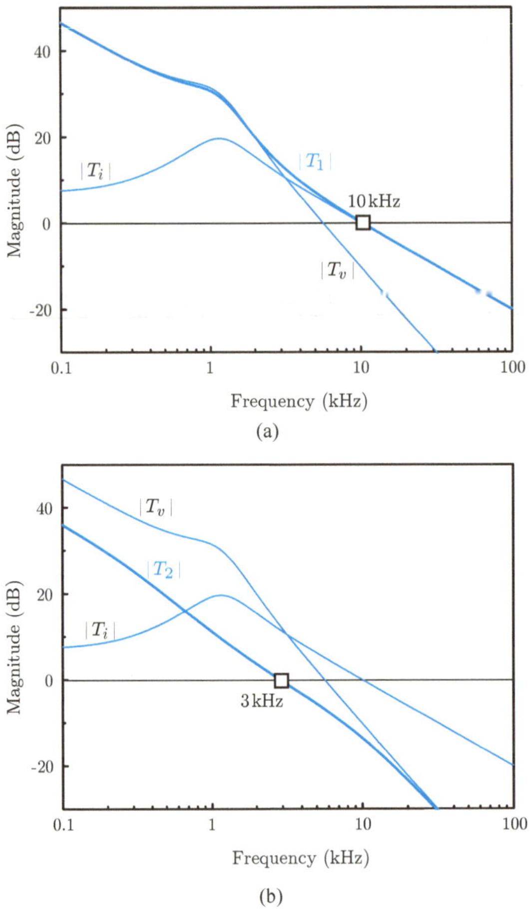

Figure 9.24 Individual feedback loops and system loop gains. (a) Overall loop gain T_1. (b) Outer loop gain T_2.

Voltage Loop Design

1) Compensation pole: $\omega_{pc} = \omega_{esr} = 2\pi \cdot 3.39 \times 10^3$ rad/s
2) Compensation zero: $\omega_{zc} = 0.8\,\omega_o = 0.8\,(2\pi \cdot 1.16 \times 10^3) = 2\pi \cdot 928$ rad/s
3) $T_2(s)$ crossover frequency: $\omega_{cr} = \omega_{esr} = 2\pi \cdot 3.39 \times 10^3$ rad/s
4) Integrator gain:

$$
\begin{aligned}
K_v &= \frac{K_{id}\,\omega_{cr}\,R_i\,\omega_{zc}}{\omega_{id}\,K_{vd}} \\
&= \frac{16\,(2\pi \cdot 3.39 \times 10^3)\,0.67\,(2\pi \cdot 928)}{(2\pi \cdot 339)16} \\
&= 3.91 \times 10^4
\end{aligned}
$$

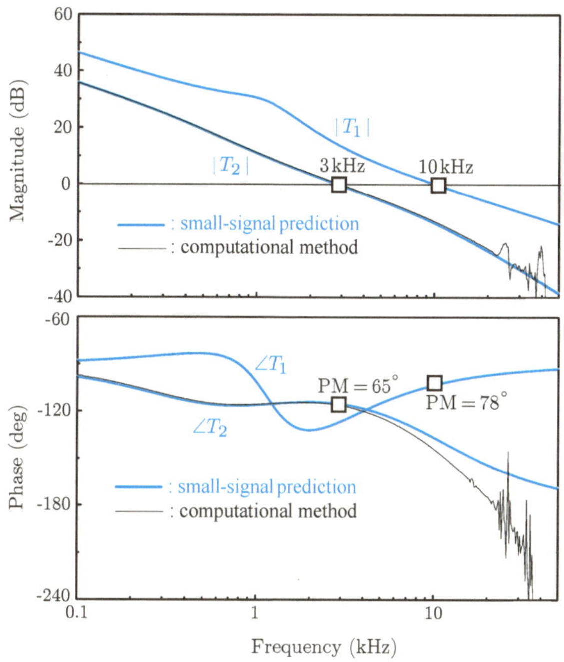

Figure 9.25 Comparison between overall loop gain T_1 and outer loop gain T_2.

5) Voltage feedback circuit: $R_1 = 10$ kΩ

$$\Rightarrow R_2 = 92.3\ k\Omega,\ C_2 = 1.86\ \text{nF, and}\ C_3 = 0.70\ \text{nF}$$

The performance of the converter is evaluated using PSpice®simulations. Whenever appropriate and informative, the theoretical predictions are compared with the results of the computational method. Figure 9.24 displays the Bode plots of the individual feedback loops and system loop gains. The Bode plots shows a close resemblance to the asymptotic plots in Fig. 9.21. Figure 9.24(a) confirms that the crossover frequency of $T_1(s)$ is located at the exact target frequency, $\omega_{ci} = 2\pi \cdot 10 \times 10^3$ rad/s. On the other hand, the $T_2(s)$ crossover frequency is placed in close proximity to the design aim, $\omega_{cr} = 2\pi \cdot 3.0 \times 10^3$ rad/s.

Figure 9.25 compares the Bode plot of $T_1(s)$ and $T_2(s)$. The theoretical forecasts of the outer loop gain $T_2(s)$ are compared with the results of the computational method. The theoretical predictions and computational results exhibit a notable disparity in the phase characteristics. This difference is due to the sampling effects of the peak current mode control, which are not considered in the classical analysis. Further discussions about this observation will be given in the next chapter. Both the loop gains have a sufficient phase margin; the phase margin of $T_1(s)$ is 78° and that of $T_2(s)$ is 65°.

The output impedance and audio-susceptibility characteristics are shown in Fig. 9.26. The time-domain performance of the converter is displayed in Fig. 9.27. Figure 9.27(a) is the transient response of the output voltage due to the step changes of

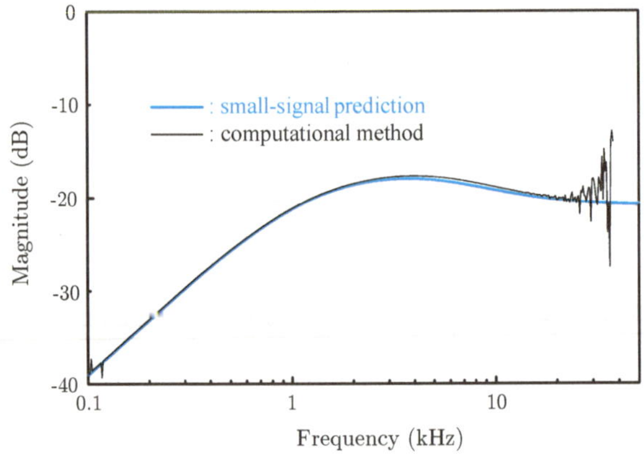

(a)

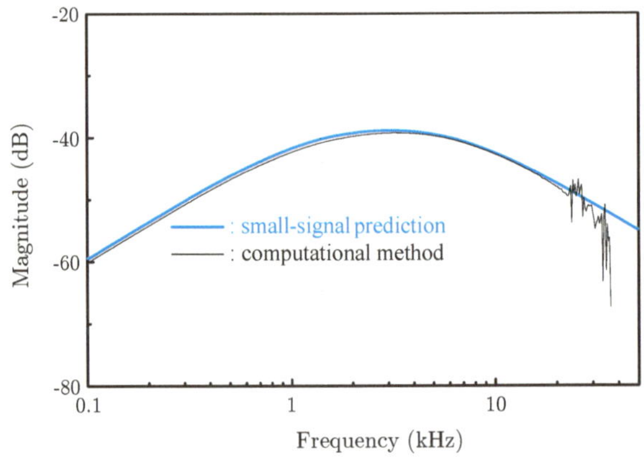

(b)

Figure 9.26 Closed-loop performance. (a) Output impedance. (b) Audio-susceptibility.

$R = 1\ \Omega \Rightarrow 0.5\ \Omega \Rightarrow 1\ \Omega$ in the load resistance, while Fig. 9.27(b) shows the transient response with the $V_S = 16\ \text{V} \Rightarrow 8\ \text{V} \Rightarrow 16\ \text{V}$ changes in the input voltage.

9.2.6 Converter Dynamics in DCM

As discussed in the earlier chapters, the power stage dynamics significantly alter when dc-to-dc converters cross the CCM/DCM boundary. Accordingly, it is needed to investigate the impacts of DCM operation on the converter performance.

In Section 8.5, we demonstrated that the control design intended for CCM operation guarantees stability in DCM operation and therefore is suited for both CCM operation and DCM operation. However, it was also found that the closed-loop performance of the converter will be deteriorated in DCM operation. The degree of the

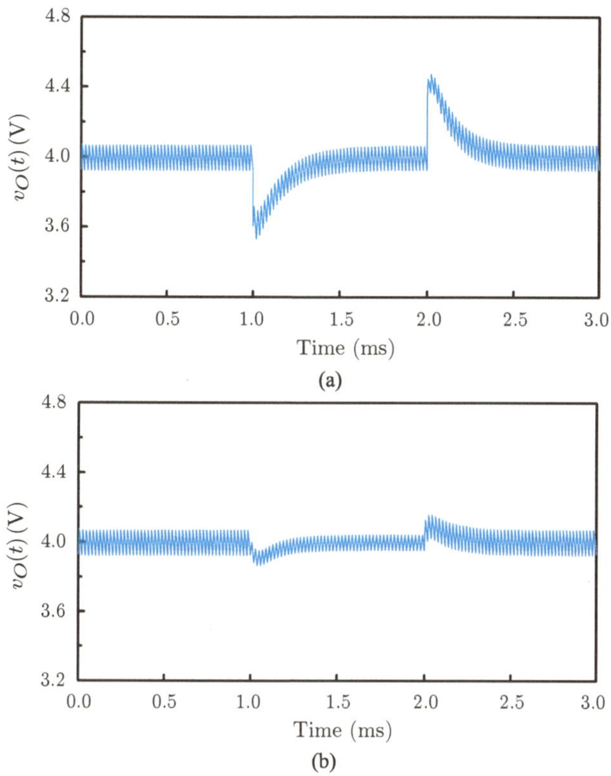

Figure 9.27 Transient response. (a) Step load response. (b) Step input response.

performance degradation is related to the control scheme of the converter. Example 8.11 in Chapter 8 analyzed the performance change of a buck converter with voltage mode control.

The current section investigates the performance change of the converter with current mode control. It will be shown that current mode control could reduce the performance change and could offer near-uniform loop gain characteristics for both CCM operation and DCM operation.

Converter Performance in DCM — EXAMPLE 9.9

The performance of the buck converter in the previous example is evaluated in DCM operation and compared with that of the CCM operation. For the given operational conditions, the critical resistance for the CCM/DCM boundary is determined as $R_{crit} = 2L/(D'T_s) = 5.33\ \Omega$. The DCM power stage model, developed earlier in Section 5.5.2, is combined with the small-signal model of peak current mode control, resulting in the complete small-signal model in DCM operation. The model predictions are presented in comparison with the results of the computational method.

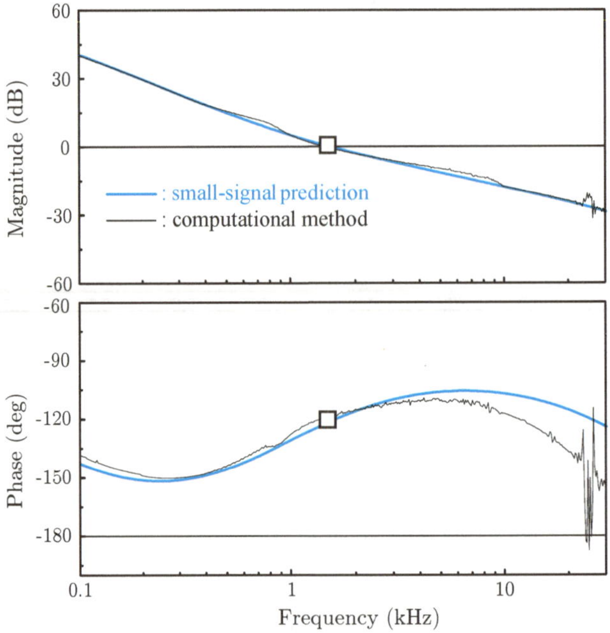

Figure 9.28 Outer loop gain in DCM with $R = 10\ \Omega$.

Figure 9.28 shows the outer loop gain of the converter evaluated with $R = 10\ \Omega$. The results confirm the accuracy of the DCM analysis.

Figure 9.29 displays the DCM outer loop gains, evaluated with $R = R_{crit} = 5.33\ \Omega$, $R = 10\ \Omega$, and $R = 20\ \Omega$, along with the CCM loop gain with $R = 1\ \Omega$. The loop gains exhibit relatively small changes in both the crossover frequency and phase margin, compared to the case of the conventional voltage mode control shown in Fig. 8.40 in Example 8.11. The reduced sensitivity is explained as follows. The change in the power stage dynamics consistently occurs in all the power stage transfer functions. In DCM operations, the split of the power stage double pole and reduction in the mid-band gain will appear in both $G_{vd}(s)$ and $G_{id}(s)$. For frequencies where $|T_i| \gg 1$, the outer loop gain is approximated as $T_2(s) \approx T_v(s)/T_i(s) = G_{vd}(s)F_v(s)F_m'/(G_{id}(s)R_iF_m')$. Accordingly, the changes occurring in both $G_{vd}(s)$ and $G_{id}(s)$ will be canceled and will not show in $T_2(s)$.†

Lastly, Fig. 9.30 compares the transitional output voltage in response to the step changes in the load resistance: $R = 0.8\ \Omega \Rightarrow 1\ \Omega \Rightarrow 0.8\ \Omega$ for CCM case, and $R = 5.3\ \Omega \Rightarrow 20\ \Omega \Rightarrow 5.3\ \Omega$ for DCM case. When compared with the voltage mode control in Fig. 8.42, the difference between the CCM and DCM transient responses is far less pronounced.

---

†For voltage mode control, this canceling does not happen and the changes in $G_{vd}(s)$ will directly propagate to the loop gain, thus showing very different loop gain characteristics in CCM operation and DCM operation.

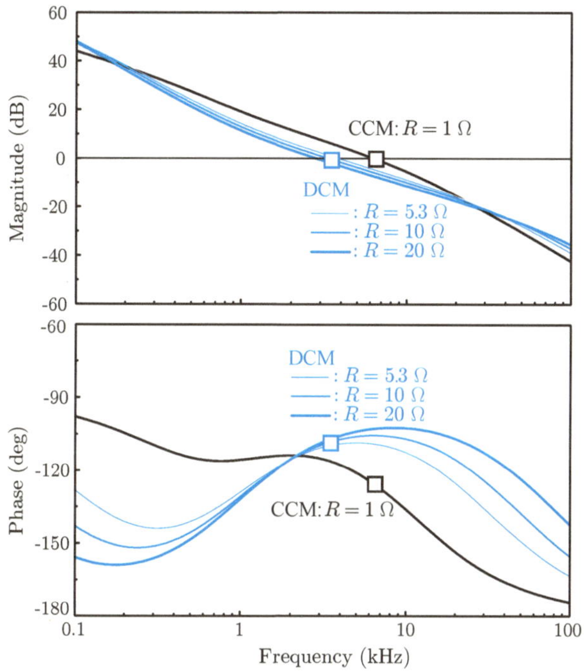

Figure 9.29 Outer loop gain in CCM and DCM operations.

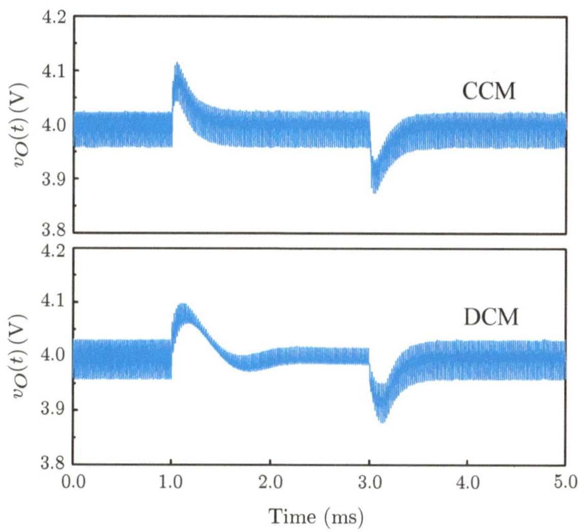

Figure 9.30 Step load response in CCM and DCM operations.

9.3 CLOSED-LOOP PERFORMANCE OF PEAK CURRENT MODE CONTROL

The previous section investigated the dynamics of the peak current mode control, focusing on the individual feedback loops and system loop gains, in consideration of their role in the stability analysis and control design. Based on these analyses, step-by-step control design procedures are established. It was found that the two-pole one-zero compensation offers both stability and good loop gain characteristics.

This section now investigates the closed-loop performance of the current-mode controlled PWM converters that employ the two-pole one-zero compensation. The audio-susceptibility and output impedance are analyzed in detail, addressing the impacts of the voltage feedback compensation. In particular, the choice of the integrator gain is discussed in conjunction with the position of the crossover frequency of the outer loop gain $T_2(s)$.

This section also analyzes the step load response of current-mode controlled PWM converters. The relationship between the output impedance and step load response is studied, leading to practical methods for predicting the step load response from the output impedance characteristics.

The analysis is presented in a general manner, as such, the results are applicable to all the three basic converters. This section first presents buck converter examples. Boost converter examples will be treated in the next section.

9.3.1 Audio-Susceptibility Analysis

The audio-susceptibility denotes the closed-loop input-to-output transfer function. The general expression of the audio-susceptibility for the three basic PWM converters is derived from Fig. 9.15

$$A_u(s) = \frac{\hat{v}_o(s)}{\hat{v}_s(s)} = \frac{G_{vs}(1 + G_{id}R_iF'_m) - G_{is}R_iF'_mG_{vd}}{1 + G_{id}R_iF'_m + G_{vd}F_vF'_m} \tag{9.49}$$

Using the definitions of the current loop, $T_i(s) = G_{id}(s)R_iF'_m$, and voltage loop, $T_v(s) = G_{vd}(s)F_v(s)F'_m$, the preceding expression is written as

$$\begin{aligned} A_u(s) &= \frac{G_{vs}(1+T_i) - G_{is}\dfrac{T_i}{G_{id}}G_{vd}}{1+T_i+T_v} \\ &= \boxed{\frac{G_{vs} + T_i\left(G_{vs} - \dfrac{G_{is}G_{vd}}{G_{id}}\right)}{1+T_i+T_v}} \end{aligned} \tag{9.50}$$

Now, the audio-susceptibility is analyzed using the expression (9.50). We first consider buck converters and later deal with boost and buck/boost converters.

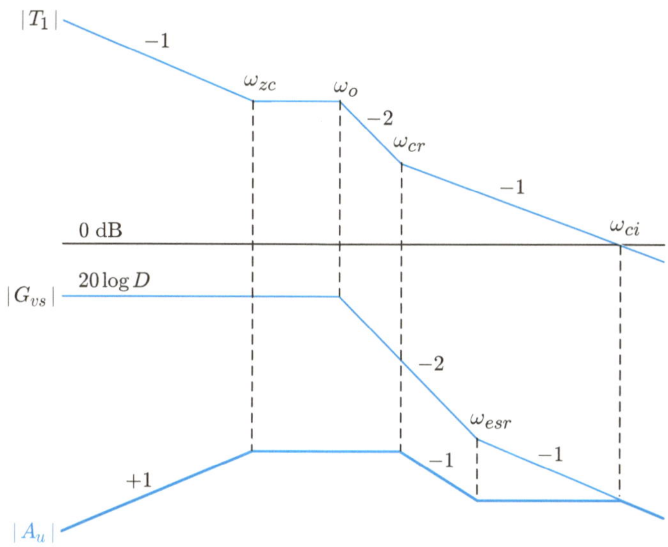

Figure 9.31 Asymptotic analysis of audio-susceptibility.

Buck Converter

The buck converter is a special case in which the audio-susceptibility analysis is performed in a simple way. For buck converters, it can be shown from Table 9.2 that

$$\frac{G_{vs}(s)}{G_{vd}(s)} = \frac{G_{is}(s)}{G_{id}(s)} = \frac{D}{V_s} \tag{9.51}$$

which implies

$$G_{vs}(s) - \frac{G_{is}(s)G_{vd}(s)}{G_{id}(s)} = 0 \tag{9.52}$$

The audio-susceptibility expression in (9.50) now reduces to

$$\boxed{A_u(s) = \frac{G_{vs}(s)}{1 + T_i(s) + T_v(s)} = \frac{G_{vs}(s)}{1 + T_1(s)}} \tag{9.53}$$

where $T_1(s) = T_i(s) + T_v(s)$ is the overall loop gain. Thus, for buck converters, the audio-susceptibility analysis is identical to that of voltage mode control. The role of the loop gain $T_m(s)$ in voltage mode control is replaced with the overall loop gain $T_1(s)$ in current mode control.

The asymptotic analysis is performed on (9.53), yielding the result shown in Fig. 9.31. The asymptotic plot is constructed using the $G_{vs}(s)$ expression in Table 9.2 and the $|T_1|$ structure shown in Fig. 9.21(a). According to the previous design procedures, the parameters of the voltage feedback compensation are selected as

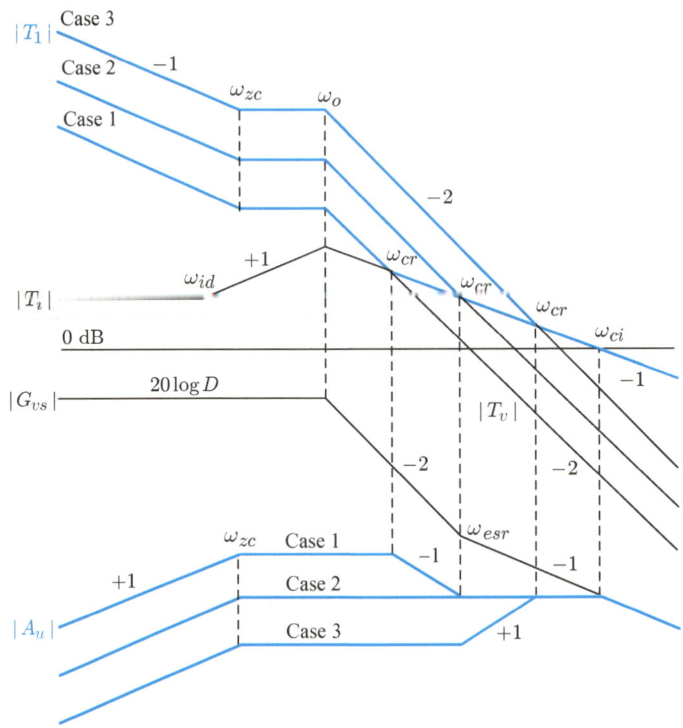

Figure 9.32 Asymptotic analysis of audio-susceptibility.

$\omega_{pc} = \omega_{esr}$ and $\omega_{zc} < \omega_o$. The asymptotic plot is converted into an equation for the audio-susceptibility

$$A_u(s) = \frac{D}{V_S F_m' K_v} s \frac{1 + \dfrac{s}{\omega_{esr}}}{\left(1 + \dfrac{s}{\omega_{zc}}\right)\left(1 + \dfrac{s}{\omega_{cr}}\right)\left(1 + \dfrac{s}{\omega_{ci}}\right)} \tag{9.54}$$

based on the analysis technique presented in Table 8.2. The parameter ω_{cr} denotes the frequency where $|T_i| = |T_v|$, which corresponds to the crossover frequency of the outer loop gain $T_2(s)$. Figure 9.31 is a specific example where ω_{cr} is located below the esr zero: $\omega_{cr} < \omega_{esr}$.

The audio-susceptibility analysis is generalized in Fig. 9.32 which shows the asymptotic plots for $|G_{vs}|$, $|T_1|$, and $|A_u|$, along with $|T_i|$ and $|T_v|$. In this general analysis, the integrator gain K_v of the voltage feedback compensation is varied to result in the following three cases:

- Case 1: $\omega_{cr} < \omega_{esr}$
- Case 2: $\omega_{cr} = \omega_{esr}$
- Case 3: $\omega_{cr} > \omega_{esr}$

The effect of K_v is clear in Fig. 9.32. The larger K_v shifts ω_{cr} towards higher frequencies and reduces the peak value of $|A_u|$, $|A_u|_{peak}$. However, this reduction only happens up to the K_v value that places ω_{cr} at ω_{esr}. Increasing K_v beyond this value does not reduce $|A_u|_{peak}$ any further. Instead, it could deteriorate the phase characteristics of $T_1(s)$; the phase margin will be lessened as ω_{cr} approaches the crossover frequency ω_{ci}. For Cases 1 and 2, the peak value of the audio-susceptibility is determined as

$$|A_u|_{peak} = 20 \log \left(\frac{D\,\omega_{zc}}{V_S F_m' K_v} \right) \tag{9.55}$$

For Case 3, the peak value of the audio-susceptibility is given by

$$\begin{aligned} |A_u|_{peak} &= 20 \log \left(\frac{D\,\omega_{zc}}{V_S F_m' K_v} \right) + 20 \log \left(\frac{\omega_{cr}}{\omega_{esr}} \right) \\ &= 20 \log \left(\frac{D\,\omega_{zc}}{V_S F_m' K_v} \frac{\omega_{cr}}{\omega_{esr}} \right) \end{aligned} \tag{9.56}$$

For all the three cases, the compensation zero ω_{zc} becomes the first pole in the audio-susceptibility. As demonstrated in Section 8.4.5, this pole determines the settling time of the step input response, $t_s = 3/\omega_{zc}$. Thus, ω_{zc} should be placed as high as possible, but without exceeding the power stage double pole ω_o. As a general guideline, it was recommended that $\omega_{zc} = (0.6-0.8)\,\omega_o$ in the previous section.

Buck Converter Example **EXAMPLE 9.10**

This example illustrates the accuracy of the previous audio-susceptibility analysis. The integrator gain of the buck converter used in Example 9.8 is now modified to result in the three cases discussed in this section, while other compensation parameters remain the same

- Case 1: $K_v = 1.97 \times 10^4 \quad \Rightarrow \omega_{cr} = 0.5\,\omega_{esr}$
- Case 2: $K_v = 3.91 \times 10^4 \quad \Rightarrow \omega_{cr} = \omega_{esr}$
- Case 3: $K_v = 7.82 \times 10^4 \quad \Rightarrow \omega_{cr} = 2.0\,\omega_{esr}$

Figure 9.33 shows the simulations of various the transfer functions of the three cases, which validate the preceding theoretical discussions.

Boost and Buck/Boost Converters

For other topologies, the expression of the audio-susceptibility does not reduce to the simple form of (9.53). For boost and buck/boost converters, the audio-susceptibility analysis proceeds from (9.50). At the frequencies below the $T_i(s)$ crossover frequency, where $|T_i| \gg 1$, the audio-susceptibility expression in (9.50) is approximated

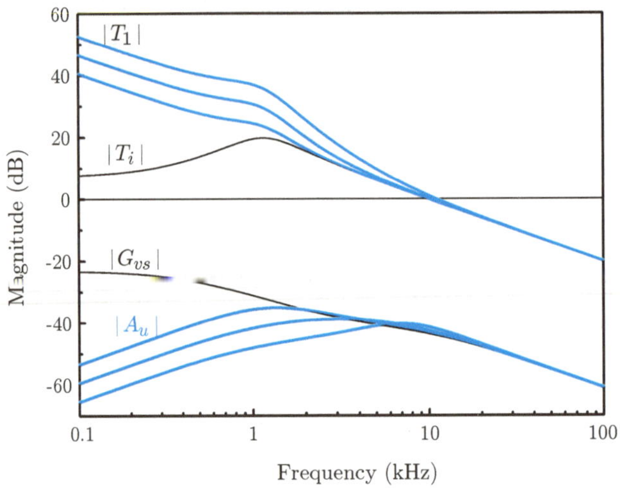

Figure 9.33 Audio-susceptibility of buck converter.

as

$$A_u(s) = \frac{G_{vs} + T_i\left(G_{vs} - \dfrac{G_{is}G_{vd}}{G_{id}}\right)}{1 + T_i + T_v}$$

$$\approx \frac{G_{vs} + (1 + T_i)\left(G_{vs} - \dfrac{G_{is}G_{vd}}{G_{id}}\right)}{1 + T_i + T_v} \tag{9.57}$$

This equation is rearranged as

$$A_u(s) = \frac{G_{vs}}{1 + T_i + T_v} + \frac{G_{vs} - \dfrac{G_{is}G_{vd}}{G_{id}}}{1 + \dfrac{T_v}{1 + T_i}}$$

$$= \frac{G_{vs}}{1 + T_1} + \frac{G_{vs} - \dfrac{G_{is}G_{vd}}{G_{id}}}{1 + T_2} \tag{9.58}$$

where $T_1(s) = T_i(s) + T_v(s)$ is the overall loop gain and $T_2(s) = T_v(s)/(1 + T_i(s))$ is the outer loop gain. From the definitions of $T_1(s)$ and $T_2(s)$, it follows that

$$|T_1| = |T_i + T_v| \gg |T_2| = \left|\frac{T_v}{1 + T_i}\right| \tag{9.59}$$

It also can be shown from Table 9.2 that

$$\left|G_{vs} - \frac{G_{is}G_{vd}}{G_{id}}\right| \gg |G_{vs}| \tag{9.60}$$

Table 9.4 Expressions for Numerator of $A_u(s)$ and $Z_o(s)$

	$A_{ui}(s) = G_{vs} - \dfrac{G_{is}G_{vd}}{G_{id}}$	$Z_{oi}(s) = Z_p - \dfrac{Z_q G_{vd}}{G_{id}}$
Buck	0	$R\dfrac{1+sCR_c}{1+sC(R+R_c)}$
Boost	$\dfrac{1}{2(1-D)}\dfrac{1+sCR_c}{1+sC(R+R_c)/2}$	$\dfrac{R}{2}\dfrac{1+sCR_c}{1+sC(R+R_c)/2}$
Buck/boost	$K_u\dfrac{1+sCR_c}{1+sC(R+R_c)/(1+D)}$	$K_z\dfrac{1+sCR_c}{1+sC(R+R_c)/(1+D)}$
	$K_u = \dfrac{D^2}{(1+D)(1-D)}$	$K_z = \dfrac{R}{(1+D)}$

The two inequalities of (9.59) and (9.60) further simplify the audio-susceptibility to

$$\boxed{A_u(s) \approx \frac{G_{vs} - \dfrac{G_{is}G_{vd}}{G_{id}}}{1+T_2(s)} = \frac{A_{ui}(s)}{1+T_2(s)}} \tag{9.61}$$

with $A_{ui}(s) = G_{vs} - G_{is}G_{vd}/G_{id}$.

The asymptotic analysis is now applied to (9.61) in order to investigate the audio-susceptibility characteristics. In this analysis, the outer loop gain $T_2(s)$ alone appears in the denominator while the overall loop gain $T_1(s)$ is not involved at all. This is in contrast to the buck converter case where $T_1(s)$ is the only loop gain associated with the audio-susceptibility analysis.

Using the power stage transfer functions in Table 9.2, the numerator of (9.61), $A_{ui}(s) = G_{vs} - G_{is}G_{vd}/G_{id}$, is expressed in a simple form. The results are shown in the left-hand side of Table 9.4. Using these expressions, along with knowledge about the outer loop gain $T_2(s)$, the audio-susceptibility is analyzed. Details about this analysis are given in a later section which handles a boost converter. It should be reminded that this analysis is valid only for the frequencies below the $T_i(s)$ crossover frequency. At higher frequencies where $|T_i| \approx 0$ and $|T_v| \approx 0$, the audio-susceptibility simply follows $G_{vs}(s)$, as can be seen from (9.50).

9.3.2 Output Impedance Analysis

The output impedance is another important performance metric for closed-loop controlled converters. The output impedance is analyzed in the same manner as that of the audio-susceptibility. The output impedance is derived from Fig. 9.15.

$$\begin{aligned} Z_o(s) &= \frac{Z_p(1+G_{id}R_iF'_m) - Z_qR_iF'_mG_{vd}}{1+G_{id}R_iF'_m+G_{vd}F_vF'_m} \\ &= \frac{Z_p(1+T_i) - Z_q\dfrac{T_i}{G_{id}}G_{vd}}{1+T_i+T_v} \end{aligned} \tag{9.62}$$

This expression is simplified to

$$Z_o(s) = \frac{Z_p + T_i\left(Z_p - \dfrac{Z_qG_{vd}}{G_{id}}\right)}{1+T_i+T_v} \approx \frac{Z_p + (1+T_i)\left(Z_p - \dfrac{Z_qG_{vd}}{G_{id}}\right)}{1+T_i+T_v} \tag{9.63}$$

in the frequency range below the $T_i(s)$ crossover frequency, and rearranged as

$$\begin{aligned} Z_o(s) &= \frac{Z_p}{1+T_i+T_v} + \frac{Z_p - \dfrac{Z_qG_{vd}}{G_{id}}}{1+\dfrac{T_v}{1+T_i}} \\ &= \frac{Z_p}{1+T_1} + \frac{Z_p - \dfrac{Z_qG_{vd}}{G_{id}}}{1+T_2} \end{aligned} \tag{9.64}$$

Due to the conditions $|T_1| \gg |T_2|$ and $|Z_p - Z_qG_{vd}/G_{id}| \gg |Z_p|$, the output impedance is further simplified to

$$\boxed{Z_o(s) \approx \frac{Z_p - \dfrac{Z_qG_{vd}}{G_{id}}}{1+T_2(s)} = \frac{Z_{oi}(s)}{1+T_2(s)}} \tag{9.65}$$

with $Z_{oi}(s) = Z_p - Z_qG_{vd}/G_{id}$.

The numerator of (9.65), $Z_{oi}(s) = Z_p - Z_qG_{vd}/G_{id}$, is interpreted as the output impedance evaluated under the condition that the current loop is only closed and the voltage loop is broken. This is because the output impedance $Z_o(s)$ reduces to $Z_{oi}(s)$ when $T_2(s) = T_v(s)/(1+T_i(s)) = 0$, which is true when $T_v(s) = 0$. The expressions for $Z_{oi}(s)$ are shown in the right-hand side of Table 9.4. For all the three basic converters, the $Z_{oi}(s)$ expression fits into the following format

$$Z_{oi}(s) = K_{oi}\frac{1+\dfrac{s}{\omega_{esr}}}{1+\dfrac{s}{\omega_{pi}}} \tag{9.66}$$

where $\omega_{esr} = 1/(CR_c)$ is the esr zero. The expressions for K_{oi} and ω_{pi} are given in Table 9.5. It should be noted that ω_{pi} is practically identical to ω_{id} in the $G_{id}(s)$

Table 9.5 Expressions for K_{oi} and ω_{pi}

	K_{oi}	ω_{pi}
Buck converter	R	$\dfrac{1}{C(R+R_c)}$
Boost converter	$\dfrac{R}{2}$	$\dfrac{2}{C(R+R_c)}$
Buck/boost converter	$\dfrac{R}{1+D}$	$\dfrac{1+D}{C(R+R_c)}$

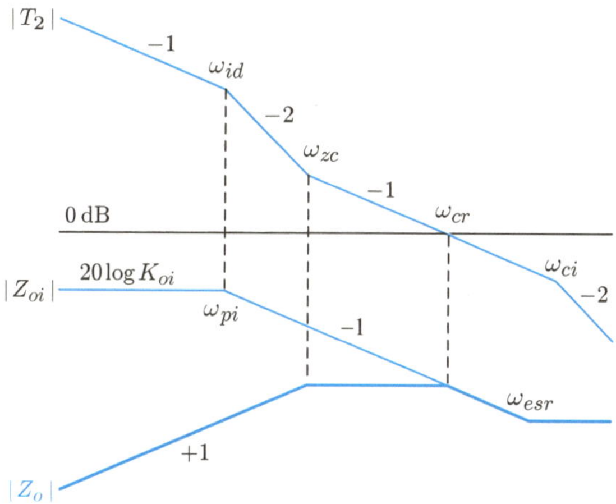

Figure 9.34 Output impedance analysis.

expression in Table 9.2. Referring to Fig. 9.21(b), the corner frequency ω_{id} appears as a pole in $|T_2|$ structure. It can be seen from Table 9.5 that the high-frequency asymptote of $|Z_{oi}|$ is the parallel combination of the load resistor and esr of the output capacitor

$$|Z_{oi}(j\infty)| = 20\log\left(K_{oi}\frac{\omega_{pi}}{\omega_{esr}}\right) = 20\log\left(\frac{R\,R_c}{R+R_c}\right) = 20\log\left(R \parallel R_c\right) \quad (9.67)$$

Now, the output impedance is investigated by performing the asymptotic analysis on (9.65). Figure 9.34 is the result of this analysis. The asymptotic plot is constructed using the expression of $Z_{oi}(s)$ in (9.66) and the structure of $|T_2|$ in Fig. 9.21(b), with the incorporation of the fact $\omega_{pi} = \omega_{id}$. This analysis assumes that the $|T_2|$ crossover occurs before the esr zero: $\omega_{cr} < \omega_{esr}$.

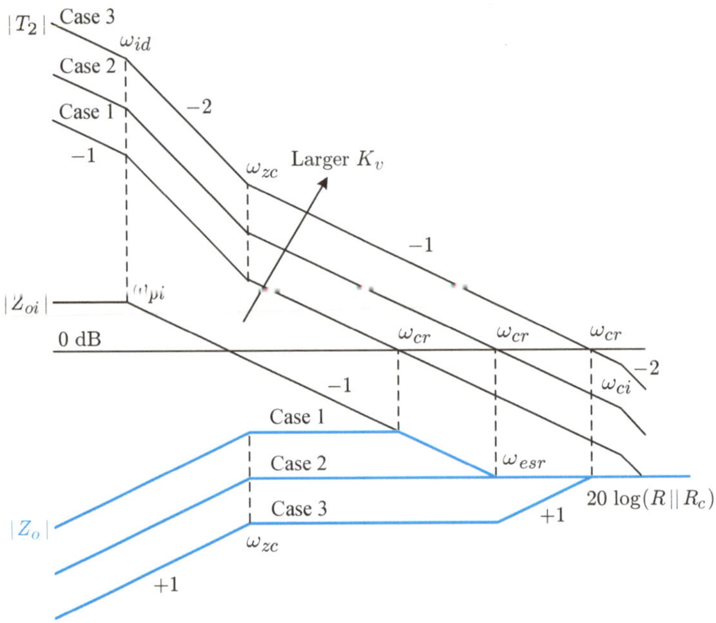

Figure 9.35 Generalization of output impedance analysis.

Figure 9.35 shows the output impedance analysis for general cases. The integrator gain K_v in the voltage feedback compensation, which controls the $T_2(s)$ crossover frequency ω_{cr}, is varied to result in the three cases: Case 1 with $\omega_{cr} < \omega_{esr}$, Case 2 with $\omega_{cr} = \omega_{esr}$, and Case 3 with $\omega_{cr} > \omega_{esr}$.

For Cases 2 and 3, the maximum value of the output impedance is reduced to $|Z(j\omega)|_{peak} = 20\log(R \parallel R_c)$, which is the theoretical minimum for the given power stage parameters. For Case 1, the peak value of the output impedance is evaluated as

$$
\begin{aligned}
|Z_o(j\omega)|_{peak} &- 20\log\left(\frac{\omega_{esr}}{\omega_{cr}}\right) = 20\log(R\,||\,R_c) \\
\Rightarrow \quad |Z_o(j\omega)|_{peak} &= 20\log(R \parallel R_c) + 20\log\left(\frac{\omega_{esr}}{\omega_{cr}}\right) \\
&= 20\log\left(R \parallel R_c\frac{\omega_{esr}}{\omega_{cr}}\right) \qquad (9.68)
\end{aligned}
$$

For all the three cases, the output impedance has the same first pole at the compensation zero ω_{zc}. As will be shown later, this pole determines the speed of the step load response. For faster response, ω_{zc} should be placed at higher frequency but should *not* exceed the power stage pole ω_o, in order *not* to become a conditionally stable system.

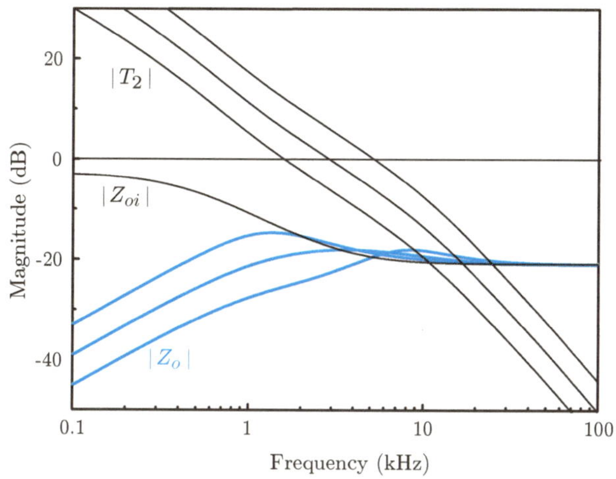

Figure 9.36 Output impedance analysis.

Figure 9.35 shows that the peak value of the output impedance can be minimized by placing the $T_2(s)$ crossover frequency at ω_{esr}. Increasing the $T_2(s)$ crossover frequency beyond ω_{esr} does not further reduce the peak value of the output impedance.

The asymptotic plots in Fig. 9.35 assume that the phase margin of $T_2(s)$ is sufficiently large. This assumption precludes the potential peaking in the output impedance at the crossover frequency of $T_2(s)$. Cases where the outer loop gain does not satisfy this assumption are addressed in the next section.

Buck Converter Example **EXAMPLE 9.11**

The buck converter used in Example 9.10 is revisited in order to substantiate the preceding output impedance analysis. The same compensation parameters as those of Example 9.10 are used to result in the three cases of the output impedance analysis: Case 1 with $\omega_{cr} = 0.5\,\omega_{esr}$, Case 2 with $\omega_{cr} = \omega_{esr}$, and Case 3 with $\omega_{cr} = 2.0\,\omega_{esr}$. Computer simulations for the three cases are shown in Fig. 9.36, which exhibit a close match with the asymptotic analysis of Fig. 9.35.

9.3.3 Step Load Response Analysis

The step load response refers to the transient waveform of the output voltage due to a step change in the load current. The analysis method for the step load response was discussed in Section 7.3.1. The step load response is investigated using the output impedance and its relationship to the output voltage

$$v_O(t) = \mathcal{L}^{-1}\left(\frac{I_{step}}{s} Z_o(s)\right) \tag{9.69}$$

where $\mathcal{L}^{-1}$ is the inverse Laplace transformation and I_{step} is the magnitude of the step change in the load current.

In this section, the step load response is investigated using the results of the previous output impedance analysis, along with the relationship (9.69). As shown in Fig. 9.35, the output impedance is classified into three cases depending on the relative location of the $T_2(s)$ crossover frequency ω_{cr} and the esr zero ω_{esr}. The three illustrative examples in Fig. 9.35 are divided into Case 1, Case 2, and Case 3 in the increasing order of the $T_2(s)$ crossover frequency. However, for smoother analytical treatment, the step load response is first discussed for the output impedance of Case 2 with $\omega_{cr} = \omega_{esr}$. The analysis then continues with Case 3 of $\omega_{cr} > \omega_{esr}$, and finally treats Case 1 with $\omega_{cr} < \omega_{esr}$. The analysis now starts with Case 2.

Case 2: $\omega_{cr} = \omega_{esr}$

This is the case in which the $T_2(s)$ crossover frequency falls at the esr zero. Figure 9.37(a) shows the output impedance of this case. The expression for the output impedance becomes

$$Z_o(s) = \frac{s}{\omega_m} \frac{1}{1 + \frac{s}{\omega_{zc}}} \tag{9.70}$$

where ω_m denotes the frequency at which the initial linear segment of $|Z_o|$ is projected to the 0 dB line. The peak magnitude of the output impedance is given by

$$|Z_o(j\omega)|_{peak} = |Z_o(j\infty)| = 20\log\left(\frac{\omega_{zc}}{\omega_m}\right) \tag{9.71}$$

The transient response of the output voltage due to the step load change of I_{step} is evaluated as

$$\begin{aligned} v_O(t) &= \mathcal{L}^{-1}\left(\frac{I_{step}}{s}\frac{s}{\omega_m}\frac{1}{1+\frac{s}{\omega_{zc}}}\right) \\ &= I_{step}\frac{\omega_{zc}}{\omega_m}e^{-\omega_{zc}t} \end{aligned} \tag{9.72}$$

The peak overshoot of the output voltage is found as

$$\Delta v_O(t)_{peak} = v_O(0) = I_{step}\frac{\omega_{zc}}{\omega_m} = I_{step}10^{\frac{|Z_o|_{peak}}{20}} \tag{9.73}$$

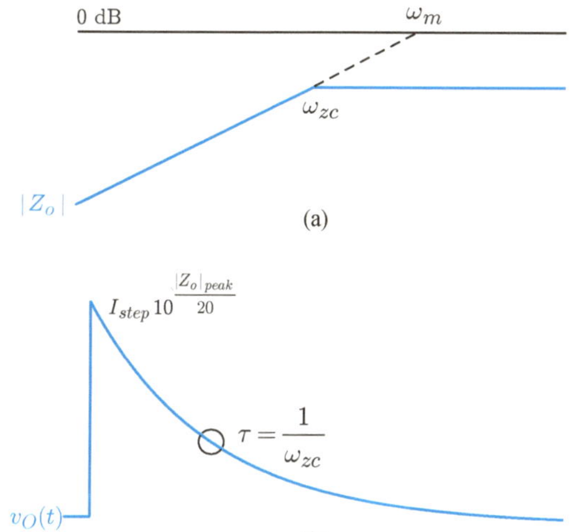

Figure 9.37 Case 2 with $\omega_{cr} = \omega_{esr}$. (a) Output impedance. (b) Step load response.

from (9.71) and (9.72).

The output voltage decays from its initial peak with a time constant of $\tau = 1/\omega_{zc}$. Thus, the settling time of v_O is given by $t_s = 3\tau = 3/\omega_{zc}$. The output voltage waveform is shown in Fig. 9.37(b).

Case 3: $\omega_{cr} > \omega_{esr}$

This case corresponds to the situation where the $T_2(s)$ crossover frequency is pushed beyond the esr zero. Figure 9.38(a) shows the output impedance for this case. The output impedance is expressed as

$$Z_o(s) = \frac{s}{\omega_m} \frac{1 + \dfrac{s}{\omega_{esr}}}{\left(1 + \dfrac{s}{\omega_{zc}}\right)\left(1 + \dfrac{s}{\omega_{cr}}\right)} \tag{9.74}$$

with $\omega_{zc} \ll \omega_{esr} \ll \omega_{cr}$. The peak magnitude of the output impedance is given by

$$|Z_o(j\omega)|_{peak} = |Z_o(j\infty)| = 20\log\left(\frac{\omega_{zc}\,\omega_{cr}}{\omega_m\,\omega_{esr}}\right) \tag{9.75}$$

The transient response of the output voltage due to the step load change of I_{step} is given by

$$v_O(t) = \mathcal{L}^{-1}\left(\frac{I_{step}}{s}\frac{s}{\omega_m}\frac{1 + \dfrac{s}{\omega_{esr}}}{\left(1 + \dfrac{s}{\omega_{zc}}\right)\left(1 + \dfrac{s}{\omega_{cr}}\right)}\right)$$

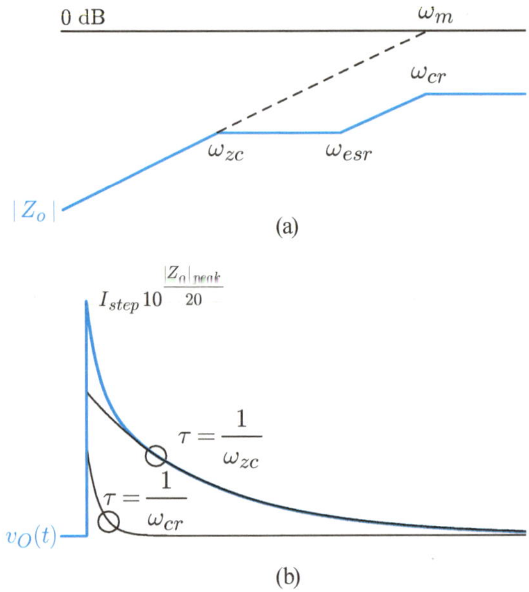

Figure 9.38 Case 3 with $\omega_{cr} > \omega_{esr}$. (a) Output impedance. (b) Step load response.

$$v_O(t) = \frac{I_{step}}{\omega_{cr} - \omega_{zc}} \frac{\omega_{zc}\omega_{cr}}{\omega_m \omega_{esr}} \left((\omega_{esr} - \omega_{zc}) e^{-\omega_{zc} t} + (\omega_{cr} - \omega_{esr}) e^{-\omega_{cr} t} \right) \tag{9.76}$$

where $(\omega_{esr} - \omega_{zc}) > 0$ and $(\omega_{cr} - \omega_{esr}) > 0$. The step load response is the sum of the two exponentially decaying terms. The second term decays much faster than the first term, due to the condition $\omega_{zc} \ll \omega_{esr} \ll \omega_{cr}$. The peak deviation of the output voltage is expressed as

$$\Delta v_O(t)_{peak} = v_O(0) = I_{step} \frac{\omega_{zc}}{\omega_m} \frac{\omega_{cr}}{\omega_{esr}} = I_{step} 10^{\frac{|Z_o|_{peak}}{20}} \tag{9.77}$$

Figure 9.38(b) shows the output voltage waveform. The output voltage decays from its initial peak with the fast rate of $1/\omega_{cr}$ in the beginning, and with the slow rate of $1/\omega_{zc}$ after the second term diminishes. The expression of the peak output voltage overshoot is the same as Case 2, but its initial decay is faster due to the condition $\omega_{zc} \ll \omega_{cr}$. Nonetheless, the settling time will be determined by the slow time constant, $t_s = 3/\omega_{zc}$.

Case 1: $\omega_{cr} < \omega_{esr}$

In this case, the $T_2(s)$ crossover frequency falls below the esr zero. Figure 9.39 shows the output impedance and the step load response for this case. The output impedance is expressed as

$$Z_o(s) = \frac{s}{\omega_m} \frac{1+\dfrac{s}{\omega_{esr}}}{\left(1+\dfrac{s}{\omega_{zc}}\right)\left(1+\dfrac{s}{\omega_{cr}}\right)} \tag{9.78}$$

with $\omega_{zc} \ll \omega_{cr} \ll \omega_{esr}$. This expression is the same as (9.74), yet the order of the corner frequencies is different. Unlike the previous two cases, the peak value of the output impedance does not coincide with the high-frequency asymptote. From Fig. 9.39(a), the peak magnitude of the output impedance is given by

$$|Z_o(j\omega)|_{peak} = \left|\frac{s}{\omega_m}\right|_{s=j\omega_{zc}} = 20\log\left(\frac{\omega_{zc}}{\omega_m}\right) \tag{9.79}$$

On the other hand, the high-frequency asymptote is expressed as

$$|Z_o(j\infty)| = 20\log\left(\frac{\omega_{zc}}{\omega_m}\frac{\omega_{cr}}{\omega_{esr}}\right) \tag{9.80}$$

The transient response due to the step load change of I_{step} is evaluated as

$$v_O(t) = \frac{I_{step}}{\omega_{cr}-\omega_{zc}}\frac{\omega_{zc}\,\omega_{cr}}{\omega_m\omega_{esr}}\left((\omega_{esr}-\omega_{zc})e^{-\omega_{zc}t} + (\omega_{cr}-\omega_{esr})e^{-\omega_{cr}t}\right) \tag{9.81}$$

with $(\omega_{esr} - \omega_{zc}) > 0$ and $(\omega_{cr} - \omega_{esr}) < 0$. The step load response is the sum of the two exponentially decaying terms, the same as Case 2. However, the signs of the two exponential terms are opposite; the first coefficient is positive while the second is negative. The second negative term decays much faster than the first positive term. The waveform for this case is shown in Fig. 9.39(b). The settling time of the output voltage is still governed by the slow time constant, $t_s = 3/\omega_{zc}$.

The initial value of the output voltage is given by

$$v_O(0) = I_{step}\frac{\omega_{zc}}{\omega_m}\frac{\omega_{cr}}{\omega_{esr}} = I_{step}10^{\frac{|Z_o(j\infty)|}{20}} \tag{9.82}$$

Unlike the previous cases, the peak deviation of the output voltage, $v_O(t)_{peak}$, cannot be explicitly described in terms of $|Z_o|_{peak}$. Instead, $|Z_o|_{peak}$ can be used to predict the upper bound of $v_O(t)_{peak}$, as will be demonstrated in the following analysis.

From the profile of $v_O(t)$ in Fig. 9.39(b), it becomes evident that

$$v_O(0) < \Delta v_O(t)_{peak} < I_{step}\frac{\omega_{zc}\,\omega_{cr}(\omega_{esr}-\omega_{zc})}{\omega_m\,\omega_{esr}(\omega_{cr}-\omega_{zc})} \tag{9.83}$$

The last term in (9.83) is the initial value of the slow-decaying first term in (9.81). With the conditions $\omega_{esr} \gg \omega_{zc}$ and $\omega_{cr} \gg \omega_{zc}$, the expression (9.83) is approximated as

$$v_O(0) < \Delta v_O(t)_{peak} < I_{step}\frac{\omega_{zc}}{\omega_m} \tag{9.84}$$

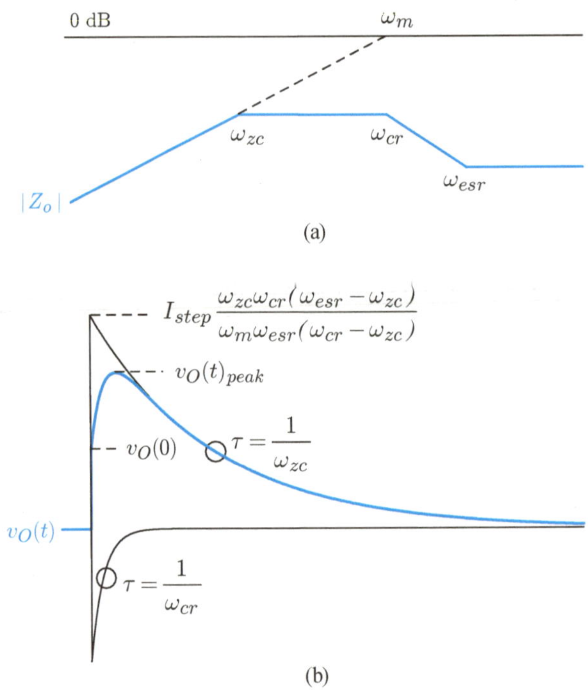

Figure 9.39 Case 1 with $\omega_{cr} < \omega_{esr}$. (a) Output impedance. (b) Step load response.

Finally, using (9.79) and (9.82), the above inequality is written as

$$I_{step}\, 10^{\frac{|Z_o(j\infty)|}{20}} < \Delta v_O(t)_{peak} < I_{step}\, 10^{\frac{|Z_o(j\omega)|_{peak}}{20}} \tag{9.85}$$

The peak deviation of the output voltage can be estimated from (9.85). The high-frequency asymptote of the output impedance, $|Z_o(j\infty)|$, predicts the lower bound of $\Delta v_O(t)_{peak}$, while the peak value of the output impedance, $|Z_o(j\omega)|_{peak}$, projects the upper bound.

EXAMPLE 9.12 Buck Converter Example

This example validates the preceding discussions about the step load response. The buck converter in Example 9.11 is used to generate the step load response for the three different cases: Case 1 with $\omega_{cr} = 0.5\,\omega_{esr}$, Case 2 with $\omega_{cr} = \omega_{esr}$, and Case 3 with $\omega_{cr} = 2.0\,\omega_{esr}$. The output impedances for these cases were shown in Fig. 9.36. A series of step changes, $I_O = 4\text{ A} \Rightarrow 8\text{ A} \Rightarrow 4\text{ A}$, is introduced to the load current and the output voltage waveforms are shown in Fig. 9.40. While the profiles of the transient responses are different, the settling time can be considered to be largely

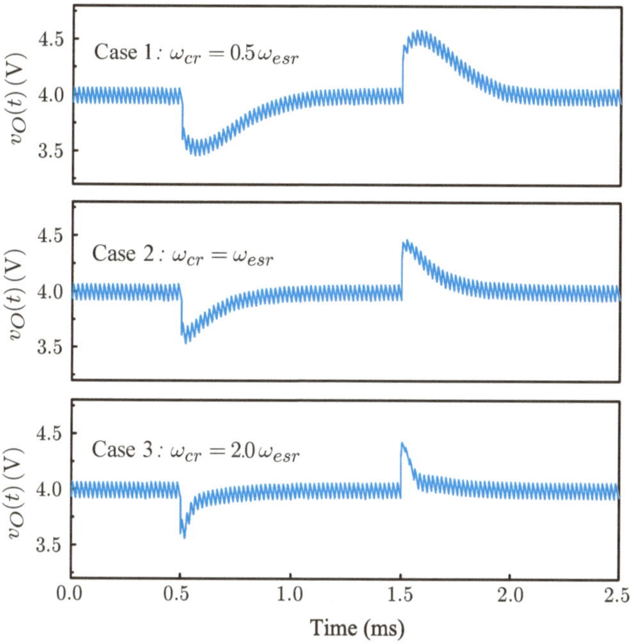

Figure 9.40 Step load response of buck converter.

the same for all the three cases

$$t_s = \frac{3}{\omega_{zc}} = \frac{3}{2\pi \cdot 928} = 0.51 \text{ ms}$$

For Cases 2 and 3, the peak deviation of the output voltage is given by

$$\Delta v_O(t)_{peak} = I_{step} 10^{\frac{|Z_o|_{peak}}{20}} = 4 \cdot 10^{-20/20} = 0.4 \text{ V}$$

For Case 1, the $\Delta v_O(t)_{peak}$ is estimated as

$$I_{step}\, 10^{\frac{|Z_o(j\infty)|}{20}} < \Delta v_O(t)_{peak} < I_{step}\, 10^{\frac{|Z_o(j\omega)|_{peak}}{20}}$$

$$\Rightarrow 4 \cdot 10^{-20/20} < \Delta v_O(t)_{peak} < 4 \cdot 10^{-15/20}$$

$$\Rightarrow 0.40 \text{ V} < \Delta v_O(t)_{peak} < 0.71 \text{ V}$$

Step Load Response and Compensation Design

As demonstrated in the previous section, the step load response is mainly determined by the voltage feedback compensation. In particular, the output voltage deviation can be reduced to the theoretical minimum by placing the $T_2(s)$ crossover frequency at the esr zero. Increasing the $T_2(s)$ crossover frequency beyond the esr zero does not reduce the peak deviation any further. Nonetheless, as shown in Example 9.12, a higher $T_2(s)$ crossover frequency results in a faster transient response at the beginning, thereby reducing the time necessary for the output voltage to reach a specified percentage of its final value.

The previous analysis implicitly assumes that the phase margin of $T_2(s)$ is sufficiently large. In real applications, however, it is not always possible to push the $T_2(s)$ crossover frequency to the esr zero, while maintaining a large phase margin. In some cases, pushing the $T_2(s)$ crossover frequency towards higher frequencies results in a smaller phase margin. In turn, a small phase margin will adversely affect the output impedance characteristics around the $T_2(s)$ crossover frequency.

The output impedance $Z_o(s)$ is related to $T_2(s)$ by the equation

$$Z_o(s) = \frac{Z_{oi}(s)}{1 + T_2(s)} \tag{9.86}$$

As discussed in Section 8.4.4, the magnitude of the output impedance at the $T_2(s)$ crossover frequency is given by

$$|Z_o| = \frac{|Z_{oi}|}{\sqrt{2 - 2\cos\phi_m}} \tag{9.87}$$

where ϕ_m represents the phase margin of $T_2(s)$. If the phase margin is less than 60°, the output impedance exhibits a peaking over the $|Z_{oi}|$ at the $T_2(s)$ crossover frequency, as a function of the phase margin ϕ_m

$$\boxed{|Z_o|_{peaking} = 20\log\left(\frac{1}{\sqrt{2 - 2\cos\phi_m}}\right)} \tag{9.88}$$

The smaller the phase margin, the larger the peaking. The peaking in the output impedance is transformed into a pair of complex poles, located near the imaginary axis. As the phase margin becomes smaller, the complex poles move closer to the imaginary axis. These complex poles would result in an oscillatory transient response, as discussed in Section 8.4.4.

The preceding discussions indicate that the phase margin should be considered in determining the $T_2(s)$ crossover frequency. As shown in Fig. 9.21(b), the outer loop gain has a pole at the $T_i(s)$ crossover frequency, ω_{ci} in Fig. 9.21(b). For an adequate phase margin, the $T_2(s)$ crossover frequency should occur well before the $T_i(s)$ crossover frequency so that the impacts of ω_{ci} on the phase margin become insignificant.

With a predetermined $T_i(s)$ crossover frequency ω_{ci}, commonly located around $10\% - 30\%$ of the switching frequency, the $T_2(s)$ crossover frequency can be increased up to the esr zero while maintaining a sufficient phase margin, only provided

Table 9.6 Four Different Control Designs

	Integrator gain	T_2 crossover frequency	T_2 phase margin
Design A	$K_v = 3.90 \times 10^4$	$\omega_{cr} = 0.2\,\omega_{esr}$	$\phi_m = 64°$
Design B	$K_v = 7.82 \times 10^4$	$\omega_{cr} = 0.4\,\omega_{esr}$	$\phi_m = 57°$
Design C	$K_v = 1.56 \times 10^5$	$\omega_{cr} = 0.6\,\omega_{esr}$	$\phi_m = 45°$
Design D	$K_v = 3.13 \times 10^5$	$\omega_{cr} = 1.0\,\omega_{esr}$	$\phi_m = 32°$

that the frequency of esr zero itself is much lower than ω_{ci}. If the esr zero does not meet this requirement, a compromise should be made; namely, the $T_2(s)$ crossover frequency can only be increased within the limit that does not reduce the phase margin less than 60°. With such a design trade-off, the $T_2(s)$ crossover frequency would occur below the esr zero, resulting in the output impedance of Case 1.[†] For this instance, the peak deviation of the output voltage is estimated from (9.85).

The settling time of the output voltage is determined by the zero of the voltage feedback compensation. For small settling time, the compensation zero ω_{zc} should be placed as high as possible without exceeding the power stage double pole ω_o. The design procedures recommended that $\omega_{zc} = (0.6 - 0.8)\,\omega_o$.

Buck Converter Example — EXAMPLE 9.13

This example illustrates the control design and closed-loop performance of a buck converter, whose esr zero is *not* sufficiently lower than the switching frequency ω_s and thus not sufficiently lower than the $T_i(s)$ crossover frequency ω_{ci} either. In the buck converter used in Example 9.11, the esr of the output capacitor is decreased to $R_c = 25$ mΩ, while other power stage parameters remain unchanged. Now, the esr zero is located at $\omega_{esr} = 2\pi \cdot 1.36 \times 10^4$ rad/s $= 0.27\,\omega_s = 0.36\,\omega_{ci}$. With this modification, the integrator gain in the voltage feedback compensation is varied between $3.90 \times 10^4 < K_v < 3.13 \times 10^5$ to yield the four different designs listed in Table 9.6. The $T_2(s)$ crossover frequency spans between $0.2\,\omega_{esr} < \omega_{cr} < 1.0\,\omega_{esr}$ with phase margin of $32° < \phi_m < 64°$.

The outer loop gains of the four designs are shown in Fig. 9.41. As the integrator gain increases, the crossover frequency shifts towards higher frequencies at the expense of the reduction in the phase margin. Figure 9.42 shows the output impedances of the four designs. The output impedance exhibits a peaking at the $T_2(s)$ crossover frequency when the phase margin reduces lower than 60°. Finally, the transient responses due to a series of step load changes, $I_O = 4\text{ A} \Rightarrow 8\text{ A} \Rightarrow 4\text{ A}$, are shown in Fig. 9.43. The output voltage shows an oscillatory behavior for Design C and Design D, where the output impedance reveals a peaking due to the small

[†]This will be particularly true for the boost and buck/boost converters in which the 0 dB crossover frequency of $T_2(s)$ is inherently narrow due to the presence of the RHP zero in the duty ratio-to-output transfer function. Details about this fact come later in Section 9.4.1.

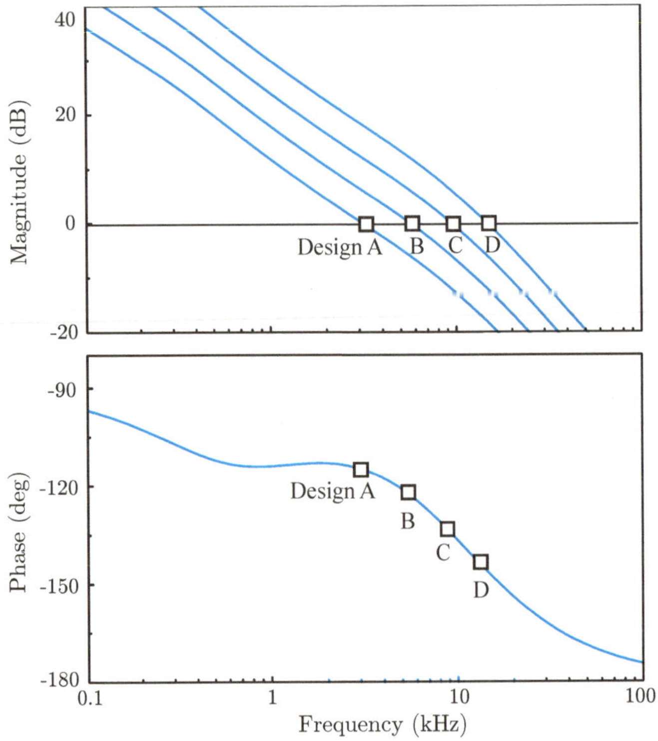

Figure 9.41 Outer loop gain of four different designs.

phase margin. Accordingly, Design A or Design B can be considered acceptable, whose output impedance characteristics belong to Case 1 in the previous analysis.

The control design procedures in Table 9.3 recommended to place the $T_2(s)$ crossover frequency in the range of $\omega_{cr} = (0.3 - 1.0)\,\omega_{esr}$. When the esr zero is sufficiently lower than the switching frequency, ω_{cr} can be placed at ω_{esr}, as is the case with Example 9.8. On the other hand, when ω_{esr} is *not* sufficiently lower than the switching frequency, ω_{cr} should be placed before ω_{esr}, for example, $\omega_{cr} = 0.4\,\omega_{esr}$ in Design B in Example 9.13.

It should be emphasized that the design strategy of $\omega_{cr} = (0.3 - 1.0)\,\omega_{esr}$ only applies to buck converters. For the boost and buck/boost converters that have a right-half plane (RHP) zero in their duty ratio-to-output transfer function, other constraints should be considered in determining the $T_2(s)$ crossover frequency. Design considerations for boost and buck/boost converters will be given in Section 9.4.

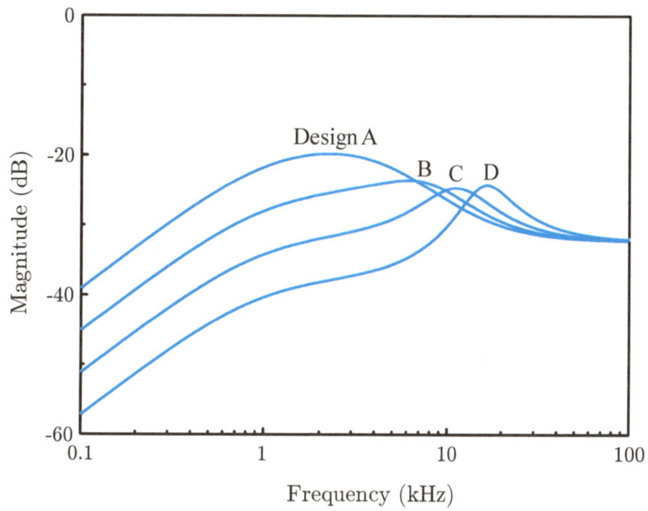

Figure 9.42 Output impedance of four different designs.

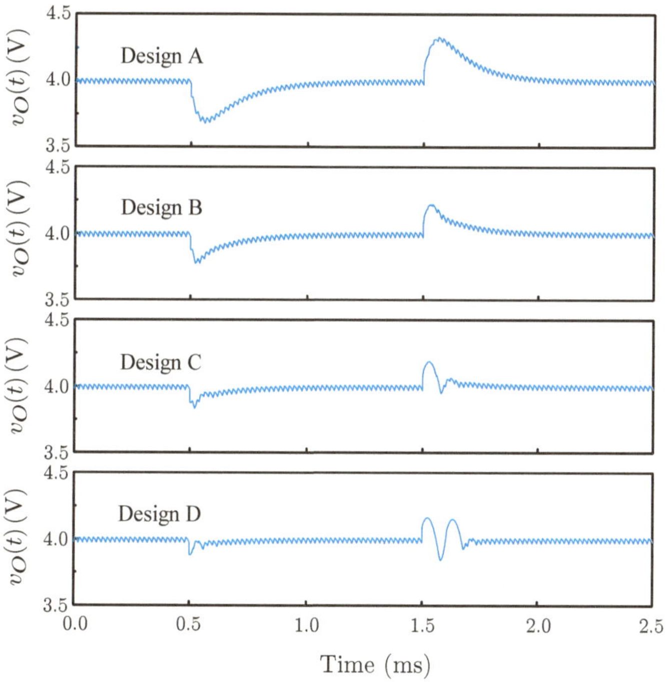

Figure 9.43 Step load response of buck converter with four different designs.

Generalization of Step Load Response

The outcomes of the previous step load response analysis are extended to all dc-to-dc converters in general.

1) The output impedance characteristics of a dc-to-dc converter can be obtained from the small-signal analysis, computational method, or experimental measurement. When the small-signal model of the converter is available, the asymptotic analysis will be performed to obtain the output impedance characteristics. If the small-signal model is not available, either the computational method or experimental measurement can be used for the acquisition of the output impedance data. The output impedance is displayed in the Bode plot format.

2) The output impedance parameters, required for the step load response analysis, are extracted from the output impedance plot.

3) The results of the previous section are then adapted to predict the step load response from the output impedance parameters.

The output impedance of practical dc-to-dc converters generally follows the pattern of Case 1, illustrated in Fig. 9.39(a). For this case, the settling time is given by $t_s = 3/\omega_{zc}$, where ω_{zc} is the first pole in Fig. 9.39(a). The peak deviation of the output voltage can be estimated from (9.85). If a more detailed prediction is desired, the expression (9.81) can be used to derive an analytical equation for the output voltage. The resulting equation is then transformed to the output voltage waveform. An example is given below to illustrate the analysis procedures described above.

EXAMPLE 9.14 Switched Capacitor Converter Example

This example demonstrates the generality of the step load response analysis. For this purpose, a switched capacitor dc-to-dc converter [8], which performs the dc-to-dc power conversion using only capacitors and switches, is used in this example. Figure 9.44(a) is the output impedance of the switched capacitor converter, which was measured from the experimental converter using an impedance analyzer. The output impedance closely resembles the profile of Case 1, and therefore the corresponding analysis results are applied here.

As shown in Fig. 9.44(a), the major output impedance parameters are measured as $\omega_{zc} = 2\pi \cdot 3 \times 10^3$ rad/s, $|Z_o|_{\text{peak}} = -23$ dB, and $|Z_o(j\infty)| = -33$ dB. The output of the converter is regulated at $V_O = 5$ V. Now, a 4 A step decrease occurs in the load current. The settling time of the output voltage is predicted as

$$t_s = \frac{3}{\omega_{zc}} = \frac{3}{2\pi \cdot 3 \times 10^3} = 0.16 \text{ ms}$$

and the peak overshoot is estimated as

$$I_{step} 10^{\frac{|Z_o(j\infty)|}{20}} < \Delta v_O(t)_{peak} < I_{step} 10^{\frac{|Z_o(j\omega)|_{peak}}{20}}$$

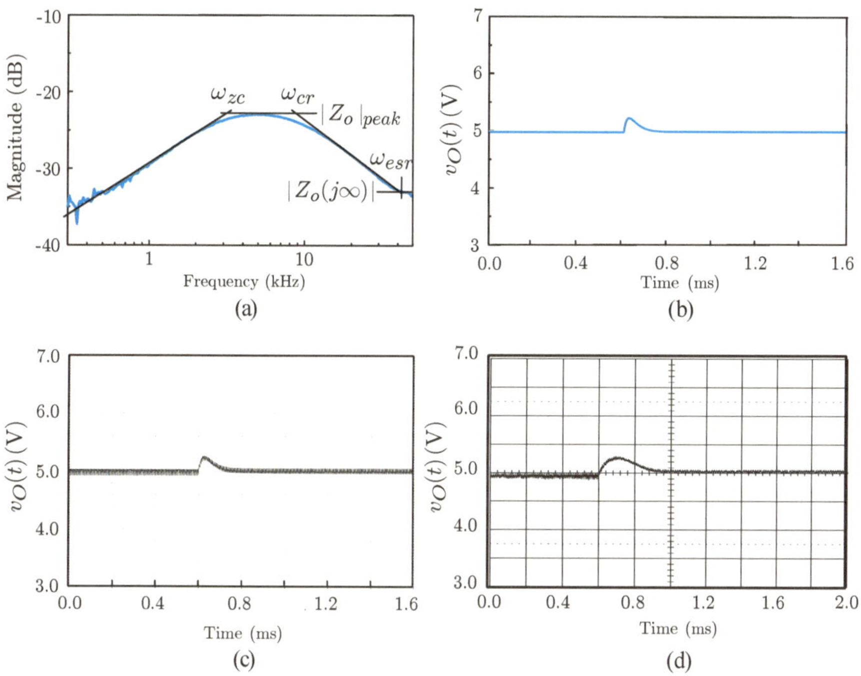

Figure 9.44 Switched capacitor converter example. (a) Output impedance. (b) Prediction of the equation (9.81). (c) Cycle-by-cycle time-domain simulation. (d) Measured step load response.

$$\Rightarrow 4 \cdot 10^{-33/20} < \quad \Delta v_O(t)_{peak} \quad < 4 \cdot 10^{-23/20}$$

$$\Rightarrow 0.09 \text{ V} < \quad \Delta v_O(t)_{peak} \quad < 0.28 \text{ V}$$

For a more detailed prediction, other output impedance parameters, ω_{cr}, ω_{esr}, and ω_m in Fig. 9.39(a) and Fig. 9.44(a), are estimated and the outcomes are put into (9.81) to obtain an analytical equation for $v_O(t)$. Figure 9.44(b) is the plot of the resulting equation.

The predictions of the step load response analysis are compared with both the time-domain simulations and experimental data. Figure 9.44(c) is the step load response generated from cycle-by-cycle time-domain simulation. Finally, Fig. 9.44(d) is the step load response measured from the experimental switched capacitor converter. The close correlation and resemblance between the output impedances and step load responses, obtained from analysis, simulation, and measurement, confirm the generality and accuracy of the step load response analysis.

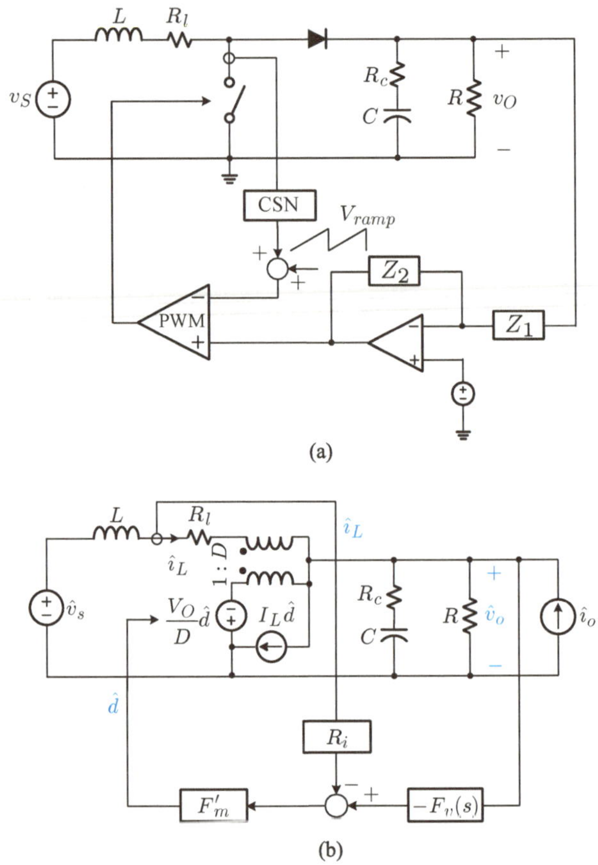

Figure 9.45 Boost converter with peak current mode control. (a) Circuit diagram. (b) Small-signal model.

9.4 CURRENT MODE CONTROL FOR BOOST AND BUCK/BOOST CONVERTERS

Chapter 8 demonstrated that voltage mode control is not suitable for the converters that have the right-half plane (RHP) zero in their power stage transfer function, such as the boost converter, the buck/boost converter, and all other isolated PWM converters derived from these two converters. Accordingly, current mode control is a viable alternative for these converters. This section presents the feedback design and dynamic analysis of the peak current mode control, adapted to the dc-to-dc converters with the RHP zero. A boost converter is used as an example to address the impact of the RHP zero on the control design and closed-loop performance.

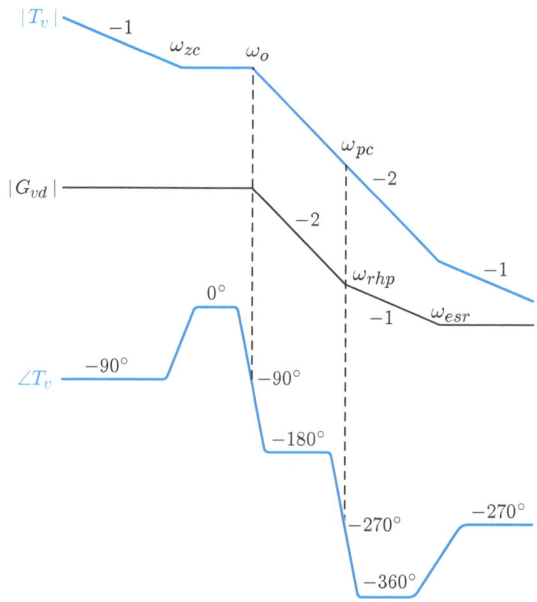

Figure 9.46 Asymptotic plots for duty ratio-to-output transfer function $G_{vd}(s)$ and voltage loop $T_v(s)$.

9.4.1 Stability Analysis and Control Design

Figure 9.45 shows the circuit diagram and small-signal model of a boost converter employing current mode control. From the small-signal model in Fig. 9.45(b), the duty ratio-to-output transfer function is derived as

$$G_{vd}(s) = \frac{\hat{v}_o(s)}{\hat{d}(s)} = K_{vd}\frac{\left(1 - \dfrac{s}{\omega_{rhp}}\right)\left(1 + \dfrac{s}{\omega_{esr}}\right)}{1 + \dfrac{s}{Q\omega_o} + \dfrac{s^2}{\omega_o^2}} \tag{9.89}$$

and the duty-ratio-to-inductor current transfer function is given by

$$G_{id}(s) = \frac{\hat{i}_L(s)}{\hat{d}(s)} = K_{id}\frac{1 + \dfrac{s}{\omega_{id}}}{1 + \dfrac{s}{Q\omega_o} + \dfrac{s^2}{\omega_o^2}} \tag{9.90}$$

The expressions for the dc gain, corner frequencies, and quality factor of the transfer functions are given in Table 9.2. Figure 9.46 shows the asymptotic plot of the duty ratio-to-output transfer function $G_{vd}(s)$ and voltage loop $T_v(s) = G_{vd}(s)F_v(s)F_m'$ of the boost converter that adopts the two-pole one-zero compensation. This example

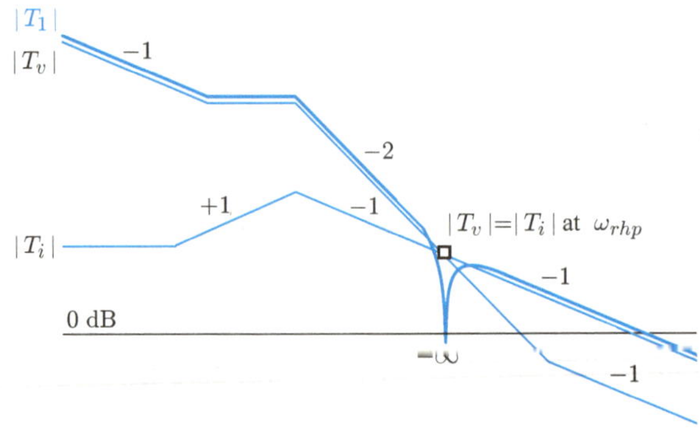

Figure 9.47 Individual feedback loops and overall loop gain with $|T_i| = |T_v|$ at ω_{rhp}.

assumes that the RHP zero comes before the esr zero, $\omega_{rhp} < \omega_{esr}$. According to the control design guidelines, the voltage feedback compensation parameters are selected as $\omega_{pc} = \omega_{rhp}$ and $\omega_{zc} < \omega_o$, resulting in the asymptotic plot shown in Fig. 9.46.

With the compensation pole ω_{pc} located at the RHP zero ω_{rhp}, the phase of the voltage loop drops by $-180°$ over ω_{rhp}. The $90°$ phase delay caused by ω_{rhp} is augmented by the additional $90°$ phase drop incurred by ω_{pc}, resulting in a $-180°$ change in $\angle T_v$ while $|T_v|$ stays in a -40 dB/dec slope. These phase characteristics are very unfavorable, mainly staying less than $-180°$ for the frequencies after the power stage double pole ω_o. In particular, $\angle T_v$ drops to $-270°$ at ω_{rhp}, as illustrated in Fig. 9.46.

Figure 9.47 shows the asymptotic plots for the current loop $|T_i|$, voltage loop $|T_v|$, and overall loop gain $|T_1|$. In order to elucidate the impact of the RHP zero on stability, it is assumed that $|T_v| = |T_i|$ at ω_{rhp} in Fig. 9.47. Under this situation, the overall loop gain at ω_{rhp}, $\overrightarrow{T_1}(j\omega_{rhp}) = \overrightarrow{T_i}(j\omega_{rhp}) + \overrightarrow{T_v}(j\omega_{rhp})$, is given by the sum of two equal-length vectors with a $180°$ phase difference: $\angle T_v = -270° = 90°$ while $\angle T_i = -90°$. Thus, $|T_1|$ dips down to negative infinity and $\angle T_1$ drops rapidly at ω_{rhp}. This indicates that the converter is unstable with a negative phase margin.

EXAMPLE 9.15 Instability with Condition $|T_v| = |T_i|$ at ω_{rhp}

This example illustrates the preceding discussions about instability. Figure 9.48 shows the Bode plots of $T_v(s)$, $T_i(s)$, and $T_1(s)$ of a boost converter which employs the two-pole one-zero compensation with the condition $|T_i| = |T_v|$ at ω_{rhp}. The overall loop gain reveals the dip in $|T_1|$ and sudden drop in $\angle T_1$ at the frequencies where $|T_v| \approx |T_i|$, thereby confirming that the converter is unstable.

The previous analysis indicates that stability of the converter is retained, only provided that the frequency at which $|T_v| = |T_i|$ comes well before the RHP zero. The frequency where $|T_v| = |T_i|$ is the crossover frequency of the outer loop gain

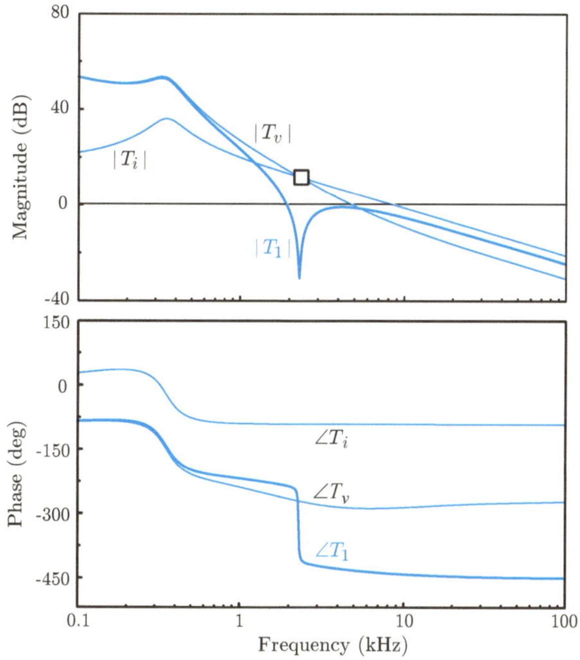

Figure 9.48 Individual feedback loops and overall loop gain with $|T_i| = |T_v|$ at ω_{rhp}.

$T_2(s)$, previously denoted as ω_{cr}. Accordingly, the $T_2(s)$ crossover frequency ω_{cr} should be placed prior to the RHP zero, ω_{rhp}.

Figure 9.49 shows the asymptotic plots for $|T_v|$, $|T_i|$, $|T_2|$, and $\angle T_2$ under the assumption that $\omega_{cr} < \omega_{rhp}$. As shown in Fig. 9.49, the phase of $T_2(s)$ at ω_{rhp} becomes $\angle T_2 = \angle T_v - \angle T_i = -270° - (-90°) = -180°$. This also validates the requirement of $\omega_{cr} < \omega_{rhp}$ for an acceptable phase margin at ω_{cr}. In the previous design procedures, it was recommended that $\omega_{cr} = (0.1\text{–}0.3)\,\omega_{rhp}$ for the boost and buck/boost converters.

The control design procedures established in the previous section are still applicable to the dc-to-dc converters with an RHP zero in their power stage transfer function. The same two-pole one-zero compensation is applied for the voltage feedback compensation. *As one critical design constraint, however, the integrator gain should be selected to place the crossover frequency of the outer loop gain $T_2(s)$ in the range of $\omega_{cr} = (0.1\text{–}0.3)\,\omega_{rhp}$.*

Current Mode Control for Boost Converter — EXAMPLE 9.16

This example illustrates the design and performance of the peak current mode control adapted to a boost converter. Figure 9.50 shows the circuit diagram of a boost converter which employs the peak current mode control. The output voltage is regu-

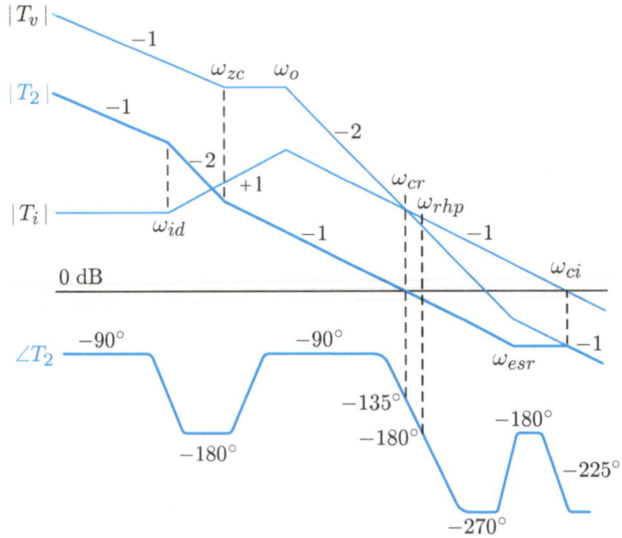

Figure 9.49 Individual feedback loops and outer loop gain with $\omega_{cr} < \omega_{rhp}$.

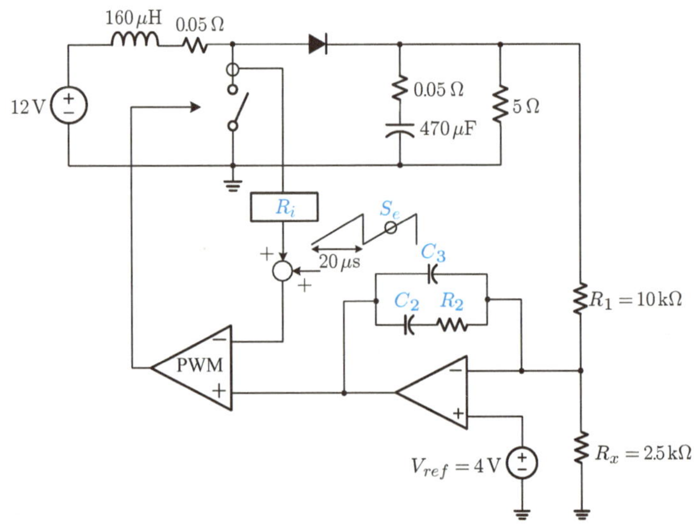

Figure 9.50 Current-mode controlled boost converter: $R_i = 0.67$, $S_e = 7.49 \times 10^4$ V/s, $R_2 = 19.6$ kΩ, $C_2 = 29.2$ nF, and $C_3 = 5.38$ nF.

lated at $V_O = V_{ref}(1+R_1/R_x) = 4\,(1+4) = 20$ V with the input voltage $V_S = 12$ V. From the power stage circuit parameters and operational conditions, the corner frequencies and dc gains of the power stage transfer functions are determined as

$$\frac{V_O}{V_S} = \frac{1}{1-D} = \frac{20}{12} \quad \Rightarrow \quad D = 0.4$$

$$\omega_o \approx \frac{1-D}{\sqrt{LC}} = \frac{1-0.4}{\sqrt{160 \times 10^{-6}\, 470 \times 10^{-6}}} = 2\pi \cdot 348 \text{ rad/s}$$

$$\omega_{rhp} \approx \frac{(1-D)^2 R}{L} = \frac{(1-0.4)^2 5}{160 \times 10^{-6}} = 2\pi \cdot 1.79 \times 10^3 \text{ rad/s}$$

$$\omega_{esr} = \frac{1}{CR_c} = \frac{1}{470 \times 10^{-6}\, 0.05} = 2\pi \cdot 6.77 \times 10^3 \text{ rad/s}$$

$$K_{vd} \approx \frac{V_S}{(1-D)^2} = \frac{12}{(1-0.4)^2} = 33.3$$

$$\omega_{id} \approx \frac{2}{CR} = \frac{2}{470 \times 10^{-6}\, 5} = 2\pi \cdot 135 \text{ rad/s}$$

$$K_{id} \approx \frac{2V_S}{(1-D)^3 R} = \frac{2 \cdot 12}{(1-0.4)^3\, 5} = 22.2$$

The switching frequency of the converter is $\omega_s = 2\pi \cdot 50 \times 10^3$ rad/s. The maximum input voltage of the PWM block is $V_{max} = 5.0$ V and the peak value of the inductor current is calculated as

$$\begin{aligned} i_{L\,peak} &= \frac{V_O}{(1-D)R} + \frac{1}{2}\frac{V_S}{L} DT_s \\ &= \frac{20}{(1-0.4)5} + \frac{1}{2}\frac{12}{160 \times 10^{-6}} 0.4 \cdot 20 \times 10^{-6} = 6.97 \text{ A} \end{aligned}$$

Now the control design is performed based on the proposed design procedures.

Current Loop Design

1) $T_i(s)$ crossover frequency: $\omega_{ci} = 0.16\,\omega_s = 2\pi \cdot 8 \times 10^3$ rad/s

2) Dc gain of $T_i(s)$:

$$K_i = \frac{\omega_{id}\,\omega_{ci}}{\omega_o^2} = \frac{(2\pi \cdot 135)(2\pi \cdot 8 \times 10^3)}{(2\pi \cdot 348)^2} = 8.92$$

3) CSN gain:

$$R_i < \frac{V_{max}}{i_{L\,peak}} = \frac{5.0}{6.97} = 0.72 \quad \Rightarrow R_i = 0.67$$

4) Modulator gain:

$$F_m' = \frac{K_i}{K_{id} R_i} = \frac{8.92}{22.2 \cdot 0.67} = 0.60$$

5) Compensation ramp:

$$
\begin{aligned}
S_e &= \frac{1}{T_s F_m'} + \frac{S_f - S_n}{2} \\
&= \frac{1}{20 \times 10^{-6}\, 0.60} + \frac{\dfrac{20-12}{160 \times 10^{-6}} 0.67 - \dfrac{12}{160 \times 10^{-6}} 0.67}{2} \\
&= 7.50 \times 10^4 \text{ V/s}
\end{aligned}
$$

Voltage Loop Design

1) Compensation pole: $\omega_{pc} = \omega_{rhp} = 2\pi \cdot 1.79 \times 10^3$ rad/s
2) Compensation zero: $\omega_{zc} = 0.8\,\omega_o = 0.8\,(2\pi \cdot 348) = 2\pi \cdot 278$ rad/s
3) $T_2(s)$ crossover frequency: $\omega_{cr} = 0.28\,\omega_{rhp} = 0.28\,(2\pi \cdot 1.79 \times 10^3)$ $= 2\pi \cdot 501$ rad/s
4) Integrator gain:

$$
\begin{aligned}
K_v &= \frac{K_{id}\,\omega_{cr}\,R_i\,\omega_{zc}}{\omega_{id}\,K_{vd}} \\
&= \frac{22.2\,(2\pi \cdot 501)\,0.67\,(2\pi \cdot 278)}{(2\pi \cdot 135)\,33.3} \\
&= 2.89 \times 10^3
\end{aligned}
$$

5) Voltage feedback circuit: $R_1 = 10$ kΩ $\Rightarrow R_2 = 19.6$ kΩ, $C_2 = 29.2$ nF, and $C_3 = 5.38$ nF

Figure 9.51 shows the Bode plots of the individual feedback loops and system loop gains. Figure 9.51(a) reveals that the crossover frequency of $T_1(s)$ is precisely located at the target frequency, $\omega_{ci} = 2\pi \cdot 8.0 \times 10^3$ rad/s. The crossover frequency of $T_2(s)$ is also placed at the exact design goal of $\omega_{cr} = 2\pi \cdot 501$ rad/s. Figure 9.52 compares the Bode plots of $T_1(s)$ and $T_2(s)$. The two loop gains have different phase and gain margins. The overall loop gain T_1 has a 90° phase margin and an ∞ dB gain margin. In contrast, the phase margin of T_2 is 45° and its gain margin is 11 dB.

As addressed in Section 9.2.3, the implications of the stability margins of the two loop gains are very different. For the overall loop gain, $T_1(s) = T_i(s) + T_v(s)$, the stability margins are defined for the sum of the two individual feedback loops. The ∞ dB gain margin implies that the converter never becomes unstable, as far as $|T_v|$ and $|T_i|$ are *simultaneously* increased, regardless of the amount of increase. However, this information is not very useful for the control design purpose because we do not increase $|T_v|$ and $|T_i|$ at the same time. Instead, $|T_i|$ is first fixed and $|T_v|$ is later adjusted for a design trade-off.

The stability margin of the outer loop gain $T_2(s) = T_v(s)/(1 + T_i(s))$ is directly related with the voltage feedback compensation. In the control design, most control parameters are theoretically determined and, at the final stage, the integrator gain K_v is calibrated for the final design refinement. The stability margins of $T_2(s)$ provide straightforward and explicit information about this design optimization. For

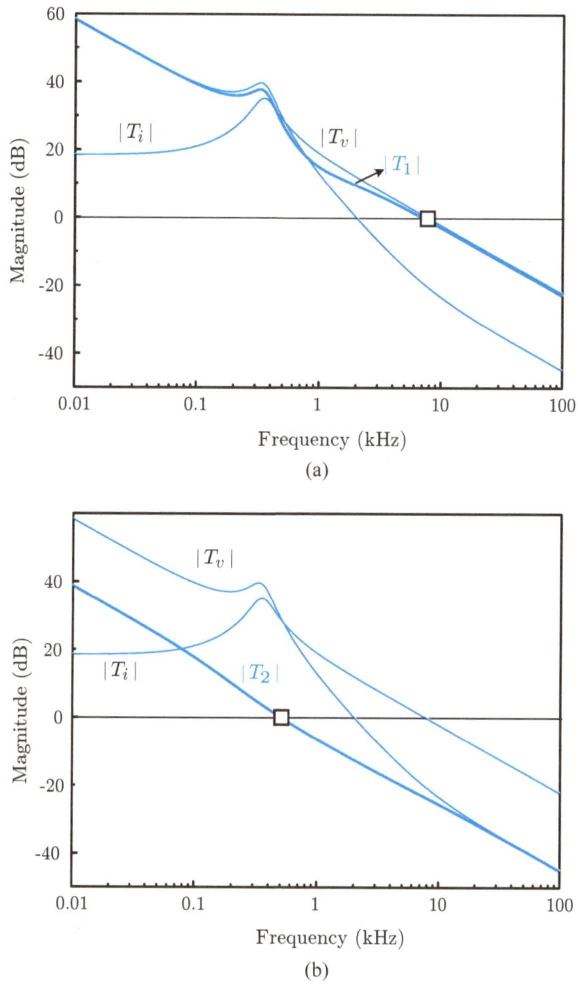

Figure 9.51 Individual feedback loops and system loop gains. (a) Overall loop gain $T_1(s)$. (b) Outer loop gain $T_2(s)$.

example, the gain margin of 11 dB indicates that the converter becomes unstable when K_v is increased more than the factor of $10^{11/20} = 3.54$. This information is critical when increasing the integrator gain to improve the closed-loop performance.

The audio-susceptibility and output impedance characteristics are displayed in Fig. 9.53. Referring to Section 9.3.1, the audio-susceptibility is approximately given by

$$A_u(s) = \frac{A_{ui}(s)}{1 + T_2(s)} \approx \frac{\dfrac{1}{2(1-D)} \dfrac{1 + sCR_c}{1 + sC(R + R_c)/2}}{1 + T_2(s)} \tag{9.91}$$

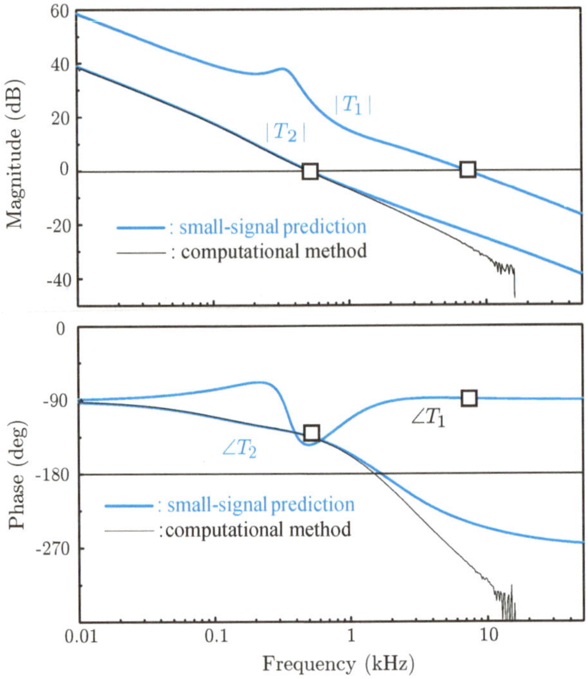

Figure 9.52 Comparison between overall loop gain $T_1(s)$ and outer loop gain $T_2(s)$.

Figure 9.53(a) shows $|A_{ui}|$, $|T_2|$, and $|A_u|$. As addressed in Section 9.3.1, the relationship (9.91) is only valid up to mid-frequencies. At high frequencies, $|A_u|$ simply follows $|G_{vs}|$.

The output impedance is approximated as

$$Z_o(s) = \frac{Z_{oi}(s)}{1+T_2(s)} \approx \frac{\frac{R}{2}\frac{1+sCR_c}{1+sC(R+R_c)/2}}{1+T_2(s)} \tag{9.92}$$

Figure 9.53(b) displays $|Z_{oi}|$, $|T_2|$, and $|Z_o|$. The first pole and peak value of the output impedance are predicted from Fig. 9.53(b) as $\omega_{zc} = 2\pi \cdot 400$ rad/s and $|Z_o|_{peak} = -2$ dB. This information will be used to predict the step load response of the output voltage.

The step input response and step load response are shown in Fig. 9.54. Figure 9.54(a) is the transient response of the output voltage due to the step changes of $V_S = 12\ \text{V} \Rightarrow 8\ \text{V} \Rightarrow 12\ \text{V}$ in the input voltage. Finally, Fig. 9.54(b) shows the output voltage waveform in response to the step changes of $I_O = 4\ \text{A} \Rightarrow 7\ \text{A} \Rightarrow 4\ \text{A}$ in the load current. The settling time of the output voltage is predicted as

$$t_s = \frac{3}{\omega_{zc}} = \frac{3}{2\pi \cdot 400} = 1.19\ \text{ms}$$

The upper limit of the output voltage deviation is predicted as

$$\Delta v_O(t)_{peak} < I_{step}\, 10^{|Z_o|_{peak}/20} = 3 \cdot 10^{-2/20} = 2.38\ \text{V}$$

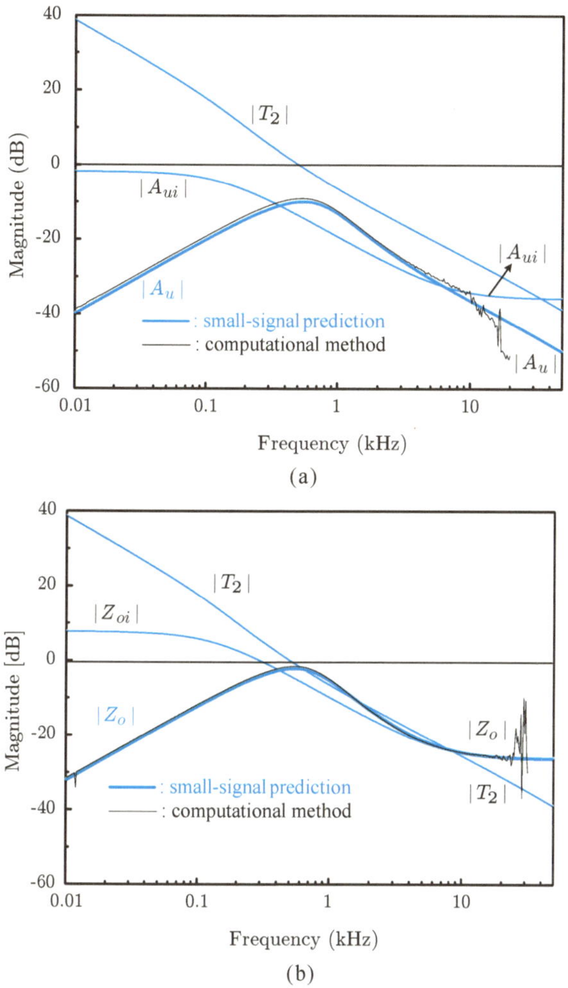

Figure 9.53 Frequency-domain performance. (a) Audio-susceptibility. (b) Output impedance.

9.4.2 Loop Gain Analysis

In the previous control design procedures, all the control parameters are systematically selected based on the power stage parameters and operational conditions. However, the integrator gain is often lastly tuned for optimal closed-loop performance. Accordingly, it would be helpful to investigate the behavior of the overall

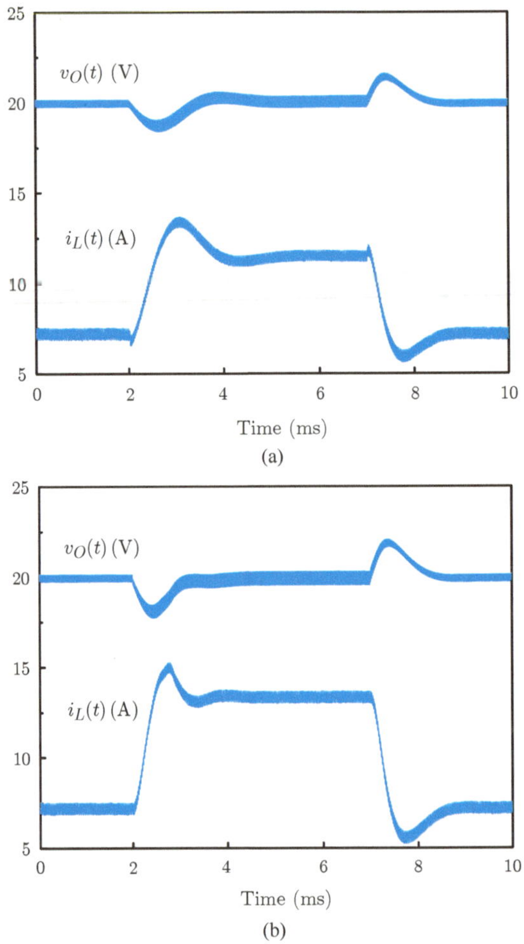

Figure 9.54 Transient response. (a) Step input response. (b) Step load response.

loop gain and outer loop gain when the integrator gain varies. This would provide insights about the implication and characteristics of the two loop gains, as well as the overall dynamics of multi-loop controlled converter systems.

EXAMPLE 9.17 Loop Gain Analysis

For the loop gain analysis purpose, the boost converter used in Example 9.16 is revisited. The overall loop gain and outer loop gain of the boost converter with the integrator gain of $K_v = 2890$ were shown in Fig. 9.52. Now, the integrator gain is successively increased from $K_v = 2890$ to 5780, 9900, and 14560. Figure 9.55 shows the overall loop gain $T_1(s)$ with the four different integrator gains. The overall loop gain exhibits rather complex behavior, demonstrating substantial changes in both magnitude and phase. This complex behavior originates from the location of the

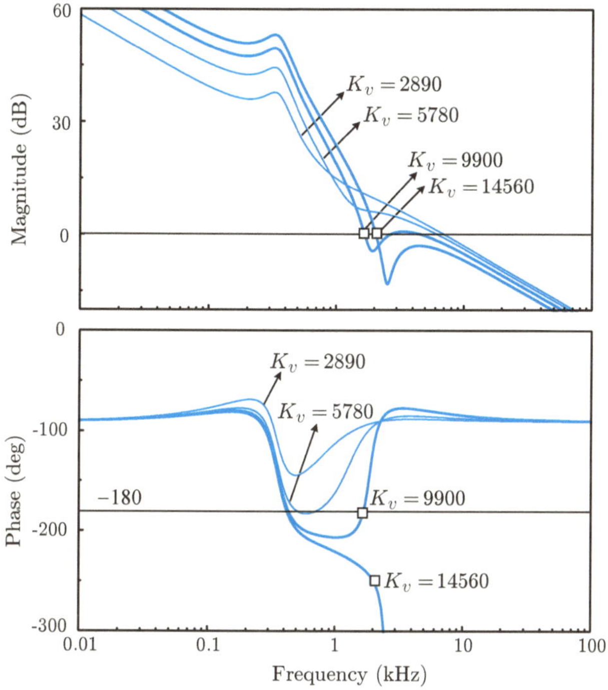

Figure 9.55 Overall loop gains with different integrator gains.

integrator gain K_v in the loop gain expression

$$T_1(s) = T_i(s) + T_v(s) = T_i(s) + G_{vd}(s)F'_m \underbrace{\frac{K_v}{s}\frac{1+\frac{s}{\omega_{zc}}}{1+\frac{s}{\omega_{pc}}}}_{F_v(s)}$$

which is rearranged as

$$T_1(s) = \boxed{T_i(s) + K_v\left(G_{vd}(s)F'_m \frac{1}{s}\frac{1+\frac{s}{\omega_{zc}}}{1+\frac{s}{\omega_{pc}}}\right)} \tag{9.93}$$

The change in K_v affects both the magnitude and phase of the loop gain, because K_v appears as a multiplication factor in the second term only.

Figure 9.56 shows the outer loop gain $T_2(s)$ with the four different integrator gains. In contrast to $T_1(s)$, the outer loop gain shows a linear change in magnitude only, while the phase remains unchanged. This transition pattern is easily inferred

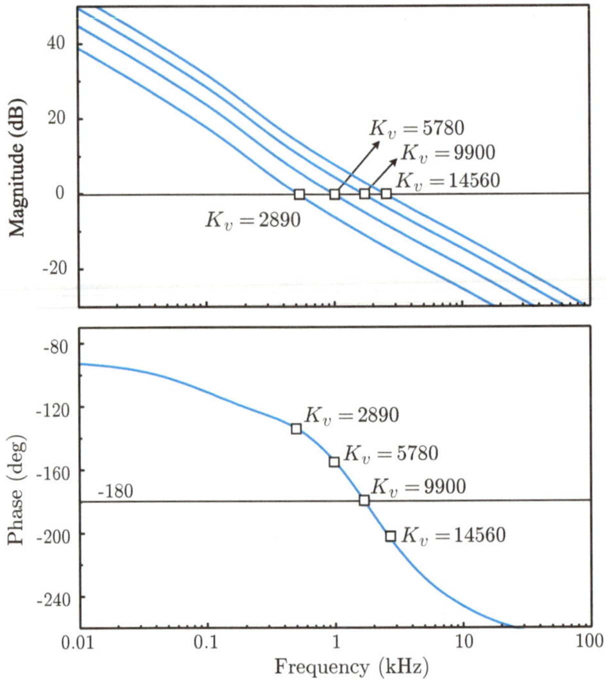

Figure 9.56 Outer loop gains with different integrator gains.

from the $T_2(s)$ expression

$$T_2(s) = \frac{T_v(s)}{1+T_i(s)} = \frac{G_{vd}(s)F'_m \dfrac{K_v}{s} \dfrac{1+\dfrac{s}{\omega_{zc}}}{1+\dfrac{s}{\omega_{pc}}}}{1+T_i(s)}$$

$$= \boxed{K_v \frac{G_{vd}(s)F'_m \dfrac{1}{s} \dfrac{1+\dfrac{s}{\omega_{zc}}}{1+\dfrac{s}{\omega_{pc}}}}{1+T_i(s)}} \tag{9.94}$$

Here, the integrator gain K_v is the common multiplication factor in the whole loop gain expression. Thus, K_v linearly increases the magnitude only, without affecting the phase. The original outer loop gain with $K_v = 2890$ exhibits a 11 dB gain margin. Accordingly, the converter is marginally stable with the integrator gain of $K_v = 2890 \cdot 10^{11/20} = 9900$, and becomes unstable for the larger gain, $K_v = 14560$. Figure 9.56 shows an exact match to this analysis.

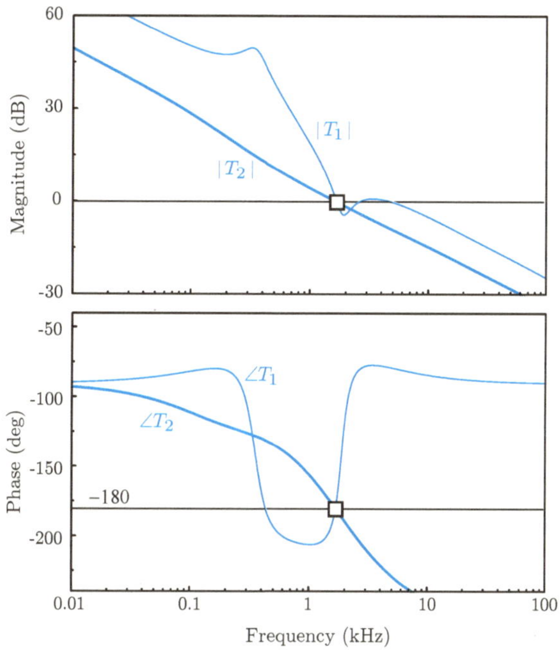

Figure 9.57 System loop gains with $K_v = 9900$.

Although the overall loop gain and outer loop gain show very different stability margins and evolution patterns, they should provide the same information about the absolute stability. As shown in Fig. 9.55 and Fig. 9.56, the two loop gains both indicate that the converter is stable with $K_v = 2890$ and 5780, and unstable with $K_v = 14560$. An important observation is made from the loop gain plots with $K_v = 9900$. The outer loop gain plot in Fig. 9.56 indicates that, when the integrator gain is increased to $K_v = 9900$, the loop gain crosses the 0 dB line at $\omega_{cr} = 2\pi \cdot 1.70 \times 10^3$ rad/s with $-180°$ phase angle. This implies

$$T_2(j\omega_{cr}) = \frac{T_v(j\omega_{cr})}{1 + T_i(j\omega_{cr})} = 1\angle - 180° = -1 \tag{9.95}$$

with $\omega_{cr} = 2\pi \cdot 1.70 \times 10^3$ rad/s. The above equation is rewritten as

$$1 + \frac{T_v(j\omega_{cr})}{1 + T_i(j\omega_{cr})} = 0 \quad \Rightarrow \quad 1 + T_i(j\omega_{cr}) + T_v(j\omega_{cr}) = 0 \tag{9.96}$$

which indicates

$$T_1(j\omega_{cr}) = T_i(j\omega_{cr}) + T_v(j\omega_{cr}) = -1 = 1\angle - 180° \tag{9.97}$$

Equations (9.95) and (9.97) imply that the overall loop gain and outer loop gain identically predict that the converter is marginally stable with the integrator gain $K_v = 9900$. This information should show up in both the Bode plot and polar plot of the two loop gains.

- The Bode plots of the overall loop gain and outer loop gain both cross the 0 dB line at the frequency $\omega_{cr} = 2\pi \cdot 1.70 \times 10^3$ rad/s with $-180°$ phase angle:

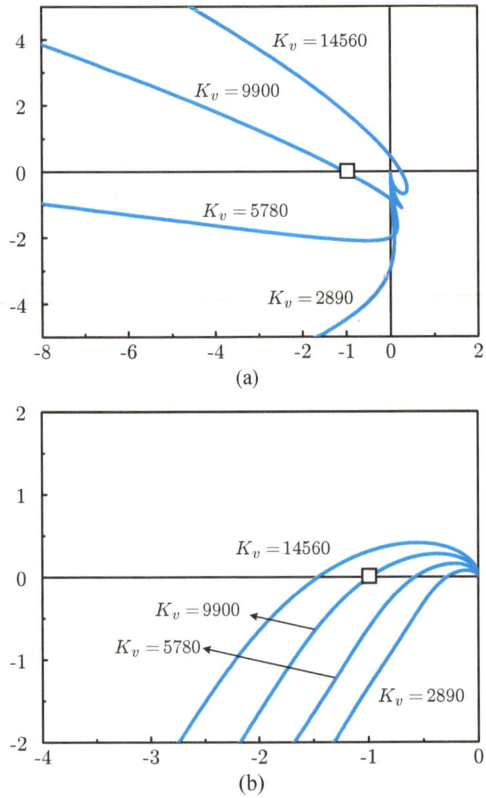

Figure 9.58 Polar plot of loop gains. (a) Overall loop gain. (b) Outer loop gain.

$T_1(j\omega_{cr}) = T_2(j\omega_{cr}) = 1\angle{-180°}$. Figure 9.57 shows the Bode plots of the two loop gains with $K_v = 9900$. The two loop gains precisely cross the 0 dB line at $\omega_{cr} = 2\pi \cdot 1.70 \times 10^3$ rad/s with a $-180°$ phase angle, thereby confirming that $T_1(j\omega_{cr}) = T_2(j\omega_{cr}) = 1\angle{-180°}$.

- The polar plots of the two loop gains hit the $(-1, 0)$ point at $\omega_{cr} = 2\pi \cdot 1.70 \times 10^3$ rad/s: $T_1(j\omega_{cr}) = T_2(j\omega_{cr}) = -1$. Figure 9.58 shows the polar plots of the loop gains evaluated with the four different integrator gains. The overall loop gain exhibits a very complex transition pattern, as predicted from (9.93). On the other hand, the polar plot of the outer loop gain expands proportionally as the integrator gain increases. In spite of the very different transition patterns, the polar plots of the two loop gains identically traverse the $(-1, 0)$ point with $K_v = 9900$.

The left-hand side of expression (9.96) also indicates that $\pm\omega_{cr} = \pm 2\pi \cdot 1.70 \times 10^3$ rad/s are the roots of the characteristic equation of the converter. Thus, a pair of the system poles lie on the imaginary axis when the integrator gain is increased to

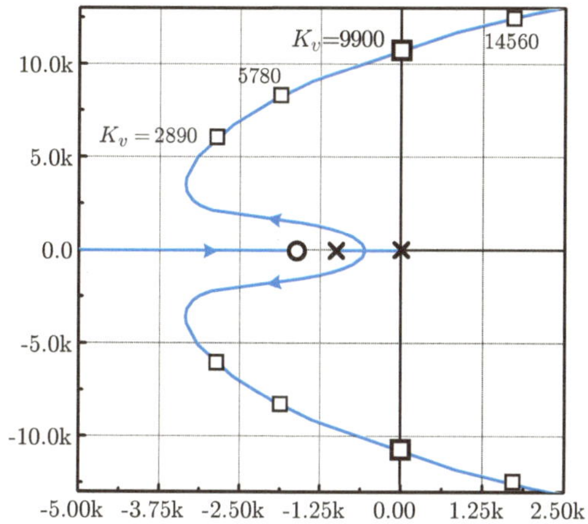

Figure 9.59 Pole trajectory with integrator gain variation.

$K_v = 9900$. Figure 9.59 shows the root locus evaluated with respect to the integrator gain K_v

$$1 + T_2(s) = 1 + K_v \frac{G_{vd}(s)F_m' \dfrac{1}{s} \dfrac{1+\dfrac{s}{\omega_{zc}}}{1+\dfrac{s}{\omega_{pc}}}}{1+T_i(s)} = 0 \tag{9.98}$$

or

$$1 + T_1(s) = 1 + T_i(s) + K_v G_{vd}(s) F_m' \frac{1}{s} \frac{1+\dfrac{s}{\omega_{zc}}}{1+\dfrac{s}{\omega_{pc}}} = 0 \tag{9.99}$$

On the root locus plot, the locations of the dominant poles are marked for $K_v =$ 2890, 5780, 9900, and 14560. As predicted, a pair of the closed-loop poles indeed cross the imaginary axis with $K_v = 9900$. The crossing points should be $\pm\omega_{cr} = \pm 2\pi \cdot 1.70 \times 10^3$ rad/s $= \pm 1.07 \times 10^4$ rad/s, which are the frequencies that satisfy $1 + T_1(j\omega_{cr}) = 1 + T_2(j\omega_{cr}) = 0$.

9.5 CHAPTER SUMMARY

Current mode control employs an additional feedback from the inductor current on top of the output voltage feedback. Using the two feedback signals, current mode control offers substantial improvements over conventional voltage mode control. Modern PWM dc-to-dc converters extensively adapt current mode control. Now, current mode control truly holds the prime position in PWM dc-to-dc power converter control schemes. This chapter presented comprehensive analysis and design of current mode control.

Peak current mode control is the most common form of current mode control. The peak current mode control senses the switch current and uses its peak value to execute the PWM function. Because the switch current is the on-time inductor current, the switch current sensing is conceptually replaced with the inductor current sensing for the purpose of dynamic analysis and control design. The analysis and design of the peak current mode control are presented using the inductor current sensing.

Current mode control is a multi-loop control scheme in which multiple system loop gains exist. Two particular system loop gains, the overall loop gain and outer loop gain, are identified and utilized for the dynamic analysis. These two loop gains are used, collectively in some cases and individually in other cases, for the analysis and design of current mode control. The respective role and implication of the two loop gains are addressed throughout this chapter.

Systematic non-iterative design procedures are formulated for peak current mode control. The current feedback circuit is designed to place the crossover frequency of the overall loop gain at high frequencies, up to $10\% - 30\%$ of the switching frequency. The voltage feedback circuit is constructed using a two-pole one-zero compensation. The compensation parameters are selected for stability and good closed-loop performance. In particular, the integrator gain is determined for the desired crossover frequency of the outer loop gain. Step-by-step design procedures for the three basic PWM converters are given in Table 9.3.

The performance of the peak current mode control adapted to the buck converter and boost converter is analyzed in detail. In particular, the step load response is thoroughly analyzed, leading to a practical method for predicting the transient response of the output voltage from the converter's output impedance characteristics. It was shown that the results of this step load analysis can be extended to all dc-to-dc converters in general. A specific example is given in Example 9.14 where the step load response of a switched capacitor converter is presented.

The use and implication of the overall loop gain and outer loop gain in the stability analysis are demonstrated using a boost converter example. The outward information of the loop gains should be interpreted based on the definition of the corresponding loop gain. Although the two loop gains have utterly different shapes, stability margins, and transition patterns, they provide a coherent message on the internal information about the converter stability. This analysis is covered in Example 9.17.

The analysis presented in this chapter is called the *classical* analysis because the analysis does not include the sampling effects of the peak current mode control and is currently less prevalent than the past. A new analysis method that includes the sampling effects has emerged as an alternative to the classical analysis. Even so, the classical analysis is still valuable because it duly and *conveniently* describes the major dynamics of current mode control using only the familiar s-domain analysis. The next chapter deals with the new analysis to include the sampling effects into the converter dynamics. Whenever appropriate and helpful, the results of the new analysis will be compared with those of the classical analysis.

REFERENCES

1. D. Sable, R. Ridley, and B. Cho, "Comparison of performance of single-loop and current-injection control for PWM converters that operate in both continuous and discontinuous modes of operation," *IEEE Trans. Power Electron.,* vol. 7, no. 1, pp. 136-142, Jan. 1992.

2. L. Dixon, "Average current mode control of switching power supplies," *UNITRODE Power Supply Seminar SEM-700,* 1990.

3. L. Dixon, "Optimizing the design of a high power factor switching regulator," *UNITRODE Power Supply Seminar SEM-700,* 1990.

4. W. Tang, F. C. Lee, R. B. Ridley, and I. Cohen, "Charge control: modeling, analysis, and design," *IEEE Trans. Power Electron.,* vol. 8, no. 4, pp. 396-403, Oct. 1993.

5. F. C. Lee, M. F. Mahmoud, and Y. Yu, *"Design handbook for a standardized control module for dc-to-dc converters,"* vol. I, NASA Contract NAS-3-20102, Apr. 1980.

6. F. C. Lee, Y. Yu, and M. F. Mahmoud, "A unified analysis and design procedure for a standardized control module for dc-dc switching regulators," in *PESC Conf. Rec.,* pp. 284-301, 1979.

7. R. B. Ridley, B. H. Cho, and F. C. Lee, "Analysis and interpretation of loop gains of multi-loop-controlled switching regulator," *IEEE Trans. Power Electron.,* vol. 3, no. 4, pp. 489-498, Oct. 1988.

8. B. Choi and W. Lim, "Control design and closed-loop analysis of a switched-capacitor dc-to-dc converter," *IEEE Trans. Aerosp. Electron. Syst.,* vol. 37, no. 3, pp. 1099-1107, Jul. 2001.

PROBLEMS

9.1* The current sensing network (CSN) discussed in Example 9.2 is employed in the buck converter operating with the following conditions:

$V_S = 24$ V $\quad V_O = 12$ V $\quad L = 80\ \mu$H $\quad R = 4\ \Omega$

$C = 400\ \mu$F $\quad R_c = 0.01\ \Omega \quad f_s = 50$ kHz

The circuit parameters of the CSN are shown in Fig. P9.1.

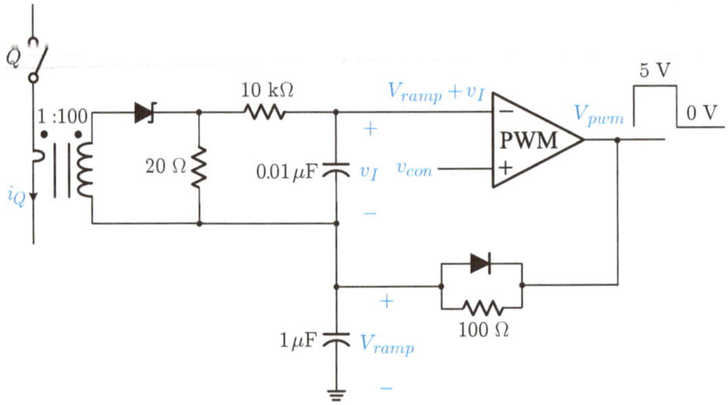

Fig. P9.1

a) Sketch the waveforms of the switch current i_Q, sensed voltage signal v_I, and compensation ramp V_{ramp} for the two operational periods. Show the maximum and minimum values of the waveforms.

b) Sketch the composite signal, $V_{ramp} + v_I$, control voltage v_{con}, and PWM output V_{pwm} for the two operational periods, in order to illustrate the control waveforms of the peak current mode control.

9.2* As discussed in Section 9.2.1, several different expressions for the modulator gain F'_m can be formulated, depending on the way of relating the average values of the on-time and/or off-time inductor currents, $\bar{i}_{on}(t)$ and $\bar{i}_{off}(t)$, to the average value of the inductor current, $\bar{i}_L(t)$.

$$\bar{i}_{on}(t) = v_{con} - S_e dT_s - \frac{1}{2} S_n dT_s$$

$$\bar{i}_{off}(t) = v_{con} - S_e dT_s - \frac{1}{2} S_f (1-d) T_s$$

a) Show that one modulator gain expression is given by

$$F'_m = \frac{2}{(2S_e + S_n)T_s}$$

by assuming $\bar{i}_L(t) = \bar{i}_{on}(t)$ and linearizing the following equation

$$\bar{i}_L(t) = v_{con} - S_e dT_s - \frac{1}{2} S_n dT_s$$

b) Prove that another modulator gain expression is formulated as

$$F'_m = \frac{2}{(2S_e - S_f)T_s}$$

by linearizing

$$\bar{\imath}_L(t) = \bar{\imath}_{off}(t) = v_{con} - S_e dT_s - \frac{1}{2}S_f(1-d)T_s$$

c) Finally, show that the other modulator gain expression is found as

$$F'_m = \frac{1}{S_e T_s}$$

by linearizing the following weighted average of $\bar{\imath}_{on}(t)$ and $\bar{\imath}_{off}(t)$

$$\begin{aligned} \bar{\imath}_L(t) &= d\,\bar{\imath}_{on}(t) + (1-d)\,\bar{\imath}_{off}(t) \\ &= v_{con} - S_e dT_s - \frac{1}{2}S_n d^2 T_s - \frac{1}{2}S_f(1-d)^2 T_s \end{aligned}$$

9.3 Derive an exact expression for the duty ratio-to-inductor current transfer function, $G_{id}(s)$, for the buck, boost, and buck/boost converters that contain the parasitic resistances of the reactive components, R_l and R_c in Fig. 9.13. Compare the exact expressions with the approximations given in Table 9.2.

9.4* Figure P9.4 shows the circuit diagram of a current-mode controlled buck converter.

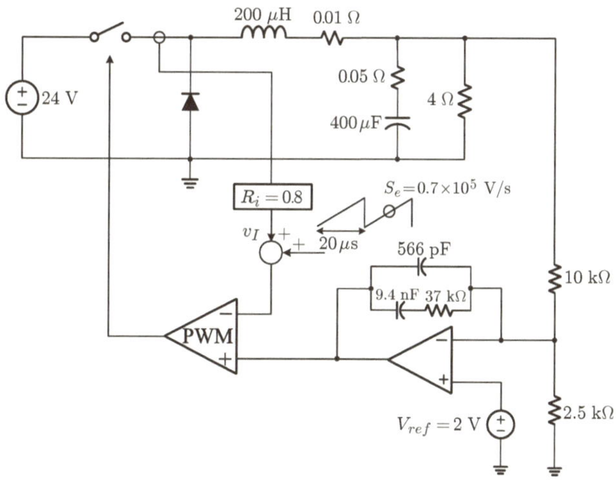

Fig. P9.4

a) Find the expressions for the current loop $T_i(s)$ and voltage loop $T_v(s)$.
b) On semi-log graph paper, sketch the asymptotic plot for $|T_i|$ and $|T_v|$ and construct the asymptotic plot for the overall loop gain $|T_1|$ and outer loop gain $|T_2|$. Predict the 0 dB crossover frequency of $T_1(s)$ and $T_2(s)$.
c) Use the results of b) to construct the factorized expressions for $T_1(s)$ and $T_2(s)$.

9.5* Shown in Fig. P9.5 is the circuit diagram of a current-mode controlled buck converter.

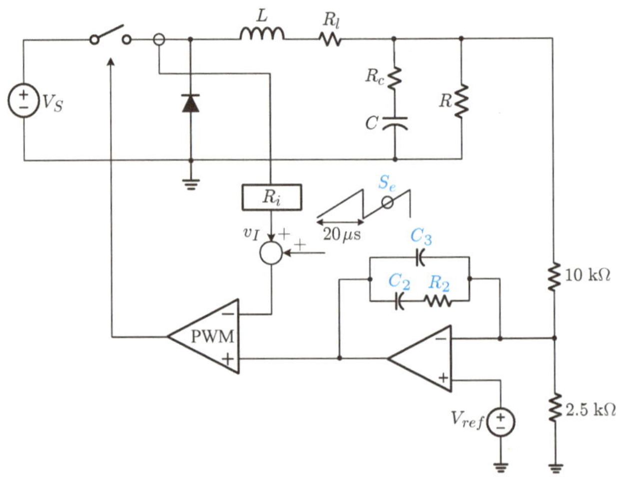

Fig. P9.5

a) Consider the following power stage parameters and operational conditions for the buck converter:

$V_S = 24$ V $\quad V_{ref} = 2$ V $\quad L = 80\ \mu$H $\quad R_l = 0.01\ \Omega$
$C = 400\ \mu$F $\quad R_c = 0.1\ \Omega$ $\quad R = 2\ \Omega$ $\quad R_i = 0.5$

Determine the slope of the compensation ramp, S_e, to place the 0 dB crossover frequency of the current loop at the 20% of the switching frequency, $\omega_{ci} = 0.2\,\omega_s$. Also, determine values for C_2, R_2, and C_3 to locate the 0 dB crossover frequency of the outer loop gain at the esr zero, $\omega_{cr} = \omega_{esr}$. Construct the asymptotic plots for $|T_i|$, $|T_v|$, $|T_1|$, and $|T_2|$ on semi-log graph paper, in order to confirm the compliance with the design requirements.

b) Repeat a) for the converter operating with

$V_S = 18$ V $\quad V_{ref} = 1.2$ V $\quad L = 60\ \mu$H $\quad R_l = 0.01\ \Omega$
$C = 600\ \mu$F $\quad R_c = 0.1\ \Omega$ $\quad R = 3\ \Omega$ $\quad R_i = 0.7$

for the design aims of $\omega_{ci} = 0.2\,\omega_s$ and $\omega_{cr} = 0.8\,\omega_{esr}$.

c) Repeat a) for the converter operating with

$$V_S = 48 \text{ V} \quad V_{ref} = 6 \text{ V} \quad L = 300\ \mu\text{H} \quad R_l = 0.01\ \Omega$$
$$C = 600\ \mu\text{F} \quad R_c = 0.05\ \Omega \quad R = 6\ \Omega \quad R_i = 0.2$$

for the design aims of $\omega_{ci} = 0.2\,\omega_s$ and $\omega_{cr} = \omega_{esr}$.

9.6* Figure P9.6 shows the four different output impedances of a current-mode controlled PWM converter. For each case, sketch the general profile of the output voltage $v_O(t)$ when a step decrease of $\Delta I_{step} = 5$ A occurs in the load current. Show all the prominent features of $v_O(t)$ including the estimations for the peak overshoot and settling time.

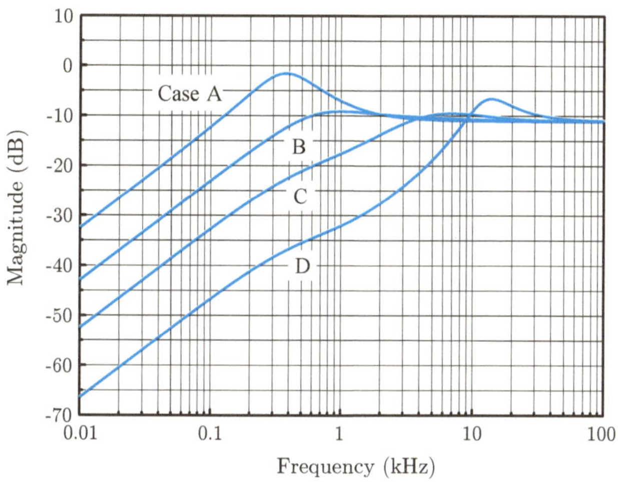

Fig. P9.6

9.7 Shown in Fig. P9.7 are the Bode plots for $|T_i|$, $|T_v|$, $|Z_{oi}|$, and $|G_{vs}|$ of a current-mode controlled buck converter.

a) Construct the asymptotic plot for $|T_1|$, $|T_2|$, $|Z_o|$, and $|A_u|$ on Fig. P9.7.

b) Use the result of a) to extract the factorized expressions for $T_1(s)$, $T_2(s)$, $Z_o(s)$, and $A_u(s)$.

9.8* Perform the asymptotic analysis to evaluate the closed-loop performance of the current-mode controlled buck converter shown in Fig. P9.8.

a) Sketch $|T_i|$, $|T_v|$, $|Z_{oi}|$, and $|G_{vs}|$ on semi-log graph paper.

b) Use the results of a) to construct the asymptotic plots for $|T_1|$, $|T_2|$, $|Z_o|$, and $|A_u|$.

c) Use the outcomes of b) to extract the factorized expressions for $T_1(s)$, $T_2(s)$, $Z_o(s)$, and $A_u(s)$.

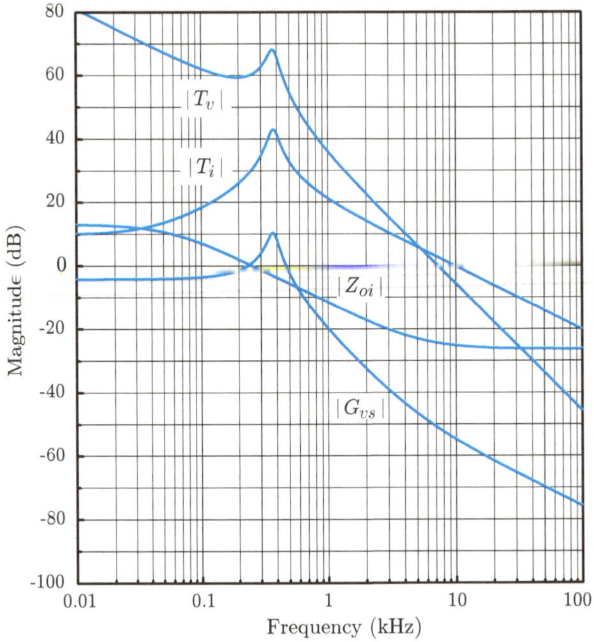

Fig. P9.7

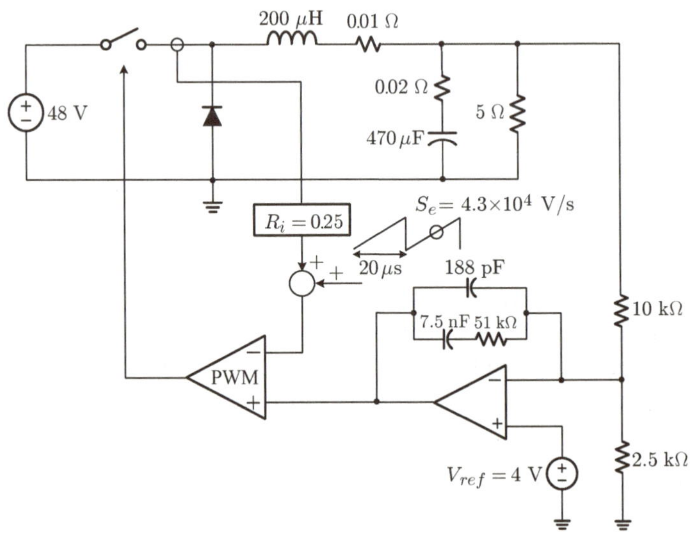

Fig. P9.8

9.9 Figure P9.9 shows the circuit diagram of a current-mode controlled boost converter.

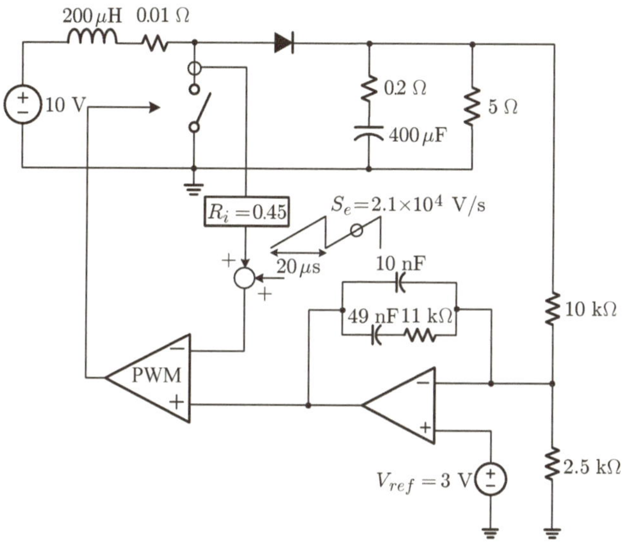

Fig. P9.9

a) Find the expressions for the current loop $T_i(s)$ and voltage loop $T_v(s)$.

b) On semi-log graph paper, sketch the asymptotic plots for $|T_i|$ and $|T_v|$ and construct the asymptotic plot for the overall loop gain $|T_1|$ and outer loop gain $|T_2|$. Estimate the 0 dB crossover frequency of the loop gains.

c) Use the results of b) to extract the factorized expressions for $T_1(s)$ and $T_2(s)$.

9.10* Shown in Fig. P9.10 is the circuit diagram of a current-mode controlled boost converter.

a) Consider the following circuit parameters and operational conditions for the boost converter:

$V_S = 24$ V $\quad V_{ref} = 9.2$ V $\quad L = 160\ \mu$H $\quad R_l = 0.01\ \Omega$

$C = 400\ \mu$F $\quad R_c = 0.05\ \Omega \quad R = 10\ \Omega \quad R_i = 0.45$

Determine the slope of the compensation ramp, S_e, to place the 0 dB crossover frequency of the current loop at the 20% of the switching frequency, $\omega_{ci} = 0.2\,\omega_s$. Also determine values for C_2, R_2, and C_3 to locate the 0 dB crossover frequency of the outer loop gain at the 20% of the RHP zero, $\omega_{cr} = 0.2\,\omega_{rhp}$. Construct the asymptotic plots for $|T_i|$, $|T_v|$, $|T_1|$, and $|T_2|$ on semi-log graph paper, in order to confirm the compliance with the design requirements.

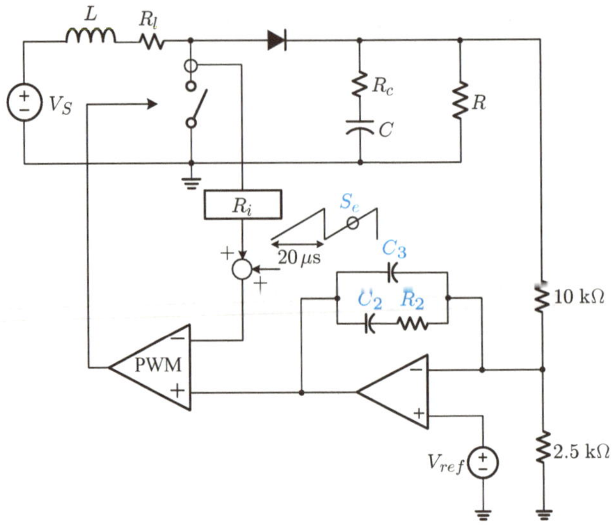

Fig. P9.10

b) Repeat a) for the converter operating with

$V_S = 30$ V $\quad V_{ref} = 9.6$ V $\quad L = 80\ \mu$H $\quad R_l = 0.01\ \Omega$

$C = 860\ \mu$F $\quad R_c = 0.05\ \Omega$ $\quad R = 10\ \Omega$ $\quad R_i = 0.15$

for the design aims of $\omega_{ci} = 0.2\,\omega_s$ and $\omega_{cr} = 0.15\,\omega_{rhp}$.

c) Repeat a) for the converter operating with

$V_S = 12$ V $\quad V_{ref} = 5.6$ V $\quad L = 240\ \mu$H $\quad R_l = 0.01\ \Omega$

$C = 400\ \mu$F $\quad R_c = 0.02\ \Omega$ $\quad R = 2\ \Omega$ $\quad R_i = 0.3$

for the design aims of $\omega_{ci} = 0.2\,\omega_s$ and $\omega_{cr} = 0.3\,\omega_{rhp}$.

CHAPTER 10

CURRENT MODE CONTROL (II) – SAMPLING EFFECTS AND NEW CONTROL DESIGN PROCEDURES

Peak current mode control is a sampled-data process because the control action is periodically executed by sampling and holding the error signal produced by the fast-varying inductor current waveform. Due to this feature, current-mode controlled converters exhibit sampled-data system characteristics, referred to as the *sampling effects* of current mode control. The dynamics of current-mode controlled converters, at the presence of the sampling effects, can be investigated via discrete-time modeling and z-domain analysis. However, the results would be too complex to expose any easy insights onto converter dynamics or control design strategy. In fact, several attempts had been made to characterize the converter dynamics in terms of z-domain expressions but the outcomes were not widely accepted due to their apparent complexity.

Traditionally, peak current mode control has been analyzed and designed using s-domain techniques, implicitly assuming that the sampling effects have negligible consequences. This classical analysis was explored in the previous chapter. We proved that the classical analysis duly predicts major dynamics of current mode control and provides correct design procedures for most cases. The classical analysis prevailed until the late 1980s.

In the early 1990s, a new class of dynamic models emerged, which accounts for the sampling effects only using conventional s-domain descriptions. These new models, which are called the *s-domain model for current mode control* in this book, rekindled interest and attention to the sampling effects of current mode control. Using this new s-domain model, several new facets of current mode control are revealed, which supplement and reinforce the outcomes of the earlier classical analysis.

In this chapter, we investigate the sampling effects with the aid of the s-domain model for current mode control. The consequence of the sampling effects on converter dynamics will be studied in detail. Based on the analysis results, new design procedures for current mode control will be developed. The predictions and performance of the new design will be compared and contrasted with those of the classical design, which does not consider the sampling effects. We will address the correlation between the new design procedures and the classical design procedures. Finally, as a practical application example, the design and evaluation of a current-mode controlled off-line flyback converter with an optocoupler-isolated feedback circuit will be discussed.

10.1 SAMPLING EFFECTS OF CURRENT MODE CONTROL

This section presents the origin, nature, and consequence of the sampling effects. It also introduces an s-domain model for current mode control, which will be used in this chapter to account for the sampling effects.

10.1.1 Origin and Consequence of Sampling Effects

Figure 10.1 shows the peak current mode control adapted to the three basic PWM converters. As discussed in Section 9.2.1, this model could represent each of the three basic converters with the respective connections of the power stage configuration.

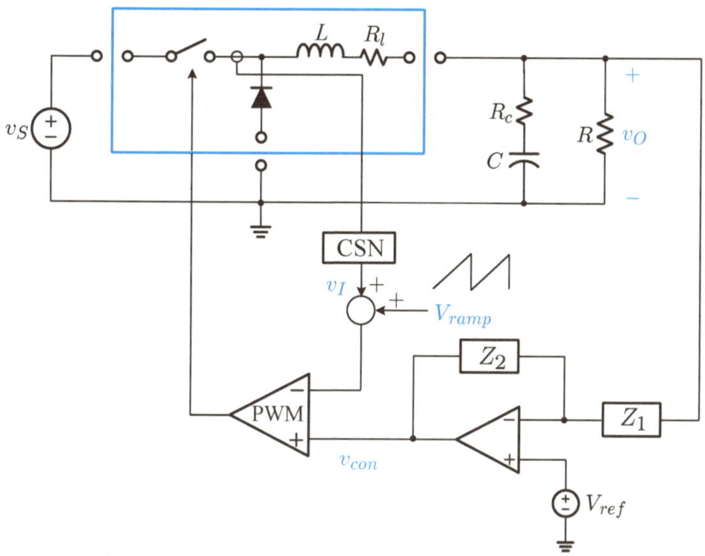

Figure 10.1 Peak current mode control adapted to three basic PWM converters.

For peak current mode control, the switch current is sensed with the current sensing network (CSN). As explained in the previous chapter, the switch current sensing is identical to the inductor current sensing, as far as the control mechanism is concerned. For the convenience of pictorial illustration and analytical treatment, the peak current mode control is described in terms of the inductor current sensing.

Origin of Sampling Effects

The sampling effects are invoked by the operation of the PWM modulator. The sampling effects of current mode control are investigated using the PWM waveforms shown in Figs. 10.2 and 10.3. Figure 10.2(a) shows the structure and waveforms of the PWM modulator, where the peak current mode control is represented by the inductor current sensing. The CSN gain is assumed unity without loss of generality, and so $v_I = i_L$ for this case. Figure 10.2(a) is transformed into the equivalent form in Fig. 10.2(b).

The control waveforms in Fig. 10.2(b) are modified into Fig. 10.3, in order to explain the evolution of the perturbed inductor current i'_L in comparison with the original inductor current i_L. The sampling effects originate from the fact that the control action is periodically exerted by sampling and holding the error signal generated by the fast-varying inductor current. The initial perturbation $\hat{i}_L(k) = i'_L(k) - i_L(k)$, occurred in the inductor current at the instant of its peak value in the kth period, propagates to the ensuing switching periods and generates the continuous error signal, $\hat{i}_L(t) = i'_L(t) - i_L(t)$. During the $(k+1)^{\text{th}}$ period, the error signal is sampled at the moment when i'_L intersects the composite signal, $v_{con} - V_{ramp}$. The sampled

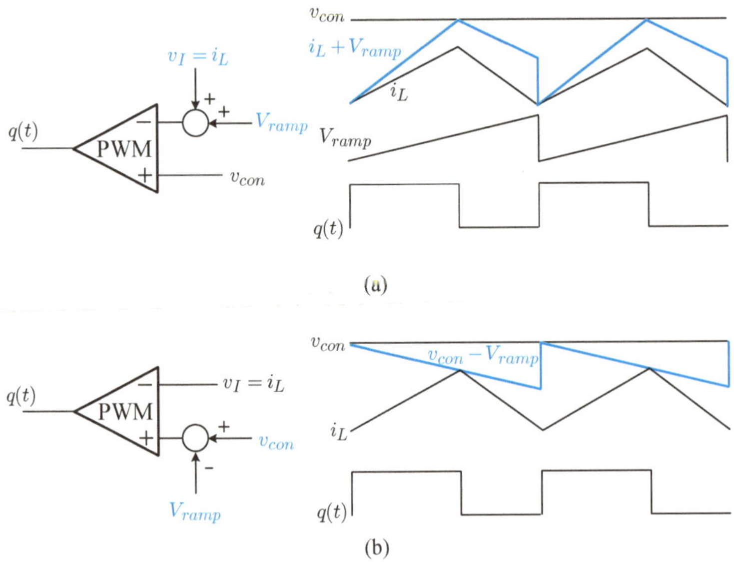

Figure 10.2 Structure and waveforms of peak current mode control. (a) Peak current mode control with unity inductor current sensing. (b) Alternative representation.

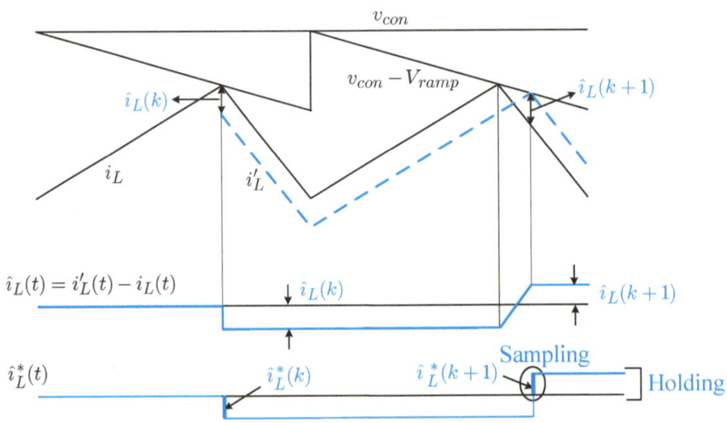

Figure 10.3 Evolution of perturbed inductor current.

error signal $\hat{\imath}_L(k+1)$ is held constant until the next sampling instant. The PWM modulator samples and holds the error signal in synchronization with the switching period. The $\hat{\imath}^*_L(t)$ at the bottom of Fig. 10.3 is a conceptual representation of the sampled-and-held error signal in the PWM modulator.

Consequence of Sampling Effects

The sampling effects penetrate into the converter dynamics and induce phenomena that the classical analysis is unable to predict. The sampling effects are mainly pronounced at higher frequencies, near or above half the switching frequency, because the PWM modulator samples and holds the error signal at the rate of the switching frequency.

Consequence of Sampling Effects EXAMPLE 10.1

This example demonstrates the consequence of the sampling effects on the loop gain characteristics. Figure 10.4 shows the Bode plots of the loop gain of a current-mode controlled boost converter, obtained from two different dynamic models. The thin curve is the prediction of the classical model that does not consider the sampling effects. The thick curve is the outcome of the new model that will be developed in this chapter to account for the sampling effects. Along with the model predictions, the loop gain curve obtained using the computational method is also shown in Fig. 10.4. The switching frequency of the converter is $f_s = 50$ kHz.

The loop gain plots do not show any differences at low frequencies. However, they exhibit notable deviations at high frequencies, particularly in the frequencies around and above half the switching frequency, $f_s/2 = 25$ kHz. When the predictions of the two models are compared with the outcome of the computational method, the new model better correlates with the computational data in both the magnitude and phase characteristics. The improved accuracy of the new model is a direct consequence of including the sampling effects of current mode control. Further discussions about the consequence of the sampling effects will be given in Sections 10.2.3 and 10.3.

10.1.2 Modeling Methodology for Sampling Effects

In current-mode controlled PWM converters, the circuit variables are classified into two categories: the slow-varying variable and fast-varying variable. The output voltage and control voltage are smoothly-filtered slow-varying variables. In contrast, the inductor current is a triangular-shaped fast-varying variable.

The sampling effects are discrete-time phenomena which need to be investigated via sampled-data modeling and z-domain analysis. The sampling effects are initially invoked by the fast-varying inductor current, and therefore the sampled-data modeling is to be applied to the inductor current feedback circuit. On the other hand, it is unnecessary to apply the sampled-data modeling for the remaining part of the converter circuit, where the slow-varying circuit variables prevail. For this part, the discrete-time analysis will not yield much different results from those of the classical analysis based on the averaging method. In fact, voltage mode control is also a sampled-data process in the sense that the duty ratio is only updated once in every switching period. Nonetheless, the averaging method offers sufficient accuracy because all the circuit variables involved with the PWM process vary slowly.

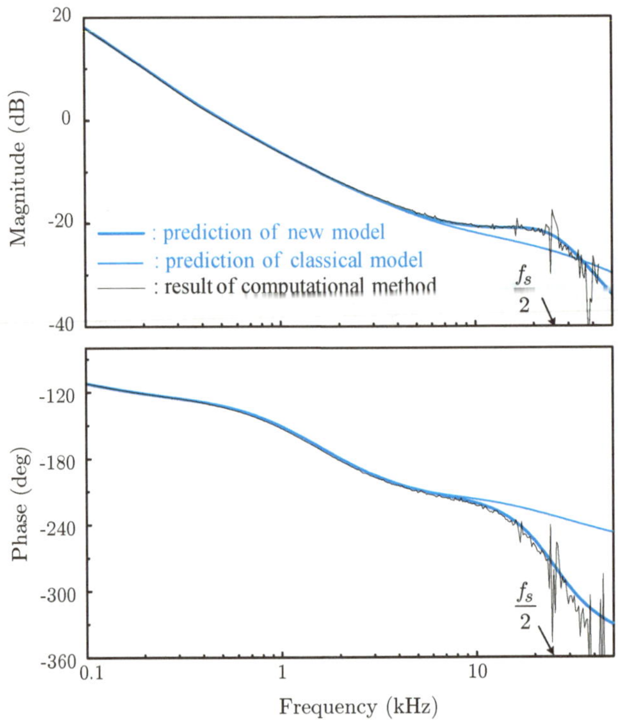

Figure 10.4 Loop gain of boost converter.

One effective modeling methodology for current mode control is to utilize the existing s-domain models as much as possible and to make only necessary modifications or additions to accommodate the sampling effects. In this practice, the sampling effects are first modeled as a discrete-time process and the resulting z-domain model is later transformed into an *equivalent s-domain representation.* The equivalent s-domain representation is then amalgamated with the existing s-domain models for slow-varying circuit variables. The final s-domain model for the entire converter provides both the simplicity of s-domain analysis and the accuracy of sampled-data modeling. The s-domain converter model accepts all the classical analysis techniques and provides the results in s-domain expressions. The analysis results will be accurate up to the Nyquist frequency, i.e., half the switching frequency of the converter, because the model duly accounts for the sample and hold function of the peak current mode control, occurring in every switching period.

10.1.3 Feedforward Gains

In addition to the sampling effects, there exists another mechanism that needs to be considered in analyzing current mode control, but is not included in the classical analysis. In peak current mode control, the changes in the input or output voltage

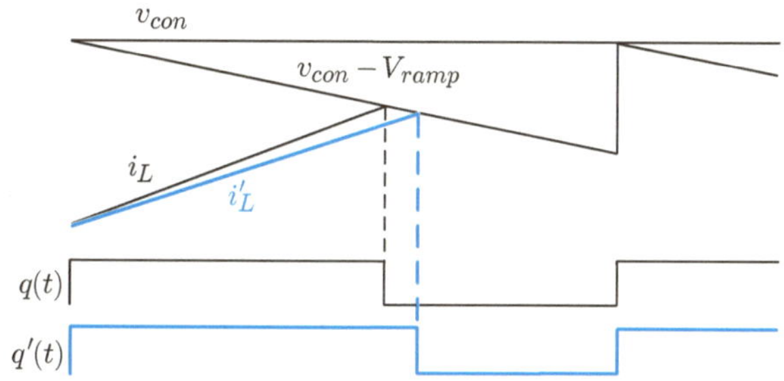

Figure 10.5 PWM waveforms with different input voltages.

affect the cycle-by-cycle duty ratio. This feature is illustrated in Fig. 10.5, which shows the inductor current and modulator waveforms of the buck converter with two different input voltages. The thin black line is the inductor current with the original input voltage. The thick blue line is the inductor current when the input voltage is decreased while the output voltage remains in regulation. The reduced input voltage lowers the on-time slope and the duty ratio is increased.

In the classical analysis in Example 9.3, the modulator gain of peak current mode control was derived assuming the slopes of the inductor current remain unchanged. This hypothesis is now discarded to improve the accuracy of the small-signal model. The slopes of the inductor current are affected by both the input voltage and output voltage, and the resulting slope changes immediately alter the duty ratio. This functional dependency is represented by the feedforward gains from the input and output voltages in the new model.

10.1.4 Complete s-Domain Model for Current Mode Control

The complete s-domain small-signal model for current-mode controlled PWM converters is obtained by adding the necessary gain blocks to the small-signal model introduced in Section 9.2.1. Figure 10.6 shows the resulting small-signal model. In this new small-signal model, four gain blocks are added or modified to represent the sampled-data feature of peak current mode control, while the voltage feedback compensation $F_v(s)$ and CSN gain R_i remain the same.

1) Sampling effects: The s-domain representation of sampling effects can be either merged into the PWM block, resulting in a new s-domain modulator gain $F_m^*(s)$, or included as a separate s-domain gain block, $H_e(s)$, located in the inductor current feedback path.

2) Feedforward gains: The gain block k_f represents the influence of the input voltage and k_r is the feedforward gain from the output voltage.

Once the four additional gain blocks are identified, the complete s-domain model for current mode control is established. With $F_m^*(s) = F'_m$, $H_e(s) = 1$, and $k_f = k_r = 0$, Fig. 10.6 reduces to the model used in the classical analysis in Chapter 9.

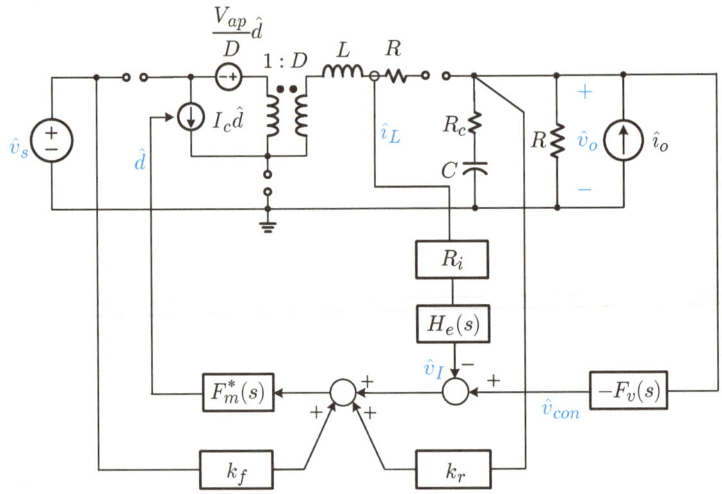

Figure 10.6 Complete s-domain small-signal model for current-mode controlled PWM converters.

10.1.5 Two Prevalent s-Domain Models for Current Mode Control

Two distinct s-domain models for peak current mode control are being widely used. The first s-domain model was proposed by R. B. Ridley [1]. In Ridley's model, the sampling effects are incorporated as a separate s-domain gain block, $H_e(s)$, while the modulator gain is modeled as a constant. The sampled-data feature of current mode control is initially described as a z-domain transfer function and later transformed into the equivalent s-domain representation $H_e(s)$. The modulator gain includes the slope of the compensation ramp, yet does not contain any frequency-dependent term. The feedforward gains from the input and output voltages are also included. The expressions for the gain blocks of Ridley's model for the three basic PWM converters are listed in Table 10.1.

Another s-domain model was soon proposed by F. D. Tan [2]. In Tan's model, the sampling effects are incorporated into the modulator gain as a frequency-dependent term, while the current feedback loop only contains the CSN gain, namely $H_e(s) = 1$. This model also contains the feedforward gains from the input and output voltages. The gain blocks for Tan's model are also shown in Table 10.1.

The two s-domain models for current mode control, Ridley's model and Tan's model, have allowed the sampled-data feature of current mode control to be analyzed using only s-domain techniques. However, these two models have subtle differences in their modeling approaches and final results, and therefore have become a subject of comparisons and clarifications. Although the exactness of these two models is still an issue of rigorous assessment, either of them can be selected as a reference model

Table 10.1 Two Prevalent s-Domain Models for Current Mode Control

		Ridley's model	Tan's model
$H_e(s)$ for all converters		$H_e(s) = 1+\frac{s}{\omega_n Q_z}+\frac{s^2}{\omega_n^2}$ $Q_z = -\frac{2}{\pi} \quad \omega_n = \frac{\pi}{T_s}$	$H_e(s) = 1$
$F_m^*(s)$ for all converters†		$F_m^*(s) = \frac{1}{(S_n + S_e)T_s}$	$F_m^*(s) = \frac{K_m^*}{1+\frac{s}{\omega_p}}$ $K_m^* = \frac{2}{(S_n - S_f + 2S_e)T_s}$ $\omega_p = \frac{\omega_s^2}{4K_m^*(S_n + S_f)}$
k_f:	buck converter	$-\frac{DT_sR_i}{L}\left(1-\frac{D}{2}\right)$	$-\frac{D(1-D)T_sR_i}{2L}$
	boost converter	$-\frac{T_sR_i}{2L}$	0
	buck/boost converter	$-\frac{DT_sR_i}{L}\left(1-\frac{D}{2}\right)$	$-\frac{D(1-D)T_sR_i}{2L}$
k_r:	buck converter	$\frac{T_sR_i}{2L}$	0
	boost converter	$\frac{(1-D)^2T_sR_i}{2L}$	$-\frac{D(1-D)T_sR_i}{2L}$
	buck/boost converter	$\frac{(1-D)^2T_sR_i}{2L}$	$-\frac{D(1-D)T_sR_i}{2L}$

† S_n is the on-time slope and S_f is the off-time slope of the sensed current feedback signal v_I, while S_e is the slope of the compensation ramp.

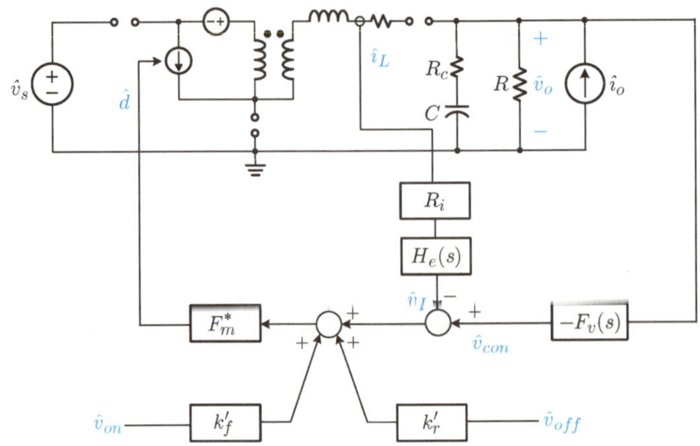

Figure 10.7 Modified small-signal model for current mode control.

in studying the sampling effects. Each model has its own theoretical developments and provides the correct design procedures for current mode control.

In this book, Ridley's model is chosen as the reference in studying the sampling effects. In the next section, the gain blocks of Ridley's model will be studied. Then, the dynamics of peak current mode control are analyzed in an attempt to establish new design procedures. Several design examples are given. The predictions and performance of the new design will be compared with those of the classical design developed in Chapter 9. The new control design procedures will be compared with the classical design procedures, to reveal the connection between the two different design approaches. Lastly, the design and performance of an off-line flyback converter with an optocoupler-based peak current mode control will be discussed.

10.2 EXPRESSIONS FOR s-DOMAIN MODEL FOR CURRENT MODE CONTROL

This section presents analytical details about Ridley's s-domain model. Expressions for the gain blocks for the sampled-data feature of current mode control are provided.

10.2.1 Modified Small-Signal Model

In order to facilitate the derivation of the gain blocks, the s-domain model in Fig. 10.6 is modified into an alternative form shown in Fig. 10.7. The modifications incorporated into Fig. 10.7 are as follows.

1) In the modified model, the modulator gain is represented by F_m^* in place of $F_m^*(s)$ because the modulator gain is a constant in Ridley's model.

2) In the original model in Fig. 10.6, the feedforward gains, k_f and k_r, are derived from the input voltage and output voltage of the converter. On the other hand, in the modified model in Fig. 10.7, the feedforward gains are derived from the small-signal sources $\hat{v}_{on}$ and $\hat{v}_{off}$. The small-signal source $\hat{v}_{on}$ represents the voltage that appears across the inductor terminals during the on-time period and $\hat{v}_{off}$ is the magnitude of the inductor voltage during the off-time period. With these changes, the feedforward gains are renamed as k'_f and k'_r.

The purpose of the above modifications is to simplify the analysis of the feedforward gains. Advantages of the modified model will be explained later in Section 10.2.4.

The modified model in Fig. 10.7 is now adapted to the buck/boost converter. The resulting model is shown in Fig. 10.8. For the buck/boost converter, the structure of the modified model is the same as the original model in Fig. 10.6, because $\hat{v}_{on} = \hat{v}_s$ and $\hat{v}_{off} = \hat{v}_o$ for this case. The voltage feedback compensation $F_v(s)$ is not included in Fig. 10.8 because $F_v(s)$ does not affect the small-signal models of current mode control. The CSN gain R_i is also known in advance.

The four gain blocks, F_m^*, $H_e(s)$, k'_f, and k'_r, are derived from Fig. 10.8. Once the four gain blocks are identified, they should be converted into the new expressions that can be put into the original small-signal model of Fig. 10.6. By comparing the two small-signal models of Fig. 10.6 and Fig. 10.7, it is apparent that the expressions for F_m^* and $H_e(s)$ remain the same. However, the feedforward gains, k'_f and k'_r, need to be converted to the expressions for k_f and k_r because $\hat{v}_{on} \neq \hat{v}_s$ and $\hat{v}_{off} \neq \hat{v}_o$ in general. The buck/boost converter is a special case where $\hat{v}_{on} = \hat{v}_s$ and $\hat{v}_{off} = \hat{v}_o$ so $k_f = k'_f$ and $k_r = k'_r$. The conversion of the feedforward gains will be discussed later in Section 10.2.4

When the four gain blocks, F_m^*, $H_e(s)$, k_f, and k_r, are inserted into Fig. 10.6, the s-domain small-signal model for the three basic converters can be obtained by appropriately arranging the connections in the power stage circuit configuration. Figure 10.8 itself is the small-signal model of the buck/boost converter with $k_f = k'_f$ and $k_r = k'_r$.

The next section presents the derivation of F_m^*, $H_e(s)$, k'_f and k'_r, using Fig. 10.8. In order to simplify forthcoming model derivations, the CSN gain is assumed to be unity, $R_i = 1$, unless stated otherwise.

10.2.2 Modulator Gain F_m^*

The modulator gain F_m^* is derived by graphically relating the perturbation in the control voltage $\hat{v}_{con}$ to the change in duty ratio $\hat{d}$. Figure 10.9 shows the modulator waveforms where a perturbation in v_{con} is incurred at the onset of the second operational period. The solid line represents the modulator waveforms with the original control voltage v_{con} and the dashed line is the waveforms with a perturbed control voltage v'_{con}. The perturbation in the control voltage, $\hat{v}_{con} = v'_{con} - v_{con}$, causes a change in the on-time period, $\hat{d}T_s$. From the geometry of the modulator waveforms, highlighted in Fig. 10.9 with an expanded view, the following relationship is seen

$$\hat{v}_{con} = S_n\hat{d}T_s + S_e\hat{d}T_s = (S_n + S_e)\hat{d}T_s \tag{10.1}$$

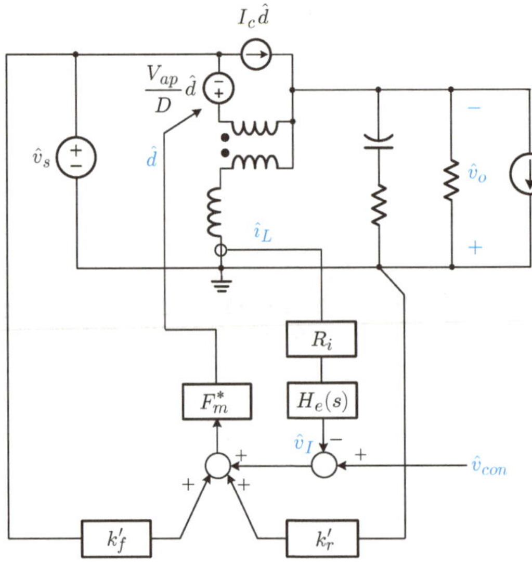

Figure 10.8 Small-signal model of buck/boost converter with current mode control.

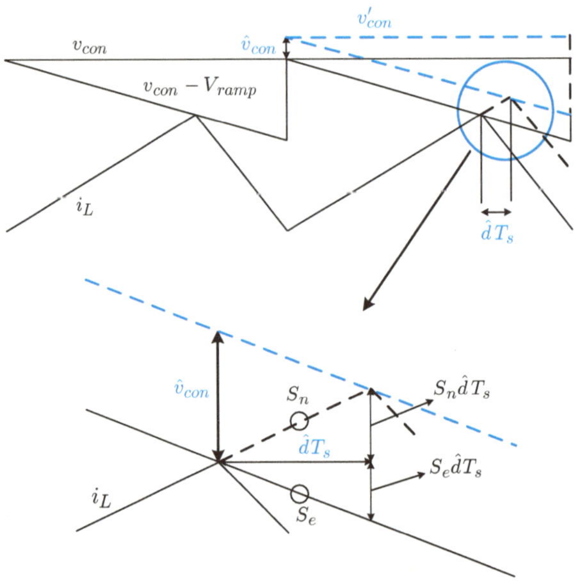

Figure 10.9 Modulator waveforms with perturbed control voltage.

where S_n is the on-time slope of the inductor current and S_e is the slope of the compensation ramp. The above equation is arranged as

$$\boxed{F_m^* = \frac{\hat{d}}{\hat{v}_{con}} = \frac{1}{(S_n + S_e)T_s}} \tag{10.2}$$

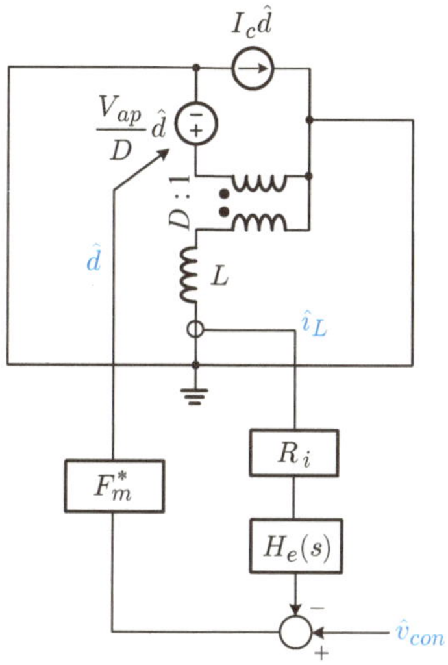

Figure 10.10 Simplified s-domain model for current mode control.

to yield the modulator gain.

10.2.3 $H_e(s)$: s-Domain Representation of Sampling Effects

The sampling effects of current mode control are modeled as a separate gain block $H_e(s)$ located in the current feedback path. For the purpose of $H_e(s)$ derivation, the input and output voltages are fixed so that $\hat{v}_s = \hat{v}_o = 0$ in Fig. 10.8. Accordingly, Fig. 10.8 is simplified to Fig. 10.10. The expression for $H_e(s)$ is now derived from Fig. 10.10 through the following three steps.

Step One: The expression for the transfer function from the control voltage to inductor current, $H_i(s) = \hat{\imath}_L(s)/\hat{v}_{con}(s)$, is derived in two different ways, thereby yielding two distinct expressions for the transfer function.The first way directly extracts the $H_i(s)$ expression from Fig. 10.10 and the resulting $H_i(s)$ explicitly contains the gain block $H_e(s)$. The second way performs the discrete-time analysis of the PWM modulator waveforms, yielding a z-domain expression of the control voltage-to-inductor current transfer function, $H_i(z) = \hat{\imath}_L(z)/\hat{v}_{con}(z)$. The z-domain transfer function is later converted into an equivalent s-domain transfer function, leading to the second s-domain $H_i(s)$ expression. The second $H_i(s)$ expression does not contain the term $H_e(s)$.

Step Two: By equating the two different $H_i(s)$ expressions derived in Step One, the gain block $H_e(s)$ is identified as $H_e(s) = sT_s/(e^{sT_s} - 1)$ where T_s is the switching period.

Step Three: The $H_e(s)$ expression is subsequently approximated to $H_e(s) = sT_s/(e^{sT_s} - 1) \approx 1 + s/(\omega_n Q_z) + s^2/\omega_n^2$ with $Q_z = -2/\pi$ and $\omega_n = \pi/T_s$, based on Taylor series expansion of the complex exponential.

1) Step One: Two Different Expressions for $H_i(s) = \hat{\imath}_L(s)/\hat{v}_{con}$

This section explains details about the derivation of the two different $H_i(s)$ expressions. The gain block $H_e(s)$ is included in one of the two $H_i(s)$ expressions.

First $H_i(s)$ expression

Direct application of Mason's gain rule to Fig. 10.10 results in the first expression for the control-to-inductor current transfer function $H_i(s)$

$$H_i(s) = \frac{\hat{\imath}_L(s)}{\hat{v}_{con}(s)} = \frac{F_m^* \dfrac{\hat{\imath}_L(s)}{\hat{d}(s)}}{1 + F_m^* \dfrac{\hat{\imath}_L(s)}{\hat{d}(s)} R_i H_e(s)} \tag{10.3}$$

This expression contains the gain block $H_e(s)$ in the denominator.

Second $H_i(s)$ expression

For the formulation of the second $H_i(s)$ expression, the dynamic behavior of the inductor current is illustrated in Fig. 10.11. Figure 10.11(a) shows the propagation of the inductor current perturbation with the assumption $R_i = 1$,[†] while Fig. 10.11(b) depicts the effects of the control voltage perturbation. The inductor current dynamics under the perturbation in both the inductor current and control voltage are described by a difference equation

$$\hat{\imath}_L(k+1) = k_1 \hat{\imath}_L(k) + k_2 \hat{v}_{con}(k+1) \tag{10.4}$$

The first term in the right-hand side of (10.4) represents the propagation of the inductor current perturbation. The inductor current perturbation at the $(k+1)^{\text{th}}$ period, $\hat{\imath}_L(k+1)$, is affected by the inductor current perturbation occurred at the previous switching period, $\hat{\imath}_L(k)$. This term is the natural response. The second term describes the impact of the control voltage perturbation on the inductor current. The disturbance in the control voltage at the $(k+1)^{\text{th}}$ period, $\hat{v}_{con}(k+1)$, affects the inductor current at the same switching period, $\hat{\imath}_L(k+1)$. This second term is the forced response. The two proportionate coefficients in (10.4), k_1 and k_2, are evaluated by analyzing

[†]For the inductor current perturbation in Fig. 10.11(a), the variable R_i is selected as the unity without loss of generality, because the CSN gain R_i appears in both the input and output variables. For the control voltage perturbation in Fig. 10.11(b), however, the CSN gain R_i should be included in the analysis, because R_i appears only in the output variable.

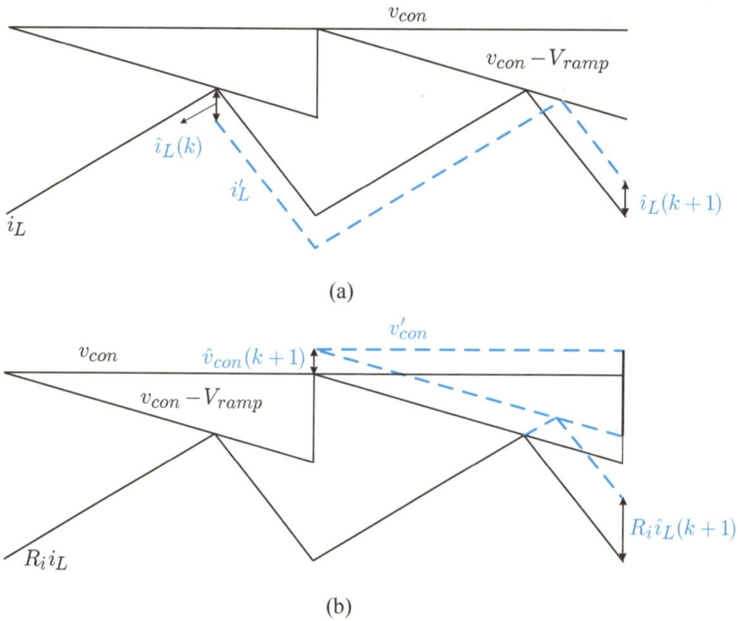

Figure 10.11 Inductor current dynamics. (a) Propagation of inductor current perturbation. (b) Effects of control voltage perturbation.

the natural and forced responses.

Natural Response: The natural response is analyzed from the waveforms of the perturbed inductor current. Figure 10.12 shows the propagation of the inductor current perturbation, $\hat{i}_L(k) = i'_L(k) - i_L(k)$, into the next sampling instant. From the geometry of the PWM waveforms, the following relationships are established

$$-\hat{i}_L(k) = S_n \hat{d} T_s + S_e \hat{d} T_s = (S_n + S_e)\hat{d} T_s \tag{10.5}$$

and

$$\hat{i}_L(k+1) = S_f \hat{d} T_s - S_e \hat{d} T_s = (S_f - S_e)\hat{d} T_s \tag{10.6}$$

From (10.5) and (10.6), it follows that

$$\frac{\hat{i}_L(k+1)}{\hat{i}_L(k)} = -\frac{S_f - S_e}{S_n + S_e} \tag{10.7}$$

which is arranged as

$$\boxed{\hat{i}_L(k+1) = -\alpha\, \hat{i}_L(k)} \tag{10.8}$$

with

$$\boxed{\alpha = \frac{S_f - S_e}{S_n + S_e}} \tag{10.9}$$

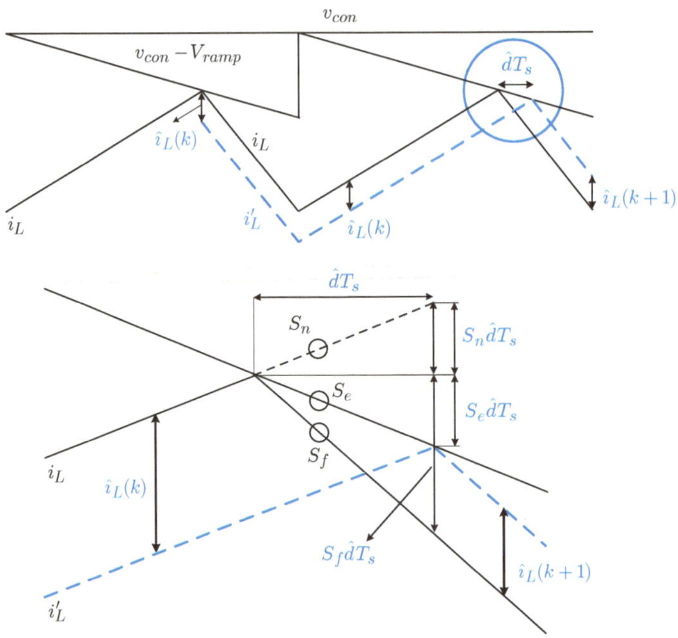

Figure 10.12 Propagation of inductor current perturbation: natural response.

Equation (10.8) is the natural response of the inductor current.

Forced Response: The forced response originates from the perturbation in the control voltage. Figure 10.13 illustrates the effect of the control voltage perturbation. From the geometry of the modulator waveforms, the following relationships are established:

$$\hat{v}_{con}(k+1) = S_n\hat{d}\,T_s + S_e\hat{d}\,T_s = (S_n + S_e)\,\hat{d}\,T_s \tag{10.10}$$

and

$$R_i\,\hat{\imath}_L(k+1) = S_n\hat{d}\,T_s + S_f\hat{d}\,T_s = (S_n + S_f)\,\hat{d}\,T_s \tag{10.11}$$

By eliminating $\hat{d}\,T_s$ from (10.10) and (10.11), it follows that

$$R_i\,\hat{\imath}_L(k+1) = \frac{S_n + S_f}{S_n + S_e}\,\hat{v}_{con}(k+1) = \left(1 + \frac{S_f - S_e}{S_n + S_e}\right)\hat{v}_{con}(k+1) \tag{10.12}$$

which is rearranged as

$$\boxed{\hat{\imath}_L(k+1) = \frac{1}{R_i}(1+\alpha)\,\hat{v}_{con}(k+1)} \tag{10.13}$$

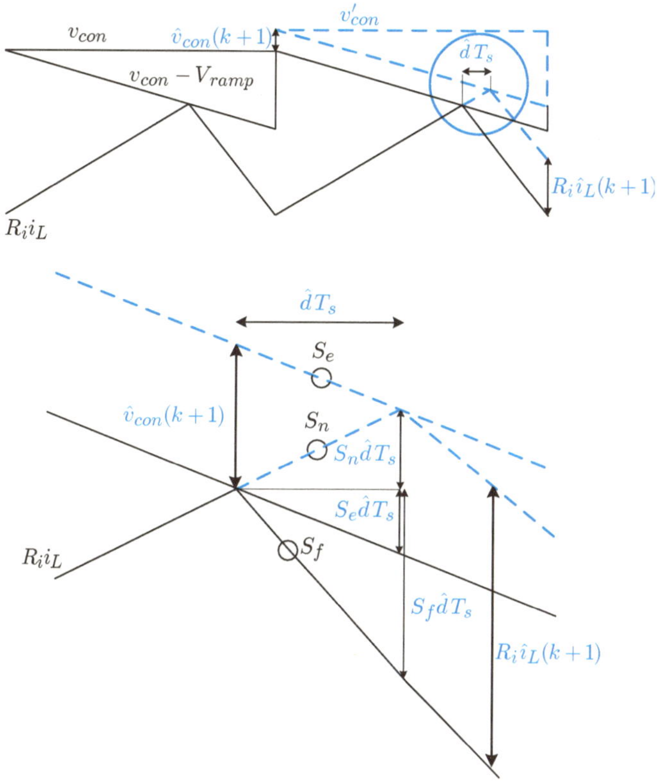

Figure 10.13 Effects of control voltage perturbation: forced response.

where $\alpha = (S_f - S_e)/(S_n + S_e)$ is defined in (10.9). The above equation is the forced response.

Complete Response: By summing the natural response and forced response, the complete response of the inductor current is given by

$$\hat{i}_L(k+1) = k_1 \hat{i}_L(k) + k_2\, \hat{v}_{con}(k+1) \tag{10.14}$$

where $k_1 = -\alpha$ and $k_2 = (1+\alpha)/R_i$ with $\alpha = (S_f - S_e)/(S_n + S_e)$. The final difference equation for the inductor current now becomes

$$\boxed{\hat{i}_L(k+1) = -\alpha\, \hat{i}_L(k) + \frac{1}{R_i}(1+\alpha)\, \hat{v}_{con}(k+1)} \tag{10.15}$$

By taking the z-transformation of (10.15)

$$z\, \hat{i}_L(z) = -\alpha\, \hat{i}_L(z) + \frac{1}{R_i}(1+\alpha)\, z\, \hat{v}_{con}(z) \tag{10.16}$$

the z-domain control voltage-to-inductor current transfer function, $H_i(z)$, is derived as

$$H_i(z) = \frac{\hat{i}_L(z)}{\hat{v}_{con}(z)} = \frac{1}{R_i}(1+\alpha)\frac{z}{z+\alpha} \tag{10.17}$$

The continuous-time representation of $H_i(z)$ is obtained from (10.17) by multiplying $(1-e^{-sT_s})/(sT_s)$ and replacing z with e^{sT_s} [3]

$$\begin{aligned} H_i(s) &= \frac{1-e^{-sT_s}}{sT_s}H_i(z=e^{sT_s}) = \frac{1-e^{-sT_s}}{sT_s}\frac{1}{R_i}(1+\alpha)\frac{e^{sT_s}}{e^{sT_s}+\alpha} \\ &= \frac{1}{R_i}\frac{1+\alpha}{sT_s}\frac{e^{sT_s}-1}{e^{sT_s}+\alpha} \end{aligned} \tag{10.18}$$

It is well known that the z-domain-to-s-domain conversion is only valid for the frequencies below the Nyquist frequency, $\omega_n = 0.5\,\omega_s = \pi/T_s$. Equation (10.18) is the second s-domain expression for the control voltage-to-inductor current transfer function. This second $H_i(s)$ expression is developed from the modulator waveforms and therefore does not include the gain block $H_e(s)$. Recall that the first $H_i(s)$ expression, given by (10.3), contains the gain block $H_e(s)$.

2) Step Two: Identification of Gain Block $H_e(s)$

Two independently-driven descriptions for $H_i(s)$, (10.3) and (10.18), are now obtained. By equating these two expressions, the expression for $H_e(s)$ is derived as

$$H_e(s) = \frac{sT_s}{e^{sT_s}-1} \tag{10.19}$$

Details about this derivation are given in Example 10.2.

EXAMPLE 10.2 Derivation of $H_e(s)$

This example illustrates the derivation of (10.19). By equating (10.3) and (10.18), it follows that

$$\frac{F_m^*\dfrac{\hat{i}_L(s)}{\hat{d}(s)}}{1+F_m^*\dfrac{\hat{i}_L(s)}{\hat{d}(s)}R_iH_e(s)} = \frac{1}{R_i}\frac{1+\alpha}{sT_s}\frac{e^{sT_s}-1}{e^{sT_s}+\alpha} \tag{10.20}$$

The objective of this analysis is to find the expression for $H_e(s)$ that satisfies (10.20). First, from the simplified s-domain model in Fig. 10.10, the duty-ratio-to-inductor current transfer function is derived as

$$\frac{\hat{i}_L(s)}{\hat{d}(s)} = \frac{V_{ap}}{D}D\frac{1}{sL} = \frac{V_{ap}}{sL} \tag{10.21}$$

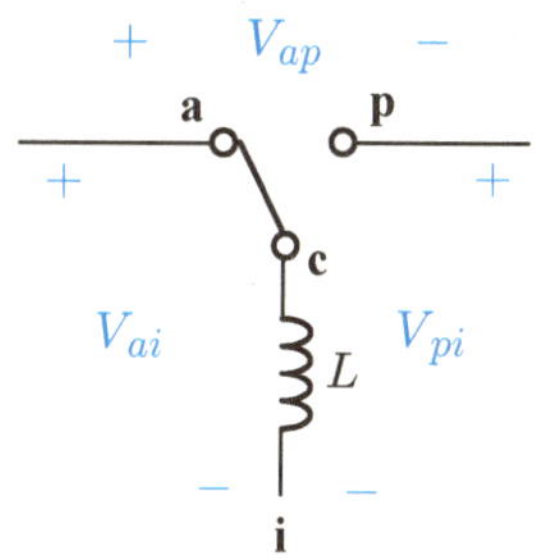

Figure 10.14 PWM switch and terminal voltage definitions.

The expression for V_{ap}, the voltage between the active and passive terminals of the PWM switch, is derived from Fig. 10.14

$$V_{ap} = V_{ai} + V_{ip} \tag{10.22}$$

The voltages V_{ai} and V_{ip} are related to the on-time and off-time slopes of the current feedback signal

$$S_n = \frac{V_{ai}}{L} R_i \tag{10.23}$$

$$S_f = \frac{V_{ip}}{L} R_i \tag{10.24}$$

From (10.22) through (10.24), it follows that

$$V_{ap} = \frac{L}{R_i}(S_n + S_f) \tag{10.25}$$

By merging (10.21) and (10.25), the duty ratio-to-inductor current transfer function is given by

$$\frac{\hat{\imath}_L(s)}{\hat{d}(s)} = \frac{1}{R_i}\frac{S_n + S_f}{s} \tag{10.26}$$

From (10.26) and $F_m^* = 1/\big((S_n + S_e)T_s\big)$, it becomes that

$$\begin{aligned} F_m^* \frac{\hat{\imath}_L(s)}{\hat{d}(s)} &= \frac{1}{(S_n + S_e)T_s}\frac{1}{R_i}\frac{S_n + S_f}{s} \\ &= \frac{1}{R_i}\frac{1+\alpha}{sT_s} \end{aligned} \tag{10.27}$$

with $\alpha = (S_f - S_e)/(S_n + S_e)$. Using (10.27), the expression (10.20) now becomes

$$\frac{\dfrac{1}{R_i}\dfrac{1+\alpha}{sT_s}}{1 + \dfrac{1}{R_i}\dfrac{1+\alpha}{sT_s}R_i H_e(s)} = \frac{1}{R_i}\frac{1+\alpha}{sT_s}\frac{e^{sT_s} - 1}{e^{sT_s} + \alpha} \tag{10.28}$$

which is solved for $H_e(s)$ to yield the expression (10.19), $H_e(s) = sT_s/(e^{sT_s} - 1)$, as shown in Problem 10.2 at the end of this chapter.

3) Step Three: Approximation of Gain Block $H_e(s)$

As the last derivation step, the s-domain representation of the sampling effects $H_e(s)$ is approximated to

$$H_e(s) = \frac{sT_s}{e^{sT_s} - 1} \approx 1 + \frac{s}{\omega_n Q_z} + \frac{s^2}{\omega_n^2} \tag{10.29}$$

with

$$Q_z = -\frac{2}{\pi} \tag{10.30}$$

and

$$\omega_n = \frac{\pi}{T_s} \tag{10.31}$$

based on the Taylor series expansion of the complex exponential. The procedures of this approximation are given in Example 10.3.

EXAMPLE 10.3 Approximation of $H_e(s)$

This example illustrates the derivation of (10.29). First, the complex exponential function is approximated as

$$e^{sT_s} = \frac{e^{s\frac{T_s}{2}}}{e^{-s\frac{T_s}{2}}} = \frac{e^{s\frac{\pi}{\omega_s}}}{e^{-s\frac{\pi}{\omega_s}}} \approx \frac{1 + s\frac{\pi}{\omega_s} + \frac{1}{2}\left(s\frac{\pi}{\omega_s}\right)^2}{1 - s\frac{\pi}{\omega_s} + \frac{1}{2}\left(s\frac{\pi}{\omega_s}\right)^2} \tag{10.32}$$

using the identity $e^x = 1 + x + x^2/2! + \cdots$ with $x \ll 1$. This approximation is accurate for frequencies $\omega \ll \omega_s/\pi$. The above expression is further modified to

$$e^{sT_s} \approx \frac{1 + s\frac{\pi}{\omega_s} + \frac{1}{2}\left(s\frac{\pi}{\omega_s}\right)^2}{1 - s\frac{\pi}{\omega_s} + \frac{1}{2}\left(s\frac{\pi}{\omega_s}\right)^2} \approx \frac{1 + \frac{1}{2/\pi}\frac{s}{\omega_s/2} + \left(\frac{s}{\omega_s/2}\right)^2}{1 - \frac{1}{2/\pi}\frac{s}{\omega_s/2} + \left(\frac{s}{\omega_s/2}\right)^2} \tag{10.33}$$

by accepting the approximation of $\pi^2/2 \approx 4$. Using (10.33), $H_e(s)$ is written as

$$H_e(s) = \frac{sT_s}{e^{sT_s} - 1} \approx \frac{s\frac{2\pi}{\omega_s}}{\left(\frac{1+\frac{1}{2/\pi}\frac{s}{\omega_s/2}+\left(\frac{s}{\omega_s/2}\right)^2}{1-\frac{1}{2/\pi}\frac{s}{\omega_s/2}+\left(\frac{s}{\omega_s/2}\right)^2}\right) - 1}$$

$$= 1 - \frac{1}{2/\pi}\frac{s}{\omega_s/2} + \left(\frac{s}{\omega_s/2}\right)^2 \tag{10.34}$$

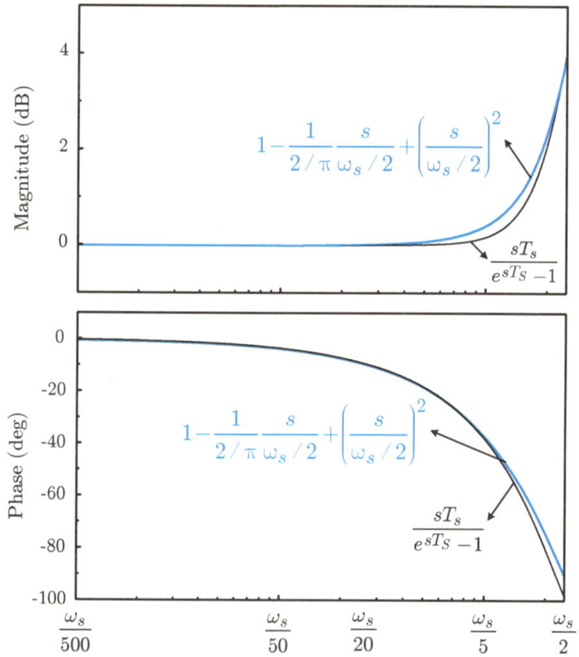

Figure 10.15 Bode plot of the original $H_e(s)$ and its low-frequency approximation.

which is put into the expressions (10.29) through (10.31) using $\omega_s = 2\pi/T_s$.

The s-domain representation of the sampling effects is valid only for the frequencies below the Nyquist frequency, due to the z-domain-to-s-domain conversion in (10.18) and the assumptions used in (10.32). Figure 10.15 compares the Bode plots of the original expression of $H_e(s)$ and its low-frequency approximation of (10.29).

The s-domain representation of the sampling effects, given by (10.29), is a double zero function with a negative quality factor, $Q_z = -2/\pi \approx -0.637$. The double zero is located at the Nyquist frequency, $\omega_n = \pi/T_s = 0.5\,\omega_s$. Accordingly, $H_e(s)$ exerts negligible effects at low frequencies and only becomes influential at high frequencies. As shown in Fig. 10.15, $H_e(s)$ does not cause any practical impacts at frequencies below $\omega_s/20$.

The maximum impact of $H_e(s)$ occurs at the Nyquist frequency where $|H_e(j\omega)|$ increases about 4 dB and $\angle H_e(j\omega)$ drops nearly 90°. This observation is consistent with Fig. 10.4 which illustrates the consequence of the sampling effects on the loop gain characteristics. The sampling effects are only pronounced at the frequencies near and above the Nyquist frequency. This fact also supports the past prevalence of the classical analysis which ignores the sampling effects. *If the control bandwidth is well*

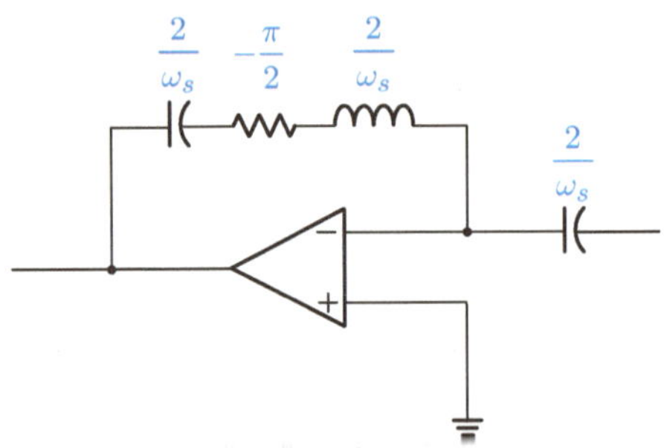

Figure 10.16 Circuit model for $H_e(s) = 1 - \frac{1}{2/\pi}\frac{s}{\omega_s/2} + \left(\frac{s}{\omega_s/2}\right)^2$.

below the Nyquist frequency, the sampling effects do not interfere with the control design and closed-loop performance.

EXAMPLE 10.4 Circuit Model for $H_e(s)$

The functional block $H_e(s)$, which represents the sampling effect of peak current mode control, can be transformed into a circuit model. Figure 10.16 shows a simple circuit model for $H_e(s)$. The unique feature of the circuit model is the existence of a negative resistance, which is accepted by most circuit simulators including PSpice®. The transfer function of the circuit is given by

$$F(s) = -\frac{\frac{1}{s\frac{2}{\omega_s}} - \frac{\pi}{2} + s\frac{2}{\omega_s}}{\frac{1}{s\frac{2}{\omega_s}}} = -\left(1 - \frac{1}{2/\pi}\frac{s}{\omega_s/2} + \left(\frac{s}{\omega_s/2}\right)^2\right) = -H_e(s) \quad (10.35)$$

This circuit model is used to simulate the small-signal dynamics of the converter with the presence of the sampling effects.

10.2.4 Feedforward Gains

Among the four small-signal gain blocks for current mode control, the modulator gain F_m^* and sampling effect $H_e(s)$ are identified. This section now presents derivations of the feedforward gains, k_f' and k_r'. Similar to the case of $H_e(s)$ development, two different expressions for the inductor voltage-to-inductor current transfer function are formulated. By equating these two alternative expressions, the feedforward gain expressions are found.

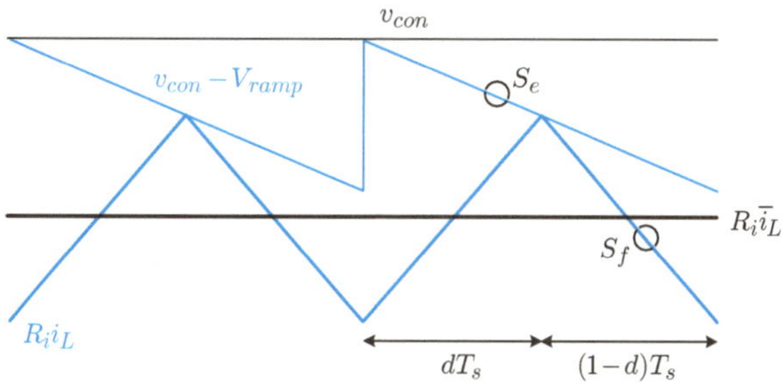

Figure 10.17 Steady-state modulator waveforms.

Feedforward Gain k'_f

The feedforward gain is derived from the modulator waveform of peak current mode control, shown in Fig. 10.17. It can be seen from Fig. 10.17 that

$$R_i\bar{i}_L = v_{con} - S_e dT_s - \frac{1}{2}S_f(1-d)T_s \tag{10.36}$$

where $\bar{i}_L$ denotes the average inductor current. The off-time slope of the sensed inductor current is given by

$$S_f = \frac{v_{off}}{L}R_i \tag{10.37}$$

where v_{off} is the magnitude of the inductor voltage during the off-time period. The flux linkage balance condition on the inductor is written as

$$v_{on}dT_s = v_{off}(1-d)T_s \tag{10.38}$$

where v_{on} is the inductor voltage during the on-time period. The above equation yields the expression for the duty ratio in terms of the inductor voltages

$$d = \frac{v_{off}}{v_{on} + v_{off}} \tag{10.39}$$

By incorporating (10.37) and (10.39) into (10.36), and linearizing the resulting equation with respect to $\bar{i}_L$ and v_{on}, the following equation is obtained

$$G_{on}(s) = \frac{\hat{\bar{i}}_L}{\hat{v}_{on}} = \frac{S_e T_s}{R_i}\frac{D}{V_{ap}} - \frac{T_s}{2L}D^2 \tag{10.40}$$

where $V_{ap} = V_{on} + V_{off}$ is the sum of the dc components of the on-time inductor voltage and off-time inductor voltage. Equation (10.40) is the first equation for $G_{on}(s) = \hat{\bar{i}}_L/\hat{v}_{on}$. Derivation of (10.40) is given in Example 10.5.

EXAMPLE 10.5 Derivation of $G_{on}(s)$

This example shows the derivation of $G_{on}(s)$ in (10.40). Incorporation of (10.37) and (10.39) into (10.36) yields

$$R_i \bar{i}_L = v_{con} - S_e T_s \frac{v_{off}}{v_{on} + v_{off}} - \frac{1}{2}\frac{v_{off}}{L} R_i T_s \frac{v_{on}}{v_{on} + v_{off}} \tag{10.41}$$

Differentiation of (10.41) with respect to v_{on} results in

$$\begin{aligned}\frac{\partial \bar{i}_L}{\partial v_{on}} &= \frac{1}{R_i} S_e T_s \frac{\partial}{\partial v_{on}}\left(\frac{v_{off}}{v_{on} + v_{off}}\right) - \frac{1}{2}\frac{v_{off}}{L} T_s \frac{\partial}{\partial v_{on}}\left(\frac{v_{on}}{v_{on} + v_{off}}\right) \\ &= \frac{1}{R_i} S_e T_s \frac{v_{off}}{(v_{on} + v_{off})^2} - \frac{T_s}{2L}\frac{v_{off}^2}{(v_{on} + v_{off})^2}\end{aligned} \tag{10.42}$$

which can be arranged in the form of (10.40) by considering the steady-state circuit variables: $D = V_{off}/(V_{on} + V_{off})$ and $V_{ap} = V_{on} + V_{off}$.

The second expression for $G_{on} = \hat{i}_L/\hat{v}_{on}$ is obtained from the simplified small-signal model of the converter, shown in Fig. 10.18. The feedforward gains only describe the low-frequency averaged dynamics of the PWM block. Therefore, the high-frequency modulator dynamics, such as the sampling effects, can be ignored. Figure 10.18 is the small-signal model constructed for the low-frequency averaged dynamics, based on the following observations.

1) The gain block $H_e(s)$ is approximated by unity: $H_e(s) = 1$.
2) The average value of the inductor voltage is zero, as such, the inductor behaves as a short circuit at low frequencies.
3) The output voltage of the converter is constant: $\hat{v}_o = \hat{v}_{off} = 0$.
4) The control voltage remains constant: $\hat{v}_{con} = 0$

From Fig. 10.18, the following relationship is established

$$(\hat{v}_{on} k_f' - R_i \hat{\bar{i}}_L) F_m^* \frac{V_{ap}}{D} = -\hat{v}_{on} \tag{10.43}$$

by noting that the transformer windings are shorted. Equation (10.43) is rearranged as

$$G_{on}(s) = \frac{\hat{\bar{i}}_L}{\hat{v}_{on}} = \frac{1}{R_i}\left(\frac{D}{F_m^* V_{ap}} + k_f'\right) \tag{10.44}$$

to yield the second expression for $G_{on}(s)$. By equating (10.40) and (10.44) with incorporation of $F_m^* = 1/((S_n + S_e)T_s)$, the feedforward gain is obtained as

$$\boxed{k_f' = -\frac{D T_s R_i}{L}\left(1 - \frac{D}{2}\right)} \tag{10.45}$$

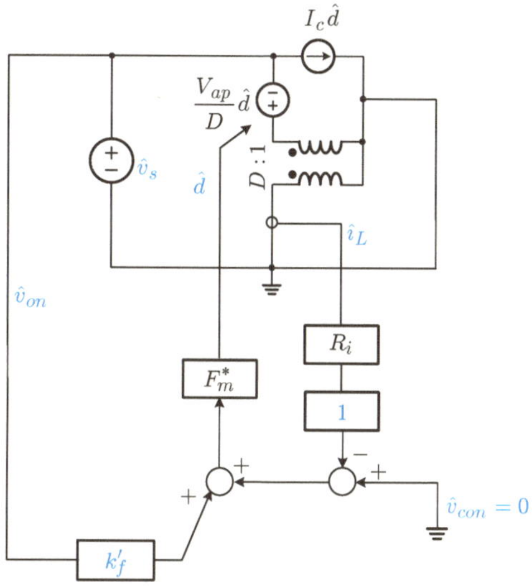

Figure 10.18 Small-signal model for derivation of feedforward gains.

Feedforward Gain k_r'

By repeating the same process for the off-time inductor voltage, the feedforward gain k_r' is evaluated as

$$\boxed{k_r' = \frac{(1-D)^2 T_s R_i}{2L}} \tag{10.46}$$

Conversion of Feedforward Gains

The feedforward gains, k_f' and k_r' derived in the previous section, should be converted to the original feedforward gains, k_f and k_r defined in Fig. 10.6. Figure 10.19 shows the small-signal model that will be used to find the relationship between $\{k_f'\, k_r'\}$ and $\{k_f\, k_r\}$. This model is derived from Fig. 10.7 with the condition $\hat{v}_I = \hat{v}_{con} = 0$.

In Fig. 10.19, the input signal to the modulator gain block F_m^* is expressed as

$$\hat{v}_{feed} = k_f'\hat{v}_{on} + k_r'\hat{v}_{off} \tag{10.47}$$

The conversion from $\{k_f'\, k_r'\}$ to $\{k_f\, k_r\}$ is explained using the buck converter. For the buck converter, the following relationships hold

$$\hat{v}_{on} = \hat{v}_s - \hat{v}_o \quad \text{and} \quad \hat{v}_{off} = \hat{v}_o \tag{10.48}$$

Equations (10.47) and (10.48) are combined

$$\hat{v}_{feed} = k_f'(\hat{v}_s - \hat{v}_o) + k_r'\hat{v}_o = k_f'\hat{v}_s + (-k_f' + k_r')\hat{v}_o \tag{10.49}$$

Table 10.2 Inductor Voltage Relationships and Feedforward Gains

	Buck converter	Boost converter	Buck/boost converter
v_{on}	$v_S - v_O$	v_S	v_S
v_{off}	v_O	$v_O - v_S$	v_O
k_f	k'_f	$k'_f - k'_r$	k'_f
k_r	$-k'_f + k'_r$	k'_r	k'_r

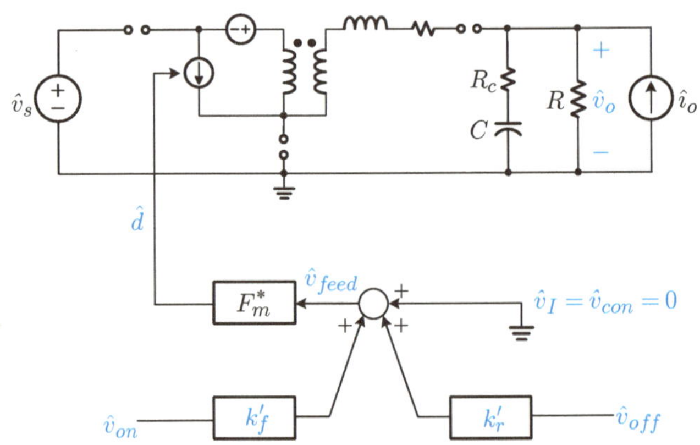

Figure 10.19 Small-signal model for feedforward gain conversion.

yielding the relationship

$$\hat{v}_{feed} = k_f \hat{v}_s + k_r \hat{v}_o \tag{10.50}$$

with

$$k_f = k'_f \quad \text{and} \quad k_r = -k'_f + k'_r \tag{10.51}$$

Expressions (10.51) are the original feedforward gains for the buck converter, shown in Fig. 10.6. The same process is applied to the boost and buck/boost converters. The results of this analysis are shown in Table 10.2. The feedforward gains, k'_f and k'_r derived in (10.45) and (10.46), are now converted into the original gains, k_f and k_r in Fig. 10.6. Table 10.2 is used for this conversion. The final expressions for the feedforward gains of the three basic converters are shown in Table 10.1.

The analysis of the small-signal gain blocks was performed using the modified model of Fig. 10.7 rather than the original model of Fig. 10.6. The advantage of Fig. 10.7 is that k_f' and k_r' are derived only once and the outcomes are adapted to the three basic converters to provide their respective k_f and k_r expressions. If Fig. 10.6 were used for this analysis, the derivation should be repeated three times for the three basic converters.

10.3 NEW CONTROL DESIGN PROCEDURES FOR CURRENT MODE CONTROL

In the previous section, we identified the expressions for the small-signal gain blocks for peak current mode control. This section presents the analysis of current mode control using these gain blocks. The objective of this analysis is to set up new design procedures that incorporate the sampling effects of current mode control.

10.3.1 New Power Stage Model

Figure 10.20(a) shows the complete small-signal model for current-mode controlled PWM dc-to-dc converters. The analysis will be facilitated by defining a *new* power stage model. The block of the small-signal model, enclosed by the solid line in Fig. 10.20(a), is considered as the new power stage model. *The new power stage model is in fact the small-signal model of power stage, amalgamated with the gain blocks of peak current model control.* Figure 10.20(a) is converted into the block diagram representation in Fig. 10.20(b) by defining the following composite gain blocks for the new power stage model

$$G_{vci}(s) \equiv \frac{\hat{v}_o(s)}{\hat{v}_{con}(s)}_{\text{current loop closed}} \tag{10.52}$$

$$A_{ui}(s) \equiv \frac{\hat{v}_o(s)}{\hat{v}_s(s)}_{\text{current loop closed}} \tag{10.53}$$

and

$$Z_{oi}(s) \equiv \frac{\hat{v}_o(s)}{\hat{i}_o(s)}_{\text{current loop closed}} \tag{10.54}$$

Each transfer function is evaluated under the condition that only the current feedback is activated and the voltage feedback path is broken. Referring to Fig. 10.20(b), the loop gain is defined as

$$T_m(s) = (-)\, G_{vci}(s)(-)\, F_v(s) = G_{vci}(s)F_v(s) \tag{10.55}$$

The $T_m(s)$ in (10.55) is the only system loop gain that can be used for the control design and performance analysis of the current-mode controlled converters. Once $G_{vci}(s)$ is identified, the voltage feedback compensation $F_v(s)$ can be designed for the loop gain offering both stability and good closed-loop performance.

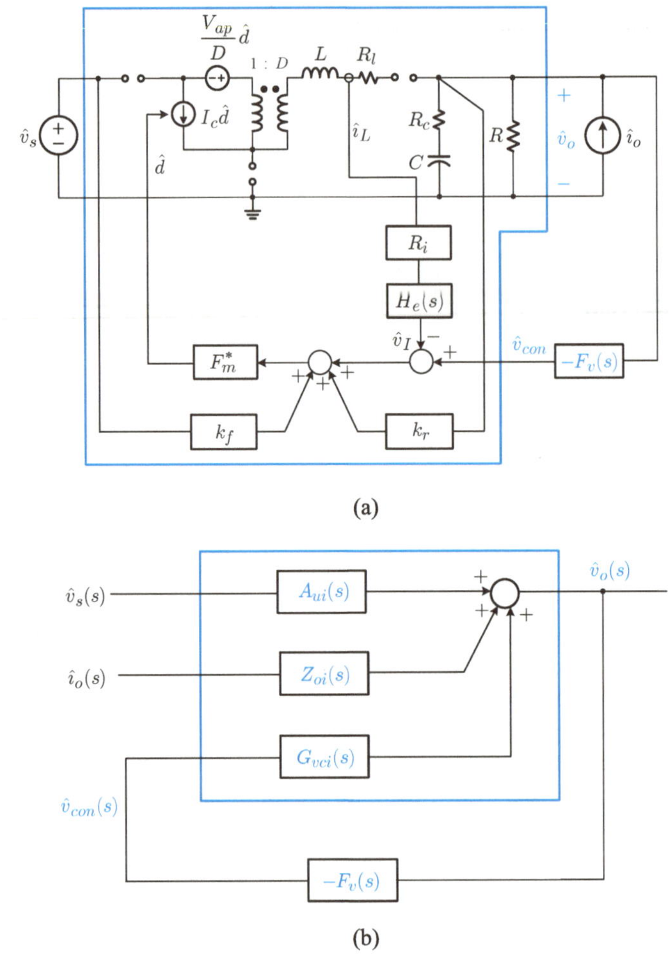

Figure 10.20 Small-signal model for current-mode controlled PWM converters. (a) Circuit model. (b) Block diagram representation.

The audio-susceptibility is given by

$$A_u(s) = \frac{A_{ui}(s)}{1 + T_m(s)} \tag{10.56}$$

and the output impedance is determined as

$$Z_o(s) = \frac{Z_{oi}(s)}{1 + T_m(s)} \tag{10.57}$$

The above descriptions can be used to investigate the closed-loop performance using the asymptotic analysis method discussed in Chapter 8. Details about such analyses can be found in [4].

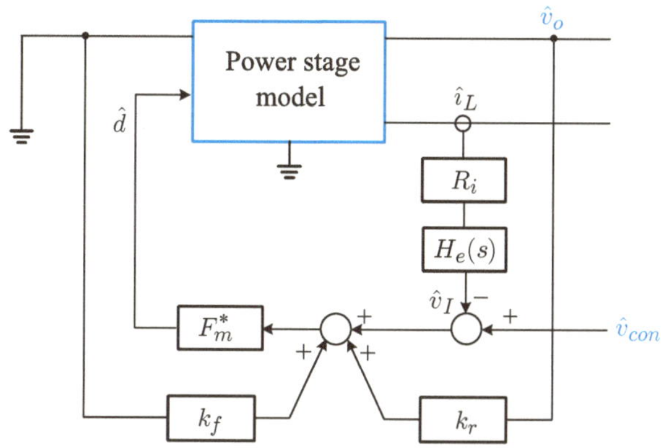

Figure 10.21 Small-signal model for $G_{vci}(s)$ evaluation.

10.3.2 Control-to-Output Transfer Function with Current Loop Closed

The most important gain block in Fig. 10.20(b) is the control-to-output transfer function evaluated with the current loop closed, $G_{vci}(s) = \hat{v}_o(s)/\hat{v}_{con}(s)$. Knowledge of this gain block is essential to the loop gain design.

Derivation of $G_{vci}(s)$

The expression for $G_{vci}(s)$ is derived from the new power stage model with the condition $\hat{v}_s(s) = \hat{i}_o(s) = 0$. Figure 10.21 shows the power stage model modified for the $G_{vci}(s)$ evaluation. From Mason's gain rule, $G_{vci}(s)$ is described as

$$G_{vci}(s) = \frac{\hat{v}_o(s)}{\hat{v}_{con}(s)} = \frac{F_m^* \dfrac{\hat{v}_o(s)}{\hat{d}(s)}}{1 - k_r F_m^* \dfrac{\hat{v}_o(s)}{\hat{d}(s)} + R_i H_e(s) F_m^* \dfrac{\hat{i}_L(s)}{\hat{d}(s)}} \tag{10.58}$$

The above description can be reduced to a simple expression with some practical assumptions on power stage parameters and operational conditions. As will be illustrated in Example 10.6, $G_{vci}(s)$ is casted into the following third-order approximation

$$G_{vci}(s) \approx K_{vc} \frac{\left(1 - \dfrac{s}{\omega_{rhp}}\right)\left(1 + \dfrac{s}{\omega_{esr}}\right)}{\left(1 + \dfrac{s}{\omega_{pl}}\right)\left(1 + \dfrac{s}{Q_p \omega_n} + \dfrac{s^2}{\omega_n^2}\right)} \tag{10.59}$$

The expressions for Q_p, ω_n, and ω_{esr} are common to all the three basic PWM converters

$$\boxed{Q_p = \frac{1}{\pi\left(\left(1+\frac{S_e}{S_n}\right)D' - 0.5\right)}} \tag{10.60}$$

$$\boxed{\omega_n = \frac{\pi}{T_s}} \tag{10.61}$$

and

$$\boxed{\omega_{esr} = \frac{1}{CR_c}} \tag{10.62}$$

The other three parameters, K_{vc}, ω_{pl}, and ω_{rhp}, vary with the converter topology. Expressions for these parameters for the three basic converters are shown in Table 10.3.

EXAMPLE 10.6 Derivation of $G_{vci}(s)$ for Buck Converter

This example provides analysis details involved with the derivation of (10.59). The derivations are illustrated using a buck converter example. The extension of the results to other converter topologies are shown in Table 10.3.

The $G_{vci}(s)$ expression for the buck converter is obtained by evaluating (10.58) with some simplifying assumptions. As the first step, the modulator gain is expressed as

$$F_m^* = \frac{1}{(S_n + S_e)T_s} = \frac{1}{m_c S_n T_s} \tag{10.63}$$

with

$$\boxed{m_c = 1 + \frac{S_e}{S_n}} \tag{10.64}$$

The modulator gain is further modified as

$$F_m^* = \frac{L}{m_c V_S (1-D) R_i T_s} \tag{10.65}$$

based on the relationship

$$S_n = \frac{V_S - V_O}{L} R_i = \frac{V_S(1-D)}{L} R_i \tag{10.66}$$

Using (10.65) and the expressions of $\hat{v}_o(s)/\hat{d}(s)$, $\hat{\imath}_L(s)/\hat{d}(s)$, and k_r for the buck converter, given in Table 9.2 and Table 10.1, $G_{vci}(s)$ is evaluated as

$$G_{vci}(s) = \frac{F_m^* \dfrac{\hat{v}_o(s)}{\hat{d}(s)}}{1 - k_r F_m^* \dfrac{\hat{v}_o(s)}{\hat{d}(s)} + R_i H_e(s) F_m^* \dfrac{\hat{\imath}_L(s)}{\hat{d}(s)}}$$

$$= \frac{\frac{L}{m_c V_S D' R_i T_s} V_S \frac{1+sCR_c}{\Delta(s)}}{1 - \frac{T_s R_i}{2L} \frac{L}{m_c V_S D' R_i T_s} V_S \frac{1+sCR_c}{\Delta(s)} + \frac{R_i L}{m_c V_S D' R_i T_s} \frac{V_S}{R} \frac{1+sCR}{\Delta(s)} H_e(s)} \tag{10.67}$$

where

$$D' = 1 - D$$

and

$$\Delta(s) = 1 + \frac{s}{Q\omega_o} + \frac{s^2}{\omega_o^2} \tag{10.68}$$

with $Q = R\sqrt{C/L}$ and $\omega_o = 1/\sqrt{LC}$.

The expression (10.67) is rearranged as

$$G_{vci}(s) = \frac{L}{R_i} \frac{1+sCR_c}{m_c T_s D' \Delta(s) + \frac{L}{R}(1+sCR) H_e(s) - \frac{T_s}{2}(1+sCR_c)} \tag{10.69}$$

Now, using the expression $H_e(s) = 1 + s/(Q_z\omega_n) + s^2/\omega_n^2$ with $Q_z = -2/\pi$ and $\omega_n = \pi/T_s$, the expression (10.69) is written as

$$G_{vci}(s) = \frac{L}{R_i} \frac{1+sCR_c}{a + bs + cs^2 + ds^3} \tag{10.70}$$

where

$$\begin{aligned}
a &= m_c T_s D' + \frac{L}{R} - \frac{T_s}{2} = T_s(m_c D' - 0.5) + \frac{L}{R} \\
b &= m_c T_s D' \frac{L}{R} + \frac{L}{R}\left(CR - \frac{T_s}{2}\right) - \frac{T_s}{2} CR_c \approx \frac{L}{R} CR - \frac{T_s}{2} CR_c \approx LC \\
c &= m_c T_s D' LC + \frac{L}{R}\left(\frac{T_s^2}{\pi^2} - \frac{CRT_s}{2}\right) \\
&\approx m_c T_s D' LC + \frac{L}{R}\left(-\frac{CRT_s}{2}\right) = m_c T_s D' LC - LC\frac{T_s}{2} \\
d &= \frac{L}{R} CR \frac{T_s^2}{\pi^2} = LC\frac{T_s^2}{\pi^2}
\end{aligned}$$

with the assumptions $CR \gg T_s/2$, $CR \gg m_c D' T_s$, $L/R_c \gg T_s/2$, and $CR \gg 2T_s/\pi^2$.

Based on the approximation of $a + bs + cs^2 + ds^3 \approx a\big(1 + (b/a)s\big)\big(1 + (c/b)s + (d/b)s^2\big)$ with the conditions $b \gg c$ and $b \gg d$, $G_{vci}(s)$ is factorized as

$$G_{vci}(s) = K_{vc} \frac{\left(1 - \frac{s}{\omega_{rhp}}\right)\left(1 + \frac{s}{\omega_{esr}}\right)}{\left(1 + \frac{s}{\omega_{pl}}\right)\left(1 + \frac{s}{Q_p\omega_n} + \frac{s^2}{\omega_n^2}\right)} \tag{10.71}$$

For the buck converter, ω_{rhp} does not exist so it becomes

$$\omega_{rhp} = \infty$$

The esr zero is given by

$$\omega_{esr} = \frac{1}{CR_c}$$

The other parameters are evaluated as

$$K_{vc} = \frac{L}{R_i}\frac{1}{a} = \frac{L}{R_i}\frac{1}{T_s(m_cD' - 0.5) + \frac{L}{R}} = \frac{R}{R_i}\frac{1}{1 + \frac{RT_s}{L}(m_cD' - 0.5)} \tag{10.72}$$

$$\omega_{pl} = \frac{a}{b} = \frac{T_s(m_cD' - 0.5) + \frac{L}{R}}{LC} = \frac{1}{CR} + \frac{T_s}{LC}(m_cD' - 0.5) \tag{10.73}$$

$$\omega_n = \sqrt{\frac{b}{d}} = \sqrt{\frac{LC}{LC\frac{T_s^2}{\pi^2}}} = \frac{\pi}{T_s} \tag{10.74}$$

and

$$Q_p\omega_n = \frac{b}{c} = \frac{LC}{m_cT_sD'LC - LC\frac{T_s}{2}} = \frac{1}{(m_cD' - 0.5)T_s} \tag{10.75}$$

yielding the expression

$$Q_p = \frac{1}{\pi(m_cD' - 0.5)} = \frac{1}{\pi\left(\left(1 + \frac{S_e}{S_n}\right)D' - 0.5\right)} \tag{10.76}$$

The assumptions used in the previous derivation, $CR \gg T_s/2$, $CR \gg m_cD'T_s$, $L/R_c \gg T_s/2$, and $CR \gg 2T_s/\pi^2$, are easily met in practice. The accuracy of the $G_{vci}(s)$ approximation will be demonstrated in a later example.

For the boost and buck/boost converters, $G_{vci}(s)$ can be derived in the same way. The expressions of $G_{vci}(s)$ for the three basic converters are shown in Table 10.3. The derivations of $G_{vci}(s)$ are treated in Problem 10.4.

Predictions of $G_{vci}(s)$

The $G_{vci}(s)$ expression is repeated below

$$G_{vci}(s) = K_{vc}\frac{\left(1 - \frac{s}{\omega_{rhp}}\right)\left(1 + \frac{s}{\omega_{esr}}\right)}{\left(1 + \frac{s}{\omega_{pl}}\right)\left(1 + \frac{s}{Q_p\omega_n} + \frac{s^2}{\omega_n^2}\right)} \tag{10.77}$$

Table 10.3 Expressions for $G_{vci}(s)$ for Three Basic Converters

Control-to-output transfer function

$$G_{vci}(s) = K_{vc}\frac{\left(1-\frac{s}{\omega_{rhp}}\right)\left(1+\frac{s}{\omega_{esr}}\right)}{\left(1+\frac{s}{\omega_{pl}}\right)\left(1+\frac{s}{Q_p\omega_n}+\frac{s^2}{\omega_n^2}\right)}$$

For all three converters:
$$Q_p = \frac{1}{\pi\left((1+S_e/S_n)D'-0.5\right)}$$
$$\omega_n = \frac{\pi}{T_s}$$
$$\omega_{esr} = \frac{1}{CR_c}$$

Buck converter[†]

$$K_{vc} = \frac{R}{R_i}\frac{1}{1+\frac{RT_s}{L}(m_cD'-0.5)} \qquad \omega_{pl} = \frac{1}{CR}+\frac{T_s}{LC}(m_cD'-0.5)$$
$$\omega_{rhp} = \infty$$

Boost converter

$$K_{vc} = \frac{D'R}{2R_i}\frac{1}{1+\frac{D'^3RT_s}{2L}(m_c-0.5)} \qquad \omega_{pl} = \frac{2}{CR}+\frac{T_sD'^3}{LC}(m_c-0.5)$$
$$\omega_{rhp} = D'^2\frac{R}{L}$$

Buck/boost converter

$$K_{vc} = \frac{D'R}{(1+D)R_i}\frac{1}{1+\frac{D'^3RT_s}{(1+D)L}(m_c-0.5)} \qquad \omega_{pl} = \frac{1+D}{CR}+\frac{T_sD'^3}{LC}(m_c-0.5)$$
$$\omega_{rhp} = \frac{D'^2}{D}\frac{R}{L}$$

[†]For the buck converter, ω_{rhp} does not exist so $\omega_{rhp} = \infty$ is used.

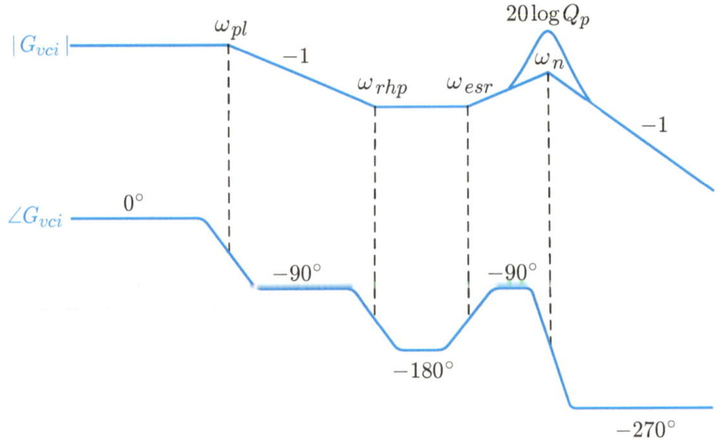

Figure 10.22 Asymptotic plot for $G_{vci}(s) of boost and buck/boost converters.$

The noticeable feature of the transfer function is the existence of the quadratic term in the denominator. The quadratic term, originated from the sampling effects of current mode control, introduces a double pole at half the switching frequency, $\omega_n = \pi/T_s$. The double pole produces a peaking by the amount of 20 log Q_p. Figure 10.22 shows the asymptotic plot for $|G_{vci}|$ of the boost and buck/boost converters with conditions $\omega_{pl} \ll \omega_{rhp} \ll \omega_{esr} \ll \omega_n{}^{\dagger}$ and $Q_p > 0.5$.

EXAMPLE 10.7 Accuracy of $G_{vci}(s)$ Analysis

This example demonstrates the accuracy of the $G_{vci}(s)$ approximation. Figure 10.23 compares the predictions of (10.77) and the exact small-signal simulations using Fig. 10.20(a). The buck converter used in Example 9.8 is revisited and three different values for Q_p are considered in this analysis: $Q_p = 1$, 4, and 16. The thick curves are the plots of the approximated $G_{vci}(s)$ expression in (10.77). The thin curves are the results of the small-signal simulation. Figure 10.24 also shows the $G_{vci}(s)$ plot of (10.77) for the case of $Q_p = 1.0$, in comparison with the result of the computational method. The close match between the Bode plots and computational data confirms the accuracy of the $G_{vci}(s)$ approximation.

10.3.3 Control Design Procedures

The most important role of $G_{vci}(s)$ lies in its use in the control design. Figure 10.25 shows the asymptotic plots for $|G_{vci}|$ and loop gain $|T_m|$. This figure assumes the

†For buck converters, ω_{esr} does not exist.

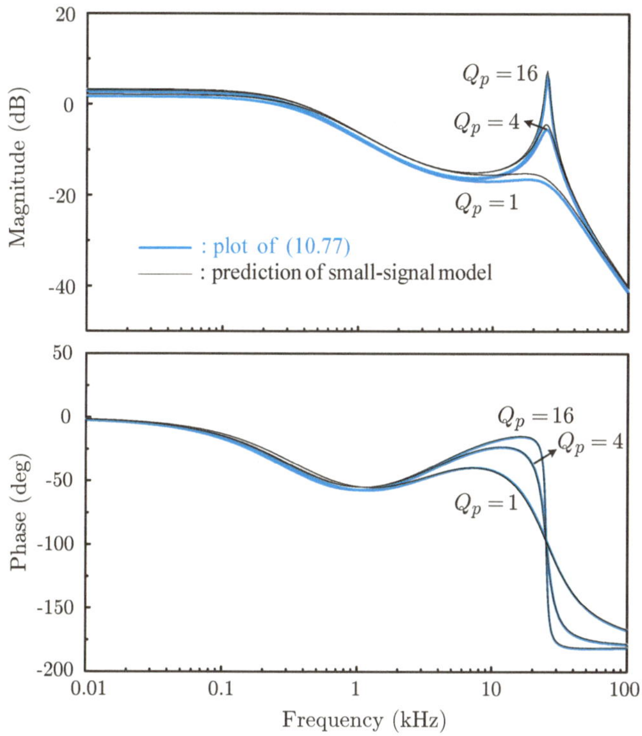

Figure 10.23 Accuracy of $G_{vci}(s)$ approximation with $Q_p = 1, 4,$ and 16.

two-pole one-zero circuit for the voltage feedback compensation

$$F_v(s) = \frac{K_v\left(1+\dfrac{s}{\omega_{zc}}\right)}{s\left(1+\dfrac{s}{\omega_{pc}}\right)} \tag{10.78}$$

The compensation pole is located at the RHP zero, $\omega_{pc} = \omega_{rhp}$.† The compensation zero is placed after the low-frequency pole of $G_{vci}(s)$, ω_{pl}, but before the power stage double pole which occurs prior to the loop gain crossover frequency. Details about the compensation parameters will be explained shortly.

The structure and circuit components for the two-pole one-zero compensation were given in Fig. 9.22, along with the equation (9.48) in Section 9.2.5. The profile of the loop gain is very similar to that of the outer loop gain in the previous classical analysis. However, *the loop gain has a double pole at half the switching frequency.*

†For the buck converter in which ω_{rhp} is not present, ω_{pc} is placed at ω_{esr} and the asymptotic plots should be accordingly modified.

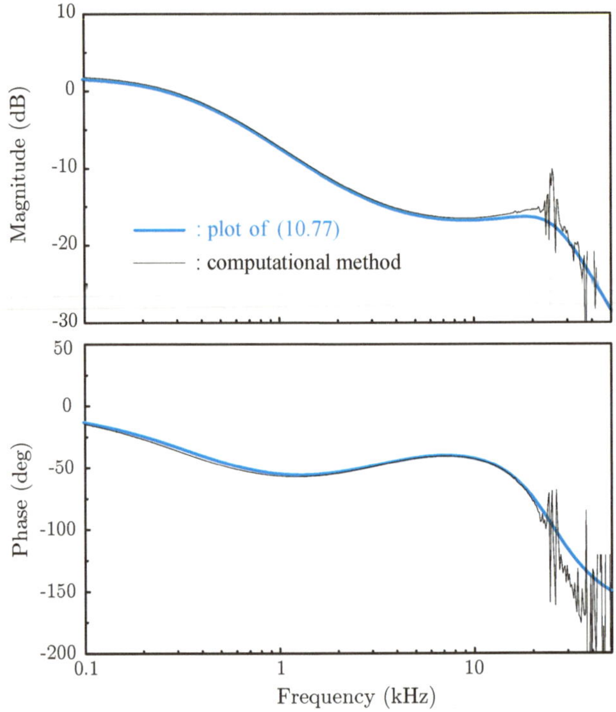

Figure 10.24 Accuracy of G_{vci} (s) approximation with $Q_p = 1$.

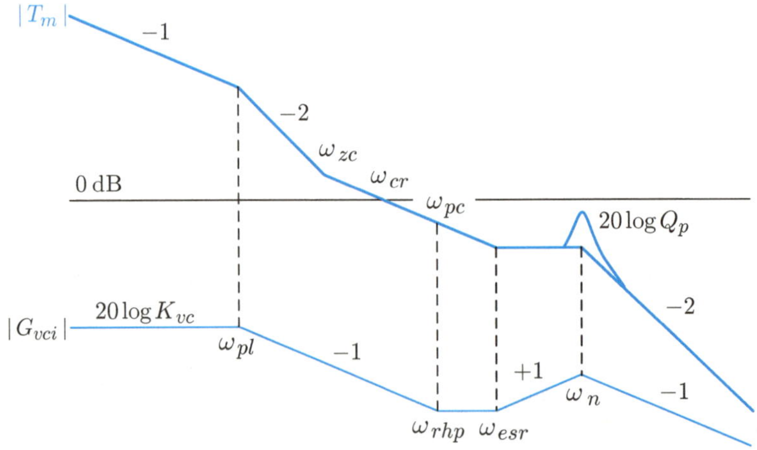

Figure 10.25 Asymptotic plot for $|G_{vci}|$ and $|T_m|$ with two-pole one-zero compensation.

The double pole produces a peaking of $20 \log Q_p$ at $\omega_n = \pi/T_s$. If the peaking is large, the loop gain will cross the 0 dB line at $\omega_n = \pi/T_s$, thereby destablizing the converter.† This becomes the cause for the high-frequency oscillations, often encountered with current mode control. As a simple example, the peaking becomes infinite

$$20 \log Q_p = 20 \log \left(\frac{1}{\pi \left((1 + S_e/S_n) D' - 0.5 \right)} \right) = \infty \text{ dB}$$

when the compensation ramp is absent, $S_e = 0$, and $D' = 0.5$, which are the conditions for the onset of the sub-harmonic oscillation.

The previous discussions naturally conclude that the quality factor of the double pole, $Q_p = 1/\left(\pi\left((1 + S_e/S_n)D' - 0.5\right)\right)$, should be properly controlled to avoid instability due to an excessive peaking. In other words, the double pole should be well damped with an adequate quality factor. The idea of properly damping the double pole at half the switching frequency leads to an easy and practical design methodology for current mode control.

Current Loop Design

The theme of the current loop design is to properly select the CSN gain and adequately control the quality factor of the double pole at half the switching frequency. The control of the quality factor is transformed into the selection of the compensation ramp slope.

1) Determine the CSN gain R_i such that $i_{L\,peak} R_i < V_{max}$ where V_{max} is the maximum allowable input voltage of the PWM block. The CSN gain R_i determines the on-time slope S_n of the current feedback signal.

2) Determine the slope of the compensation ramp S_e to provide a quality factor between $0.3 < Q_p < 1.3$ for the double pole at $\omega_n = \pi/T_s$

$$0.3 < \frac{1}{\pi \left((1 + S_e/S_n) D' - 0.5 \right)} < 1.3$$

This prevents an excessive peaking at half the switching frequency, which could trigger high-frequency oscillations.

† The 0 dB crossing due to the peaking would cause an abrupt phase decrease so that the polar plot violates the Nyquist stability criterion.

Voltage Feedback Compensation Design

The voltage feedback compensation is designed to achieve the desirable loop gain characteristics. With the two-pole one-zero compensation, the loop gain is given by

$$T_m(s) = G_{vci}(s)\,F_v(s)$$
$$= \underbrace{K_{vc}\frac{\left(1-\frac{s}{\omega_{rhp}}\right)\left(1+\frac{s}{\omega_{esr}}\right)}{\left(1+\frac{s}{\omega_{pl}}\right)\left(1+\frac{s}{Q_p\omega_n}+\frac{s^2}{\omega_n^2}\right)}}_{G_{vci}(s)}\;\underbrace{\frac{K_v\left(1+\frac{s}{\omega_{zc}}\right)}{s\left(1+\frac{s}{\omega_{pc}}\right)}}_{F_v(s)} \qquad (10.79)$$

The design procedures for the voltage feedback compensation are formulated as follows.

1) Place the compensation pole ω_{pc} at the lowest frequency among the RHP zero, esr zero, and half the switching frequency: $\omega_{pc} = \min\{\omega_{rhp}\ \omega_{esr}\ 0.5\,\omega_s\}$.

2) Locate the compensation zero ω_{zc} at the frequencies higher than ω_{pl}, but lower than the power stage double pole ω_o: $\omega_{zc} = (0.6 - 0.8)\omega_o$.

3) Select the 0 dB crossover frequency of the loop gain, denoted as ω_{cr} in Fig. 10.25, and determine the integrator gain K_v required for the selected ω_{cr}. The position of ω_{cr} is chosen based on the guidelines established in Chapter 9: $\omega_{cr} = (0.3 - 1.0)\,\omega_{esr}$ for buck converters and $\omega_{cr} = (0.1 - 0.3)\,\omega_{rhp}$ for boost and buck/boost converters. From Fig. 10.25, the following relationship is formulated

$$20\log\left(K_{vc}\frac{K_v}{\omega_{pl}}\right) - 40\log\left(\frac{\omega_{zc}}{\omega_{pl}}\right) - 20\log\left(\frac{\omega_{cr}}{\omega_{zc}}\right) = 0\ \text{dB} \qquad (10.80)$$

The first term in (10.80) is the magnitude of the loop gain $|T_m|$, evaluated at ω_{pl}: $|T_m(j\omega_{pl})| = 20\log\left(K_{vc}K_v/\omega_{pl}\right)$. The above equation is converted into the design equation

$$\frac{K_{vc}K_v}{\omega_{pl}}\left(\frac{\omega_{pl}}{\omega_{zc}}\right)^2\frac{\omega_{zc}}{\omega_{cr}} = 1 \qquad (10.81)$$

from which the integrator gain is determined as

$$\boxed{K_v = \frac{\omega_{zc}\,\omega_{cr}}{K_{vc}\,\omega_{pl}}} \qquad (10.82)$$

The above design procedures are summarized in Table 10.4.

Table 10.4 New Design Procedures for Peak Current Mode Control

Current loop design

$$Q_p = \frac{1}{\pi\left(\left(1+\dfrac{S_e}{S_n}\right)D' - 0.5\right)}$$

1) Select the CSN gain R_i such that $i_{L\,peak}\,R_i < V_{max}$, where V_{max} is the maximum allowable input voltage for the PWM block.

2) Determine the quality factor Q_p: $0.3 < Q_p < 1.3$

3) Evaluate the compensation ramp slope:

$$S_e = S_n\left(\frac{\dfrac{1}{\pi\,Q_p} + 0.5}{D'} - 1\right)$$

Voltage loop design

$$T_m(s) = \underbrace{K_{vc}\frac{\left(1-\dfrac{s}{\omega_{rhp}}\right)\left(1+\dfrac{s}{\omega_{esr}}\right)}{\left(1+\dfrac{s}{\omega_{pl}}\right)\left(1+\dfrac{s}{Q_p\omega_n}+\dfrac{s^2}{\omega_n^2}\right)}}_{G_{vci}(s)}\;\underbrace{\frac{K_v\left(1+\dfrac{s}{\omega_{zc}}\right)}{s\left(1+\dfrac{s}{\omega_{pc}}\right)}}_{F_v(s)}$$

1) Set the compensation pole: $\omega_{pc} = \min\{\omega_{rhp}\;\omega_{esr}\;\omega_s/2\}^{\dagger}$.

2) Select the compensation zero: $\omega_{zc} = (0.6 - 0.8)\,\omega_o$.

3) Set the loop gain crossover frequency ω_{cr}:

$\omega_{cr} = (0.3 - 1.0)\,\omega_{esr}$ for buck converters

$\omega_{cr} = (0.1 - 0.3)\,\omega_{rhp}$ for boost and buck/boost converters.

4) Evaluate the integrator gain:

$$K_v = \frac{\omega_{zc}\,\omega_{cr}}{K_{vc}\,\omega_{pl}}$$

5) Check the phase margin and adjust K_v to secure a 45°–70° phase margin.

6) Evaluate the circuit components for voltage feedback compensation using (9.48) in Chapter 9.

†For buck converters, ω_{rhp} does not exist so $\omega_{rhp} = \infty$ is used.

EXAMPLE 10.8 Buck Converter Design Example

This example illustrates the adaptation of the new design procedures developed in this section. For this purpose, the peak current mode control employed to the buck converter in Example 9.8, is redesigned, based on the design procedures developed in this section. The corner frequencies and operational conditions of the buck converter are $\omega_o = 2\pi \cdot 1.16 \times 10^3$ rad/s, $\omega_{esr} = 2\pi \cdot 3.39 \times 10^3$ rad/s, and $D = 0.25$. The switching frequency is $\omega_s = 2\pi \cdot 50 \times 10^3$ rad/s. The maximum input voltage for the PWM block is assumed as $V_{max} = 5.0$ V and the peak value of the inductor current is calculated as $i_{L\,peak} = 4.75$ A. Based on the new design procedures, the control design is performed.

Current Loop Design

The current loop design is the process of selecting the CSN gain and compensation ramp slope for a desired quality factor Q_p.

1) CSN gain:

$$R_i < \frac{V_{max}}{i_{L\,max}} = \frac{5.0}{4.75} = 1.05 \quad \Rightarrow R_i = 0.67$$

2) Quality factor of double pole and compensation ramp slope:

$$Q_p = \frac{1}{\pi\left(\left(1+\frac{S_e}{S_n}\right)D' - 0.5\right)} = 1 \quad \text{with}$$

$$S_n = \frac{V_S - V_O}{L}R_i = \frac{16-4}{40\times10^{-6}}0.67 = 2.02\times10^5 \text{ V/s}$$

$$\Rightarrow \quad S_e = S_n\left(\frac{\frac{1}{\pi\, Q_p}+0.5}{D'} - 1\right)$$

$$= 2.02\times10^5\left(\frac{\frac{1}{\pi\, 1}+0.5}{0.75} - 1\right) = 1.84\times10^4 \text{ V/s}$$

$$\Rightarrow \quad m_c = 1+\frac{S_e}{S_n} = 1+\frac{1.84\times10^4}{2.02\times10^5} = 1.09$$

The slope of the compensation ramp, $S_e = 1.84 \times 10^4$ V/s, is noticeably smaller than that of the conventional design in Example 9.8, $S_e = 1.46 \times 10^5$ V/s. Detailed comparisons between the current design and the earlier design will be given later.

Voltage Loop Design

The voltage loop design is the process of selecting the compensation parameters for voltage feedback circuit.

1) Compensation pole: $\omega_{pc} = \omega_{esr} = 2\pi \cdot 3.39 \times 10^3$ rad/s
2) Compensation zero: $\omega_{zc} = 0.8\,\omega_o = 2\pi \cdot 928$ rad/s

3) Loop gain crossover frequency: $\omega_{cr} = \omega_{esr} = 2\pi \cdot 3.39 \times 10^3$ rad/s

4) Integrator gain:

$$K_v = \frac{\omega_{zc}\,\omega_{cr}}{K_{vc}\,\omega_{pl}} \quad \text{with}$$

$$K_{vc} = \frac{R}{R_i}\frac{1}{1+\dfrac{RT_s}{L}(m_c D' - 0.5)}$$

$$= \frac{1}{0.67}\frac{1}{1+\dfrac{1\cdot 20\times 10^{-6}}{40\times 10^{-6}}(1.09\cdot 0.75 - 0.5)} = 1.29$$

$$\omega_{pl} = \frac{1}{CR} + \frac{T_s}{LC}(m_c D' - 0.5)$$

$$= \frac{1}{470\times 10^{-6}\,1} + \frac{20\times 10^{-6}}{40\times 10^{-6}\,470\times 10^{-6}}$$

$$\cdot (1.09\cdot 0.75 - 0.5) = 2\pi \cdot 392 \text{ rad/s}$$

$$\text{Thus,} \quad K_v = \frac{(2\pi\cdot 928)(2\pi\cdot 3.39\times 10^3)}{1.29(2\pi\cdot 392)} = 3.91\times 10^4$$

5) Voltage feedback circuit: $R_1 = 10$ kΩ

$$\Rightarrow R_2 = 92.3 \text{ k}\Omega,\quad C_2 = 1.86 \text{ nF},\quad \text{and} \quad C_3 = 0.70 \text{ nF}$$

The design results are the same as those of the classical design in Example 9.8, except for the reduced compensation ramp slope. The theoretical background for this sameness will be discussed in the next section.

Performance of Current Mode Control

The performance of the buck converter with the newly-designed peak current mode control is evaluated. Figure 10.26 shows the loop gain of the converter. The prediction of the new model is compared with the result of the computational method. The new model prediction is also contrasted with the outcome of the classical model which does not include the sampling effects but has the *same control parameters as those of the new model* – that is, the control parameters of the new design.

The existence of the double pole at half the switching frequency, which was theoretically predicted by the new model, is well supported by the computational data. The new model accurately correlates with the computational data and thus demonstrates notable improvements over the classical model.

The loop gain crossover frequency is located at the exact target frequency of $\omega_{cr} = 2\pi \cdot 3.39 \times 10^3$ rad/s with a sufficient phase margin. Figure 10.27(a) shows the output impedance characteristics of the converter. The predictions of the new model and classical model both show a good match with the computational data. Figure 10.27(b) depicts the audio-susceptibility of the converter. The audio-susceptibility curves show the impacts of the feedforward gains in the new model. The new model shows improved audio-susceptibility characteristics compared to the classical model which assumes the *zero* feedforward gains.

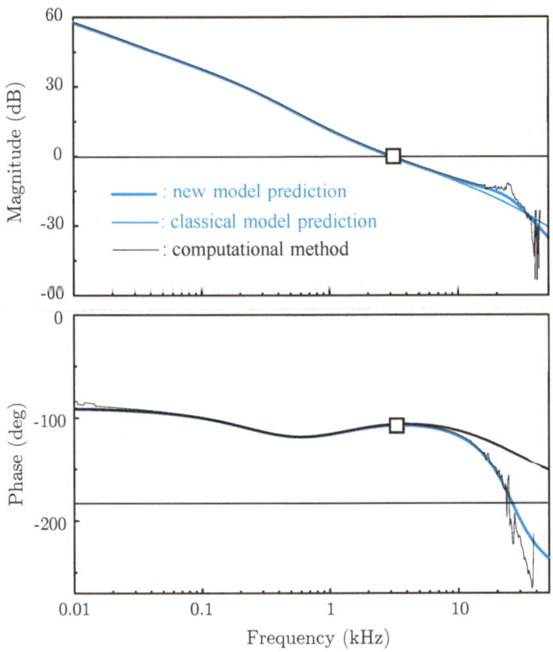

Figure 10.26 Loop gain of buck converter: predictions of new and classical models.

Comparison between New Design and Classical Design

In this part, the performance of the new design will be compared with that of the previous classical design, which was done in Example 9.8 without considering the sampling effects. For this comparative study, the new model developed in this chapter is used for the theoretical predictions of the two designs. The theoretical predictions are compared with the results of the computational method. Figure 10.28 compares the loop gain of the two designs. Although the new design and classical design both predict stability with a sufficient phase margin, they show discernible disparities at high frequencies.

As will be demonstrated in the next section, the classical design is in fact an *exact match* to the new design that is executed with the target of $Q_p = 0.4$. Accordingly, the classical design corresponds to the new design with $Q_p = 0.4$, while the new design was performed with the design goal of $Q_p = 1.0$. The difference in Q_p is well reflected in the loop gain characteristics. The classical design shows more gradual high-frequency phase decay with a small quality factor of $Q_p = 0.4$, while the new design exhibits a gain boost with $Q_p = 1$ at the frequencies around half the switching frequency.

Figure 10.29 shows the output impedance and audio-susceptibility of the two designs. As shown in Fig. 10.29(a), the two designs do not show any noticeable difference in the output impedance characteristics. However, the new design offers improved audio-susceptibility characteristics. This improvement is due to the reduction in the compensation ramp slope. The compensation ramp slope of the

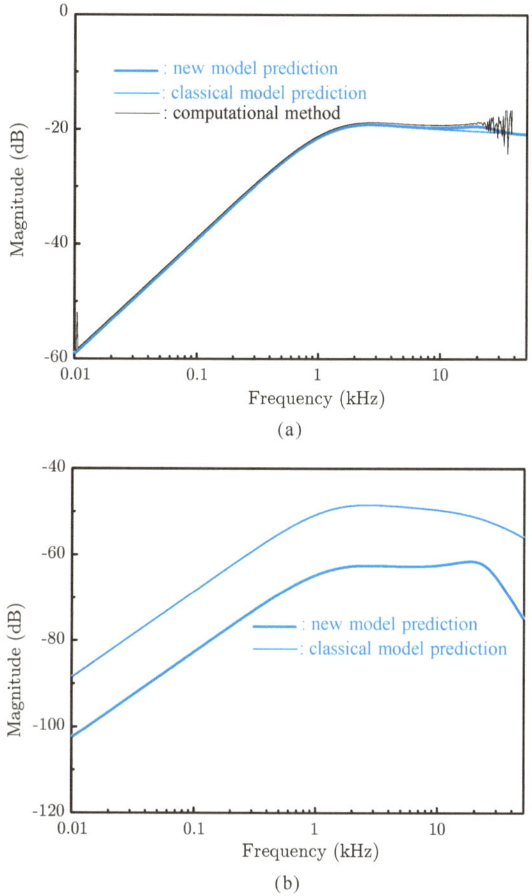

Figure 10.27 Frequency-domain performance of buck converter. (a) Output impedance. (b) Audio-susceptibility.

new design, $S_e = 9.2 \times 10^3$ V/s, is much smaller than the classical design case, $S_e = 1.46 \times 10^5$ V/s. Referring to Section 9.2.2, the smaller ramp slope increases the magnitude of the current loop. The enhanced current loop in turn provides more attenuation for the audio-susceptibility. However, this improvement only occurs in the buck converter. It was shown in Section 9.3.1 that, in the case of the buck converter, the audio-susceptibility is governed by the overall loop gain which is directly affected by the current loop. In contrast, for the boost and buck/boost converters, the audio-susceptibility is dictated by the outer loop gain which is not largely influenced by the current loop.

Figure 10.30 compares the time-domain performance. For this analysis, the same time-domain model is used for both the new design and classical design. The only difference is the slope of the compensation ramp. Figure 10.30(a) is the transient response of the output voltage due to the $I_O = 4$ A $\Rightarrow$ 8 A $\Rightarrow$ 4 A changes in the

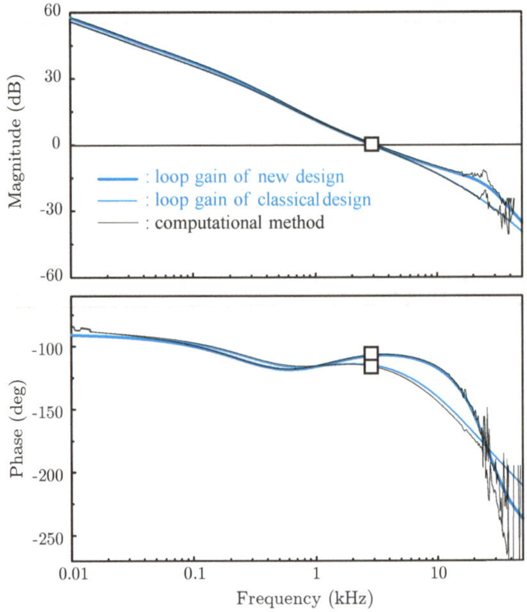

Figure 10.28 Loop gain of buck converter: comparison of new and classical designs.

load current, while Fig. 10.30(b) shows the transient waveform of the output voltage in response to the $V_S = 16$ V $\Rightarrow 8$ V $\Rightarrow 16$ V changes in the input voltage. The step load responses are identical in both designs, as predicted from the output impedance analysis. On the other hand, the new design shows an improved step input response. This result is consistent with the audio-susceptibility analysis and demonstrates the benefits of the enhanced current loop magnitude.

EXAMPLE 10.9 Boost Converter Design Example

The new design procedures are now applied to the boost converter that was used in Example 9.16. The small-signal parameters and operational conditions are $\omega_o = 2\pi \cdot 348$ rad/s, $\omega_{rhp} = 2\pi \cdot 1.79 \times 10^3$ rad/s, $\omega_{esr} = 2\pi \cdot 6.77 \times 10^3$ rad/s, and $D = 0.4$. The switching frequency is $\omega_s = 2\pi \cdot 50 \times 10^3$ rad/s and the maximum input voltage of PWM block is $V_{max} = 5.0$ V. The peak value of the inductor current is $i_{L\,peak} = 6.97$ A. The new design procedures are adapted to these conditions.

Current Loop Design

1) CSN gain:

$$R_i < \frac{V_{max}}{i_{L\,peak}} = \frac{5.0}{6.97} = 0.72 \quad \Rightarrow R_i = 0.67$$

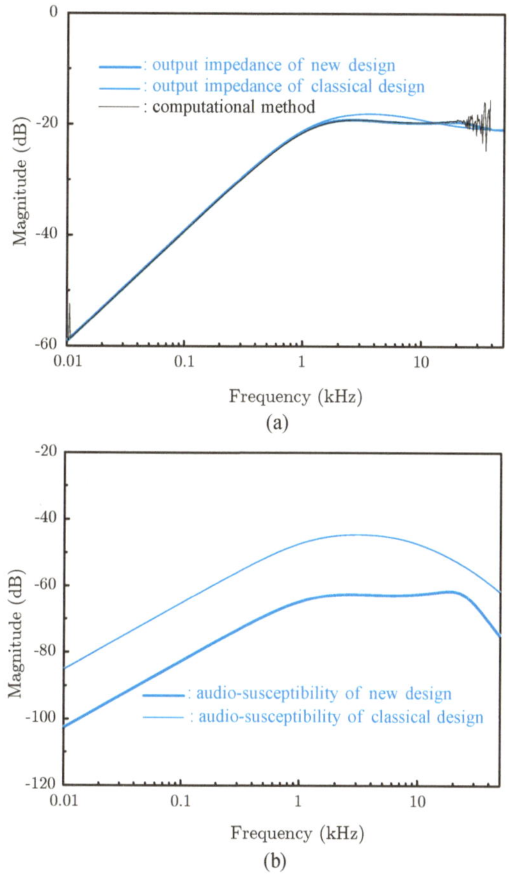

Figure 10.29 Frequency-domain performance of buck converter. (a) Output impedance. (b) Audio-susceptibility.

2) Damping ratio of double pole:

$$Q_p = \frac{1}{\pi\left(\left(1+\dfrac{S_e}{S_n}\right)D' - 0.5\right)} = 1 \quad \text{with}$$

$$S_n = \frac{V_S}{L}R_i = \frac{12}{160\times 10^{-6}}0.67 = 5.03\times 10^4 \text{ V/s}$$

$$\Rightarrow \quad S_e = S_n\left(\frac{\dfrac{1}{\pi\, Q_p}+0.5}{D'}-1\right) = 5.03\times 10^4\left(\frac{\dfrac{1}{\pi\, 1}+0.5}{0.6}-1\right) = 1.83\times 10^4 \text{ V/s}$$

$$\Rightarrow \quad m_c = 1+\frac{S_e}{S_n} = 1+\frac{1.83\times 10^4}{5.03\times 10^4} = 1.36$$

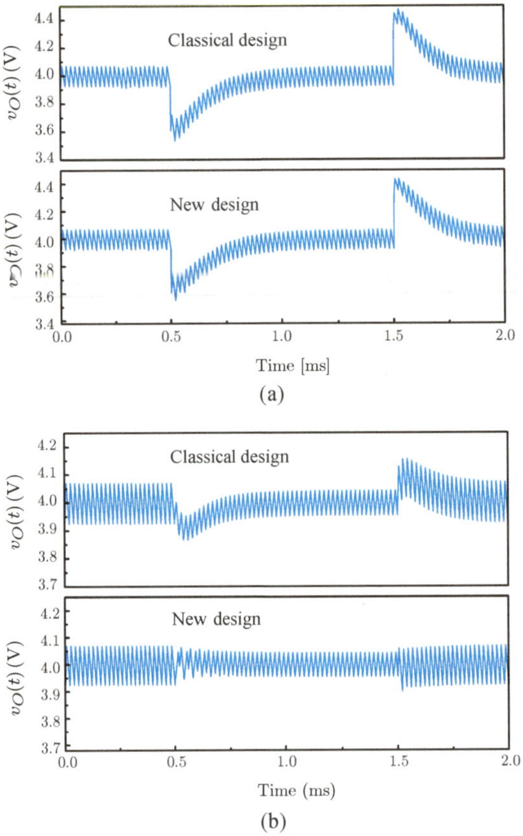

Figure 10.30 Time-domain performance of buck converter. (a) Step load response. (b) Step input response.

Voltage Loop Design

1) Compensation pole: $\omega_{pc} = \omega_{rhp} = 2\pi \cdot 1.79 \times 10^3$ rad/s
2) Compensation zero: $\omega_{zc} = 0.8\,\omega_o = 2\pi \cdot 278$ rad/s
3) Loop gain crossover frequency: $\omega_{cr} = 0.28\,\omega_{rhp} = 2\pi \cdot 501$ rad/s
4) Integrator gain:

$$K_v = \frac{\omega_{zc}\,\omega_{cr}}{K_{vc}\,\omega_{pl}}$$

with

$$K_{vc} = \frac{D'R}{2\,R_i}\frac{1}{1 + \dfrac{D'^3 RT_s}{2\,L}(m_c - 0.5)}$$

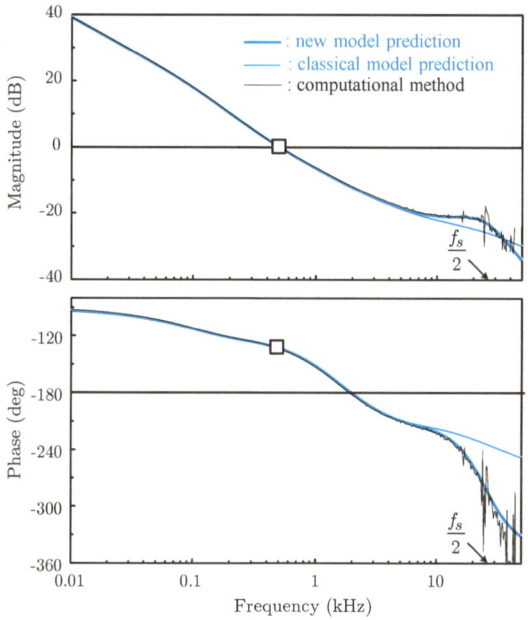

Figure 10.31 Loop gain of boost converter: predictions of new and classical models.

$$= \frac{0.6 \cdot 5}{2 \cdot 0.67} \frac{1}{1 + \dfrac{0.6^3\, 5 \cdot 20 \times 10^{-6}}{2 \cdot 160 \times 10^{-6}}(1.36 - 0.5)} = 2.12$$

$$\omega_{pl} = \frac{2}{CR} + \frac{T_s D'^3}{LC}(m_c - 0.5)$$

$$= \frac{2}{470 \times 10^{-6}\, 5} + \frac{20 \times 10^{-6}\, 0.6^3}{(160 \times 10^{-6})\,(470 \times 10^{-6})}$$

$$\cdot (1.36 - 0.5) = 2\pi \cdot 143 \text{ rad/s}$$

$$\text{Thus,} \quad K_v = \frac{(2\pi \cdot 278)(2\pi \cdot 501)}{2.12\,(2\pi \cdot 143)} = 2.88 \times 10^3$$

5) Voltage feedback circuit: $R_1 = 10$ kΩ

$$\Rightarrow R_2 = 19.6 \text{ k}\Omega, \quad C_2 = 29.2 \text{ nF}, \quad \text{and } C_3 = 5.38 \text{ nF}$$

The design results are identical to those of the classical design in Example 9.16, except for the reduced compensation ramp slope. The compensation ramp slope of the new design is $S_e = 1.83 \times 10^4$ V/s, while that of the classical design is $S_e = 7.50 \times 10^4$ V/s.

Performance of Current Mode Control

The loop gain of the boost converter is shown in Fig. 10.31. The loop gain is displayed using the new model and the classical model, together with the results of the computational method. The loop gain exhibits an exact match to the design target

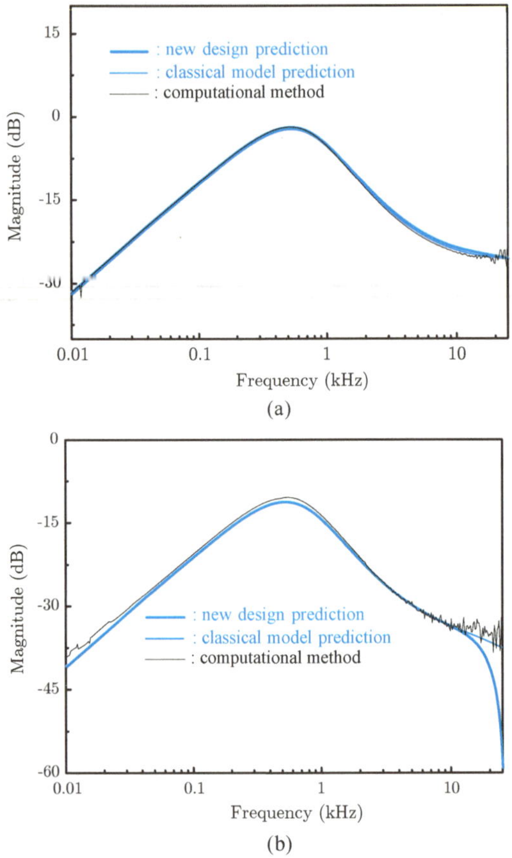

Figure 10.32 Frequency-domain performance of boost converter. (a) Output impedance. (b) Audio-susceptibility.

of $\omega_{cr} = 2\pi \cdot 501$ rad/s with a sufficient phase margin. The new model also shows good correlations to the computational data and demonstrates improved model accuracy over the classical model. The double pole at half the switching frequency is evident in the new model prediction and computational result. Figure 10.32 shows the output impedance and audio-susceptibility of the boost converter. Here, both the classical model and new model show good correspondence to the computational data.

Comparison between New Design and Classical Design

In Fig. 10.33, the loop gain characteristics of the new design are compared with those of the classical design, which was done in Example 9.16 without considering the sampling effects. The new model incorporating the sampling effects is used to assess the loop gain characteristics of the two designs. Both the designs show good loop

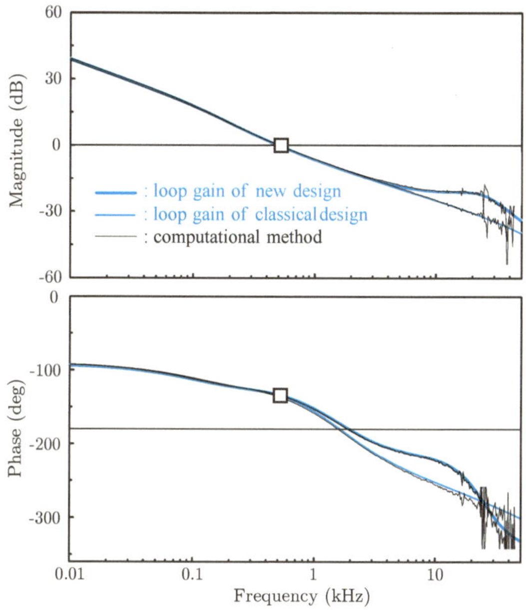

Figure 10.33 Loop gain of boost converter: comparison of new and classical designs.

gain characteristics and also exhibit close agreement with the computational data. It will be shown that the classical design is an exact match to the new design with $Q_p = 0.32$. On the other hand, the current design is done with the goal of $Q_p = 1.0$. This difference in Q_p values is well reflected in the loop gain characteristics.

10.3.4 Correlation between New and Classical Design Procedures

The new control design procedures developed in this chapter are different from those of Chapter 9. The new design procedures explicitly incorporate the sampling effects of the peak current mode control. In contrast, the classical design does not consider the sampling effects at all. Nonetheless, the two design procedures showed remarkable similarity, as illustrated in Examples 10.8 and 10.9. Theoretical backgrounds behind this similarity are explored in this section. The current loop designs are first investigated and the voltage loop designs are subsequently compared.

Current Loop Design

The goal of the new current loop design is to provide a predetermined value for Q_p. On the other hand, the classical design targets to place the crossover frequency of the current loop at the desired frequency. Although the objectives of the two designs are different, the means of obtaining their goal is the same. Once the power stage and CSN parameters are determined, the only remaining design freedom is the slope of the compensation ramp.

The new design attempts to provide a desired Q_p value by adjusting the compensation ramp slope S_e in the following expression

$$Q_p = \frac{1}{\pi\left(\left(1+\frac{S_e}{S_n}\right)D' - 0.5\right)} \tag{10.83}$$

The classical design aims to adjust the ratio of the current-loop crossover frequency to the switching frequency, ω_{ci}/ω_s, by controlling S_e in the following equation

$$\frac{\omega_{ci}}{\omega_s} = \frac{K_{id}R_i\frac{\omega_o^2}{\omega_{id}}\frac{2}{(S_n - S_f + 2S_e)T_s}}{\frac{2\pi}{T_s}} \tag{10.84}$$

The above design equation is obtained from (9.39) and (9.41) in Chapter 9. An explicit correlation between the new and classical designs can be established using (10.83) and (10.84), as illustrated in the following example.

EXAMPLE 10.10 Buck and Boost Converter Examples

The buck converter in Example 10.8 is used as the first example to illustrate the connection between the Q_p value in the new design and the ω_{ci}/ω_s ratio in the classical design. Figure 10.34(a) is the $Q_p - \omega_{ci}/\omega_s$ curve, obtained by relating (10.83) and (10.84) with the given power stage parameters and operational conditions. First, the required S_e value for a specific Q_p is calculated using (10.83). The ratio of ω_{ci}/ω_s is then evaluated from (10.84) using the calculated S_e value. By repeating this process, the continuous $Q_p - \omega_{ci}/\omega_s$ curve is obtained. The curve is evaluated for $0.3 < Q_p < 1.3$. Figure 10.34(a) reveals a linear relationship between Q_p and ω_{ci}/ω_s. For one example, the goal of $Q_p = 0.5$ in the new design requires the S_e value that offers the condition $\omega_{ci}/\omega_s = 0.25$ in the classical design. As another example, the design aim of $Q_p = 1$ in the new design translates to the condition $\omega_{ci}/\omega_s = 0.5$ in the classical design: that is, placing the current loop crossover frequency at half the switching frequency. Furthermore, selecting the damping ratio between $0.3 < Q_p < 1.3$ in the new design is equivalent to placing the constraint $0.15 < \omega_{ci}/\omega_s < 0.65$ in the classical design.

Figure 10.34(b) shows the $Q_p - \omega_{ci}/\omega_s$ curve for the boost converter used in Example 10.9. Interestingly, the curve is identical to that of Fig. 10.34(a). The sameness in Fig. 10.34(a) and Fig. 10.34(b) is actually the expected result of a general relationship that links Q_p and ω_{ci}/ω_s. The exact description of this relationship is given next.

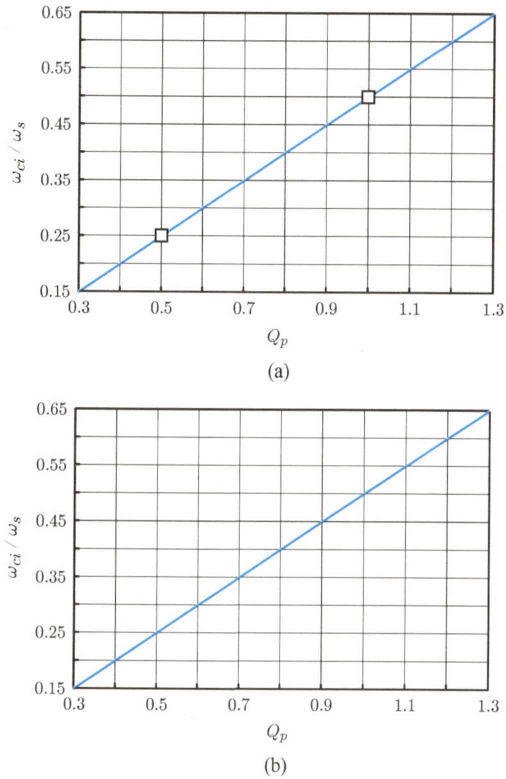

Figure 10.34 Relationship between Q_p and ω_{ci}/ω_s. (a) Buck converter in Example 10.8. (b) Boost converter in Example 10.9.

As demonstrated in the previous example, a direct correlation exists between the new and classical designs. In fact, it can be proved that the following relationship holds true

$$\frac{\omega_{ci}}{\omega_s} = \frac{Q_p}{2} \tag{10.85}$$

for all the three basic dc-to-dc converters, regardless of the operational conditions and power stage parameters. The equality between Figs. 10.34(a) and 10.34(b) is now self-evident from (10.85). The verification of (10.85) is given in Example 10.11.

Derivation of $\omega_{ci}/\omega_s = Q_p/2$ for Buck Converter — EXAMPLE 10.11

This example shows the validity of (10.85) using the buck converter. First, the following two identities are derived for the buck converter

$$K_{id}R_i\frac{\omega_o^2}{\omega_{id}} = \frac{V_S}{R}R_i\frac{\frac{1}{LC}}{\frac{1}{CR}} = \frac{V_S}{L}R_i \tag{10.86}$$

and

$$\begin{aligned}\frac{2}{(S_n - S_f + 2S_e)T_s} &= \frac{2}{(S_n(2m_c - 1) - S_f)T_s} \\ &= \frac{2}{\left(\frac{V_S - V_O}{L}R_i(2m_c - 1) - \frac{V_O}{L}R_i\right)T_s} \\ &= \frac{L}{2R_iV_S}\frac{2}{(m_cD' - 0.5)T_s} \end{aligned} \tag{10.87}$$

with

$$m_c = 1 + \frac{S_e}{S_n}$$

Using (10.86) and (10.87), the equation (10.84) is written as

$$\begin{aligned}\frac{\omega_{ci}}{\omega_s} &= \frac{K_{id}R_i\frac{\omega_o^2}{\omega_{id}}\frac{2}{(S_n - S_f + 2S_e)T_s}}{\frac{2\pi}{T_s}} \\ &= \frac{\left(\frac{V_S}{L}R_i\right)\left(\frac{L}{2R_iV_S}\frac{2}{(m_cD' - 0.5)T_s}\right)}{\frac{2\pi}{T_s}} \\ &= \frac{1}{2\,\pi(m_cD' - 0.5)} = \frac{1}{2\,\pi\left(\left(1 + \frac{S_e}{S_n}\right)D' - 0.5\right)} = \frac{Q_P}{2}\end{aligned} \tag{10.88}$$

The same procedures can be applied to the boost and buck/boost converters to show that the relationship $\omega_{ci}/\omega_s = Q_p/2$ is valid for all the three basic PWM converters.

Voltage Loop Design

The objective of the voltage loop design is to select the three parameters, ω_{pc}, ω_{zc}, and K_v, for the two-pole one-zero compensation. The selection of ω_{pc} and ω_{zc} is identical in both the new design and the classical design. The only difference exists in the selection of K_v.

In the new design, the integrator gain K_v in (10.82) is selected from

$$K_v = \frac{1}{K_{vc}\,\omega_{pl}}\omega_{zc}\,\omega_{cr} \tag{10.89}$$

where ω_{cr} is the desired crossover frequency of the loop gain $T_m(s)$. In the classical design, K_v in (9.47) is determined from

$$K_v = \frac{1}{\frac{K_{vd}}{K_{id}R_i}\omega_{id}}\omega_{zc}\,\omega_{cr} \tag{10.90}$$

where ω_{cr} is the desired crossover frequency of the outer loop gain $T_2(s)$. For the three basic PWM converters, the following relationship holds true

$$K_{vc}\,\omega_{pl} = \frac{K_{vd}}{K_{id}R_i}\omega_{id} \tag{10.91}$$

as will be shown in Example 10.12. Accordingly, if the same ω_{cr} is selected in the two designs, the required K_v is always identical. Now, all the three parameters, ω_{pc}, ω_{zc}, and K_v, are the same in the two design approaches.

Buck Converter Example — EXAMPLE 10.12

This example shows the validity of (10.91). For the buck converter, it follows that

$$K_{vc} = \frac{R}{R_i}\frac{1}{1+\dfrac{RT_s}{L}(m_cD'-0.5)} \tag{10.92}$$

and

$$\omega_{pl} = \frac{1}{CR} + \frac{T_s}{LC}(m_cD'-0.5) \tag{10.93}$$

The product of K_{vc} and ω_{pl} is given by

$$\begin{aligned} K_{vc}\omega_{pl} &= \frac{R}{R_i}\frac{1}{1+\dfrac{RT_s}{L}(m_cD'-0.5)}\left(\frac{1}{CR}+\frac{T_s}{LC}(m_cD'-0.5)\right) \\ &= \frac{R}{R_i}\frac{1}{1+\dfrac{RT_s}{L}(m_cD'-0.5)}\frac{1}{CR}\left(1+\frac{RT_s}{L}(m_cD'-0.5)\right) \\ &= \frac{1}{R_iC} \end{aligned} \tag{10.94}$$

On the other hand, from the power stage transfer functions, it can be shown that

$$\frac{K_{vd}}{K_{id}R_i} = \frac{R}{R_i} \tag{10.95}$$

and

$$\omega_{id} = \frac{1}{CR} \tag{10.96}$$

yielding

$$\frac{K_{vd}}{K_{id}R_i}\omega_{id} = \frac{1}{R_iC} \tag{10.97}$$

From the expressions (10.92) through (10.97), the relationship (10.91) is clear

$$K_{vc}\omega_{pl} = \frac{K_{vd}}{K_{id}R_i}\omega_{id} = \frac{1}{R_iC} \tag{10.98}$$

Final Remarks

The analysis of the sampling effects has led to the new design procedures for peak current mode control. This alternative design controls the quality factor in the control-to-output transfer function so that the sampling effects neither cause stability problems nor interfere with the closed-loop performance. The rationale of the new design is utterly different from that of the previous classical design which does not consider the sampling effects at all. Nonetheless, the results of the two designs markedly resemble each other.

The current loop designs are intimately connected by the relationship, $\omega_{ci}/\omega_s = Q_p/2$. If this information is incorporated in the current loop design, the results of the two designs should be the same. More importantly, *if the conservative design constraint,* $0.1 < \omega_{ci}/\omega_s < 0.3$*, is imposed on the classical design as recommended in Chapter* 9*, the performance of the converter is not adversely affected by the sampling effect of current mode control.* This is because the design constraint of $0.1 < \omega_{ci}/\omega_s < 0.3$ is the same as the requirement of $0.2 < Q_p < 0.6$, which avoids the detrimental consequence of the sampling effects. Now, the conservative design guideline of $0.1 < \omega_{ci}/\omega_s < 0.3$ is relaxed to a less restrictive condition of $0.1 < \omega_{ci}/\omega_s < 0.65$ with the knowledge of the sampling effects of current mode control.

The voltage loop designs in the two approaches yield identical results, given that the loop gain crossover frequency of the new design is the same as the crossover frequency of the outer loop gain in the classical design.

In conclusion, either the classical design or the new design can be adapted for the three basic PWM converters, as long as the design constraint $0.1 < \omega_{ci}/\omega_s < 0.65$ is used in the classical design, or the equivalent condition $0.2 < Q_p < 1.3$ is incorporated in the new design. For this case, both the new and classical designs offer stability and good closed-loop performance. Furthermore, there is an exact one-to-one correspondence, matched by the condition $\omega_{ci}/\omega_s = Q_p/2$, between the two designs. For example, the classical design with $\omega_{ci}/\omega_s = 0.5$ is interpreted as the new design with $Q_p = 1$ because they are in fact the same design.

10.4 OFF-LINE FLYBACK CONVERTER WITH OPTOCOUPLER ISOLATED CURRENT MODE CONTROL

This section deals with the design and analysis of an optocoupler-isolated peak current mode control. A flyback converter, which functions as a rear-end dc-to-dc converter in an off-line power supply, is used to illustrate the details involved with practical applications of the peak current mode control.

10.4.1 Off-Line Power Supplies

Off-line power supplies are the power converters that receive an ac voltage from the utility line and provide a dc voltage to the load. Figure 10.35 is the functional diagram of a typical off-line power supply. The system consists of an ac-to-dc rectifier, bulk

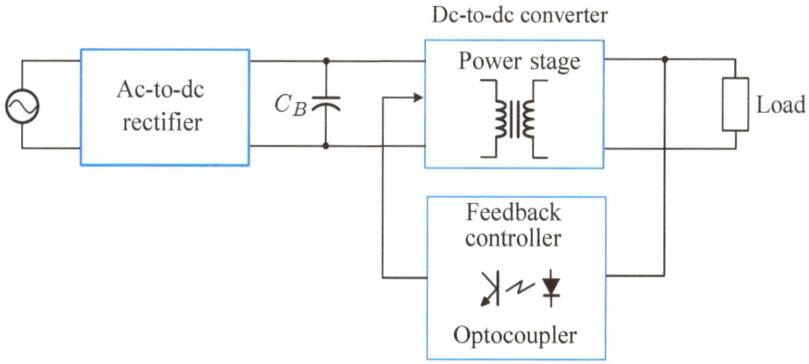

Figure 10.35 Structure of off-line power supplies.

capacitor C_B, and dc-to-dc converter. The front-end ac-to-dc rectifier converts the ac utility voltage into an intermediate dc voltage. The rear-end dc-to-dc converter then alters the intermediate voltage into the voltage level required by the load.

The ac-to-dc rectifier comes in a wide variety in structure and function, ranging from simple line-frequency rectifiers to sophisticated power factor corrected (PFC) ac-to-dc converters. An introduction to a PFC ac-to-dc converter was given in Section 9.1.3. A large capacitor C_B is connected at the output of the ac-to-dc rectifier to stabilize the intermediate dc voltage.

The rear-end dc-to-dc converter should offer galvanic isolation for practical reasons such as safety. The isolation should be provided in both the power stage and feedback controller. For the power stage isolation, one of the transformer-isolated topologies is employed. For the feedback isolation, the controller typically employs an optocoupler which couples the infrared diode and photo-transistor in order to transmit the control signal with isolation. This optocoupler-isolated feedback control is widely adapted to cost-sensitive applications such as off-line power supplies for consumer electronics.

10.4.2 Current Mode Control for Flyback Converter with Optocoupler-Isolated Feedback

The flyback converter is widely used as a rear-end dc-to-dc converter for off-line power supplies, because the converter provides the power stage isolation with minimal component count. The off-line flyback converter usually adapts the peak current mode control using the optocoupler-isolated feedback technique.

Figure 10.36 shows the conceptual circuit diagram of an off-line power supply configured using a flyback converter with an optocoupler-isolated feedback control. A simple bridge rectifier converts the ac line voltage to an intermediate dc voltage for the flyback converter downstream. An EMI filter is placed in front of the bridge rectifier and a bulk capacitor C_B is employed at the output of the rectifier. The flyback

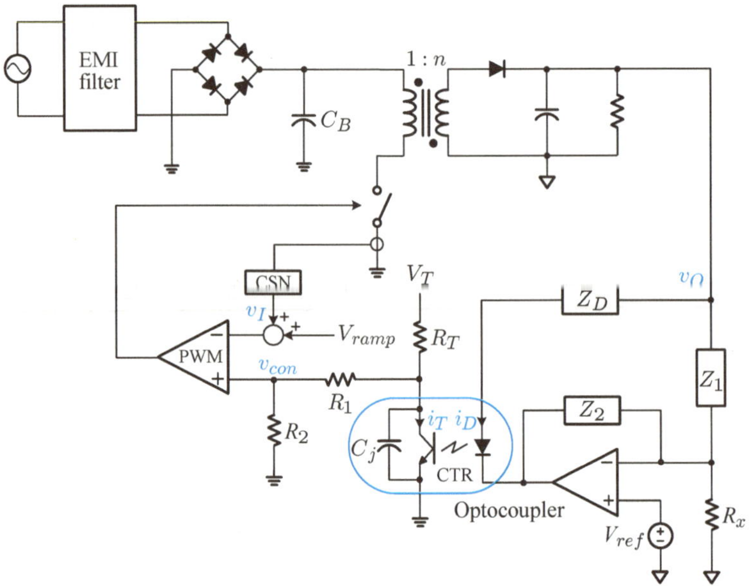

Figure 10.36 Optocoupler-isolated current-mode controlled flyback converter.

converter employs the peak current mode control, implemented with an optocoupler-isolated feedback controller.

Optocoupler-Isolated Feedback Circuit

The flyback converter adopts the standard optocoupler-isolated feedback circuit, illustrated in Fig. 10.36. The two gain blocks, $Z_1(s)$ and $Z_2(s)$, denote the impedances of the passive circuit components in the voltage feedback circuit. The resistor R_x is employed to adjust the output voltage V_O with a fixed reference voltage V_{ref}: $V_O = V_{ref}(1 + Z_1(j0)/R_x)$. The infrared diode in the optocoupler is powered by the converter's output voltage v_O through the current limiting circuit, represented by its impedance $Z_D(s)$ in Fig. 10.36. The $Z_D(s)$ also functions as an element of the voltage feedback circuit. The dc voltage V_T and resistance R_T allow the phototransistor in the optocoupler to deliver the current i_T in proportion to the infrared diode current i_D.

Peak Current Mode Control

The peak current mode control is adapted to the optocoupler-isolated feedback circuit, as shown in Fig. 10.36. The switch current is sensed with the current sensing network (CSN) and combined with the compensation ramp V_{ramp}. The resulting signal is compared with the output of the optocoupler-isolated feedback circuit, v_{con}, in order to generate the PWM output. The techniques for the analysis and design of

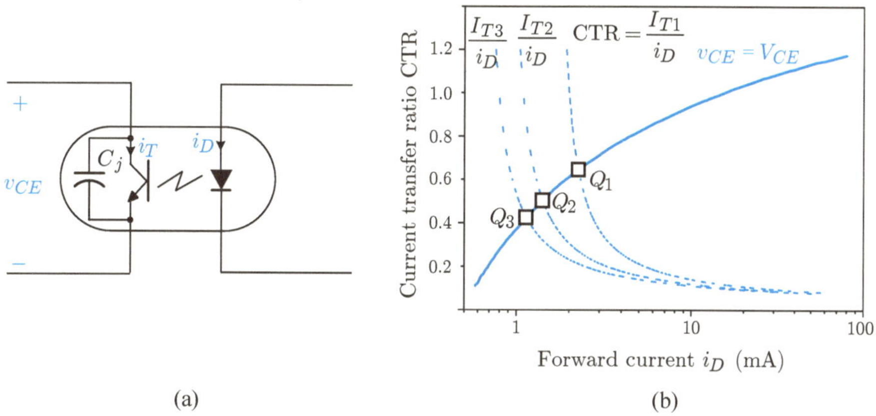

Figure 10.37 Optocoupler-isolated current mode control. (a) Optocoupler. (b) Operating point.

the conventional peak current mode control are extended to Fig. 10.36 with minor modifications, as will be illustrated in later sections.

The electrical characteristics of the optocoupler are specified by two parameters. The first is the ratio of the photo-transistor current i_T to the infrared diode current i_D. This ratio is referred to as the current transfer ratio, $\mathrm{CTR} \equiv i_T/i_D$. The other parameter is the parasitic capacitance between the collector and emitter terminals of the photo-transistor, denoted as C_j. These two parameters are largely affected by the operational conditions. Thus, the actual values for CTR and C_j need to be evaluated at the given bias point of the optocoupler.

Operating Point and CTR of Optocoupler As the prerequisite to assess the electrical parameters, the operating point of an optocoupler should be first identified. The two optocoupler parameters depend on the collector-emitter voltage v_{CE} and the collector current i_T of the photo transistor. Figure 10.37(a) shows v_{CE} and i_T on the circuit symbol of an optocoupler.

In most optocoupler-isolated feedback circuits, the dc values of v_{CE} and i_T are prefixed at the specific values, required to produce the steady-state duty ratio of the converter. Once V_{CE} and I_T are fixed, the values of CTR and C_j are then determined from the current transfer characteristics of the optocoupler.

First, we determine the CTR of the optocoupler, based on the given V_{CE}, I_T, and the transfer characteristics of the optocoupler. Figure 10.37(b) shows the current transfer curve of an optocoupler, given by the diode current vs the current transfer ratio (CTR) curve at the given V_{CE}. The current transfer curve is very sensitive to the V_{CE} value. Thus, the current transfer curve at the given V_{CE} should be obtained from the data sheet or experimentally measured. Figure 10.37(b) is such an example.

The transistor current i_T and the diode current i_D is related by the equation of $i_T = \mathrm{CTR}\, i_D$. Figure 10.37(b) illustrates how the operating point is determined

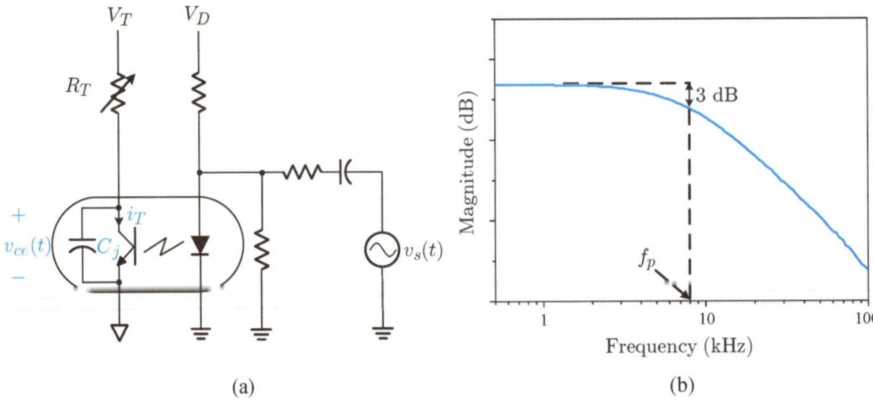

Figure 10.38 Measurement of optocoupler parameters. (a) Measurement circuit. (b) Measurement result.

when i_T and i_D – CTR curve at the given V_{CE} are known. From a given value of $i_T = I_T$, the operating point of the optocoupler is found by over-plotting the relation of

$$\text{CTR} = \frac{I_T}{i_D} \tag{10.99}$$

on the i_D – CTR curve. Examples of the operating points, Q_1, Q_2, and Q_3, with the three different i_T values, I_{T1}, I_{T2}, and I_{T3}, are illustrated in Fig. 10.37(b). Each operating point simultaneously determines the required diode current I_D and the resulting current transfer ratio CTR for the predetermined transistor current $i_T = I_{T1,2,3}$ at the given V_{CE}.

Measurement of C_j Figure 10.38(a) is an experimental circuit to measure the parasitic capacitor C_j of the optocoupler at the given operating point of $i_T = I_T$ and $v_{CE} = V_{CE}$. The resistance R_T and the source voltage V_T are selected to result in the predetermined transistor current I_T and collector-emitter voltage V_{CE}. Now, the small-signal source $v_s(t)$ is activated and the frequency response of $|v_{ce}(j\omega)|/|v_s(j\omega)|$ is recorded using an impedance analyzer. Figure 10.38(b) shows an example of such measurements. The frequency response exhibits the first-order low-pass filter characteristics, whose corner frequency is given by

$$f_p = \frac{1}{2\pi C_j R_T} \tag{10.100}$$

under the assumption that $\omega_T = 1/(C_j R_T)$ is a dominant pole in the circuit. This equation can be used to determine the value of C_j with a predetermined R_T.

Small-Signal Model

As discussed in Chapter 5, the flyback converter is transformed into an equivalent buck/boost converter. Figure 10.39 shows the steps of the circuit transformation described in Section 5.5.3.

Figure 10.40 shows the small-signal model of the flyback converter, obtained by merging the small-signal model of the equivalent buck/boost converter and s-domain model for current mode control. In the small-signal model, the CSN gain is given by

$$R_i' = nR_i \tag{10.101}$$

where R_i is the CSN gain of the original flyback circuit and n is the turns ratio of the transformer. This modification is necessary because the current feedback signal is extracted from the secondary side of the transformer in the small-signal model, while the current feedback is actually employed to the primary side of the transformer in the original flyback circuit. The gain block $F_v(s)$ represents the transfer function of the optocoupler feedback circuit.

Optocoupler-Isolated Feedback Circuit

The optocoupler feedback circuit in Fig. 10.36 is redrawn in Fig. 10.41(a). This circuit can be configured into the two-pole one-zero circuit

$$F_v(s) = \frac{K_v\left(1+\dfrac{s}{\omega_{zc}}\right)}{s\left(1+\dfrac{s}{\omega_{pc}}\right)} \tag{10.102}$$

or three-pole two-zero circuit

$$F_v'(s) = \frac{K_v\left(1+\dfrac{s}{\omega_{zc}}\right)\left(1+\dfrac{s}{\omega_{zc}'}\right)}{s\left(1+\dfrac{s}{\omega_{pc}}\right)\left(1+\dfrac{s}{\omega_{pc}'}\right)} \tag{10.103}$$

by appropriately selecting the impedances, $Z_D(s)$, $Z_1(s)$, and $Z_2(s)$. Figure 10.41(b) shows the circuit implementation of the two-pole one-zero circuit, while Fig. 10.41(c) depicts the three-pole two-zero circuit. The expressions for the dc gain and corner frequencies of the two transfer functions are shown in Table 10.5. Derivations of the equations (10.102) and (10.103) are given in the following example.

Analysis of Optocoupler-Isolated Feedback Circuit — EXAMPLE 10.13

This example shows the derivation of the transfer function of the optocoupler-isolated feedback circuit. Referring to Fig. 10.41(a), the voltage feedback compensation is given by

$$F_v(s) = -\frac{\hat{v}_{con}(s)}{\hat{v}_o(s)} = -\frac{\hat{\imath}_D(s)}{\hat{v}_o(s)}\frac{\hat{\imath}_T(s)}{\hat{\imath}_D(s)}\frac{\hat{v}_{con}(s)}{\hat{\imath}_T(s)} \tag{10.104}$$

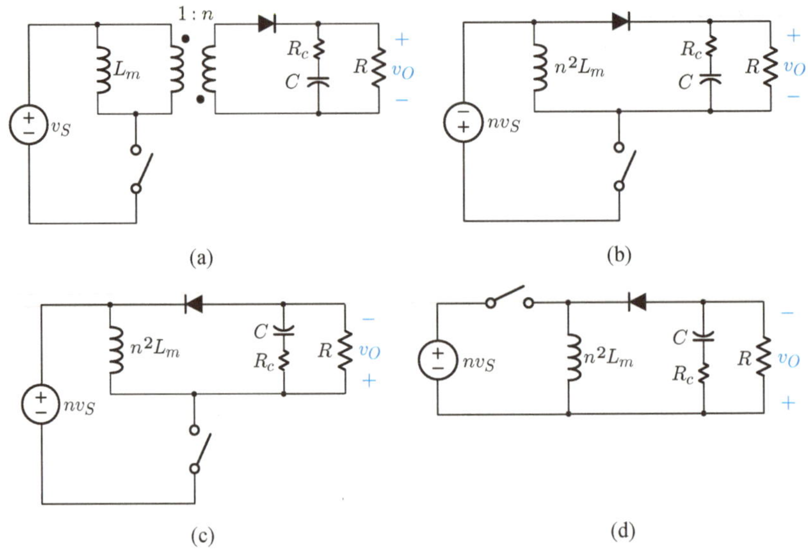

Figure 10.39 Simplification of power stage circuit. (a) Flyback converter. (b) Reflection of input voltage and magnetizing inductance. (c) Modified circuit. (d) Equivalent buck/boost converter.

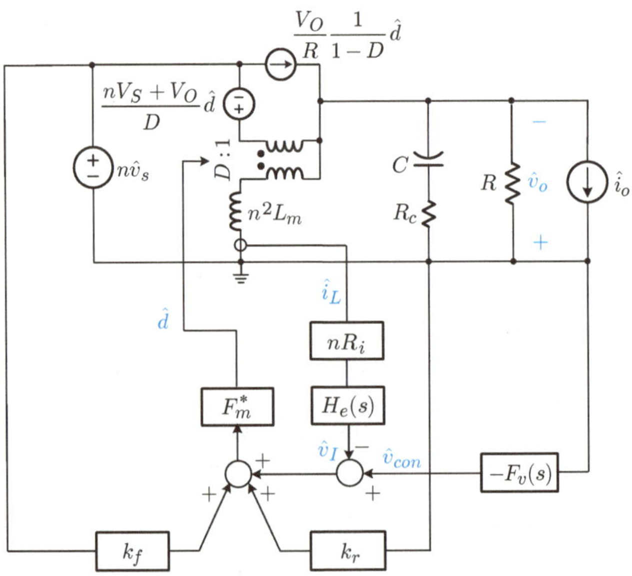

Figure 10.40 Small-signal model of flyback converter.

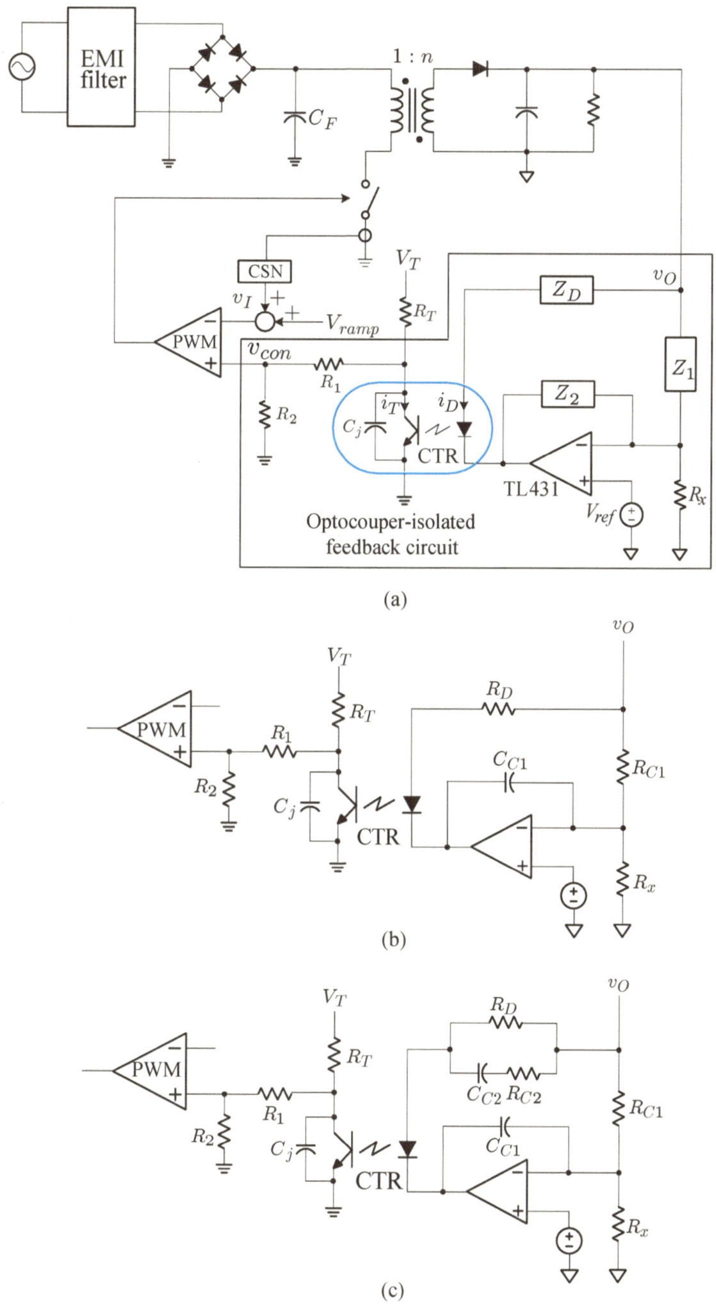

Figure 10.41 Optocoupler-isolated feedback circuit. (a) Voltage feedback circuit. (b) Two-pole one-zero circuit. (c) Three-pole two-zero circuit.

The first term in the right-hand side of (10.104) is evaluated as follows. By recognizing that the inverting terminal of the op amp is the virtual ground for the ac signal, the relationship between $\hat{v}_o$ and $\hat{\imath}_D$ is given by

$$\hat{\imath}_D(s) = \frac{\hat{v}_o(s) - \left(-\frac{Z_2(s)}{Z_1(s)}\right)\hat{v}_o(s)}{Z_D(s)} \tag{10.105}$$

which is rearranged as

$$\frac{\hat{\imath}_D(s)}{v_o(s)} = \frac{1}{Z_D(s)}\left(1 + \frac{Z_2(s)}{Z_1(s)}\right) \tag{10.106}$$

The second term in (10.104) is the current transfer ratio CTR of the optocoupler.

By considering the photo-transistor in Fig. 10.41(a) as a small-signal current source $\hat{\imath}_T$, the following relationship is formulated

$$\hat{v}_{con}(s) = -\left(\frac{1}{sC_j} \parallel R_T \parallel (R_1 + R_2)\right)\left(\frac{R_2}{R_1 + R_2}\right)\hat{\imath}_T(s) \tag{10.107}$$

yielding the description

$$\frac{\hat{v}_{con}(s)}{\hat{\imath}_T(s)} = -\left(\frac{R_T \parallel (R_1 + R_2)}{1 + sC_j(R_T \parallel (R_1 + R_2))}\right)\left(\frac{R_2}{R_1 + R_2}\right) \tag{10.108}$$

By combining (10.106) and (10.108), the transfer function of the feedback circuit is given by

$$\begin{aligned} F_v(s) &= \frac{R_2}{R_1 + R_2}\frac{R_T \parallel (R_1 + R_2)}{Z_D(s)}\text{CTR} \\ &\quad \cdot \left(1 + \frac{Z_2(s)}{Z_1(s)}\right)\frac{1}{1 + sC_j(R_T \parallel (R_1 + R_2))} \end{aligned} \tag{10.109}$$

The above transfer function can be converted into either the two-pole one-zero or three-pole two-zero circuit. Figure 10.41(b) becomes the two-pole one-zero compensation with the selection of $Z_D(s) = R_D$, $Z_1(s) = R_{C1}$, and $Z_2(s) = 1/(sC_{C1})$. The resulting transfer function is denoted as $F_v(s)$ in Table 10.5. In the circuit in Fig. 10.41(c), where $Z_D(s) = R_D \parallel \left(R_{C2} + 1/(sC_{C2})\right)$, $Z_1(s) = R_{C1}$, and $Z_2(s) = 1/(sC_{C1})$, a pole-zero pair is added to the previous two-pole one-zero compensation, thus yielding the three-pole two-zero compensation. This transfer function is described as $F_v'(s)$ in Table 10.5.

Control Design Procedures

The control design procedures established in the previous section are now applied to the small-signal model of Fig. 10.40. Detailed design steps are illustrated in the following example.

EXAMPLE 10.14 Current Mode Control for Off-Line Flyback Converter

Figure 10.42 shows the circuit diagram of a flyback converter which employs peak current mode control using the optocoupler-isolated feedback.

Table 10.5 Dc Gain and Corner Frequencies of Optocoupler-Isolated Feedback Circuit

Transfer function	
$F_v(s) = \dfrac{K_v\left(1+\dfrac{s}{\omega_{zc}}\right)}{s\left(1+\dfrac{s}{\omega_{pc}}\right)}$	$F_v'(s) = \dfrac{K_v\left(1+\dfrac{s}{\omega_{zc}}\right)\left(1+\dfrac{s}{\omega_{zc}'}\right)}{s\left(1+\dfrac{s}{\omega_{pc}}\right)\left(1+\dfrac{s}{\omega_{pc}'}\right)}$
$K_v = \dfrac{R_2}{R_1+R_2}\dfrac{R_T \parallel (R_1+R_2)}{R_D}\text{CTR}\dfrac{1}{C_{C1}R_{C1}}$	
$\omega_{pc} = \dfrac{1}{C_j\big(R_T \parallel (R_1+R_2)\big)}$	$\omega_{pc}' = \dfrac{1}{C_{C2}R_{C2}}$
$\omega_{zc} = \dfrac{1}{C_{C1}R_{C1}}$	$\omega_{zc}' = \dfrac{1}{C_{C2}(R_{C2}+R_D)}$

The feedback controller is implemented using a PWM chip, optocoupler, and op amp, along with other passive components. A simple resistive sensing is used for a low-cost CSN implementation. The switching frequency of the flyback converter is $\omega_s = 2\pi \cdot 65 \times 10^3$ rad/s and the turns ratio of the transformer is $n = 6/62 = 0.097$. From the information given in Fig. 10.42, the small-signal parameters and operational conditions are determined as follows.

$$
\begin{aligned}
\frac{V_O}{V_S} &= \frac{D}{1-D}n \\
&\Rightarrow \quad \frac{10}{220\sqrt{2}} = \frac{D}{1-D}0.097 \quad \Rightarrow \quad D = 0.25 \\
\omega_o &= \frac{1-D}{\sqrt{n^2 L_m C}} \\
&= \frac{1-0.25}{\sqrt{0.097^2\, 1.5\times10^{-3}\, 910\times10^{-6}}} = 2\pi\cdot 1.05\times10^3 \text{ rad/s} \\
\omega_{rhp} &= \frac{(1-D)^2}{D}\frac{R}{n^2 L_m} \\
&= \frac{(1-0.25)^2}{0.25}\frac{2}{0.097^2\, 1.5\times10^{-3}} = 2\pi\cdot 5.07\times10^4 \text{ rad/s} \\
\omega_{esr} &= \frac{1}{CR_c} = \frac{1}{910\times10^{-6}\, 0.04} = 2\pi\cdot 4.37\times10^3 \text{ rad/s}
\end{aligned}
$$

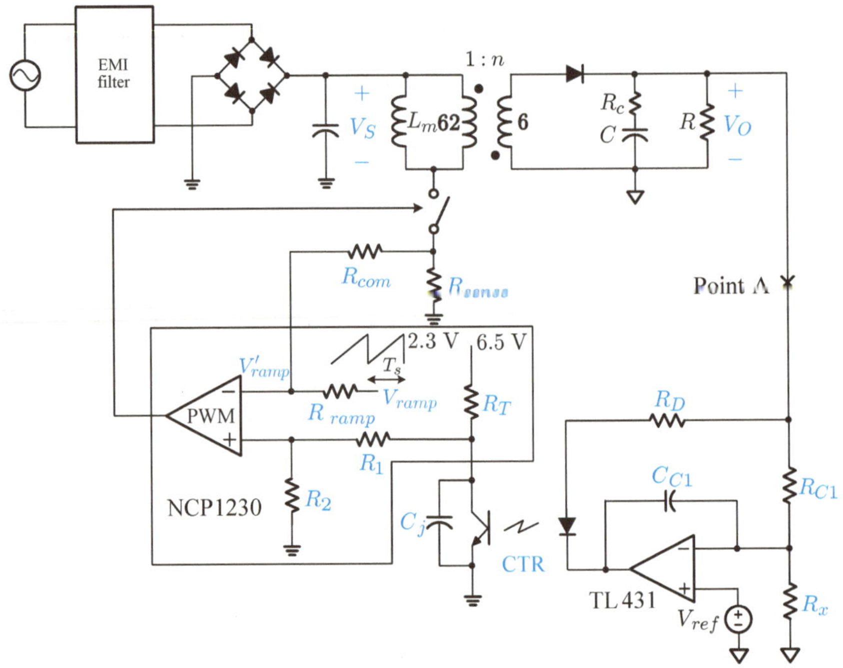

Figure 10.42 Off-line flyback converter with optocoupler-isolated peak current mode control: $V_S = 220\sqrt{2}$ V, $V_O = 10$ V, $L_m = 1.5$ mH, $n = 6/62 = 0.097$, $C = 910\ \mu$F, $R_c = 0.04\ \Omega$, $R = 2\ \Omega$, $R_D = 393\ \Omega$, $R_{C1} = 2$ kΩ, $C_{C1} = 0.1\ \mu$F, $V_{ref} = 2.5$ V, $R_x = 667\ \Omega$, CTR $= 0.236$, $C_j = 2.3$ nF, $R_T = 20$ kΩ, $R_1 = 55$ kΩ, $R_2 = 25$ kΩ, $R_{com} = 6.4$ kΩ, $R_{sense} = 0.67\ \Omega$, $R_{ramp} = 18$ kΩ, $V_{ramp} = 2.3$ V, $T_s = (65 \times 10^3)^{-1}$ s, PWM chip: NCP1230® from On Semiconductor, and op amp: TL431.

The maximum value of the switch current is evaluated as

$$
\begin{aligned}
i_{Q\,max} &= n\frac{V_O}{R}\frac{1}{1-D} + \frac{1}{2}\frac{V_S}{L_m}DT_s \\
&= 0.097\frac{10}{2}\frac{1}{1-0.25} + \frac{1}{2}\frac{220\sqrt{2}}{1.5\times 10^{-3}}0.25\frac{1}{65\times 10^3} \\
&= 1.05\ \text{A}
\end{aligned}
$$

The maximum input voltage for the PWM block is given as $V_{max} = 1.0$ V. The four resistances inside the PWM chip are prefixed: $R_{ramp} = 18$ kΩ, $R_T = 20$ kΩ, $R_1 = 55$ kΩ, and $R_2 = 25$ kΩ. In addition, the PWM chip provides a ramp signal whose magnitude is fixed at $V_{ramp} = 2.3$ V. The ramp signal is divided by the three resistors, R_{ramp}, R_{com}, and R_{sense}, and the resulting signal is applied to the PWM block as the compensation ramp V'_{ramp}.

Current Loop Design

The current loop design involves the selection of resistances for R_{com} and R_{sense} in compliance with the design specification, while using the predetermined conditions, $R_{ramp} = 18\ \text{k}\Omega$ and $V_{ramp} = 2.3$ V.

1) CSN gain:

$$R_i < \frac{V_{max}}{i_{Q\,max}} = \frac{1.0}{1.05} = 0.95 \quad \Rightarrow R_i = 0.5$$

2) Damping ratio of double pole:

$$Q_p = \frac{1}{\pi\left(\left(1+\dfrac{S_e}{S_n}\right)D' - 0.5\right)} = 0.6$$

$$S_n = \frac{nV_S}{n^2 L_m} nR_i = \frac{0.097 \cdot 220\sqrt{2}}{0.097^2\ 1.5 \times 10^{-3}} 0.097 \cdot 0.5 = 1.04 \times 10^5\ \text{V/s}$$

$$\Rightarrow \quad S_e = S_n\left(\frac{\dfrac{1}{\pi\ Q_p} + 0.5}{D'} - 1\right)$$

$$= 1.04 \times 10^5\left(\frac{\dfrac{1}{\pi\ 0.6} + 0.5}{0.75} - 1\right) = 3.89 \times 10^4\ \text{V/s}$$

$$\Rightarrow \quad V'_{ramp} = S_e T_s = 3.89 \times 10^4 \frac{1}{65 \times 10^3} = 0.6\ \text{V}$$

$$m_c = 1 + \frac{S_e}{S_n} = 1 + \frac{3.89 \times 10^4}{1.04 \times 10^5} = 1.37$$

where V'_{ramp} denotes the magnitude of the actual compensation ramp applied to the input of the PWM block.

3) Selection of R_{com} and R_{sense} with the assumptions of $R_{sense} \ll R_{ramp}$ and $R_{sense} \ll R_{com}$:

$$V'_{ramp} = V_{ramp}\frac{R_{com} + R_{sense}}{R_{ramp} + R_{com} + R_{sense}} \approx V_{ramp}\frac{R_{com}}{R_{ramp} + R_{com}}$$

$$\Rightarrow 0.6 = 2.3\frac{R_{com}}{18 \times 10^3 + R_{com}}$$

$$\Rightarrow R_{com} = 6.4\ \text{k}\Omega$$

$$R_i = \frac{R_{ramp} + R_{com}}{R_{ramp} + R_{com} + R_{sense}} R_{sense} \frac{R_{ramp}}{R_{ramp} + R_{com}}$$

$$= \frac{R_{sense}}{R_{ramp} + R_{com} + R_{sense}} R_{ramp} \approx \frac{R_{sense}}{R_{ramp} + R_{com}} R_{ramp}$$

$$\Rightarrow \quad 0.5 = \frac{R_{sense}}{18 \times 10^3 + 6.4 \times 10^3} 18 \times 10^3$$

$$\Rightarrow \quad R_{sense} = 0.68\ \Omega$$

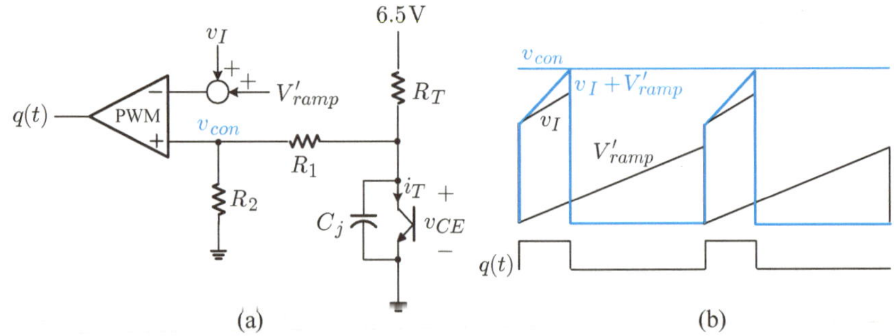

Figure 10.43 PWM block and waveform. (a) PWM block diagram. (b) PWM waveforms.

Optocoupler Parameter Assessment

As previously addressed, the optocoupler parameters need to be calculated in consideration of the operational conditions of the converter. Figure 10.43(a) is a functional representation of the PWM block. The current feedback signal v_I is summed with the compensation ramp V'_{ramp}. The resulting signal is then compared with the control voltage v_{con} to yield the predetermined duty ratio. The waveforms of the PWM block are illustrated in Fig. 10.43(b).

We first need to evaluate the steady-state values of the collector-emitter voltage V_{CE} and the collector current I_T of the photo transistor. Once the V_{CE} and I_T are determined, the current transfer ratio (CTR) and the parasitic capacitance C_j of the optocoupler are evaluated as shown in Figs. 10.37 and 10.38. Referring to Fig. 10.43, the values of V_{CE} and I_T, along with the optocoupler parameters, CTR *and* C_j, are evaluated as described below.

1) Steady-state control voltage V_{con}:

$$V_{con} = D\,V'_{ramp} + R_i\, i_{Q\,max} = 0.25 \cdot 0.6 + 0.5 \cdot 1.05 = 0.68\text{V}$$

2) Transistor collector-emitter voltage V_{CE}:

$$\begin{aligned} V_{con} &= \frac{R_2}{R_1 + R_2} V_{CE} \\ &\Rightarrow \quad 0.68 = \frac{25 \times 10^3}{55 \times 10^3 + 25 \times 10^3} V_{CE} \qquad \Rightarrow V_{CE} = 2.16 \text{ V} \end{aligned}$$

3) Transistor current I_T:

$$I_T = \frac{6.5 - V_{CE}}{R_T} - \frac{V_{con}}{R_2} = \frac{6.5 - 2.16}{20 \times 10^3} - \frac{0.68}{25 \times 10^3} = 1.9 \times 10^{-4} \text{ A}$$

4) Optocoupler parameters: With the collector-emitter voltage of $V_{CE} = 2.16$ V and the transistor current of $I_T = 1.9 \times 10^{-4}$, the optocoupler parameters are found as CTR $= 0.236$ and $C_j = 1.2$ nF, based on the procedures described in Figs. 10.37 and 10.38.

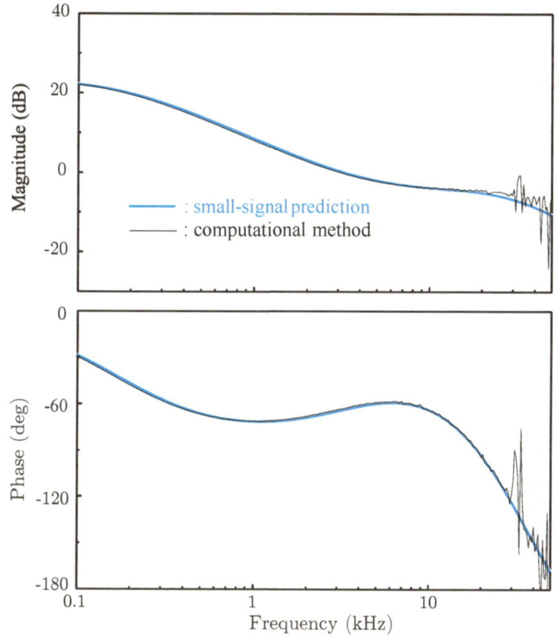

Figure 10.44 Control-to-output transfer function with current loop closed.

Control-to-Output Transfer Function with Current Loop Closed

Referring to Table 10.3, the control-to-output transfer function of the flyback converter with the current loop closed is given by

$$G_{vci}(s) = K_{vc}\frac{\left(1-\dfrac{s}{\omega_{rhp}}\right)\left(1+\dfrac{s}{\omega_{esr}}\right)}{\left(1+\dfrac{s}{\omega_{pl}}\right)\left(1+\dfrac{s}{Q_p\omega_n}+\dfrac{s^2}{\omega_n^2}\right)} \qquad (10.110)$$

where

$$\omega_n = \frac{\omega_s}{2} = 2\pi\cdot 32.5\times 10^3 \text{ rad/s}$$

$$\begin{aligned} K_{vc} &= \frac{D'R}{(1+D)nR_i}\frac{1}{1+\dfrac{D'^3RT_s}{(1+D)n^2L_m}(m_c-0.5)} \\ &= \frac{0.75\cdot 2}{(1+0.25)0.097\cdot 0.5} \\ &\quad\times\frac{1}{1+\dfrac{0.75^3\,2\,(65\times 10^3)^{-1}}{(1+0.25)0.097^2\,1.5\times 10^{-3}}(1.37-0.5)} = 15.1 \end{aligned}$$

$$
\begin{aligned}
\omega_{pl} &= \frac{1+D}{CR} + \frac{T_s D'^3}{n^2 L_m C}(m_c - 0.5) \\
&= \frac{1+0.25}{910 \times 10^{-6}\, 2} \\
&\quad + \frac{(65 \times 10^3)^{-1}\, 0.75^3}{0.097^2\, 1.5 \times 10^{-3}\, 910 \times 10^{-6}}(1.37 - 0.5) \\
&= 2\pi \cdot 179 \text{ rad/s}
\end{aligned}
$$

Figure 10.44 shows the control-to-output transfer function evaluated with only the current loop closed. The theoretical prediction is shown in parallel with the result of the computational method.

Voltage Loop Design

The voltage feedback circuit adopts the two-pole one-zero compensation shown in Fig. 10.41(b)

$$
F_v(s) = \frac{K_v\left(1 + \dfrac{s}{\omega_{zc}}\right)}{s\left(1 + \dfrac{s}{\omega_{pc}}\right)} \tag{10.111}
$$

where

$$
\omega_{pc} = \frac{1}{C_j\,(R_T \parallel (R_1 + R_2))} \tag{10.112}
$$

$$
\omega_{zc} = \frac{1}{C_{C1} R_{C1}} \tag{10.113}
$$

$$
K_v = \frac{R_2}{R_1 + R_2}\frac{R_T \parallel (R_1 + R_2)}{R_D}\,\text{CTR}\,\frac{1}{C_{C1} R_{C1}} \tag{10.114}
$$

1) Compensation pole: $\omega_{pc} = \min\{\omega_{rhp}\ \omega_{esr}\ 0.5\,\omega_s\} = \omega_{esr} = 2\pi \cdot 4.37 \times 10^3$ rad/s. By evaluating (10.112) with the given resistive parameters, the desired capacitance is determined as $C_j = 2.3$ nF. The collector-emitter parasitic capacitance of the optocoupler was measured as 1.2 nF. Thus, an external capacitance $C_{ext} = 1.1$ nF is added across the collector and emitter terminals in order to obtain the required capacitance $C_j = 2.3$ nF.
2) Compensation zero: $\omega_{zc} = 0.76\,\omega_o = 2\pi \cdot 800$ rad/s. One circuit parameter is initially selected as $R_{C1} = 2$ kΩ. Then, $C_{C1} = 0.1$ μF is determined from (10.113). Finally, $R_x = 667$ Ω is selected to produce the desired output voltage $V_O = 10$ V with $V_{ref} = 2.5$ V: $V_O = V_{ref}(1 + R_{C1}/R_x) \Rightarrow 10 = 2.5\,(1 + 2 \times 10^3/667)$.
3) T_m crossover frequency: $\omega_{cr} = 0.16\,\omega_{rhp} = 2\pi \cdot 8.12 \times 10^3$ rad/s
4) Integrator gain: $K_v = \dfrac{\omega_{zc}\,\omega_{cr}}{K_{vc}\,\omega_{pl}} = \dfrac{(2\pi \cdot 800)\,(2\pi \cdot 8.12 \times 10^3)}{15.1\,(2\pi \cdot 179)} = 1.51 \times 10^4$

The current transfer ratio of the optocoupler was found as CTR = 0.236. Accordingly, the last compensation component is determined as $R_D = 393$ Ω from (10.114).

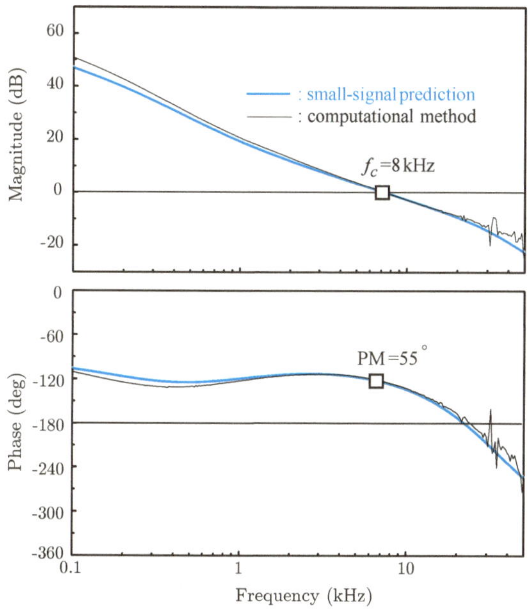

Figure 10.45 Loop gain characteristics of flyback converter.

Performance of Current Mode Control

Figure 10.45 compares the theoretical and computational loop gains. The loop gain should be evaluated at Point A in Fig. 10.42. Point A is the only place that provides the correct loop gain information. The loop gain closely meets the design target of $\omega_{cr} = 2\pi \cdot 8.12 \times 10^3$ rad/s with a 55° phase margin.

Although the loop gain magnitude does not show any sign of the peaking at the Nyquist frequency due to the design goal of $Q_p = 0.6$, the phase characteristics support the existence of the double pole. Figure 10.46 compares the simulated and measured step load responses due to the step changes between $I_o = 10$ A and 5 A in the load current.

Performance in DCM Operation

When the load resistance is increased beyond the critical value

$$R_{crit} = \frac{2n^2 L_M}{(1-D)^2 T_s} = \frac{2 \cdot 0.097^2 \; 1.5 \times 10^{-3}}{(1-0.25)^2 \, (65 \times 10^3)^{-1}} = 3.3 \ \Omega$$

the flyback converter enters DCM operation. Figure 10.47 shows the loop gain characteristics of the converter in DCM operation, in comparison with those of the CCM operation. The DCM loop gains with $R = 3.3\ \Omega$, $10\ \Omega$, and $20\ \Omega$ are compared with the CCM loop gain with $R = 2\ \Omega$. Although the mid-band gain and 0 dB crossover

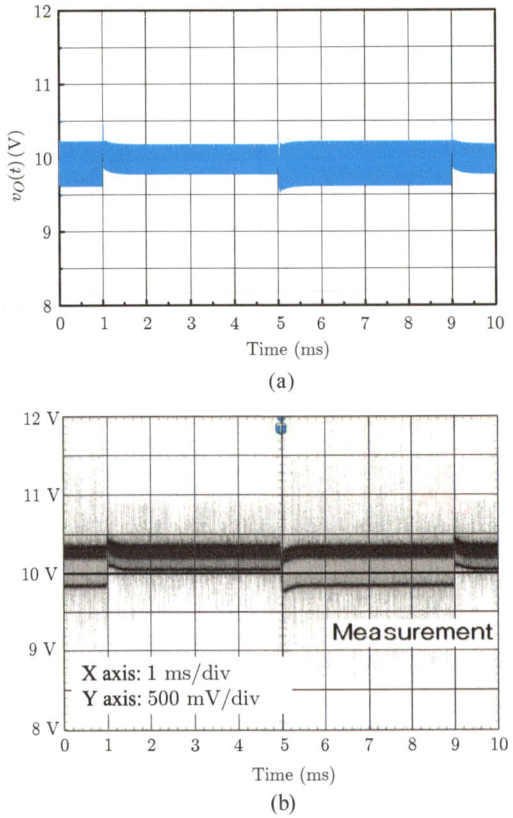

Figure 10.46 Step load response of flyback converter. (a) Simulated response. (b) Measured response.

frequency are reduced, the loop gain curves confirm that the converter maintains stability in DCM operation with sufficient phase margin.

10.5 CHAPTER SUMMARY

The sampling effects of current mode control originate from the fact that the control action is periodically executed by sampling and holding the error signal produced by the fast-varying inductor current. In other words, the PWM modulator samples and holds the error signal at the rate of the switching frequency. The sampling effects, ignored in the preceding classical analysis, need to be duly addressed to accurately describe the high-frequency dynamics of current mode control. This chapter presented comprehensive analyses of the sampling effects using Ridley's s-domain model for current mode control.

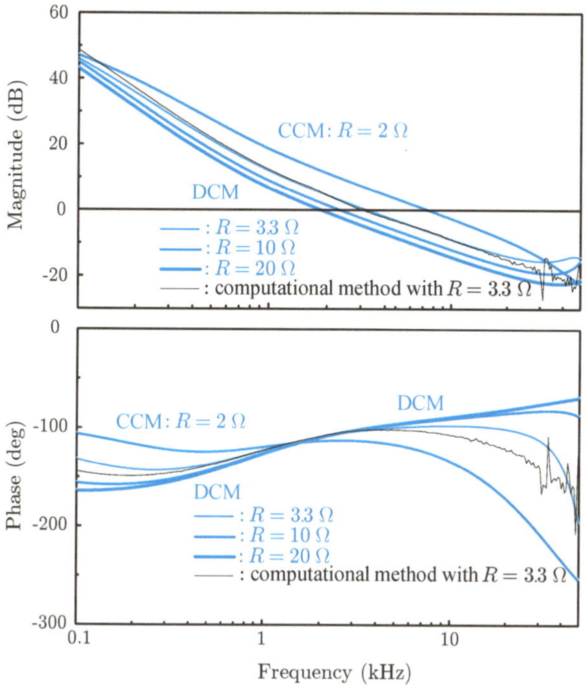

Figure 10.47 Loop gain characteristics of flyback converter in CCM and DCM operations.

The analysis of the sampling effects culminates with the derivation of the control-to-output transfer function evaluated with the current loop closed

$$G_{vci}(s) = K_{vc}\frac{\left(1-\dfrac{s}{\omega_{rhp}}\right)\left(1+\dfrac{s}{\omega_{esr}}\right)}{\left(1+\dfrac{s}{\omega_{pl}}\right)\left(1+\dfrac{s}{Q_p\omega_n}+\dfrac{s^2}{\omega_n^2}\right)}$$

The most distinctive feature of the transfer function is the presence of the quadratic term in the denominator. As an ultimate upshot of the sampling effects, the quadratic term introduces a double pole at half the switching frequency. This term is common to all the three basic PWM converters and offers the foundation for the current loop design for all PWM converters.

The control-to-output transfer function produces the peaking of $20\log Q_p$ at half the switching frequency, $\omega_n = \pi/T_s$. This peaking could cause the loop gain to violate the Nyquist stability criterion, thereby destabilizing the converter. To prevent such a consequence, the quality factor of the quadratic term needs to be controlled between $0.2 < Q_p < 1.3$. The idea of *adequately damping the double pole at half the switching frequency* leads to a simple and practical design methodology for current mode control. These new design procedures are illustrated in Section 10.3.3.

The goal of the new design procedures is to provide a predetermined value for Q_p. On the other hand, the classical design covered in Chapter 9 targets to place the crossover frequency of the current loop ω_{ci} at the desired frequency, thereby achieving a prefixed value for the ratio of the current loop crossover frequency to the switching frequency, ω_{ci}/ω_s. These two apparently different design goals are intimately linked by the equation

$$\frac{\omega_{ci}}{\omega_s} = \frac{Q_p}{2}$$

Therefore, there exists an exact one-to-one match between the new design and classical design. For example, selecting $Q_p = 1$ in the new design is identical to placing the current loop crossover frequency at half the switching frequency in the classical design. This rather surprising fact was also verified in [2] using an entirely different approach.

Details of the new design were addressed in this chapter However, it should be reminded that the classical design is also a valid design approach and yields the same results as those of the new design.

This chapter presented several analysis and design examples of the peak current mode control adapted various of PWM converters, including the off-line flyback converter with an optocoupler-isolated feedback control. The outcomes of these examples can readily be adapted to all types of isolated or non-isolated PWM dc-to-dc converters for real applications.

REFERENCES

1. R. B. Ridley, "A new, continuous-time model for current-mode control," *IEEE Trans. Power Electron.,* vol. 6, no. 2, pp. 271- 280, Apr. 1991.
2. F. D. Tan and R. D. Middlebrook, "A unified model for current-programmed converters," *IEEE Trans. Power Electron.,* vol. 10, no. 4, pp. 397-408, Jul. 1995.
3. G. F. Franklin, J. D. Power, and M. Workman, *Digital Control of Dynamic Systems,* 3rd Ed., Addison-Wesley Longman, Inc., 1998, Chapter 6.
4. R. B. Ridley, *A new small-signal model for current mode control,* Ph. D. Dissertation, Virginia Polytechnic Institute and State University, Blacksburg, VA, Nov. 1990.

PROBLEMS

10.1* In Example 9.3 in Chapter 9, the modulator gain of the peak current mode control was derived from the expression

$$\bar{i}_L(t) = v_{con} - S_e d T_s - \frac{1}{2}(S_n + S_f) d(1-d) T_s$$

with the assumption that the slopes of the inductor current remain unchanged. Now, discard this assumption and introduce the perturbation to both S_n and S_f, in order to derive the feedforward gains from the input voltage/output voltage to duty ratio. Show that the feedforword gains are given by

Buck converter: $k_f = -\dfrac{D(1-D)T_s R_i}{2L}$ $\quad k_r = 0$

Boost converter: $k_f = 0$ $\quad k_r = -\dfrac{D(1-D)T_s R_i}{2L}$

Buck/boost converter: $k_f = -\dfrac{D(1-D)T_s R_i}{2L}$

$$k_r = -\frac{D(1-D)T_s R_i}{2L}$$

The above expressions differ from those of Ridley's model, but they coincide with the feedforward gains of Tan's model, as can be confirmed in Table 10.1.

10.2 Solve the expression (10.28)

$$\frac{\frac{1}{R_i}\frac{1+\alpha}{sT_s}}{1+\frac{1}{R_i}\frac{1+\alpha}{sT_s}R_i H_e(s)} = \frac{1}{R_i}\frac{1+\alpha}{sT_s}\frac{e^{sT_s}-1}{e^{sT_s}+\alpha} \quad \text{with} \quad \alpha = \frac{S_f - S_e}{S_n + S_e}$$

in order to derive the expression for $H_e(s)$

$$H_e(s) = \frac{sT_s}{e^{sT_s}-1}$$

10.3 Modify the procedures given in Example 10.5 to derive the expression for the feedforward gain from the output voltage, given in (10.46)

$$k'_r = \frac{(1-D)^2 T_s R_i}{2L}$$

10.4 The procedures for casting the control-to-output transfer function of current-mode controlled converters into the third-order expression

$$G_{vci}(s) = K_{vc}\frac{\left(1-\dfrac{s}{\omega_{rhp}}\right)\left(1+\dfrac{s}{\omega_{esr}}\right)}{\left(1+\dfrac{s}{\omega_{pl}}\right)\left(1+\dfrac{s}{Q_p\omega_n}+\dfrac{s^2}{\omega_n^2}\right)}$$

are illustrated in Example 10.6.

a) Follow the procedures of Example 10.6 to derive the control-to-output transfer function of the boost converter given in Table 10.3

$$K_{vc} = \frac{D'R}{2R_i}\frac{1}{1+\dfrac{D'^3RT_s}{2L}(m_c-0.5)}$$

$$\omega_{pl} = \frac{2}{CR}+\frac{T_sD'^3}{LC}(m_c-0.5)$$

$$\omega_{rhp} = D'^2\frac{R}{L}$$

Specify the restrictions for the power stage parameters and operational conditions, which will improve the accuracy of the approximations.

b) Repeat a) for the buck/boost converter

$$K_{vc} = \frac{D'R}{(1+D)R_i}\frac{1}{1+\dfrac{D'^3RT_s}{(1+D)L}(m_c-0.5)}$$

$$\omega_{pl} = \frac{1+D}{CR}+\frac{T_sD'^3}{LC}(m_c-0.5)$$

$$\omega_{rhp} = \frac{D'^2}{D}\frac{R}{L}$$

10.5* Adaption of the new design procedures to a buck converter was illustrated in Example 10.8. Redesign the feedback controller for $Q_p = 0.4$, while keeping other design criteria unchanged. Compare the results of your design with those of Example 9.8.

10.6* Redesign the boost converter used in Example 10.9 with $Q_p = 0.32$, while keeping other design criteria the same. Compare the outcomes of your design with those of Example 9.16.

10.7 Example 10.11 proved the relationship

$$\frac{\omega_{ci}}{\omega_s} = \frac{Q_p}{2}$$

for the buck converter. Show that the above relationship also holds true for the boost and buck/boost converters.

10.8 Example 10.12 illustrated the relationship

$$K_{vc}\omega_{pl} = \frac{K_{vd}}{K_{id}R_i}\omega_{id}$$

for the buck converter. Verify that this relationship is also valid for the boost and buck/boost converters.

10.9* Figure P10.9 shows the off-line flyback converter with an optocoupler-isolated peak current mode control. The switching frequency of the converter is $\omega_s = 2\pi \cdot 65 \times 10^3$ rad/s. Perform the control design to meet the specifications of $Q_p = 0.8$ and $\omega_{cr} = 0.1\omega_{rhp}$, while adapting the design guidelines practiced in Example 10.14. Assume CTR = 0.236. Specify values for R_{sense}, R_{com}, C_{C1}, R_D, and C_j.

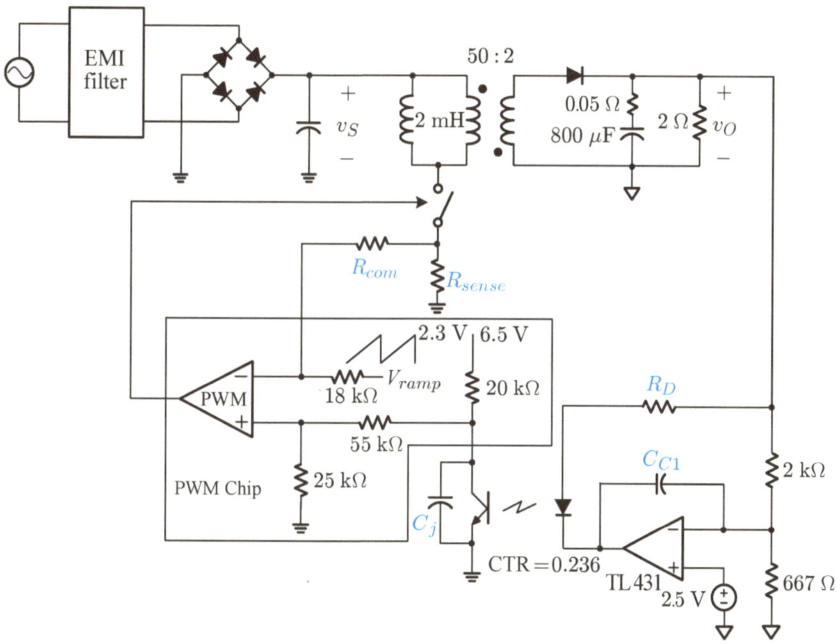

Fig. P10.9

10.10* Shown in Fig. P10.10 is a current-mode controlled buck converter. The switching frequency of the converter is 50 kHz and the output voltage is regulated at 10 V. The output of the PWM block alters between 15 V and 0 V.

a) Draw the small-signal model of the converter. Show the details about the small-signal model including the parameters of the PWM switch model, the numerical values/expressions of the gain blocks of current mode control, and a symbolic expression of the voltage feedback compensation, $F_v(s)$.

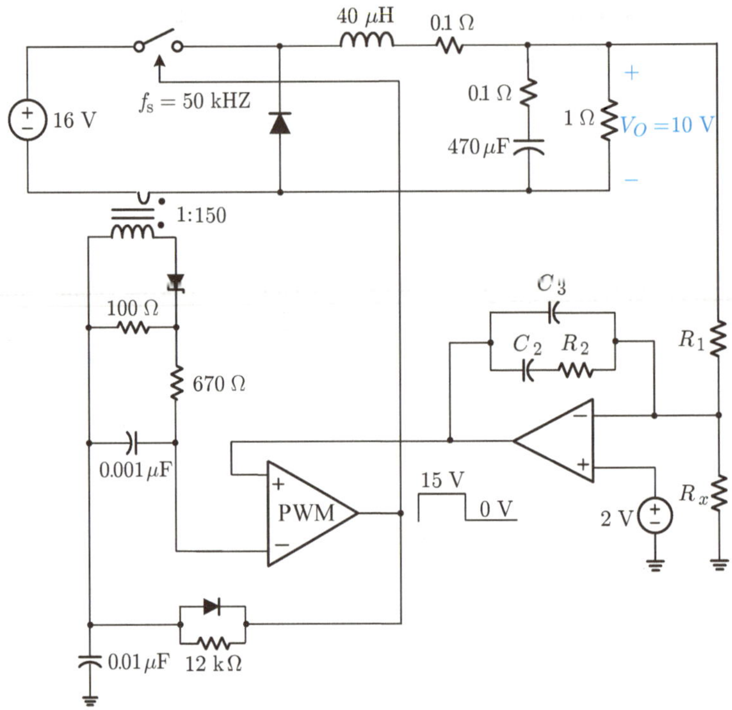

Fig. P10.10

b) Sketch the asymptotic plot of the control-to-output transfer function $|G_{vci}|$. Show the corner frequencies, slopes, asymptotic values, and other important features of $|G_{vci}|$. Also, derive the expression of $G_{vci}(s)$

c) Design the voltage feedback compensation for $\omega_{cr} = 0.5\,\omega_{esr}$. Specify the values for R_1, R_2, C_2, C_3, and R_x.

d) Sketch the asymptotic plot of the loop gain $|T_m|$. Show the frequencies, slopes, and other prominent features of $|T_m|$.

e) Sketch the asymptotic plot of the output impedance $|Z_o|$. Show the corner frequencies, slopes, and asymptotic values of $|Z_o|$. Derive the expression of $Z_o(s)$.

f) Sketch the output voltage v_O in response to a 5 A step decrease in the load current. Assume the loop gain secures a sufficient phase margin and show the prominent features of the waveform.

10.11* The circuit shown in Fig. P10.11 is a voltage feedback circuit for an optocoupler-isolated current mode control.

a) Convert the buck converter in Problem 10.10 into an optocoupler-isolated version using the circuit shown in Fig. P10.11. Show the circuit diagram of the modified converter.

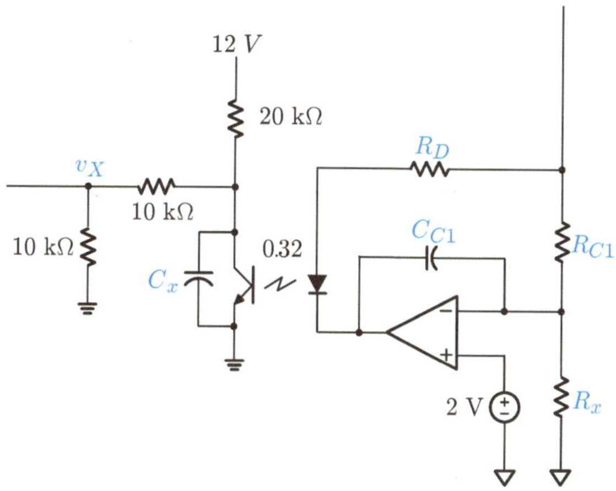

Fig. P10.11

b) Assume CTR=0.32 for the optocoupler. Also assume that the parasitic capacitance of the photo-transistor is so small that an external capacitor C_x is added as a circuit component for the voltage feedback circuit. Determine values of R_{C1}, C_{C1}, R_D, C_x, and R_x so that the modified converter produces the performance you predicted in Problem 10.10.

c) Find the average value of v_X.

APPENDIX A

ANSWERS TO END-OF-CHAPTER PROBLEMS

Note: Some design problems and graphical problems do not have a unique solution. For these problems, the answers in this list are a typical illustration and other solutions are equally acceptable.

CHAPTER 2

2.1 **a)** $i_{L\,max} = 10$ A $i_{L\,min} = -10$ A **b)** $v_{C\,max} = 8$ V $v_{C\,min} = -4$ V

2.2 **a)** Converter A: $V_S = 12$ V Converter B: $V_S = 50$V Converter C: $V_S = 20$V
b) Converter A: $i_{L\,max} = 7.2$ A Converter B: $i_{L\,max} = 12$A Converter C: $i_{L\,max} = 12$ A

2.4 **a)** $L = 40$ mH **b)** $i_{L\,max} = 20$ A **c)** $t = 130$ ms **d)** $19.2 \text{ V} \leq V_X$

2.5 **a)** i) $i_{L\,max} = 1.5$ A $i_{L\,min} = 0$ A ii) $i_{L\,max} = 0.5$ A $i_{L\,min} = 0$ A
b) i) $v_{C\,max} = 1.5$ V $v_{C\,min} = 0$ V ii) $v_{C\,max} = 0.5$ V $v_{C\,min} = 0$ V

2.6 **a)** (a) $v_2/v_1 = 1 - D$ (b) $v_2/v_1 = 1/D$ (c) $v_2/v_1 = -(1 - D)/D$
b) (d) $i_2/i_1 = (1 - D)/D$

2.7 **a)** $D = 0.25$ $T_s = 120\ \mu$s **b)** $D = 0.2$ $T_s = 150\ \mu$s

2.8 **a)** $\bar{v}_C = 50$ V **b)** $V_2 = 20$ V **c)** $\bar{\imath}_{L1} = 4/3$ A
d) $i_{L1\,max} = (4/3 + 1/8)$ A $i_{L1\,min} = (4/3 - 1/8)$ A $i_{L2\,max} = 2.5$ A $i_{L2\,min} = 1.5$ A $i_{C\,max} = (4/3 + 1/8)$ A $i_{C\,min} = -2.5$ A $v_{C\,max} = (50 + 1/3)$ V $v_{C\,min} = (50 - 1/3)$ V $\Delta v_C = 2/3$ V

2.9 **a)** $\bar{v}_C = 60\text{ V}$ $V_2 = 40\text{ V}$
b) $i_{L1\,max} = 5\text{ A}$ $i_{L1\,min} = 3\text{ A}$ $i_{L2\,max} = 4\text{ A}$ $i_{L2\,min} = 0\text{ A}$
c) $i_{C\,max} = 5\text{ A}$ $i_{C\,min} = -4\text{ A}$ $v_{C\,max} = 62\text{ V}$ $v_{C\,min} = 58\text{ V}$

2.10 **a)** $\bar{v}_C = 40\text{ V}$ $V_2 = 20\text{ V}$ **b)** $i_{L1\,max} = 3\text{ A}$ $i_{L1\,min} = -3\text{ A}$ $i_{L2\,max} = 3\text{ A}$
$i_{L2\,min} = -3\text{ A}$ $i_{C\,max} = 3\text{ A}$ $i_{C\,min} = -3\text{ A}$ $v_{C\,max} = 40.56\text{ V}$
$v_{C\,min} = 39.44\text{ V}$

2.12 **a)** $i_P = 1.75\cos t$ $v_S = 2\sin t$
b) $i_P = 2.75\cos t$ $i_m = 1.5\cos t$ $v_m = 2\sin t$ $v_O = 4\sin t$

2.13 **a)** $L_m = 0.5\text{ H}$ $n = 0.25$
b) (a) $i_p = -40\cos t$ (b) $i_p = -42.5\cos t$ (c) $i_p = -40\cos t$
(d) $i_p = (32/3)\sin t$ (e) $i_p = 32\sin t$

2.14 **a)** i) $L = 2.5\text{ mH}$ ii) $n = 5$
b) i) $V_{S\,max} = 120\text{ V}$ ii) $T_{s\,max} = 4\text{ ms}$ iii) $i_{P\,max} = 3\text{ A}$ $i_{P\,min} = -3\text{ A}$
c) $i_{S\,max} = 1\text{ A}$ $i_{S\,min} = -1\text{ A}$ $i_{P\,max} = 3.5\text{ A}$ $i_{P\,min} = -3.5\text{ A}$

2.15 **a)** $L_1 = 2\text{ mH}$ $L_2 = 5\text{ mH}$ $n = 7/5$ **b)** $i'_p = -0.5 \times 10^3 \cos t$

2.16 **a)** $i_{P\,max} = 1.25\text{ A}$ $i_{P\,min} = -1.25\text{ A}$ $i_{S\,max} = 0.5\text{ A}$ $i_{S\,min} = -0.5\text{ A}$
b) $i_{P\,max} = 1.25\text{ A}$ $i_{P\,min} = -1.25\text{ A}$ $i_{S\,max} = 0.5\text{ A}$ $i_{S\,min} = -0.5\text{ A}$

2.17 **a)** $i_{P\,max} = 12\text{ A}$ $i_{P\,min} = 0\text{ A}$ $i_{Q\,max} = 12\text{ A}$ $i_{Q\,min} = 0\text{ A}$
$i_{D\,max} = 12\text{ A}$ $i_{D\,min} = 0\text{ A}$ $v_{S\,max} = 0\text{ V}$ $v_{S\,min} = -48\text{ V}$
b) $i_{P\,max} = 6\text{ A}$ $i_{P\,min} = 0\text{ A}$ $i_{D\,max} = 6\text{ A}$ $i_{D\,min} = 0\text{ A}$ $v_{S\,max} = 48\text{ V}$ $v_{S\,min} = -48\text{ V}$
c) $i_{P\,max} = 6\text{ A}$ $i_{P\,min} = 0\text{ A}$ $i_{D\,max} = 6\text{ A}$ $i_{D\,min} = 0\text{ A}$ $v_{S\,max} = 96\text{ V}$ $v_{S\,min} = -48\text{ V}$

2.18 $i_{S\,max} = 2\text{ A}$ $i_{S\,min} = 0\text{ A}$ $i_{P\,max} = 6\text{ A}$ $i_{P\,min} = 0\text{ A}$
$i_{Q\,max} = 6\text{ A}$ $i_{Q\,min} = 0\text{ A}$ $i_{D\,max} = 6\text{ A}$ $i_{D\,min} = 0\text{ A}$

2.19 **a)** $L = 48\text{ mH}$ $R_D = 5\,\Omega$ **b)** i) 0.6 J ii) 0.6 J iii) 10 W
c) i) $i_{L\,max} = 7.5\text{ A}$ $i_{L\,min} = 0\text{ A}$ $v_{L\,max} = 15\text{ V}$ $v_{L\,min} = -10\text{ V}$
ii) $i_{L\,max} = 10\text{ A}$ $i_{L\,min} = 0\text{ A}$ $v_{L\,max} = 15\text{ V}$ $v_{L\,min} = -5\text{ V}$
Not workable due to saturation of the solenoid inductance

2.20 **a)** i) $i_L = 4e^{-(t-0.01)/0.01}$ ii) 16 W
b) i) $i_{L\,max} = 4\text{ A}$ $i_{L\,min} = 0\text{ A}$ ii) Transferred to the voltage source V_{S2}
c) i) $v_C = 126.5\text{ V}$ ii) $i_{L\,max} = 4\text{ A}$ $v_C\,(200\text{ ms}) = 178.8\text{ V}$
iii) v_C continues to grow until the capacitor explodes.

2.21 (a) $i_{T\,max} = 10\text{ A}$ $i_{T\,min} = 0\text{ A}$ $i_{S\,max} = 10\text{ A}$ $i_{S\,min} = -10\text{ A}$
(b) $i_{T\,max} = 10\text{ A}$ $i_{T\,min} = 5\text{ A}$ $i_{S\,max} = 10\text{ A}$ $i_{S\,min} = -10\text{ A}$

2.22 **a)** $i_{S\,max} = 2\text{ A}$ $i_{S\,min} = -2\text{ A}$ $v_{L\,max} = 40\text{ V}$ $v_{L\,min} = -80\text{ V}$
b) i) 0.4 mJ ii) 0.2 mJ iii) 0.2 mJ

2.23 $i_{L\,max} = 2\,\text{A}$ $i_{L\,min} = 0\,\text{A}$ $i_{D1\,max} = 2\,\text{A}$ $i_{D1\,min} = 0\,\text{A}$ $i_{D2\,max} = 2\,\text{A}$ $i_{D2\,min} = 0\,\text{A}$ $i_{Q1\,max} = 6\,\text{A}$ $i_{Q1\,min} = 0\,\text{A}$ $v_{X\,max} = 24\,\text{V}$ $v_{X\,min} = 0\,\text{V}$

2.24 **a)** i) $L_m = 200\,\mu\text{H}$ ii) $n = 0.5$ iii) $D = 0.4$ iv) $C = 4.5\,\mu\text{F}$
b) i) $0.114\,\text{J}$ ii) $v_C = 253\,\text{V}$ iii) $t = 1.4\,\text{ms}$
c) Turn-on the switch as soon as i_D is reduced to zero, while keeping the on-time constant.

2.25 **a)** $i_{Q\,max} = 2\,\text{A}$ $i_{Q\,min} = 0\,\text{A}$ $v_{C\,max} = 4\sqrt{5}\,\text{V}$ $v_{C\,min} = 0\,\text{A}$
b) $v_C = 20\sqrt{10}\,\text{V}$

2.26 **a)** $i_{P\,max} = 6\,\text{A}$ $i_{P\,min} = 0\,\text{A}$ $v_{S\,max} = 0\,\text{V}$ $v_{S\,min} = -48\,\text{V}$
b) $i_{P\,max} = 3\,\text{A}$ $i_{P\,min} = 0\,\text{A}$ $v_{S\,max} = 48\,\text{V}$ $v_{S\,min} = -96\,\text{V}$
c) $i_{P\,max} = 5.4\,\text{A}$ $i_{P\,min} = 0\,\text{A}$ $v_{S\,max} = 48\,\text{V}$ $v_{S\,min} = -48\,\text{V}$
d) $i_{P\,max} = 3\,\text{A}$ $i_{P\,min} = 0\,\text{A}$ $v_{S\,max} = 4.25\,\text{V}$ $v_{S\,min} = 0\,\text{V}$

2.27 **a)** i) $1.35\,\text{mJ}$ ii) $1.01\,\text{mJ}$ iii) $0\,\text{mJ}$ iv) $0.34\,\text{mJ}$
b) i) $0.9\,\text{mJ}$ ii) $0.45\,\text{mJ}$ iii) $0.45\,\text{mJ}$ iv) $0\,\text{mJ}$

2.28 **a)** i) $6\,\text{mJ}$ ii) $5\,\text{mJ}$ iii) $1\,\text{mJ}$ iv) $v_C = 100\,\text{V}$
b) i) $6\,\text{mJ}$ ii) $1\,\text{mJ}$ iii) $5\,\text{mJ}$ iv) $v_C = 44.7\,\text{V}$

2.29 **a)** $v_{C\,max} = 800\,\text{V}$ $v_{C\,min} = 0\,\text{V}$ $i_{C\,max} = 40\,\text{A}$ $i_{C\,min} = -400\,\text{A}$
b) $P = 640\,\text{W}$

2.30 **a)** $L = 2\,\text{mH}$ **b)** $V_Z = 4\,\text{V}$ **c)** $i_{L1\,max} = 12\,\text{A}$ $i_{L1\,min} = 0\,\text{A}$ $i_{D2\,max} = 6\,\text{A}$ $i_{D2\,min} = 0\,\text{A}$ $v_{C\,max} = 58.8\,\text{V}$ $v_{C\,min} = 0\,\text{V}$

2.31 **a)** i) $7.2\,\text{mJ}$ ii) $2.16\,\text{mJ}$ iii) $1.44\,\text{mJ}$ iv) $3.6\,\text{mJ}$ iv) $v_C = 65.7\,\text{V}$
b) $i_{D1\,max} = 12\text{A}$ $i_{D1\,min} = 0\text{A}$ $i_{D2\,max} = 6\text{A}$ $i_{D2\,min} = 0\text{A}$ $i_{D3\,max} = 12\,\text{A}$ $i_{D3\,min} = 0\,\text{A}$ **c)** $\bar{\imath}_{D1} = 1.5\,\text{A}$ $\bar{\imath}_{D3} = 0.6\,\text{A}$

2.32 **a)** $i_{L1\,max} = 6\,\text{A}$ $i_{L1\,min} = 0\,\text{A}$ $i_{D1\,max} = 6\,\text{A}$ $i_{D1\,min} = 0\,\text{A}$ $i_{T\,max} = 6\,\text{A}$ $i_{T\,min} = 0\,\text{A}$ $i_{D2\,max} = 3\,\text{A}$ $i_{D2\,min} = 0\,\text{A}$
$v_{C\,max} = 26.8\,\text{V}$ $v_{C\,min} = 0\,\text{V}$
b) i) $L = 2\,\text{mH}$ ii) $V_z = 4.8\,\text{V}$ iii) $i_{L2\,max} = 4.8\,\text{A}$ $i_{L2\,min} = 0\,\text{A}$ $i_{D3\,max} = 4.8\,\text{A}$ $i_{D3\,min} = 0\,\text{A}$

CHAPTER 3

3.2 **a)** $\Delta v_O = 7.64\,\mu\text{V}$ **b)** $L_1 = 3.18\,\text{mH}$ $C_1 = 3.18\,\text{mF}$ **c)** The two-stage filter requires only 11% of the total inductance and capacitance of the single-stage filter for the same attenuation.

3.3 **a)** $V_O = 4\text{V}$ **b)** $v_{L\,max} = 6\text{V}$ $v_{L\,min} = -4\text{V}$ $i_{D\,max} = 5\text{A}$ $i_{D\,min} = 0\text{A}$
c) $L = 60\,\mu\text{H}$ $R = 1\,\Omega$

3.4 **a)** $\bar{\imath}_{Q1} = 1.6$ A $\bar{\imath}_{D1} = 2.4$ A $\bar{\imath}_{Q2} = 2.4$ A $\bar{\imath}_{D2} = 1.6$ A
b) $P_{Q1} = 3.2$ W $P_{D1} = 1.2$ W $P_{Q2} = 4.8$ W $P_{D2} = 0.8$ W **c)** $\eta = 0.89$

3.5 **a)** $\Delta v_O = 0.064$ V **b)** $\Delta v_O = 0.062$ V

3.6 **a)** $L = 200\ \mu$H $C = 25\ \mu$F **b)** $L = 2$ mH

3.7 **a)** $\Delta v_O = 57$ mV
b) First circuit: $\Delta v_O = 5.09$ V Second circuit: $\Delta v_O = 0.81$ V

3.8 **a)** Converter A: 0.6 V Converter B: 0.06 V Converter C: 0.005 V
b) Converter A: 0.43 V Converter B: 0.043 V Converter C: 0.0043 V

3.10 **c)** $i_{L\,max} = 1.5$ A $i_{L\,min} = -0.5$ A

3.11 **a)** $V_O = 16$ V $I_O = 5$ A $R = 3.2\,\Omega$ $L = 40\ \mu$F

3.12 **a)** $v_{L\,max} = 3$ V $v_{L\,min} = -9$ V **b)** $L = 40\ \mu$H $R = 5\,\Omega$
c) $i_{C\,max} = 2.7$ A $i_{C\,min} = -1.8$ A **d)** $\Delta v_O = 0.54$ V **e)** $\Delta v_O = 25.9$ mV

3.13 **a)** $i_{L\,max} = 10$ A $i_{L\,min} = 0$ A $\bar{\imath}_L = 5$ A **b)** $i_{L\,max} = 7$ A $i_{L\,min} = 0$ A $\bar{\imath}_L = 2.97$ A **c)** $i_{L\,max} = 3.6$ A $i_{L\,min} = 0$ A $\bar{\imath}_L = 1.8$ A

3.14 **a)** $L = 40\ \mu$H **b)** $L > 100\ \mu$H **c)** $D_{crit} = 0.4$

3.15 **a)** i) $i_{L\,max} = (4 + 5.3/2)$ A $i_{L\,min} = (4 - 5.3/2)$ A $v_{con} = 2$ V ii) $i_{L\,max} = 4$ A $i_{L\,min} = 0$ A $v_{con} = 1$ V
b) i) $i_{L\,max} = 4$ A $i_{L\,min} = 0$ A $v_{con} = 1$ V ii) $i_{L\,max} = (4 + 5.3/2)$ A $i_{L\,min} = (4 - 5.3/2)$ A $v_{con} = 2$ V

3.16 Converter A: $\bar{v}_{con} = 3$ V $V_O = 6$ V Converter B: $\bar{v}_{con} = 1.5$ V $V_O = 3$ V
Converter C: $\bar{v}_{con} = 1.5$ V $V_O = 3$ V Converter D: $\bar{v}_{con} = 2$ V $V_O = 4$ V

3.17 **a)** $R = 5\,\Omega : \bar{\imath}_L = 1$ A slope: 1.25×10^5 $i_{L\,peak} = 1.625$A
$R = 8\,\Omega : \bar{\imath}_L = 0.625$ A slope: 1.25×10^5
$R = 12\,\Omega : \bar{\imath}_L = 0.47$ A slope: 1.08×10^5
b) $R = 5\,\Omega : \bar{\imath}_L = 0.4$ A slope: 2×10^5 $i_{L\,peak} = 0.8$A
$R = 8\,\Omega : \bar{\imath}_L = 0.25$ A slope: 2×10^5
$R = 12\,\Omega : \bar{\imath}_L = 0.17$ A slope: 2×10^5

3.18 **a)** $i_{L\,max} = 6.5$ A $i_{L\,min} = 3.5$ A $v_{con} = 1.25$ V **b)** $1\text{ V} < v_{con} < 5/3$ V
c) $|Z_2(j\,0)/Z_1(j\,0)| = 1 \neq \infty$ **d)** $V_{con} = 2$ V $V_O = 8$ V

3.19 **a)** $0.16 < D < 0.9$ $0.64\text{ V} < v_{con} < 3.6$ V **b)** $L = 200\ \mu$H
c) $V_S = 20\text{ V} : i_{L\,max} = 1.35$ A $i_{L\,min} = 0.45$ A
$V_S = 22.5\text{ V} : i_{L\,max} = 1.8$ A $i_{L\,min} = 0$ A
$V_S = 60\text{ V} : i_{L\,max} = 3.36$ A $i_{L\,min} = 0$ A

3.20 **a)** i) $8\text{ V} < V_S < 10$ V $1.6\text{ V} < v_{con} < 2$ V
ii) $10\text{ V} < V_S < 20$ V $0.692\text{ V} < v_{con} < 1.6$ V

b) i) $V_O = 4\,\text{V}$ $V_{con} = 1.6\,\text{V}$ ii) No change because $|Z_2(j\,0)/Z_1(j\,0)| = \infty$ iii) $V_{con} = 2.29\,\text{V}$ $V_O = 5.71\,\text{V}$

3.21 **a)** $i_{L\,max} = 6\,\text{A}$ $i_{L\,min} = 0\,\text{A}$ $\tilde{v}_{O\,max} = 0.6\,\text{V}$ $\tilde{v}_{O\,min} = -0.6\,\text{V}$
b) i) $0.25 < D < 0.5$ ii) $1.25\,\text{V} < v_{con} < 2.5\,\text{V}$ iii) $3\,\text{A} < \Delta i_L < 6\,\text{A}$
iv) $0.06\,\text{V} < \Delta v_O < 0.12\,\text{V}$

3.22 A: energy deficit with a smaller input voltage
B: reduced input voltage and increased duty ratio
C: energy surplus with a larger input voltage
D: larger duty ratio with a reduced input voltage
F: smaller duty ratio in DCM operation

CHAPTER 4

4.1 **a)** $i_{L\,max} = 2\,\text{A}$ $i_{L\,min} = 0\,\text{A}$ $\tilde{v}_{O\,max} = 0.02\,\text{V}$ $\tilde{v}_{O\,min} = -0.02\,\text{V}$
b) $i_{L\,max} = 2.5\,\text{A}$ $i_{L\,min} = 1.5\,\text{A}$ $\tilde{v}_{O\,max} = 0.03\,\text{V}$ $\tilde{v}_{O\,min} = -0.02\,\text{V}$

4.2 **a)** $i_{L\,max} = 3.5\,\text{A}$ $i_{L\,min} = 0.5\,\text{A}$ $I_L = 2\,\text{A}$ **b)** $i_{L\,max} = 6.875\,\text{A}$ $i_{L\,min} = -0.625\,\text{A}$ $I_L = 3.125\,\text{A}$

4.3 **a)** For both MOSFET and diode: voltage stress: 77.8 V current stress: 15 A
b) For both MOSFET and diode: voltage stress: 200 V current stress: 51.3 A

4.4 **a)** i) DCM operation ii) $I_L = 0.57\,\text{A}$ iii) $\Delta i_L = 1.5\,\text{A}$ iv) $\Delta v_O = 0.075\,\text{V}$
v) $\bar{v}_{con} = 2\,\text{V}$
b) i) CCM operation ii) $I_L = 16.7\,\text{A}$ iii) $\Delta i_L = 2.4\,\text{A}$ iv) $\Delta v_O = 0.895\,\text{V}$
v) $\bar{v}_{con} = 3.2\,\text{V}$

4.5 **a)** $L = 30\,\mu\text{H}$ **b)** $i_{L\,max} = 8\,\text{A}$ $i_{L\,min} = 0\,\text{A}$ $i_{D\,max} = 8\,\text{A}$ $i_{D\,min} = 0\,\text{A}$ $\bar{v}_{con} = 2.4\,\text{V}$ $\tilde{v}_{O\,max} = 0.08\,\text{V}$ $\tilde{v}_{O\,min} = -0.08\,\text{V}$
c) $L = 25\,\mu\text{H}$ $i_{L\,max} = 20\,\text{A}$ $i_{L\,min} = 0\,\text{A}$ $i_{D\,max} = 20\,\text{A}$ $i_{D\,min} = 0\,\text{A}$ $\bar{v}_{con} = 3\,\text{V}$ $\tilde{v}_{O\,max} = 0.3\,\text{V}$ $\tilde{v}_{O\,min} = -0.1\,\text{V}$

4.6 **a)** $L = 9.6\,\mu\text{H}$ **b)** $D_{crit} = 0.6$ $D_{max} = 0.8\,(V_S = 6\,\text{V})$ $D_{min} = 0\,(V_S = 30\,\text{V})$ **c)** CCM $(D = 0.8)$ $i_{L\,max} = 20\,\text{A}$ $i_{L\,min} = 10\,\text{A}$ $\bar{v}_{con} = 4.8\,\text{V}$

4.7 **a)** $\bar{v}_{con} = 3\,\text{V}$ $V_O = 8\,\text{V}$ **b)** $\bar{v}_{con} = 8/3\,\text{V}$ $V_O = 16/3\,\text{V}$ **c)** $\bar{v}_{con} = 1.0\,\text{V}$ $V_O = 4\,\text{V}$ **d)** $\bar{v}_{con} = 1.5\,\text{V}$ $V_O = 4\,\text{V}$

4.8 (a) $V_{ref} = 2\,\text{V}$ (b) $V_{ref} = 11\,\text{V}$ (c) $V_{ref} = 20\,\text{V}$ (d) $V_{ref} = 8\,\text{V}$

4.9 **a)** $\bar{i}_{D1} = 1.88\,\text{A}$ $\bar{i}_{D2} = 6.67\,\text{A}$ $\bar{i}_{D3} = 1.67\,\text{A}$
b) $\bar{i}_{Q1} = 0.63\,\text{A}$ $\bar{i}_{Q2} = 2.22\,\text{A}$ $\bar{i}_{Q3} = 0.56\,\text{A}$
c) $v_{Q1\,max} = 20\,\text{V}$ $v_{Q2\,max} = 26.7\,\text{V}$ $v_{Q3\,max} = 26.7\,\text{V}$

4.10 (a) $\bar{v}_{con} = 2.5\,\text{V}$ $\bar{v}_O = 6.2\,\text{V}$ (b) $\bar{v}_{con} = 5/3\,\text{V}$ $\bar{v}_O = 10/3\,\text{V}$
(c) $\bar{v}_{con} = 0.625\,\text{V}$ $\bar{v}_O = 2.5\,\text{V}$ (d) $\bar{v}_{con} = 1.0\,\text{V}$ $\bar{v}_O = 2.5\,\text{V}$

4.11 **a)** Converter A: $v_{Q\,max} = 50$ V $I_L = 25$ A Converter B: $v_{Q\,max} = 50$ V $I_L = 15$ A **b)** $\eta = 0.95$

4.12 (a) $V_O = 9$ V $R = 1.875\ \Omega$ $L = 37.5\ \mu$H (b) $V_O = 38.4$ V $R = 12.8\ \Omega$ $L = 60\ \mu$H (c) $V_O = 14.4$ V $R = 4.8\ \Omega$ $L = 60\ \mu$H

4.13 Converter A: $V_O = 40$ V $I_O = 2$ A $L = 60\ \mu$H $R = 20\ \Omega$
Converter B: $V_O = 16$ V $I_O = 2$ A $L = 60\ \mu$H $R = 8\ \Omega$

4.14 $\bar{\imath}_{Q1} = 0.83$ A $\bar{\imath}_{D1} = 1.24$ A $\bar{\imath}_{Q2} = 0.4$ A $\bar{\imath}_{D2} = 0.6$ A $\bar{\imath}_{Q3} = 0.67$ A $\bar{\imath}_{D3} = 1$ A

4.15 $\bar{\imath}_{L1} = 22$ A $\bar{\imath}_{L2} = 13.3$ A $\bar{\imath}_{L3} = 16.7$ A

4.16 **a)** $v_{Q1\,max} = 60$ V $v_{Q2\,max} = 150$ V $v_{Q3\,max} = 60$ V
b) $P_{D1} = 8.7$ W $P_{D2} = 5$ W $P_{D3} = 0.8$ W

4.17 **a)** Circuit A: boost mode: $\bar{v}_1 = 3.5$ V Circuit B: buck mode: $\bar{v}_1 = 56$ V
Circuit C: buck/boost mode: $\bar{v}_1 = 42$ V
b) Circuit A: boost mode: $\bar{v}_1 = 10.5$ V Circuit B: buck mode: $\bar{v}_1 = 18.7$ V
Circuit C: buck/boost mode: $\bar{v}_1 = 4.67$ V

4.19 **a)** v_O continues to grow until the capacitor explodes with an overvoltage condition.
b) i_L continues to grow until the inductor saturates

4.20 **a)** $\bar{v}_{O1} = 3$ V $\bar{v}_{O2} = 6$ V
b) $i_{L1\,max} = 4.25$ A $i_{L1\,min} = 3.75$ A $i_{L2\,max} = 4.25$ A $i_{L2\,min} = 3.75$ A
c) $i_{C1\,max} = 0.5$ A $i_{C1\,min} = -0.5$ A $i_{C2\,max} = 2.25$ A $i_{C2\,min} = -2$ A

4.21 **a)** $\bar{v}_{O1} = 48$ V $\bar{v}_{O2} = 24$ V
b) $i_{L1\,max} = 8.5$ A $i_{L1\,min} = 3.5$ A $i_{L2\,max} = 8.5$ A $i_{L2\,min} = 3.5$ A
c) $i_{C1\,max} = 5$ A $i_{C1\,min} = -5$ A $i_{C2\,max} = 2.5$ A $i_{C2\,min} = -2.5$ A

4.22 i) D_c open: no reset for the magnetizing inductance
ii) D_2 open: no freewheeling path for the inductor current
iii) C open: new steady-state with an increased current ripple
iv) $D = 0.15$: new steady-state with a reduced output voltage
v) $D = 0.65$: saturation of the magnetizing inductance

4.23 **a)** i) (b) ii) (d) iii) (c)
b) (a) $0 < D < 48/72$ (b) $0 < D < 8/32$ (c) $0 < D < 1$ (d) $0 < D < 1/2$
c) (a) 72 V (b) 32 V (c) 28.8 V (d) 24 V

4.24 **a)** To obtain a controllable magnetizing inductance
b) Flyback converter: use of magnetizing inductance as a functional inductor
Bridge-type converter: automatic magnetizing inductance reset
c) (a) $v_{O\,max} = 9.3$ V $v_{Q\,max} = 84$ V (b) $v_{O\,max} = 4.7$ V $v_{Q\,max} = 42$ V
(c) $v_{O\,max} = \infty$ $v_{Q\,max} = \infty$ (d) $v_{O\,max} = 7$ V $v_{Q\,max} = 28$ V
(e) $v_{O\,max} = 7$ V $v_{Q\,max} = 28$ V (f) $v_{O\,max} = 14$ V $v_{Q\,max} = 56$ V

4.25 **a)** $V_O = 12.5\text{ V}$ $v_{Q\,max} = 125\text{ V}$ $\bar{i}_D = 6.25\text{ A}$
b) $V_O = 10\text{ V}$ $v_{Q\,max} = 150\text{ V}$ $\bar{i}_D = 2.5\text{ A}$ $\bar{i}'_D = 1\text{ A}$
c) $V_O = 10\text{ V}$ $v_{Q\,max} = 300\text{ V}$ $\bar{i}_D = 2.5\text{ A}$ $\bar{i}'_D = 0.25\text{ A}$
d) $V_O = 10\text{ V}$ $v_{Q\,max} = 100\text{ V}$ $\bar{i}_D = 2.5\text{ A}$ $\bar{i}'_D = 0.5\text{ A}$
e) $V_O = 5\text{ V}$ $v_{Q\,max} = 100\text{ V}$ $\bar{i}_D = 1.25\text{ A}$

CHAPTER 5

5.2 **b)** $i_A = D\,\hat{i}_c$ $v_B = V_{ap}\hat{d}$

5.4 **b)** $V_O/V_S = 1/(1-D)$ $I_O/I_S = 1-D$

5.6 **b)** $$\frac{V_O}{V_S} = \frac{D}{1-D}\frac{1}{1+\dfrac{1}{(1-D)^2}\dfrac{R_l}{R}}$$

5.7 **a)** $\hat{i}/\hat{v} = 0.001\text{ A/V}$ **b)** $\hat{v} = 0.2\text{ V}$
c) $i(t) = (2 + 0.1\sin 20t)\text{ mA}$ **d)** $1.5\text{ mA} < |i(t)| < 7.5\text{ mA}$

5.8 **a)** $F_m = (1-D)/V_{con}$ **b)** $F_m = 0.05$

5.9 i) $\hat{z} = 4\hat{x} + 4\hat{y}$ $ii)\hat{z} = 2\hat{x} - 4\hat{y}$

5.11 $$G_{vs}(s) = D\frac{R}{R+R_l}\frac{1+\dfrac{s}{\omega_{esr}}}{1+\dfrac{s}{Q\omega_o}+\dfrac{s^2}{{\omega_o}^2}}$$

$$\omega_{esr} = \frac{1}{CR_c} \qquad \omega_o = \sqrt{\frac{1}{LC}\frac{R+R_l}{R+R_c}}$$

$$Q = \frac{1}{\omega_o}\frac{R+R_l}{L + C(RR_l + RR_c + R_lR_c)}$$

$$G_{vd}(s) = V_S\frac{R}{R+R_l}\frac{1+\dfrac{s}{\omega_{esr}}}{1+\dfrac{s}{Q\omega_o}+\dfrac{s^2}{{\omega_o}^2}}$$

$$Z_p(s) = R \parallel R_l\frac{\left(1+\dfrac{s}{\omega_z}\right)\left(1+\dfrac{s}{\omega_{esr}}\right)}{1+\dfrac{s}{Q\omega_o}+\dfrac{s^2}{{\omega_o}^2}} \qquad \omega_z = \frac{R_l}{L}$$

5.12 **a)** $\hat{v}_{ac} = (1-D)\,\hat{v}_{bc} - V_{bc}\hat{d} \quad \hat{\imath}_b = (1-D)\,\hat{\imath}_a - I_a\hat{d}$

d) $G_{vs}(s) = \dfrac{1}{1-D}\dfrac{1}{1+\dfrac{s}{Q\omega_o}+\dfrac{s^2}{\omega_o^2}} \quad \omega_o = \sqrt{\dfrac{(1-D)^2}{LC}}$

$Q = (1-D)R\sqrt{\dfrac{C}{L}}$

5.14 i) Buck converter: $r_i = \dfrac{V_O}{\mu I_S} \quad r_o = \dfrac{\mu V_O}{I_S} \quad g_f = \dfrac{2I_S}{V_O} \quad k_o = \dfrac{2I_S}{\mu D}$
$k_i = \dfrac{2I_S}{D}$

with $I_S = \dfrac{MV_O}{R} \quad \mu = \dfrac{M}{1-M} \quad M = \dfrac{2D}{D+\sqrt{D^2+\dfrac{8L}{RT_s}}}$

ii) Boost converter: $r_i = \dfrac{V_S}{\mu I_O} \quad r_o = \dfrac{\mu V_S}{I_O} \quad g_f = \dfrac{2I_O}{V_S} \quad k_o = -\dfrac{2I_O}{D} \quad k_i = -\dfrac{2\mu I_O}{D}$

with $I_O = \dfrac{V_O}{R} \quad \mu = M-1 \quad M = \dfrac{1}{2}\left(1+\sqrt{1+\dfrac{2D^2RT_s}{L}}\right)$

5.15 **a)** i) $i_{D1\,max} = 12.5\text{ A} \quad i_{D1\,min} = 0\text{ A} \quad \bar{\imath}_{D1} = 3\text{ A}$
$i_{D2\,max} = 12.5\text{ A} \quad i_{D2\,min} = 0\text{ A} \quad \bar{\imath}_{D2} = 9\text{ A}$
$i_{C\,max} = 0.5\text{ A} \quad i_{C\,min} = 0\text{ A} \quad \bar{\imath}_C = 0.125\text{ A}$
b) i) Same as **a)** except for $i_C(t) = 0$

5.16 **a)** $i_{D1\,max} = 4.75\text{ A} \quad i_{D1\,min} = 0\text{ A} \quad i_{C\,max} = 3.34\text{ A} \quad i_{C\,min} = 0\text{ A}$
$v_{S\,max} = 60\text{ V} \quad v_{S\,min} = -60\text{ V} \quad v_{Q\,max} = 60\text{ V} \quad v_{Q\,min} = 0\text{ V}$

5.18 **a)** $i_{D1\,max} = 6\text{ A} \quad i_{D1\,min} = 0\text{ A} \quad v_{S\,max} = 24\text{ V} \quad v_{S\,min} = -36\text{ V}$
$v_{Q\,max} = 60\text{ V} \quad v_{Q\,min} = 0\text{ V}$

5.19 **a)** $F_m = 1/V_m$ **b)** $F_m = 2/V_m$ **c)** $F_m = -1/V_m$

CHAPTER 6

Note: Some design problems and graphical problems do not have a unique solution. For these problems, the answers in this list are a typical illustration and other solutions are equally acceptable.

6.1 **a)** $\dfrac{100}{1+\dfrac{s}{2}}$ **b)** $0.01\left(1+\dfrac{s}{2}\right)$ **c)** $\dfrac{1}{2}s\dfrac{1}{1+\dfrac{s}{200}}$ **d)** $\dfrac{50}{s}\left(1+\dfrac{s}{5}\right)$

e) $\dfrac{2000}{s\left(1+\dfrac{s}{20}\right)}$ **f)** $\dfrac{1}{2}s\left(1+\dfrac{s}{20}\right)$

6.2 i) $|v_O(t)| = \dfrac{\sqrt{2}\left|1-\dfrac{\omega_z^2}{\omega_{o1}^2}\right|}{\sqrt{1+\dfrac{\omega_z^2}{\omega_p^2}}\left|1-\dfrac{\omega_z^2}{\omega_{o2}^2}\right|}V_m$ ii) $|v_O(t)| = 0$

iii) $|v_O(t)| = \dfrac{\sqrt{1+\dfrac{\omega_p^2}{\omega_z^2}}\left|1-\dfrac{\omega_p^2}{\omega_{o1}^2}\right|}{\sqrt{2}\left|1-\dfrac{\omega_p^2}{\omega_{o2}^2}\right|}V_m$ iv) $|v_O(t)| = \infty$

6.3 **a)** $\dfrac{10^{30/20}}{20}s\dfrac{1}{\left(1+\dfrac{s}{20}\right)\left(1+\dfrac{s}{800}\right)}$ **b)** $\dfrac{20\cdot 10^{-30/20}}{s}\left(1+\dfrac{s}{20}\right)\left(1+\dfrac{s}{800}\right)$

c) $10^{-30/20}\dfrac{1+\dfrac{s}{20}}{1+\dfrac{s}{20\times 10^3}}$ **d)** $10^{30/20}\dfrac{1+\dfrac{s}{800}}{1+\dfrac{s}{800\times 10^{-3}}}$

6.4 **a)** $|T|_{max} = 20\log K_t \quad |T|_{min} = 20\log\left(K_t\dfrac{\omega_{z2}}{\omega_{p2}}\right)$

b) $|T|_{max} = 20\log K_t \quad |T|_{min} = 20\log\left(K_t\dfrac{\omega_{z1}\omega_{z2}}{\omega_{p1}\omega_{p2}}\right)$

6.5 **a)** $|T|_{max} = 20\log(K_t\omega_p)$ **b)** $|T|_{max} = 20\log(K_t\omega_{p1})$

c) $|T|_{max} = 20\log\left(K_t\dfrac{\omega_{p1}\omega_{p2}}{\omega_z}\right)$ **d)** $|T|_{max} = 20\log(K_t\omega_{p1})$

6.6 **a)** X = 63.2 rad/s Y = −46.5 dB **b)** X = 11.2×10^3 rad/s Y = 10.7×10^3 rad/s **c)** X = 13.4 rad/s Y = −43.5 dB

6.8 **a)** X = 46 dB Y = 1.42×10^3 rad/s **b)** X = 4.5×10^3 rad/s Y = 15.8×10^3 rad/s

6.11 $L = 40\,\mu\text{H}$ $R_l = 0.01\,\Omega$ $C = 375\,\mu\text{F}$ $R_c = 0.056\,\Omega$ $R = 0.65\,\Omega$ $V_S = 10\,\text{V}$

6.12 i) Boost converter : $G_{vd}(s) = K_{vd}\dfrac{\left(1 - \dfrac{s}{\omega_{rhp}}\right)\left(1 + \dfrac{s}{\omega_{esr}}\right)}{1 + \dfrac{s}{Q\omega_o} + \dfrac{s^2}{\omega_o^2}}$

$$K_{vd} = \frac{RV_S}{RD'^2 + R_l}\left(1 - \frac{R_l}{RD'^2}\right) \quad \omega_{rhp} = \frac{RD'^2 - R_l}{L} \quad \omega_{esr} = \frac{1}{CR_c}$$

$$\omega_o = \sqrt{\frac{1}{LC}\frac{RD'^2 + R_l}{R + R_c}} \quad Q = \frac{1}{\omega_o}\frac{RD'^2 + R_l}{L + C(R_lR_c + R_lR + R_cRD'^2)}$$

ii) Buck/boost converter: Same as the boost converter case except for

$$\omega_{rhp} = \frac{RD'^2 - R_lD}{LD}$$

6.14 **a)** $K_t = 2\pi \cdot 1 \times 10^3$ $\omega_z = 2\pi \cdot 60$ rad/s $Q = 3.98$ $\omega_o = 2\pi \cdot 400$ rad/s
b) $K_t = \dfrac{1}{2\pi \cdot 200}$ $\omega_{p1} = 2\pi \cdot 50$ rad/s $\omega_{p2} = 2\pi \cdot 1.2 \times 10^3$ rad/s
c) $K_t = 2\pi \cdot 1000$ $\omega_z = 2\pi \cdot 15$ rad/s $\omega_p = 2\pi \cdot 1.2 \times 10^3$ rad/s

6.15 **a)** $L = 25\,\mu\text{H}$ $C = 400\,\mu\text{F}$ $R = 0.56\,\Omega$ $R_c = 0.1\Omega$
b) $v_O(t) = 22.4 + 0.56 \sin 2\pi \cdot 200t$ $v_{O\,max} = 23$ V $v_{O\,min} = 21.9$ V
c) $v_O(t) = 22.4 + 0.032 \sin(2\pi \cdot 200t + 90°)$ $v_{O\,max} = 22.43$ V $v_{O\,min} = 22.37$ V

6.16 **a)** $G_{vd}(s) = K_{vd}\dfrac{1 + \dfrac{s}{\omega_{esr}}}{1 + \dfrac{s}{Q\omega_o} + \dfrac{s^2}{\omega_o^2}}$ $K_{vd} = 56.5$

$\omega_{esr} = 2\pi \cdot 5 \times 10^3$ rad/s $\omega_o = 2\pi \cdot 600$ rad/s $Q = 2$

b) $G_{vs}(s) = K_{vs}\dfrac{1 + \dfrac{s}{\omega_{esr}}}{1 + \dfrac{s}{Q\omega_o} + \dfrac{s^2}{\omega_o^2}}$ $K_{vs} = 0.25$

c) $V_O = 14.13$ V

CHAPTER 7

Note: Some design problems and graphical problems do not have a unique solution. For these problems, the answers in this list are a typical illustration and other solutions are equally acceptable.

7.1 **a)** i) $T_m(s) = K_t \dfrac{1}{1 + \dfrac{s}{Q\omega_o} + \dfrac{s^2}{\omega_o^2}}$ $K_t = 40$ $Q = 3.43$ $\omega_o = 2\pi \cdot 1.16 \times 10^3$ rad/s iii) $\bar{v}_O = 4.29$ V $\bar{v}_{con} = 1.07$ V

b) i) $T_m(s) = \dfrac{K_t}{s} \dfrac{1}{1 + \dfrac{s}{Q\omega_o} + \dfrac{s^2}{\omega_o^2}}$ $K_t = 1.25 \times 10^4$ iii) $\bar{v}_O = 4$V $\bar{v}_{con} = 1$ V

c) i) $T_m(s) = \dfrac{K_t}{1 + \dfrac{s}{\omega_p}} \dfrac{1}{1 + \dfrac{s}{Q\omega_o} + \dfrac{s^2}{\omega_o^2}}$ $K_t = 40$ $\omega_p = 2\pi \cdot 49.7$ rad/s

iii) $\bar{v}_O = 4.29$ V $\bar{v}_{con} = 1.07$ V

7.2 **a)** $v_O(t) = \dfrac{V_{step} K_a \omega_{p1} \omega_{p2}}{\omega_{p2} - \omega_{p1}} \left(e^{-\omega_{p1} t} - e^{-\omega_{p2} t}\right)$ $\omega_{p1} \ll \omega_{p2}$

b) $K_a = \dfrac{1}{2\pi \cdot 2 \times 10^3}$ $\omega_{p1} = 2\pi \cdot 70$ rad/s $\omega_{p2} = 2\pi \cdot 800$ rad/s

c) $v_O(t) = 0.077 \left(e^{-2\pi \cdot 70t} - e^{-2\pi \cdot 800t}\right)$

7.3 $v_O(t) = 3 + 0.01 \sin(2\pi \cdot 20t + 90°)$

7.4 **a)** $v_O(t) = I_{step} K_z \omega_p e^{-\omega_p t}$ **b)** $K_z = \dfrac{1}{2\pi \cdot 7 \times 10^3}$ $\omega_p = 2\pi \cdot 500$ rad/s

c) $v_O(t) = 0.143 e^{-2\pi \cdot 500t}$

7.5 i) $v_O(t) = I_{step} K_z \omega_p e^{-\omega_p t}$

ii) $v_O(t) = I_{step} K_z \dfrac{\omega_{p1} \omega_{p2}}{\omega_{p2} - \omega_{p1}} \dfrac{1}{\omega_z} \left((\omega_z - \omega_{p1}) e^{-\omega_{p1} t} + (\omega_{p2} - \omega_z) e^{-\omega_{p2} t}\right)$

$(\omega_z - \omega_{p1}) > 0$ $(\omega_{p2} - \omega_z) > 0$

iii) same as ii) except for $(\omega_z - \omega_{p1}) > 0$ $(\omega_{p2} - \omega_z) < 0$

7.6 **a)** $|Z_o|_{peak} = -20$ dB **b)** $\dfrac{s}{3 \times 10^3} \dfrac{1}{1 + \dfrac{s}{300}}$ **c)** $v_O(t) = 0.5 e^{-300t}$

7.8 **a)** $\dfrac{v_O(s)}{v_S(s)} = F_A(s)\dfrac{Z_i(s)}{Z_i(s)+Z_o(s)}F_B(s)$

b) $\dfrac{v_O(s)}{v_S(s)} = F_A(s)F_B(s)\dfrac{1}{1+T_m(s)}$ with $T_m(s) = \dfrac{Z_o(s)}{Z_i(s)}$

c) $\left|\dfrac{Z_o(j\omega)}{Z_i(j\omega)}\right| < 1$ for all frequencies

f) i) When $|Z_o|$ doubles. ii) When $|Z_O|$ increases by 2/3.

7.9 i) Passing $(-1, 0)$ point at $\omega = \omega_c$ ii) $|T_m(j\omega_c)| = 0$ dB with $\angle T_m(j\omega_c) = -180°$ iii) Poles at $\pm j\omega_c$ iv) Oscillation at $t_{os} = 2\pi/\omega_c$ v) $|G(j\omega_c)| = \infty$

7.10 **a)** Stable PM $= 45°$ GM $= 12$ dB $\omega_c = 2\pi \cdot 200$ rad/s
c) 12 dB gain increase **d)** $45°$ phase delay

7.11 **a)** Unstable PM $= -20°$ GM $= -10$ dB $\omega_c = 2\pi \cdot 55 \times 10^3$ rad/s
b) i) 20 dB gain decrease ii) 40 dB gain decrease iii) 30 dB gain decrease

7.12 **a)** Conditionally stable **b)** $1 < K_v < 10$ $20 < K_v < 400$
c) $K_v = 10$: oscillation at $\omega = 200$ rad/s
$K_v = 20$: oscillation at $\omega = 500$ rad/s
$K_v = 400$: oscillation at $\omega = 2 \times 10^3$ rad/s

7.13 **a)** Conditionally stable **c)** $1 < K_v < 10$ $10^2 < K_v < 10^4$

CHAPTER 8

Note: Some design problems and graphical problems do not have a unique solution. For these problems, the answers in this list are a typical illustration and other solutions are equally acceptable.

8.1 **a)** $F(s)_{\text{asym}} = 2.5 \times 10^{-3} s \dfrac{1+\dfrac{s}{400}}{\left(1+\dfrac{s}{40}\right)\left(1+\dfrac{s}{4\times 10^3}\right)}$

b) $F(s)_{\text{exac}} = F(s)_{\text{asym}}$

8.2 **a)** $K_f = \dfrac{K_g}{\omega_c}$ **b)** $K_f = K_g\dfrac{\omega_p}{\omega_z\omega_c'}$ **c)** $K_f = K_g\dfrac{\omega_p}{\omega_z\omega_c'} = \dfrac{K_g}{\omega_c}$

8.3 **a)** $|G_{vd}(j\,0)| = 24$ dB $|G_{vs}(j\,0)| = -12$ dB $|Z_p(j\infty)| = -26$ dB
b) $K_v = 1.34 \times 10^3$
c) i) $\omega_{z1} = 0.8\,\omega_0 = 2\pi\cdot 681$ rad/s $\omega_{z2} = 1.5\omega_o = 2\pi\cdot 1.28\times 10^3$ rad/s
$\omega_{p1} = \omega_{esr} = 2\pi\cdot 4.55\times 10^3$ rad/s $\omega_{p2} = 0.8\,\omega_s = 2\pi\cdot 3.2\times 10^4$ rad/s
ii) $K_v = 8.57 \times 10^3$

iii) $A_u(s) = K_a s \dfrac{1}{\left(1+\dfrac{s}{\omega_{z1}}\right)\left(1+\dfrac{s}{\omega_{z2}}\right)}$ $K_a = 7.29 \times 10^{-6}$

$|A_u|_{peak} = -30.12$ dB

8.4 **a)** $\omega_o = 2\pi \cdot 700$ rad/s **b)** $\omega_{z1} = 2\pi \cdot 350$ rad/s
c) $\omega_c = 2\pi \cdot 5 \times 10^3$ rad/s PM $= 33°$ **e)** $F_m = 4.19$
f) 1.5 kHz $< f_c <$ 30 kHz $15° < \phi_m < 65°$

8.5 **a)** $|T_m(j\,0)| = 34$ dB $\omega_c = 2\pi \cdot 50 \cdot 10^{34/20}$ rad/s
b) $\bar{v}_{con} = 1.08$ V $\bar{v}_O = 5.39$ V
c) $|A_u(j\,0)| = -39.4$ dB $|A_u|_{peak} = -5.4$ dB **d)** $R_1 = 500\ \Omega$

8.6 **a)** To avoid being a conditionally stable system
b) Speed of the step input response
c) Speed of the step load response and phase margin
d) To extend the frequencies where the slope of $|T_m|$ remains with a -20 dB/dec.
e) $\omega_{z1} \ll \omega_{z2}$
f)

	(Conditionally stable system)
(Fast step load response)	(Reduced phase margin)
(Better noise attenuation)	(Reduced phase margin)
(Higher crossover frequency)	(Less accurate model prediction)

8.7 **a)** Unstable **c)** Sustained oscillation at $\omega_c = 2\pi \cdot 5 \times 10^3$ rad/s
d) Sustained oscillation at $\omega_c = 2\pi \cdot 1 \times 10^3$ rad/s
e) None of the two approaches is desirable
i) Increasing K_v more than 20 dB makes the converter conditionally stable.
ii) Decreasing K_v more than 20 dB makes the converter have marginal stability margins.

8.9 **a)** Conditionally stable
c) $2500 \times 10^{-3} < K_v < 2500 \times 10^{-1}$ $2500 \times 10^{1} < K_v$
d) $K_v = 2500 \times 10^{1}$

8.10 **a)** $A_u(s)_{\text{asym}} = 5 \times 10^{-3} s \dfrac{1}{\left(1+\dfrac{s}{40}\right)\left(1+\dfrac{s}{4\times 10^3}\right)}$
b) $A_u(s)_{\text{exac}} = A_u(s)_{\text{asym}}$ **c)** $v_O(t) = 2\left(e^{-40t} - e^{-4\times 10^3 t}\right)$

8.11 **a)** $|Z_o|_{\text{peak}} = -40$ dB
b) $Z_o(s) = \dfrac{s}{1\times 10^4}\dfrac{1}{\left(1+\dfrac{s}{100}\right)\left(1+\dfrac{s}{1\times 10^3}\right)}$ **c)** $v_O(t) = \dfrac{1}{900}\left(e^{-100t} - e^{-1\times 10^3 t}\right)$

8.12 **a)** i) $\omega_{z1} = 4.29 \times 10^3$ rad/s ii) $\omega_{z2} = 10^4$ rad/s iii) $\omega_{p1} = \omega_{esr} = 2.86 \times 10^4$ rad/s $K_v = 1.07 \times 10^4$ iv) $\omega_{z2} = 3.45 \times 10^4$ rad/s

b) $A_u(s) = K_a s \dfrac{1}{(1+s/\omega_{z1})(1+/\omega_{z2})}$ $\quad K_a = 5.84 \times 10^{-6} \quad |A_u|_{peak} =$ -32 dB

8.13 **a)** $T_1(s) = \dfrac{\omega_{cm}}{s}\dfrac{1}{1+\dfrac{s}{\omega_{ci}}}$ $\quad \omega_c = \omega_{cm} \quad$ PM $= 90°$ **b)** $T_2(s) = \dfrac{s}{\omega_{cm}}\dfrac{1}{1+\dfrac{s}{Q\omega_{cv}}+\dfrac{s^2}{\omega_{cv}^2}}$

c) $1+T_1(s) = 0 \;\Rightarrow\; 1+\dfrac{T_v(s)}{1+T_i(s)} = 0 \;\Rightarrow\; 1+T_i(s)+T_v(s) = 0$

$1+T_2(s) = 0 \Rightarrow 1+\dfrac{T_i(s)}{1+T_v(s)} = 0 \;\Rightarrow\; 1+T_v(s)+T_i(s) = 0$

d): Both the converters are stable.

e) $T_3(s) = \dfrac{\omega_{ci}}{s}\dfrac{\left(1+\dfrac{s}{\omega_{zc}}\right)\left(1+\dfrac{s}{\omega_{cm}}\right)}{1+\dfrac{s}{Q\omega_o}+\dfrac{s^2}{\omega_o^2}}$ $\quad \omega_c = \omega_{ci} \quad$ PM $= 90°$

GM $= \infty$ dB

8.14 **a)** $|Z_o| = \dfrac{|Z_p|}{\sqrt{1+|T_m|^2-2|T_m|\cos\theta}}$

b) i) $|Z_o| = \dfrac{|Z_p|}{\sqrt{2-2\cos\theta}}$ $\quad$ ii) $|Z_o| = \dfrac{|Z_p|}{1-e^{-\mathrm{GM}/20}}$

c) i) $|Z_o|$ peaking at the frequency where $|T_m| = 0$ dB as a function of the phase margin θ.

ii) $|Z_o|$ peaking at the frequency where $\angle T_m = -180°$ as a function of the gain margin GM.

8.15 **a)** $R_l = 0.1\,\Omega \quad L = 53\,\mu\mathrm{H} \quad R_c = 0.04\,\Omega \quad C = 613\,\mu\mathrm{F}$

b) $v_O(t) = 2.4 + 0.1\sin 2\pi\cdot 100t + 0.1\sin(2\pi\cdot 4\times 10^3 t - 90°)$

c) $\omega_c = 2\pi\cdot 4\times 10^3$ r/s $\quad$ PM $= 10.2°$

d) $\left.\dfrac{v_o(s)}{i_o(s)}\right|_B = \dfrac{s}{2\pi\cdot 2\times 10^4}\dfrac{1+\dfrac{s}{2\pi\cdot 6.5\times 10^3}}{1+\dfrac{s}{Q\omega_o}+\dfrac{s^2}{\omega_o^2}}$ $\quad \omega_o = 2\pi\cdot 4\times 10^3 \quad Q =$ $10^{15/20} = 5.6$

e) $v_O(t) = 2 + 0.001\sin(2\pi 100t + 90°) + 0.56\sin(2\pi\cdot 4\times 10^3 t)$

f) $v_O(t)_{min} = 1.8$ V $\quad t_{os} = 2.5\times 10^{-4}$ s

8.16 **a)** PM $= 20°$ $\quad$ GM $= 3$ dB $\quad$ **b)** 9.2 dB $\quad$ **c)** $20\log\left(\dfrac{1}{1-10^{-3/20}}\right) =$ 10.7 dB

8.17 **a)** PM $= 90°$ GM $= 60$ dB **b)** $F_v(s) = \dfrac{4 \times 10^3}{s} \dfrac{1 + \dfrac{s}{400}}{1 + \dfrac{s}{2 \times 10^4}}$

c) $Z_o(s) = \dfrac{s}{4 \times 10^3} \dfrac{1 + \dfrac{s}{8 \times 10^4}}{\left(1 + \dfrac{s}{400}\right)\left(1 + \dfrac{s}{4 \times 10^4}\right)}$ $|Z_o|_{peak} = -26$ dB

8.18 **a)** $\omega_c = 2\pi \cdot 5 \times 10^3$ rad/s PM $= 20°$
b) $|Z_o|_{peak} = -9.18$ dB at $\omega_c = 2\pi \cdot 5 \times 10^3$ rad/s
c) Underdamped oscillation with $t_{os} = 0.2$ ms
d) $K'_v = 47.43$qquad Unacceptable duo to narrow crossover frequency
e) $K''_v = 1.5 \times 10^4$ **f)** $K'''_v = 0.8K''_v = 1.2 \times 10^4$ in anticipation of an additional phase delay to high frequencies.

8.19 **a)** $\omega_c = 2\pi \cdot 3 \times 10^3$ r/s PM $= 40°$ **c)** $|A_u(j2\pi \cdot 100)| = -52$ dB
d) $V_m = 5.09$ V **e)** $2\pi \cdot 1 \times 10^3 < \omega_c < 2\pi \cdot 7 \times 10^3$ $20° <$ PM $< 53°$
f) Move the second compensation zero ω_{p2} towards higher frequencies or move the first compensation zero ω_{z1} towards lower frequencies.

8.20 **a)** $\omega_c = 2\pi \cdot 10 \times 10^3$ rad/s **b)** $D = 0.22$
c) $v_O(t) = 2.2 + 0.02\sin(2\pi \cdot 400t + 90°)$ **d)** $t_s = 0.8$ ms

8.21 **a)** $Z_p(s) = 0.1 \dfrac{1 + \dfrac{s}{400}}{1 + \dfrac{s}{40}}$ $T_m(s) = \dfrac{4 \times 10^4}{s} \dfrac{1 + \dfrac{s}{400}}{1 + \dfrac{s}{40}}$

b) $Z_o(s) = 2.5 \times 10^{-6} s \dfrac{1}{1 + \dfrac{s}{4 \times 10^3}}$ **c)** $v_O(t) = 0.1e^{-4\times10^3 t}$

8.22 **a)** i) PM $= 10.2°$ ii) $f_c = 4$ kHz **b)** i) $f_c = 5 - 10$ kHz ii) $|T_m(j2\pi \cdot 10^3)| = 32$ dB iii) $t_s = 0.12$ ms $\Delta v_{O\,\text{peak}} = 0.12$ V

8.23 **a)** $\omega_c = 2\pi \cdot 5 \times 10^3$ rad/s **b)** PM $= 8°$ **c)** $\omega_{z1} = 2\pi \cdot 350$ rad/s
d) $D = 0.25$ **e)** $t_s = 1.4$ ms

CHAPTER 9

Note: Some design problems and graphical problems do not have a unique solution. For these problems, the answers in this list are a typical illustration and other solutions are equally acceptable.

9.1 **a)** $i_{Q\,max} = 3.75$ A $i_{Q\,min} = 0$ A $v_{I\,max} = 0.75$ V $v_{I\,min} = 0$ V

$V_{ramp\,max} = 0.5$ V $V_{ramp\,min} = 0$ V

b) $(v_I + V_{ramp})_{max} = 1.25$ V $(v_I + V_{ramp})_{min} = 0$ V $v_{con} = 1.25$ V

9.3 $$G_{id}(s) = K_{id}\frac{1+\dfrac{s}{\omega_{id}}}{1+\dfrac{s}{Q\omega_o}+\dfrac{s^2}{\omega_o^2}}$$

i) Buck converter

$$K_{id} = \frac{V_S}{R+R_l} \quad \omega_{id} = \frac{1}{C(R+R_c)} \quad \omega_o = \sqrt{\frac{1}{LC}\frac{R+R_l}{R+R_c}}$$

$$Q = \frac{1}{\omega_o}\frac{R+R_l}{L+C(R_lR_c+R_lR+R_cR)}$$

ii) Boost converter

$$K_{id} = \frac{2V_S}{D'\left(RD'^2+R_l\right)} \quad \omega_{id} = \frac{2}{C(R+2R_c)} \quad \omega_o = \sqrt{\frac{1}{LC}\frac{RD'^2+R_l}{R+R_c}}$$

$$Q = \frac{1}{\omega_o}\frac{RD'^2+R_l}{L+C\left(R_lR_c+R_lR+R_cRD'^2\right)}$$

iii) Buck/boost converter

$$K_{id} = \frac{(1+D)V_S}{D'\left(RD'^2+R_l\right)} \quad \omega_{id} = \frac{1+D}{C(R+(1+D)R_c)} \quad \omega_o = \sqrt{\frac{1}{LC}\frac{RD'^2+R_l}{R+R_c}}$$

$$Q = \frac{1}{\omega_o}\frac{RD'^2+R_l}{L+C\left(R_lR_c+R_lR+R_cRD'^2\right)}$$

9.4 **a)** $$T_i(s) = K_i\frac{1+\dfrac{s}{\omega_{id}}}{1+\dfrac{s}{Q\omega_o}+\dfrac{s^2}{\omega_o^2}}$$

$K_i = 3.076 \quad \omega_{id} = 2\pi \cdot 100\ \text{rad/s} \quad \omega_o = 2\pi \cdot 562\ \text{rad/s} \quad Q = 5.65$

$$T_v(s) = K_v\frac{1}{s}\frac{1+\dfrac{s}{\omega_{zc}}}{1+\dfrac{s}{Q\omega_o}+\dfrac{s^2}{\omega_o^2}} \qquad K_v = 1.54\times10^5 \quad \omega_{zc} = 2\pi\cdot450\ \text{rad/s}$$

b) $T_1(s)$ crossover frequency: $\omega_{ci} = 2\pi \cdot 1 \times 10^4$ rad/s
$T_2(s)$ crossover frequency: $\omega_{cr} = 2\pi \cdot 1.8 \times 10^3$ rad/s

c) $$T_1(s) = K_1\frac{\left(1+\dfrac{s}{\omega_{zc}}\right)\left(1+\dfrac{s}{\omega_{cr}}\right)}{s\left(1+\dfrac{s}{Q\omega_o}+\dfrac{s^2}{\omega_o^2}\right)} \qquad K_1 = 1.54 \times 10^5 \qquad \omega_{cr} =$$
$2\pi \cdot 1.8 \times 10^3$ rad/s

$$T_2(s) = K_2 \frac{1+\dfrac{s}{\omega_{zc}}}{s\left(1+\dfrac{s}{\omega_{id}}\right)\left(1+\dfrac{s}{\omega_{ci}}\right)} \qquad K_2 = 5\times 10^4 \quad \omega_{ci} = 2\pi\cdot 1\times$$

10^4 rad/s

9.5 **a)** $S_e = 1.06\times 10^5$ V/s $C_2 = 3.67$ nF $R_2 = 60.89$ kΩ $C_3 = 0.8$ nF
b) $S_e = 1.31\times 10^5$ V/s $C_2 = 3.2$ nF $R_2 = 74$ kΩ $C_3 = 1.1$ nF
c) $S_e = 2.95\times 10^4$ V/s $C_2 = 12.5$ nF $R_2 = 42$ kΩ $C_3 = 0.75$ nF

9.6 Case A: $5\cdot 10^{-11/20}\text{ V} < \Delta v_O(t)_{peak} < 5\cdot 10^{-1.5/20}\text{ V} \quad t_s = 1.5\text{ ms}$

Case B: $\Delta v_O(t)_{peak} = 5\cdot 10^{-9/20}\text{ V} \quad t_s = 1.2\text{ ms}$

Case C: $\Delta v_O(t)_{peak} = 5\cdot 10^{-9.5/20}\text{ V} \quad t_s = 1.2\text{ ms}$

Case D: $\Delta v_O(t)_{peak} = 5\cdot 10^{-6.5/20}\text{ V} \quad t_s = 1.2\text{ ms}$

9.7 $\omega_{id} = 2\pi\cdot 100$ rad/s $\omega_{zc} = 2\pi\cdot 300$ rad/s $\omega_o = 2\pi\cdot 375$ rad/s $Q = 5.65$
$\omega_{esr} = 2\pi\cdot 5\times 10^3$ rad/s $\omega_{cr} = 2\pi\cdot 5\times 10^3$ rad/s $\omega_{ci} = 2\pi\cdot 1\times 10^4$ rad/s

b) $$T_1(s) = K_1 \frac{\left(1+\dfrac{s}{\omega_{zc}}\right)\left(1+\dfrac{s}{\omega_{cr}}\right)}{s\left(1+\dfrac{s}{Q\omega_o}+\dfrac{s^2}{\omega_o^2}\right)} \qquad K_1 = 7.05\times 10^5$$

$$T_2(s) = K_2 \frac{1+\dfrac{s}{\omega_{zc}}}{s\left(1+\dfrac{s}{\omega_{id}}\right)\left(1+\dfrac{s}{\omega_{ci}}\right)} \qquad K_2 = 2.23\times 10^5$$

$$A_u(s) = K_a s \frac{1+\dfrac{s}{\omega_{esr}}}{\left(1+\dfrac{s}{\omega_{zc}}\right)\left(1+\dfrac{s}{\omega_{cr}}\right)\left(1+\dfrac{s}{\omega_{ci}}\right)} \qquad K_a = 7.98\times 10^{-7}$$

$$Z_o(s) = K_z s \frac{1+\dfrac{s}{\omega_{esr}}}{\left(1+\dfrac{s}{\omega_{zc}}\right)\left(1+\dfrac{s}{\omega_{cr}}\right)} \qquad K_z = 1.79\times 10^{-5}$$

9.8 **a)** $$T_i(s) = K_i \frac{1+\dfrac{s}{\omega_{id}}}{1+\dfrac{s}{Q\omega_o}+\dfrac{s^2}{\omega_o^2}} \qquad K_i = 2.5 \quad \omega_{id} = 2\pi\cdot 68\text{ rad/s}$$

$\omega_o = 2\pi\cdot 520$ rad/s $Q = 5.67$

$$T_v(s) = \frac{K_v}{s}\frac{1+\frac{s}{\omega_{zc}}}{1+\frac{s}{Q\omega_o}+\frac{s^2}{\omega_o^2}} \qquad K_v = 6.5\times 10^5 \quad \omega_{zc} = 2\pi\cdot 416 \text{ rad/s}$$

$$Z_{oi}(s) = K_{oi}\frac{1+\frac{s}{\omega_{esr}}}{1+\frac{s}{\omega_{pi}}} \qquad K_{oi} = 5 \quad \omega_{pi} = 2\pi\cdot 67.4 \text{ rad/s}$$

$\omega_{esr} = 2\pi\cdot 1.69\times 10^4$ rad/s

$$G_{vs}(s) = K_{vs}\frac{1+\frac{s}{\omega_{esr}}}{1+\frac{s}{Q\omega_o}+\frac{s^2}{\omega_o^2}} \qquad K_{vs} = 0.417$$

c) $T_1(s) = K_1\dfrac{\left(1+\frac{s}{\omega_{zc}}\right)\left(1+\frac{s}{\omega_{cr}}\right)}{s\left(1+\frac{s}{Q\omega_o}+\frac{s^2}{\omega_o^2}\right)} \qquad K_1 = 6.5\times 10^5 \qquad \omega_{cr} =$ $2\pi\cdot 6\times 10^3$ rad/s

$T_2(s) = K_2\dfrac{1+\frac{s}{\omega_{zc}}}{s\left(1+\frac{s}{\omega_{id}}\right)\left(1+\frac{s}{\omega_{ci}}\right)} \qquad K_2 = 2.6\times 10^5 \qquad \omega_{ci} =$ $2\pi\cdot 1\times 10^4$ rad/s

$$A_u(s) = K_a s\frac{1+\frac{s}{\omega_{esr}}}{\left(1+\frac{s}{\omega_{zc}}\right)\left(1+\frac{s}{\omega_{cr}}\right)\left(1+\frac{s}{\omega_{ci}}\right)} \qquad K_a = 6.42\times 10^{-7}$$

$$Z_o(s) = K_z s\frac{1+\frac{s}{\omega_{esr}}}{\left(1+\frac{s}{\omega_{zc}}\right)\left(1+\frac{s}{\omega_{cr}}\right)} \qquad K_z = 1.92\times 10^{-5}$$

9.9 **a)** $T_i(s) = K_i\dfrac{1+\frac{s}{\omega_{id}}}{1+\frac{s}{Q\omega_o}+\frac{s^2}{\omega_o^2}} \qquad K_i = 11.36 \quad \omega_{id} = 2\pi\cdot 80 \text{ rad/s}$

$\omega_o = 2\pi\cdot 375$ rad/s $\quad Q = 4.71$

$$T_v(s) = K_v \frac{\left(1+\frac{s}{\omega_{zc}}\right)\left(1-\frac{s}{\omega_{rhp}}\right)\left(1+\frac{s}{\omega_{esr}}\right)}{s\left(1+\frac{s}{Q\omega_o}+\frac{s^2}{\omega_o^2}\right)\left(1+\frac{s}{\omega_{rhp}}\right)} \qquad K_v = 7.15\times10^4$$

$\omega_{zc} = 2\pi{\cdot}300\,\text{rad/s}$ $\omega_{rhp} = 2\pi{\cdot}1.77{\times}10^3\,\text{rad/s}$ $\omega_{esr} = 2\pi{\cdot}2{\times}10^3\,\text{rad/s}$

b) $T_1(s)$ crossover frequency: $\omega_{ci} = 2\pi\cdot 1\times10^4$ rad/s
$T_2(s)$ crossover frequency: $\omega_{cr} = 2\pi\cdot 550$ rad/s

c) $$T_1(s) = K_1 \frac{\left(1+\frac{s}{\omega_{zc}}\right)\left(1+\frac{s}{\omega_{cr}}\right)\left(1-\frac{s}{\omega_{rhp}}\right)}{s\left(1+\frac{s}{Q\omega_o}+\frac{s^2}{\omega_o^2}\right)\left(1+\frac{s}{\omega_{rhp}}\right)}$$

$K_1 = 7.15\times10^4$ $\omega_{cr} = 2\pi\cdot 550$ rad/s

$$T_2(s) = K_2 \frac{\left(1+\frac{s}{\omega_{zc}}\right)\left(1-\frac{s}{\omega_{rhp}}\right)\left(1+\frac{s}{\omega_{esr}}\right)}{s\left(1+\frac{s}{\omega_{id}}\right)\left(1+\frac{s}{\omega_{ci}}\right)\left(1+\frac{s}{\omega_{rhp}}\right)} \qquad K_2 = 6.3\times10^3$$

$\omega_{ci} = 2\pi\cdot 1\times10^4$ rad/s

9.10 **a)** $S_e = 1.01\times10^5$ V/s $C_2 = 47$ nF $R_2 = 12.8$ kΩ $C_3 = 5$ nF
b) $S_e = 6.07\times10^4$ V/s $C_2 = 32$ nF $R_2 = 16.5$ kΩ $C_3 = 2.8$ nF
c) $S_e = 3.04\times10^3$ V/s $C_2 = 0.194$ μF $R_2 = 4.7$ kΩ $C_3 = 0.5$ μF

CHAPTER 10

Note: Some design problems and graphical problems do not have a unique solution. For these problems, the answers in this list are a typical illustration and other solutions are equally acceptable.

10.5 $S_e = 1.46\times10^5$ V/s $R_1 = 10$ kΩ $R_2 = 92.3$ kΩ $C_2 = 1.86$ nF $C_3 =$ 0.70 nF

10.6 $S_e = 7.5\times10^4$ V/s $R_1 = 10$ kΩ $R_2 = 19.6$ kΩ $C_2 = 29.2$ nF $C_3 =$ 5.38 nF

10.9 $R_{sense} = 0.58\,\Omega$ $R_{com} = 2.9$ kΩ $C_{C1} = 0.1$ μF $R_D = 851\,\Omega$ $C_j =$ 2.28 nF

10.10 **c)** $R_1 = 10$ kΩ $R_x = 2.5$ kΩ $R_2 = 45.98$ kΩ $C_2 = 3.73$ nF $C_3 = 1.41$ nF

e) $Z_o(s) = \frac{s}{\omega_m} \frac{1 + \frac{s}{\omega_{esr}}}{\left(1 + \frac{s}{\omega_{zc}}\right)\left(1 + \frac{s}{\omega_{cr}}\right)}$

$\omega_{zc} = 2\pi \cdot 298$ rad/s $\omega_{cr} = 2\pi \cdot 1.69 \times 10^3$ rad/s $\omega_{esr} = 2\pi \cdot 3.39 \times 10^3$ rad/s $\omega_m = 2\pi \cdot 4.65 \times 10^3$ rad/s

f) $v_O(t) = 1.6e^{-2\pi \cdot 928t} - 1.1e^{-2\pi \cdot 1.69 \times 10^3 t}$

10.11 **b)** $R_{c1} = 10\,\mathrm{k\Omega}$ $C_{c1} = 17.15\,\mathrm{nF}$ $R_D = 239.8\,\Omega$ $C_x = 4.7\,\mathrm{nF}$ $R_x = 2.5\,\mathrm{k\Omega}$ **c)** $v_x(t) = 8.866\,\mathrm{V}$

Index